Symbol Table

$\overline{A}$	complement of event A
H_0	null hypothesis
H_1	alternative hypothesis
α	alpha; probability of a type I error or the area of the critical region
β	beta; probability of a type II error
r	sample linear correlation coefficient
ρ	rho; population linear correlation coefficient
r^2	coefficient of determination
R^2	multiple coefficient of determination
r_s	Spearman's rank correlation coefficient
b_1	point estimate of the slope of the regression line
b_0	point estimate of the y-intercept of the regression line
$\hat{y}$	predicted value of y
d	difference between two matched values
$\overline{d}$	mean of the differences d found from matched sample data
s_d	standard deviation of the differences d found from matched sample data
s_e	standard error of estimate

T	rank sum; used in the Wilcoxon signed-ranks test
H	Kruskal-Wallis test statistic
R	sum of the ranks for a sample; used in the Wilcoxon rank-sum test
μ_R	expected mean rank; used in the Wilcoxon rank-sum test
σ_R	expected standard deviation of ranks; used in the Wilcoxon rank-sum test
$\mu_{\overline{x}}$	mean of the population of all possible sample means $\overline{x}$
$\sigma_{\overline{x}}$	standard deviation of the population of all possible sample means $\overline{x}$
E	margin of error of the estimate of a population parameter, or expected value
Q_1, Q_2, Q_3	quartiles
$D_1, D_2, \ldots, D_9$	deciles
$P_1, P_2, \ldots, P_{99}$	percentiles
x	data value

Symbol Table

f	frequency with which a value occurs	t	t distribution
Σ	capital sigma; summation	$t_{\alpha/2}$	critical value of t
Σx	sum of the values	df	number of degrees of freedom
Σx^2	sum of the squares of the values	F	F distribution
$(\Sigma x)^2$	square of the sum of all values	χ^2	chi-square distribution
Σxy	sum of the products of each x value multiplied by the corresponding y value	χ^2_R	right-tailed critical value of chi-square
		χ^2_L	left-tailed critical value of chi-square
n	number of values in a sample	p	probability of an event or the population proportion
$n!$	n factorial		
N	number of values in a finite population; also used as the size of all samples combined	q	probability or proportion equal to $1 - p$
k	number of samples or populations or categories	$\hat{p}$	sample proportion
		$\hat{q}$	sample proportion equal to $1 - \hat{p}$
$\bar{x}$	mean of the values in a sample	$\bar{p}$	proportion obtained by pooling two samples
μ	mu; mean of all values in a population	$\bar{q}$	proportion or probability equal to $1 - \bar{p}$
s	standard deviation of a set of sample values	$P(A)$	probability of event A
σ	lowercase sigma; standard deviation of all values in a population	$P(A\|B)$	probability of event A, assuming event B has occurred
s^2	variance of a set of sample values	$_nP_r$	number of permutations of n items selected r at a time
σ^2	variance of all values in a population	$_nC_r$	number of combinations of n items selected r at a time
z	standard score		
$z_{\alpha/2}$	critical value of z		

Biostatistics

for the Biological
and Health Sciences

Biostatistics

for the Biological
and Health Sciences

Marc M. Triola, M.D.
New York University School of Medicine

Mario F. Triola

PEARSON
Addison
Wesley

Boston San Francisco New York
London Toronto Sydney Tokyo Singapore Madrid
Mexico City Munich Paris Cape Town Hong Kong Montreal

Publisher: Greg Tobin
Executive Editor: Deirdre Lynch
Project Editor: Katie Nopper
Assistant Editor: Sara Oliver
Managing Editor: Ron Hampton
Senior Production Supervisor: Peggy McMahon
Senior Designer: Barbara T. Atkinson
Photo Researcher: Beth Anderson
Digital Assets Manager: Jason Miranda
Executive Media Producer: Michelle Neil
Media Producer: Cecilia Fleming
Software Development: Mary Durnwald and Marty Wright
Marketing Manager: Phyllis Hubbard
Marketing Assistant: Celena Carr
Senior Author Support/Technology Specialist: Joe Vetere
Senior Prepress Supervisor: Caroline Fell
Permissions Administrator: Shannon Barbe
Senior Manufacturing Buyer: Evelyn Beaton
Cover Design: Night & Day Design
Production Services, Composition, and Illustration: Nesbitt Graphics, Inc.
Printer: Courier Kendallville
Cover photo: CORBIS and GettyImages

For permission to use copyrighted material, grateful acknowledgment is made to the copyright holders on pages 687–688 in the back of the book, which is hereby made part of this copyright page.

Many of the designations used by manufacturers and sellers to distinguish their products are claimed as trademarks. Where those designations appear in this book, and Addison-Wesley was aware of a trademark claim, the designations have been printed in initial caps or all caps.

Library of Congress Cataloging-in-Publication Data
Triola, Marc M.
  Biostatistics for the biological and health sciences/Marc M. Triola, Mario F. Triola
    p. cm.
  ISBN 0-321-19436-5
  1. Biometry. 2. Medical statistics. I. Triola, Mario F. II. Title.

QH323.5.T75 2006
570'.1'5195—dc22

2005045835

15 16 17 18 19—V0UD—19 18 17 16 15

To Dushana
To Ginny

About the Authors

Marc M. Triola, M.D. is a graduate of Johns Hopkins University, NYU School of Medicine and the NYU/Mount Sinai Research Fellowship in Medical Informatics. He is Chief of the Section of Medical Informatics and Director of Research for the Advanced Educational Systems laboratory, both at NYU School of Medicine. His research has focused on factors affecting the usability and impact of clinical information systems and the assessment of change in knowledge and attitudes resulting from computer-based education programs. Dr. Triola uses biostatistics as an integral part of his research. He completely reprogrammed the Statdisk statistical software package, and he has included functions particularly applicable to biostatistics.

Mario F. Triola is a Professor Emeritus of Mathematics at Dutchess Community College, where he has taught statistics for over 30 years. Marty is the author of *Elementary Statistics,* now in its tenth edition. He is also the author of *Essentials of Statistics, Elementary Statistics Using Excel, Elementary Statistics Using the Graphing Calculator,* and he is a co-author of *Statistical Reasoning for Everyday Life* and *Business Statistics.* He has written several manuals and workbooks for technology supporting statistics education. Outside of the classroom, Marty's consulting work includes the design of casino slot machines and fishing rods, and he has worked with attorneys in determining probabilities in paternity lawsuits, identifying salary inequities based on gender, and analyzing disputed election results. Marty has testified as an expert witness in New York State Supreme Court for an election dispute. The Text and Academic Authors Association has awarded Mario F. Triola a "Texty" for Excellence for his work on *Elementary Statistics*.

Preface

About This Book

The major objective of this book is to provide the best possible introduction to statistics for students in the biological, life, and health sciences. This goal is realized through features that include a friendly writing style, content that reflects the important features of a modern introductory statistics course, an abundance of interesting and real data, and a variety of helpful pedagogical components. This text reflects recommendations and guidelines from the American Statistical Association.

Biostatistics for the Biological and Health Sciences differs from traditional statistics textbooks in some very important ways. Most topics and methods in this book are the same as those found in traditional introductory statistics textbooks, but this book emphasizes applications in the fields of biology and the health sciences. Most examples and exercises are biological or health related, although the text also includes a few interesting applications from other fields. This book also includes some topics not usually found in traditional introductory statistics textbooks, such as relative risk, odds ratio, rates of fatality, rates of fertility, rates of morbidity, and life tables. Finally, some statistical methods have been included in this book because they are particularly applicable to the biological and health sciences, such as the construction of confidence interval estimates of odds ratios, McNemar's test for matched pairs, and logistic regression.

Audience/Prerequisites

Biostatistics for the Biological and Health Sciences is written for students majoring in the biological and health sciences, and it is designed for a wide variety of students taking their first statistics course, including those with limited mathematical backgrounds. Readers should have completed at least high school algebra, but calculus is not necessary. No prior statistics course is required. In many cases, underlying theory is included, but this book does not stress mathematical rigor that would be more suitable for mathematics majors or for students with extensive mathematical backgrounds that include calculus courses.

Technology

Biostatistics for the Biological and Health Sciences can be used easily without reference to any specific technology. This book is designed so that it can be used successfully with students using nothing more than one of a variety of scientific calculators. For those who choose to supplement the course with a specific technology, supplemental materials are available.

One important supplement is STATDISK, which is a statistics software package designed for this book and included free with new copies. It includes modules for almost all of the major procedures included in the text. STATDISK can be downloaded from the CD included with new copies of this book, or it can be downloaded from the Web site at www.aw-bc.com/triola.

Flexible Syllabus

The organization of this book reflects the preferences of most statistics instructors, but there are two common variations that can be easily used with this book:

- **Early coverage of correlation/regression:** Some professors prefer to cover the basics of correlation and regression early in the course, such as immediately following the topics of Chapter 2. *Sections 9-2 (Correlation) and 9-3 (Regression) can be covered early in the course.* In Section 9-2, cover only the subsection labeled as "Part 1: Basic Concepts of Correlation" and omit the subsection labeled "Part 2: Beyond the Basic Concept of Correlation." In Section 9-3, cover only the subsection labeled as "Part 1: Basic Concepts of Regression" and omit the subsection labeled "Part 2: Beyond the Basics of Regression."
- **Minimum probability:** Some professors feel strongly that coverage of probability should be extensive, while others feel just as strongly that coverage should be kept to a bare minimum. Professors preferring minimum coverage can include Section 3-2 while skipping the remaining sections of Chapter 3, as they are not essential for the chapters that follow. Many professors prefer to cover the fundamentals of probability along with the basics of the addition rule and multiplication rule, and the coverage of the multiplication rule (Sections 3-4 and 3-5) offers that flexibility.

Features

Beyond an interesting and accessible (and sometimes humorous) writing style, great care has been taken to ensure that each chapter of *Biostatistics for the Biological and Health Sciences* will help students understand the concepts presented. The following features are designed to help meet that objective:

- **Exercises:** There are 1271 exercises. Because the use of real data is such an important consideration for students, much effort has been devoted to finding real, meaningful, and interesting data. Among the 1271 exercises, there are 752 (59%) that involve *real data*. Some exercises, denoted by an asterisk *, are more difficult or involve new concepts. These exercises are located at the end of the exercise sets.

- **Real Data Sets:** These are used extensively throughout the entire book. Appendix B lists 12 large data sets. These data sets are provided in printed form in Appendix B, and in electronic form on the Web site and the CD bound in the back of new copies of the book.

- **Chapter-opening features:** A list of chapter sections previews the chapter for the student; a chapter-opening problem, using real data, then motivates the chapter material; and the first section is a chapter overview that provides a statement of the chapter's objectives.

- **End-of-chapter features:**
  A **Chapter Review** summarizes the key concepts and topics of the chapter.
  Review Exercises offer practice on the chapter concepts and procedures.
  Cumulative Review Exercises reinforce earlier material.
  From Data to Decision: Critical Thinking is a capstone problem that requires critical thinking and a writing component.
  Cooperative Group Activities encourage active learning in groups.
  Technology Projects are for use with SPSS, SAS, STATDISK, Minitab, Excel, or a TI-83/84 Plus calculator.

- **Margin Essays:** The text includes 53 margin essays, which illustrate uses and abuses of statistics in real, practical, and interesting applications. Topics include "Do Boys or Girls Run in the Family?" and "Test of Touch Therapy."

- **Flowcharts:** These appear throughout the text to simplify and clarify more complex concepts and procedures.

- **Statistical Software:** SPSS, SAS, STATDISK, Minitab, Excel and TI-83/84 Plus output appear throughout the text.

- **CD-ROM:** Prepared by the authors and packaged with every new copy of the text, the CD includes the data sets from Appendix B in the textbook. These data sets are stored as Minitab worksheets, SPSS files, SAS files, SAS Transport files, JMP files, Excel workbooks, and a TI-83 Plus and TI-84 Plus calculator APP (application), and as text files that can be used with other applications. The CD also includes programs for the TI-83 Plus and TI-84 Plus graphing calculators, STATDISK statistical software, and DDXL, an Excel Add-in, which enhances the capabilities of Excel's statistics programs.

- *Addison-Wesley Tutor Center:* Free tutoring is available to students who purchase a copy of *Biostatistics for the Biological and Health Sciences* bundled with an access code. The Addison-Wesley Math Tutor Center is staffed by qualified statistics and mathematics instructors who provide students with tutoring on text examples and any exercise with an answer in the back of the book. Tutoring assistance is provided by toll-free telephone, fax, e-mail, and whiteboard technology—which allows tutors and students to actually see the problems worked while they "talk" in real time over the Internet. This service is available five days a week, seven hours a day. For more information, please contact your Addison-Wesley sales representative.

Supplements

The student and instructor supplements packages are intended to be the most complete and helpful learning system available for the introductory biostatistics course. Instructors should contact their local Addison-Wesley sales representative, or e-mail the company directly at exam@aw.com for examination copies.

For the Instructor

- *Instructor's Solutions Manual,* by J. Jackson Barnette (University of Alabama at Birmingham) and Ian C. Walters (D'Youville College) contains detailed solutions to all the exercises. ISBN: 0-321-28688-X.

- **Testing System:** Great care has been taken to create the most comprehensive testing system possible for *Biostatistics for the Biological and Health Sciences.* Not only is there a printed test bank, there is also a computerized test generator, **TestGen,** that allows instructors to view and edit test bank questions, transfer them to tests, and print in a variety of formats. The program also offers many options for sorting, organizing, and displaying test banks and tests. A built-in random number and test generator makes **TestGen** ideal for creating multiple versions of tests and provides more possible test items than printed test bank questions. Users can export tests to be compatible with a variety of course management systems, or even just to display in a web browser. Additionally, tests created with TestGen can be used with **QuizMaster,** which enables students to take exams on a computer network. **The Printed Test Bank,** by Karin DeAmicis (Kennesaw State University), contains several tests for each chapter of the main text, along with answer keys. ISBN: 0-321-35830-9; **TestGen** for Mac and Windows ISBN: 0-321-35829-5.

For the Student

- **Triola Web Site:** This Web site may be accessed at http://www.aw-bc.com/triola_biostatistics, and provides the Appendix B data sets in formats for SPSS, SAS, Excel, Minitab, JMP, and TI-83 Plus or TI-84 Plus calculators. The STATDISK statistical software package is also inlcuded along with programs for TI-83 Plus and TI-84 Plus calculators.

- *Student's Solutions Manual,* by J. Jackson Barnette (University of Alabama), provides detailed, worked-out solutions to all odd-numbered text exercises. ISBN: 0-321-28689-8.

The following technology supplements are suitable for most of the topics in *Biostatistics for the Biological and Health Sciences.*

- *Excel® Student Laboratory Manual and Workbook,* ISBN: 0-321-36909-2.
- *Minitab® Student Laboratory Manual and Workbook,* ISBN: 0-321-36919-X.
- *SAS Student Laboratory Manual and Workbook,* ISBN: 0-321-36910-6.
- *SPSS® Student Laboratory Manual and Workbook,* ISBN: 0-321-36911-4.

- *STATDISK Student Laboratory Manual and Workbook,* ISBN: 0-321-36912-2.

- *TI-83/84 Plus® Companion to Elementary Statistics,* ISBN: 0-321-36920-3.

- **Triola Version of** *ActivStats®,* developed by Paul Velleman and Data Description, Inc., provides complete coverage of introductory statistics topics on CD-ROM, using a full range of multimedia. *ActivStats* integrates video, simulation, animation, narration, text, and interactive experiments, World Wide Web access, and Data Desk®, a statistical software package. Homework problems and data sets from the text are included on the CD-ROM. *ActivStats* for Windows and Macintosh ISBN: 0-321-30364-4. Also available in versions for Excel, JMP, Minitab and SPSS. See your Addison-Wesley sales representative for details or check the Web site at www.aw-bc.com/activstats.

- *The Student Edition of MINITAB, Release 14 for Windows* is an educational version of the professional version MINITAB Release 14 for Windows. It includes all but a limited number of the features and functions of the professional version. MINITAB is an easy-to-use general-purpose statistical computing package for analyzing data. It is a flexible and powerful tool that was designed from the beginning to be used by students and researchers new to statistics. It is now one of the most widely used statistics packages in the world. MINITAB quickly performs tedious computations and produces accurate and professional quality graphs almost instantly. This power frees the user to focus on the exploration of the structure of the data and the interpretation of the output. MINITAB 14 software is available for purchase only when bundled with an Addison-Wesley textbook. ISBN: 0-321-11313-6. It can also be purchased with *The Student Guide to MINITAB, Release 14.* ISBN: 0-201-77469-0.

- *The Student Guide to MINITAB, Release 14,* by John McKenzie and Robert Goldman, includes both a Getting Started with MINITAB section and hands-on, self-paced Tutorials designed to teach students how to use the software capabilities through a variety of approaches. The Tutorials cover all of the primary features and capabilities of MINITAB, including graphical and numerical methods for one and two or more variables, bivariate analysis, total quality management tools, time series analysis, and taking data from the web, to name a few. ISBN: 0-321-11312-8 (Guide only). ISBN: 0-201-77469-0 (Guide plus Student Edition of MINITAB, Release 14 for Windows software.

Any of these products can be purchased separately, or bundled with Addison-Wesley texts. Instructors can contact local sales representatives for details on purchasing and bundling supplements with the textbook or contact the company at exam@aw.com for examination copies of any of these items.

Acknowledgments

The success of *Biostatistics for the Biological and Health Sciences* is due to the efforts of many dedicated professionals. We wish to express our most sincere thanks to Deirdre Lynch, Greg Tobin, Peggy McMahon, Barbara Atkinson, Katie Nopper, Sara Oliver, Yolanda Cossio, Phyllis Hubbard, Celena Carr, Joe Vetere, Ceci Fleming, Janet Nuciforo, and the entire Addison-Wesley staff.

We are grateful for the many suggestions and comments provided by instructors, and to the hundreds of researchers who did studies, conducted surveys, and compiled interesting and meaningful data used in examples and exercises throughout this book.

We would also like to thank the following for their help with this text:

Text Accuracy Reviewers
Anita L. Davelos Baines, University of Texas–Pan American
Chad Cross, University of Nevada–Las Vegas
Eric Howington, Coastal Carolina University
Peggy Gallup, Southern Connecticut State University
Sharon Marie Homan, Saint Louis University School of Public Health
Sunny Kim, Florida International University
Julie Norton, California State University–Hayward
Sandra Woolson, University of North Carolina at Chapel Hill School of Medicine

Reviewers
Ahmed A. Arif, Texas Tech University Health Sciences Center
Priya Banerjee, State University of New York–Brockport
Hokwon A. Cho, University of Nevada–Las Vegas
Erica Corbett, Southeastern Oklahoma State University
Robert Downer, Louisiana State University
Karen Gaines, University of South Dakota
Tyler Haynes, Saginaw Valley State University
John C. Higginbotham, University of Alabama
Jeremiah N. Jarrett, Central Connecticut State University
James Leeper, University of Alabama
Deborah Lurie, St. Joseph's University
Katrina Mangin, University of Arizona
Osvaldo Mendez, University of Texas at El Paso
Jim Novak, University of South Dakota
Jason Roy, University of Rochester

Lisa Sullivan, Boston University
John W. Wilson, University of Pittsburgh
Robert F. Woolson, Medical University of South Carolina
Ken Yasukawa, Beloit College
George Zimmermann, Richard Stockton College of New Jersey

Contents

Biostatistics

for the Biological
and Health Sciences

1

Introduction

What can we learn from this health care survey?

USA Today ran a "Health Care Survey" that occupied 3/4 page in one of its issues. Readers were asked to "please take a moment to fill out this survey and return it to us." Readers could mail or fax their responses. The first question asked how often they visited a doctor in a year. The second question asked for a check-off of health conditions for the past year, including colds, fevers, hemorrhoids, motion sickness, and warts. Most questions asked about health conditions, tobacco use, and prescription drugs. Question 17 of the survey was this: "May we contact you again to participate in other surveys for *USA Today*?" Readers wishing to participate in other surveys for *USA Today* were asked to provide an address, daytime phone number, evening phone number, and e-mail address.

Consider the way in which the data are collected for this survey. How does it affect what we can con-clude about the general population based on the results obtained from such surveys? Can we use the reported numbers of doctor visits to estimate the mean (average) number of doctor visits for the whole population? The answers to such questions are critically important for the evaluation of such survey results. The issue being considered here is the single most important issue of this entire chapter, and it may well be the single most important issue of this entire book.

In this chapter, we will address the validity of such surveys. We will see that we can often form important conclusions by applying simple common sense. After completing this chapter, we should be able to identify the key issue affecting the validity of the above survey, and we should have an important understanding of data collection methods in general.

The State of Statistics

1-1   Overview

The Chapter Problem on the previous page involves a survey, which is one of many tools that can be used for collecting data. A common goal of a survey is to collect data from a small part of a larger group so that we can learn something about the larger group. This is a common and important goal of the subject of statistics: Learn about a large group by examining data from some of its members. In this context, the terms *sample* and *population* become important. Formal definitions for these and other basic terms are given here.

Definitions

Data are observations (such as measurements, genders, survey responses) that have been collected.

Statistics is a collection of methods for planning experiments, obtaining data, and then organizing, summarizing, analyzing, interpreting, presenting, and drawing conclusions based on the data.

A **population** is the complete collection of all elements (scores, people, measurements, and so on) to be studied. The collection is complete in the sense that it includes *all* subjects to be studied.

A **census** is the collection of data from *every* member of the population.

A **sample** is a *subcollection* of members selected from part of a population.

For example, a Reuters/Zogby poll asked 1000 adults if they believe that life exists elsewhere in the universe. The 1000 survey subjects constitute a *sample*, whereas the *population* consists of the entire collection of all 202,682,345 adult Americans. Every 10 years, the United States government attempts to obtain a *census* of every citizen, but fails because it is impossible to reach everyone.

An important activity of this book is to demonstrate how we can use sample data to form conclusions about populations. We will see that it is *extremely* critical to obtain sample data that are representative of the population from which the data are drawn. As we proceed through this chapter, we should focus on these key concepts:

- **Sample data must be collected in an appropriate way, such as through a process of *random* selection.**
- **If sample data are not collected in an appropriate way, the data may be so completely useless that no amount of statistical torturing can salvage them.**

1-2   Types of Data

In Section 1-1 we defined the terms *population* and *sample*. The following two terms are used to distinguish between cases in which we have data for an entire population, and cases in which we have data for a sample only.

Definitions

A **parameter** is a measurement describing some characteristic of a *population*.
A **statistic** is a measurement describing some characteristic of a *sample*.

EXAMPLES

1. **Parameter:** After a freshwater pond is excavated and filled, it is stocked with a population of 500 rainbow trout with a total weight of 2100 lb. If we divide the total weight by the number of trout, we get a mean (average) of 4.2 lb. If we consider the collection of all 500 trout to be the population of the lake, then the 4.2 lb is a *parameter,* not a statistic.
2. **Statistic:** Based on a sample of 877 surveyed executives, it is found that 45% of them would not hire someone with a typographic error on their job application. That figure of 45% is a *statistic* because it is based on a sample, not the entire population of all executives.

Some data sets consist of numbers (such as heights of 66 in. and 72 in.), while others are non-numerical (such as eye colors of green and brown). The terms *quantitative data* and *qualitative data* are often used to distinguish between these types.

Definitions

Quantitative data consist of numbers representing counts or measurements.
Qualitative (or **categorical** or **attribute**) **data** can be separated into different categories that are distinguished by some non-numeric characteristic.

EXAMPLES

1. **Quantitative data:** The weights of manatees.
2. **Qualitative data:** The genders (male/female) of bears.

When working with quantitative data, it is important to use the appropriate units of measurement, such as dollars, hours, feet, meters, and so on. We should be especially careful to observe such references as "all amounts are in *thousands of dollars*" or "all times are in *hundredths of a second*" or "units are in *kilograms*." To ignore such units of measurement could lead to very wrong conclusions. NASA lost its $125 million Mars Climate Orbiter when it crashed because the controlling software had acceleration data in *English* units, but they were incorrectly assumed to be in *metric* units.

Quantitative data can be further described by distinguishing between *discrete* and *continuous* types.

Definitions

Discrete data result when the number of possible values is either a finite number or a "countable" number. (That is, the number of possible values is 0 or 1 or 2 and so on.)

Continuous (numerical) data result from infinitely many possible values that correspond to some continuous scale that covers a range of values without gaps, interruptions or jumps.

EXAMPLES

1. **Discrete data:** The numbers of eggs that hens lay are *discrete* data because they represent counts.

2. **Continuous data:** The amounts of milk from cows are *continuous* data because they are measurements that can assume any value over a continuous span. During a given time interval, a cow might yield an amount of milk that can be any value between 0 gallons and 5 gallons. It would be possible to get 2.343115 gallons because the cow is not restricted to the discrete amounts of 0, 1, 2, 3, 4, or 5 gallons.

When describing relatively smaller amounts, correct grammar dictates that we use "fewer" for discrete amounts, and "less" for continuous amounts. For example, it is correct to say that we drank *fewer* cans of cola and, in the process, we drank *less* cola. The numbers of cans of cola are discrete data, whereas the actual volume amounts of cola are continuous data.

Another common way of classifying data is to use four levels of measurement: nominal, ordinal, interval, and ratio. In applying statistics to real problems, the level of measurement of the data is an important factor in determining which procedure to use. There will be some references to these levels of measurement in this book, but the important point here is based on common sense: Don't do computations and don't use statistical methods with data that are not appropriate. For example, it would not make sense to compute an average of social security numbers, because those numbers are data that are used for identification, and they don't represent measurements or counts of anything. For the same reason, it would make no sense to compute an average of the numbers entered on the identification tags placed on homing pigeons.

Definition

The **nominal level of measurement** is characterized by data that consist of names, labels, or categories only. The data cannot be arranged in an ordering scheme (such as low to high).

EXAMPLES

Here are examples of sample data at the nominal level of measurement.

1. **Yes/no/undecided:** Survey responses of yes, no, and undecided.
2. **Colors:** The colors of pea pods (green, yellow) used in a genetics experiment.

Because nominal data lack any ordering or numerical significance, they should not be used for calculations. Numbers are sometimes assigned to the different categories (especially when data are coded for computers), but these numbers have no real computational significance and any average calculated with them is meaningless.

Definition

Data are at the **ordinal level of measurement** if they can be arranged in some order, but *differences* between data values either cannot be determined or are meaningless.

EXAMPLES

Here are examples of sample data at the ordinal level of measurement.

1. **Course grades:** A college professor assigns grades of A, B, C, D, or F. These grades can be arranged in order, but we can't determine differences between such grades. For example, we know that A is higher than B (so there is an ordering), but we cannot subtract B from A (so the difference cannot be found).
2. **Ranks:** Based on several criteria, a biologist ranks all bears in one region according to their aggressiveness. Those ranks (first, second, third, and so on) determine an ordering. However, the differences between ranks are meaningless. For example, a difference of "second minus first" might suggest $2 - 1 = 1$, but this difference of 1 is meaningless because it is not an exact quantity that can be compared to other such differences. The difference between the aggressiveness of the first bear and the second bear is not necessarily the same as the difference between the aggressiveness of the second bear and the third bear.

Ordinal data provide information about relative comparisons, but not the magnitudes of the differences. Usually, ordinal data should not be used for calculations such as an average, but this guideline is sometimes violated (such as when we use letter grades to calculate a grade-point average).

Should You Believe a Statistical Study?

In *Statistical Reasoning for Everyday Life*, 2nd edition, authors Jeff Bennett, William Briggs, and Mario Triola list the following eight guidelines for critically evaluating a statistical study: (1) Identity the goal of the study, the population considered, and the type of study. (2) Consider the source, particularly with regard to a possibility of bias. (3) Analyze the sampling method. (4) Look for problems in defining or measuring the variables of interest. (5) Watch out for confounding variables that could invalidate conclusions. (6) Consider the setting and wording of any survey. (7) Check that graphs represent data fairly, and conclusions are justified. (8) Consider whether the conclusions achieve the goals of the study, whether they make sense, and whether they have practical significance.

Definition

The **interval level of measurement** is like the ordinal level, with the additional property that the difference between any two data values is meaningful. However, data at this level do not have a *natural* zero starting point (where *none* of the quantity is present).

EXAMPLES

The following examples illustrate the interval level of measurement.

1. **Temperatures:** Body temperatures of 98.2°F and 98.6°F are examples of data at this interval level of measurement. Those values are ordered, and we can determine their difference of 0.4°F. However, there is no natural starting point. The value of 0°F might seem like a starting point, but it is arbitrary and does not represent the total absence of heat. Because 0°F is not a natural zero starting point, it is wrong to say that 50°F is *twice* as hot as 25°F.

2. **Years of cicada emergence:** The years 1936, 1953, 1970, 1987, and 2004. (Time did not begin in the year 0, so the year 0 is arbitrary instead of being a natural zero starting point representing "no time.")

Definition

The **ratio level of measurement** is the interval level with the additional property that there is also a natural zero starting point (where zero indicates that *none* of the quantity is present). For values at this level, *differences* and *ratios* are both meaningful.

EXAMPLES

The following are examples of data at the ratio level of measurement. Note the presence of the natural zero value, and note the use of meaningful ratios of "twice" and "three times."

1. **Weights:** Weights (in kg) of bald eagles (0 kg does represent no weight, and 4 kg is twice as heavy as 2 kg).

2. **Ages:** Ages (in days) of bald eagles (0 does represent a newborn with no aging, and an age of 60 days is three times as old as an age of 20 days).

This level of measurement is called the ratio level because the zero starting point makes ratios meaningful. Among the four levels of measurement, most difficulty arises with the distinction between the interval and ratio levels. *Hint:* To simplify that distinction, use a simple "ratio test:" Consider two quantities where one number is twice the other, and ask whether "twice" can be used to correctly describe the quantities. Because a 4-kg weight is *twice* as heavy as a 2-kg weight, but 50°F is *not twice* as hot as 25°F, weights are at the ratio level while Fahrenheit temperatures are at the interval level. For a concise comparison and review, study Table 1-1 for the differences among the four levels of measurement.

Table 1-1	Levels of Measurement of Data	
Level	Summary	Example
Nominal	Categories only. Data cannot be arranged in an ordering scheme.	Bear encounter states: 5 New York 20 Idaho 40 Wyoming } Categories or names only
Ordinal	Categories are ordered, but differences can't be found or are meaningless.	Bears according to aggressiveness: 5 not aggressive 20 somewhat aggressive 40 highly aggressive } An order is determined by "not," "somewhat," "highly."
Interval	Differences are meaningful, but there is no natural starting point, and ratios are meaningless.	Bear den temperatures: 5°F 20°F 40°F } 0°F doesn't mean "no heat." 40°F is not twice as hot as 20°F.
Ratio	There is a natural zero starting point and ratios are meaningful.	Bear migration distances: 5 miles 20 miles 40 miles } 40 miles is twice as far as 20 miles.

1-2 Exercises

In Exercises 1–4, determine whether the given value is a statistic or a parameter.

1. In studying the effects of geese near an airport, a random sample of Canada geese includes 12 males.

2. A sample of Canada geese who mate is observed and the average (mean) number of offspring is 3.8.

3. A study involves attaching altimeters to individual frigate birds, and their average (mean) altitude is found to be 226 m.

4. In a study of all cloned sheep it is found that their average (mean) age is 2.7 years.

In Exercises 5–8, determine whether the given values are from a discrete or continuous data set.

5. In a study of birds in Buldir Island, Alaska, 312 breeding adult red-legged kittiwakes are banded.

6. A sample of broad-billed hummingbirds is observed and their mean length is 3.25 in.

7. In a survey of 1059 adults, it is found that 39% of them have guns in their homes (based on a Gallup poll).

8. The bears in Data Set 6 in Appendix B have an average (mean) head width of 6.19 in.

In Exercises 9–16, determine which of the four levels of measurement (nominal, ordinal, interval, ratio) is most appropriate.

9. Lengths of broad-billed hummingbirds.

10. Ratings of excellent, good, average, or poor for the racing ability of thoroughbred horses.

11. Body temperatures of sandhill cranes.

12. Case numbers on files of manatees killed by boats.

13. Blood types of A, B, AB, and O.

14. Social security numbers used to identify subjects in a clinical trial.

15. The numbers of manatees killed by boats in each of the last 10 years.

16. Zip codes of locations of bald eagle sightings.

In Exercises 17–20, identify the (a) sample and (b) population. Also, determine whether the sample is likely to be representative of the population.

17. A marine biologist obtains the weights of rainbow trout that she catches in a net.

18. A Florida newspaper runs a health survey and obtains 750 responses.

19. In a Gallup poll of 1059 randomly selected adults, 39% answered "yes" when asked "Do you have a gun in your home?"

20. A graduate student at the University of Newport conducts a research project about how adult Americans communicate. She begins with a survey mailed to 500 of the adults that she knows. She asks them to mail back a response to this question: "Do you prefer to use e-mail or snail mail (the U.S. Postal Service)?" She gets back 65 responses, with 42 of them indicating a preference for snail mail.

1-3 Design of Experiments

Successful use of statistics typically requires more *common sense* than mathematical expertise (despite Voltaire's warning that "common sense is not so common"). Because we now have access to calculators and computers, modern applications of statistics no longer require us to master complex algorithms of mathematical manipulations. Instead, we can focus on *interpretation* of data and results. For an extremely important example illustrating the role of common sense, consider the following definition.

> ### Definition
>
> A **voluntary response sample** (or **self-selected sample**) is one in which the respondents themselves decide whether to be included.

For a good example of a voluntary response sample, see the Chapter Problem involving a *USA Today* health survey in which readers were asked to respond by way of mail or fax. When individuals decide themselves whether to participate, it often happens that people with strong interests or opinions are more likely to participate, so the responses are not representative of the whole population. This suggests that mail-in or fax surveys, Internet surveys, and telephone call-in polls are seriously flawed in the sense that we should not make conclusions about a population based on such biased samples. With such voluntary response samples, valid conclusions can be made only about the specific group of people who chose to participate, but a common practice is to incorrectly state or imply conclusions about a larger population. From a statistical viewpoint, such a sample is fundamentally flawed and should not be used for making general statements about a larger population. This example leads naturally to a more general principle that is critically important throughout this entire book:

If sample data are not collected in an appropriate way, the data may be so completely useless that no amount of statistical analysis can salvage them.

Statistical methods are driven by data. We typically obtain data from two distinct sources: *observational studies* and *experiments*.

Definitions

In an **observational study,** we observe and measure specific characteristics, but we don't attempt to *modify* the subjects being studied.

In an **experiment,** we apply some *treatment* and then proceed to observe its effects on the subjects.

A Gallup poll is a good example of an observational study. A good example of an experiment is a **clinical trial,** which is a well-planned and well-designed experiment with a **treatment group** (in which some subjects are given a special treatment) and a **control group** (in which some subjects are not given the special treatment or are given a placebo). Drugs, such as Lipitor, are tested in such clinical trials. The Gallup poll is observational in the sense that we merely observe people (often through interviews) without modifying them in any way. But the clinical trial of Lipitor involves treating some people with the drug, so the treated people are modified (and those in the control group may also be modified). There are different types of observational studies, as illustrated in Figure 1-1. These terms, commonly used in many different professional journals, are defined below.

Definitions

In a **cross-sectional study,** data are observed, measured, and collected at one point in time.

In a **retrospective** (or **case-control**) **study,** data are collected from the past by going back in time (through examination of records, interviews, and so on).

In a **prospective** (or **longitudinal** or **cohort**) **study,** data are collected in the future from groups sharing common factors (called *cohorts*).

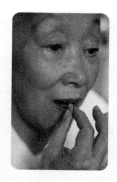

Clinical Trials vs. Observational Studies

In a *New York Times* article about hormone therapy for women, reporter Denise Grady wrote about a report of treatments tested in randomized controlled trials. She stated that "Such trials, in which patients are assigned at random to either a treatment or a placebo, are considered the gold standard in medical research. By contrast, the observational studies, in which patients themselves decide whether to take a drug, are considered less reliable. . . . Researchers say the observational studies may have painted a falsely rosy picture of hormone replacement because women who opt for the treatments are healthier and have better habits to begin with than women who do not."

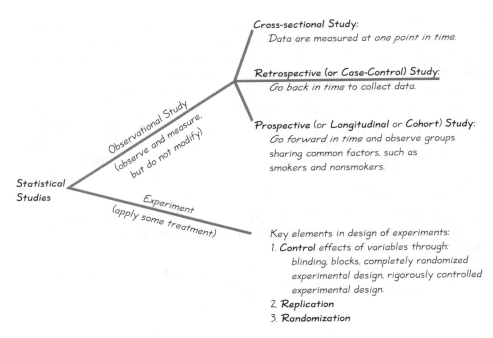

FIGURE 1-1 **Elements of Statistical Studies**

There is an important distinction between the sampling done in retrospective and prospective studies. In retrospective studies we go back in time to collect data about the resulting characteristic that is of concern. For example, a retrospective study of a disease might work backward to determine which conditions could be linked to the disease. Such a retrospective study requires subjects with the disease and subjects without the disease, so that the effects of the prior conditions are linked to the disease, not the general population. In *The Lady Tasting Tea,* author David Salsburg cites the case of a retrospective study showing that artificial sweeteners were linked to bladder cancer. However, almost all of the diseased subjects were from low socioeconomic classes, while almost all of the disease-free subjects were from higher socioeconomic classes. Consequently, the two groups were not comparable, and this retrospective study was flawed.

In prospective studies we go forward in time by following groups with a potentially causative factor and those without it, such as a group of drivers who use cell phones and a group of drivers who do not use cell phones.

The above three definitions apply to observational studies, but we now shift our focus to experiments. Results of experiments are sometimes ruined because of *confounding.*

Definition

Confounding occurs when effects of variables are somehow mixed so that the *individual* effects of the variables cannot be identified. (That is, confounding is basically confusion of variable effects.)

Try to plan the experiment so that confounding does not occur.

For example, suppose we treat 1000 people with a vaccine designed to prevent Lyme disease caused by ticks. If an early onset of cold weather causes the ticks to hibernate and the 1000 vaccinated subjects subsequently experience an unusually low incidence of Lyme disease, we don't know if the lower disease rate is the result of an effective vaccine or the early onset of cold weather. Confounding has occurred because the effects of the vaccine treatment and the effects of the cold weather have been mixed. We cannot distinguish between the effects of the vaccine and the effects of the cold weather, and we cannot determine whether the vaccine is an effective treatment. A better experimental design would somehow account for the vaccine treatment and the cold weather so that their individual effects can be identified or controlled.

Controlling Effects of Variables

Figure 1-1 shows that one of the key elements in the design of experiments is controlling effects of variables. We can gain that control by using such devices as blinding, blocks, a completely randomized experimental design, or a rigorously controlled experimental design, described as follows.

Blinding   In 1954, a massive experiment was designed to test the effectiveness of the Salk vaccine in preventing the polio that killed or paralyzed thousands of children. In that experiment, a treatment group was given the actual Salk vaccine, while a second group was given a placebo that contained no vaccine. In experiments involving placebos, there is often a **placebo effect** that occurs when an untreated subject reports an improvement in symptoms. (The reported improvement in the placebo group may be real or imagined.) This placebo effect can be minimized or accounted for through the use of **blinding**, a technique in which the subject doesn't know whether he or she is receiving a treatment or a placebo. Blinding allows us to determine whether the treatment effect is significantly different from the placebo effect. In a **single-blind** experiment, the subjects don't know whether they are getting the treatment or a placebo. The polio experiment was **double-blind**, meaning that blinding occurred at two levels: (1) The children being injected didn't know whether they were getting the Salk vaccine or a placebo, and (2) the doctors who gave the injections and evaluated the results did not know either.

Blocks   When designing an experiment to test the effectiveness of one or more treatments, it is important to put the subjects (often called *experimental units*) in different groups (or *blocks*) in such a way that those groups are very similar. A **block** is a group of subjects that are known (prior to the experiment) to be similar in the ways that might affect the outcome of the experiment.

> **When conducting an experiment of testing one or more different treatments, form blocks (or groups) of subjects with similar characteristics.**

Randomization in the Design of Experiments   When assigning subjects to different treatments, it is a common practice to somehow include random selection. One approach is to use a **completely randomized design,** whereby the treatments are assigned to the subjects by using a completely random assignment process. Here is an example of a completely randomized experimental design: Children are assigned to a vaccine treatment group or placebo group through a process of random coin flips. For each child, an outcome of heads results in a vaccine and an outcome of tails results in a placebo. In a **randomized block design,** we first form blocks so that within each block, the subjects have similar characteristics, then we use randomization to assign subjects to treatments separately within each block. Here is an example of a randomized block design: We form a block of male subjects and another block of female subjects, then within each block we flip a coin to determine whether the subject is given a vaccine or placebo. A completely randomized design has no restrictions on the randomization, whereas the randomized block design has no restrictions on randomization within each block. If the vaccine really does affect males and females differently, the randomized block design has a much better chance of identifying that difference, whereas the completely randomized design might not allow us to see such a gender difference.

Replication and Sample Size

In addition to controlling effects of variables, another key element of experimental design is the size of the samples. Samples should be large enough so that the erratic behavior that is characteristic of very small samples will not disguise the true effects of different treatments. Repetition of an experiment is called **replication,** and replication is used effectively when we have enough subjects to recognize differences from different treatments. (In another context, *replication* refers to the repetition or duplication of an experiment so that results can be confirmed or verified.) With replication, the large sample sizes increase the chance of recognizing different treatment effects. However, a large sample is not necessarily a good sample. Although it is important to have a sample that is sufficiently large, it is more important to have a sample in which data have been chosen in some appropriate way, such as random selection (described below).

> **Use a sample size that is large enough so that we can see the true nature of any effects, and obtain the sample using an appropriate method.**

In the experiment designed to test the Salk vaccine, 200,000 children were given the actual Salk vaccine and 200,000 other children were given a placebo. Because the actual experiment used sufficiently large sample sizes, the effectiveness of the vaccine became evident. Nevertheless, even though the treatment and placebo groups were very large, the experiment would have been a failure if subjects had not been assigned to the two groups in a way that made both groups similar in the ways that were important to the experiment.

Randomization and Other Sampling Strategies

In statistics, one of the worst mistakes is to collect data in a way that is inappropriate. We cannot overstress this very important point:

If sample data are not collected in an appropriate way, the data may be so completely useless that no amount of statistical analysis can salvage them.

Early in this section we saw that a voluntary response sample is one in which the subjects decide themselves whether to respond. Such samples are very common, but their results are generally useless for making valid inferences about larger populations.

We now define some of the more common methods of sampling.

Definitions

In a **random sample,** members from the population are selected in such a way that each *individual member* has the same chance of being selected.

A **simple random sample** of size *n* subjects is selected in such a way that every possible *sample of the same size n* has the same chance of being chosen.

EXAMPLE **Random Sample and Simple Random Sample**
Picture a classroom with 36 students arranged in six rows of 6 students each. Assume that the professor selects a sample of 6 students by rolling a die and selecting the row corresponding to the outcome. Is the result a random sample? Simple random sample?

SOLUTION   The sample is a random sample because each individual student has the same chance (one chance in six) of being selected.

However, the sample is not a simple random sample because not all samples of size 6 have the same chance of being chosen. This use of a die to select a row makes it *impossible* to select 6 students who are in different rows. There is one chance in six of selecting the students in the first row, but there is *no* chance of selecting the six students sitting in the first seat of each row. Because all samples of size six do *not* have the same chance of being selected, we do not have a simple random sample.

Important: Throughout this book, we will use a variety of different statistical procedures, and there is often a requirement that we have a *simple random sample*, as defined above. With random sampling we expect all components of the population to be (approximately) proportionately represented. Random samples are selected by many different methods, including the use of computers to generate random numbers. (Before computers, tables of random numbers were often used instead. For truly exciting reading, see this book consisting of one million digits that were randomly generated: *A Million Random Digits* published by Free Press. The *Cliff Notes* summary of the plot is not yet available.) Unlike careless or haphazard sampling, random sampling usually requires very careful planning and execution.

In addition to random sampling, there are other sampling techniques in use, and we describe the common ones here. See Figure 1-2 for an illustration depicting the different sampling approaches. Keep in mind that only random sampling and simple random sampling will be used throughout the remainder of the book.

Definitions

In **systematic sampling,** we randomly select a starting point and then select every *k*th (such as every 50th) element in the population.

With **convenience sampling,** we simply collect results that are very easy (convenient) to get.

With **stratified sampling,** we subdivide the population into at least two different subgroups (or strata) that share the same characteristics (such as gender or age bracket), then we draw a sample from each subgroup (or stratum).

In **cluster sampling,** we first divide the population area into sections (or clusters), then randomly select some of those clusters, and then choose *all* the members from those selected clusters.

It is easy to confuse stratified sampling and cluster sampling, because they both involve the formation of subgroups. But cluster sampling uses *all* members from a *sample* of clusters, whereas stratified sampling uses a *sample* of members from *all* strata. An example of cluster sampling can be found in a pre-election poll, whereby we randomly select 30 election precincts from a large number of precincts, then survey all the people from each of those precincts. This is much faster and much less expensive than selecting one person from each of the many precincts in the population area. Results from stratified or cluster sampling are often adjusted or weighted to correct for any disproportionate representations of groups.

For a fixed sample size, if you randomly select subjects from different strata, you are likely to get more consistent (and less variable) results than by simply selecting a random sample from the general population. For that reason, stratified sampling is often used to reduce the variation in the results. Many of the methods discussed later in this book have a requirement that sample data are a *simple random sample,* and neither stratified sampling nor cluster sampling satisfies that requirement.

Figure 1-2 illustrates the five common methods of sampling. Professionals often collect data by using some combination of the five methods. Here is one typical example of what is called a *multistage sample design*: First, randomly select a sample of counties from all 50 states, then randomly select cities and towns in those counties, then randomly select residential blocks in each city or town, then randomly select households in each block, then randomly select someone from each household. We will not use such a sample design in this book. We should again stress that the methods of this book typically require that we have a *simple random sample*.

Sampling Errors

No matter how well you plan and execute the sample collection process, there is likely to be some error in the results. For example, randomly select 1000 adults, ask them if they graduated from high school, and record the sample percentage of "yes" responses. If you randomly select another sample of 1000 adults, it is likely that you will obtain a *different* sample percentage.

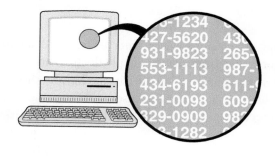

Random Sampling:

Each member of the population has an equal chance of being selected. Computers are often used to generate random telephone numbers.

Simple Random Sampling:

A sample of size n subjects is selected in such a way that every possible sample of the same size n has the same chance of being chosen.

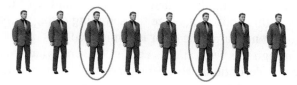

Systematic Sampling:

Select some starting point, then select every kth (such as every 50th) element in the population.

Convenience Sampling:

Use results that are easy to get.

Stratified Sampling:

Subdivide the population into at least two different subgroups (or strata) that share the same characteristics (such as gender or age bracket), then draw a sample from each subgroup.

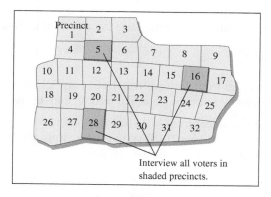

Cluster Sampling:

Divide the population area into sections (or clusters), then randomly select some of those clusters, and then choose all members from those selected clusters.

FIGURE 1-2 **Common Sampling Methods**

> ### Definitions
>
> A **sampling error** is the difference between a sample result and the true population result; such an error results from chance sample fluctuations.
>
> A **nonsampling error** occurs when the sample data are incorrectly collected, recorded, or analyzed (such as by selecting a biased sample, using a defective measure instrument, or recording the data incorrectly).

If we carefully collect a sample so that it is representative of the population, we can use methods in this book to analyze the sampling error, but we must exercise extreme care so that nonsampling error is minimized.

After reading through this section, it is easy to be somewhat overwhelmed by the variety of different definitions. But remember this main point: The method used to collect data is absolutely and critically important, and we should know that *randomness* is particularly important. If sample data are not collected in an appropriate way, the data may be so completely useless that no statistical analysis can salvage them.

1-3 Exercises

In Exercises 1–4, determine whether the given description corresponds to an observational study or an experiment.

1. Drug Testing Patients are given Lipitor to determine whether this drug has the effect of lowering cholesterol levels.

2. Treating Syphilis Much controversy arose over a study of patients with syphilis who were *not* given a treatment that could have cured them. Their health was followed for years after they were found to have syphilis.

3. Quality Control The Federal Drug Administration randomly selects a sample of Bayer aspirin tablets. The amount of aspirin in each tablet is measured for accuracy.

4. Magnetic Bracelets Cruise ship passengers are given magnetic bracelets, which they agree to wear in an attempt to eliminate or diminish the effects of motion sickness.

In Exercises 5–8, identify the type of observational study (cross-sectional, retrospective, prospective).

5. Medical Research A researcher from the New York University School of Medicine obtains data about head injuries by examining hospital records from the past five years.

6. Psychology of Trauma A researcher from Mt. Sinai Hospital in New York City plans to obtain data by following (to the year 2015) siblings of victims who perished in the World Trade Center terrorist attack of September 11, 2001.

7. Flu Incidence The Centers for Disease Control obtains current flu data by polling 3000 people this month.

8. Deer Encounters A state Fish and Game official investigates the size of the deer population by collecting data from reports of damage caused by vehicles hitting deer in each of the last ten years.

In Exercises 9–20, identify which of these types of sampling is used: random, systematic, convenience, stratified, or cluster.

 9. Aspirin Usage A medical student researches the prevalence of aspirin usage by questioning all of the patients who enter her clinic for treatment.

10. Health Survey Selection The Connecticut Department of Health obtains an alphabetical list of 2,537,243 adult Connecticut residents and constructs a pool of survey subjects by selecting every 1000th name on that list.

11. Telephone Polls In a Gallup poll of 1059 adults, the interview subjects were selected by using a computer to randomly generate telephone numbers that were then called.

12. Diet A dietitian has partitioned people into age categories of under 18, 18–49, 50–69, and over 69. She is surveying 200 members from each category.

13. Student Drinking Motivated by a student who died from binge drinking, the College of Newport conducts a study of student drinking by randomly selecting 10 different classes and interviewing all of the students in each of those classes.

14. Clinical Trial of Blood Treatment In phase two testing of a new drug designed to increase the red blood cell count, a researcher creates envelopes with the names and addresses of all treated subjects. She wants to increase the dosage in a subsample of 12 subjects, so she thoroughly mixes all of the envelopes in a bin, then pulls 12 of those envelopes to identify the subjects to be given the increased dosage.

15. Sobriety Checkpoint One of the authors observed a sobriety checkpoint at which every 5th driver was stopped and interviewed.

16. Orange You Sweet A farmer has 5267 orange trees. She randomly selects 20 trees, then picks all of the oranges on those trees and measures the sugar content in each orange.

17. Education and Health A researcher is studying the effect of education on health and conducts a survey of 150 randomly selected workers from each of these categories: Less than a high school degree; high school degree; more than a high school degree.

18 Anthropometrics A statistics student obtains height/weight data by interviewing family members.

19. Medical Research A Johns Hopkins University researcher surveys all cardiac patients in each of 30 randomly selected hospitals.

20. AIDS Survey In studying the incidence of AIDS, an epidemiologist is planning a survey in which 500 people will be randomly selected from each of these groups: unmarried adult men, married adult men, unmarried adult women, married adult women.

Exercises 21–26 relate to random samples and simple random samples.

21. Sampling Aspirin Tablets A pharmacist thoroughly mixes a container of 1000 Bufferin tablets, then scoops a sample of 50 tablets that are to be measured for the exact aspirin content. Does this sampling plan result in a random sample? Simple random sample? Explain.

22. Sampling Students A classroom consists of 30 students seated in six different rows, with five students in each row. The instructor rolls a die and the outcome is used to select a sample consisting of all students in a particular row. Does this sampling plan result in a random sample? Simple random sample? Explain.

23. Convenience Sample A nurse plans to obtain a sample of patients by selecting the first 36 patients that he encounters when he begins his work tomorrow. Does this sampling plan result in a random sample? Simple random sample? Explain.

24. Systematic Sample A quality control engineer selects every 100th thermometer that is produced by her company. Does this sampling plan result in a random sample? Simple random sample? Explain.

25. Stratified Sample SmithKline Beecham plans to conduct a marketing survey of 100 men and 100 women in Orange County, which consists of an equal number of men and women. Does this sampling plan result in a random sample? Simple random sample? Explain.

26. Type of Sampling Refer to Data Set 12 in Appendix B. After reviewing the gender and race composition of the sample, identify the type of sampling that is used (random, systematic, convenience, stratified, or cluster). What is it about the sample that enables you to identify the sampling type?

27. Cluster Sample A researcher randomly selects ten blocks in the Village of Newport, then asks all adult residents of the selected blocks whether they have ever had a blood transfusion. Does this sampling plan result in a random sample? Simple random sample? Explain.

28. Random Selection Among the 50 states, one state is randomly selected. Then, a statewide voter registration list is obtained and one name is randomly selected. Does this procedure result in a randomly selected voter?

29. Sample Design In "Cardiovascular Effects of Intravenous Triiodothyronine in Patients Undergoing Coronary Artery Bypass Graft Surgery" (*Journal of the American Medical Association,* Vol. 275, No. 9), the authors explain that patients were assigned to one of three groups: (1) a group treated with triiodothyronine, (2) a group treated with normal saline bolus and dopamine, and (3) a placebo group given normal saline. The authors summarize the sample design as a "prospective, randomized, double-blind, placebo-controlled trial." Describe the meaning of each of those terms in the context of this study.

30. Drivers with Cell Phones What are two major problems likely to be encountered in a prospective study in which some drivers have no cell phones while others are asked to use their cell phones while driving?

31. Motorcycle Helmets The Hawaii State Senate held hearings when it was considering a law requiring that motorcyclists wear helmets. Some motorcyclists testified that they had been in crashes in which helmets would not have been helpful. Which important group was not able to testify? (See "A Selection of Selection Anomalies" by Wainer, Palmer, and Bradlow in *Chance,* Vol. 11, No. 2.)

Review

This chapter presented some important basics. There were fundamental definitions, such as *sample* and *population*, along with some very basic principles. Section 1-2 discussed different types of data. Section 1-3 introduced important elements in the design of experiments. On completing this chapter, you should be able to do the following:

- Distinguish between a population and a sample and distinguish between a parameter and a statistic.
- Identify the level of measurement (nominal, ordinal, interval, ratio) of a set of data.

- Understand the importance of good experimental design, including the control of variable effects, replication, and randomization.

- Recognize the importance of good sampling methods in general, and recognize the importance of a *simple random sample* in particular. Understand that if sample data are not collected in an appropriate way, the data may be so completely useless that no amount of statistical analysis can salvage them.

Review Exercises

1. Sampling Shortly after the World Trade Center towers were destroyed by terrorists, America Online ran a poll of its Internet subscribers and asked this question: "Should the World Trade Center towers be rebuilt?" Among the 1,304,240 responses, 768,731 answered yes, 286,756 answered no, and 248,753 said that it was too soon to decide. Given that this sample is extremely large, can the responses be considered to be representative of the population of the United States? Explain.

2. Sampling Design You have been hired by Merck & Company, Inc. to conduct a survey to determine the percentage of adult Americans who use its Prilosec drug. Describe a procedure for obtaining a sample of each type: random, systematic, convenience, stratified, cluster.

3. Level of Measurement Identify the level of measurement (nominal, ordinal, interval, ratio) used in each of the following.
   a. The weights of people in a sample of airplane passengers
   b. A physician's descriptions of "abstains from alcohol, light drinker, moderate drinker, heavy drinker."
   c. Tree classifications of "oak, maple, elm."
   d. Bob, who is different in many ways, measures time in days, with 0 corresponding to his birth date. The day before his birth is -1, the day after his birth is $+1$, and so on. Bob has converted the dates of major historical events to his numbering system. What is the level of measurement of these numbers?

4. Salk Vaccine The clinical trial of a polio vaccine involved 200,000 children treated with the Salk vaccine and 200,000 other children given a placebo. Only 33 of the children in the Salk treatment group later developed polio, but 115 of the children in the placebo group later developed polio.
   a. Are the given values discrete or continuous?
   b. Identify the level of measurement (nominal, ordinal, interval, ratio) for the numbers 33 and 115.
   c. If 50 children from the treatment group are randomly selected and their average (mean) age is calculated, is the result a statistic or a parameter?
   d. What would be wrong with gauging opinions about this clinical trial by mailing a questionnaire that parents could complete and mail back?

5. Smokers Identify the type of sampling (random, systematic, convenience, stratified, cluster) used when a sample of people who smoke is obtained as described below. Then determine whether the sampling scheme is likely to result in a sample that is representative of the population of all people who smoke.
   a. A complete list of all smokers is somehow compiled and every 1000th name is selected.
   b. At the next New York City marathon, a survey is conducted of all runners.
   c. Fifty different grocery store owners are randomly selected, and all of their customers are surveyed.

    d. A computer file of all smokers is compiled and they are all numbered consecutively, then random numbers generated by computer are used to select a sample of 1000 smokers.

    e. All of the smokers' zip codes are collected, and 5 smokers are randomly selected from each zip code.

6. Design of Experiment You plan to conduct an experiment to test the effectiveness of Sleepeze, a new drug that is supposed to reduce insomnia. You will use a sample of subjects that are treated with the drug and another sample of subjects that are given a placebo.

    a. What is "blinding" and how might it be used in this experiment?

    b. Why is it important to use blinding in this experiment?

    c. What is a completely randomized design?

    d. What is a rigorously controlled design?

    e. What is replication, and why is it important?

Cumulative Review Exercises

The Cumulative Review Exercises in this book are designed to include topics from preceding chapters. For Chapters 2–13, the cumulative review exercises include topics from preceding chapters. The exercises given here relate to percentages and scientific notation.

In Exercises 1–6, answer the given questions that relate to percentages.

1. Percentages

    a. What is 57% of 1500?

    b. What is 26% of 950?

2. Percentages in a Gallup Poll

    a. In a Gallup poll, 52% of 1038 surveyed adults said that secondhand smoke is "very harmful." What is the actual number of adults who said that secondhand smoke is "very harmful"?

    b. Among the 1038 surveyed adults, 52 said that secondhand smoke is "not at all harmful." What is the percentage of people who chose "not at all harmful"?

3. Percentages in a Study of Lipitor

    a. In a study of the cholesterol drug Lipitor, 270 patients were given a placebo, and 19 of those 270 patients reported headaches. What percentage of this placebo group reported headaches?

    b. Among the 270 patients in the placebo group, 3.0% reported back pains. What is the actual number of patients who reported back pains?

4. Percentages in Campus Crime In a study of college campus crimes committed by students high on alcohol or drugs, a mail survey of 1875 students was conducted. A *USA Today* article noted, "Eight percent of the students responding anonymously say they've committed a campus crime. And 62% of that group say they did so under the influence of alcohol or drugs." Assuming that the number of students responding anonymously is 1875, how many actually committed a campus crime while under the influence of alcohol or drugs?

5. Percentages in the Media A *New York Times* editorial criticized a chart caption that described a dental rinse as one that "reduces plaque on teeth by over 300%." What is wrong with that statement?

6. Phony Data A researcher at the Sloan-Kettering Cancer Research Center was once criticized for falsifying data. Among his data were figures obtained from 6 groups of mice, with 20 individual mice in each group. These values were given for the percentage of successes in each group: 53%, 58%, 63%, 46%, 48%, 67%. What is the major flaw?

In Exercises 7–10, the given expressions are designed to yield results expressed in a form of scientific notation. For example, the computer-displayed result of 1.23E5 can be expressed as 123,000, and the result of 4.56E − 4 can be expressed as 0.000456. Use a calculator or computer to perform the indicated operation and express the result as an ordinary number that is not in scientific notation.

7. 0.95^{500} **8.** 8^{14} **9.** 9^{12} **10.** 0.25^{17}

Cooperative Group Activities

1. *In-class activity:* From the cafeteria, obtain 18 straws. Cut 6 of them in half, cut 6 of them into quarters, and leave the other 6 as they are. There should now be 42 straws of different lengths. Put them in a bag, mix them up, then select one straw, find its length, then replace it. Repeat this until 20 straws have been selected. Important: Select the straws without looking into the bag and select the first straw that is touched. Find the average (mean) of the sample of 20 straws. Now remove all of the straws and find the mean of the population. Did the sample provide an average that was close to the true population average? Why or why not?

2. *In-class activity:* Identify the problems with a televised report on CNN Headline News that included a comment that crime in the United States fell in the 1980s because of the growth of abortions in the 1970s, which resulted in fewer unwanted children.

Technology Project

The objective of this project is to introduce the technology resources that you will be using in your statistics course. Refer to Data Set 1 in Appendix B and use the 40 pulse rates of males. Using a statistics software package or a TI-83/84 Plus calculator, enter those 40 values, then obtain a printout of them.

from DATA *to* DECISION

Critical Thinking

The Swiss physician H. C. Lombard once compiled longevity data for different professions. He used death certificates that included name, age at death, and profession. He then proceeded to compute the average (mean) length of life for the different professions, and he found that students were lowest with a mean of only 20.7 years! (See "A Selection of Selection Anomalies" by Wainer, Palmer, and Bradlow in *Chance*, Volume 11, No. 2.) Similar results would be obtained if the same data were collected today in the United States.

Analyzing the Results

Is being a student really more dangerous than being a police officer, a taxicab driver, or a postal employee? Explain.

2

Describing, Exploring, and Comparing Data

Are nonsmoking people really affected by others who are smoking cigarettes, or is the effect of *secondhand smoke* a myth?

Data Set 5 in Appendix B includes some of the latest available data from the National Institutes of Health. The data, reproduced below in Table 2-1, were obtained as part of the National Health and Nutrition Examination Survey. The data values consist of the measured levels of serum cotinine (in ng/ml) in people selected as study subjects. (The data were rounded to the nearest whole number, so a value of zero does not necessarily indicate the total absence of cotinine. In fact, all of the original values were greater than zero.) Cotinine is a metabolite of nicotine, meaning that when nicotine is absorbed by the body, cotinine is produced. Because it is known that nicotine is absorbed through cigarette smoking, we have an effective way of measuring the presence of cigarette smoke indirectly through cotinine.

Here are important issues: Should nonsmokers be concerned about their health because they are in the presence of smokers? In recent years, many regulations have been enacted to restrict smoking in public places. Are those regulations justified based on reasons of health, or are they simply causing unnecessary hardship for smokers?

Critical Thinking: A visual comparison of the numbers in the three groups in Table 2-1 might provide some insight, but in this chapter we will present methods for gaining much greater understanding. We will be able to make intelligent and productive comparisons. We will learn techniques for describing, exploring, and comparing data sets such as the three groups of data in Table 2-1.

Table 2-1	Measured Cotinine Levels in Three Groups

Smoker: Subjects reporting tobacco use.

ETS: (Environmental Tobacco Smoke) Subjects are nonsmokers who are exposed to environmental tobacco smoke ("secondhand smoke") at home or work.

NOETS: (No Environmental Tobacco Smoke) Subjects are nonsmokers who are not exposed to environmental tobacco smoke at home or work. That is, the subjects do not smoke and are not exposed to secondhand smoke.

Smoker:	1	0	131	173	265	210	44	277	32	3
	35	112	477	289	227	103	222	149	313	491
	130	234	164	198	17	253	87	121	266	290
	123	167	250	245	48	86	284	1	208	173
ETS:	384	0	69	19	1	0	178	2	13	1
	4	0	543	17	1	0	51	0	197	3
	0	3	1	45	13	3	1	1	1	0
	0	551	2	1	1	1	0	74	1	241
NOETS:	0	0	0	0	0	0	0	0	0	0
	0	9	0	0	0	0	0	0	244	0
	1	0	0	0	90	1	0	309	0	0
	0	0	0	0	0	0	0	0	0	0

2-1 Overview

This chapter is extremely important because it presents the basic tools for measuring and describing different characteristics of a set of data. When describing, exploring, and comparing data sets, the following characteristics are usually extremely important.

Important Characteristics of Data

1. **Center:** A representative or average value that indicates where the middle of the data set is located.

2. **Variation:** A measure of the amount that the data values vary among themselves.

3. **Distribution:** The nature or shape of the distribution of the data (such as bell-shaped, uniform, or skewed).

4. **Outliers:** Sample values that lie very far away from the vast majority of the other sample values.

5. **Time:** Changing characteristics of the data over time.

Hint: To help remember the above five characteristics, use a mnemonic for their first letters of CVDOT, such as "**C**omputer **V**iruses **D**estroy **O**r **T**erminate."

Critical Thinking and Interpretation: Going Beyond Formulas

Technology has allowed us to enjoy this principle of modern statistics usage: It is not so important to memorize formulas or manually perform complex arithmetic calculations. Instead, we can focus on obtaining results by using some form of technology (calculator or software), then making practical sense of the results through critical thinking. Keep this in mind as you proceed through this chapter. For example, when studying the extremely important *standard deviation* in Section 2-5, try to see how the key formula serves as a measure of variation, then learn how to find values of standard deviations, but really work on *understanding* and *interpreting* values of standard deviations.

Although this chapter includes some detailed steps for important procedures, it is not necessary to master those steps in all cases. However, we recommend that in each case you perform a few manual calculations before using your calculator or computer. Your understanding will be enhanced, and you will acquire a better appreciation of the results obtained from the technology.

The methods of this chapter are often called methods of **descriptive statistics,** because the objective is to summarize or *describe* the important characteristics of a set of data. Later in this book we will use methods of **inferential statistics** when we use sample data to make inferences (or generalizations) about a population. With inferential statistics, we are making an inference that goes beyond the known data. Descriptive statistics and inferential statistics are the two general divisions of the subject of statistics, and this chapter deals with basic concepts of descriptive statistics.

2-2 Frequency Distributions

When working with large data sets, it is often helpful to organize and summarize the data by constructing a table that lists the different possible data values (either individually or by groups) along with the corresponding frequencies, which represent the number of times those values occur.

Definition

A **frequency distribution** lists data values (either individually or by groups of intervals), along with their corresponding frequencies (or counts).

Table 2-2 is a frequency distribution summarizing the measured cotinine levels of the 40 smokers listed in Table 2-1. The **frequency** for a particular class is the number of original values that fall into that class. For example, the first class in Table 2-2 has a frequency of 11, indicating that 11 of the original data values are between 0 and 99 inclusive.

We will first present some standard terms used in discussing frequency distributions, and then we will describe how to construct and interpret them.

Table 2-2 Frequency Distribution of Cotinine Levels of Smokers	
Serum Cotinine Level (ng/ml)	Frequency (Number of Smokers)
0–99	11
100–199	12
200–299	14
300–399	1
400–499	2

Definitions

Lower class limits are the smallest numbers that can belong to the different classes. (Table 2-2 has lower class limits of 0, 100, 200, 300, and 400.)

Upper class limits are the largest numbers that can belong to the different classes. (Table 2-2 has upper class limits of 99, 199, 299, 399, and 499.)

Class boundaries are the numbers used to separate classes, but without the gaps created by class limits. They are obtained as follows: Find the size of the gap between the upper class limit of one class and the lower class limit of the next class. Add half of that amount to each upper class limit to find the upper class boundaries; subtract half of that amount from each lower class limit to find the lower class boundaries. (Table 2-2 has gaps of exactly 1 unit, so 0.5 is added to the upper class limits and subtracted from the lower class limits. The first class has boundaries of −0.5 and 99.5, the second class has boundaries of 99.5 and 199.5, and so on. The complete list of boundaries used for all classes is −0.5, 99.5, 199.5, 299.5, 399.5, and 499.5.)

Class midpoints are the midpoints of the classes. (Table 2-2 has class midpoints of 49.5, 149.5, 249.5, 349.5, and 449.5.) Each class midpoint can be found by adding the lower class limit to the upper class limit and dividing the sum by 2.

Class width is the difference between two consecutive lower class limits or two consecutive lower class boundaries. (Table 2-2 uses a class width of 100.)

The definitions of class width and class boundaries are tricky. Be careful to avoid the easy mistake of making the class width the difference between the lower class limit and the upper class limit. See Table 2-2 and note that the class width is 100, not 99. You can simplify the process of finding class boundaries by understanding that they basically fill the gaps between classes by splitting the difference between the end of one class and the beginning of the next class.

The following procedure can be used to manually construct a frequency table, but many statistics software packages and some calculators can be used to largely automate that procedure.

Procedure for Constructing a Frequency Distribution

Frequency distributions are constructed for these reasons: (1) Large data sets can be summarized, (2) we can gain some insight into the nature of data, and (3) we have a basis for constructing important graphs (such as *histograms*, introduced in the next section). Many uses of technology allow us to automatically obtain frequency distributions without manually constructing them, but here is the basic procedure:

1. Decide on the number of classes you want.   The number of classes should be between 5 and 20, and the number you select might be affected by the convenience of using round numbers.

2. Calculate

$$\text{Class width} \approx \frac{(\textit{Maximum value}) - (\textit{Minimum value})}{\textit{Number of classes}}$$

Round this result to get a convenient number. (Usually round *up*.) You might need to change the number of classes, but the priority should be to use values that are easy to understand.

3. Starting point: Begin by choosing a number for the lower limit of the first class. Choose either the lowest data value or a convenient value that is a little smaller.

4. Using the lower limit of the first class and the class width, proceed to list the other lower class limits. (Add the class width to the starting point to get the second lower class limit. Add the class width to the second lower class limit to get the third, and so on.)

5. List the lower class limits in a vertical column and proceed to enter the upper class limits, which can be easily identified.

6. Go through the data set putting a tally in the appropriate class for each data value. Use the tally marks to find the total frequency for each class.

When constructing a frequency distribution, be sure that classes do not overlap so that each of the original values must belong to exactly one class. Include all classes, even those with a frequency of zero. Try to use the same width for all classes, although it is sometimes impossible to avoid open-ended intervals, such as "65 years or older."

Measuring Disobedience

How are data collected about something that doesn't seem to be measurable, such as people's level of disobedience? Psychologist Stanley Milgram devised the following experiment: A researcher instructed a volunteer subject to operate a control board that gave increasingly painful "electrical shocks" to a third person. Actually, no real shocks were given, and the third person was an actor. The volunteer began with 15 volts and was instructed to increase the shocks by increments of 15 volts. The disobedience level was the point at which the subject refused to increase the voltage. Surprisingly, two-thirds of the subjects obeyed orders even though the actor screamed and faked a heart attack.

EXAMPLE **Cotinine Levels of Smokers** Using the 40 cotinine levels for the *smokers* in Table 2-1, follow the above procedure to construct the frequency distribution shown in Table 2-2. Assume that you want 5 classes.

SOLUTION

Step 1:   Begin by selecting 5 as the number of desired classes.

Step 2:   Calculate the class width. In the following calculation, 98.2 is rounded up to 100, which is a more convenient number.

$$\text{Class width} \approx \frac{(\text{Maximum value}) - (\text{Minimum value})}{\text{Number of classes}}$$

$$= \frac{491 - 0}{5} = 98.2 \approx 100$$

Step 3:   We choose a starting point of 0, which is the lowest value in the list and is also a convenient number.

Step 4:   Add the class width of 100 to the starting point of 0 to determine that the second lower class limit is 100. Continue to add the class width of 100 to get the remaining lower class limits of 200, 300, and 400.

Step 5:   List the lower class limits vertically as shown in the margin. From this list, we can easily identify the corresponding upper class limits as 99, 199, 299, 399, and 499.

Step 6:   After identifying the lower and upper limits of each class, proceed to work through the data set by entering a tally mark for each value. When the tally marks are completed, add them to find the frequencies shown in Table 2-2.

0	–
100	–
200	–
300	–
400	–

Relative Frequency Distribution

An important variation of the basic frequency distribution uses **relative frequencies,** which are easily found by dividing each class frequency by the total of all frequencies. A **relative frequency distribution** includes the same class limits as a frequency distribution, but relative frequencies are used instead of actual frequencies. The relative frequencies are often expressed as percents.

$$\text{relative frequency} = \frac{\text{class frequency}}{\text{sum of all frequencies}}$$

In Table 2-3 the actual frequencies from Table 2–2 are replaced by the corresponding relative frequencies expressed as percents. The first class has a relative frequency of 11/40 = 0.275, or 27.5%, which is often rounded to 28%. The second class has a relative frequency of 12/40 = 0.3, or 30.0%, and so on. If constructed correctly, the sum of the relative frequencies should total 1 (or 100%), with some small discrepancies allowed for rounding errors. Because 27.5% was rounded to 28% and 2.5% was rounded to 3%, the sum of the relative frequencies in Table 2-3 is 101% instead of 100%.

Table 2-3

Relative Frequency Distribution of Cotinine Levels in Smokers

Serum Cotinine Level (ng/ml)	Relative Frequency
0–99	28%
100–199	30%
200–299	35%
300–399	3%
400–499	5%

Table 2-4

Cumulative Frequency Distribution of Cotinine Levels in Smokers

Serum Cotinine Level (ng/ml)	Cumulative Frequency
Less than 100	11
Less than 200	23
Less than 300	37
Less than 400	38
Less than 500	40

Table 2-5

Last Digits of Male Pulse Rates

Last Digit	Frequency
0	7
1	0
2	6
3	0
4	11
5	0
6	9
7	0
8	7
9	0

Because they use simple percentages, relative frequency distributions make it easier for us to understand the distribution of the data and to compare different sets of data.

Cumulative Frequency Distribution

Another variation of the standard frequency distribution is used when cumulative totals are desired. The **cumulative frequency** for a class is the sum of the frequencies for that class and all previous classes. Table 2-4 is the cumulative frequency distribution based on the frequency distribution of Table 2-2. Using the original frequencies of 11, 12, 14, 1, and 2, we add 11 + 12 to get the second cumulative frequency of 23, then we add 11 + 12 + 14 = 37 to get the third, and so on. See Table 2-4 and note that in addition to using cumulative frequencies, the class limits are replaced by "less than" expressions that describe the new range of values.

Critical Thinking: Interpreting Frequency Distributions

The transformation of raw data to a frequency distribution is typically a means to some greater end. The following examples illustrate how frequency distributions can be used to describe, explore, and compare data sets. (The following section shows how the construction of a frequency distribution is often the first step in the creation of a graph that visually depicts the nature of the distribution.)

EXAMPLE **Describing Data** Refer to Data Set 1 in Appendix B for the pulse rates of 40 randomly selected adult males. Table 2-5 summarizes the *last digits* of those pulse rates. If the pulse rates are measured by counting the number of heartbeats in one minute, we expect that those last digits should occur with frequencies that are roughly the same. But note that the frequency distribution shows that the last digits are all *even* numbers; there are *no* odd numbers present. This suggests that the pulse rates were not counted for one minute. Perhaps they were counted for 30 seconds and the values were then doubled. (Upon further examination of the *original* pulse rates, we can see that every original value is a multiple of four, suggesting that the number of heartbeats was counted for 15 seconds, then that count was multiplied by four.) It's fascinating to learn something about the method of data collection by simply describing some characteristics of the data.

EXAMPLE **Exploring Data** It is well known that adult men have heights that are generally greater than heights of adult women. We might expect that if we construct a table describing the frequency distribution of men and women combined, the frequencies would begin and end with low numbers (for exceptionally short or tall subjects). We might also expect that there would be a dip in frequency values near the middle, illustrating the differences in heights between men and women. For example, we might expect to see a pattern that starts low, then rises, then falls, then rises again, then falls, as in this pattern: 1, 2, 45, 675, 5247, 4364, 5773, 1448, 131, 3. However, see Table 2-6, which depicts real data obtained from the last national health survey. Note that the frequencies start low, then reach a *single* peak of 6364, then decrease! If we were to analyze heights of adult men and women combined by analyzing only the single frequency distribution, we would not see the fundamental difference between heights of men and women.

EXAMPLE Comparing Data Sets The Chapter Problem given at the beginning of this chapter includes data sets consisting of measured cotinine levels from smokers, nonsmokers exposed to tobacco smoke, and nonsmokers not exposed to tobacco smoke. Table 2-7 shows the relative frequencies for the three groups. By comparing those relative frequencies, it should be obvious that the frequency distribution for smokers is very different from the frequency distributions for the other two groups. Because the two groups of nonsmokers (exposed and not exposed) have such high frequency amounts for the first class, it might be helpful to further compare those data sets with a closer examination of their values.

Table 2-6

Heights of Adult Men and Women

Height (cm)	Frequency
110.0–119.9	1
120.0–129.9	2
130.0–139.9	45
140.0–149.9	675
150.0–159.9	4247
160.0–169.9	6364
170.0–179.9	4773
180.0–189.9	1448
190.0–199.9	131
200.0–209.9	3

Open-Ended Intervals The frequency distributions discussed in this section are tidy in the sense that they all involve classes with the same width. As we noted earlier, it is often necessary to use open-ended intervals, such as the age category of "65 years or older." It is often better to use such an open-ended interval that captures a relatively small proportion of the sample data than to use several classes (such as 65–74, 75–84, 85–94, 95–104), with each containing a tiny proportion of the sample data. However, an open-ended interval does introduce a vague element that might become a nuisance when trying to perform calculations or construct graphs such as those that are introduced in the next section.

Table 2-7   Cotinine Levels for Three Groups

Serum Cotinine Level (ng/ml)	Smokers	Nonsmokers Exposed to Smoke	Nonsmokers Not Exposed to Smoke
0–99	28%	85%	95%
100–199	30%	5%	0%
200–299	35%	3%	3%
300–399	3%	3%	3%
400–499	5%	0%	0%
500–599	0%	5%	0%

2-2 Exercises

In Exercises 1–4, identify the class width, class midpoints, and class boundaries for the given frequency distribution based on Data Set 1 in Appendix B.

1. Systolic Blood Pressure of Men	Frequency
90–99	1
100–109	4
110–119	17
120–129	12
130–139	5
140–149	0
150–159	1

2. Systolic Blood Pressure of Women	Frequency
80–99	9
100–119	24
120–139	5
140–159	1
160–179	0
180–199	1

3. Cholesterol of Men	Frequency
0–199	13
200–399	11
400–599	5
600–799	8
800–999	2
1000–1199	0
1200–1399	1

4. Body Mass Index of Women	Frequency
15.0–20.9	10
21.0–26.9	15
27.0–32.9	11
33.0–38.9	2
39.0–44.9	2

In Exercises 5–8, construct the relative frequency distribution that corresponds to the frequency distribution in the exercise indicated.

5. Exercise 1   **6.** Exercise 2   **7.** Exercise 3   **8.** Exercise 4

In Exercises 9–12, construct the cumulative frequency distribution that corresponds to the frequency distribution in the exercise indicated.

9. Exercise 1   **10.** Exercise 2   **11.** Exercise 3   **12.** Exercise 4

13. Bears Refer to Data Set 6 in Appendix B and construct a frequency distribution of the weights of bears. Use 11 classes beginning with a lower class limit of 0 and use a class width of 50 lb.

14. Body Temperatures Refer to Data Set 2 in Appendix B and construct a frequency distribution of the body temperatures for midnight on the second day. Use 8 classes beginning with a lower class limit of 96.5 and use a class width of 0.4°F. Describe two different notable features of the result.

15. Head Circumferences Refer to Data Set 4 in Appendix B. Construct a frequency distribution for the head circumferences of baby boys and construct a separate frequency distribution for the head circumferences of baby girls. In both cases, use the classes of 34.0–35.9, 36.0–37.9, and so on. Then compare the results and determine whether there appears to be a significant difference between the two genders.

16. Very Poplar Trees Data Set 9 in Appendix B includes data from a sample of poplar trees. Using the 40 values of the dry weights (in kilograms) of the poplar trees from year 2, construct a frequency distribution summarizing those weights. Use classes of 0.00–0.49, 0.50–0.99, and so on.

17. Yeast Cell Counts Refer to Data Set 11 in Appendix B. The listed yeast cell counts are whole numbers between 1 and 12 inclusive, so it is natural to use a frequency distribution with the first class representing 1, the second class representing 2, and so on. Construct the frequency distribution. What is the most common cell count?

*18. Number of Classes   In constructing a frequency distribution, Sturges' guideline suggests that the ideal number of classes can be approximated by $1 + (\log n)/(\log 2)$, where n is the number of data values. Use this guideline to complete the table for determining the ideal number of classes.

Table for Exercise 18

Number of Values	Ideal Number of Classes
16–22	5
23–45	6
	7
	8
	9
	10
	11
	12

2-3   Visualizing Data

The main objective of this chapter is to learn basic techniques for investigating the important characteristics of data sets: center, variation, distribution, outliers, and changes over time ("CVDOT"). In Section 2-2 we introduced frequency distributions as a tool for describing, exploring, or comparing *distributions* of data sets. In this section we continue the study of distributions by introducing graphs that are pictures of distributions. As you read through this section, keep in mind that the objective is not simply to construct graphs, but rather to learn something about data sets—that is, to understand the nature of their distributions.

Histograms

Among the different types of graphs presented in this section, the histogram is particularly important.

> **Definition**
>
> A **histogram** is a bar graph in which the horizontal scale represents classes of data values and the vertical scale represents frequencies. The heights of the bars correspond to the frequency values, and the bars are drawn adjacent to each other (without gaps).

We can construct a histogram after we have first completed a frequency distribution table for a data set. The cotinine levels of smokers are depicted in the histogram of Figure 2-1, which corresponds directly to the frequency distribution in Table 2-2 given in the preceding section. Each bar of the histogram is marked with its lower class boundary at the left and its upper class boundary at the right. Instead of using class boundaries along the horizontal scale, it is often more practical to use class midpoint values centered below their corresponding bars. The use of class midpoint values is very common in software packages that automatically generate histograms.

Before constructing a histogram from a completed frequency distribution, we must give some thought to the scales used on the vertical and horizontal axes. The maximum frequency (or the next higher convenient number) should suggest a value for the top of the vertical scale; 0 should be at the bottom. In Figure 2-1 we designed the vertical scale to run from 0 to 15. The horizontal scale should be subdivided in a way that allows all the classes to fit well. Ideally, we should try to follow the rule of

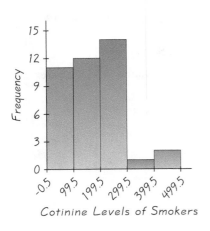

FIGURE 2-1 **Histogram**

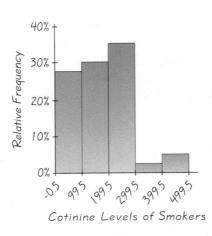

FIGURE 2-2 **Relative Frequency Histogram**

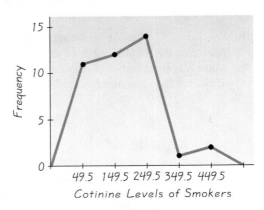

FIGURE 2-3 **Frequency Polygon**

thumb that the vertical height of the histogram should be about three-fourths of the total width. Both axes should be clearly labeled.

Interpreting a Histogram   Remember that the objective is not simply to construct a histogram, but rather to learn something about the data. Analyze the histogram to see what can be learned about the center of the data, the variation (which will be discussed at length in Section 2-5), the shape of the distribution, and whether there are any outliers (values far away from the other values). The histogram is not suitable for determining whether there are changes over time. Examining Figure 2-1, we see that the histogram is centered around 175, the values vary from around zero to 500, and the shape of the distribution is heavier on the left.

Relative Frequency Histogram

A **relative frequency histogram** has the same shape and horizontal scale as a histogram, but the vertical scale is marked with relative frequencies instead of actual frequencies, as in Figure 2-2.

Frequency Polygon

A **frequency polygon** uses line segments connected to points located directly above class midpoint values. See Figure 2-3 for the frequency polygon corresponding to Table 2-2. The heights of the points correspond to the class frequencies, and the line segments are extended to the right and left so that the graph begins and ends on the horizontal axis.

Ogive

An **ogive** (pronounced "oh-jive") is a line graph that depicts *cumulative* frequencies, just as the cumulative frequency distribution (see Table 2-4 in the preceding section) lists cumulative frequencies. Figure 2-4 is an ogive corresponding to

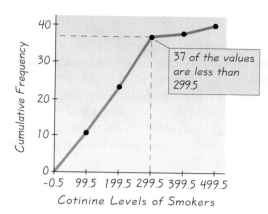

FIGURE 2-4 **Ogive**

Table 2-4. Note that the ogive uses class boundaries along the horizontal scale, and the graph begins with the lower boundary of the first class and ends with the upper boundary of the last class. Ogives are useful for determining the number of values below some particular value. For example, see Figure 2-4 where it is shown that 37 of the cotinine values are less than 299.5.

Dotplots

A **dotplot** consists of a graph in which each data value is plotted as a point (or dot) along a scale of values. Dots representing equal values are stacked. See Figure 2-5, which includes two separate dotplots generated by Minitab with the same scale. The top dotplot representing fertilizer and irrigation shows dots corresponding to values of 3.2, 3.9, 4.4, 5.4, 6.3, 6.3, 6.4, 6.7, 6.8, 6.8, 6.9, 6.9, 6.9, 7.1, 7.3, 7.3, 7.3, 7.6, 7.7, and 8.0. The values are heights (in meters) of poplar trees. Note that the two values of 6.3 m are represented by the two dots that are stacked vertically. The two values of 6.8 m are also stacked, as are the three heights of 6.9 m. The top dotplot represents 20 poplar trees treated with both fertilizer and irrigation, and the bottom dotplot represents 20 similar trees treated with irrigation only. (The data are from a study conducted by researchers at Pennsylvania State University; the data were obtained from Minitab, Inc.) We can see that the inclusion of fertilizer as part of the treatment appears to have the effect of resulting in trees that tend to be somewhat taller.

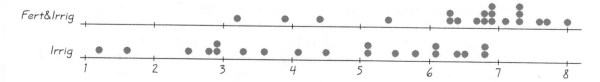

FIGURE 2-5 **Dotplots of Poplar Tree Heights**

Stem-and-Leaf Plots

A **stem-and-leaf plot** (or stemplot) represents data by separating each value into two parts: the stem (such as the leftmost digit) and the leaf (such as the rightmost digit). The illustration below shows a stem-and-leaf plot for the same heights (in meters) of the poplar trees represented in the top dotplot from the preceding illustration. It is easy to see how the first height of 3.2 m is separated into its stem of 3 and leaf of 2. Each of the remaining values is broken up in a similar way. The leaves are arranged in increasing order, not necessarily the order in which they occur in the original list.

Stem-and-Leaf Plot

Stem (units)	Leaves (tenths)	
3.	29	← Values are 3.2 and 3.9.
4.	4	
5.	4	
6.	334788999	
7.	133367	
8.	0	← Value is 8.0.

By turning the page on its side, we can see a distribution of these data. A great advantage of the stem-and-leaf plot is that we can see the distribution of data and yet retain all the information in the original list. If necessary, we could reconstruct the original list of values. Another advantage is that construction of a stem-and-leaf plot is a quick and easy way to *sort* data (arrange them in order), and sorting is required for some statistical procedures (such as finding medians, percentiles, or ranks).

The rows of digits in a stem-and-leaf plot are similar in nature to the bars in a histogram. One of the guidelines for constructing histograms is that the number of classes should be between 5 and 20, and the same guideline applies to stem-and-leaf plots for the same reasons. Better stem-and-leaf plots are often obtained by first rounding the original data values. Also, stem-and-leaf plots can be *expanded* to include more rows and can be *condensed* to include fewer rows (by combining stems). For example, the above stem-and-leaf plot can be expanded by subdividing rows into those with leaves having digits of 0 through 4 and those with digits 5 through 9.

Pareto Charts

In a recent year, there were 13,800 accidental deaths among United States residents between the ages of 15 and 24 (based on data from the National Safety Council). Here is the breakdown by category: firearms (150), poison (870), motor vehicle (10,500), fires and burns (240), drowning (700), falls (210), and other causes (1130). Although the preceding sentence correctly describes the data, better understanding can be achieved through an effective graph. A graph suitable for this data set is a **Pareto chart,** which is a bar graph for qualitative data, with the bars arranged in order according to frequencies. As in histograms, vertical scales in Pareto charts can represent frequencies or relative frequencies. The tallest bar is

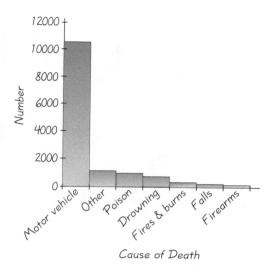

FIGURE 2-6 **Pareto Chart of Causes of Accidental Deaths Among U.S. Residents Aged 15–24**

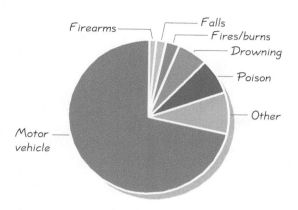

FIGURE 2-7 **Pie Chart of Types of Accidental Deaths Among U.S. Residents Aged 15–24**

at the left, and the smaller bars are farther to the right. By arranging the bars in order of frequency, the Pareto chart focuses attention on the more important categories. Figure 2-6 is a Pareto chart clearly showing that the category of motor vehicle accidents is most serious by far.

Pie Charts

Pie charts are also used to visually depict qualitative data. Figure 2-7 is an example of a **pie chart,** which is a graph depicting qualitative data as slices of a pie. Figure 2-7 represents the same data as Figure 2-6. Construction of a pie chart involves slicing up the pie into the proper proportions. The category of motor vehicles represents 76% of the total, so the wedge representing motor vehicles should be 76% of the total (with a central angle of $0.76 \times 360° = 274°$).

The Pareto chart (Figure 2-6) and the pie chart (Figure 2-7) depict the same data in different ways, but a comparison will probably show that the Pareto chart does a better job of showing the relative sizes of the different categories. Pie charts are not recommended as an effective tool for representing data.

Scatter Diagrams

A **scatter diagram** (or **scatterplot**) is a plot of paired (x, y) data with a horizontal x-axis and a vertical y-axis. The data are paired in a way that matches each value from one data set with a corresponding value from a second data set. To manually construct a scatter diagram, construct a horizontal axis for the values of the first variable, construct a vertical axis for the values of the second variable, then plot the points. The pattern of the plotted points is often helpful in determining whether there is some relationship between the two variables. (This issue is discussed at

FIGURE 2-8
Scatter Diagram

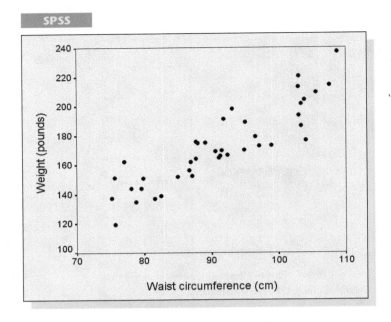

length when the topic of correlation is considered in Section 9-2.) Using the weight (in lb) and waist circumference (in cm) for the males in Data Set 1 of Appendix B, we used SPSS to generate the scatter diagram shown here (Figure 2-8). On the basis of that graph, there does appear to be a relationship between weight and waist circumference, as shown by the pattern of the points.

Time-Series Graph

Time-series data are data that have been collected at different points in time. For example, Figure 2-9 shows the numbers of manatee deaths resulting from power-boat encounters over the past decade (based on data from the Florida Marine Research Institute). We can see that for this time period, there is a trend of increasing values. It is often critically important to know when population values change over time. Companies have gone bankrupt because they failed to monitor the

FIGURE 2-9 Manatee Deaths from Powerboat Encounters

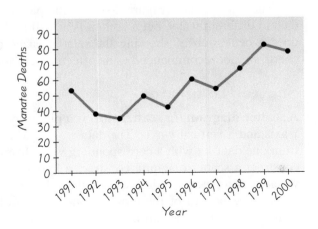

quality of their goods or services and incorrectly believed that they were dealing with stable data. They did not realize that their products were becoming seriously defective as important population characteristics were changing.

Other Graphs

Numerous pictorial displays other than the ones just described can be used to represent data dramatically and effectively. In Section 2-7 we present boxplots, which are very useful for revealing the spread of data. Pictographs depict data by using pictures of objects, such as soldiers, tanks, airplanes, stacks of coins, or moneybags.

A notable graph of historical importance is one developed by the world's most famous nurse, Florence Nightingale. This graph, shown in Figure 2-10, is particularly interesting because it actually saved lives when Nightingale used it to convince British officials that military hospitals needed to improve sanitary conditions, treatment, and supplies. It is drawn somewhat like a pie chart, except that the central angles are all the same and different radii are used to show changes in the numbers of deaths each month. The outermost regions of Figure 2-10 represent deaths due to preventable diseases, the innermost regions represent deaths from wounds, and the middle regions represent deaths from other causes.

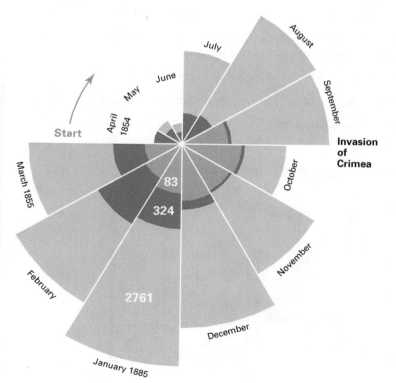

FIGURE 2-10 Deaths in British Military Hospitals During the Crimean War

Outermost regions, deaths due to preventable diseases
Middle regions, deaths from causes other than diseases or wounds
Innermost regions, deaths from wounds

Conclusion

The effectiveness of Florence Nightingale's graph illustrates well this important point: A graph is not in itself an end result; it is a tool for describing, exploring, and comparing data, as described below.

Describing data: In a histogram, for example, consider center, variation, distribution, and outliers (CVDOT without the last element of time). What is the approximate value of the center of the distribution, and what is the approximate range of values? Consider the overall shape of the distribution. Are the values evenly distributed? Is the distribution skewed (lopsided) to the right or left? Does the distribution peak in the middle? Identify any extreme values and any other notable characteristics.

Exploring data: Look for features of the graph that reveal some useful and/or interesting characteristics of the data set. In Figure 2-10, for example, we see that more soldiers were dying from inadequate hospital care than were dying from battle wounds.

Comparing data: Construct similar graphs that make it easy to compare data sets. For example, see the dotplots included in this section to see that the poplar trees treated with fertilizer and irrigation tended to be taller than those treated with irrigation only.

2-3 Exercises

In Exercises 1–4, answer the questions by referring to the SPSS-generated histogram given below. The histogram represents the lengths (mm) of cuckoo eggs found in the nests of other birds (based on data from O. M. Latter and Data and Story Library).

1. **Center** What is the approximate value of the center? That is, what egg length appears to be near the center of all lengths included in the graph?

2. **Variation** What are the lowest and highest possible egg lengths?

3. **Percentage** What percentage of the 120 egg lengths are less than 21.125 mm?

4. **Class Width** What is the class width?

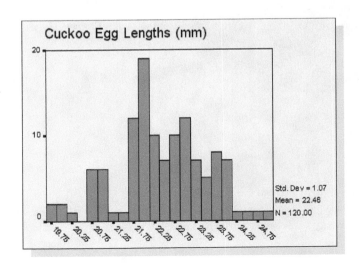

In Exercises 5 and 6, refer to the accompanying pie chart of blood groups for a large sample of people (based on data from the Greater New York Blood Program).

5. Interpreting Pie Chart What is the approximate percentage of people with Group A blood? Assuming that the pie chart is based on a sample of 500 people, approximately how many of those 500 people have Group A blood?

6. Interpreting Pie Chart What is the approximate percentage of people with Group B blood? Assuming that the pie chart is based on a sample of 500 people, approximately how many of those 500 people have Group B blood?

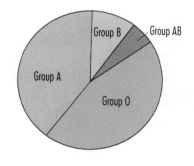

7. Bears Exercise 13 in Section 2-2 referred to Data Set 6 in Appendix B. Using the frequency distribution of the weights of bears (with 11 classes beginning with a lower class limit of 0 and a class width of 50 lb), construct the corresponding histogram. What is the approximate weight that is at the center?

8. Body Temperatures Exercise 14 in Section 2-2 referred to Data Set 2 in Appendix B. Using the frequency distribution of the body temperatures for midnight on the second day (with 8 classes beginning with a lower class limit of 96.5 and a class width of 0.4°F), construct the corresponding histogram. What does the distribution suggest about the common belief that the average body temperature is 98.6°F? If the subjects are randomly selected, the distribution should be approximately bell-shaped. Is it?

9. Yeast Cell Counts Refer to Data Set 11 in Appendix B and construct a histogram. Some statistical procedures discussed later in the book require a distribution with a shape that is approximately bell-shaped. Does this distribution appear to meet that requirement?

In Exercises 10 and 11, use graphs to compare the two indicated data sets.

10. Head Circumferences Exercise 15 in Section 2-2 referred to Data Set 4 in Appendix B. Using the frequency distributions of the head circumferences of baby boys and baby girls, use the same set of axes to construct the corresponding frequency polygons, then compare the results to determine whether there appears to be a significant difference between the two genders.

11. Poplar Trees Figure 2-5 shows two dotplots of tree heights. Those dotplots suggest that when poplar trees are treated with fertilizer and irrigation, they tend to be taller than the poplar trees treated with irrigation only. Use any suitable graphs to compare the heights of poplar trees in a control group with no treatment to poplar trees in a group treated with irrigation only. The heights (in meters) are listed below.

Control Group: 3.2 1.9 3.6 4.6 4.1 5.1 5.5 4.8 2.9 4.9
                  3.9 6.9 6.8 6.0 6.9 6.5 5.5 5.5 4.1 6.3

Irrigation Treatment: 4.1 2.5 4.5 3.6 2.8 5.1 2.9 5.1 1.2 1.6
                       6.4 6.8 6.5 6.8 5.8 2.9 5.5 3.3 6.1 6.1

In Exercises 12 and 13, list the original data represented by the given stem-and-leaf plots.

12.

Stem (tens)	Leaves (units)
20	0005
21	69999
22	2233333
23	
24	1177

13.

Stem (hundreds)	Leaves (tens and units)
50	12 12 12 55
51	
52	00 00 00 00
53	27 27 35
54	72

In Exercises 14 and 15, construct the dotplot for the data represented by the stem-and-leaf plot in the given exercise.

14. Exercise 12   **15.** Exercise 13

16. Stem-and-Leaf Plot Refer to Exercise 11 and construct a stem-and-leaf plot using the heights of the 20 trees treated with irrigation.

In Exercises 17–20, use the given paired data from Appendix B to construct a scatter diagram.

17. Parent/Child Height In Data Set 3, use the mother's height for the horizontal scale and use the child's height for the vertical scale. Determine whether there appears to be a relationship between the height of a mother and the height of her child. If so, describe the relationship.

18. Bear Neck/Weight In Data Set 6, use the distances around bear necks for the horizontal scale and use the bear weights for the vertical scale. Based on the result, what is the relationship between a bear's neck size and its weight?

19. Sepal/Petal Lengths of Irises In Data Set 7, use the sepal lengths and petal lengths of the irises listed in the class of Setosa. Determine whether there appears to be a relationship between the sepal lengths and petal lengths.

20. Red and White Blood Counts In Data Set 10, use the white blood cell counts and the red blood cell counts. Is there a relationship between them?

2-4 Measures of Center

The main objective of this chapter is to master basic tools for measuring and describing different characteristics of a set of data. In Section 2-1 we noted that when describing, exploring, and comparing data sets, these characteristics are usually extremely important: center, variation, distribution, outliers, and changes over time. The mnemonic of CVDOT (*C*omputer *V*iruses *D*estroy *O*r *T*erminate) is helpful for remembering those characteristics. In Sections 2-2 and 2-3 we saw that frequency distributions and graphs, such as histograms, are helpful in investigating distribution. The focus of this section is the characteristic of *center*.

Part 1 of this section includes core concepts that should be understood before considering Part 2.

PART 1   Basic Concepts

Definition

A **measure of center** is a value at the center or middle of a data set.

There are several different ways to determine the center, so we have different definitions of measures of center, including the mean, median, mode, and midrange. We begin with the mean.

Mean

The (arithmetic) mean is generally the most important of all numerical measurements used to describe data, and it is what most people call an *average*.

Definition

The **arithmetic mean** of a set of values is the measure of center found by adding the values and dividing the total by the number of values. This measure of center will be used often throughout the remainder of this text, and it will be referred to simply as the **mean.**

This definition can be expressed as Formula 2-1, which uses the Greek letter Σ (uppercase Greek sigma) to indicate that the data values should be added. That is, Σx represents the sum of all data values. The symbol n denotes the **sample size,** which is the number of values in the data set.

Formula 2-1

$$\text{Mean} = \frac{\Sigma x}{n}$$

The mean is denoted by $\bar{x}$ (pronounced "x-bar") if the data set is a *sample* from a larger population; if all values of the population are used, then we denote the mean by μ (lowercase Greek mu). (Sample statistics are usually represented by English letters, such as $\bar{x}$, and population parameters are usually represented by Greek letters, such as μ.)

Notation

Σ	denotes the *addition* of a set of values.
x	is the *variable* usually used to represent the individual data values.
n	represents the *number of values* in a *sample*.
N	represents the *number of values* in a *population*.
$\bar{x} = \dfrac{\Sigma x}{n}$	is the mean of a set of *sample* values.
$\mu = \dfrac{\Sigma x}{N}$	is the mean of all values in a *population*.

Class Size Paradox

There are at least two ways to obtain the mean class size, and they can have very different results. At one college, if we take the numbers of students in 737 classes, we get a mean of 40 students. But if we were to compile a list of the class sizes for each student and use this list, we would get a mean class size of 147. This large discrepancy is due to the fact that there are many students in large classes, while there are few students in small classes. Without changing the number of classes or faculty, we could reduce the mean class size experienced by students by making all classes about the same size. This would also improve attendance, which is better in smaller classes.

EXAMPLE Monitoring Lead in Air Lead is known to have some serious adverse affects on health. Listed below are measured amounts of lead (in micrograms per cubic meter, or $\mu\text{g/m}^3$) in the air. The Environmental Protection Agency has established an air quality standard for lead: a maximum of 1.5 $\mu\text{g/m}^3$. The measurements shown below were recorded at Building 5 of the World Trade Center site on different days immediately following the destruction caused by the terrorist attacks of September 11, 2001. Find the mean for this sample of measured levels of lead in the air.

5.40   1.10   0.42   0.73   0.48   1.10

continued

SOLUTION   The mean is computed by using Formula 2-1. First add the values, then divide by the number of values:

$$\bar{x} = \frac{\Sigma x}{n} = \frac{5.40 + 1.10 + 0.42 + 0.73 + 0.48 + 1.10}{6} = \frac{9.23}{6} = 1.538$$

The mean lead level is 1.538 $\mu g/m^3$. Apart from the value of the mean, it is also notable that the data set includes one value (5.40) that is very far away from the others. It would be wise to investigate such an "outlier." In this case, the lead level of 5.40 $\mu g/m^3$ was measured the day after the collapse of the two World Trade Center towers, and there were excessive levels of dust and smoke. Also, some of the lead may have come from the emissions from the large number of vehicles that rushed to the site. These factors provide a reasonable explanation for such an extreme value.

Median

One disadvantage of the mean is that it is sensitive to every value, so one exceptional value can affect the mean dramatically. The median largely overcomes that disadvantage. The median can be thought of as a "middle value" in the sense that about half of the values in a data set are below the median and half are above it. The following definition is more specific.

Definition

The **median** of a data set is the measure of center that is the *middle value* when the original data values are arranged in order of increasing (or decreasing) magnitude. The median is often denoted by $\tilde{x}$ (pronounced "x-tilde").

To find the median, first sort the values (arrange them in order), then follow one of these two procedures:

1. If the number of values is odd, the median is the number located in the exact middle of the list.
2. If the number of values is even, the median is found by computing the mean of the two middle numbers.

EXAMPLE Monitoring Lead in Air Listed below are measured amounts of lead (in $\mu g/m^3$) in the air. Find the median for this sample.

$$5.40 \quad 1.10 \quad 0.42 \quad 0.73 \quad 0.48 \quad 1.10$$

SOLUTION   First sort the values by arranging them in order:

$$0.42 \quad 0.48 \quad 0.73 \quad 1.10 \quad 1.10 \quad 5.40$$

Because the number of values is an even number (6), the median is found by computing the mean of the two middle values of 0.73 and 1.10.

$$\text{Median} = \frac{0.73 + 1.10}{2} = \frac{1.83}{2} = 0.915$$

Because the number of values is an even number (6), the median is found by computing the mean of the two middle numbers (0.73 and 1.10), so the median is 0.915 $\mu g/m^3$. Note that the median is very different from the mean of 1.538 $\mu g/m^3$ that was found from the same set of sample data in the preceding example. The reason for this large discrepancy is the effect that 5.40 had on the mean. If this extreme value were reduced to 1.20, the mean would drop from 1.538 $\mu g/m^3$ to 0.838 $\mu g/m^3$, but the median would not change.

EXAMPLE **Monitoring Lead in Air** Repeat the preceding example after including the measurement of 0.66 $\mu g/m^3$ recorded on another day. That is, find the median of these lead measurements:

5.40   1.10   0.42   0.73   0.48   1.10   0.66

SOLUTION   First arrange the values in order:

0.42   0.48   0.66   0.73   1.10   1.10   5.40

Because the number of values is an odd number (7), the median is the value in the exact middle of the sorted list: 0.73 $\mu g/m^3$.

After studying the preceding two examples, the procedure for finding the median should be clear. Also, it should be clear that the mean is dramatically affected by extreme values, whereas the median is not dramatically affected. Because the median is not so sensitive to extreme values, it is often used for data sets with a relatively small number of extreme values. For example, the U.S. Census Bureau recently reported that the *median* household income is $36,078 annually. The median was used because there is a small number of households with really high incomes.

Mode

Definition

The **mode** of a data set is the value that occurs *most frequently*.

- When two values occur with the same greatest frequency, each one is a mode and the data set is **bimodal.**

- When more than two values occur with the same greatest frequency, each is a mode and the data set is said to be **multimodal.**
- When no value is repeated, we say that there is no mode.

EXAMPLE **Find the modes of the following data sets.**

a. 5.40   1.10   0.42   0.73   0.48   1.10

b. 27   27   27   55   55   55   88   88   99

c. 1   2   3   6   7   8   9   10

continued

SOLUTION

a. The number 1.10 is the mode because it is the value that occurs most often.

b. The numbers 27 and 55 are both modes because they occur with the same greatest frequency. This data set is bimodal because it has two modes.

c. There is no mode because no value is repeated.

In reality, the mode isn't used much with numerical data. But among the different measures of center we are considering, the mode is the only one that can be used with data at the nominal level of measurement. (Recall that the nominal level of measurement applies to data that consist of names, labels, or categories only.)

Midrange

Definition

The **midrange** is the measure of center that is the value midway between the highest and lowest values in the original data set. It is found by adding the highest data value to the lowest data value and then dividing the sum by 2, as in the following formula.

$$\text{midrange} = \frac{\text{maximum value} + \text{minimum value}}{2}$$

The midrange is rarely used. Because it uses only the maximum and minimum values, it is too sensitive to those extremes. However, the midrange does have three redeeming features: (1) It is easy to compute; (2) it helps to reinforce the important point that there are several different ways to define the center of a data set; (3) it is sometimes incorrectly used for the median, so confusion can be reduced by clearly defining the midrange along with the median.

EXAMPLE **Monitoring Lead in Air** Listed below are measured amounts of lead (in $\mu g/m^3$) in the air from the site of the World Trade Center on different days after September 11, 2001. Find the midrange for this sample:

> 5.40   1.10   0.42   0.73   0.48   1.10

SOLUTION   The midrange is found as follows:

$$\frac{\text{maximum value} + \text{minimum value}}{2} = \frac{5.40 + 0.42}{2} = 2.910$$

The midrange is 2.910 $\mu g/m^3$.

Unfortunately, the term *average* is sometimes used for any measure of center and is sometimes used for the mean. Because of this ambiguity, the term *average*

Table 2-8	Comparison of Cotinine Levels in Smokers, Nonsmokers Exposed to Environmental Tobacco Smoke (ETS), and Nonsmokers Not Exposed to Environmental Tobacco Smoke (NOETS)		
	Smokers	ETS	NOETS
Mean	172.5	60.6	16.4
Median	170.0	1.5	0.0
Mode	1 and 173	1	0
Midrange	245.5	275.5	154.5

should not be used when referring to a particular measure of center. Instead, we should use the specific term, such as mean, median, mode, or midrange.

In the spirit of describing, exploring, and comparing data, we provide Table 2-8, which summarizes the different measures of center for the cotinine levels listed in Table 2-1 in the Chapter Problem. Recall that cotinine is a metabolite of nicotine, so that when nicotine is absorbed by the body, cotinine is produced. A comparison of the measures of center suggests that cotinine levels are highest in smokers. Also, the cotinine levels in nonsmokers exposed to smoke are higher than nonsmokers not exposed. This suggests that "secondhand smoke" does have an effect. There are methods for determining whether such apparent differences are statistically significant, and we will consider some of those methods later in this book.

Round-Off Rule

A simple rule for rounding answers is this:

> **Carry one more decimal place than is present in the original set of values.**

When applying this rule, round only the final answer, *not intermediate values that occur during calculations.* Thus the mean of 2, 3, 5, is 3.333333 . . . , which is rounded to 3.3. Because the original values are whole numbers, we rounded to the nearest tenth. As another example, the mean of 80.4 and 80.6 is 80.50 (one more decimal place than was used for the original values).

PART 2   Beyond the Basics

Weighted Mean

In some cases, the values vary in their degree of importance, so we may want to weight them accordingly. We can then proceed to compute a **weighted mean** of

Just an Average Guy

Men's Health magazine published statistics describing the "average guy," who is 34.4 years old, weighs 175 pounds, is about 5 ft 10 in. tall, and is named Mike Smith. The age, weight, and height are all mean values, but the name of Mike Smith is the mode that corresponds to the most common first and last names. Other notable statistics: The average guy sleeps about 6.9 hours each night, drinks about 3.3 cups of coffee each day, and consumes 1.2 alcoholic drinks per day. He earns about $36,100 per year, and has debts of $2563 from two credit cards. He has banked savings of $3100.

the x values, which is a mean computed with the different values assigned different weights denoted by w in Formula 2-2.

Formula 2-2   Weighted mean: $\bar{x} = \dfrac{\Sigma(w \cdot x)}{\Sigma w}$

Formula 2-2 tells us to multiply each weight w by the corresponding value x, then add the products, then divide that total by the sum of the weights w. For example, suppose we need a mean of 3 test scores (85, 90, 75), but the first test counts for 20%, the second test counts for 30%, and the third test counts for 50% of the final grade. We can assign weights of 20, 30, and 50 to the test scores, then proceed to calculate the mean by using Formula 2-2 with x = 85, 90, 75 and the corresponding weights of w = 20, 30, 50, as shown here:

$$\bar{x} = \frac{\Sigma(w \cdot x)}{\Sigma w}$$
$$= \frac{(20 \times 85) + (30 \times 90) + (50 \times 75)}{20 + 30 + 50} = \frac{8150}{100} = 81.5$$

As another example, college grade-point averages can be computed by assigning each letter grade the appropriate number of points (A = 4, B = 3, etc.), then assigning to each number a weight equal to the number of credit hours. Again, Formula 2-2 can be used to compute the grade-point average.

The Best Measure of Center

So far, we have considered the mean, median, mode, and midrange as measures of center. Which one of these is best? Unfortunately, there is no single best answer to that question because there are no objective criteria for determining the most representative measure for all data sets. The different measures of center have different advantages and disadvantages, some of which are summarized in Table 2-9. An important advantage of the mean is that it takes every value into account, but an important disadvantage is that it is sometimes dramatically affected by a few extreme values. This disadvantage can be overcome by using a trimmed mean, as described in Exercise 20.

Skewness

A comparison of the mean, median, and mode can reveal information about the characteristic of skewness, defined below and illustrated in Figure 2-11.

FIGURE 2-11 **Skewness**

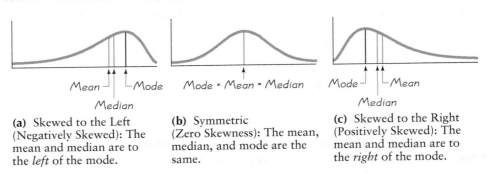

(a) Skewed to the Left (Negatively Skewed): The mean and median are to the *left* of the mode.

(b) Symmetric (Zero Skewness): The mean, median, and mode are the same.

(c) Skewed to the Right (Positively Skewed): The mean and median are to the *right* of the mode.

				Takes Every	Affected by	
Measure of Center	Definition	How Common?	Existence	Value into Account?	Extreme Values?	Advantages and Disadvantages
Mean	$\bar{x} = \dfrac{\Sigma x}{n}$	most familiar "average"	always exists	yes	yes	used through-out this book; works well with many statistical methods
Median	middle value	commonly used	always exists	no	no	often a good choice if there are some ex-treme values
Mode	most frequent data value	sometimes used	might not exist; may be more than one mode	no	no	appropriate for data at the nominal level
Midrange	$\dfrac{\text{high} + \text{low}}{2}$	rarely used	always exists	no	yes	very sensitive extreme values

Table 2-9   Comparison of Mean, Median, Mode, and Midrange

General comments:
- For a data collection that is approximately symmetric with one mode, the mean, median, mode, and midrange tend to be about the same.
- For a data collection that is obviously asymmetric, it would be good to report both the mean and median.
- The mean is relatively *reliable*. That is, when samples are drawn from the same population, the sample means tend to be more con-sistent than the other measures of center (consistent in the sense that the means of samples drawn from the same population don't vary as much as the other measures of center).

Definition

A distribution of data is **skewed** if it is not symmetric and extends more to one side than the other. (A distribution of data is **symmetric** if the left half of its his-togram is roughly a mirror image of its right half.)

Data **skewed to the left** (also called *negatively skewed*) have a longer left tail, and the mean and median are to the left of the mode. [Although not always pre-dictable, data skewed to the left generally have a mean less than the median, as in Figure 2-11(a)]. Data **skewed to the right** (also called *positively skewed*) have a longer right tail, and the mean and median are to the right of the mode. [Again, al-though not always predictable, data skewed to the right generally have the mean to the right of the median, as in Figure 2-11(c)].

If we examine the histogram in Figure 2-1 for the cotinine levels of smokers, we see a graph that appears to be skewed to the right. In practice, many distributions of data are symmetric and without skewness. Distributions skewed to the right are more common than those skewed to the left because it's often easier to get exceptionally large values than values that are exceptionally small. With annual incomes, for example, it's impossible to get values below the lower limit of zero, but there are a few people who earn millions of dollars in a year. Annual incomes therefore tend to be skewed to the right, as in Figure 2-11(c).

2-4 Exercises

In Exercises 1–8, find the (a) mean, (b) median, (c) mode, and (d) midrange for the given sample data.

1. Tobacco Use in Children's Movies   In "Tobacco and Alcohol Use in G-Rated Children's Animated Films," by Goldstein, Sobel and Newman (*Journal of the American Medical Association*, Vol. 281, No. 12), the lengths (in seconds) of scenes showing tobacco use were recorded for animated movies from Universal Studios. Six of those times are listed below.

   0   223   0   176   0   548

2. Cereal   A dietitian obtains the amounts of sugar (in grams) from one gram in each of 16 different cereals, including Cheerios, Corn Flakes, Fruit Loops, Trix, and 12 others. Those values are listed below. Is the mean of those values likely to be a good estimate of the mean amount of sugar in each gram of cereal consumed by the population of all Americans who eat cereal? Why or why not?

   0.03 0.24 0.30 0.47 0.43 0.07 0.47 0.13 0.44 0.39 0.48 0.17 0.13 0.09 0.45 0.43

3. Body Mass Index   As part of the National Health Survey, the body mass index is measured for a random sample of women. Some of the values included in Data Set 1 from Appendix B are listed below. Is the mean of this sample reasonably close to the mean of 25.74, which is the mean for all 40 women included in Data Set 1?

   19.6 23.8 19.6 29.1 25.2 21.4 22.0 27.5 33.5 20.6 29.9 17.7 24.0 28.9 37.7

4. Drunk Driving   The blood alcohol concentrations of a sample of drivers involved in fatal crashes and then convicted with jail sentences are given below (based on data from the U.S. Department of Justice). Given that current state laws prohibit driving with levels above 0.08 or 0.10, does it appear that these levels are significantly above the maximum that is allowed?

   0.27 0.17 0.17 0.16 0.13 0.24 0.29 0.24 0.14 0.16 0.12 0.16 0.21 0.17 0.18

5. Motorcycle Fatalities   Listed below are ages of motorcyclists when they were fatally injured in traffic crashes (based on data from the U.S. Department of Transportation). Do the results support the common belief that such fatalities are incurred by a greater proportion of younger drivers?

   17   38   27   14   18   34   16   42   28   24   40   20   23   31   37   21   30   25

6. Fruit Flies   Listed below are the thorax lengths (in millimeters) of a sample of male fruit flies. [Fruit flies (*Drosophila*) are a favorite subject of researchers because they have a simple chromosome composition, they reproduce quickly, they have large

numbers of offspring, and they are easy to care for. (These are characteristics of the fruit flies, not necessarily the researchers.)]

0.72   0.90   0.84   0.68   0.84   0.90   0.92   0.84   0.64   0.84   0.76

7. Blood Pressure Measurements Fourteen different second-year medical students at Bellevue Hospital measured the blood pressure of the same person. The systolic readings (in mmHg) are listed below. What is notable about this data set?

138   130   135   140   120   125   120   130   130   144   143   140   130   150

8. Phenotypes of Peas An experiment was conducted to determine whether a deficiency of carbon dioxide in soil affects the phenotypes of peas. Listed below are the phenotype codes, where 1 = smooth-yellow, 2 = smooth-green, 3 = wrinkled-yellow, and 4 = wrinkled-green. What is the level of measurement of these data (nominal, ordinal, interval, ratio)? Can the measures of center be obtained for these values? Do the results make sense?

| 2 | 1 | 1 | 1 | 1 | 1 | 1 | 4 | 1 | 2 | 2 | 1 | 2 |
| 3 | 3 | 2 | 3 | 1 | 3 | 1 | 3 | 1 | 3 | 2 | 2 | |

In Exercises 9–12, find the mean, median, mode, and midrange for each of the samples, then compare the two sets of results.

9. Patient Waiting Times The Newport Health Clinic experiments with two different configurations for serving patients. In one configuration, all patients enter a single waiting line that feeds three different physicians. In another configuration, patients wait in individual lines at three different physician stations. Waiting times (in minutes) are recorded for ten patients from each configuration. Compare the results.

Single line: 65   66   67   68   71   73   74   77   77   77

Multiple lines: 42   54   58   62   67   77   77   85   93   100

Interpret the results by determining whether there is a difference between the two data sets that is not apparent from a comparison of the measures of center. If so, what is it?

10. Skull Breadths Maximum breadth of samples of male Egyptian skulls from 4000 B.C. and 150 A.D. (based on data from *Ancient Races of the Thebaid* by Thomson and Randall-Maciver):

4000 B.C.:   131   119   138   125   129   126   131   132   126   128   128   131

150 A.D.:   136   130   126   126   139   141   137   138   133   131   134   129

Compare the results. Changes in head sizes over time suggest interbreeding with people from other regions. Do the head sizes appear to have changed from 4000 B.C. to 150 A.D.?

11. Poplar Trees Listed below are heights (in meters) of trees included in an experiment. Compare the results.

Control Group:	3.2	1.9	3.6	4.6	4.1	5.1	5.5	4.8	2.9	4.9
	3.9	6.9	6.8	6.0	6.9	6.5	5.5	5.5	4.1	6.3
Irrigation Treatment:	4.1	2.5	4.5	3.6	2.8	5.1	2.9	5.1	1.2	1.6
	6.4	6.8	6.5	6.8	5.8	2.9	5.5	3.3	6.1	6.1

12. Poplar Trees The given stem-and-leaf plot represents heights (in meters) of poplar trees treated with fertilizer and irrigation. Compare the measures of center for these trees to the control group of trees listed in Exercise 11.

Stem (units)	Leaves (tenths)
3.	29
4.	4
5.	4
6.	334788999
7.	133367
8.	0

In Exercises 13–16, refer to the data set in Appendix B. Use computer software or a calculator to find the means and medians, then compare the results as indicated.

13. Head Circumference While studying the disorder of hydrocephalus, a pediatrician investigates head circumferences of 2-year-old males and females. Use the sample results listed in Data Set 4 in Appendix B. Does there appear to be a difference between the two genders?

14. Body Mass Index It is well known that men tend to be taller than women and men tend to weigh more than women. The body mass index (BMI) of an individual is computed from height and weight. Refer to Data Set 1 in Appendix B and use the BMI values of men and the BMI values of women. Does there appear to be a difference between the two genders?

15. Petal Lengths of Irises In Data Set 7 from Appendix B, use the petal lengths of the irises from each of the three different classes. Do the three different classes appear to be about the same?

16. Sepal Widths of Irises In Data Set 7 from Appendix B, use the sepal widths of the irises from each of the three different classes. Do the three different classes appear to be about the same?

In Exercises 17 and 18, find the mean of the data summarized in the given frequency distribution.

17. Mean from a Frequency Distribution When data are summarized in a frequency distribution, we might not know the exact values falling in a particular class. To make calculations possible, we can pretend that in each class, all sample values are equal to the class midpoint. Because each class midpoint is repeated a number of times equal to the class frequency, the sum of all sample values is $\Sigma(f \cdot x)$, where f denotes frequency and x represents the class midpoint. The total number of sample values is the sum of frequencies Σf. The formula below is then used to compute the mean when the sample data are summarized in a frequency distribution. Using the frequency distribution given in Table 2-2, find the mean cotinine level of the sample of smokers. How does the result compare to the value of $\bar{x} = 172.5$ that is obtained by using the original list of 40 values listed in Table 2-1?

$$\bar{x} = \frac{\Sigma(f \cdot x)}{\Sigma f} \qquad \text{(mean from frequency distribution)}$$

18. Body Temperatures The accompanying frequency distribution summarizes a sample of human body temperatures. (See the temperatures for midnight on the second day, as listed in Data Set 2 in Appendix B.) Find the mean. (See Exercise 17.) How does the mean compare to the value of 98.6°F, which is the value commonly assumed to be the mean?

Table for Exercise 18

Temperature	Frequency
96.5–96.8	1
96.9–97.2	8
97.3–97.6	14
97.7–98.0	22
98.1–98.4	19
98.5–98.8	32
98.9–99.2	6
99.3–99.6	4

19. Mean of Means Using an almanac, a researcher finds the mean physician's salary for each state. He adds those 50 values, then divides by 50 to obtain their mean. Is the result equal to the national mean physician's salary? Why or why not?

*20. Trimmed Mean Because the mean is very sensitive to extreme values, we say that it is not a *resistant* measure of center. The **trimmed mean** is more resistant. To find the 10% trimmed mean for a data set, first arrange the data in order, delete the bottom 10% of the values and the top 10% of the values, then calculate the mean of the remaining values. For the weights of the bears in Data Set 6 from Appendix B, find (a) the mean; (b) the 10% trimmed mean; (c) the 20% trimmed mean. How do the results compare?

*21. Censored Data An experiment is conducted to study longevity of trees treated with fertilizer. The experiment is run for a fixed time of five years. (The test is said to be *censored* at five years.) The sample results (in years) are 2.5, 3.4, 1.2, 5+, 5+ (where 5+ indicates that the tree was still alive at the end of the experiment). What can you conclude about the mean tree life?

2-5   Measures of Variation

Study Hint: Because this section introduces the concept of variation, which is so important in statistics, this is one of the most important sections in the entire book. First read through Part 1 of this section quickly and gain a general understanding of the characteristic of variation. Next, learn how to obtain measures of variation, especially the standard deviation. Finally, consider Part 2 and try to understand the reasoning behind the formula for standard deviation, but do not spend much time memorizing formulas or doing arithmetic calculations. Instead, place a high priority on learning how to *interpret* values of standard deviation.

PART 1   Basic Concepts

For a visual illustration of variation, see Figure 2-12, which shows Excel-generated bar graphs for six patient waiting times at two different clinics. In the first clinic, three patients all wait in a single line that feeds three different

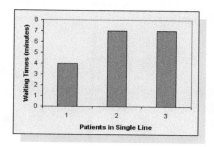
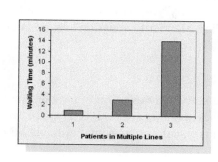

FIGURE 2-12 **Waiting Times (min) of Clinic Patients**

physicians. In the second clinic, three patients wait in three different lines for each of the three different physicians. Here are the specific patient waiting times (in minutes):

Single waiting line	4	7	7
Multiple waiting lines	1	3	14

If we consider only the mean, we would not recognize any difference between the two samples, because they both have a mean of $\bar{x} = 6.0$ min. However, it should be obvious that the samples are very different in the amounts that the waiting times *vary*. Although both samples of patients have a mean waiting time of 6.0 minutes, the waiting times of the patients in multiple lines vary much more than those in the single line. In this section, we want to develop the ability to measure and understand such variation.

Let's now develop some specific ways of actually measuring variation so that we can use specific numbers instead of subjective judgments. We begin with the range.

Range

Definition

The **range** of a set of data is the difference between the maximum value and the minimum value.

$$\text{Range} = (\text{Maximum value}) - (\text{Minimum value})$$

To compute the range, simply subtract the lowest value from the highest value. For the single-line waiting times, the range is $7 - 4 = 3$ min. The waiting times with multiple lines have a range of 13 min, and this larger value suggests greater variation.

The range is very easy to compute, but because it depends on only the maximum and the minimum values, it isn't as useful as the other measures of variation that use every value.

Standard Deviation of a Sample

The standard deviation is the measure of variation that is generally the most important and useful. We define the standard deviation now, but in order to understand it fully you will need to study the subsection of "Interpreting and Understanding Standard Deviation" found later in this section.

Definition

The **standard deviation** of a set of sample values is a measure of variation of values about the mean. It is a type of average deviation of values from the mean that is calculated by using Formulas 2-3 or 2-4.

Formula 2-3

$$s = \sqrt{\frac{\Sigma(x - \bar{x})^2}{n - 1}}$$

sample standard deviation

Formula 2-4

$$s = \sqrt{\frac{n(\Sigma x^2) - (\Sigma x)^2}{n(n - 1)}}$$

shortcut formula for
sample standard deviation

Later in this section we will discuss the rationale for these formulas, but for now we recommend that you use Formula 2-3 for a few examples, then learn how to find standard deviation values using your calculator and by using a software program. (Most scientific calculators are designed so that you can enter a list of values and automatically get the standard deviation.) For now, we cite important properties that are consequences of the way in which the standard deviation is defined.

- The standard deviation is a measure of variation of all values from the *mean*.

- The value of the standard deviation s is usually positive. It is zero only when all of the data values are the same number. Also, larger values of s indicate greater amounts of variation.

- The value of the standard deviation s can increase dramatically with the inclusion of one or more outliers (data values that are very far away from all of the others).

- The units of the standard deviation s (such as minutes, feet, pounds, and so on) are the same as the units of the original data values.

Procedure for Finding the Standard Deviation with Formula 2-3

Step 1:   Compute the mean $\bar{x}$.

Step 2:   Subtract the mean from each individual value to get a list of deviations of the form $(x - \bar{x})$.

Step 3:   Square each of the differences obtained from Step 2. This produces numbers of the form $(x - \bar{x})^2$.

Step 4:   Add all of the squares obtained from Step 3. This is the value of $\Sigma(x - \bar{x})^2$.

Step 5:   Divide the total from Step 4 by the number $(n - 1)$, which is 1 less than the total number of values present.

Step 6:   Find the square root of the result of Step 5.

EXAMPLE Using Formula 2-3 Use Formula 2-3 to find the standard deviation of the waiting times from the multiple lines. Those times (in minutes) are 1, 3, 14.

SOLUTION   We will use the six steps in the procedure just given. Refer to those steps and refer to Table 2-10, which shows the detailed calculations.

continued

Table 2-10	Calculating Standard Deviation	
x	$x - \bar{x}$	$(x - \bar{x})^2$
1	-5	25
3	-3	9
14	8	64
Totals: 18		98

$$\bar{x} = \frac{18}{3} = 6.0 \text{ min} \quad \Big| \quad s = \sqrt{\frac{98}{3-1}} = \sqrt{49} = 7.0 \text{ min}$$

Step 1: Obtain the mean of 6.0 by adding the values and then dividing by the number of values:

$$\bar{x} = \frac{\Sigma x}{n} = \frac{18}{3} = 6.0 \text{ min}$$

Step 2: Subtract the mean of 6.0 from each value to get these values of $(x - \bar{x})$: $-5, -3, 8$.

Step 3: Square each value obtained in Step 2 to get these values of $(x - \bar{x})^2$: 25, 9, 64.

Step 4: Sum all of the preceding values to get the value of

$$\Sigma(x - \bar{x})^2 = 98.$$

Step 5: With $n = 3$ values, divide by 1 less than 3:

$$\frac{98}{2} = 49.0$$

Step 6: Find the square root of 49.0. The standard deviation is

$$\sqrt{49.0} = 7.0 \text{ min}$$

Ideally, we would now interpret the meaning of the result, but such interpretations will be discussed later in this section.

EXAMPLE **Using Formula 2-4** The preceding example used Formula 2-3 for finding the standard deviation of patient waiting times. Using the same data set, find the standard deviation by using Formula 2-4.

SOLUTION   Formula 2-4 requires that we first find values for n, Σx, and Σx^2.

$n = 3$ (because there are 3 values in the sample)

$\Sigma x = 18$ (found by adding the 3 sample values)

$\Sigma x^2 = 206$ (found by adding the squares of the sample values, as in $1^2 + 3^2 + 14^2$)

Using Formula 2-4, we get

$$s = \sqrt{\frac{n(\Sigma x^2) - (\Sigma x)^2}{n(n-1)}} = \sqrt{\frac{3(206) - (18)^2}{3(3-1)}} = \sqrt{\frac{294}{6}} = 7.0 \text{ min}$$

A good activity is to stop here and calculate the standard deviation of the waiting times of 4 min, 7 min, and 7 min (for the single line). Follow the same procedures used in the preceding two examples and verify that $s = 1.7$ min. (It will also become important to develop an ability to obtain values of standard deviations by using a calculator and software.) Although the interpretations of these standard deviations will be discussed later, we can now compare them to see that the standard deviation of the times for the single line (1.7 min) is much lower than the standard deviation for multiple lines (7.0 min). This supports our subjective conclusion that the waiting times with the single line have much less variation than the times from multiple lines.

Standard Deviation of a Population

In our definition of standard deviation, we referred to the standard deviation of *sample* data. A slightly different formula is used to calculate the standard deviation σ (lowercase Greek sigma) of a population: Instead of dividing by $n - 1$, divide by the population size N, as in the following expression.

$$\sigma = \sqrt{\frac{\Sigma(x - \mu)^2}{N}} \qquad \text{population standard deviation}$$

Because we generally deal with sample data, we will usually use Formula 2-3, in which we divide by $n - 1$. Many calculators give both the sample standard deviation and the population standard deviation, but they use a variety of different notations. Be sure to identify the notation used by your calculator, so that you get the correct result.

Variance of a Sample and Population

We are using the term *variation* as a general description of the amount that values vary among themselves. (The terms *variability, dispersion, spread* are sometimes used instead of *variation*.) The term *variance* refers to a specific definition.

Definition

The **variance** of a set of values is a measure of variation equal to the square of the standard deviation.

Sample variance: Square of the standard deviation s.

Population variance: Square of the population standard deviation σ.

The sample variance s^2 is said to be an **unbiased estimator** of the population variance σ^2, which means that values of s^2 tend to target the value of σ^2 instead of systematically tending to overestimate or underestimate σ^2. For example, consider an IQ test designed so that the variance is 225. If you repeat the process of randomly selecting 100 subjects, giving them IQ tests, and calculating the sample variance s^2 in each case, the sample variances that you obtain will tend to center around 225, which is the population variance.

> **EXAMPLE** **Finding Variance** In the preceding example, we used the patient waiting times of 1 min, 3 min, and 14 min to find that the standard deviation is given by $s = 7.0$ min. Find the variance of that same sample.
>
> **SOLUTION**   Because the variance is the square of the standard deviation, we get the result shown below. Note that the original data values are in units of minutes, the standard deviation is 7.0 min, but the variance is given in units of min^2.
>
> $$\text{Sample variance} = s^2 = 7.0^2 = 49.0 \text{ min}^2$$

The variance is an important statistic used in some important statistical methods, such as analysis of variance discussed in Chapter 11. For our present purposes, the variance has this serious disadvantage: *The units of variance are different than the units of the original data set.* For example, if the original waiting times are in minutes, the units of the variance are in min^2. What is a square minute? Because the variance uses different units, it is extremely difficult to understand variance by relating it to the original data set. Because of this property, we will focus on the standard deviation when we try to develop an understanding of variation later in this section.

We now present the notation and round-off rule we are using.

Notation

s = *sample* standard deviation
s^2 = *sample* variance

σ = *population* standard deviation
σ^2 = *population* variance

Note: Articles in professional journals and reports often use SD for standard deviation and VAR for variance.

Round-Off Rule

We use the same round-off rule given in Section 2-4:

Carry one more decimal place than is present in the original set of data.

Round only the final answer, not values in the middle of a calculation. (If it becomes absolutely necessary to round in the middle, carry at least twice as many decimal places as will be used in the final answer.)

PART 2   Beyond the Basics

Comparing Variation in Different Populations

We stated earlier that because the units of the standard deviation are the same as the units of the original data, it is easier to understand the standard deviation than the variance. However, that same property makes it difficult to compare variation for values taken from different populations. By resulting in a value that is free of specific units of measure, the *coefficient of variation* overcomes this disadvantage.

Definition

The **coefficient of variation** (or **CV**) for a set of non-negative sample or population data, expressed as a percent, describes the standard deviation relative to the mean, and is given by the following:

$$\text{Sample} \qquad\qquad \text{Population}$$

$$CV = \frac{s}{\bar{x}} \cdot 100\% \qquad CV = \frac{\sigma}{\mu} \cdot 100\%$$

EXAMPLE **Heights and Weights of Men** Using the sample height and weight data for the 40 males included in Data Set 1 in Appendix B, we find the statistics given in the table below. Find the coefficient of variation for heights, then find the coefficient of variation for weights, then compare the two results.

	Mean ($\bar{x}$)	Standard Deviation (s)
Height	68.34 in.	3.02 in.
Weight	172.55 lb	26.33 lb

SOLUTION   Because we have sample statistics, we find the two coefficients of variation as follows.

Heights: $CV = \dfrac{s}{\bar{x}} \cdot 100\% = \dfrac{3.02 \text{ in.}}{68.34 \text{ in.}} \cdot 100\% = 4.42\%$

Weights: $CV = \dfrac{s}{\bar{x}} \cdot 100\% = \dfrac{26.33 \text{ lb}}{172.55 \text{ lb}} \cdot 100\% = 15.26\%$

Although the difference in the units makes it impossible to compare the standard deviation of 3.02 in. to the standard deviation of 26.33 lb, we can compare the coefficients of variation, which have no units. We can see that heights (with $CV = 4.42\%$) have considerably less variation than weights (with $CV = 15.26\%$). This makes intuitive sense, because we routinely see that weights among men vary much more than heights. For example, it is very rare to see two adult men with one of them being twice as tall as the other, but it is much more common to see two men with one of them weighing twice as much as the other.

Interpreting and Understanding Standard Deviation

This subsection is extremely important, because we will now try to make some intuitive sense of the standard deviation. First, we should clearly understand that the standard deviation measures the *variation* among values. Values close together will yield a small standard deviation, whereas values spread farther apart will yield a larger standard deviation.

One crude but simple tool for understanding standard deviation is the **range rule of thumb,** which is based on the principle that for many data sets, the vast majority (such as 95%) of sample values lie within two standard deviations of the mean. (We could improve the accuracy of this rule by taking into account such factors as the size of the sample and the nature of the distribution, but we prefer to sacrifice accuracy for the sake of simplicity. Also, we could use three or even four standard deviations instead of two standard deviations, which is a somewhat arbitrary choice. But we want a simple rule that will help us interpret values of standard deviations; methods you will learn later produce more accurate results.)

Range Rule of Thumb

For Estimating a Value of the Standard Deviation *s*: To roughly estimate the standard deviation, use

$$s \approx \frac{\text{range}}{4}$$

where range = (maximum value) − (minimum value).

For Interpreting a Known Value of the Standard Deviation *s*: If the standard deviation *s* is known, use it to find rough estimates of the minimum and maximum "usual" sample values by using

minimum "usual" value = (mean) − 2 × (standard deviation)

maximum "usual" value = (mean) + 2 × (standard deviation)

When calculating a standard deviation using Formula 2-3 or 2-4, you can use the range rule of thumb as a check on your result, but you must realize that although the approximation will get you in the general vicinity of the answer, it can be off by a fairly large amount.

EXAMPLE Cotinine Levels of Smokers Use the range rule of thumb to find a rough estimate of the standard deviation of the sample of 40 cotinine levels of smokers, as listed in Table 2-1.

SOLUTION   In using the range rule of thumb to estimate the standard deviation of sample data, we find the range and divide by 4. By scanning the list of cotinine levels, we can see that the lowest is 0 and the highest is 491, so the range is 491. The standard deviation *s* is estimated as follows:

$$s \approx \frac{\text{range}}{4} = \frac{491}{4} = 122.75 \approx 123$$

INTERPRETATION This result is quite close to the correct value of 119.5 that is found by calculating the exact value of the standard deviation with Formula 2-3 or 2-4. We should not expect the range rule of thumb to work this well in all other cases.

The next example is particularly important as an illustration of one way to *interpret* the value of a standard deviation.

EXAMPLE **Head Circumferences of Girls** Past results from the National Health Survey suggest that the head circumferences of two-month-old girls have a mean of 40.05 cm and a standard deviation of 1.64 cm. Use the range rule of thumb to find the minimum and maximum "usual" head circumferences. (These results could be used by a physician who can identify "unusual" circumferences that might be the result of a disorder such as hydrocephalus.) Then determine whether a circumference of 42.6 cm would be considered "unusual."

SOLUTION   With a mean of 40.05 cm and a standard deviation of 1.64 cm, we use the range rule of thumb to find the minimum and maximum usual heights as follows:

$$\text{minimum "usual" value} = (\text{mean}) - 2 \times (\text{standard deviation})$$
$$= 40.05 - 2(1.64) = 36.77 \text{ cm}$$
$$\text{maximum "usual" value} = (\text{mean}) + 2 \times (\text{standard deviation})$$
$$= 40.05 + 2(1.64) = 43.33 \text{ cm}$$

INTERPRETATION Based on these results, we expect that typical two-month-old girls have head circumferences between 36.77 cm and 43.33 cm. Because 42.6 cm falls within those limits, it would be considered usual or typical, not unusual.

Empirical (or 68–95–99.7) Rule for Data with a Bell-Shaped Distribution

Another rule that is helpful in interpreting values for a standard deviation is the **empirical rule.** This rule states that *for data sets having a distribution that is approximately bell-shaped*, the following properties apply. (See Figure 2-13.)

- About 68% of all values fall within 1 standard deviation of the mean.
- About 95% of all values fall within 2 standard deviations of the mean.
- About 99.7% of all values fall within 3 standard deviations of the mean.

EXAMPLE **Heights of Women** Heights of women have a bell-shaped distribution with a mean of 163 cm and a standard deviation of 6 cm. What percentage of women have heights between 145 cm and 181 cm?

continued

FIGURE 2-13
The Empirical Rule

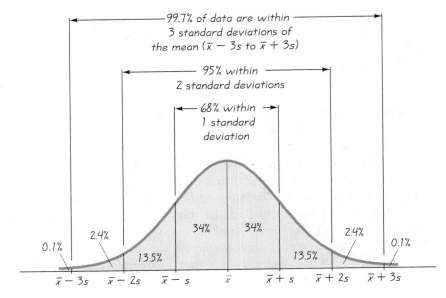

SOLUTION   The key to solving this problem is to recognize that 145 cm and 181 cm are each exactly 3 standard deviations away from the mean of 163, as shown below.

$$3 \text{ standard deviations} = 3s = 3(6) = 18$$

Therefore, 3 standard deviations from the mean is

$$163 - 18 = 145$$

or

$$163 + 18 = 181$$

The empirical rule tells us that about 99.7% of all values are within 3 standard deviations of the mean, so about 99.7% of all heights of women are between 145 cm and 181 cm.

Hint: Difficulty with applying the empirical rule usually stems from confusion with interpreting phrases such as "within 3 standard deviations of the mean." Stop here and review the preceding example until the meaning of that phrase becomes clear. Also, see the following general interpretations of such phrases.

Phrase	Meaning
Within 1 standard deviation of the mean	Between $(\bar{x} - s)$ and $(\bar{x} + s)$
Within 2 standard deviations of the mean	Between $(\bar{x} - 2s)$ and $(\bar{x} + 2s)$
Within 3 standard deviations of the mean	Between $(\bar{x} - 3s)$ and $(\bar{x} + 3s)$

A third concept that is helpful in understanding or interpreting a value of a standard deviation is **Chebyshev's theorem.** The preceding empirical rule applies only to data sets with a bell-shaped distribution. Instead of being limited to data sets with bell-shaped distributions, Chebyshev's theorem applies to *any* data set, but its results are very approximate.

Chebyshev's Theorem

The proportion (or fraction) of any set of data lying within K standard deviations of the mean is always *at least* $1 - 1/K^2$, where K is any positive number greater than 1. For $K = 2$ and $K = 3$, we get the following statements:

- At least 3/4 (or 75%) of all values lie within 2 standard deviations of the mean.
- At least 8/9 (or 89%) of all values lie within 3 standard deviations of the mean.

EXAMPLE **Heights of Women** Heights of women have a mean of 163 cm and a standard deviation of 6 cm. What can we conclude from Chebyshev's theorem?

SOLUTION   Applying Chebyshev's theorem with a mean of 163 cm and a standard deviation of 6 cm, we can reach the following conclusions:

- At least 3/4 (or 75%) of all heights of women are within 2 standard deviations of the mean (between 151 cm and 175 cm).
- At least 8/9 (or 89%) of all heights of women are within 3 standard deviations of the mean (between 145 cm and 181 cm).

When trying to make sense of a value of a standard deviation, we should use one or more of the preceding three concepts. To gain additional insight into the nature of the standard deviation, we now consider the underlying rationale leading to Formula 2-3, which is the basis for its definition. (Formula 2-4 is simply another version of Formula 2-3, derived so that arithmetic calculations can be simplified.)

Rationale for Standard Deviation

The standard deviation of a set of sample data is defined by Formulas 2-3 and 2-4, which are equivalent in the sense that they will always yield the same result. Formula 2-3 has the advantage of reinforcing the concept that the standard deviation is a type of average deviation. Formula 2-4 has the advantage of being easier to use when you must calculate standard deviations on your own. Formula 2-4 also eliminates the intermediate rounding errors introduced in Formula 2-3 when the exact value of the mean is not used. Formula 2-4 is used in calculators and programs because it requires only three memory locations (for n, Σx, and Σx^2) instead of a memory location for every value in the data set.

Why define a measure of variation in the way described by Formula 2-3? In measuring variation in a set of sample data, it makes sense to begin with the individual amounts by which values deviate from the mean. For a particular value x, the amount of **deviation** is $x - \bar{x}$, which is the difference between the individual x value and the mean. For the waiting times of 1, 3, 14, the mean is 6.0 so the deviations away from the mean are -5, -3, and 8. It would be good to somehow combine those deviations into a single collective value. Simply adding the deviations doesn't work, because the sum will always be zero. To get a statistic that measures variation (instead of always being zero), we need to avoid the canceling out of negative and positive numbers. One approach is to add absolute values, as in $\Sigma|x - \bar{x}|$. If we find the mean of that sum, we get the **mean absolute deviation** (or **MAD**), which is the mean distance of the data from the mean.

$$\text{mean absolute deviation} = \frac{\Sigma|x - \bar{x}|}{n}$$

Because the waiting times of 1, 3, 14 have deviations of -5, -3, and 8, the mean absolute deviation is $(5 + 3 + 8)/3 = 16/3 = 5.3$.

Why Not Use the Mean Absolute Deviation?

Because the mean absolute deviation requires that we use absolute values, it uses an operation that is not algebraic. (The algebraic operations include addition, multiplication, extracting roots, and raising to powers that are integers or fractions, but absolute value is not included.) The use of absolute values creates algebraic difficulties in inferential methods of statistics. For example, Section 8-3 presents a method for making inferences about the means of two populations, and that method is built around an additive property of variances, but the mean absolute deviation has no such additive property. (Here is a simplified version of the additive property of variances: If you have two independent populations and you randomly select one value from each population and add them, such sums will have a variance equal to the sum of the variances of the two populations.) That same additive property underlies the rationale for regression discussed in Chapter 9 and analysis of variance discussed in Chapter 11. The mean absolute deviation lacks this important additive property. Also, the mean absolute value is *biased*, meaning that when you find mean absolute values of samples, you do not tend to target the mean absolute value of the population. In contrast, the standard deviation uses only algebraic operations. Because it is based on the square root of a sum of squares, the standard deviation closely parallels distance formulas found in algebra. There are many instances where a statistical procedure is based on a similar sum of squares. Therefore, instead of using absolute values, we get a better measure of variation by making all deviations $(x - \bar{x})$ non-negative by squaring them, and this approach leads to the standard deviation. For these reasons, scientific calculators typically include a standard deviation function, but they almost never include the mean absolute deviation.

Why Divide by $n - 1$?

After finding all of the individual values of $(x - \bar{x})^2$, we combine them by finding their sum, then we get an average by dividing by $n - 1$. We divide by $n - 1$ because there are only $n - 1$ independent values. That is, with a given mean, only $n - 1$ values can be freely assigned any number before the last value is determined. See Exercise 27, which provides concrete numbers illustrating that division by $n - 1$ yields a better result than division by n. That exercise shows that if s^2 were defined with division by n, it would systematically underestimate the value of σ^2, so we compensate by increasing its overall value by making its denominator smaller (by using $n - 1$ instead of n). Exercise 27 shows how division by $n - 1$ causes sample variances s^2 to target the value of the population variance σ^2, whereas division by n causes sample variances s^2 to underestimate the value of the population variance σ^2.

Step 6 in the above procedure for finding a standard deviation is to find a square root. We take the square root to compensate for the squaring that took place in Step 3. An important consequence of taking the square root is that *the standard deviation has the same units of measurement as the original values*. For example, if patient waiting times are in minutes, the standard deviation of those times will also be in minutes. If we were to stop at Step 5, the result would be in units of "square minutes," which is an abstract concept having no direct link to reality.

After studying this section, you should understand that the standard deviation is a measure of variation among values. Given sample data, you should be able to compute the value of the standard deviation. You should be able to interpret the values of standard deviations that you compute. You should know that for typical data sets, it is unusual for a value to differ from the mean by more than 2 or 3 standard deviations.

2-5 Exercises

In Exercises 1–8, find the range, variance, and standard deviation for the given sample data. (The same data were used in Section 2-4 where we found measures of center. Here we find measures of variation.)

1. Tobacco Use in Children's Movies In "Tobacco and Alcohol Use in G-Rated Children's Animated Films," by Goldstein, Sobel and Newman (*Journal of the American Medical Association,* Vol. 281, No. 12), the lengths (in seconds) of scenes showing tobacco use were recorded for animated movies from Universal Studios. Six of those times are listed below.

$$0 \quad 223 \quad 0 \quad 176 \quad 0 \quad 548$$

2. Cereal A dietitian obtains the amounts of sugar (in grams) from one gram in each of 16 different cereals, including Cheerios, Corn Flakes, Fruit Loops, Trix, and 12 others. Those values are listed below. Based on these sample values, would an amount of 0.95 be considered "unusual"? Why or why not? (*Hint*: See the range rule of thumb.)

0.03	0.24	0.30	0.47	0.43	0.07	0.47	0.13
0.44	0.39	0.48	0.17	0.13	0.09	0.45	0.43

3. Body Mass Index As part of the National Health Examination, the body mass index (BMI) is measured for a random sample of women. Some of the values included in Data Set 1 from Appendix B are listed below. Based on these sample data, would a BMI of 34.0 be considered "unusual"? Why or why not? (*Hint*: See the range rule of thumb.)

19.6	23.8	19.6	29.1	25.2	21.4	22.0	27.5
33.5	20.6	29.9	17.7	24.0	28.9	37.7	

4. Drunk Driving The blood alcohol concentrations of a sample of drivers involved in fatal crashes and then convicted with jail sentences are given below (based on data

from the U.S. Department of Justice). When a state wages a campaign to reduce drunk driving, is the campaign intended to lower the standard deviation?

0.27	0.17	0.17	0.16	0.13	0.24	0.29	0.24
0.14	0.16	0.12	0.16	0.21	0.17	0.18	

5. Motorcycle Fatalities Listed below are ages of motorcyclists when they were fatally injured in traffic crashes (based on data from the U.S. Department of Transportation). How does the variation of these ages compare to the variation of ages among licensed drivers in the general population?

17	38	27	14	18	34	16	42	28
24	40	20	23	31	37	21	30	25

6. Fruit Flies Listed below are the thorax lengths (in millimeters) of a sample of male fruit flies. Based on these sample values, is a thorax length of 0.63 mm unusual? Why or why not? (*Hint*: Use the range rule of thumb.)

0.72	0.90	0.84	0.68	0.84	0.90
0.92	0.84	0.64	0.84	0.76	

7. Blood Pressure Measurements Fourteen different second-year medical students at Bellevue Hospital measured the blood pressure of the same person. The systolic readings (in mmHg) are listed below. What do the measures of variation suggest about the accuracy of the readings?

138	130	135	140	120	125	120
130	130	144	143	140	130	150

8. Phenotypes of Peas An experiment was conducted to determine whether a deficiency of carbon dioxide in soil affects the phenotypes of peas. Listed below are the phenotype codes, where 1 = smooth-yellow, 2 = smooth-green, 3 = wrinkled-yellow, and 4 = wrinkled-green. What is the level of measurement of these data (nominal, ordinal, interval, ratio)? Can the measures of variation be obtained for these values? Do the results make sense?

2	1	1	1	1	1	1	4	1	2	2	1	2
3	3	2	3	1	3	1	3	1	3	2	2	

In Exercises 9–12, find the range, variance, and standard deviation for each of the two samples, then compare the two sets of results. (The same data were used in Section 2-4.)

9. Patient Waiting Times The Newport Health Clinic experiments with two different configurations for serving patients. In one configuration, all patients enter a single waiting line that feeds three different physicians. In another configuration, patients wait in individual lines at three different physician stations. Waiting times (in minutes) are recorded for ten patients from each configuration. Compare the results.

Single line:	65	66	67	68	71	73	74	77	77	77
Multiple lines:	42	54	58	62	67	77	77	85	93	100

10. Skull Breadths Maximum breadth of samples of male Egyptian skulls from 4000 B.C. and 150 A.D. (based on data from *Ancient Races of the Thebaid* by Thomson and Randall-Maciver):

4000 B.C.:	131	119	138	125	129	126	131	132	126	128	128	131
150 A.D.:	136	130	126	126	139	141	137	138	133	131	134	129

11. Poplar Trees Listed below are heights (in meters) of trees included in an experiment.

Control group:

| 3.2 | 1.9 | 3.6 | 4.6 | 4.1 | 5.1 | 5.5 | 4.8 | 2.9 | 4.9 |
| 3.9 | 6.9 | 6.8 | 6.0 | 6.9 | 6.5 | 5.5 | 5.5 | 4.1 | 6.3 |

Irrigation treatment:

| 4.1 | 2.5 | 4.5 | 3.6 | 2.8 | 5.1 | 2.9 | 5.1 | 1.2 | 1.6 |
| 6.4 | 6.8 | 6.5 | 6.8 | 5.8 | 2.9 | 5.5 | 3.3 | 6.1 | 6.1 |

12. Poplar Trees The given stem-and-leaf plot represents heights (in meters) of poplar trees treated with fertilizer and irrigation. Compare the measures of variation for these trees to the control group of trees listed in Exercise 11.

Stem (units)	Leaves (tenths)
3.	29
4.	4
5.	4
6.	334788999
7.	133367
8.	0

In Exercises 13–16, refer to the data set in Appendix B. Use computer software or a calculator to find the standard deviations, then compare the results.

13. Head Circumference In order to correctly diagnose the disorder of hydrocephalus, a pediatrician investigates head circumferences of 2-year-old males and females. Use the sample results listed in Data Set 4 in Appendix B. Does there appear to be a difference in variation between the two genders?

14. Body Mass Index It is well known that men tend to be taller than women and men tend to weigh more than women. The body mass index (BMI) of an individual is computed from height and weight. Refer to Data Set 1 in Appendix B and use the BMI values of men and the BMI values of women. Does there appear to be a difference in variation between the two genders?

15. Petal Lengths of Irises In Data Set 7, use the petal lengths of the irises from each of the three different classes. Do the three different classes appear to have the same amount of variation?

16. Sepal Widths of Irises In Data Set 7, use the sepal widths of the irises from each of the three different classes. Do the three different classes appear to have the same amount of variation?

In Exercises 17 and 18, find the standard deviation of the data summarized in the given frequency distribution. (The same frequency distributions were used in Section 2-4.)

17. Finding Standard Deviation from a Frequency Distribution To find the standard deviation of sample data summarized in a frequency distribution table, use the formula below, where x represents the class midpoint and f represents the class frequency. Using the frequency distribution given in Table 2-2, find the standard deviation of the cotinine levels for the sample of smokers. How does the result compare to the value of $s = 119.5$ that is obtained by using the original list of 40 values listed in Table 2-1?

$$s = \sqrt{\frac{n[\Sigma(f \cdot x^2)] - [\Sigma(f \cdot x)]^2}{n(n-1)}}$$ Standard deviation for frequency distribution

Table for Exercise 18

Temperature	Frequency
96.5–96.8	1
96.9–97.2	8
97.3–97.6	14
97.7–98.0	22
98.1–98.4	19
98.5–98.8	32
98.9–99.2	6
99.3–99.6	4

18. Body Temperatures The accompanying frequency distribution summarizes a sample of human body temperatures. (See the temperatures for midnight on the second day, as listed in Data Set 2 in Appendix B.) Find the standard deviation. (See Exercise 17.)

19. Range Rule of Thumb Use the range rule of thumb to estimate the standard deviation of ages of all faculty members at your college.

20. Range Rule of Thumb Using the sample data in Data Set 1 from Appendix B, the sample of 40 women have upper leg lengths with a mean of 38.86 cm and a standard deviation of 3.78 cm. Use the range rule of thumb to estimate the minimum and maximum "usual" upper leg lengths for women. Is a length of 47.0 cm considered unusual in this context?

21. Empirical Rule Heights of men have a bell-shaped distribution with a mean of 176 cm and a standard deviation of 7 cm. Using the empirical rule, what is the approximate percentage of men between
   a. 169 cm and 183 cm?
   b. 155 cm and 197 cm?

22. Coefficient of Variation Refer to the data in Exercise 9. Find the coefficient of variation for each of the two samples, then compare the results.

23. Coefficient of Variation Use the sample data listed below to find the coefficient of variation for each of the two samples, then compare the results.
   Heights (in.) of men: 71   66   72   69   68   69
   Lengths (mm) of cuckoo eggs: 19.7 21.7 21.9 22.1 22.1 22.3 22.7 22.9 23.9

24. Equality for All What do you know about the values in a data set having a standard deviation of $s = 0$?

25. Understanding Units of Measurement If a data set consists of longevity times (in days) of fruit flies, what are the units used for standard deviation? What are the units used for variance?

26. Interpreting Outliers A data set consists of 20 values that are fairly close together. Another value is included, but this new value is an outlier (very far away from the other values). How is the standard deviation affected by the outlier? No effect? A small effect? A large effect?

*27. Why Divide by $n - 1$? Let a population consist of the values 3, 6, 9. Assume that samples of two values are randomly selected *with replacement*.
   a. Find the variance σ^2 of the population {3, 6, 9}.
   b. List the nine different possible samples of two values selected with replacement, then find the sample variance s^2 (which includes division by $n - 1$) for each of them. If you repeatedly select two sample values, what is the mean value of the sample variances s^2?
   c. For each of the nine samples, find the variance by treating each sample as if it is a population. (Be sure to use the formula for population variance, which includes division by n.) If you repeatedly select two sample values, what is the mean value of the population variances?
   d. Which approach results in values that are better estimates of σ^2: Part (b) or part (c)? Why? When computing variances of samples, should you use division by n or $n - 1$?
   e. The preceding parts show that s^2 is an unbiased estimator of σ^2. Is s an unbiased estimator of σ?

2-6 Measures of Relative Standing

This section introduces measures that can be used to compare values from different data sets, or to compare values within the same data set. We introduce z scores (for comparing values from different data sets) and quartiles and percentiles (for comparing values within the same data set).

z Scores

A z score (or standardized score) is found by converting a value to a standardized scale, as given in the following definition. We will use z scores extensively in Chapter 5 and later chapters, so they are extremely important.

Definition

A **standardized score,** or **z score,** is the number of standard deviations that a given value x is above or below the mean. It is found using the following expressions.

Sample   Population

$$z = \frac{x - \bar{x}}{s} \quad \text{or} \quad z = \frac{x - \mu}{\sigma}$$

(Round z to two decimal places.)

The following example illustrates how z scores can be used to compare values, even though they might come from different populations.

EXAMPLE **Comparing Heights** Former NBA superstar Michael Jordan is 78 in. tall and WNBA basketball player Rebecca Lobo is 76 in. tall. Jordan is obviously taller by 2 in., but which player is *relatively* taller? Does Jordan's height among men exceed Lobo's height among women? Men have heights with a mean of 69.0 in. and a standard deviation of 2.8 in.; women have heights with a mean of 63.6 in. and a standard deviation of 2.5 in. (based on data from the National Health Survey).

SOLUTION To compare the heights of Michael Jordan and Rebecca Lobo relative to the populations of men and women, we need to standardize those heights by converting them to z scores.

$$\text{Jordan:} \quad z = \frac{x - \mu}{\sigma} = \frac{78 - 69.0}{2.8} = 3.21$$

$$\text{Lobo:} \quad z = \frac{x - \mu}{\sigma} = \frac{76 - 63.6}{2.5} = 4.96$$

INTERPRETATION Michael Jordan's height is 3.21 standard deviations above the mean, but Rebecca Lobo's height is a whopping 4.96 standard deviations above the mean. Rebecca Lobo's height among women is relatively greater than Michael Jordan's height among men.

Process of Drug Approval

Gaining FDA approval for a new drug is expensive and time consuming. It begins with a **phase one study** in which the safety of the drug is tested with a small (20–100) group of volunteers. In **phase two,** the drug is tested for effectiveness in randomized trials involving a larger (100–300) group of subjects. This phase often has subjects randomly assigned to either a treatment group or a placebo group. In **phase three,** the goal is to better understand the effectiveness of the drug as well as its adverse reactions. Phase three typically involves 1000 to 3000 subjects and this phase alone might require several years of testing. Lisa Gibbs wrote in *Money* magazine that "the (drug) industry points out that for every 5,000 treatments tested, only five make it to clinical trials and only one ends up in drugstores." Total cost estimates vary from a low of $40 million to as much as $1.5 billion.

z Scores and Unusual Values

In Section 2-5 we used the range rule of thumb to conclude that a value is "unusual" if it is more than 2 standard deviations away from the mean. It follows that unusual values have z scores less than −2 or greater than +2. (See Figure 2-14). Using this criterion, both Michael Jordan and Rebecca Lobo are unusually tall because they both have heights with z scores greater than 2.

While considering professional basketball players with exceptional heights, another player is Mugsy Bogues, who was successful even though he is only 5 ft 3 in. tall. (We again use the fact that men have heights with a mean of 69.0 in. and a standard deviation of 2.8 in.) After converting 5 ft 3 in. to 63 in., we standardize his height by converting it to a z score as follows:

$$\text{Bogues: } z = \frac{x - \mu}{\sigma} = \frac{63 - 69.0}{2.8} = -2.14$$

Let's be grateful to Mugsy Bogues for his many years of inspired play and for illustrating this principle:

Whenever a value is less than the mean, its corresponding z score is negative.

Ordinary values: **−2 ≤ z score ≤ 2**

Unusual values: **z score < −2 *or* z score > 2**

z scores are measures of position in the sense that they describe the location of a value (in terms of standard deviations) relative to the mean. A z score of 2 indicates that a value is 2 standard deviations *above* the mean, and a z score of −3 indicates that a value is 3 standard deviations *below* the mean. Quartiles and percentiles are also measures of position, but they are defined differently than z scores and they are useful for comparing values within the same data set or between different sets of data.

Quartiles and Percentiles

From Section 2-4 we know that the median of a data set is the middle value, so that 50% of the values are equal to or less than the median and 50% of the values are greater than or equal to the median. Just as the median divides the data into two equal parts, the three **quartiles,** denoted by Q_1, Q_2, and Q_3, divide the sorted values into four equal parts. (Values are *sorted* when they are arranged in order.)

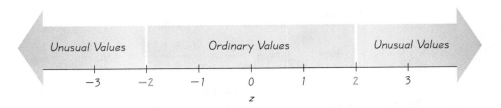

FIGURE 2-14 Interpreting z Scores
Unusual values are those with z scores less than −2.00 or greater than 2.00.

Here are descriptions of the three quartiles:

Q_1 (First quartile):   Separates the bottom 25% of the sorted values from the top 75%. (To be more precise, at least 25% of the sorted values are less than or equal to Q_1, and at least 75% of the values are greater than or equal to Q_1.)

Q_2 (Second quartile):   Same as the median; separates the bottom 50% of the sorted values from the top 50%.

Q_3 (Third quartile):   Separates the bottom 75% of the sorted values from the top 25%. (To be more precise, at least 75% of the sorted values are less than or equal to Q_3, and at least 25% of the values are greater than or equal to Q_3.)

We will describe a procedure for finding quartiles after we discuss percentiles. There is not universal agreement on a single procedure for calculating quartiles, and different computer programs often yield different results. For example, if you use the data set of 1, 3, 6, 10, 15, 21, 28, and 36, you will get these results:

	Q_1	Q_2	Q_3
STATDISK	4.5	12.5	24.5
Minitab	3.75	12.5	26.25
SPSS	3.75	12.5	26.25
Excel	5.25	12.5	22.75
SAS	4.5	12.5	24.5
TI-83/84 Plus	4.5	12.5	24.5

If you use a calculator or computer software for exercises involving quartiles, you may get results that differ slightly from the answers given in the back of the book.

Just as there are three quartiles separating a data set into four parts, there are also 99 **percentiles,** denoted $P_1, P_2, \ldots, P_{99}$, which partition the data into 100 groups with about 1% of the values in each group. (Quartiles and percentiles are examples of *quantiles*—or *fractiles*—which partition data into groups with roughly the same number of values.)

The process of finding the percentile that corresponds to a particular value x is fairly simple, as indicated in the following expression.

$$\text{percentile of value } x = \frac{\text{number of values less than } x}{\text{total number of values}} \cdot 100$$

EXAMPLE Cotinine Levels of Smokers Table 2-11 lists the 40 sorted cotinine levels of smokers included in Table 2-1. Find the percentile corresponding to the cotinine level of 112.

continued

Table 2-11		*Sorted* Cotinine Levels of 40 Smokers							
0	1	1	3	17	32	35	44	48	86
87	103	112	121	123	130	131	149	164	167
173	173	198	208	210	222	227	234	245	250
253	265	266	277	284	289	290	313	477	491

SOLUTION   From Table 2-11 we see that there are 12 values less than 112, so

$$\text{percentile of } 112 = \frac{12}{40} \cdot 100 = 30$$

INTERPRETATION   The cotinine level of 112 is the 30th percentile.

The preceding example shows how to convert from a given sample value to the corresponding percentile. There are several different methods for the reverse procedure of converting a given percentile to the corresponding value in the data set. The procedure we will use is summarized in Figure 2-15, which uses the following notation.

Notation

n total number of values in the data set

k percentile being used (Example: For the 25th percentile, $k = 25$.)

L locator that gives the *position* of a value (Example: For the 12th value in the sorted list, $L = 12$.)

P_k kth percentile (Example: P_{25} is the 25th percentile.)

 EXAMPLE **Cotinine Levels of Smokers** Refer to the sorted cotinine levels of smokers in Table 2-11 and use Figure 2-15 to find the value of the 68th percentile, P_{68}.

SOLUTION   Referring to Figure 2-15, we see that the sample data are already sorted, so we can proceed to find the value of the locator L. In this computation we use $k = 68$ because we are trying to find the value of the 68th percentile. We use $n = 40$ because there are 40 data values.

$$L = \frac{k}{100} \cdot n = \frac{68}{100} \cdot 40 = 27.2$$

Next, we are asked if L is a whole number and we answer no, so we proceed to the next lower box where we change L by rounding it up from 27.2 to 28. (In this book we typically round off the usual way, but this is one of two cases where we round *up* instead of rounding *off*.) Finally, the bottom box shows that the value of P_{68} is the 28th value, counting up from the lowest. In Table 2-11, the 28th value is 234. That is, $P_{68} = 234$.

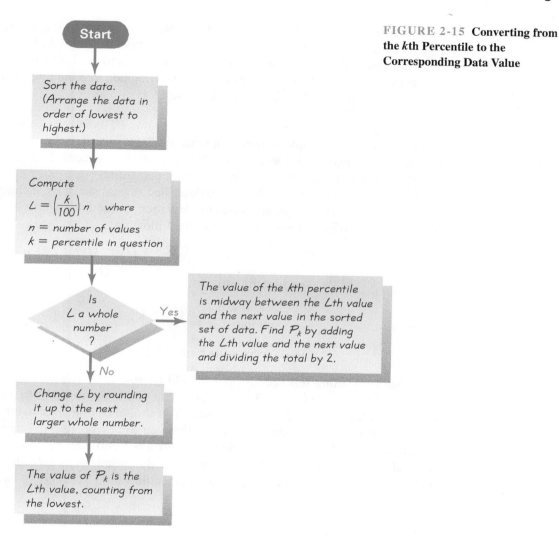

FIGURE 2-15 **Converting from the kth Percentile to the Corresponding Data Value**

Start

Sort the data. (Arrange the data in order of lowest to highest.)

Compute
$$L = \left(\frac{k}{100}\right) n \quad \text{where}$$
$n = $ number of values
$k = $ percentile in question

Is L a whole number?

Yes → The value of the kth percentile is midway between the Lth value and the next value in the sorted set of data. Find P_k by adding the Lth value and the next value and dividing the total by 2.

No ↓

Change L by rounding it up to the next larger whole number.

The value of P_k is the Lth value, counting from the lowest.

EXAMPLE **Cotinine Levels of Smokers** Refer to the sample of cotinine levels of smokers given in Table 2-11. Use Figure 2-15 to find the value of Q_1, which is the first quartile.

SOLUTION First we note that Q_1 is the same as P_{25}, so we can proceed with the objective of finding the value of the 25th percentile. Referring to Figure 2-15, we see that the sample data are already sorted, so we can proceed to compute the value of the locator L. In this computation, we use $k = 25$ because we are attempting to find the value of the 25th percentile, and we use $n = 40$ because there are 40 data values.

$$L = \frac{k}{100} \cdot n = \frac{25}{100} \cdot 40 = 10$$

Next, we are asked if L is a whole number and we answer yes, so we proceed to the box located at the right. We now see that the value of the kth (25th) per-

continued

centile is midway between the Lth (10th) value and the next value in the original set of data. That is, the value of the 25th percentile is midway between the 10th value and the 11th value. The 10th value is 86 and the 11th value is 87, so the value midway between them is 86.5. We conclude that the 25th percentile is $P_{25} = 86.5$. The value of the first quartile Q_1 is also 86.5.

$Q_1 = P_{25}$
$Q_2 = P_{50}$
$Q_3 = P_{75}$

The preceding example showed that when finding a quartile value (such as Q_1), we can use the equivalent percentile value (such as P_{25}) instead. See the margin for relationships relating quartiles to equivalent percentiles.

In earlier sections of this chapter we described several statistics, including the mean, median, mode, range, and standard deviation. Some other statistics are defined using quartiles and percentiles, as in the following:

$$\text{interquartile range (or IQR)} = Q_3 - Q_1$$

$$\text{semi-interquartile range} = \frac{Q_3 - Q_1}{2}$$

$$\text{midquartile} = \frac{Q_3 + Q_1}{2}$$

$$\text{10–90 percentile range} = P_{90} - P_{10}$$

After completing this section, you should be able to convert a value into its corresponding z score (or standard score) so that you can compare it to other values, which may be from different data sets. You should be able to convert a value into its corresponding percentile value so that you can compare it to other values in some data set. You should be able to convert a percentile to the corresponding data value. And finally, you should understand the meanings of quartiles and be able to relate them to their corresponding percentile values (as in $Q_3 = P_{75}$).

2-6 Exercises

In Exercises 1–4, express all z scores with two decimal places.

1. Darwin's Height Men have heights with a mean of 176 cm and a standard deviation of 7 cm. Charles Darwin had a height of 182 cm.
   a. What is the difference between Darwin's height and the mean?
   b. How many standard deviations is that [the difference found in part (a)]?
   c. Convert Darwin's height to a z score.
   d. If we consider "usual" heights to be those that convert to z scores between -2 and 2, is Darwin's height usual or unusual?

2. Heights of Men Adult males have heights with a mean of 176 cm and a standard deviation of 7 cm. Find the z scores corresponding to the following.
   a. Actor Danny DeVito, who is 152 cm tall
   b. NBA basketball player Shaquille O'Neal, who is 216 cm tall

3. Pulse Rates of Adults Assume that adults have pulse rates (beats per minute) with a mean of 72.9 and a standard deviation of 12.3 (based on data from the National

Health Survey). When this exercise question was written, one of the authors had a pulse rate of 48.

  a. What is the difference between this author's pulse rate and the mean?

  b. How many standard deviations is that [the difference found in part (a)]?

  c. Convert a pulse rate of 48 to a z score.

  d. If we consider "usual" pulse rates to be those that convert to z scores between -2 and 2, is a pulse rate of 48 usual or unusual? Can you identify one reason for an unusually low pulse rate?

4. Body Temperatures Human body temperatures have a mean of 98.20°F and a standard deviation of 0.62°F. Convert the given temperatures to z scores.

  a. 100°F

  b. 96.96°F

  c. 98.20°F

In Exercises 5–8, express all z scores with two decimal places. Consider a score to be unusual if its z score is less than -2.00 or greater than 2.00.

5. Heights of Women The Beanstalk Club is limited to women and men who are very tall. The minimum height requirement for women is 70 in. Women's heights have a mean of 63.6 in. and a standard deviation of 2.5 in. Find the z score corresponding to a woman with a height of 70 in. and determine whether that height is unusual.

6. Length of Pregnancy A woman wrote to *Dear Abby* and claimed that she gave birth 308 days after a visit from her husband, who was in the Navy. Lengths of pregnancies have a mean of 268 days and a standard deviation of 15 days. Find the z score for 308 days. Is such a length unusual? What do you conclude?

7. Body Temperature Human body temperatures have a mean of 98.20°F and a standard deviation of 0.62°F. An emergency room patient is found to have a temperature of 101°F. Convert 101°F to a z score. Is that temperature unusually high? What does it suggest?

8. Cholesterol Levels For men aged between 18 and 24 years, serum cholesterol levels (in mg/100 mL) have a mean of 178.1 and a standard deviation of 40.7 (based on data from the National Health Survey). Find the z score corresponding to a male, aged 18–24 years, who has a serum cholesterol level of 259.0 mg/100 mL. Is this level unusually high?

9. Comparing Test Scores Which is relatively better: A score of 85 on a biology test or a score of 45 on an economics test? Scores on the biology test have a mean of 90 and a standard deviation of 10. Scores on the economics test have a mean of 55 and a standard deviation of 5.

10. Comparing Scores Three students take equivalent stress tests. Which is the highest relative score?

  a. A score of 144 on a test with a mean of 128 and a standard deviation of 34.

  b. A score of 90 on a test with a mean of 86 and a standard deviation of 18.

  c. A score of 18 on a test with a mean of 15 and a standard deviation of 5.

In Exercises 11–14, use the 40 sorted cotinine levels of smokers listed in Table 2-11. Find the percentile corresponding to the given cotinine level.

11. 149   **12.** 210   **13.** 35   **14.** 250

In Exercises 15–22, use the 40 sorted cotinine levels of smokers listed in Table 2-11. Find the indicated percentile or quartile.

15. P_{20} **16.** Q_3 **17.** P_{75} **18.** Q_2

19. P_{33} **20.** P_{21} **21.** P_1 **22.** P_{85}

*23. Cotinine Levels of Smokers Use the sorted cotinine levels of smokers listed in Table 2-11.

    a. Find the interquartile range.
    b. Find the midquartile.
    c. Find the 10–90 percentile range.
    d. Does $P_{50} = Q_2$? If so, does P_{50} *always* equal Q_2?
    e. Does $Q_2 = (Q_1 + Q_3)/2$? If so, does Q_2 *always* equal $(Q_1 + Q_3)/2$?

2-7 Exploratory Data Analysis (EDA)

This chapter presents the basic tools for describing, exploring, and comparing data, and the focus of this section is the exploration of data. We begin this section by first defining exploratory data analysis, then we introduce outliers, 5-number summaries, and boxplots.

> **Definition**
>
> **Exploratory data analysis** is the process of using statistical tools (such as graphs, measures of center, measures of variation) to investigate data sets in order to understand their important characteristics.

Recall that in Section 2-1 we listed five important characteristics of data, and we began with (1) *center*, (2) *variation*, and (3) the nature of the *distribution*. These characteristics can be investigated by calculating the values of the mean and standard deviation, and by constructing a histogram. It is generally important to further investigate the data set to identify any notable features, especially those that could strongly affect results and conclusions. One such feature is the presence of outliers.

Outliers

An **outlier** is a value that is located very far away from almost all of the other values. Relative to the other data, an outlier is an *extreme* value. When exploring a data set, outliers should be considered because they may reveal important information, and they may strongly affect the value of the mean and standard deviation, as well as seriously distorting a histogram. The following example uses an incorrect entry as an example of an outlier, but not all outliers are errors; some outliers are correct values.

EXAMPLE **Cotinine Levels of Smokers** When using computer software or a calculator, it is often easy to make keying errors. Refer to the cotinine levels of smokers listed in Table 2-1 with the Chapter Problem and assume that the first entry of 1 is incorrectly entered as 11111 because you were distracted by a meteorite landing on your porch. The incorrect entry of 11111 is an outlier because it is located very far away from the other values. How does that outlier affect the mean, standard deviation, and histogram?

SOLUTION   When the entry of 1 is replaced by the outlier value of 11111, the mean changes from 172.5 to 450.2, so the effect of the outlier is very substantial. The incorrect entry of 11111 causes the standard deviation to change from 119.5 to 1732.7, so the effect of the outlier here is also substantial. Figure 2-1 in Section 2-3 depicts the histogram for the correct values of cotinine levels of smokers in Table 2-1, but the STATDISK display below shows the histogram that results from using the same data with the value of 1 replaced by the incorrect value of 11111. Compare this STATDISK histogram to Figure 2-1 and you can easily see that the presence of the outlier dramatically affects the shape of the distribution.

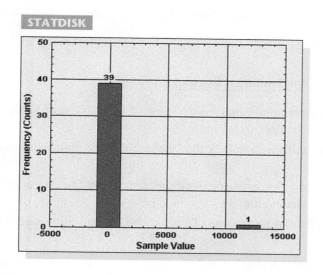

The preceding example illustrates these important principles:

1. **An outlier can have a dramatic effect on the mean.**
2. **An outlier can have a dramatic effect on the standard deviation.**
3. **An outlier can have a dramatic effect on the scale of the histogram so that the true nature of the distribution is totally obscured.**

An easy procedure for finding outliers is to examine a *sorted* list of the data. In particular, look at the minimum and maximum sample values and determine whether they are very far away from the other typical values. Some outliers are correct values and some are errors, as in the preceding example. If we are sure that an outlier is an error, we should correct it or delete it. If we include an outlier because we know that it is correct, we might study its effects by constructing graphs and calculating statistics with and without the outlier included.

Boxplots

In addition to the graphs presented in Section 2-3, a boxplot is another graph that is used often. Boxplots are useful for revealing the center of the data, the spread of the data, the distribution of the data, and the presence of outliers. The construction of a boxplot requires that we first obtain the minimum value, the maximum value, and quartiles, as defined in the *5-number summary*.

Definition

For a set of data, the **5-number summary** consists of the minimum value, the first quartile Q_1, the median (or second quartile Q_2), the third quartile Q_3, and the maximum value.

A **boxplot** (or **box-and-whisker diagram**) is a graph of a data set that consists of a line extending from the minimum value to the maximum value, and a box with lines drawn at the first quartile Q_1, the median, and the third quartile Q_3. (See Figure 2-16.)

Procedure for Constructing a Boxplot

1. Find the 5-number summary consisting of the minimum value, Q_1, the median, Q_3, and the maximum value.

2. Construct a scale with values that include the minimum and maximum data values.

3. Construct a box (rectangle) extending from Q_1 to Q_3, and draw a line in the box at the median value.

4. Draw lines extending outward from the box to the minimum and maximum data values.

Boxplots don't show as much detailed information as histograms or stem-and-leaf plots, so they might not be the best choice when dealing with a single data set. They are often great for comparing two or more data sets. When using two or more boxplots for comparing different data sets, it is important to use the same scale so that correct comparisons can be made.

EXAMPLE Cotinine Levels of Smokers Refer to the 40 cotinine levels of smokers in Table 2-1 (without the error of 11111 used in place of 1, as in the preceding example).

a. Find the values constituting the 5-number summary.

b. Construct a boxplot.

SOLUTION

a. The 5-number summary consists of the minimum, Q_1, median, Q_3, and maximum. To find those values, first sort the data (by arranging them in order from lowest to highest). The minimum of 0 and the maximum of 491 are easy to identify from the sorted list. Now proceed to find the quartiles. Using the flowchart of Figure 2-15, we get $Q_1 = P_{25} = 86.5$, which is located by calculating the locator $L = (25/100)40 = 10$ and finding the value midway between the 10th value and the 11th value in the sorted list. The median is 170, which is the value midway between the 20th and 21st values. We also find that $Q_3 = 251.5$ by using Figure 2-15 for the 75th percentile. The 5-number summary is therefore 0, 86.5, 170, 251.5, and 491.

b. In Figure 2-16 we graph the boxplot for the data. We use the minimum (0) and the maximum (491) to determine a scale of values, then we plot the values from the 5-number summary as shown.

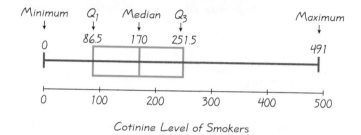

FIGURE 2-16 **Boxplot**

In Figure 2-17 we show some generic boxplots along with common distribution shapes. It appears that the cotinine levels of smokers have a skewed distribution.

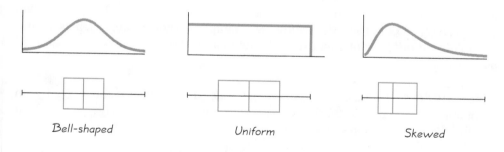

FIGURE 2-17 **Boxplots Corresponding to Bell-Shaped, Uniform, and Skewed Distributions**

To illustrate the use of boxplots to compare data sets, see the accompanying Minitab and SPSS displays of cholesterol levels for a sample of males and a sample of females, based on the National Health Survey data included in Data Set 1 of Appendix B. Based on the sample data, it appears that males have cholesterol levels that are generally higher than females, and the cholesterol levels of males appear to vary more than those of females.

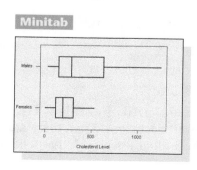

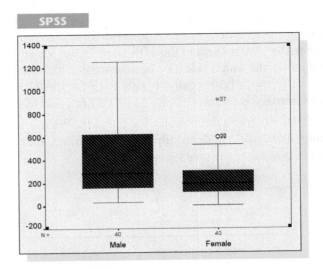

The preceding description of boxplots describes **skeletal** (or **regular**) **boxplots,** but some statistical software packages provide **modified boxplots,** which show outliers as special points (as in the displayed SPSS-generated boxplot for the cholesterol levels of females). For example, Minitab uses asterisks to identify points that are exceptional because they are greater than almost all other points or less than almost all other points. Minitab's specific criterion is to use asterisks for points representing values less than $Q_1 - 1.5(Q_3 - Q_1)$ or greater than $Q_3 + 1.5(Q_3 - Q_1)$. (See the following example.) Another approach is to use solid dots for *mild outliers* and small hollow circles for *extreme outliers,* defined as follows:

Mild Outliers   (plotted as *solid dots*): Values below Q_1 or above Q_3 by an amount that is greater than $1.5(Q_3 - Q_1)$ but not greater than $3(Q_3 - Q_1)$. That is, mild outliers are values of x such that

$$Q_1 - 3(Q_3 - Q_1) \le x < Q_1 - 1.5(Q_3 - Q_1)$$

or

$$Q_3 + 1.5(Q_3 - Q_1) < x \le Q_3 + 3(Q_3 - Q_1)$$

Extreme Outliers   (plotted as small *hollow circles*): Values that are either below Q_1 by more than $3(Q_3 - Q_1)$ or above Q_3 by more than $3(Q_3 - Q_1)$. That is, extreme outliers are values of x such that

$$x < Q_1 - 3(Q_3 - Q_1)$$

or

$$x > Q_3 + 3(Q_3 - Q_1)$$

EXAMPLE **Do Males and Females Have the Same Pulse Rates?** It has been well documented that there are important physiological differences between males and females. Males tend to weigh more and they tend to be taller than females. But is there a difference in pulse rates of males and females? Data Set 1 in Appendix B lists pulse rates for a sample of 40 males and a sample of 40 females. Later in this book we will describe important statistical methods that can be used to formally test for differences, but for now, let's explore the data to see what can be learned. (Even if we already know how to apply those formal statistical methods, it would be wise to first explore the data before proceeding with the formal analysis.)

SOLUTION   Let's begin with an investigation into the key elements of center, variation, distribution, outliers, and characteristics over time (the same "CVDOT" list introduced in Section 2-1). The accompanying displays show Minitab-generated boxplots for each of the two samples (with asterisks denoting values that are outliers), an SPSS-generated histogram of the male pulse rates, and an SAS-generated relative-frequency histogram of the female pulse rates. Listed below are measures of center (mean), measures of variation (standard deviation), and the 5-number summary for the pulse rates listed in Data Set 1.

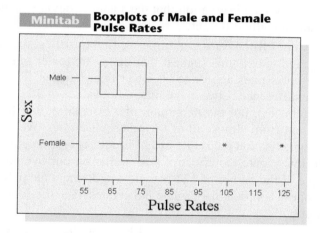

Minitab   **Boxplots of Male and Female Pulse Rates**

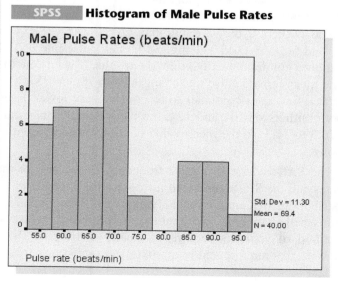

SPSS   **Histogram of Male Pulse Rates**

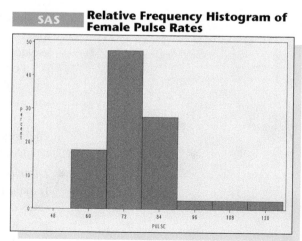

SAS   **Relative Frequency Histogram of Female Pulse Rates**

continued

	Mean	Standard Deviation	Minimum	Q_1	Median	Q_3	Maximum
Males	69.4	11.3	56	60.0	66.0	76.0	96
Females	76.3	12.5	60	68.0	74.0	80.0	124

INTERPRETATION Examining and comparing the statistics and graphs, we make the following important observations.

- *Means:* The mean pulse rates of 69.4 for males and 76.3 for females appear to be very different. The boxplots are based on minimum and maximum values along with the quartiles, so the boxplots do not depict the values of the two means, but they do suggest that the pulse rates of males are somewhat lower than those of females, so it follows that the mean pulse rates for males do appear to be lower than the mean for females.

- *Variation:* The standard deviations 11.3 and 12.3 do not appear to be dramatically different. Also, the boxplots depict the spread of the data. The widths of the boxplots do not appear to be very different, further supporting the observation that the standard deviations do not appear to be dramatically different.

- The values listed for the minimums, first quartiles, medians, third quartiles, and maximums suggest that males are lower in each case, so that male pulse rates appear to be lower than females. There is a very dramatic difference between the maximum values of 96 (for males) and 124 (for females), but the maximum of 124 is an outlier because it is very different from almost all of the other pulse rates. We should now question whether that pulse rate of 124 is correct or is an error, and we should also investigate the effect of that outlier on our overall results. For example, if the outlier of 124 is removed, do the male pulse rates continue to appear less than the female pulse rates? See the following comments about outliers.

- *Outliers:* The Minitab-generated boxplots include two asterisks corresponding to two female pulse rates considered to be outliers. [Minitab identifies points as outliers if they are less than $Q_1 - 1.5(Q_3 - Q_1)$ or greater than $Q_3 + 1.5(Q_3 - Q_1)$.] The highest female pulse rate of 124 is an outlier that does not dramatically affect the results. If we delete the value of 124, the mean female pulse rate changes from 76.3 to 75.08, so the mean pulse rate for males continues to be considerably lower. Even if we delete both outliers of 104 and 124, the mean female pulse rate changes from 76.3 to 74.3, so the mean value of 69.4 for males continues to appear lower.

- *Distributions:* The SAS-generated relative frequency histogram for the female pulse rates and the SPSS-generated histogram for the male pulse rates depict distributions that are not radically different. Both graphs appear to be roughly bell-shaped, as we might have expected. If the use of a particular method of statistics requires normally distributed (bell-shaped) populations, that requirement is approximately satisfied for both sets of sample data.

We now have considerable insight into the nature of the pulse rates for males and females. Based on our exploration, we can conclude that males appear to have pulse rates with a mean that is less than that of females. There are more advanced methods we could use later (such as the methods of Section 8-3), but the tools presented in this chapter give us considerable insight.

Critical Thinking

Armed with a list of tools for investigating center, variation, distribution, outliers, and characteristics of data over time, we might be tempted to develop a rote and mindless procedure, but critical thinking is critically important. In addition to using the tools presented in this chapter, we should consider any other relevant factors that might be crucial to the conclusions we form. We might pose questions such as these: Is the sample likely to be representative of the population, or is the sample somehow biased? What is the source of the data, and might the source be someone with an interest that could affect the quality of the data? Suppose, for example, that we want to estimate the mean income of biologists. Also suppose that we mail questionnaires to 200 biologists and receive 20 responses. We could calculate the mean, standard deviation, construct graphs, identify outliers, and so on, but the results will be what statisticians refer to as hogwash. The sample is a voluntary response sample, and it is not likely to be representative of the population of all biologists. In addition to the specific statistical tools presented in this chapter, we should also *think!*

2-7 Exercises

1. Testing Corn Seeds In 1908, William Gosset published the article "The Probable Error of a Mean" under the pseudonym of "Student" (*Biometrika*, Vol. 6, No. 1). He included the data listed below for two different types of corn seed (regular and kiln dried) that were used on adjacent plots of land. The listed values are the yields of head corn in pounds per acre. Using the yields from *regular seed*, find the 5-number summary and construct a boxplot.

Regular	1903	1935	1910	2496	2108	1961	2060	1444	1612	1316	1511
Kiln Dried	2009	1915	2011	2463	2180	1925	2122	1482	1542	1443	1535

2. Testing Corn Seeds Using the yields from the *kiln dried seed* listed in Exercise 1, find the 5-number summary and construct a boxplot. Do the results appear to be substantially different from those obtained in Exercise 1?

3. Fruit Flies Listed below are the thorax lengths (in millimeters) of a sample of male fruit flies. Find the 5-number summary and construct a boxplot.

0.72	0.90	0.84	0.68	0.84	0.90
0.92	0.84	0.64	0.84	0.76	

4. Bear Lengths Refer to Data Set 6 for the lengths (in inches) of the 54 bears that were anesthetized and measured. Find the 5-number summary and construct a boxplot. Does the distribution of the lengths appear to be symmetric or does it appear to be skewed?

5. Body Temperatures Refer to Data Set 2 in Appendix B for the 106 body temperatures for 12 A.M. on day 2. Find the 5-number summary and construct a boxplot, then determine whether the sample values support the common belief that the mean body temperature is 98.6°F.

6. Cuckoo Egg Lengths Refer to Data Set 8 in Appendix B and use the lengths (in mm) of cuckoo eggs found in nests of meadow pipits. Find the 5-number summary and construct a boxplot.

In Exercises 7–16, find 5-number summaries, construct boxplots, and compare the data sets.

7. Cuckoo Egg Lengths Refer to Data Set 8 in Appendix B. Exercise 6 uses the lengths (in mm) of cuckoo eggs found in nests of meadow pipits. Also use the lengths of cuckoo eggs found in nests of tree pipits and robins. Compare the three sets of results.

8. Poplar Trees Refer to Data Set 9 in Appendix B. Use the second-year weights from the poplar trees given no treatment and those treated with fertilizer and irrigation. Are there any substantial differences?

9. Head Circumferences Refer to Data Set 4 in Appendix B and use the head circumferences of boys and the head circumferences of girls. Is there a substantial difference?

10. Body Mass Indexes Refer to Data Set 1 in Appendix B and use the body mass indexes of males and the body mass indexes of females. Do they differ?

11. Ages In "Ages of Oscar-Winning Best Actors and Actresses" (*Mathematics Teacher* magazine) by Richard Brown and Gretchen Davis, the authors compare the ages of actors and actresses at the time they won Oscars. Results for winners from both categories are listed in the following table. Use boxplots to compare the two data sets.

Actors:	32	37	36	32	51	53	33	61	35	45	55	39
	76	37	42	40	32	60	38	56	48	48	40	43
	62	43	42	44	41	56	39	46	31	47	45	60
	46	40	36									
Actresses:	50	44	35	80	26	28	41	21	61	38	49	33
	74	30	33	41	31	35	41	42	37	26	34	34
	35	26	61	60	34	24	30	37	31	27	39	34
	26	25	33									

12. Cotinine Levels Refer to Table 2-1 located in the Chapter Problem. We have already found that the 5-number summary for the cotinine levels of smokers is 0, 86.5, 170, 251.5, and 491. Find the 5 number summaries for the other two groups, then construct the three boxplots using the same scale. Are there any apparent differences?

13. Petal Widths of Irises Refer to Data Set 7 in Appendix B and use the petal widths of the irises from each of the three different classes. Find the 5-number summaries and construct boxplots. Do the three data sets appear to be approximately the same?

14. Sepal Lengths of Irises Refer to Data Set 7 in Appendix B and use the sepal lengths of the irises from each of the three different classes. Find the 5-number summaries and construct boxplots. Do the three data sets appear to be approximately the same?

15. Hemoglobin Counts Refer to Data Set 10 in Appendix B and use the hemoglobin counts for males and the hemoglobin counts for females. Find the 5-number summaries and construct boxplots. Do the hemoglobin counts appear to be the same for both genders?

16. White Blood Cell Counts Refer to Data Set 10 in Appendix B and use the white blood cell counts for males and the white blood cell counts for females. Find the 5-number summaries and construct boxplots. Do the white blood cell counts appear to be the same for both genders?

Review

In this chapter we considered methods for describing, exploring, and comparing data sets. When investigating a data set, these characteristics are generally very important:

1. *Center:* A representative or average value.

2. *Variation:* A measure of the amount that the values vary.

3. *Distribution:* The nature or shape of the distribution of the data (such as bell-shaped, uniform, or skewed).

4. *Outliers:* Sample values that lie very far away from the vast majority of the other sample values.

5. *Time:* Changing characteristics of the data over time.

After completing this chapter we should be able to do the following:

- Summarize data by constructing a frequency distribution or relative frequency distribution (Section 2-2).
- Visually display the nature of the distribution by constructing a histogram, dotplot, stem-and-leaf plot, pie chart, or Pareto chart (Section 2-3).
- Calculate measures of center by finding the mean, median, mode, and midrange (Section 2-4).
- Calculate measures of variation by finding the standard deviation, variance, and range (Section 2-5).
- Compare individual values by using z scores, quartiles, or percentiles (Section 2-6).
- Investigate and explore the spread of data, the center of the data, and the range of values by constructing a boxplot (Section 2-7).

In addition to creating these tables, graphs, and measures, you should be able to understand and interpret those results. For example, you should clearly understand that the standard deviation is a measure of how much data vary, and you should be able to use the standard deviation to distinguish between values that are usual and those that are unusual.

Review Exercises

1. Tree Heights In a study of the relationship between heights and trunk diameters of trees, botany students collected sample data. Listed below are the tree circumferences (in feet). The data are based on results in "Tree Measurements" by Stanley Rice, *American Biology Teacher*, Vol. 61, No. 9. Using the circumferences listed below,

find the (a) mean; (b) median; (c) mode; (d) midrange; (e) range; (f) standard deviation; (g) variance; (h) Q_1; (i) Q_3; (j) P_{10}.

| 1.8 | 1.9 | 1.8 | 2.4 | 5.1 | 3.1 | 5.5 | 5.1 | 8.3 | 13.7 |
| 5.3 | 4.9 | 3.7 | 3.8 | 4.0 | 3.4 | 5.2 | 4.1 | 3.7 | 3.9 |

2. a. Using the results from Exercise 1, convert the circumference of 13.7 ft to a z score.
   b. In the context of these sample data, is the circumference of 13.7 ft "unusual"? Why or why not?
   c. Using the range rule of thumb, identify any other listed circumferences that are unusual.

3. Frequency distribution Using the same tree circumferences listed in Exercise 1, construct a frequency distribution. Use seven classes with 1.0 as the lower limit of the first class, and use a class width of 2.0.

4. Histogram Using the frequency distribution from Exercise 3, construct a histogram and identify the general nature of the distribution (such as uniform, bell-shaped, skewed).

5. Boxplot Using the same circumferences listed in Review Exercise 1, construct a boxplot and identify the values constituting the 5-number summary.

Cumulative Review Exercises

1. Tree Measurements Refer to the sample of tree circumferences (in feet) listed in Review Exercise 1.
   a. Are the given values from a population that is discrete or continuous?
   b. What is the level of measurement of these values? (Nominal, ordinal, interval, ratio)

2. a. A set of data is at the nominal level of measurement and you want to obtain a representative data value. Which of the following is most appropriate: mean, median, mode, or midrange? Why?
   b. A botanist wants to obtain data about the plants being grown in homes. If a sample is obtained by telephoning the first 250 people listed in the local telephone directory, what type of sampling is being used? (Random, stratified, systematic, cluster, convenience)
   c. An exit poll is conducted by surveying everyone who leaves the polling booth at 50 randomly selected election precincts. What type of sampling is being used? (Random, stratified, systematic, cluster, convenience)
   d. A manufacturer makes fertilizer sticks to be used for growing plants. A manager finds that the amounts of fertilizer placed in the sticks are not very consistent, so that some fertilization lasts longer than claimed, but others don't last long enough. She wants to improve quality by making the amounts of fertilizer more consistent. When analyzing the amounts of fertilizer, which of the following statistics is most relevant: mean, median, mode, midrange, standard deviation, first quartile, third quartile? Should the value of that statistic be raised, lowered, or left unchanged?

Cooperative Group Activities

1. *Out-of-class activity:* Are estimates influenced by anchoring numbers? In the article "Weighing Anchors" in *Omni* magazine, author John Rubin observed that when people estimate a value, their estimate is often "anchored" to (or influenced by) a preceding number, even if that preceding number is totally unrelated to the quantity being estimated. To demonstrate this, he asked people to give a quick estimate of the value of $8 \times 7 \times 6 \times 5 \times 4 \times 3 \times 2 \times 1$. The average answer given was 2250, but when the order of the numbers was reversed, the average became 512. Rubin explained that when we begin calculations with larger numbers (as in $8 \times 7 \times 6$), our estimates tend to be larger. He noted that both 2250 and 512 are far below the correct product, 40,320. The article suggests that irrelevant numbers can play a role in influencing real estate appraisals, estimates of car values, and estimates of the likelihood of nuclear war.

Conduct an experiment to test this theory. Select some subjects and ask them to quickly estimate the value of

$$8 \times 7 \times 6 \times 5 \times 4 \times 3 \times 2 \times 1$$

Then select other subjects and ask them to quickly estimate the value of

$$1 \times 2 \times 3 \times 4 \times 5 \times 6 \times 7 \times 8$$

Record the estimates along with the particular order used. Carefully design the experiment so that conditions are uniform and the two sample groups are selected in a way that minimizes any bias. Don't describe the theory to subjects until after they have provided their estimates. Compare the two sets of sample results by using the methods of this chapter. Provide a printed report that includes the data collected, the detailed methods used, the method of analysis, any relevant graphs and/or statistics, and a statement of conclusions. Include a critique of reasons why the results might not be correct and describe ways in which the experiment could be improved.

A variation of the preceding experiment is to survey people about their knowledge of the population of Kenya. First ask half of the subjects whether they think the population is above 5 million or below 5 million, then ask them to estimate the population with an actual number. Ask the other half of the subjects whether they think the population is above 80 million or below 80 million, then ask them to estimate the population. (Kenya's population is 28 million.) Compare the two sets of results and identify the "anchoring" effect of the initial number that the survey subjects are given.

2. *Out-of-class activity:* In each group of three or four students, collect an original data set of values at the interval or ratio levels of measurement. Provide the following: (1) a list of sample values, (2) printed computer results of descriptive statistics and graphs, and (3) a written description of the nature of the data, the method of collection, and important characteristics.

3. *In-class activity:* Given below are the ages of motorcyclists at the time they were fatally injured in traffic accidents (based on data from the U.S. Department of Transportation). If your objective is to dramatize the dangers of motorcycles for young people, which would be most effective: histogram, Pareto chart, pie chart, dotplot, mean, median, . . . ? Construct the graph and find the statistic that best meets that objective. Is it okay to deliberately distort data if the objective is saving lives of motorcyclists?

17	38	27	14	18	34	16
42	28	24	40	20	23	31
37	21	30	25	17	28	33
25	23	19	51	18	29	

4. *Out-of-class activity:* In each group of three or four students, select one of the following items and construct a graph that is effective in addressing the question:
   a. Is there a difference between the body mass index (BMI) values for men and for women? (See Data Set 1 in Appendix B.)
   b. Is there a relationship between the heights of sons (or daughters) and the heights of their fathers (or mothers)? (See Data Set 3 in Appendix B.)

Technology Project

When dealing with larger data sets, manual construction of graphs and manual calculations of data can become quite tedious. Select a technology, such as Minitab, Excel, SPSS, SAS, STATDISK, or a TI-83/84 Plus graphing calculator. Load the data consisting of the 54 weights of bears in Data Set 6 of Appendix B. Instead of manually entering the 54 weights, try to retrieve the data from the CD included with this book. Proceed to generate a histogram and find appropriate statistics that allow you to better understand this data set. Are there any notable features? Are there any outliers? Describe the key elements of center, variation, distribution, and outliers. Write a brief report including your conclusions and supporting graphs.

from DATA to DECISION

Critical Thinking

Car crash fatalities are devastating to the families involved, and they often involve lawsuits and large insurance payments. Listed below are the ages of 100 randomly selected drivers who were killed in car crashes. Also given is a frequency distribution of licensed drivers by age.

Ages (in years) of Drivers Killed in Car Crashes

37	76	18	81	28	29	18	18	27	20
18	17	70	87	45	32	88	20	18	28
17	51	24	37	24	21	18	18	17	40
25	16	45	31	74	38	16	30	17	34
34	27	87	24	45	24	44	73	18	44
16	16	73	17	16	51	24	16	31	38
86	19	52	35	18	18	69	17	28	38
69	65	57	45	23	18	56	16	20	22
77	18	73	26	58	24	21	21	29	51
17	30	16	17	36	42	18	76	53	27

Age	Licensed Drivers (millions)
16–19	9.2
20–29	33.6
30–39	40.8
40–49	37.0
50–59	24.2
60–69	17.5
70–79	12.7
80–89	4.3

Analysis

Convert the given frequency distribution to a relative frequency distribution, then create a relative frequency distribution for the ages of drivers killed in car crashes. Compare the two relative frequency distributions. Which age categories appear to have substantially greater proportions of fatalities than the proportions of licensed drivers? If you were responsible for establishing the rates for auto insurance, which age categories would you select for higher rates? Construct a graph that is effective in identifying age categories that are more prone to fatal car crashes.

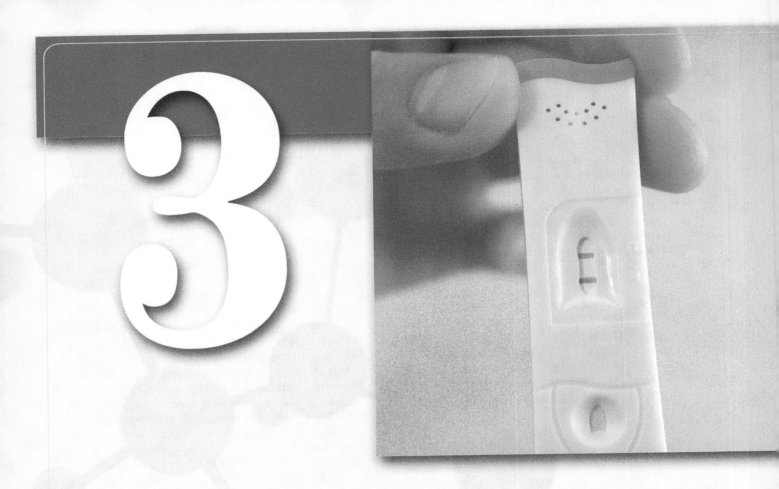

3

Probability

False positives and false negatives

All of us undergo a variety of health tests at different times in our lives. Some health tests are as simple as using a thermometer to determine whether body temperature is too high or low, or using a sphygmomanometer to determine whether blood pressure is too high or low. Some other tests involve the analysis of blood samples for identifying the presence of disease. In this Chapter Problem we consider results from a clinical study of a test for pregnancy. It is important for a woman to know if she is pregnant so that she can discontinue any activities, medications, exposure to toxicants, smoking, or alcohol consumption that could be potentially harmful to the baby. Pregnancy tests, like almost all health tests, do not yield results that are 100% accurate. In clinical trials of a blood test for pregnancy, 99 women are randomly selected from a population of women who seek medical help in determining whether they are pregnant. See results shown in Table 3-1 for the Abbot blood test (based on data from "Specificity and Detection Limit of Ten Pregnancy Tests," by Tiitinen and Stenman, *Scandinavian Journal of Clinical Laboratory Investigation*, Volume 53, Supplement 216). There are factors that affect the accuracy of these tests, such as timing. Pregnancy tests are typically most reliable when they are taken at least two weeks after conception. Other tests are more reliable than the test with results given in Table 3-1. For example, the Abbot Testpack Plus has a 0.2% false positive rate and a 0.6% false negative rate. The terms *false positive* and *false negative* are included among the following terms commonly used with health tests or screening procedures. The "condition" in the following definitions can be pregnancy, presence of some disease, or any other factor being considered.

- False positive: Test *incorrectly* indicates the presence of a condition when the subject does not actually have that condition.

- False negative: Test *incorrectly* indicates that the subject does not have a condition when the subject actually does have that condition.

- True positive: Test *correctly* indicates that a condition is present when it really is present.

- True negative: Test *correctly* indicates that a condition is not present when it really is not present.

- Test sensitivity: The probability of a true positive.

- Test specificity: The probability of a true negative.

- Positive predictive value: Probability that a subject is a true positive given that the test yields a positive result (indicating that the condition is present).

- Negative predictive value: Probability that the subject is a true negative given that the test yields a negative result (indicating that the condition is not present).

- Prevalence: Proportion of subjects having some condition.

Based on the results in Table 3-1, what is the probability of a woman being pregnant if the test indicates a negative result? What is the probability of a false positive? We will address these questions in this chapter.

Table 3-1	Pregnancy Test Results	
	Positive Test Result (Pregnancy is indicated)	Negative Test Result (Pregnancy is not indicated)
Subject is pregnant	80 (True positive)	5 (False negative)
Subject is not pregnant	3 (False positive)	11 (True negative)

3-1 Overview

Probability is the underlying foundation on which the important methods of inferential statistics are built. As a simple example, suppose that you have developed a gender-selection procedure and you claim that it greatly increases the likelihood of a baby being a girl. Suppose that independent test results from 100 couples result in 98 girls and only 2 boys. Even though there is a chance of getting 98 girls in 100 births with no special treatment, that chance is so incredibly low that it would be rejected as a reasonable explanation. Instead, it would be generally recognized that the results provide strong support for the claim that the gender-selection technique is effective. This is exactly how statisticians think: They reject explanations based on very low probabilities. Statisticians use the *rare event rule*.

Rare Event Rule for Inferential Statistics

If, under a given assumption, the probability of a particular observed event is extremely small, we conclude that the assumption is probably not correct.

The main objective in this chapter is to develop a sound understanding of probability values, which will be used in following chapters. A secondary objective is to develop the basic skills necessary to determine probability values in a variety of important circumstances.

3-2 Fundamentals

In considering probability, we deal with procedures (such as rolling a die, answering a multiple-choice test question, or undergoing a test for pregnancy) that produce outcomes.

Definitions

An **event** is any collection of results or outcomes of a procedure.

A **simple event** is an outcome or an event that cannot be further broken down into simpler components.

The **sample space** for a procedure consists of all possible *simple* events. That is, the sample space consists of all outcomes that cannot be broken down any further.

EXAMPLES

In the following display, we use f to denote a female baby and we use m to denote a male baby.

Procedure	Example of Event	Sample Space
Birth	female (simple event)	{f, m}
3 births	2 females and a male (ffm, fmf, mff are all simple events resulting in 2 females and a male)	{fff, ffm, fmf, fmm, mff, mfm, mmf, mmm}

With one birth, the result of a female is a *simple event* because it cannot be broken down any further. With three births, the event of "2 females and a male" is *not a simple event* because it can be broken down into simpler events, such as ffm, fmf, or mff. With three births, the *sample space* consists of the 8 simple events listed above. With three births, the outcome of ffm is considered a simple event, because it is an outcome that cannot be broken down any further. We might incorrectly think that ffm can be further broken down into the individual results of f, f, and m, but f, f, and m are not individual outcomes from three births. With three births, there are exactly 8 outcomes that are simple events: fff, ffm, fmf, fmm, mff, mfm, mmf, and mmm.

There are different ways to define the probability of an event, and we will present three approaches. First, however, we list some basic notation.

Notation for Probabilities

P denotes a probability.
A, B, and C denote specific events.
$P(A)$ denotes the probability of event A occurring.

Rule 1: Relative Frequency Approximation of Probability

Conduct (or observe) a procedure, and count the number of times that event A actually occurs. Based on these actual results, $P(A)$ is *estimated* as follows:

$$P(A) = \frac{\text{number of times } A \text{ occurred}}{\text{number of times trial was repeated}}$$

Rule 2: Classical Approach to Probability (Requires Equally Likely Outcomes)

Assume that a given procedure has n different simple events and that *each of those simple events has an equal chance of occurring*. If event A can occur in s of these n ways, then

$$P(A) = \frac{\text{number of ways } A \text{ can occur}}{\text{number of different simple events}} = \frac{s}{n}$$

Probabilities That Challenge Intuition

In certain cases, our subjective estimates of probability values are dramatically different from the actual probabilities. Here is a classic example: If you take a deep breath, there is better than a 99% chance that you will inhale a molecule that was exhaled in dying Caesar's last breath. In that same morbid and unintuitive spirit, if Socrates' fatal cup of hemlock was mostly water, then the next glass of water you drink will likely contain one of those same molecules. Here's another less morbid example that can be verified: In classes of 25 students, there is better than a 50% chance that at least two students will share the same birthday.

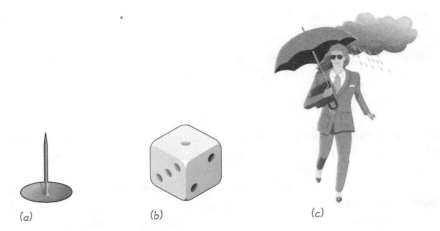

(a)   *(b)*   *(c)*

FIGURE 3-1 **Three Approaches to Finding Probability**

(a) Relative Frequency Approach (Rule 1): When trying to determine: *P*(tack lands point up), we must repeat the procedure of tossing the tack many times and then find the ratio of the number of times the tack lands with the point up to the number of tosses.

(b) Classical Approach (Rule 2): When trying to determine *P*(2) with a balanced and fair die, each of the six faces has an equal chance of occurring.

$$P(2) = \frac{\text{number of ways 2 can occur}}{\text{number of simple events}} = \frac{1}{6}$$

(c) Subjective Probability (Rule 3): When trying to estimate the probability of rain tomorrow, meteorologists use their expert knowledge of weather conditions to develop an estimate of the probability.

Rule 3: Subjective Probabilities

$P(A)$, the probability of event A, is estimated by using knowledge of the relevant circumstances.

It is very important to note that the classical approach (*Rule 2*) requires *equally likely outcomes*. If the outcomes are not equally likely, we must use the relative frequency estimate or we must rely on our knowledge of the circumstances to make an *educated guess*. Figure 3-1 illustrates the three approaches.

When finding probabilities with the relative frequency approach (Rule 1), we obtain an *approximation* instead of an exact value. As the total number of observations increases, the corresponding approximations tend to get closer to the actual probability. This property is stated as a theorem commonly referred to as the *law of large numbers*.

Law of Large Numbers

As a procedure is repeated again and again, the relative frequency probability (from Rule 1) of an event tends to approach the actual probability.

The law of large numbers tells us that the relative frequency approximations from Rule 1 tend to get better with more observations. This law reflects a simple notion supported by common sense: A probability estimate based on only a few trials can be off by substantial amounts, but with a very large number of trials, the estimate tends to be much more accurate. For example, suppose that we want to estimate some population value based on data obtained from a clinical trial. If the clinical trial uses only a dozen randomly selected people, the estimate could easily be in error by large amounts, but a clinical trial with thousands of randomly selected people will yield an estimate that will be fairly close to the true population value.

EXAMPLE **Secondhand Smoke** Find the probability that a randomly selected adult American believes that secondhand smoke is not at all harmful. Here are results from a Gallup poll: Among 1038 randomly selected adults, 52 said that secondhand smoke is not at all harmful.

SOLUTION   The sample space consists of two simple events: The selected person believes that secondhand smoke is not at all harmful, or the person does not have that belief. Because the sample space consists of events that are not equally likely, we can't use the classical approach (Rule 2). We can use the relative frequency approach (Rule 1) with the given results from the Gallup poll. We get the following result:

$$P(\text{person believes that secondhand smoke is not at all harmful})$$
$$= \frac{52}{1038} \approx 0.0501$$

EXAMPLE **Genotypes** As part of a study of the genotypes AA, Aa, aA, and aa, you write each individual genotype on an index card, then you shuffle the four cards and randomly select one of them. What is the probability that you select the genotype Aa?

SOLUTION   Because the sample space (AA, Aa, aA, aa) in this case includes equally likely outcomes, we use the classical approach (Rule 2) to get

$$P(\text{Aa}) = \frac{1}{4}$$

EXAMPLE **Crashing Meteorites** What is the probability that your car will be hit by a meteorite this year?

SOLUTION   In the absence of historical data on meteorites hitting cars, we cannot use the relative frequency approach of Rule 1. There are two possible outcomes (hit or no hit), but they are not equally likely, so we cannot use the classical approach of Rule 2. That leaves us with Rule 3, whereby we make a subjective estimate. In this case, we all know that the probability in question is very, very small. Let's estimate it to be, say, 0.000000000001 (equivalent to

continued

1 in a trillion). That subjective estimate, based on our general knowledge, is likely to be in 4 general ballpark of the true probability.

In basic probability problems of the type we are now considering, it is very important to examine the available information carefully and to identify the total number of possible outcomes correctly. In some cases, the total number of possible outcomes is given, but in other cases it must be calculated, as in the following example.

EXAMPLE **Cloning of Humans** Adults are randomly selected for a Gallup poll, and they are asked if they think that cloning of humans should or should not be allowed. Among the randomly selected adults surveyed, 91 said that cloning of humans should be allowed, 901 said that it should not be allowed, and 20 had no opinion. Based on these results, estimate the probability that a randomly selected person believes that cloning of humans should be allowed.

SOLUTION   *Hint:* Instead of trying to formulate an answer directly from the written statement, summarize the given information in a format that allows you to better understand it. For example:

91	cloning of humans should be allowed
901	cloning of humans should not be allowed
20	no opinion
1012	total

We can now use the relative frequency approach (Rule 1) as follows:

P(cloning of humans should be allowed)

$$= \frac{\text{number believing that cloning of humans should be allowed}}{\text{total}}$$

$$= \frac{91}{1012} = 0.0899$$

We estimate that there is a 0.0899 probability that when an adult is randomly selected, he or she believes that cloning of humans should be allowed. As with all surveys, the accuracy of this result depends on the quality of the sampling method and the survey procedure. Because the poll was conducted by the Gallup organization, the results are likely to be reasonably accurate. Chapter 6 will include more advanced procedures for analyzing such survey results.

EXAMPLE **Gender of Children** Find the probability that when a couple has 3 children, they will have exactly 2 boys. Assume that boys and girls are equally likely and that the gender of any child is not influenced by the gender of any other child.

SOLUTION   The biggest obstacle here is correctly identifying the sample space. It involves more than working only with the numbers 2 and 3 that were given in the statement of the problem. The sample space consists of 8 different ways that 3 children can occur, and we list them in the margin. Those 8 outcomes are equally likely, so we use Rule 2. Of those 8 different possible outcomes, 3 correspond to exactly 2 boys, so

$$P(\text{2 boys in 3 births}) = \frac{3}{8} = 0.375$$

INTERPRETATION   There is a 0.375 probability that if a couple has 3 children, exactly 2 will be boys.

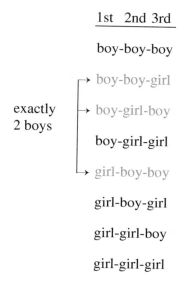

The statements of the three rules for finding probabilities and the preceding examples might seem to suggest that we should always use Rule 2 when a procedure has equally likely outcomes. In reality, many procedures are so complicated that the classical approach (Rule 2) is impractical to use. In the game of solitaire, for example, the outcomes (hands dealt) are all equally likely, but it is extremely frustrating to try to use Rule 2 to find the probability of winning. In such cases we can more easily get good estimates by using the relative frequency approach (Rule 1). Simulations are often helpful when using this approach. (A **simulation** of a procedure is a process that behaves in the same ways as the procedure itself, so that similar results are produced. See the Technology Project at the end of this chapter.) For example, it's much easier to use Rule 1 for estimating the probability of winning at solitaire—that is, to play the game many times (or to run a computer simulation)—than to perform the extremely complex calculations required with Rule 2.

EXAMPLE   **Thanksgiving Day** If a year is selected at random, find the probability that Thanksgiving Day will be on a (a) Wednesday, (b) Thursday.

SOLUTION

a. Thanksgiving Day always falls on the fourth Thursday in November. It is therefore impossible for Thanksgiving to be on a Wednesday. When an event is impossible, we say that its probability is 0.

b. It is certain that Thanksgiving will be on a Thursday. When an event is certain to occur, we say that its probability is 1.

Because any event imaginable is impossible, certain, or somewhere in between, it follows that the mathematical probability of any event is 0, 1, or a number between 0 and 1 (see Figure 3-2).

- The probability of an impossible event is 0.
- The probability of an event that is certain to occur is 1.
- For any event A, the probability of A is between 0 and 1 inclusive. That is, $0 \le P(A) \le 1$.

In Figure 3-2, the scale of 0 through 1 is shown, and the more familiar and common expressions of likelihood are included.

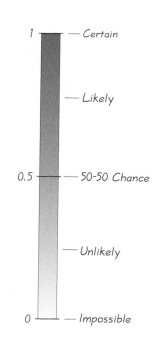

FIGURE 3-2 **Possible Values for Probabilities**

Complementary Events

Sometimes we need to find the probability that an event A does *not* occur.

Definition

The **complement** of event A, denoted by $\overline{A}$, consists of all outcomes in which event A does *not* occur.

EXAMPLE **Birth Genders** In reality, more boys are born than girls. In one typical group, there are 205 newborn babies, 105 of whom are boys. If one baby is randomly selected from the group, what is the probability that the baby is *not* a boy?

SOLUTION   Because 105 of the 205 babies are boys, it follows that 100 of them are girls, so

$$P(\text{not selecting a boy}) = P(\overline{\text{boy}}) = P(\text{girl}) = \frac{100}{205} = 0.488$$

Although it is difficult to develop a universal rule for rounding off probabilities, the following guide will apply to most problems in this text.

Rounding Off Probabilities

When expressing the value of a probability, either give the *exact* fraction or decimal or round off final decimal results to three significant digits. (*Suggestion:* When a probability is not a simple fraction such as 2/3 or 5/9, express it as a decimal so that the number can be better understood.)

All digits in a number are significant except for the zeros that are included for proper placement of the decimal point.

EXAMPLES

- The probability of 0.021491 has five significant digits (21491), and it can be rounded to three significant digits as 0.0215.
- The probability of 1/3 can be left as a fraction, or rounded to 0.333. Do *not* round to 0.3.
- The probability of heads in a coin toss can be expressed as 1/2 or 0.5; because 0.5 is exact, there's no need to express it as 0.500.
- The fraction 432/7842 is exact, but its value isn't obvious, so express it as the decimal 0.0551.

An important concept in this section is the mathematical expression of probability as a number between 0 and 1. This type of expression is fundamental and common in statistical procedures, and we will use it throughout the remainder of this text. A typical computer output, for example, may include a "P-value" expression such as "significance less than 0.001." We will discuss the meaning of P-values later, but they are essentially probabilities of the type discussed in this section. For now, you should recognize that a probability of 0.001 (equivalent to 1/1000) corresponds to an event so rare that it occurs an average of only once in a thousand trials.

3-2 Exercises

In Exercises 1 and 2, express the indicated degree of likelihood as a probability value.

1. Identifying Probability Values
   a. "You have a 50-50 chance of choosing the correct answer."
   b. "There is a 20% chance of rain tomorrow."
   c. "You have absolutely no chance of marrying my daughter."

2. Identifying Probability Values
   a. "There is a 90% chance of snow tomorrow."
   b. "It will definitely become dark tonight."
   c. "You have one chance in ten of being correct."

3. Identifying Probability Values Which of the following values *cannot* be probabilities?

$$0, \quad 1, \quad -1, \quad 2, \quad 0.0123, \quad 3/5, \quad 5/3, \quad \sqrt{2}$$

4. Identifying Probability Values
   a. What is the probability of an event that is certain to occur?
   b. What is the probability of an impossible event?
   c. A sample space consists of 10 separate events that are equally likely. What is the probability of each?
   d. On a true/false test, what is the probability of answering a question correctly if you make a random guess?
   e. On a multiple-choice test with five possible answers for each question, what is the probability of answering a question correctly if you make a random guess?

5. Gender of Children In this section, we gave an example that included a list of the eight outcomes that are possible when a couple has three children. Refer to that list, and find the probability of each event.
   a. Among three children, there is exactly one girl.
   b. Among three children, there are exactly two girls.
   c. Among three children, all are girls.

6. Cell Phones and Brain Cancer In a study of 420,095 cell phone users in Denmark, it was found that 135 developed cancer of the brain or nervous system. Estimate the probability that a randomly selected cell phone user will develop such a cancer. Is the result very different from the probability of 0.000340 that was found for the general population? What does the result suggest about cell phones as a cause of such cancers, as has been claimed?

7. Mendelian Genetics When Mendel conducted his famous genetics experiments with peas, one sample of offspring consisted of 428 green peas and 152 yellow peas. Based on those results, estimate the probability of getting an offspring pea that is green. Is the result reasonably close to the value of 3/4 that was expected?

8. Being Struck by Lightning In a recent year, 389 of the 281,421,906 people in the United States were struck by lightning. Estimate the probability that a randomly selected person in the United States will be struck by lightning this year.

Using Probability to Identify Unusual Events. In Exercises 9–12, consider an event to be "unusual" if its probability is less than or equal to 0.05. (This is equivalent to the same criterion commonly used in inferential statistics, but the value of 0.05 is not absolutely rigid, and other values such as 0.01 are sometimes used instead.)

9. Probability of a Wrong Result Table 3-1 shows that among 85 women who were pregnant, the test for pregnancy yielded the wrong conclusion 5 times.
  a. Based on the available results, find the probability of a wrong test conclusion for a woman who is pregnant.
  b. Is it "unusual" for the test conclusion to be wrong for women who are pregnant?

10. Probability of a Wrong Result Table 3-1 shows that among 14 women who were *not* pregnant, the test for pregnancy yielded the wrong conclusion 3 times.
  a. Based on the available results, find the probability of a wrong test conclusion for a woman who is not pregnant.
  b. Is it "unusual" for the test conclusion to be wrong for women who are not pregnant?

11. Cholesterol-Reducing Drug In a clinical trial of Lipitor, a common drug used to lower cholesterol, one group of patients was given a treatment of 10 mg atorvastatin tablets. (Lipitor consists of atorvastatin calcium.) That group consists of 19 patients who experienced flu symptoms and 844 who did not (based on data from Pfizer, Inc.).
  a. Estimate the probability that a patient taking the drug will experience flu symptoms.
  b. Is it "unusual" for a patient taking the drug to experience flu symptoms?

12. Guessing Birthdays On their first date, Kelly asks Mike to guess the date of her birth, not including the year.
  a. What is the probability that Mike will guess correctly? (Ignore leap years.)
  b. Would it be "unusual" for him to guess correctly on his first try?
  c. If you were Kelly, and Mike did guess correctly on his first try, would you believe his claim that he made a lucky guess, or would you be convinced that he already knew when you were born?
  d. If Kelly asks Mike to guess her age, and Mike's guess is too high by 15 years, what is the probability that Mike and Kelly will have a second date?

13. Probability of a Birthday
  a. If a person is randomly selected, find the probability that his or her birthday is October 18, which is National Statistics Day in Japan. Ignore leap years.
  b. If a person is randomly selected, find the probability that his or her birthday is in October. Ignore leap years.
  c. If a person is randomly selected, find the probability that he or she was born on a day of the week that ends with the letter *y*.

14. Probability of a Car Crash Among 400 randomly selected drivers in the 20–24 age bracket, 136 were in a car accident during the last year (based on data from the National Safety Council). If a driver in that age bracket is randomly selected, what is the approximate probability that he or she will be in a car accident during the next year? Is the resulting value high enough to be of concern to those in the 20–24 age bracket?

15. Probability of an Adverse Drug Reaction When the drug Viagra was clinically tested, 117 patients reported headaches and 617 did not (based on data from Pfizer, Inc.). Use this sample to estimate the probability that a Viagra user will experience a headache. Is the probability high enough to be of concern to Viagra users?

16. Gender of Children: Constructing Sample Space This section included a table summarizing the gender outcomes for a couple planning to have three children.
   a. Construct a similar table for a couple planning to have *two* children.
   b. Assuming that the outcomes listed in part (a) are equally likely, find the probability of getting two girls.
   c. Find the probability of getting exactly one child of each gender.

17. Genetics: Constructing Sample Space Both parents have the brown/blue pair of eye-color genes, and each parent contributes one gene to a child. Assume that if the child has at least one brown gene, that color will dominate and the eyes will be brown. (The actual determination of eye color is somewhat more complicated.)
   a. List the different possible outcomes. Assume that these outcomes are equally likely.
   b. What is the probability that a child of these parents will have the blue/blue pair of genes?
   c. What is the probability that the child will have brown eyes?

18. Genetics: Constructing Sample Space Repeat Exercise 17 assuming that one parent has a brown/brown pair of eye-color genes while the other parent has a brown/blue pair of eye-color genes.

19. Genetics: Constructing Sample Space Repeat Exercise 17 assuming that one parent has a brown/brown pair of eye-color genes while the other parent has a blue/blue pair of eye-color genes.

20. Interpreting Effectiveness of a Treatment A double-blind experiment is designed to test the effectiveness of the drug Statisticzene as a treatment for number blindness. When treated with Statisticzene, subjects seem to show improvement. Researchers calculate that there is a 0.04 probability that the treatment group would show improvement if the drug has no effect. What should you conclude about the effectiveness of Statisticzene?

3-3 Addition Rule

The main objective of this section is to introduce the *addition rule* as a device for finding probabilities that can be expressed as $P(A \text{ or } B)$, the probability that either event A occurs or event B occurs (or they both occur) as the single outcome of a procedure. The key word to remember is *or*. Throughout this text we use the *inclusive or*, which means either one or the other or both. (Except for Exercise 24, we will not consider the *exclusive or*, which means either one or the other but not both.)

In the previous section we presented the fundamentals of probability and considered events categorized as *simple*. In this and the following section we consider *compound events*.

Definition

A **compound event** is any event combining two or more simple events.

Notation for Addition Rule

$$P(A \text{ or } B) = P(\text{in a single trial, event } A \text{ occurs or event } B \\ \text{occurs or they both occur})$$

Table 3-2			
Peas Used in Hybridization Experiments			
		Flower	
		Purple	White
Pod	Green	5	3
	Yellow	4	2

Probabilities play a very prominent role in genetics. See Figure 3-3, which depicts a sample of peas, like those Mendel used in his famous hybridization experiments. The peas shown have green or yellow pods and they have purple or white flowers. Sometimes, the visual display of data can help or hinder our understanding. Figure 3-3 might be a bit difficult to work with, but Table 3-2 summarizes the same key data in a format that might be much easier.

In the sample of 14 peas summarized in Figure 3-3 and Table 3-2, how many of them have "green pods *or* purple flowers"? (Remember, "green pod or purple flower" really means "green pod, or purple flower, or both.") Examination of Figure 3-3 should show that a total of 12 peas have green pods or purple flowers.

FIGURE 3-3 **Peas Used in a Genetics Study**

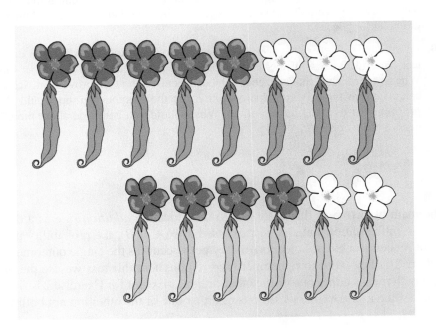

(*Important note*: It is *wrong* to add the 8 peas with green pods to the 9 peas with purple flowers, because this total of 17 would have counted 5 of the peas twice, but they are individuals that should be counted once each.) Because 12 of the 14 peas have "green pods or purple flowers," the probability of randomly selecting a pea that has a green pod or a purple flower can be expressed as P(green pod or purple flower) $= 12/14 = 6/7$.

This example suggests a general rule whereby we add the number of outcomes corresponding to each of the events in question:

> **When finding the probability that event *A* occurs or event *B* occurs, find the total of the number of ways *A* can occur and the number of ways *B* can occur, but *find that total in such a way that no outcome is counted more than once.***

One approach is to combine the number of ways event *A* can occur with the number of ways event *B* can occur and, if there is any overlap, compensate by subtracting the number of outcomes that are counted twice, as in the following rule.

Formal Addition Rule

$$P(A \text{ or } B) = P(A) + P(B) - P(A \text{ and } B)$$

where $P(A \text{ and } B)$ denotes the probability that *A* and *B* both occur at the same time as an outcome in a trial of a procedure.

The formal addition rule is presented as a formula, but the blind use of formulas is not recommended. It is generally better to understand the spirit of the rule and apply it intuitively, as follows.

Intuitive Addition Rule

To find $P(A \text{ or } B)$, find the sum of the number of ways event *A* can occur and the number of ways event *B* can occur, *adding in such a way that every outcome is counted only once*. $P(A \text{ or } B)$ is equal to that sum, divided by the total number of outcomes in the sample space.

Figure 3-4 shows a Venn diagram that provides a visual illustration of the formal addition rule. In this figure we can see that the probability of *A* or *B* equals the probability of *A* (left circle) plus the probability of *B* (right circle) minus the probability of *A* and *B* (football-shaped middle region). This figure shows that the addition of the areas of the two circles will cause double-counting of the football-shaped middle region. This is the basic concept that underlies the addition rule. Because of the relationship between the addition rule and the Venn diagram shown in Figure 3-4, the notation $P(A \cup B)$ is sometimes used in place of

Convicted by Probability

A witness described a Los Angeles robber as a Caucasian woman with blond hair in a ponytail who escaped in a yellow car driven by an African-American male with a mustache and beard. Janet and Malcolm Collins fit this description, and they were convicted based on testimony that there is only about 1 chance in 12 million that any couple would have these characteristics. It was estimated that the probability of a yellow car is 1/10, and the other probabilities were estimated to be 1/4, 1/10, 1/3, 1/10, and 1/1000. The convictions were later overturned when it was noted that no evidence was presented to support the estimated probabilities or the independence of the events. However, because the couple was not randomly selected, a serious error was made in not considering the probability of *other* couples being in the same region with the same characteristics.

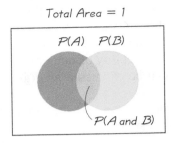

FIGURE 3-4 Venn Diagram Showing Overlapping Events

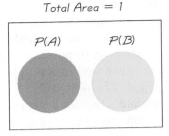

FIGURE 3-5 Venn Diagram Showing Nonoverlapping Events

$P(A$ or $B)$. Similarly, the notation $P(A \cap B)$ is sometimes used in place of $P(A$ and $B)$ so the formal addition rule can be expressed as

$$P(A \cup B) = P(A) + P(B) - P(A \cap B)$$

The addition rule is simplified whenever A and B cannot occur simultaneously, so $P(A$ and $B)$ becomes zero. Figure 3-5 illustrates that when there is no overlapping of A and B, we have $P(A$ or $B) = P(A) + P(B)$. The following definition formalizes the lack of overlapping shown in Figure 3-5.

Definition

Events A and B are **disjoint** (or **mutually exclusive**) if they cannot occur at the same time.

The flowchart of Figure 3-6 shows how disjoint events affect the addition rule.

FIGURE 3-6 Applying the Addition Rule

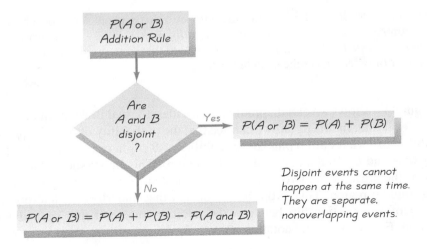

EXAMPLE **Clinical Trials of Pregnancy Test** Refer to Table 3-1, reproduced here for your convenience. Assuming that 1 person is randomly selected from the 99 people included in the study, apply the addition rule to find the probability of selecting a subject who is pregnant or had a positive test result.

Table 3-1	Pregnancy Test Results	
	Positive Test Result (Pregnancy is indicated)	Negative Test Result (Pregnancy is not indicated)
Subject is pregnant	80	5
Subject is not pregnant	3	11

SOLUTION   From the table we can easily see that there are 88 subjects who were pregnant or tested positive. We obtain that total of 88 by adding the pregnant subjects to the subjects who tested positive, being careful to count the 80 pregnant subjects who tested positive only once. It would be very wrong to add the 85 pregnant subjects to the 83 subjects who tested positive, because the total of 168 would count some subjects twice, even though they are individuals that should be counted only once. Dividing the correct total of 88 by the overall total of 99, we get this result: P(pregnant or positive) = 88/99 = 8/9 or 0.889.

In the preceding example, there are several strategies you could use for counting the subjects who were pregnant or tested positive. Any of the following would work.

- Color the cells representing subjects who are pregnant or tested positive, then add the numbers in those colored cells, being careful to add each number only once. This approach yields

$$3 + 80 + 5 = 88$$

- Add the 85 pregnant subjects to the 83 subjects who tested positive, but compensate for the double-counting by subtracting the 80 pregnant subjects who tested positive. This approach yields a result of

$$85 + 83 - 80 = 88$$

- Start with the total of 85 pregnant subjects, then add those subjects who tested positive and were not yet included in that total, to get a result of

$$85 + 3 = 88$$

Carefully study the preceding example, because it makes clear this essential feature of the addition rule: "Or" suggests addition, and the addition must be done without double-counting.

We can summarize the key points of this section as follows:

1. To find $P(A$ or $B)$, begin by associating "or" with addition.

2. Consider whether events A and B are disjoint; that is, can they happen at the same time? If they are not disjoint (that is, they can happen at the same time), be sure to avoid (or at least compensate for) double-counting when adding the relevant probabilities. If you understand the importance of not double-counting when you find $P(A$ or $B)$, you don't necessarily have to calculate the value of $P(A) + P(B) - P(A$ and $B)$.

Errors made when applying the addition rule often involve double-counting; that is, events that are not disjoint are treated as if they were. One indication of such an error is a total probability that exceeds 1; however, errors involving the addition rule do not always cause the total probability to exceed 1.

Complementary Events

In Section 3-2 we defined the complement of event A and denoted it by $\overline{A}$. We said that $\overline{A}$ consists of all the outcomes in which event A does *not* occur. Events A and $\overline{A}$ must be disjoint, because it is impossible for an event and its complement to occur at the same time. Also, we can be absolutely certain that A either does or does not occur, which implies that either A or $\overline{A}$ must occur. These observations let us apply the addition rule for disjoint events as follows:

$$P(A \text{ or } \overline{A}) = P(A) + P(\overline{A}) = 1$$

We justify $P(A$ or $\overline{A}) = P(A) + P(\overline{A})$ by noting that A and $\overline{A}$ are disjoint; we justify the total of 1 by our certainty that A either does or does not occur. This result of the addition rule leads to the following three equivalent expressions.

Rule of Complementary Events

$$P(A) + P(\overline{A}) = 1$$
$$P(\overline{A}) = 1 - P(A)$$
$$P(A) = 1 - P(\overline{A})$$

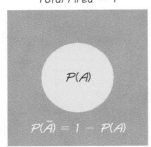

Total Area = 1

$P(A)$

$P(\overline{A}) = 1 - P(A)$

FIGURE 3-7 Venn Diagram for the Complement of Event A

Figure 3-7 visually displays the relationship between $P(A)$ and $P(\overline{A})$.

EXAMPLE In reality, when a baby is born, $P(\text{boy}) = 0.5121$. Find $P(\overline{\text{boy}})$.

SOLUTION   Using the rule of complementary events, we get
$$P(\overline{\text{boy}}) = 1 - P(\text{boy}) = 1 - 0.5121 = 0.4879.$$

That is, the probability of not getting a boy, which is the probability of a girl, is 0.4879.

A major advantage of the *rule of complementary events* is that its use can greatly simplify certain problems. We will illustrate this particular advantage in Section 3-5.

3-3 Exercises

Determining Whether Events Are Disjoint. For each part of Exercises 1 and 2, are the two events disjoint for a single trial? (Hint: Consider "disjoint" to be equivalent to "separate" or "not overlapping.")

1. **a.** Randomly selecting a cardiac surgeon
       Randomly selecting a female physician
   b. Randomly selecting a female college student
       Randomly selecting a college student who drives a motorcycle
   c. Randomly selecting someone treated with the cholesterol-reducing drug Lipitor
       Randomly selecting someone in a control group given no medication

2. **a.** Randomly selecting a fruit fly with red eyes
       Randomly selecting a fruit fly with sepia (dark brown) eyes
   b. Receiving a phone call from a volunteer survey subject who opposes all cloning
       Receiving a phone call from a volunteer survey subject who approves of cloning of sheep
   c. Randomly selecting a nurse
       Randomly selecting a male

3. Finding Complements
   a. If $P(A) = 0.05$, find $P(\overline{A})$.
   b. Women have a 0.25% rate of red/green color blindness. If a woman is randomly selected, what is the *probability* that she does *not* have red/green color blindness? (*Hint:* Be careful expressing 0.25% as a probability value, which is *not* 0.25.)

4. Finding Complements
   a. Find $P(\overline{A})$ given that $P(A) = 0.0175$.
   b. A Reuters/Zogby poll showed that 61% of Americans say they believe that life exists elsewhere in the galaxy. What is the probability of randomly selecting someone not having that belief?

5. Peas on Earth Refer to Figure 3-3. Find the probability of randomly selecting one of the peas and getting one with a green pod or a white flower.

6. Peas on Earth Refer to Figure 3-3. Find the probability of randomly selecting one of the peas and getting one with a yellow pod or a purple flower.

7. National Statistics Day If someone is randomly selected, find the probability that his or her birthday is not October 18, which is National Statistics Day in Japan. Ignore leap years.

8. Birthday and Complement If someone is randomly selected, find the probability that his or her birthday is not in October. Ignore leap years.

In Exercises 9–12, use the data in the following table, which summarizes results from the sinking of the Titanic.

	Men	Women	Boys	Girls
Survived	332	318	29	27
Died	1360	104	35	18

9. *Titanic* Passengers If one of the *Titanic* passengers is randomly selected, find the probability of getting someone who is a woman or child.

10. *Titanic* Passengers If one of the *Titanic* passengers is randomly selected, find the probability of getting a man or someone who survived the sinking.

11. *Titanic* Passengers If one of the *Titanic* passengers is randomly selected, find the probability of getting a child or someone who survived the sinking.

12. *Titanic* Passengers If one of the *Titanic* passengers is randomly selected, find the probability of getting a woman or someone who did not survive the sinking.

In Exercises 13–20, use the data in the following table, which summarizes blood groups and Rh types for 100 typical people. These values may vary in different regions according to the ethnicity of the population.

		Group			
		O	A	B	AB
	Positive	39	35	8	4
Rh Type	Negative	6	5	2	1

13. Blood Groups and Types If one person is randomly selected, find the probability of getting someone who is not group A.

14. Blood Groups and Types If one person is randomly selected, find the probability of getting someone who is type Rh^-.

15. Blood Groups and Types If one person is randomly selected, find the probability of getting someone who is group A or type Rh^-.

16. Blood Groups and Types If one person is randomly selected, find the probability of getting someone who is group A or group B.

17. Blood Groups and Types If one person is randomly selected, find $P(\text{not type } Rh^+)$.

18. Blood Groups and Types If one person is randomly selected, find $P(\text{group B or type } Rh^+)$.

19. Blood Groups and Types If one person is randomly selected, find $P(\text{group AB or type } Rh^+)$.

20. Blood Groups and Types If one person is randomly selected, find $P(\text{group A or O or type } Rh^+)$.

21. Poll Resistance Pollsters are concerned about declining levels of cooperation among persons contacted in surveys. A pollster contacts 84 people in the 18–21 age bracket and finds that 73 of them respond and 11 refuse to respond. When 275 people in the 22–29 age bracket are contacted, 255 respond and 20 refuse to respond (based on data

from "I Hear You Knocking but You Can't Come In," by Fitzgerald and Fuller, *Sociological Methods and Research,* Vol. 11, No. 1). Assume that 1 of the 359 people is randomly selected. Find the probability of getting someone in the 18–21 age bracket or someone who refused to respond.

22. Poll Resistance Refer to the same data set as in Exercise 21. Assume that 1 of the 359 people is randomly selected, and find the probability of getting someone who is in the 18–21 age bracket or someone who responded.

23. Disjoint Events If events *A* and *B* are disjoint and events *B* and *C* are disjoint, must events *A* and *C* be disjoint? Give an example supporting your answer.

*24. Exclusive *Or* How is the addition rule changed if the *exclusive or* is used instead of the *inclusive or*? In this section it was noted that the *exclusive or* means either one or the other, but not both.

3-4 Multiplication Rule: Basics

In Section 3-3 we presented the addition rule for finding $P(A$ or $B)$, the probability that a trial has an outcome of A or B or both. The objective of this section is to develop a rule for finding $P(A$ and $B)$, the probability that event A occurs in a first trial and event B occurs in a second trial.

Notation

$$P(A \text{ and } B) = P(\text{event } A \text{ occurs in a first trial and event } B \text{ occurs in a second trial})$$

In Section 3-3 we associated the word "or" with addition; in this section we will associate the word "and" with multiplication. We will see that $P(A$ and $B)$ involves multiplication of probabilities and that we must sometimes adjust the probability of event B to reflect the outcome of event A.

Probability theory is used extensively in the analysis and design of standardized tests, such as the SAT, ACT, MCAT (for medicine), and LSAT (for law). For ease of grading, such tests typically use true/false or multiple-choice questions. Let's assume that the first question on a test is a true/false type, while the second question is a multiple-choice type with five possible answers (a, b, c, d, e). We will use the following two questions. Try them!

1. True or false: A pound of goose feathers is heavier than a pound of gold.

2. Which one of the following has had the most influence on our understanding of genetics?

   a. Gene Hackman

   b. Gene Simmons

   c. Gregor Mendel

   d. jeans

   e. Jean-Jacques Rousseau

The answers to the two questions are T (for "true") and c. (The answer to the first question is true. Weights of feathers are expressed in Avoirdupois pounds, but weights of gold are expressed in Troy pounds.) Let's find the probability that if someone makes random guesses for both answers, the first answer will be correct *and* the second answer will be correct. One way to find that probability is to list the sample space as follows.

$$T,a \quad T,b \quad T,c \quad T,d \quad T,e$$
$$F,a \quad F,b \quad F,c \quad F,d \quad F,e$$

If the answers are random guesses, then the 10 possible outcomes are equally likely, so

$$P(\text{both correct}) = P(T \text{ and } c) = \frac{1}{10} = 0.1$$

Now note that $P(T \text{ and } c) = 1/10$, $P(T) = 1/2$, and $P(c) = 1/5$, from which we see that

$$\frac{1}{10} = \frac{1}{2} \cdot \frac{1}{5}$$

so that

$$P(T \text{ and } c) = P(T) \times P(c)$$

This suggests that, in general, $P(A \text{ and } B) = P(A) \cdot P(B)$, but let's consider another example before making that generalization.

First, note that tree diagrams are sometimes helpful in determining the number of possible outcomes in a sample space. A **tree diagram** is a picture of the possible outcomes of a procedure, shown as line segments emanating from one starting point. These diagrams are sometimes helpful if the number of possibilities is not too large. The tree diagram shown in Figure 3-8 summarizes the outcomes of the true/false and multiple-choice questions. From Figure 3-8 we see that if both answers are random guesses, all 10 branches are equally likely and the probability of getting the correct pair (T, c) is 1/10. For each response to the first question, there are 5 responses to the second. The total number of outcomes is 5 taken 2 times, or 10. The tree diagram in Figure 3-8 illustrates the reason for the use of multiplication.

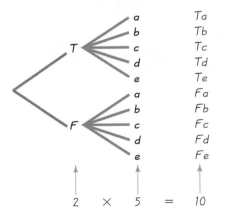

FIGURE 3-8 Tree Diagram of Test Answers

Our first example of the true/false and multiple-choice questions suggested that $P(A \text{ and } B) = P(A) \cdot P(B)$, but the next example will introduce another important element.

EXAMPLE **Genetics Experiment** Mendel's famous hybridization experiments involved peas, like those illustrated in Figure 3-3 and summarized in Table 3-2 from Section 3-3. Table 3-2 is reproduced here. If two of the peas included in Table 3-2 are randomly selected *without replacement*, find the probability that the first selection has a green pod and the second selection has a yellow pod.

Table 3-2	Peas Used in Hybridization Experiments		
		Flower	
		Purple	White
Pod	Green	5	3
	Yellow	4	2

SOLUTION

First selection: $P(\text{green pod}) = 8/14$ (because there are 14 peas, 8 of which have green pods)

Second selection: $P(\text{yellow pod}) = 6/13$ (after the first selection of a pea with a green pod, there are 13 peas remaining, 6 of which have yellow pods)

With $P(\text{first pea with green pod}) = 8/14$ and $P(\text{second pea with yellow pod}) = 6/13$, we have

$$P(\text{1st pea with green pod and 2nd pea with yellow pod}) = \frac{8}{14} \cdot \frac{6}{13} \approx 0.264$$

The key point is that *we must adjust the probability of the second event to reflect the outcome of the first event.* Because selection of the second pea is made without replacement of the first pea, the second probability must take into account the result of a pea with a green pod for the first selection. After a pea with a green pod has been selected on the first trial, only 13 peas remain and 6 of them have yellow pods, so the second selection yields this: $P(\text{pea with yellow pod}) = 6/13$.

This example illustrates the important principle that *the probability for the second event B should take into account the fact that the first event A has already occurred.* This principle is often expressed using the following notation.

Notation for Conditional Probability

$P(B|A)$ represents the probability of event B occurring after it is assumed that event A has already occurred. (We can read $B|A$ as "B given A" or as "event B occurring after event A has already occurred.")

Definitions

Two events A and B are **independent** if the occurrence of one does not affect the probability of the occurrence of the other. (Several events are similarly independent if the occurrence of any does not affect the probabilities of the occurrence of the others.) If A and B are not independent, they are said to be **dependent.**

For example, playing the California lottery and then playing the New York lottery are *independent* events because the result of the California lottery has absolutely no effect on the probabilities of the outcomes of the New York lottery. In contrast, the event of having your car start and the event of getting to your doctor's appointment on time are *dependent* events, because the outcome of trying to start your car does affect the probability of getting to the appointment on time.

Using the preceding notation and definitions, along with the principles illustrated in the preceding examples, we can summarize the key concept of this section as the following *formal multiplication rule*, but it is recommended that you work with the *intuitive multiplication rule*, which is more likely to reflect understanding instead of blind use of a formula.

Formal Multiplication Rule

$$P(A \text{ and } B) = P(A) \cdot P(B|A)$$

FIGURE 3-9 Applying the Multiplication Rule

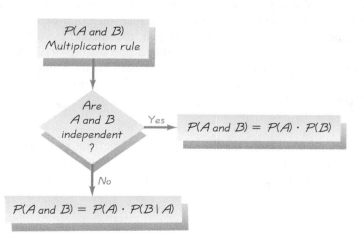

If *A* and *B* are independent events, $P(B|A)$ is really the same as $P(B)$. See the following *intuitive multiplication rule*. (Also see Figure 3-9.)

Intuitive Multiplication Rule

When finding the probability that event *A* occurs in one trial and event *B* occurs in the next trial, multiply the probability of event *A* by the probability of event *B*, but be sure that the probability of event *B* takes into account the previous occurrence of event *A*.

EXAMPLE   Plants A biologist experiments with a sample of two vascular plants (denoted here by V) and four nonvascular plants (denoted here by N). Listed below are the codes for the six plants being studied. She wants to randomly select two of the plants for further experimentation. Find the probability that the first selected plant is nonvascular (N) and the second plant is also nonvascular (N). Assume that the selections are made (a) with replacement; (b) without replacement.

V V N N N N

SOLUTION

a. If the two plants are selected *with replacement*, the two selections are independent because the second event is not affected by the first outcome. In each of the two selections there are four nonvascular (N) plants among the six plants, so we get

$$P(\text{first plant is N and second plant is N}) = \frac{4}{6} \cdot \frac{4}{6} = \frac{16}{36} \text{ or } 0.444$$

b. If the two plants are selected *without replacement*, the two selections are dependent because the probability of the second event is affected by the first outcome. In the first selection, four of the six plants are nonvascular (N). After selecting a nonvascular plant on the first selection, we are left with five plants including three that are nonvascular. We therefore get

$$P(\text{first plant is N and second plant is N}) = \frac{4}{6} \cdot \frac{3}{5} = \frac{12}{30} = \frac{2}{5} \text{ or } 0.4$$

Note that in this case, we adjust the second probability to take into account the selection of a nonvascular plant (N) in the first outcome. After selecting N the first time, there would be three Ns among the five plants that remain.

So far we have discussed two events, but the multiplication rule can be easily extended to several events. In general, the probability of any sequence of independent events is simply the product of their corresponding probabilities. For example, the probability of tossing a coin three times and getting all heads is $0.5 \cdot 0.5 \cdot 0.5 = 0.125$. We can also extend the multiplication rule so that it applies to several dependent events; simply adjust the probabilities as you go along.

Independent Jet Engines

Soon after departing from Miami, Eastern Airlines Flight 855 had one engine shut down because of a low oil pressure warning light. As the L-1011 jet turned to Miami for landing, the low pressure warning lights for the other two engines also flashed. Then an engine failed, followed by the failure of the last working engine. The jet descended without power from 13,000 ft to 4000 ft when the crew was able to restart one engine, and the 172 people on board landed safely. With independent jet engines, the probability of all three failing is only 0.001^3, or about one chance in a trillion. The FAA found that the same mechanic who replaced the oil in all three engines failed to replace the oil plug sealing rings. The use of a single mechanic caused the operation of the engines to become dependent, a situation corrected by requiring that the engines be serviced by different mechanics.

Monkey Typists

A classic claim is that a monkey randomly hitting a keyboard would eventually produce the complete works of Shakespeare, assuming that it continues to type century after century. The multiplication rule for probability has been used to find such estimates. One result of 1,000,000,000,000,000,000,000, 000,000,000,000 years is considered by some to be too short. In the same spirit, Sir Arthur Eddington wrote this poem. "There once was a brainy baboon, who always breathed down a bassoon. For he said, 'It appears that in billions of years, I shall certainly hit on a tune.'"

Treating Dependent Events as Independent   Part (b) of the last example involved selecting items without replacement, and we therefore treated the events as being dependent. However, it is a common practice to treat events as independent when *small samples* are drawn from *large populations*. In such cases, it is rare to select the same item twice. Here is a common guideline:

> **If a sample size is no more than 5% of the size of the population, treat the selections as being *independent* (even if the selections are made without replacement, so they are technically dependent).**

Pollsters use this guideline when they survey roughly 1000 adults from a population of millions. They assume independence, even though they sample without replacement.

The following example gives us some insight into the important procedure of *hypothesis testing* that is introduced in Chapter 7.

EXAMPLE **Gender Selection** A geneticist develops a procedure for increasing the likelihood that offspring of fruit flies will be females. In an initial test, the parents are treated and the results consist of 20 females among 20 offspring. Assuming that the gender-selection procedure has no effect, find the probability of getting 20 females among 20 offspring. Based on the result, is there strong evidence to support the geneticist's claim that the procedure is effective in increasing the likelihood that offspring will be females?

SOLUTION   We want to find P(all 20 offspring are female) with the assumption that the procedure has no effect, so that the probability of any individual offspring being a female is 0.5. Because separate pairs of parents were used, we will treat the events as if they are independent. We get this result:

P(all 20 offspring are female)

$= $(1st is female and 2nd is female and 3rd is female . . . and 20th is female)

$= P$(female) $\cdot$ P(female) $\cdot$ $\cdot$ $\cdot$ $\cdot$ P(female)

$= 0.5 \cdot 0.5 \cdot \cdot \cdot \cdot 0.5$

$= 0.5^{20} = 0.000000954$

The low probability of 0.000000954 indicates that instead of getting 20 females by chance, a more reasonable explanation is that females appear to be more likely with the gender-selection procedure. Because there is such a small chance (0.000000954) of getting 20 females in 20 births, we do have sufficient evidence to conclude that the gender-selection procedure appears to be effective in increasing the likelihood that an offspring is female. That is, the procedure does appear to be effective.

We can summarize the fundamentals of the addition and multiplication rules as follows:

- In the addition rule, the word "or" in $P(A \text{ or } B)$ suggests addition. Add $P(A)$ and $P(B)$, being careful to add in such a way that every outcome is counted only once.

- In the multiplication rule, the word "and" in $P(A$ and $B)$ suggests multiplication. Multiply $P(A)$ and $P(B)$, but be sure that the probability of event B takes into account the previous occurrence of event A.

3-4 Exercises

Identifying Events as Independent or Dependent. In Exercises 1 and 2, for each given pair of events, classify the two events as independent or dependent.

1. **a.** Randomly selecting a plant and getting a tree
   Randomly selecting a plant and getting one with needles
   b. Randomly selecting a plant and getting a rose
   Randomly selecting a plant and getting one with red petals
   c. Wearing plaid shorts with black sox and sandals
   Asking someone on a date and getting a positive response

2. **a.** Finding that your calculator is not working
   Finding that your refrigerator is not working
   b. Finding that your kitchen light is not working
   Finding that your refrigerator is not working
   c. Drinking until your driving ability is impaired
   Being involved in a car crash

3. Coin and Die Find the probability of getting the outcome of tails and 3 when a coin is tossed and a single die is rolled.

4. Letter and Digit A new computer owner creates a password consisting of two characters. She randomly selects a letter of the alphabet for the first character and a digit (0, 1, 2, 3, 4, 5, 6, 7, 8, 9) for the second character. What is the probability that her password is "K9"? Would this password be effective as a deterrent against someone trying to gain access to her computer?

Pregnancy Test Results. In Exercises 5–8, use the data in Table 3-1, reproduced here.

Table 3-1	Pregnancy Test Results	
	Positive Test Result (Pregnancy is indicated)	Negative Test Result (Pregnancy is not indicated)
Subject is pregnant	80	5
Subject is not pregnant	3	11

5. Positive Test Result If two *different* subjects are randomly selected, find the probability that they both test positive for pregnancy.

6. Pregnant If one of the subjects is randomly selected, find the probability of getting someone who tests negative or someone who is not pregnant.

7. Pregnant If two *different* subjects are randomly selected, find the probability that they are both pregnant.

8. Negative Test Result If three *different* people are randomly selected, find the probability that they all test negative.

In Exercises 9–13, use the data in the following table, which summarizes blood groups and Rh types for 100 typical people. These values may vary in different regions according to the ethnicity of the population.

		Group O	A	B	AB
Rh Type	Positive	39	35	8	4
	Negative	6	5	2	1

9. Blood Groups and Types Two of the 100 people are randomly selected. Find the probability that they both have group O blood.
  a. Assume that the first selected subject is replaced so that the second selection is made from the same pool of 100 people.
  b. Assume that the first selected subject is not replaced before the second selection is made.

10. Blood Groups and Types Two of the 100 people are randomly selected. Find the probability that they both have blood types of Rh$^+$.
  a. Assume that the first selected subject is replaced so that the second selection is made from the same pool of 100 people.
  b. Assume that the first selected subject is not replaced before the second selection is made.

11. Blood Groups and Types Three of the 100 people are randomly selected. Find the probability that all have group B blood.
  a. Assume that each selected subject is replaced before the subsequent selections are made.
  b. Assume that none of the selected subjects are replaced.

12. Blood Groups and Types Four of the 100 people are randomly selected. Find the probability that they all have blood types of Rh$^-$.
  a. Assume that each selected subject is replaced before the subsequent selections are made.
  b. Assume that none of the selected subjects are replaced.

13. Blood Groups and Types The given data suggest that 40% of the population consists of people with group A blood. If 10 people are randomly selected from the general population (instead of the sample of 100 people represented in the given table), find the probability that they all have group A blood.

14. Wearing Hunter Orange A study of hunting injuries and the wearing of "hunter" orange clothing showed that among 123 hunters injured when mistaken for game, 6 were wearing orange (based on data from the Centers for Disease Control). If a follow-up study begins with the random selection of hunters from this sample of 123, find the probability that the first two selected hunters were both wearing orange.
  a. Assume that the first hunter is replaced before the next one is selected.
  b. Assume that the first hunter is not replaced before the second one is selected.
  c. Given a choice between selecting with replacement and selecting without replacement, which choice makes more sense in this situation? Why?

15. Probability and Guessing A biology professor gives a surprise quiz consisting of 10 true/false questions, and she states that passing requires at least 7 correct responses. Assume that an unprepared student adopts the questionable strategy of guessing for each answer.
  a. Find the probability that the first 7 responses are correct and the last 3 are wrong.
  b. Is the probability from part (a) equal to the probability of passing? Why or why not?

16. Poll Confidence Level It is common for public opinion polls to have a "confidence level" of 95%, meaning that there is a 0.95 probability that the poll results are accurate within the claimed margins of error. If five different organizations conduct independent polls, what is the probability that all five of them are accurate within the claimed margins of error? Does the result suggest that with a confidence level of 95%, we can expect that almost all polls will be within the claimed margin of error?

17. Testing Effectiveness of Gender-Selection Method Recent developments appear to make it possible for couples to dramatically increase the likelihood that they will conceive a child with the gender of their choice. In a test of a gender-selection method, 10 couples try to have baby girls. If this gender-selection method has no effect, what is the probability that the 10 babies will be all girls? If there are actually 10 girls among 10 children, does this gender-selection method appear to be effective? Why?

18. Flat Tire Excuse A classic excuse for a missed test is offered by four students who claim that their car had a flat tire. On the makeup test, the instructor asks the students to identify the particular tire that went flat. If they really didn't have a flat tire and randomly select one that supposedly went flat, what is the probability that they will all select the same tire?

19. Quality Control A production manager for the Medtyme Company claims that her new process for manufacturing blood pressure monitors is better because the rate of defects is lower than 2%, which had been the rate of defects in the past. To support her claim, she manufactures a batch of 5000 blood pressure monitors, then randomly selects 15 of them for testing, with the result that there are no defects among the 15 selected units. Assuming that the new method has the same 2% defect rate as in the past, find the probability of getting no defects among the 15 selected blood pressure monitors. Based on the result, is there strong evidence to support the manager's claim that her new process is better?

20. Redundancy The principle of redundancy is used when system reliability is improved through redundant or backup components. Assume that a surgeon's alarm clock has a 0.975 probability of working on any given morning.
  a. What is the probability that the surgeon's alarm clock will not work on the morning of an important operation?
  b. If the surgeon has two such alarm clocks, what is the probability that they both fail on the morning of an important operation?
  c. With one alarm clock, the surgeon has a 0.975 probability of being awakened. What is the probability of being awakened if she uses two alarm clocks?

21. Stocking Fish One of the authors captured three bass fish from one pond and placed them in another pond that had no bass fish. Find the probability that the three fish include at least one male and at least one female so that reproduction is possible. Assume that the three fish were captured from a population in which males and females are equally likely.

3-5 Multiplication Rule: Beyond the Basics

Section 3-4 introduced the basic concept of the multiplication rule, but in this section we extend our use of that rule to other special applications. We begin with situations in which we want to find the probability that among several trials, *at least one* will result in some specified outcome.

Complements: The Probability of "At Least One"

The multiplication rule and the rule of complements can be used together to greatly simplify the solution to this type of problem: Find the probability that among several trials, *at least one* will result in some specified outcome. In such cases, it is critical that the meaning of the language be clearly understood:

- "At least one" is equivalent to "one or more."
- The complement of getting at least one item of a particular type is that you get *no* items of that type.

Suppose a couple plans to have three children and they want to know the probability of getting at least one girl. See the following interpretations:

At least 1 girl among 3 children = 1 or more girls.

The complement of "at least 1 girl" = no girls = all 3 children are boys.

We could easily find the probability from a list of the entire sample space of eight outcomes, but we want to illustrate the use of complements, which can be used in many other problems that cannot be solved so easily.

Redundancy

Reliability of systems can be greatly improved with redundancy of critical components. Race cars in the NASCAR Winston Cup series have two ignition systems so that if one fails, the other can be used. Airplanes have two independent electrical systems, and aircraft used for instrument flight typically have two separate radios. The following is from a *Popular Science* article about stealth aircraft: "One plane built largely of carbon fiber was the Lear Fan 2100 which had to carry two radar transponders. That's because if a single transponder failed, the plane was nearly invisible to radar." Such redundancy is an application of the multiplication rule in probability theory. If one component has a 0.001 probability of failure, the probability of two independent components both failing is only 0.000001.

EXAMPLE **Gender of Children** Find the probability of a couple having at least 1 girl among 3 children. Assume that boys and girls are equally likely and that the gender of a child is independent of the gender of any brothers or sisters.

SOLUTION

Step 1: Use a symbol to represent the event desired. In this case, let A = at least 1 of the 3 children is a girl.

Step 2: Identify the event that is the complement of A.

$$\overline{A} = \textit{not} \text{ getting at least 1 girl among 3 children}$$
$$= \text{all 3 children are boys}$$
$$= \text{boy and boy and boy}$$

Step 3: Find the probability of the complement.

$$P(\overline{A}) = P(\text{boy and boy and boy})$$
$$= \frac{1}{2} \cdot \frac{1}{2} \cdot \frac{1}{2} = \frac{1}{8}$$

Step 4:   Find $P(A)$ by evaluating $1 - P(\overline{A})$.

$$P(A) = 1 - P(\overline{A}) = 1 - \frac{1}{8} = \frac{7}{8}$$

INTERPRETATION There is a 7/8 probability that if a couple has 3 children, at least 1 of them is a girl.

The principle used in this example can be summarized as follows:

> **To find the probability of *at least one* of something, calculate the probability of *none*, then subtract that result from 1. That is,**
>
> $$P(\text{at least one}) = 1 - P(\text{none}).$$

Conditional Probability

Next we consider the second major point of this section, which is based on the principle that the probability of an event is often affected by knowledge of circumstances. For example, if you randomly select someone from the general population, the probability of getting a male is 0.5, but if you then learn that the selected person smokes cigars, there is a dramatic increase in the probability that the selected person is a male (because 85% of cigar smokers are males). A *conditional probability* of an event occurs when the probability is affected by the knowledge of other circumstances. The conditional probability of event B occurring, given that event A has already occurred, can be found by using the multiplication rule $[P(A \text{ and } B) = P(A) \cdot P(B|A)]$ and solving for $P(B|A)$ by dividing both sides of the equation by $P(A)$.

Definition

A **conditional probability** of an event is a probability obtained with the additional information that some other event has already occurred. $P(B|A)$ denotes the conditional probability of event B occurring, given that event A has already occurred, and it can be found by dividing the probability of events A and B both occurring by the probability of event A:

$$P(B|A) = \frac{P(A \text{ and } B)}{P(A)}$$

This preceding formula is a formal expression of conditional probability, but blind use of formulas is not recommended. We recommend the following intuitive approach:

Intuitive Approach to Conditional Probability

The conditional probability of B given A can be found by assuming that event A has occurred and, working under that assumption, calculating the probability that event B will occur.

EXAMPLE **Clinical Trials of Pregnancy Test** Refer to Table 3-1, reproduced here for your convenience. Find the following:

a. If 1 of the 99 subjects is randomly selected, find the probability that the person tested positive, given that she was pregnant.

b. If 1 of the 99 subjects is randomly selected, find the probability that she was pregnant, given that she tested positive.

Table 3-1	Pregnancy Test Results	
	Positive Test Result (Pregnancy is indicated)	Negative Test Result (Pregnancy is not indicated)
Subject is pregnant	80	5
Subject is not pregnant	3	11

SOLUTION

a. We want $P(\text{positive} \mid \text{pregnant})$, the probability of getting someone who tested positive, *given that the selected person was pregnant.* Here is the key point: If we assume that the selected person was pregnant, we are dealing only with the 85 subjects in the first row of Table 3-1. Among those 85 subjects, 80 tested positive, so

$$P(\text{positive} \mid \text{pregnant}) = \frac{80}{85} = 0.941$$

The same result can be found by using the formula given with the definition of conditional probability. In the following calculation, we use the fact that 80 of the 99 subjects were both pregnant and tested positive. Also, 85 of the 99 subjects were pregnant. We get

$$P(\text{positive} \mid \text{pregnant}) = \frac{P(\text{pregnant and positive})}{P(\text{pregnant})}$$

$$= \frac{80/99}{85/99} = 0.941$$

b. Here we want $P(\text{pregnant} \mid \text{positive})$. If we assume that the person selected tested positive, we are dealing with the 83 subjects in the first column of Table 3-1. Among those 83 subjects, 80 were pregnant, so

$$P(\text{pregnant} \mid \text{positive}) = \frac{80}{83} = 0.964$$

Again, the same result can be found by applying the formula for conditional probability:

$$P(\text{pregnant}|\text{positive}) = \frac{P(\text{positive and pregnant})}{P(\text{positive})}$$

$$= \frac{80/99}{83/99} = 0.964$$

By comparing the results from parts (a) and (b), we see that $P(\text{positive}|\text{pregnant})$ is not the same as $P(\text{pregnant}|\text{positive})$.

INTERPRETATION The first result of $P(\text{positive}|\text{pregnant}) = 0.941$ indicates that a pregnant woman has a 0.941 probability of testing positive. This suggests that if a woman does not test positive, she cannot be confident that she is not pregnant, so she should pursue additional testing. The second result of $P(\text{pregnant}|\text{positive}) = 0.964$ indicates that for a woman who tests positive, there is a 0.964 probability that she is actually pregnant. A woman who tests positive would be wise to pursue additional testing.

Confusion of the Inverse

In the preceding example, $P(\text{positive}|\text{pregnant}) \neq P(\text{pregnant}|\text{positive})$. Although the two values of 0.941 and 0.964 are not too far apart in this example, values of $P(B|A)$ and $P(A|B)$ can be very far apart in other cases. To incorrectly believe that $P(B|A)$ and $P(A|B)$ are the same, or to incorrectly use one value for the other, is often called *confusion of the inverse*. Studies have shown that physicians often give very misleading information when they have confusion of the inverse. Based on real studies, they tended to confuse $P(\text{cancer}|\text{positive test})$ with $P(\text{positive test}|\text{cancer})$. About 95% of physicians estimated $P(\text{cancer}|\text{positive test})$ to be about 10 times too high, with the result that patients were given diagnoses that were very misleading, and patients were unnecessarily distressed by the incorrect information.

Bayes' Theorem

With *Bayes' theorem* (or *Bayes' rule*), we revise a probability value based on additional information that is later obtained. One key to understanding the essence of Bayes' theorem is to recognize that we are dealing with *sequential* events, whereby new additional information is obtained for a subsequent event, and that new information is used to revise the probability of the initial event. In this context, the terms *prior probability* and *posterior probability* are commonly used.

Definitions

A **prior probability** is an initial probability value originally obtained before any additional information is obtained.

A **posterior probability** is a probability value that has been revised by using additional information that is later obtained.

EXAMPLE 1 Using New Information to Revise Probability

The Gallup organization randomly selects an adult American for a survey. Use subjective probabilities to estimate the following:

a. What is the probability that the selected subject is a male?

b. After selecting a subject, it is later learned that this person was smoking a cigar during the interview. What is the probability that the selected subject is a male?

c. Which of the preceding two results is a prior probability? Which is a posterior probability?

SOLUTION

a. Roughly half of all Americans are males, so we estimate the probability of selecting a male subject to be 0.5. Denoting a male by M, we can express this probability as follows: $P(M) = 0.5$.

b. Although some women smoke cigars, the vast majority of cigar smokers are males. A reasonable estimate is that 85% of cigar smokers are males. Based on this additional subsequent information that the survey respondent was smoking a cigar, we estimate the probability of this person being a male as 0.85. Denoting a male by M and denoting a cigar smoker by C, we can express this result as follows: $P(M|C) = 0.85$.

c. In part (a), the value of 0.5 is the initial probability, so we refer to it as the prior probability. Because the probability of 0.85 in part (b) is a revised probability based on the additional information that the survey subject was smoking a cigar, this value of 0.85 is referred to as a posterior probability.

The Reverend Thomas Bayes [1701 (approximately)–1761] was an English minister and mathematician. Although none of his work was published during his lifetime, later publications included the following theorem (or rule) that he developed for determining probabilities of events by incorporating information about subsequent events.

Bayes' Theorem

According to **Bayes' theorem** (or **Bayes' rule**), the probability of event A, given that event B has subsequently occurred, is

$$P(A|B) = \frac{P(A) \cdot P(B|A)}{[P(A) \cdot P(B|A)] + [P(\overline{A}) \cdot P(B|\overline{A})]}$$

That's a formidable expression, but we will simplify its calculation. See the following example, which illustrates the above expression, but also see the alternative method based on a more intuitive application of Bayes' theorem.

EXAMPLE 2 Using New Information to Revise Probability

In Orange County, 51% of the adults are males. (It follows that the other 49% are females.) One adult is randomly selected for a survey involving credit card usage.

a. Find the prior probability that the selected person is a male.

b. It is later learned that the selected survey subject was smoking a cigar. Also, 9.5% of males smoke cigars, whereas 1.7% of females smoke cigars (based on data from the Substance Abuse and Mental Health Services Administration). Use this additional information to find the probability that the selected subject is a male.

SOLUTION   Let's use the following notation:

M = male $\overline{M}$ = female (or not male)

C = cigar smoker $\overline{C}$ = not a cigar smoker

a. Before using the information given in part (b), we know only that 51% of the adults in Orange County are males, so the probability of randomly selecting an adult and getting a male is given by $P(M) = 0.51$.

b. Based on the additional given information, we have the following:

$P(M) = 0.51$ because 51% of the adults are males.

$P(\overline{M}) = 0.49$ because 49% of the adults are females (not males).

$P(C|M) = 0.095$ because 9.5% of the males smoke cigars. (That is, the probability of getting someone who smokes cigars, given that the person is a male, is 0.095.)

$P(C|\overline{M}) = 0.017$ because 1.7% of the females smoke cigars. (That is, the probability of getting someone who smokes cigars, given that the person is a female, is 0.017.)

Let's now apply Bayes' theorem by using the preceding formula with M in place of A and C in place of B. We get the following result:

$$P(M|C) = \frac{P(M) \cdot P(C|M)}{[P(M) \cdot P(C|M)] + [P(\overline{M}) \cdot P(C|\overline{M})]}$$

$$= \frac{0.51 \cdot 0.095}{[0.51 \cdot 0.095] + [0.49 \cdot 0.017]}$$

$$= 0.85329341$$

$$= 0.853 \text{ (rounded)}$$

Before we knew that the survey subject smoked a cigar, there is a 0.51 probability that the survey subject is male (because 51% of the adults in Orange County are males). However, after learning that the subject smoked a cigar, we revised the probability to 0.853. There is a 0.853 probability that the cigar-smoking respondent is a male. This makes sense, because the likelihood of a male increases dramatically with the additional information that the subject smokes cigars (because so many more males smoke cigars than females).

Intuitive Bayes' Theorem

The preceding solution illustrates the application of Bayes' theorem with its calculation using the formula. Unfortunately, that calculation is complicated enough to create an abundance of opportunities for errors and/or incorrect substitution of the involved probability values. Here is another approach that is much more intuitive and easier:

Assume some convenient value for the total of all items involved, then construct a table of rows and columns with the individual cell frequencies based on the known probabilities.

For the preceding example, simply assume some value for the adult population of Orange County, such as 100,000, then use the given information to construct a table, such as the one shown below.

Finding the number of males who smoke cigars: If 51% of the 100,000 adults are males, then there are 51,000 males. If 9.5% of the males smoke cigars, then the number of cigar-smoking males is 9.5% of 51,000, or $0.095 \times 51,000 = 4845$. See the entry of 4845 in the table. The other males who do *not* smoke cigars must be $51,000 - 4845 = 46,155$. See the value of 46,155 in the table.

Finding the number of females who smoke cigars: Using similar reasoning, 49% of the 100,000 adults are females, so the number of females is 49,000. Given that 1.7% of the females smoke cigars, the number of cigar-smoking females is $0.017 \times 49,000 = 833$. The number of females who do *not* smoke cigars is $49,000 - 833 = 48,167$. See the entries of 833 and 48,167 in the table.

	C (Cigar Smoker)	$\overline{C}$ (Not a Cigar Smoker)	Total
M (male)	4845	46,155	**51,000**
$\overline{M}$ (female)	833	48,167	**49,000**
Total	**5678**	**94,322**	**100,000**

The above table involves relatively simple arithmetic. Simply partition the assumed population into the different cell categories by using suitable percentages.

Now we can easily address the key question as follows: To find the probability of getting a male subject, given that the subject smokes cigars, simply use the same conditional probability described earlier in this section. To find the probability of getting a male given that the subject smokes, restrict the table to the column of cigar smokers, then find the probability of getting a male in that column. Among the 5678 cigar smokers, there are 4845 males, so the probability we seek is $4845/5678 = 0.85329341$. That is, $P(M|C) = 4845/5678 = 0.85329341 = 0.853$ (rounded).

3-5 Exercises

Describing Complements. In Exercises 1–4, provide a written description of the complement of the given event.

1. Blood Testing When 10 students are tested for blood group, at least one of them has group A blood.

Composite Sampling

The U.S. Army once tested for syphilis by giving each inductee an individual blood test that was analyzed separately. One researcher suggested mixing pairs of blood samples. After the mixed pairs were tested, syphilitic inductees could be identified by retesting the few blood samples that were in the pairs that tested positive. The total number of analyses was reduced by pairing blood specimens, so why not put them in groups of three or four or more? Probability theory was used to find the most efficient group size, and a general theory was developed for detecting the defects in any population. This technique is known as *composite sampling.*

2. Quality Control When 50 electrocardiograph units are shipped, all of them are free of defects.

3. X-Linked Disorder When 12 males are tested for a particular X-linked recessive gene, none of them are found to have the gene.

4. Type O Blood When five different blood samples are obtained from donors, at least one of them has type O blood.

5. Subjective Conditional Probability Use subjective probability to estimate the probability that a credit card is being used fraudulently, given that today's charges were made in several different countries.

6. Subjective Conditional Probability Use subjective probability to estimate the probability of randomly selecting an adult and getting a male, given that the selected person owns a motorcycle. If a criminal investigator finds that a motorcycle is registered to Pat Ryan, is it reasonable to believe that Pat is a male?

7. Probability of At Least One Girl If a couple plans to have five children, what is the probability that they will have at least one girl? Is that probability high enough for the couple to be very confident that they will get at least one girl in five children?

8. Probability of At Least One Girl If a couple plans to have 12 children (it could happen), what is the probability that there will be at least one girl? If the couple eventually has 12 children and they are all boys, what can the couple conclude?

9. Probability of a Girl Find the probability of a couple having a baby girl when their third child is born, given that the first two children were both boys. Is the result the same as the probability of getting three girls among three children?

10. Mendelian Genetics Refer to Figure 3-3 and/or Table 3-2 in Section 3-3 for the peas used in a genetics experiment. If one of the peas is randomly selected and is found to have a green pod, what is the probability that it has a purple flower?

11. Clinical Trials of Pregnancy Test Refer to Table 3-1 and assume that one of the subjects is randomly selected. Find the probability of a negative test result given that the selected subject is not pregnant. What should a woman do if she uses this pregnancy test and obtains a negative result?

12. Clinical Trials of Pregnancy Test Refer to Table 3-1 and assume that one of the subjects is randomly selected. Find the probability that the selected subject is not pregnant, given that the test was negative. Is the result the same as the probability of a negative test result given that the selected subject is not pregnant?

13. Redundancy in Alarm Clocks A surgeon wants to ensure that she is not late for an early operation because of a malfunctioning alarm clock. Instead of using one alarm clock, she decides to use three. What is the probability that at least one of her alarm clocks works correctly if each individual alarm clock has a 99% chance of working correctly? Does the surgeon really gain much by using three alarm clocks instead of only one?

14. Acceptance Sampling With one method of the procedure called *acceptance sampling*, a sample of items is randomly selected without replacement, and the entire batch is rejected if there is at least one defect. The Medtyme Company has just manufactured 5000 blood pressure monitors, and 3% are defective. If 10 of them are selected and tested, what is the probability that the entire batch will be rejected?

15. Using Composite Blood Samples When doing blood testing for HIV infections, the procedure can be made more efficient and less expensive by combining samples of blood specimens. If samples from three people are combined and the mixture tests negative, we know that all three individual samples are negative. Find the probability of a positive result for three samples combined into one mixture, assuming the probability of an individual blood sample testing positive is 0.1 (the probability for the "at-risk" population, based on data from the New York State Health Department).

16. Using Composite Water Samples The Orange County Department of Public Health tests water for contamination due to the presence of *E. coli* (*Escherichia coli*) bacteria. To reduce laboratory costs, water samples from six public swimming areas are combined for one test, and further testing is done only if the combined sample fails. Based on past results, there is a 2% chance of finding *E. coli* bacteria in a public swimming area. Find the probability that a combined sample from six public swimming areas will reveal the presence of *E. coli* bacteria.

Conditional Probabilities. In Exercises 17–20, use the Titanic *mortality data in the accompanying table.*

	Men	Women	Boys	Girls
Survived	322	318	29	27
Died	1360	104	35	18

17. If we randomly select someone who was aboard the *Titanic*, what is the probability of getting a man, given that the selected person died?

18. If we randomly select someone who died, what is the probability of getting a man?

19. What is the probability of getting a boy or girl, given that the randomly selected person is someone who survived?

20. What is the probability of getting a man or woman, given that the randomly selected person is someone who died?

Applying Bayes' Theorem. In Exercises 21–24, assume that the Altigauge Manufacturing Company makes 80% of all electrocardiograph machines, the Bryant Company makes 15% of them, and the Cardioid Company makes the other 5%. The electrocardiograph machines made by Altigauge have a 4% rate of defects, the Bryant machines have a 6% rate of defects, and the Cardioid machines have a 9% rate of defects.

21. a. If a particular electrocardiograph machine is randomly selected from the general population of all such machines, find the probability that it was made by the Altigauge Manufacturing Company.
   b. If a randomly selected electrocardiograph machine is then tested and is found to be defective, find the probability that it was made by the Altigauge Manufacturing Company.

22. a. Find the probability of randomly selecting an electrocardiograph machine and getting one manufactured by the Bryant Company.
   b. If an electrocardiograph machine is randomly selected and tested, find the probability that it was manufactured by the Bryant Company if the test indicates that the machine is defective.

23. **a.** Find the probability of randomly selecting an electrocardiograph machine and getting one manufactured by the Cardioid Company.
    b. An electrocardiograph machine is randomly selected and tested. If the test indicates that it is defective, find the probability that it was manufactured by the Cardioid Company.

24. An electrocardiograph machine is randomly selected and tested. If the test indicates that it is *not* defective, find the probability that it is from the Altigauge Company.

25. *HIV* The New York State Health Department reports a 10% rate of the HIV virus for the "at-risk" population. Under certain conditions, a preliminary screening test for the HIV virus is correct 95% of the time. (Subjects are not told that they are HIV infected until additional tests verify the results.) One person is randomly selected from the at-risk population.
    a. What is the probability that the selected person has the HIV virus if it is known that this person has tested positive in the initial screening?
    b. What is the probability that the selected person tests positive in the initial screening if it is known that this person has the HIV virus?

26. *HIV* Use the same data from the preceding exercise.
    a. What is the probability that the selected person has the HIV virus if it is known that he or she has tested negative in the initial screening?
    b. What is the probability that the selected person tests negative in the initial screening if it is known that he or she has the HIV virus?

3-6   Risks and Odds

One simple way to measure risk is to use a probability value. For example, in one of the largest medical experiments ever conducted, it was found that among 200,745 children injected with the Salk vaccine, 33 developed paralytic polio (poliomyelitis). It follows that for this treatment group, $P(\text{polio}) = 33/200,745 = 0.000164$. However, that single measure is deficient, because no information is given about the rate of polio for those children who were injected with a placebo. The risk of polio for children treated with the Salk vaccine should be somehow compared to the risk of polio for those children given a placebo. Let's consider the data summarized in Table 3-3.

Table 3-3	Prospective Study of Polio and the Salk Vaccine		
	Polio	No Polio	**Total**
Salk Vaccine	33	200,712	200,745
Placebo	115	201,114	201,229

Based on the data in Table 3-3, we can identify the following probabilities:

Polio rate for treatment group:

$$P(\text{polio} \mid \text{Salk vaccine}) = \frac{33}{200{,}745} = 0.000164$$

Polio rate for placebo group:

$$P(\text{polio} \mid \text{placebo}) = \frac{115}{201{,}229} = 0.000571$$

Informal comparison of the preceding two probabilities likely suggests that there is a substantial difference between the two polio rates. Later chapters will use more effective methods for determining whether the apparent difference is actually significant, but in this section we introduce some simple measures for comparing the two rates.

The preceding table can be generalized with the following format, shown in Table 3-4:

Table 3-4	Generalized Table Summarizing Results of a Prospective Study	
	Disease	No Disease
Treatment (or Exposed)	a	b
Placebo (or Not Exposed)	c	d

We noted that in this section we introduce some simple measures for comparing two rates, such as the polio rate for the Salk vaccine treatment group and the polio rate for the placebo group, as summarized in Table 3-3. We begin with the *absolute risk reduction*.

Definition

When comparing two probabilities or rates, the **absolute risk reduction** is simply their difference, found by computing the following:

absolute risk reduction = $|P$ (event occurring in treatment group)

$- P$ (event occurring in control group)$|$

If the data are in the generalized format of Table 3-4, we can express the absolute risk reduction as follows:

absolute risk reduction = $|P$ (event occurring in treatment group)

$- P$ (event occurring in control group)$|$

$$= \left| \frac{a}{a+b} - \frac{c}{c+d} \right|$$

(In the above expression, "treatment" might be replaced by "the presence of some condition" or "exposed to disease" or some other equivalent description.)

EXAMPLE **Computing Absolute Risk Reduction for Salk Vaccine** Using the data summarized in Table 3-3, find the absolute risk reduction, which can be used to measure the effectiveness of the Salk vaccine.

SOLUTION   Based on the data in Table 3-3, we have already found that $P(\text{polio} \mid \text{Salk vaccine}) = 0.000164$ and $P(\text{polio} \mid \text{placebo}) = 0.000571$. It follows that

$$\text{absolute risk reduction} = \mid P(\text{polio} \mid \text{Salk vaccine}) - P(\text{polio} \mid \text{placebo}) \mid$$

$$= \left| \frac{33}{33 + 200{,}712} - \frac{115}{115 + 201{,}114} \right| = 0.000407$$

One commonly used measure of comparison is the *relative risk*. We introduce the following notation, then define relative risk.

Notation

p_t = proportion (or incidence rate) of some characteristic in a *treatment group*

p_c = proportion (or incidence rate) of some characteristic in a *control group*

Definition

In a *prospective study*, the **relative risk** (or **risk ratio**) of a characteristic is the ratio P_t / P_c, where P_t is the proportion of the characteristic in the treatment group and p_c is the proportion in the control group. If the data are in the same format as the generalized Table 3-4, then the relative risk is found by evaluating

$$p_t / p_c = \frac{\dfrac{a}{a + b}}{\dfrac{c}{c + d}}$$

EXAMPLE **Computing Relative Risk** Using the data in Table 3-3, find the relative risk.

SOLUTION   For the sample data in Table 3-3, we will consider the treatment group to be the group of children given the Salk vaccine, and the control group is the group of children given a placebo. Using the preceding notation, we have

$$p_t = \text{proportion of polio in } \textit{treatment group} = \frac{33}{33 + 200{,}712} = 0.000164$$

p_c = proportion of polio in *control (placebo) group*

$$= \frac{115}{115 + 201{,}114} = 0.000571$$

continued

Using the above values, we can now find the relative risk as follows:

$$\text{relative risk} = \frac{p_t}{p_c} = \frac{0.000164}{0.000571} = 0.287$$

INTERPRETATION We can interpret this result as follows: The polio rate for children given the Salk vaccine is 0.287 of the polio rate for children given a placebo. (A relative risk less than 1 indicates that the treatment results in a reduced risk.) If we were to consider the reciprocal value of $0.000571/0.000164 = 3.48$, we see that children in the placebo group are 3.48 times more likely to get polio.

One problem with relative risk is that it may be misleading by suggesting that a treatment is superior or inferior, even when the absolute difference between rates is not very large. For example, if 3 out of 10,000 aspirin users were to experience an immediate cure of a cold compared to only 1 out of 10,000 placebo users, the relative risk of 3.00 correctly indicates that the incidence of immediate cold cures is *three times* as high for aspirin users, but the cure rates of 0.0003 and 0.0001 are so close that, for all practical purposes, aspirin should not be considered as a factor affecting the immediate cure of a cold. With cure rates of 0.0003 and 0.0001, the absolute risk reduction is 0.0002. Because the absolute risk reduction is so small, the effectiveness of the aspirin treatment would be negligible. In such a situation, the *number needed to treat* would be an effective measure that is not so misleading.

Definition

The **number needed to treat** is the number of subjects that must be treated in order to prevent *one* event, such as a disease or adverse reaction. It is calculated by dividing 1 by the absolute risk reduction.

$$\text{number needed to treat} = \frac{1}{\text{absolute risk reduction}}$$

Round-Off Rule

If the calculated value of the number needed to treat is not a whole number, round it *up* to the next larger whole number.

If the sample data are in the format of the generalized Table 3-4, then

$$\text{number needed to treat} = \frac{1}{\left| \dfrac{a}{a+b} - \dfrac{c}{c+d} \right|} \qquad \text{(rounded } up \text{ to the next larger whole number)}$$

If 3 out of 10,000 aspirin users were to experience an immediate cure of a cold compared to only 1 out of 10,000 placebo users, the absolute risk reduction is $|0.0003 - 0.0001| = 0.0002$, and the number needed to treat is $1/0.0002 = 5000$. This means that we would need to treat 5000 subjects with colds to get *one* person who experiences an immediate cure.

EXAMPLE **Computing the Number Needed to Treat**   Using the polio data in Table 3-3, find the number needed to treat, then interpret the result.

SOLUTION   We have already obtained these results:

p_t = proportion of polio in *treatment group*   = 0.000164

p_c = proportion of polio in *control (placebo) group* = 0.000571

absolute risk reduction = $|0.000164 - 0.000571|$ = 0.000407

It is now easy to find the number needed to treat.

$$\text{number needed to treat} = \frac{1}{\text{absolute risk reduction}}$$

$$= \frac{1}{0.000407} = 2457.002457$$

$$= 2458 \text{ (rounded } up)$$

INTERPRETATION   The result of 2458 can be interpreted as follows: We would need to vaccinate 2458 children with the Salk vaccine (instead of a placebo) to prevent *one* of the children from getting polio.

So far in this chapter, we have used probability values to express likelihood of various events. Probability values are numbers between 0 and 1 inclusive. However, expressions of likelihood are often given as *odds,* such as 50:1 (or "50 to 1").

Definitions

The **actual odds against** event A occurring are the ratio $P(\overline{A})/P(A)$, usually expressed in the form of *m:n* (or "*m* to *n*"), where *m* and *n* are integers having no common factors.

The **actual odds in favor** of event A can be expressed as $P(A)/P(\overline{A})$, which is the reciprocal of the actual odds against that event. If the odds against A are *m:n*, then the odds in favor of A are *n:m*.

Note that in the two preceding definitions, the *actual odds against* and the *actual odds in favor* describe the actual likelihood of some event. (Gambling situations typically use *payoff odds*, which describe the amount of profit relative to the amount of a bet. For example, if you bet on the number 7 in roulette, the actual odds against winning are 37:1, but the payoff odds are 35:1. Racetracks and casinos are in business to make a profit, so the payoff odds will usually differ from the actual odds.)

EXAMPLE Consider the data in Table 3-5.

a. For those babies discharged early, find the probability of being rehospitalized within a week.

b. For those babies discharged early, find the odds in favor of being rehospitalized early.

Table 3-5	Retrospective Study of Newborn Discharge and Rehospitalization		
	Rehospitalized within a week	Not rehospitalized within a week	**Total**
Early discharge (<30 hours)	457	3199	3656
Late discharge (30+ hours)	260	2860	3120

SOLUTION

a. There were 3656 babies discharged early, and 457 of them were rehospitalized within a week, so

$$P(\text{rehospitalized}) = \frac{457}{3656} = \frac{1}{8}$$

b. Because $P(\text{rehospitalized}) = 1/8$, it follows that $P(\overline{\text{rehospitalized}}) = 1 - 1/8 = 7/8$. We can now find the odds in favor of rehospitalization as follows:

odds in favor of rehospitalization for the early discharge group

$$\frac{P(\text{rehospitalized})}{P(\overline{\text{rehospitalized}})} = \frac{1/8}{7/8} = \frac{1}{7}$$

This result is often expressed as 1:7.

With odds of 1:7 in favor of rehospitalization for babies discharged early, it follows that the odds *against* rehospitalization for early discharge are 7:1.

For the data in Table 3-5, how does the likelihood of rehospitalization differ between the early discharge group and the late discharge group? One way to address that question is to use the *odds ratio*.

Definition

In a *retrospective* or *prospective study*, the **odds ratio** (or **OR** or **relative odds**) is a measure of risk found from the ratio of the odds for the treatment group (or case group exposed to the risk factor) to the odds for the control group, evaluated as follows:

$$\text{odds ratio} = \frac{\text{odds in favor of event for treatment (or exposed) group}}{\text{odds in favor of event for control group}}$$

If the data are in the format of the generalized Table 3-4, then the odds ratio can be computed as follows:

$$\text{odds ratio} = \frac{ad}{bc}$$

EXAMPLE **Computing Odds Ratio** Using the data in Table 3-5, find the odds ratio for rehospitalization.

SOLUTION   For this example, we consider the case group to be the babies discharged early, and we consider the control group to be the babies discharged late. The preceding example showed that for the early discharge group, the odds in favor of rehospitalization are 1:7. Using similar calculations for the late discharge group, we get odds in favor of rehospitalization of 1/11 or 1:11. We can now find the odds ratio:

$$\text{odds ratio} = \frac{\text{odds for rehospitalization in early discharge group}}{\text{odds for rehospitalization in late discharge group}} = \frac{457/3199}{260/2860}$$

$$= \frac{1/7}{1/11} = 1.571$$

If we take advantage of the fact that Table 3-5 does correspond to the generalized Table 3-4, then the odds ratio can also be calculated as follows:

$$\text{odds ratio} = \frac{ad}{bc} = \frac{(457)(2860)}{(3199)(260)} = 1.571$$

INTERPRETATION This result indicates that the odds in favor of rehospitalization are 1.571 times higher for babies discharged early when compared to those discharged late. This suggests that newborns discharged early are at substantially increased risk of rehospitalization.

Why Not Use Relative Risk for *Retrospective Studies?*

Note this distinction made in the preceding definitions: Relative risk is defined for *prospective* studies, whereas odds ratio is defined for *prospective* and *retrospective* studies.

Prospective study: Use relative risk and/or odds ratio.

Retrospective study: Use odds ratio *only*.

A risk ratio makes sense only if the involved probabilities are good estimates of the actual incidence rates, as in a prospective study. In a prospective study, we know the results required to estimate risks. For example, Table 3-3 includes the results from the prospective study conducted to test the effectiveness of the Salk vaccine. We know that among the 200,745 children given the vaccine, 33 developed polio, so the risk of polio for vaccinated children can be estimated to be 33/200,745. Similarly, the risk of polio for children given a placebo can be estimated to be 115/201,229. The relative risk is simply the ratio of those two known estimates. But in a retrospective study, the researcher might decide to choose a sampling plan that does not yield probabilities that are good estimates of incidence rates. For example, suppose a researcher is studying hypertension and heart disease, and decides to choose subjects in such a way that 50% of them have hypertension while 50% do not. The incidence rate of 50% is the result of the researcher's decision, not the actual rate of hypertension. Here, the risk ratio of 1 would be wrong.

Consider retrospective studies with the data in Tables 3-6 and 3-7. In Table 3-6, the researcher tried to select subjects in proportion to the general population. In Table 3-7, note that the frequencies in the "Other" column are *doubled*, which could be the result of a decision by a researcher during the design of the sampling plan. Note that the relative risk values are different, but the odds ratios are the same, indicating that for retrospective studies, the odds ratio is a reliable measure, but the relative risk is not a reliable measure. Also, see the following results:

From Table 3-6: $P(\text{lung cancer} \mid \text{smoker}) = 140/672 = 0.208$

From Table 3-7: $P(\text{lung cancer} \mid \text{smoker}) = 140/1204 = 0.116$

The above probabilities are very different, so both of them cannot be good estimates of the incidence rate of death from lung cancer for smokers. Such incidence rates are good estimates in prospective studies, but not in retrospective studies, which include values that might depend on the design plan of the researcher. Consequently, relative risk should not be used with retrospective studies.

Table 3-6	Retrospective Study of Causes of Death (RR = 17.1; OR = 21.4)	
	Lung Cancer	Other
Smoker	140	532
Nonsmoker	21	1707

Table 3-7	Retrospective Study of Causes of Death (RR = 19.0; OR = 21.4)	
	Lung Cancer	Other
Smoker	140	1064
Nonsmoker	21	3414

3-6 Exercises

Headaches and Viagra. In Exercises 1–9, use the data in the accompanying table (based on data from Pfizer, Inc.). That table describes results from a clinical trial of the well-known drug Viagra.

	Viagra Treatment	Placebo
Headache	117	29
No Headache	617	696

1. Is the study retrospective or prospective?

2. For those in the Viagra treatment group, find the probability that the subject experienced a headache.

3. Compare $P(\text{headache} \mid \text{Viagra treatment})$ and $P(\text{headache} \mid \text{placebo})$.

4. Find the value of the absolute risk reduction for headaches in the treatment and placebo groups.

5. Find the number of Viagra users that would need to stop using Viagra in order to prevent a single headache.

6. For those in the Viagra treatment group, find the odds in favor of a headache, then find the odds against a headache.

7. Find the relative risk of a headache for those in the treatment group compared to those in the placebo group. Interpret the result.

8. Find the odds ratio for headaches in the treatment group compared to the placebo group, then interpret the result.

9. Based on the preceding results, should Viagra users be concerned about headaches as an adverse reaction?

Facial Injuries and Bicycle Helmets. In Exercises 10–19, use the data in the accompanying table (based on data from "A Case-Control Study of the Effectiveness of Bicycle Safety Helmets in Preventing Facial Injury," by Thompson, Rivara, and Wolf, American Journal of Public Health, Vol. 80, No. 12).

	Helmet Worn	No Helmet
Facial injuries received	30	182
All injuries nonfacial	83	236

10. Is the study retrospective or prospective?

11. For the group that did not wear helmets, find the probability of randomly selecting a subject and getting someone who had facial injuries. Can that probability be used as an estimate of the incidence of facial injuries among those who do not wear helmets? Why or why not?

12. Is the calculation of a risk ratio justified for the given data?

13. Find the value of the absolute risk reduction for facial injuries in the group that wore helmets and the group that did not wear helmets.

14. For those who did not wear helmets, find the odds in favor of facial injuries.

15. Find the odds ratio for facial injuries in the group that did not wear helmets compared to the group that did wear helmets, then interpret the result.

16. After doubling the frequencies in the second row of the table, calculate the odds ratio for facial injuries in the group that did not wear helmets compared to the group that did wear helmets. Compare the result to the odds ratio obtained in the preceding exercise.

17. Based on the preceding results, does wearing a helmet appear to decrease the risk of facial injuries? Why or why not?

18. Assuming that calculation of a risk ratio is justified for the given data, find the risk ratio of facial injuries for those not wearing helmets compared to those wearing helmets. Does the result change if the frequencies in the second row are doubled?

19. Modify the table by tripling the frequencies in the first row. When comparing facial injuries in the group that wore helmets and the group that did not wear helmets, how does the change in frequencies affect the odds ratio? Risk ratio? What do these results suggest about the use of odds ratios in retrospective studies? What do these results suggest about risk ratios in retrospective studies?

20. *Design of Experiments* You would like to conduct a study to determine the effectiveness of seat belts in saving lives in car crashes.
    a. What would be wrong with randomly selecting 2000 drivers, then randomly assigning half of them to a group that uses seat belts and another group that does not wear seat belts?
    b. If 2000 drivers are randomly selected and separated into two groups according to whether they use seat belts, what is a practical obstacle in conducting a prospective study of the effectiveness of seat belts in car crashes?

3-7   Rates of Mortality, Fertility, and Morbidity

Sections 3-2 through 3-5 dealt with principles of probability values, and Section 3-6 introduced odds as another tool for describing the likelihood of an event. In the biological and health sciences, *rates* are often used to describe the likelihood of an event. Rates are used by researchers and health professionals to monitor the health status of a community. Although any specific time interval could be used, we assume a time interval of *one year* throughout this section. We begin with the following formal definition.

Definition

A **rate** describes the frequency of occurrence of some event. It is the relative frequency of an event, multiplied by some number, typically a value such as 1000 or 100,000. A rate can be expressed as

$$\left(\frac{a}{b}\right)k$$

where
a = frequency count of the number of people for whom the event occurred
b = total number of people exposed to the risk of the event occurring
k = multiplier number, such as 1000 or 10,000

The above general definition is commonly applied to measures of mortality, fertility, and morbidity. For the following rates, mortality refers to deaths, fertility refers to births, morbidity refers to diseases, *infants* are babies who were born alive, and *neonates* are infants under the age of 28 days. The following common and important rates use some multiplier k, such as 1000.

Mortality Rates

Crude (or unadjusted) mortality rate $= \left(\dfrac{\text{number of deaths}}{\text{total number in population}}\right)k$

Infant mortality rate $= \left(\dfrac{\text{number of deaths of infants under 1 year of age}}{\text{total number of live births}}\right)k$

Neonatal mortality rate $=$
$\left(\dfrac{\text{number of deaths of infants under 28 days of age}}{\text{total number of live births}}\right)k$

Fetal mortality rate $=$
$\left(\dfrac{\text{total number of fetuses delivered without life after 20 weeks of gestation}}{\text{total number of live births} + \text{fetuses delivered without life after 20 weeks of gestation}}\right)k$

Perinatal mortality rate $=$
$\left(\dfrac{\text{number of fetal deaths} + \text{number of neonatal deaths}}{\text{number of live births} + \text{number of fetal deaths}}\right)k$

where a fetal death occurs when a fetus is delivered without life after 20 weeks of gestation, and a neonatal death occurs when an infant dies under 28 days of age.

Fertility Rates

Crude birthrate $= \left(\dfrac{\text{number of live births}}{\text{total number in population}}\right)k$

General fertility rate $= \left(\dfrac{\text{number of live births}}{\text{number of women aged } 15-44}\right)k$

Morbidity (Disease) Rates

Incidence rate $= \left(\dfrac{\text{number of reported cases of disease}}{\text{total number in population}}\right)k$

Prevalence rate $= \left(\dfrac{\text{total number of people with disease at a given time}}{\text{number in population at the given point in time}}\right)k$

EXAMPLE Crude Mortality Rate For a recent year in the United States, there were 2,416,000 deaths in a population of 285,318,000 people. Use those values with a multiplier of 1000 to find the crude mortality rate.

SOLUTION With 2,416,000 people who died, with 285,318,000 people in the total population, and letting $k = 1000$, we compute the crude mortality rate as follows:

$$\left(\frac{2,416,00}{285,318,000}\right)1000 = 8.5 \text{ (rounded)}$$

continued

INTERPRETATION: For this particular year, the death rate is 8.5 people for each 1000 people in the population. Using the relative frequency definition of probability given in Section 3-2, we might also say that for a randomly selected person, the probability of death in this year is 2,416,000/ 285,318,000 = 0.0085. One advantage of the mortality rate of 8.5 people (per 1000 people in the population) is that it results in a value that uses fewer decimal places and is generally easier to use and understand.

EXAMPLE **Infant Mortality Rate** The infant mortality rate is a very important measure of the health of a region. (According to the World Bank, the infant mortality rate is 5 per 1000 live births in high-income countries, but it is 88 per 1000 live births in developing countries.) For a recent year, there were 4,026,000 live births in the United States, and there were 27,500 deaths of infants under 1 year of age. Using a multiplying factor of $k = 1000$, find the infant mortality rate of the United States and compare it to the rate of 36.7 for Egypt.

SOLUTION   The infant mortality rate is computed as shown below.

$$\left(\frac{\text{number of deaths of infants under 1 year of age}}{\text{total number of live births}}\right)k =$$

$$\left(\frac{27,500}{4,026,000}\right)1000 = 6.8 \text{ (rounded)}$$

INTERPRETATION The infant mortality rate of 6.8 deaths per 1000 infants under 1 year of age is substantially less than the infant mortality rate of 36.7 in Egypt.

A *crude rate*, as defined and illustrated above, is a single value based on crude totals. When comparing two different regions, such as Florida and Colorado, a comparison of rates can be misleading because of differences in factors (such as age) that might affect the rates. In a recent year, the crude mortality rates (per 1000 population) were 11.8 in Florida and 6.4 in Colorado. This is not too surprising considering that in Florida, roughly 20% of the population is over the age of 65, compared to only 10% for Colorado. The higher mortality rate for Florida does not mean that Florida is less healthy; in this case, it appears that Florida has a higher death rate largely because it has a higher proportion of older residents. Instead of using crude rates, we might use either *specific rates* or *adjusted rates*. **Specific rates** are rates specific for some particular group, such as people aged 18-24, or rates specific for some particular cause of death, such as deaths due to myocardial infarction. **Adjusted rates** involve calculations that can be quite complicated, but they basically make adjustments for important factors, such as age, gender, or race. Because age is the characteristic that typically affects mortality the most, it is the most common factor used as the basis for adjustment. Calculations of adjusted rates involve the creation of a theoretical *standardized popula-*

tion that is used for the regions being compared. (See Exercise 17.) A population of 1,000,000 people with the same composition as the United States is often used as the standardized population. Adjusted rates are valuable for comparing different regions, but they do not necessarily reflect the true crude death rates. Adjusted rates should not be used as death rates and they should not be compared to crude rates.

When assessing the accuracy of rates, we should consider the source. Mortality rates found in a source such as the *Statistical Abstract of the United States* (compiled by the U.S. Bureau of the Census) are likely to be quite accurate, because each state now has a mandatory death reporting system. However, morbidity rates are likely to be less accurate, because some diseases are known to be underreported or not reported at all. Some morbidity rates might be the result of very questionable surveys. However, some surveys, such as the annual National Health Survey, involve large samples of people who are very carefully chosen, so that the results are likely to be very accurate.

3-7 Exercises

Finding Rates. In Exercises 1–8, use the data given below (based on data from the U.S. Census Bureau) to find the indicated rates. Round results to one decimal place, and use a multiplying factor of k = 1000 unless indicated otherwise.

Vital Statistics for the United States in One Year

Population: 285,318,000
Deaths: 2,416,000
Live births: 4,026,000
Deaths of infants under 1 year of
    age: 27,500
Deaths of infants under 28 days of
    age: 18,000
Motor vehicle deaths: 43,900

Women aged 15-44: 61,811,000
Fetuses delivered without life after
    20 weeks of gestation: 39,900
HIV infected persons: 900,000
Deaths from HIV infections: 17,402

1. Find the neonatal mortality rate.

2. Find the fetal mortality rate.

3. Find the perinatal mortality rate.

4. Find the crude birthrate.

5. Find the general fertility rate.

6. Using a multiplier of $k = 100,000$, find the motor vehicle death incidence rate.

7. Find the HIV incidence rate.

8. Find the HIV mortality rate for HIV-infected persons.

9. Finding Probability An example in this section involved the crude mortality rate, which was found to be 8.5 persons per 1000 population. Find the *probability* of randomly selecting someone and getting a person who died within the year. What advantage does the crude mortality rate have over the probability value?

10. Finding Probability The crude death rate for France was recently 9.1, and that rate was computed using a multiplier of $k = 1000$.

   a. Find the probability that a randomly selected French person died within the year.

   b. If two French people are randomly selected, find the probability that they both died within the year, and express the result using three significant digits.

   c. If two French people are randomly selected, find the probability that neither of them died within the year, and express the result using seven decimal places.

11. Finding Probability The crude death rate for Canada was recently 7.5, and that rate was computed using a multiplier of $k = 1000$.

   a. Find the probability that a randomly selected Canadian died within the year.

   b. If two Canadians are randomly selected, find the probability that they both died within the year, and express the result using three significant digits.

   c. If two Canadians are randomly selected, find the probability that neither of them died within the year, and express the result using seven decimal places.

12. Finding Probability In a recent year in the United States, there were 700,142 deaths due to heart disease, and the population was 285,318,000.

   a. Find the crude mortality rate for heart disease. (This result is sometimes called the *cause-specific death rate.*)

   b. Find the probability that a randomly selected person died of heart disease.

   c. Find the probability that when two people are randomly selected, neither of them died because of heart disease.

13. Cause-of-Death Ratio In a recent year in the United States, there were 2,416,000 deaths, and 700,142 of them were due to heart disease. The **cause-of-death ratio** is expressed as follows:

$$\left(\frac{\text{number of deaths due to specific disease}}{\text{total number of deaths}} \right) k$$

where $k = 100$.

   a. Find the cause-of-death ratio for heart disease.

   b. If three deaths are randomly selected, find the probability that none of them are due to heart disease.

14. Crude Mortality Rates The table below lists numbers of deaths and population sizes for different age groups for Florida and the United States for a recent year.

   a. Find the crude mortality rate for Florida and the crude mortality rate for the United States. Although we should not compare crude mortality rates, what does the comparison suggest in this case?

   b. Using only the age group of 65 and older, find the mortality rates for Florida and the United States. Compare the results.

   c. What percent of the Florida population is made up of people aged 65 and older? What is the percent for the United States? What do the results suggest about the crude mortality rate for Florida compared to the United States?

	0–24	25–64	65 and older
Florida deaths	4085	35,458	128,119
Florida population	4,976,942	6,436,092	2,807,598
U.S. deaths	75,011	464,655	1,876,334
U.S. population	100,464,000	149,499,000	35,355,000

15. Number of Deaths The number of deaths in the United States has been steadily increasing each year. Does this mean that the health of the nation is declining? Why or why not?

16. Comparing Rates In a recent year, the crude mortality rate of the United States was 8.5 (per 1000 population), and the corresponding crude mortality rate for China was 6.7. What is a major problem with comparing the crude mortality rates of the United States and China?

*17. Adjusted Mortality Rate Refer to the data listed in Exercise 14. Change the Florida population sizes for the three age categories so that they fit the same distribution as the United States. Next, adjust the corresponding numbers of deaths proportionately. (Use the same Florida mortality rates for the individual age categories, but apply those rates to the adjusted population sizes.) Finally, compute the mortality rate using the adjusted values. The result is a mortality rate adjusted for the variable of age. (Better results could be obtained by using more age categories.) How does this adjusted mortality rate for Florida compare to the mortality rate for the United States? (There are other methods for computing adjusted rates.)

3-8 Counting

In some probability problems, the major obstacle is finding the total number of outcomes. For example, suppose a test of a method of gender selection is designed to determine whether the method increases the likelihood of a female, as claimed. Suppose also that a test of this method results in 10 females among 10 births. Our conclusion about the method depends on the probability of getting 10 females in 10 births by chance. What is the probability of getting 10 females in 10 births by chance? Because we are dealing with equally likely outcomes, we can use the classical definition (see Rule 2 in Section 3-2) with $P(A) = s/n$. The number s in this case is 1, because there is only one way to get 10 females in 10 births: FFFFFFFFFF. The denominator n is the total number of possible outcomes, and that is more difficult to determine. Instead of listing the possible outcomes, we will use a much more efficient method for counting them. This section introduces different methods for finding such numbers without directly listing and counting the possibilities. We begin with the *fundamental counting rule*.

Fundamental Counting Rule

For a sequence of two events in which the first event can occur m ways and the second event can occur n ways, the events together can occur a total of $m \cdot n$ ways.

The fundamental counting rule easily extends to situations involving more than two events, as illustrated in the following examples.

EXAMPLE **Gender Selection** In a test of a method of gender selection, 10 births result in 10 females. Find the probability of getting 10 females in 10 births by chance, assuming that males and females are equally likely. Based on the result, does it appear that the method of gender selection is effective?

SOLUTION Each of the 10 births has two possible outcomes: male, female. By applying the fundamental counting rule, we get

$$2 \cdot 2 \cdot 2 \cdot 2 \cdot 2 \cdot 2 \cdot 2 \cdot 2 \cdot 2 \cdot 2 = 1024$$

Only one of those 1024 outcomes consists of 10 females, so $P(10 \text{ females in 10 births}) = 1/1024$. Because that probability is so small, it is very unlikely that the 10 females occurred by chance. A more reasonable explanation is that the method of gender selection is effective so that the probability of a female is considerably greater than 0.5.

EXAMPLE **DNA Nucleotides** DNA (deoxyribonucleic acid) is made of nucleotides, and each nucleotide can contain any one of these nitrogenous bases: A (adenine), G (guanine), C (cytosine), T (thymine). If one of those four bases (A, G, C, T) must be selected three times to form a linear triplet, how many different triplets are possible? Note that all four bases can be selected for each of the three components of the triplet.

SOLUTION There are 4 possibilities for each of the three components making up the triplet, so the number of different possibilities is $4 \cdot 4 \cdot 4 = 64$. The 64 cases could be listed as AAA, AAG, AAC, . . . , TTT.

EXAMPLE **Cotinine in Smokers** Data Set 5 in Appendix B lists measured cotinine levels for a sample of people from each of three groups: smokers (denoted here by S), nonsmokers who were exposed to tobacco smoke (denoted by E), and nonsmokers not exposed to tobacco smoke (denoted by N). When nicotine is absorbed by the body, cotinine is produced. If we calculate the mean cotinine level for each of the three groups, then arrange those means in order from low to high, we get the sequence NES. An antismoking lobbyist claims that this is evidence that tobacco smoke is unhealthy, because the presence of cotinine increases as exposure to and use of tobacco increase. How many ways can the three groups denoted by N, E, and S be arranged? If an arrangement is selected at random, what is the probability of getting the sequence of NES? Is the probability low enough to conclude that the sequence of NES indicates that the presence of cotinine increases as exposure to and use of tobacco increase?

SOLUTION In arranging sequences of the groups N, E, and S, there are 3 possible choices for the first group, 2 remaining choices for the second group, and only 1 choice for the third group. The total number of possible arrangements is therefore

$$3 \cdot 2 \cdot 1 = 6$$

There are six different ways to arrange the N, E, and S groups. (They can be listed as NES, NSE, ESN, ENS, SNE, and SEN.) If we randomly select one of the six possible sequences, there is a probability of 1/6 that the sequence NES is obtained. Because that probability of 1/6 is relatively high, we know that the sequence of NES could easily occur by chance. The probability is not low enough to conclude that the sequence of NES indicates that the presence of cotinine increases as exposure to and use of tobacco increase. We would need a smaller probability, such as 0.01.

In the preceding example, we found that 3 groups can be arranged $3 \cdot 2 \cdot 1 = 6$ different ways. This particular solution can be generalized by using the following notation for the symbol ! and the following *factorial rule*.

Notation

The **factorial symbol !** denotes the product of decreasing positive whole numbers. For example, $4! = 4 \cdot 3 \cdot 2 \cdot 1 = 24$. By special definition, $0! = 1$.

Factorial Rule

A collection of n different items can be arranged in order $n!$ different ways. (This **factorial rule** reflects the fact that the first item may be selected n different ways, the second item may be selected $n - 1$ ways, and so on.)

Routing problems often involve application of the factorial rule. Verizon wants to route telephone calls through the shortest networks. Federal Express wants to find the shortest routes for its deliveries. American Airlines wants to find the shortest route for returning crew members to their homes. See the following example.

EXAMPLE **Routes to All 50 Capitals** You have been hired to personally visit each of the 50 state capitals and interview the head of the state health department. As you plan your route of travel, you want to determine the number of different possible routes. How many different routes are possible?

SOLUTION   By applying the factorial rule, we know that 50 items can be arranged in order 50! different ways. That is, the 50 state capitals can be arranged 50! ways, so the number of different routes is 50!, or

30,414,093,201,713,378,043,612,608,166,064,768,844,377,641,568,960,512, 000,000,000,000

Now there's a large number.

The preceding example is a variation of a classical problem called the *traveling salesman problem*. It is especially interesting because the large number of possibilities means that we can't use a computer to calculate the distance of each route. The time it would take even the fastest computer to calculate the shortest possible route is about

$$1,000,000,000,000,000,000,000,000,000,000,000,000,000 \text{ centuries}$$

Considerable effort is currently being devoted to finding efficient ways of solving such problems.

According to the factorial rule, *n* different items can be arranged *n*! different ways. Sometimes we have *n* different items, but we need to select *some* of them instead of all of them. If we must conduct surveys in state capitals, as in the preceding example, but we have time to visit only four capitals, the number of different possible routes is $50 \cdot 49 \cdot 48 \cdot 47 = 5,527,200$. Another way to obtain this same result is to evaluate

$$\frac{50!}{46!} = 50 \cdot 49 \cdot 48 \cdot 47 = 5,527,200$$

In this calculation, note that the factors in the numerator divide out with the factors in the denominator, except for the factors of 50, 49, 48, and 47 that remain. We can generalize this result by noting that if we have *n* different items available and we want to select *r* of them, the number of different arrangements possible is $n!/(n - r)!$ as in $50!/46!$. This generalization is commonly called the *permutations rule*.

Permutations Rule (When Items Are All Different)

The number of **permutations** (or sequences) of *r* items selected from *n* available items (without replacement) is

$$_nP_r = \frac{n!}{(n - r)!}$$

Many calculators can evaluate expressions of $_nP_r$.

It is very important to recognize that the permutations rule requires the following conditions:

- We must have a total of *n* *different* items available. (This rule does not apply if some of the items are identical to others.)
- We must select *r* of the *n* items (without replacement).
- We must consider rearrangements of the same items to be different sequences.

When we use the term *permutations, arrangements,* or *sequences,* we imply that *order is taken into account* in the sense that different orderings of the same items are counted separately. The letters *ABC* can be arranged six different ways: *ABC, ACB, BAC, BCA, CAB, CBA.* (Later, we will refer to *combinations,* which do not count such arrangements separately.) In the following example, we are asked to find the total number of different sequences that are possible. That suggests use of the permutations rule.

EXAMPLE **Clinical Trial of New Drug** When testing a new drug, phase 1 involves only 20 volunteers, and the objective is to assess the drug's safety. To be very cautious, you plan to treat the 20 subjects in sequence, so that any particularly adverse effect can stop the treatments before any other subjects are treated. If 30 volunteers are available, how many different sequences of 20 subjects are possible?

SOLUTION   We need to select $r = 20$ subjects from $n = 30$ volunteers that are available. The number of different sequences of arrangements is found as shown:

$$_nP_r = \frac{n!}{(n-r)!} = \frac{30!}{(30-20)!} = 7.31 \times 10^{25}$$

There are 7.31×10^{25} (or 73,100,000,000,000,000,000,000,000) different possible arrangements of 20 subjects selected from the 30 that are available.

We sometimes need to find the number of permutations, but some of the items are identical to others. The following variation of the permutations rule applies to such cases.

Permutations Rule (When Some Items Are Identical to Others)

If there are n items with n_1 alike, n_2 alike, . . . , n_k alike, the number of permutations of all n items is

$$\frac{n!}{n_1! n_2! \ldots, n_k!}$$

EXAMPLE **Gender Selection** The classic examples of the permutations rule are those showing that the letters of the word *Mississippi* can be arranged 34,650 different ways and that the letters of the word *statistics* can be arranged 50,400 ways. We will consider a different application.

In the first example of this section, we found that the probability of getting 10 females in 10 births by chance is 1/1024. Suppose that instead of getting 10 females, a test of a method of gender selection results in 8 females and 2 males. Find the probability of getting 8 females and 2 males among 10 births. Is that probability useful for assessing the effectiveness of the method of gender selection?

The Random Secretary

One classic problem of probability goes like this: A secretary addresses 50 different letters and envelopes to 50 different people, but the letters are randomly mixed before being put into envelopes. What is the probability that at least one letter gets into the correct envelope? Although the probability might seem like it should be small, it's actually 0.632. Even with a million letters and a million envelopes, the probability is 0.632. The solution is beyond the scope of this text—way beyond.

continued

SOLUTION   We have $n = 10$ items, with $n_1 = 8$ alike and $n_2 = 2$ others that are alike. The number of permutations is computed as follows:

$$\frac{n!}{n_1!n_2!} = \frac{10!}{8!2!} = \frac{3,628,800}{80,640} = 45$$

There are 45 different ways that 8 females and 2 males can be arranged. Because there are 1024 different possible arrangements and 45 ways of getting 8 females and 2 males, the probability of 8 females and 2 males is given by $P(\text{8 females and 2 males}) = 45/1024 = 0.0439$. This probability of 0.0439 is *not* the probability that should be used in assessing the effectiveness of the gender selection method. Instead of the probability of 8 females in 10 births, we should consider the probability of 8 *or more* females in 10 births, which is 0.0547. (In Section 4-2 we will clarify the reason for using the probability of 8 or more females instead of the probability of exactly 8 females in 10 births.)

The preceding example involved n items, each belonging to one of two categories. When there are only two categories, we can stipulate that x of the items are alike and the other $n - x$ items are alike, so the permutations formula simplifies to

$$\frac{n!}{(n - x)!x!}$$

This particular result will be used for the discussion of binomial probabilities, which are introduced in Section 4-3.

When we intend to select r items from n different items but *do not take order into account,* we are really concerned with possible combinations rather than permutations. That is, **when different orderings of the same items are counted separately, we have a permutation problem, but when different orderings of the same items are not counted separately, we have a combination problem** and may apply the following rule.

Combinations Rule

The number of **combinations** of r items selected from n different items is

$$_nC_r = \frac{n!}{(n - r)!r!}$$

It is very important to recognize that in applying the combinations rule, the following conditions apply:

- We must have a total of n different items available.
- We must select r of the n items (without replacement).
- We must consider rearrangements of the same items to be the same. (The combination *ABC* is the same as *CBA*.)

Because choosing between the permutations rule and the combinations rule can be confusing, we provide the following example, which is intended to emphasize the difference between them.

EXAMPLE **Phase 1 of a Clinical Trial** When testing a new drug on humans, a clinical test is normally done in three phases. Phase 1 is conducted with a relatively small number of healthy volunteers. Let's assume that we want to treat 20 healthy humans with a new drug, and we have 30 suitable volunteers available.

a. If the subjects are selected and treated *in sequence*, so that the trial is discontinued if anyone presents with a particularly adverse reaction, how many different sequential arrangements are possible if 20 people are selected from the 30 that are available?

b. If 20 subjects are selected from the 30 that are available, and the 20 selected subjects are all treated at the same time, how many different treatment groups are possible?

SOLUTION   Note that in part (a), order is relevant because the subjects are treated sequentially and the trial is discontinued if anyone exhibits a particularly adverse reaction. However, in part (b) the order of selection is irrelevant because all of the subjects are treated at the same time.

a. Because order does count, we want the number of *permutations* of $r = 20$ people selected from the $n = 30$ available people. In a preceding example in this section, we found that the number of permutations is 7.31×10^{25} (or 73,100,000,000,000,000,000,000,000).

b. Because order does *not* count, we want the number of *combinations* of $r = 20$ people selected from the $n = 30$ available people. We get

$$_nC_r = \frac{n!}{(n - r)!r!} = \frac{30!}{(30 - 20)!20!} = 30{,}045{,}015$$

With order taken into account, there are 7.31×10^{25} permutations, but without order taken into account, there are 30,045,015 combinations.

3-8 Exercises

Calculating Factorials, Combinations, Permutations. In Exercises 1–8, evaluate the given expressions and express all results using the usual format for writing numbers (instead of scientific notation).

1. 6!   **2.** 15!   **3.** $_{25}P_2$ **4.** $_{100}P_3$

5. $_{25}C_2$ **6.** $_{100}C_3$ **7.** $_{52}C_5$ **8.** $_{52}P_5$

9. Gender Sequences In a test of a method of gender selection, a trial is designed so that five couples give birth to five children. How many different gender sequences are possible?

10. Probability of Six Females What is the probability that the next six babies that are born are all females?

11. Tree Growth Experiment When designing an experiment to study tree growth, the following four treatments are used: none, irrigation only, fertilization only, irrigation and fertilization. A row of 10 trees extends from a moist creek bed to a dry land area. If one of the four treatments is randomly assigned to each of the 10 trees, how many different treatment arrangements are possible?

12. Probability of Defective Pills A batch of pills consists of 7 that are good and 3 that are defective (because they contain the wrong amount of the drug). Three pills are randomly selected and tested.
  a. How many different samples of pills are possible when three pills are randomly selected (without replacement) from the 10 that are available?
  b. Find the probability that all three of the defective pills are selected.

13. Age Discrimination The Newport General Hospital reduced its emergency room staff from 32 physicians to 28. The chief executive officer claimed that four physicians were randomly selected for job termination. However, the four physicians chosen are the four oldest among the original staff of 32. Find the probability that when four physicians are randomly selected from a group of 32, the four oldest are selected. Is that probability low enough to charge that instead of using random selection, the Newport General Hospital actually fired the oldest employees?

14. Computer Design In designing a computer, if a *byte* is defined to be a sequence of 8 bits and each bit must be a 0 or 1, how many different bytes are possible? (A byte is often used to represent an individual character, such as a letter, digit, or punctuation symbol. For example, one coding system represents the letter *A* as 01000001.) Are there enough different bytes for the characters that we typically use, including lower-case letters, capital letters, digits, punctuation symbols, dollar sign, and so on?

15. Selection of Treatment Group Walton Pharmaceuticals wants to test the effectiveness of a new drug designed to relieve allergy symptoms. The initial test will be conducted by treating six people chosen from a pool of 15 volunteers. If the treatment group is randomly selected, what is the probability that it consists of the six youngest people in the pool? If the six youngest are selected, is there sufficient evidence to conclude that instead of being random, the selection was based on age?

16. Air Routes You have just started your own medical transport company and you have one plane for a route connecting Austin, Boise, and Chicago. One route is Austin–Boise–Chicago and a second route is Chicago–Boise–Austin. How many total routes are possible? How many different routes are possible if service is expanded to include a total of 8 cities?

17. Social Security Numbers Each social security number is a sequence of nine digits. What is the probability of randomly generating nine digits and getting *your* social security number?

18. Electrifying When testing for electrical current in a cable with five color-coded wires, a meter is used to test two wires at a time. How many tests are required for every possible pairing of two wires?

19. Elected Board of Directors There are 12 members on the board of directors for the Newport General Hospital.
  a. If they must elect a chairperson, first vice chairperson, second vice chairperson, and secretary, how many different slates of candidates are possible?

b. If they must form an ethics subcommittee of four members, how many different subcommittees are possible?

20. Probabilities of Gender Sequences
    a. If a couple plans to have eight children, how many different gender sequences are possible?
    b. If a couple has four boys and four girls, how many different gender sequences are possible?
    c. Based on the results from parts (a) and (b), what is the probability that when a couple has eight children, the result will consist of four boys and four girls?

21. Is the Researcher Cheating? You become suspicious when a genetics researcher randomly selects groups of 20 newborn babies and seems to consistently get 10 girls and 10 boys. The researcher explains that it is common to get 10 boys and 10 girls in such cases.
    a. If 20 newborn babies are randomly selected, how many different gender sequences are possible?
    b. How many different ways can 10 boys and 10 girls be arranged in sequence?
    c. What is the probability of getting 10 boys and 10 girls when 20 babies are born?
    d. Based on the preceding results, do you agree with the researcher's explanation that it is common to get 10 boys and 10 girls when 20 babies are randomly selected?

22. Cracked Eggs A carton contains 12 eggs, 3 of which are cracked. If we randomly select 5 of the eggs for hard boiling, what is the probability of the following events?
    a. All of the cracked eggs are selected.
    b. None of the cracked eggs are selected.
    c. Two of the cracked eggs are selected.

Review

We began this chapter with the basic concept of probability, which is so important for methods of inferential statistics introduced later in this book. We should know that a probability value, which is expressed as a number between 0 and 1, reflects the likelihood of some event. We should know that a value such as 0.01 represents an event that is very unlikely to occur. In Section 3-1 we introduced the rare event rule for inferential statistics: If, under a given assumption, the probability of a particular event is extremely small, we conclude that the assumptions are probably not correct. As an example of the basic approach used, consider a test of a method of gender selection. If we conduct a trial of a gender-selection technique and get 10 girls in 10 births, we can make one of two inferences from these sample results:

1. The technique of gender selection is not effective, and the string of 10 consecutive girls is a fluke that happened by chance.

2. The technique of gender selection is effective (or there is some other explanation for why boys and girls are not occurring with the same frequencies).

Statisticians use the rare event rule when deciding which inference is correct: In this case, the probability of getting 10 consecutive girls is so small (1/1024) that the inference of an effective technique of gender selection is the better choice. Here we can see the important role played by probability in the standard methods of statistical inference.

In Section 3-2 we presented the basic definitions and notation, including the representation of events by letters such as A. We defined probabilities of simple events as

$$P(A) = \frac{\text{number of times that } A \text{ occurred}}{\text{number of times trial was repeated}} \qquad \text{(relative frequency)}$$

$$P(A) = \frac{\text{number of ways } A \text{ can occur}}{\text{number of different simple events}} = \frac{s}{n} \qquad \text{(for equally likely outcomes)}$$

We noted that the probability of any impossible event is 0, the probability of any certain event is 1, and for any event A, $0 \leq P(A) \leq 1$. Also, $\overline{A}$ denotes the complement of event A. That is, $\overline{A}$ indicates that event A does *not* occur.

In Sections 3-3, 3-4, and 3-5 we considered compound events, which are events combining two or more simple events. We associate the word "or" with addition and associate the word "and" with multiplication. Always keep in mind the following key considerations:

- When conducting one trial, do we want the probability of event *A or B?* If so, use the addition rule, but be careful to avoid counting any outcomes more than once.

- When finding the probability that event A occurs on one trial *and* event B occurs on a second trial, use the multiplication rule. Multiply the probability of event A by the probability of event B. *Caution:* When calculating the probability of event B, be sure to take into account the fact that event A has already occurred.

In Section 3-6 we considered measures of risk and odds. We defined absolute risk reduction, relative risk, odds in favor and odds against, and odds ratio. We noted that relative risk should not be used with data from retrospective studies. Section 3-7 discussed important rates, including mortality rates, fertility rates, and morbidity rates.

In some probability problems, the biggest obstacle is finding the total number of possible outcomes. Section 3-8 was devoted to the following counting techniques:

- Fundamental counting rule
- Factorial rule
- Permutations rule (when items are all different)
- Permutations rule (when some items are identical to others)
- Combinations rule

Review Exercises

Clinical Test of Lipitor. In Exercises 1–8, use the data in the accompanying table (based on data from Parke-Davis). The cholesterol-reducing drug Lipitor consists of atorvastatin calcium.

	10 mg Atorvastatin	Placebo
Headache	15	65
No headache	17	3

1. If 1 of the 100 subjects is randomly selected, find the probability of getting someone who had a headache.

2. If 1 of the 100 subjects is randomly selected, find the probability of getting someone who was treated with 10 mg of atorvastatin.

3. If 1 of the 100 subjects is randomly selected, find the probability of getting someone who had a headache or was treated with 10 mg of atorvastatin.

4. If 1 of the 100 subjects is randomly selected, find the probability of getting someone who was given a placebo or did not have a headache.

5. If two different subjects are randomly selected, find the probability that they both used placebos.

6. If two different subjects are randomly selected, find the probability that they both had a headache.

7. If one subject is randomly selected, find the probability that he or she had a headache, given that the subject was treated with 10 mg of atorvastatin.

8. If one subject is randomly selected, find the probability that he or she was treated with 10 mg of atorvastatin, given that the subject had a headache.

9. Acceptance Sampling With one method of acceptance sampling, a sample of items is randomly selected without replacement and the entire batch is rejected if there is at least one defect. The Medtyme Pharmaceutical Company has just manufactured 2500 aspirin tablets, and 2% are defective because they contain too much or too little aspirin. If 4 of the tablets are selected and tested, what is the probability that the entire batch will be rejected?

10. Testing a Claim The Biogene Research Company claims that it has developed a technique for ensuring that a baby will be a girl. In a test of that technique, 12 couples all have baby girls. Find the probability of getting 12 baby girls by chance, assuming that boys and girls are equally likely and that the gender of any child is independent of the others. Does that result appear to support the company's claim?

11. Selecting Members The board of directors for the Genetics Technology Corporation has 10 members.
   a. If 3 members are randomly selected to oversee the auditors, find the probability that the three wealthiest members are selected.
   b. If members are elected to the positions of chairperson, vice chairperson, and treasurer, how many different slates are possible?

12. Life Insurance The New England Life Insurance Company issues one-year policies to 12 men who are all 27 years of age. Based on data from the Department of Health and Human Services, each of these men has a 99.82% chance of living through the year. What is the probability that they all survive the year?

13. Chlamydia Rate For a recent year, the rate of chlamydia infection was reported as 278.32 per 100,000 population.
   a. Find the probability that a randomly selected person has chlamydia.
   b. If two people are randomly selected, find the probability that they both have chlamydia, and express the result using three significant digits.
   c. If two people are randomly selected, find the probability that neither of them have chlamydia, and express the result using seven decimal places.

Cumulative Review Exercises

1. Treating Chronic Fatigue Syndrome Patients suffering from chronic fatigue syndrome were treated with medication, then their change in fatigue was measured on a scale from -7 to $+7$, with positive values representing improvement and 0 representing no change. The results are listed below (based on data from "The Relationship Between Neurally Mediated Hypotension and the Chronic Fatigue Syndrome," by Bou-Holaigah, Rowe, Kan, and Calkins, *Journal of the American Medical Association*, Volume 274, Number 12).

$$6\ 5\ 0\ 5\ 6\ 7\ 3\ 3\ 2\ 4\ 4\ 0\ 7\ 3\ 4\ 3\ 6\ 0\ 5\ 5\ 6$$

   a. Find the mean.
   b. Find the median.
   c. Find the standard deviation.
   d. Find the variance.
   e. Based on the results, does it appear that the treatment was effective?
   f. If one value is randomly selected from this sample, find the probability that it is positive.
   g. If two different values are randomly selected from this sample, find the probability that they are both positive.
   h. Ignore the three values of 0 and assume that only positive or negative values are possible. Assuming that the treatment is ineffective and that positive and negative values are equally likely, find the probability that 18 subjects all have positive values (as in this sample group). Is that probability low enough to justify rejection of the assumption that the treatment is ineffective? Does the treatment appear to be effective?

2. Women's Heights The accompanying boxplot depicts heights (in inches) of a large collection of randomly selected adult women.
   a. What is the mean height of adult women?
   b. If one of these women is randomly selected, find the probability that her height is between 56.1 in. and 62.2 in.
   c. If one of these women is randomly selected, find the probability that her height is below 62.2 in. or above 63.6 in.
   d. If two women are randomly selected, find the probability that they both have heights between 62.2 in. and 63.6 in.
   e. If five women are randomly selected, find the probability that three of them are taller than the mean and the other two are shorter than the mean.

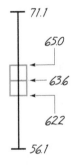

71.1
65.0
63.6
62.2
56.1

Cooperative Group Activities

1. *In-class activity:* Divide into groups of three or four and use coin tossing to develop a simulation that emulates the kingdom that abides by this decree: After a mother gives birth to a son, she will not have any other children. If this decree is followed, does the proportion of girls increase?

2. *Out-of-class activity:* Marine biologists often use the *capture-recapture* method as a way to estimate the size of a population, such as the number of fish in a lake. This method involves capturing a sample from the population, tagging each member in the sample, then returning them to the population. A second sample is later captured and the tagged members are counted along with the total size of this second sample. The results can be used to estimate the size of the population.

   Instead of capturing real fish, simulate the procedure using some uniform collection of items such as BB's, colored beads, M&Ms, Fruit Loop cereal pieces, or index cards. Start with a large collection of such items. Collect a sample of 50 and use a magic marker to "tag" each one. Replace the tagged items, mix the whole population, then select a second sample and proceed to estimate the population size. Compare the result to the actual population size obtained by counting all of the items.

3. *Out-of-class activity:* Divide into groups of two for the purpose of doing an experiment designed to show one approach to dealing with sensitive survey questions, such as those related to drug use, sexual activity (or inactivity), stealing, or cheating. Instead of actually using a controversial question that would reap wrath upon the authors, we will use this innocuous question: "Were you born in a month that has the letter *r* in it?" About 2/3 of all responses should be "yes," but let's pretend that the question is very sensitive and that survey subjects are reluctant to answer honestly. Survey people by asking them to flip a coin and respond as follows:

   - Answer "yes" if the coin turns up tails *or* you were born in a month containing the letter *r*.
   - Answer "no" if the coin turns up heads *and* you were born in a month not containing the letter *r*.

Supposedly, respondents tend to be more honest because the coin flip protects their privacy. Survey people and analyze the results to determine the proportion of people born in a month containing the letter *r*. The accuracy of the results could be checked against their actual birth dates, which can be obtained from a second question. The experiment could be repeated with a question that is more sensitive, but such a question is not given here because the authors already receive enough mail.

Technology Project

Probabilities Through Simulations

Students typically find that the topic of probability is the most difficult topic in a statistics course. Some probability problems might sound simple while their solutions are incredibly complex. In this chapter we have identified several basic and important rules commonly used for finding probabilities, but in this project we use a very different approach that can overcome much of the difficulty encountered with the application of formal rules. This alternative approach consists of developing a **simulation**, which is a process that behaves the same way as the procedure, so that similar results are produced. For example, when testing techniques of gender selection, medical researchers need to know probability values of different outcomes, such as the probability of getting at least 60 girls among 100 children. Assuming that male and female births are equally likely, we can develop a simulation that results in the genders of 100 newborn babies. One approach is simply to flip a coin 100 times, with heads representing females and tails representing males. Another approach is to use a calculator or computer to randomly generate 0s and 1s, with 0 representing a male and 1 representing a female. The numbers must be generated in such a way that they are equally likely. There are several ways of obtaining randomly generated numbers, including the following:

- **A table of random digits:** Refer, for example, to the *CRC Standard Probability and Statistics Tables and Formulae,* which contains a table of 14,000 digits.

- **Statistics software:** Minitab, Excel, STATDISK, SPSS, SAS, and the TI-83/84 Plus calculator are among the many technologies capable of generating various types of random data.

Simulating Hybridization

When Mendel conducted his famous hybridization experiments, he used peas with green pods and yellow pods. One experiment involved crossing peas in such a way that 25% of the offspring peas were expected to have yellow pods.

a. Use the random digits in the margin to develop a simulation for finding the probability that when two offspring peas are produced, at least one of them has yellow pods. How does the result compare to the correct probability of 7/16, which can be found by using the probability rules of this chapter? (*Hint*: Because 25% of the offspring are expected to have yellow pods and the other 75% are expected to have green pods, let the digit 1 represent yellow pods and let the digits 2, 3, 4 represent green pods, and ignore any other digits.)

b. Use a technology, such as Minitab, Excel, STATDISK, SPSS, SAS, or a TI-83/84 Plus calculator to develop a simulation for finding the probability that when two offspring peas are produced, at least one of them has yellow pods. How does the result compare to the correct probability of 7/16, which can be found by using the probability rules of this chapter?

46196
99438
72113
44044
86763
00151
64703
78907
19155
67640
98746
29910
82855
25259
14752
85446
75260
92532
87333
55848

from DATA to DECISION

Critical Thinking: When you apply for a job, should you be concerned about drug testing?

According to the American Management Association, about 70% of U.S. companies now test at least some employees and job applicants for drug use. The U.S. National Institute on Drug Abuse claims that about 15% of people in the 18–25 age bracket use illegal drugs. Quest Diagnostics estimates that 3% of the general workforce in the United States uses marijuana. Let's assume that you applied for a job with excellent qualifications, you took a test for marijuana usage, and you were not offered a job. You might suspect that you failed the test, even if you do not use marijuana.

Analyzing the Results

The accompanying table shows data from a test of the 1-Panel-THC test for the screening of marijuana usage. This test device costs $5.95 and is provided by the company Drug Test Success. The test results were confirmed using gas chromatography/mass spectrometry, which the company describes as "the preferred confirmation method." These results are based on using 50 ng/ml as the cutoff level for determining the presence of marijuana. Based on the results given in the table, find each of the following:

a. P(false positive)   b. P(false negative)
c. P(true positive)   d. P(true negative)
e. Test sensitivity   f. Test specificity
g. Positive predictive   h. Negative predictive
   value   value

Are the probabilities of wrong results low enough so that job applicants need not be concerned?

1-Panel-THC Test for Marijuana Use

	Confirmed Presence of Marijuana	Confirmed Absence of Marijuana
Positive test result	119	24
Negative test result	3	154

4

Discrete Probability Distributions

Is this gender selection method effective?

Recent advances in medicine, genetics, and technology have been astounding. It now seems possible that a human being could be cloned. Once considered risky and dangerous, heart bypass operations are now routine. Laser surgery can correct vision so that many people can throw out their eyeglasses or contact lenses. Instead of relying solely on random chance, couples planning to have babies can use techniques for determining the gender of their children. Such advances are sometimes accompanied by considerable controversy. Some people believe that techniques of cloning or gender selection have serious moral ramifications and should be strictly avoided, regardless of the reason. Lisa Belkin wrote this in "Getting the Girl" (an article in the *New York Times Magazine*): "If we allow parents to choose the sex of their child today, how long will it be before they order up eye color, hair color, personality traits, and IQ?" There are some convincing arguments in favor of at least limited use of gender selection. One such argument asserts that some couples carry X-linked recessive genes with this result: Any male children have a 50% chance of inheriting a serious disorder, but none of the female children will inherit the disorder. Such couples may want to use gender selection as a way to ensure that they have baby girls, thereby guaranteeing that the serious disorder will not be inherited by any of their children.

The Genetics and IVF Institute in Fairfax, Virginia developed a technique called MicroSort, which supposedly increases the chances of a couple having a baby girl. In a preliminary test, 14 couples who wanted baby girls were found. After using the MicroSort technique, 13 of them had girls and one couple had a boy. These results provide us with an interesting question: Given that 13 out of 14 couples had girls, can we conclude that the MicroSort technique is effective, or might we explain the outcome as just a chance sample result? In answering that question, we will use principles of probability to determine whether the observed births differ significantly from results that we would expect from random chance. This example raises an issue that is at the very core of inferential statistics: How do we determine whether results are attributable to random chance or to some other factor, such as the use of the MicroSort technique of gender selection?

4-1 Overview

In this chapter we combine the methods of *descriptive statistics* presented in Chapter 2 and those of *probability* presented in Chapter 3. Figure 4-1 presents a visual summary of what we will accomplish in this chapter. As the figure shows, using the methods of Chapter 2, we would repeatedly roll the die to collect sample data, which then can be described with graphs (such as a histogram or boxplot), measures of center (such as the mean), and measures of variation (such as the standard deviation). Using the methods of Chapter 3, we could find the probability of each possible outcome. In this chapter we will combine those concepts as we develop probability distributions that describe what will *probably* happen instead of what actually *did* happen. In Chapter 2 we constructed frequency tables and histograms using *observed* sample values that were actually collected, but in this chapter we will construct probability distributions by presenting possible outcomes along with the relative frequencies we *expect*.

The table at the extreme right in Figure 4-1 represents a probability distribution that serves as a model of a theoretically perfect population frequency distribution. In essence, we can describe the relative frequency table for a die rolled an infinite number of times. With this knowledge of the population of outcomes, we are able to find its important characteristics, such as the mean and standard deviation. The remainder of this book and the very core of inferential statistics are based on some knowledge of probability distributions. We begin by examining the concept of a random variable, and then we consider important distributions that have many real applications.

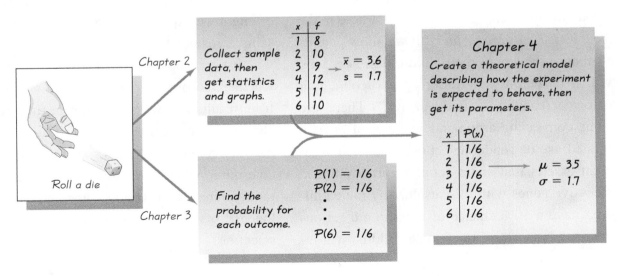

FIGURE 4-1 Combining Descriptive Methods and Probabilities to Form a Theoretical Model of Behavior

4-2 Random Variables

In this section we discuss random variables, probability distributions, procedures for finding the mean and standard deviation for a probability distribution, and methods for distinguishing between outcomes that are likely to occur by chance and outcomes that are "unusual." We begin with the related concepts of *random variable* and *probability distribution*.

Definitions

A **random variable** is a variable (typically represented by x) that has a single numerical value, determined by chance, for each outcome of a procedure.

A **probability distribution** is a graph, table, or formula that gives the probability for each value of the random variable.

EXAMPLE Gender of Children A study consists of randomly selecting 14 newborn babies and counting the number of girls (as in the Chapter Problem). If we assume that boys and girls are equally likely and if we let

$$x = \text{number of girls among 14 babies}$$

then x is a random variable because its value depends on chance. The possible values of x are 0, 1, 2, . . . , 14. Table 4-1 lists the values of x along with the corresponding probabilities. (In Section 4-3 we will see how to find the probability values, such as those listed in Table 4-1.) Because Table 4-1 gives the probability for each value of the random variable x, that table describes a probability distribution.

In Section 1-2 we made a distinction between discrete and continuous data. Random variables may also be discrete or continuous, and the following two definitions are consistent with those given in Section 1-2.

Definitions

A **discrete random variable** has either a finite number of values or a countable number of values, where "countable" refers to the fact that there might be infinitely many values, but they can be associated with a counting process.

A **continuous random variable** has infinitely many values, and those values can be associated with measurements on a continuous scale without gaps or interruptions.

Table 4-1
Probability Distribution

x (girls)	P(x)
0	0.000
1	0.001
2	0.006
3	0.022
4	0.061
5	0.122
6	0.183
7	0.209
8	0.183
9	0.122
10	0.061
11	0.022
12	0.006
13	0.001
14	0.000

This chapter deals exclusively with discrete random variables, but the following chapters will deal with continuous random variables.

(a) *Discrete Random Variable: Count of the number of movie patrons.*

(b) *Continuous Random Variable: The measured voltage of a smoke detector battery.*

FIGURE 4-2 Devices Used to Count and Measure Discrete and Continuous Random Variables

EXAMPLES The following are examples of discrete and continuous random variables.

1. Let x = the number of eggs that a hen lays in a day. This is a *discrete* random variable because its only possible values are 0, or 1, or 2, and so on. No hen can lay 2.343115 eggs, which would have been possible if the data had come from a continuous scale.

2. The count of the number of emergency room patients in one day at New York City's Bellevue Hospital is a whole number and is therefore a discrete random variable. The counting device shown in Figure 4-2(a) is capable of indicating only a finite number of values, so it is used to obtain values for a *discrete* random variable.

3. Let x = the amount of milk a cow produces in one day. This is a *continuous* random variable because it can have any value over a continuous span. During a single day, a cow might yield an amount of milk that can be any value between 0 gallons and 5 gallons. It would be possible to get 4.123456 gallons, because the cow is not restricted to the discrete amounts of 0, 1, 2, 3, 4, or 5 gallons.

4. The measure of voltage for a particular smoke detector can be any value between 0 volts and 9 volts. It is therefore a continuous random variable. The voltmeter shown in Figure 4-2(b) is capable of indicating values on a continuous scale, so it can be used to obtain values for a *continuous* random variable.

Graphs

There are various ways to graph a probability distribution, but we will consider only the **probability histogram.** Figure 4-3 is a probability histogram that is very similar to the relative frequency histogram discussed in Chapter 2, but the vertical scale shows *probabilities* instead of relative frequencies based on actual sample results.

In Figure 4-3, note that along the horizontal axis, the values of 0, 1, 2, . . . , 14 are located at the centers of the rectangles. This implies that the rectangles are each 1 unit wide, so the areas of the rectangles are 0.000, 0.001, 0.006, and so on. The areas of these rectangles are the same as the *probabilities* in Table 4-1. We will see in Chapter 5 and future chapters that such a correspondence between area and probability is very useful in statistics.

Every probability distribution must satisfy each of the following two requirements.

Requirements for a Probability Distribution

1. $\Sigma P(x) = 1$ where x assumes all possible values. (That is, the sum of all probabilities must be 1.)

2. $0 \leq P(x) \leq 1$ for every individual value of x. (That is, each probability value must be between 0 and 1 inclusive.)

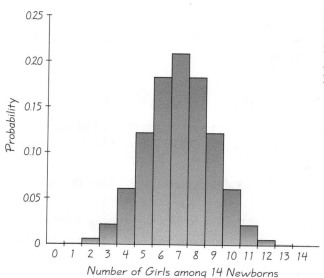

The first requirement makes sense when we realize that the values of the random variable x represent all possible events in the entire sample space, so we are certain (with probability 1) that one of the events will occur. In Table 4-1, the sum of the probabilities is 0.999; that sum is exactly 1 if we carry more decimal places and thereby eliminate the small rounding error. Also, the probability rule stating $0 \leq P(x) \leq 1$ for any event A (given in Section 3-2) implies that $P(x)$ must be between 0 and 1 for any value of x. Again, refer to Table 4-1 and note that each individual value of $P(x)$ does fall between 0 and 1 for any value of x. Because Table 4-1 does satisfy both of the requirements, it is an example of a probability distribution. A probability distribution may be described by a table, such as Table 4-1, or a graph, such as Figure 4-3, or a formula.

EXAMPLE Does Table 4-2 describe a probability distribution?

SOLUTION   To be a probability distribution, $P(x)$ must satisfy the preceding two requirements. But

$$\Sigma P(x) = P(0) + P(1) + P(2) + P(3)$$
$$= 0.2 + 0.5 + 0.4 + 0.3$$
$$= 1.4 \text{ [showing that } \Sigma P(x) \neq 1]$$

Because the first requirement is not satisfied, we conclude that Table 4-2 does not describe a probability distribution.

EXAMPLE Does $P(x) = x/3$ (where x can be 0, 1, or 2) determine a probability distribution?

SOLUTION   For the given function we find that $P(0) = 0/3$, $P(1) = 1/3$ and $P(2) = 2/3$, so that

1. $\Sigma P(x) = \dfrac{0}{3} + \dfrac{1}{3} + \dfrac{2}{3} = \dfrac{3}{3} = 1$

continued

Table 4-2	
Probabilities for a Random Variable	
x	$P(x)$
0	0.2
1	0.5
2	0.4
3	0.3

2. Each of the $P(x)$ values is between 0 and 1.

Because both requirements are satisfied, the $P(x)$ function given in this example is a probability distribution.

Mean, Variance, and Standard Deviation

In Chapter 2 we described the following important characteristics of data (which can be remembered with the mnemonic of CVDOT for "Computer Viruses Destroy Or Terminate"):

1. *Center:* A representative or average value that indicates where the middle of the data set is located.

2. *Variation:* A measure of the amount that the values vary among themselves.

3. *Distribution:* The nature or shape of the distribution of the data (such as bell-shaped, uniform, or skewed).

4. *Outliers:* Sample values that lie very far away from the vast majority of the other sample values.

5. *Time:* Changing characteristics of the data over time.

The probability histogram can give us insight into the nature or shape of the distribution. Also, we can often find the mean, variance, and standard deviation of data, which provide insight into the other characteristics. The mean, variance, and standard deviation for a probability distribution can be found by applying Formulas 4-1, 4-2, 4-3, and 4-4.

Formula 4-1 $\quad \mu = \Sigma[x \cdot P(x)]$ mean for a probability distribution

Formula 4-2 $\quad \sigma^2 = \Sigma[(x - \mu)^2 \cdot P(x)]$ variance for a probability distribution

Formula 4-3 $\quad \sigma^2 = \Sigma[x^2 \cdot P(x)] - \mu^2$ variance for a probability distribution

Formula 4-4 $\quad \sigma = \sqrt{\Sigma[x^2 \cdot P(x)] - \mu^2}$ standard deviation for a probability distribution

Caution: Evaluate $\Sigma[x^2 \cdot P(x)]$ by first squaring each value of x, then multiplying each square by the corresponding probability $P(x)$, then adding.

Rationale for Formulas 4-1 through 4-4

Why do Formulas 4-1 through 4-4 work? Formula 4-1 accomplishes the same task as the formula for the mean of a frequency table. (Recall that f represents class frequency and N represents population size.) Rewriting the formula for the mean of a frequency table so that it applies to a population and then changing its form, we get

$$\mu = \frac{\Sigma(f \cdot x)}{N} = \Sigma\left[\frac{f \cdot x}{N}\right] = \Sigma\left[x \cdot \frac{f}{N}\right] = \Sigma[x \cdot P(x)]$$

In the fraction f/N, the value of f is the frequency with which the value x occurs and N is the population size, so f/N is the probability for the value of x.

Similar reasoning enables us to take the variance formula from Chapter 2 and apply it to a random variable for a probability distribution; the result is

Formula 4-2. Formula 4-3 is a shortcut version that will always produce the same result as Formula 4-2. Although Formula 4-3 is usually easier to work with, Formula 4-2 is easier to understand directly. Based on Formula 4-2, we can express the standard deviation as

$$\sigma = \sqrt{\Sigma (x - \mu)^2 \cdot P(x)}$$

or as the equivalent form given in Formula 4-4.

When using Formulas 4-1 through 4-4, use this rule for rounding results.

Round-off Rule for μ, σ, and σ^2

Round results by carrying one more decimal place than the number of decimal places used for the random variable x. If the values of x are integers, round μ, σ, and σ^2 to one decimal place.

It is sometimes necessary to use a different rounding rule because of special circumstances, such as results that require more decimal places to be meaningful. For example, with four-engine jets the mean number of jet engines working successfully throughout a flight is 3.999714286, which becomes 4.0 when rounded to one more decimal place than the original data. Here, 4.0 would be misleading because it suggests that all jet engines always work successfully. We need more precision to correctly reflect the true mean, such as the precision in the number 3.999714.

Identifying Unusual Results with the Range Rule of Thumb

The range rule of thumb (discussed in Section 2-5) may also be helpful in interpreting the value of a standard deviation. According to the range rule of thumb, most values should lie within 2 standard deviations of the mean; it is unusual for a value to differ from the mean by more than 2 standard deviations. (The use of 2 standard deviations is not an absolutely rigid value, and other values such as 3 could be used instead.) We can therefore identify "unusual" values by determining that they lie outside of these limits:

$$\text{maximum usual value} = \mu + 2\sigma$$
$$\text{minimum usual value} = \mu - 2\sigma$$

EXAMPLE Table 4-1 describes the probability distribution for the number of girls among 14 randomly selected newborn babies. Assuming that we repeat the study of randomly selecting 14 newborn babies and counting the number of girls each time, find the mean number of girls (among 14), the variance, and the standard deviation. Use those results and the range rule of thumb to find the maximum and minimum usual values.

SOLUTION   In Table 4-3, the two columns at the left describe the probability distribution given earlier in Table 4-1, and we create the three columns at the right for the purposes of the calculations required.

continued

Table 4-3	Calculating μ, σ, and σ^2 for a Probability Distribution			
x	$P(x)$	$x \cdot P(x)$	x^2	$x^2 \cdot P(x)$
0	0.000	0.000	0	0.000
1	0.001	0.001	1	0.001
2	0.006	0.012	4	0.024
3	0.022	0.066	9	0.198
4	0.061	0.244	16	0.976
5	0.122	0.610	25	3.050
6	0.183	1.098	36	6.588
7	0.209	1.463	49	10.241
8	0.183	1.464	64	11.712
9	0.122	1.098	81	9.882
10	0.061	0.610	100	6.100
11	0.022	0.242	121	2.662
12	0.006	0.072	144	0.864
13	0.001	0.013	169	0.169
14	0.000	0.000	196	0.000
Total		6.993		52.467
		$\uparrow$		$\uparrow$
		$\Sigma[x \cdot P(x)]$		$\Sigma[x^2 \cdot P(x)]$

Using Formulas 4-1 and 4-3 and the table results, we get

$$\mu = \Sigma[x \cdot P(x)] = 6.993 = 7.0 \qquad \text{(rounded)}$$

$$\sigma^2 = \Sigma[x^2 \cdot P(x)] - \mu^2$$
$$= 52.467 - 6.993^2 = 3.564951 = 3.6 \qquad \text{(rounded)}$$

The standard deviation is the square root of the variance, so

$$\sigma = \sqrt{3.564951} = 1.9 \qquad \text{(rounded)}$$

We now know that among groups of 14 newborn babies, the mean number of girls is 7.0, the variance is 3.6 "girls squared," and the standard deviation is 1.9 girls.

Using the range rule of thumb, we can now find the maximum and minimum usual values as follows:

maximum usual value: $\mu + 2\sigma = 7.0 + 2(1.9) = 10.8$

minimum usual value: $\mu - 2\sigma = 7.0 - 2(1.9) = 3.2$

INTERPRETATION Based on these results, we conclude that for groups of 14 randomly selected babies, the number of girls should usually fall between 3.2 and 10.8.

Identifying Unusual Results with Probabilities

Strong recommendation: The following discussion includes some difficult concepts, but it also includes an extremely important approach used often in statistics. You should make every effort to understand this discussion, even if it requires several readings. Keep in mind that this discussion is based on the *rare event rule* introduced in Section 3-2:

> **If, under a given assumption (such as the assumption that boys and girls are equally likely), the probability of a particular observed event (such as 13 girls in 14 births) is extremely small, we conclude that the assumption is probably not correct.**

In the Chapter Problem, we noted that with the MicroSort technique, there were 13 girls among 14 babies. Is this result unusual? Does this result really suggest that the technique is effective, or could we get 13 girls among 14 babies just by chance? To address this issue, we could use the range rule of thumb to find the minimum and maximum likely outcomes, but here we will consider another approach. We will find the probability of getting *13 or more* girls (not the probability of getting *exactly* 13 girls). It might be difficult to see why this probability of *13 or more girls* is the relevant probability, so let's try to clarify it with a more obvious example.

Suppose you were flipping a coin to determine whether it favors heads, and suppose 1000 tosses resulted in 501 heads. This is not evidence that the coin favors heads, because it is very easy to get a result like 501 heads in 1000 tosses just by chance. Yet, the probability of getting *exactly* 501 heads in 1000 tosses is actually quite small: 0.0252. This low probability reflects the fact that with 1000 tosses, any *specific* number of heads will have a very low probability. However, we do not consider 501 heads among 1000 tosses to be *unusual*, because the probability of getting *at least* 501 heads is high: 0.487. This principle can be generalized as follows:

Using Probabilities to Determine When Results Are Unusual

- **Unusually high:** x successes among n trials is an *unusually high* number of successes if $P(x$ or more$)$ is very small (such as 0.05 or less)

- **Unusually low:** x successes among n trials is an *unusually low* number of successes if $P(x$ or fewer$)$ is very small (such as 0.05 or less)

EXAMPLE **Gender Selection** Using the preceding two criteria based on probabilities, is it unusual to get 13 girls among 14 births? Does the MicroSort technique of gender selection appear to be effective?

SOLUTION   Thirteen girls among 14 births is unusually high if $P(13$ or more girls$)$ is very small. If we refer to Table 4-1, we get this result:

$$P(13 \text{ or more girls}) = P(13) + P(14)$$
$$= 0.001 + 0.000$$
$$= 0.001$$

continued

INTERPRETATION Because the probability 0.001 is so low, we conclude that it is unusual to get 13 girls among 14 births. This suggests that the MicroSort technique of gender selection appears to be effective, because it is highly unlikely that the result of 13 girls among 14 births happened by chance.

Expected Value

The mean of a discrete random variable is the theoretical mean outcome for infinitely many trials. We can think of that mean as the *expected value* in the sense that it is the average value that we would expect to get if the trials could continue indefinitely. The uses of expected value (also called *expectation,* or *mathematical expectation*) are extensive and varied, and they play a very important role in an area of application called *decision theory.*

Definition

The **expected value** of a discrete random variable is denoted by *E,* and it represents the average value of the outcomes. It is obtained by finding the value of $\Sigma[x \cdot P(x)]$.

$$E = \Sigma[x \cdot P(x)]$$

From Formula 4-1 we see that $E = \mu$. That is, the mean of a discrete random variable is the same as its expected value. For example, repeat the procedure of randomly generating five births, and the *mean* number of males is 2.5; when children are born, the *expected value* of the number of males is also 2.5.

In this section we learned that a random variable has a numerical value associated with each outcome of some random procedure, and a probability distribution has a probability associated with each value of a random variable. We examined methods for finding the mean, variance, and standard deviation for a probability distribution. We saw that the expected value of a random variable is really the same as the mean. Finally, an extremely important concept of this section is the use of probabilities for determining when outcomes are *unusual.*

4-2 Exercises

Identifying Discrete and Continuous Random Variables. In Exercises 1 and 2, identify the given random variable as being discrete or continuous.

1. **a.** The height of a randomly selected giraffe living in Kenya
    b. The number of bald eagles located in New York State
    c. The exact gestation time of a randomly selected pregnant panda
    d. The number of blue whales currently living
    e. The number of manatees killed by Florida boats in one year

2. **a.** The cost of conducting a genetics experiment
    b. The number of eukaryotic cells in an ant
    c. The exact life span of a koala bear

d. The number of monkeys in Gibraltar
e. The weight of an elephant

Identifying Probability Distributions. *In Exercises 3–8, determine whether a probability distribution is given. In those cases where a probability distribution is not described, identify the requirements that are not satisfied. In those cases where a probability distribution is described, find its mean and standard deviation.*

3. Gender Selection In a study of the MicroSort gender-selection method, couples in a control group are not given a treatment, and they each have three children. The probability distribution for the number of girls is given in the accompanying table.

x	P(x)
0	0.125
1	0.375
2	0.375
3	0.125

4. Numbers of Girls A researcher reports that when groups of four children are randomly selected from a population of couples meeting certain criteria, the probability distribution for the number of girls is as given in the accompanying table.

x	P(x)
0	0.502
1	0.365
2	0.098
3	0.011
4	0.001

5. Genetics Experiment A genetics experiment involves offspring peas in groups of four. A researcher reports that for one group, the number of peas with white flowers has a probability distribution as given in the accompanying table.

x	P(x)
0	0.04
1	0.26
2	0.36
3	0.20
4	0.08

6. Mortality Study For a group of four men, the probability distribution for the number x who live through the next year is as given in the accompanying table.

x	P(x)
0	0.0000
1	0.0001
2	0.0006
3	0.0387
4	0.9606

7. Genetic Disorder Three males with an X-linked genetic disorder have one child each. The random variable x is the number of children among the three who inherit the X-linked genetic disorder.

x	P(x)
0	0.4219
1	0.4219
2	0.1406
3	0.0156

8. Diseased Seedlings An experiment involves groups of four seedlings grown under controlled conditions. The random variable x is the number of seedlings in a group that meet specific criteria for being classified as "diseased."

x	P(x)
0	0.805
1	0.113
2	0.057
3	0.009
4	0.002

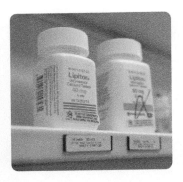

9. **Determining Whether Gender-Selection Technique Is Effective** Assume that in a test of a gender-selection technique, a clinical trial results in 9 girls in 14 births. Refer to Table 4-1 and find the indicated probabilities.
   a. Find the probability of exactly 9 girls in 14 births.
   b. Find the probability of 9 or more girls in 14 births.
   c. Which probability is relevant for determining whether 9 girls in 14 births is unusually high: the result from part (a) or part (b)?
   d. Does 9 girls in 14 births suggest that the gender-selection technique is effective? Why or why not?

10. **Determining Whether Gender-Selection Technique Is Effective** Assume that in a test of a gender-selection technique, a clinical trial results in 12 girls in 14 births. Refer to Table 4-1 and find the indicated probabilities.
   a. Find the probability of exactly 12 girls in 14 births.
   b. Find the probability of 12 or more girls in 14 births.
   c. Which probability is relevant for determining whether 12 girls in 14 births is unusually high: the result from part (a) or part (b)?
   d. Does 12 girls in 14 births suggest that the gender-selection technique is effective? Why or why not?

11. **Determining Whether Gender-Selection Technique Is Effective** Assume that in a test of a gender-selection technique, a clinical trial results in 11 girls in 14 births. Refer to Table 4-1.
   a. What is the probability value that should be used for determining whether the result of 11 girls in 14 births is unusually high?
   b. Does 11 girls in 14 births suggest that the gender-selection technique is effective? Why or why not?

12. **Determining Whether Gender-Selection Technique Is Effective** Assume that in a test of a gender-selection technique, a clinical trial results in 10 girls in 14 births. Refer to Table 4-1.
   a. What is the probability value that should be used for determining whether the result of 10 girls in 14 births is unusually high?
   b. Does 10 girls in 14 births suggest that the gender-selection technique is effective? Why or why not?

13. **Finding Mean and Standard Deviation** Let the random variable x represent the number of girls in a family of four children. Construct a table describing the probability distribution, then find the mean and standard deviation. (*Hint:* List the different possible outcomes.)

4-3 Binomial Probability Distributions

In Section 4-2 we discussed several different discrete probability distributions, but in this section we will focus on one specific type: the binomial probability distribution. Binomial probability distributions are important because they allow us to deal with circumstances in which the outcomes belong to *two* relevant categories, such as acceptable/defective or yes/no responses in a survey question. The Chapter Problem involves counting the number of girls in 14 births. The problem involves the two categories of boy/girl, so it has the required key element of "twoness." Other requirements are given in the following definition.

Definition

A **binomial probability distribution** results from a procedure that meets all the following requirements:

1. The procedure has a *fixed number of trials.*
2. The trials must be *independent.* (The outcome of any individual trial doesn't affect the probabilities in the other trials.)
3. Each trial must have all outcomes classified into *two categories.*
4. The probabilities must remain *constant* for each trial.

If a procedure satisfies these four requirements, the distribution of the random variable x is called a *binomial probability distribution* (or *binomial distribution*). The following notation is commonly used.

Notation for Binomial Probability Distributions

S and F (success and failure) denote the two possible categories of all outcomes; p and q will denote the probabilities of S and F, respectively, so

$$P(S) = p \qquad\qquad (p = \text{probability of a success})$$
$$P(F) = 1 - p = q \qquad\qquad (q = \text{probability of a failure})$$

n denotes the fixed number of trials.

x denotes a specific number of successes in n trials, so x can be any whole number between 0 and n, inclusive.

p denotes the probability of *success* in *one* of the n trials.

q denotes the probability of *failure* in *one* of the n trials.

$P(x)$ denotes the probability of getting exactly x successes among the n trials.

The word *success* as used here is arbitrary and does not necessarily represent something good. Either of the two possible categories may be called the success S as long as its probability is identified as p. Once a category has been designated as the success S, be sure that p is the probability of a success and x is the number of successes. That is, *be sure that the values of p and x refer to the same category designated as a success.* (The value of q can always be found by subtracting p from 1; if $p = 0.95$, then $q = 1 - 0.95 = 0.05$.) Here is an important hint for working with binomial probability problems:

Be sure that x and p both refer to the *same* category being called a success.

When selecting a sample for some statistical analysis, we usually sample without replacement. For example, when testing manufactured items or conducting surveys, we usually design the sampling process so that selected items cannot be selected a second time. Strictly speaking, sampling without replacement

Do Boys or Girls Run in the Family?

The authors of this book, their siblings, and their siblings' children consist of 11 males and only one female. Is this an example of a phenomenon whereby one particular gender runs in a family? This issue was studied by examining a random sample of 8770 households in the United States. The results were reported in the *Chance* magazine article "Does Having Boys or Girls Run in the Family?" by Joseph Rodgers and Debby Doughty. Part of their analysis involves use of the binomial probability distribution discussed in this section. Their conclusion is that "We found no compelling evidence that sex bias runs in the family."

involves dependent events, which violates the second requirement in the above definition. However, the following rule of thumb is based on the fact that if the sample is very small relative to the population size, we can treat the trials as being independent (even though they are actually dependent) because the difference in results will be negligible.

When sampling without replacement, the events can be treated as if they were independent if the sample size is no more than 5% of the population size. (That is, $n \leq 0.05N$.)

EXAMPLE Analysis of Multiple-Choice Answers Because they are so easy to correct, multiple-choice questions are commonly used for class tests, SAT tests, MCAT tests for medical schools, and many other circumstances. A professor of anatomy and physiology plans to give a surprise quiz consisting of 4 multiple-choice questions, each with 5 possible answers (a, b, c, d, e), one of which is correct. Let's assume that an unprepared student makes random guesses and we want to find the probability of exactly 3 correct responses to the 4 questions.

a. Does this procedure result in a binomial distribution?

b. If this procedure does result in a binomial distribution, identify the values of n, x, p, and q.

SOLUTION

a. This procedure does satisfy the requirements for a binomial distribution, as shown below.

  1. The number of trials (4) is fixed.

  2. The 4 trials are independent because a correct or wrong response for any individual question does not affect the probability of being correct or wrong on another question.

  3. Each of the 4 trials has two categories of outcomes: The response is either correct or wrong.

  4. For each response, there are 5 possible answers (a, b, c, d, e), one of which is correct, so the probability of a correct response is 1/5 or 0.2. That probability remains constant for each of the 4 trials.

b. Having concluded that the given procedure does result in a binomial distribution, we now proceed to identify the values of n, x, p, and q.

  1. With 4 test questions, we have $n = 4$.

  2. We want the probability of exactly 3 correct responses, so $x = 3$.

  3. The probability of success (correct response) for one question is 0.2, so $p = 0.2$.

  4. The probability of failure (wrong response) is 0.8, so $q = 0.8$.

Again, it is very important to be sure that x and p both refer to the same concept of "success." In this example, we use x to count the *correct* responses, so p must be the probability of a *correct* response. Therefore, x and p do use the same concept of success (correct response) here.

We will now discuss three methods for finding the probabilities correspond-ing to the random variable x in a binomial distribution. The first method involves calculations using the *binomial probability formula* and is the basis for the other two methods. The second method involves the use of Table A-1, and the third method involves the use of statistical software or a calculator. If you are using software or a calculator that automatically produces binomial probabilities, we recommend that you solve one or two exercises using Method 1 to ensure that you understand the basis for the calculations. Understanding is always infinitely better than blind application of formulas.

Method 1: Using the Binomial Probability Formula   In a binomial distri-bution, probabilities can be calculated by using the binomial probability formula.

Formula 4-5 $$P(x) = \frac{n!}{(n - x)!x!} \cdot p^x \cdot q^{n-x} \qquad \text{for } x = 0, 1, 2, \ldots, n$$

where n = number of trials

x = number of successes among n trials

p = probability of success in any one trial

q = probability of failure in any one trial ($q = 1 - p$)

The factorial symbol !, introduced in Section 3-7, denotes the product of de-creasing factors. Two examples of factorials are $3! = 3 \cdot 2 \cdot 1 = 6$ and $0! = 1$ (by definition). Many calculators have a factorial key, as well as a key labeled $_nC_r$ that can simplify the computations. For calculators with the $_nC_r$ key, use this ver-sion of the binomial probability formula (where n, x, p, and q are the same as in Formula 4-5):

$$P(x) = {_nC_r} \cdot p^x \cdot q^{n-x}$$

EXAMPLE **Analysis of Multiple-Choice Answers** Use the bino-mial probability formula to find the probability of getting exactly 3 correct an-swers when random guesses are made for 4 multiple choice questions. That is, find $P(3)$ given that $n = 4$, $x = 3$, $p = 0.2$, and $q = 0.8$.

SOLUTION   Using the given values of n, x, p, and q in the binomial proba-bility formula (Formula 4-5), we get

$$P(x) = \frac{4!}{(4 - 3)!3!} \cdot 0.2^3 \cdot 0.8^{4-3}$$

$$= \frac{4!}{1!3!} \cdot 0.008 \cdot 0.8$$

$$= (4)(0.008)(0.8) = 0.0256$$

The probability of getting exactly 3 correct answers out of 4 is 0.0256.

Calculation hint: When computing a probability with the binomial probability formula, it's helpful to get a single number for $n!/[(n - x)!x!]$, a single number for p^x, and a single number for q^{n-x}, then simply multiply the three factors together as shown at the end of the calculation for the preceding example. Don't round too much when you find those three factors; round only at the end.

Method 2: Using Table A-1 in Appendix A

In some cases, we can easily find binomial probabilities by simply referring to Table A-1 in Appendix A. First locate n and the corresponding value of x that is desired. At this stage, one row of numbers should be isolated. Now align that row with the proper probability of p by using the column across the top. The isolated number represents the desired probability. A very small probability, such as 0.000000345, is indicated by 0+.

Part of Table A-1 is shown in the margin. With $n = 4$ and $p = 0.2$ in a binomial distribution, the probabilities of 0, 1, 2, 3, and 4 successes are 0.410, 0.410, 0.154, 0.026, and 0.002, respectively.

From Table A-1:

		p
n	x	0.2
4	0	0.410
	1	0.410
	2	0.154
	3	0.026
	4	0.002

↓

Binomial probability distribution for $n = 4$ and $p = 0.2$

x	$P(x)$
0	0.410
1	0.410
2	0.154
3	0.026
4	0.002

EXAMPLE Use the portion of Table A-1 (for $n = 4$ and $p = 0.2$) shown in the margin to find the following.

a. The probability of exactly 3 successes

b. The probability of *at least* 3 successes

SOLUTION

a. The display from Table A-1 shows that when $n = 4$ and $p = 0.2$, the probability of $x = 3$ is given by $P(3) = 0.026$, which is the same value (except for rounding) computed with the binomial probability formula in the preceding example.

b. "At least" 3 successes means that the number of successes is 3 or 4.

$$P(\text{at least } 3) = P(3 \text{ or } 4)$$
$$= P(3) + P(4)$$
$$= 0.026 + 0.002$$
$$= 0.028$$

In part (b) of the preceding solution, if we wanted to find P(at least 3) by using the binomial probability formula, we would need to apply that formula twice to compute two different probabilities, which would then be added. Given this choice between the formula and the table, it makes sense to use the table. Unfortunately, Table A-1 includes only limited values of n as well as limited values of p, so the table doesn't always work, and we must then find the probabilities by using the binomial probability formula, software, or a calculator, as in the following method.

Method 3: Using Technology

STATDISK, Minitab, Excel, SPSS, SAS, and the TI-83/84 Plus calculator are all examples of technologies that can be used to find binomial probabilities. (Instead of directly providing probabilities for individual values of x, SPSS and SAS are more difficult to use because they provide *cumulative* probabilities of x or fewer successes.) We will not present details for using each of those technologies, but see the typical screen displays that list binomial probabilities for $n = 4$ and $p = 0.2$.

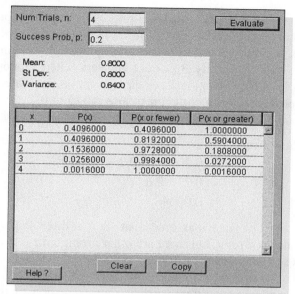

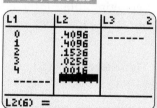

Given that we now have three different methods for finding binomial probabilities, here is an effective and efficient strategy:

1. Use computer software or a TI-83/84 Plus calculator, if available.

2. If neither software nor the TI-83/84 Plus calculator is available, use Table A-1, if possible.

3. If neither software nor the TI-83/84 Plus calculator is available and the probabilities can't be found using Table A-1, use the binomial probability formula.

Rationale for the Binomial Probability Formula

The binomial probability formula is the basis for all three methods presented in this section. Instead of accepting and using that formula blindly, let's see why it works.

Earlier in this section, we used the binomial probability formula for finding the probability of getting exactly 3 correct answers when random guesses are made for 4 multiple-choice questions. For each question there are 5 possible answers so that the probability of a correct response is 1/5 or 0.2. If we use the multiplication rule from Section 3-4, we get the following result:

P(3 correct answers followed by 1 wrong answer)
$$= 0.2 \cdot 0.2 \cdot 0.2 \cdot 0.8$$
$$= 0.2^3 \cdot 0.8$$
$$= 0.0064$$

This result isn't correct because it assumes that the *first* three responses are correct and the *last* response is wrong, but there are other arrangements possible for three correct responses and one wrong response.

In Section 3-8 we saw that with three items identical to each other (such as *correct* responses) and one other item (such as a *wrong* response), the total number of arrangements (permutations) is $4!/[(4-3)!3!]$ or 4. Each of those 4 different arrangements has a probability of $0.2^3 \cdot 0.8$, so the total probability is as follows:

$$P(3 \text{ correct among } 4) = \frac{4!}{(4-3)!3!} \cdot 0.2^3 \cdot 0.8^1$$

Generalize this result as follows: Replace 4 with n, replace x with 3, replace 0.2 with p, replace 0.8 with q, and express the exponent of 1 as $4-3$, which can be replaced with $n-x$. The result is the binomial probability formula. That is, the binomial probability formula is a combination of the multiplication rule of probability and the counting rule for the number of arrangements of n items when x of them are identical to each other and the other $n-x$ are identical to each other. (See Exercises 9 and 10.)

The number of outcomes with exactly x successes among n trials

The probability of x successes among n trials for any one particular order

$$P(x) = \frac{n!}{(n-x)!x!} \cdot p^x \cdot q^{n-x}$$

4-3 Exercises

Identifying Binomial Distributions. In Exercises 1–8, determine whether the given procedure results in a binomial distribution. For those that are not binomial, identify at least one requirement that is not satisfied.

1. Surveying people by asking them what they think of methods of gender selection.

2. Surveying 1012 people and recording whether there is a "should not" response to this question: "Do you think the cloning of humans should or should not be allowed?"

3. Treating 50 smokers with Nicorette and asking them how their mouth and throat feel.

4. Treating 50 smokers with Nicorette and asking them if they experience any mouth or throat soreness.

5. Recording the genders of 250 newborn babies.

6. Determining whether each of 3000 heart pacemakers is acceptable or defective.

7. Surveying 500 married couples by asking them how many children they have.

8. Surveying 500 married couples by asking them if they have any children.

9. Finding Probabilities When Guessing Answers Multiple-choice questions each have five possible answers, one of which is correct. Assume that you guess the answers to three such questions.
   a. Use the multiplication rule to find the probability that the first two guesses are wrong and the third is correct. That is, find $P(WWC)$, where C denotes a correct answer and W denotes a wrong answer.
   b. Beginning with WWC, make a complete list of the different possible arrangements of two wrong answers and one correct answer, then find the probability for each entry in the list.
   c. Based on the preceding results, what is the probability of getting exactly one correct answer when three guesses are made?

10. Finding Probabilities When Guessing Answers A test consists of multiple-choice questions, each having four possible answers, one of which is correct. Assume that you guess the answers to six such questions.
   a. Use the multiplication rule to find the probability that the first two guesses are wrong and the last four guesses are correct. That is, find $P(WWCCCC)$, where C denotes a correct answer and W denotes a wrong answer.
   b. Beginning with WWCCCC, make a complete list of the different possible arrangements of two wrong answers and four correct answers, then find the probability for each entry in the list.
   c. Based on the preceding results, what is the probability of getting exactly four correct answers when six guesses are made?

Using Table A-1. In Exercises 11–16, assume that a procedure yields a binomial distribution with a trial repeated n times. Use Table A-1 to find the probability of x successes given the probability p of success on a given trial.

11. $n = 2, x = 0, p = 0.01$ 12. $n = 7, x = 2, p = 0.1$

13. $n = 4, x = 3, p = 0.95$ 14. $n = 6, x = 5, p = 0.99$

15. $n = 10, x = 4, p = 0.95$ 16. $n = 11, x = 7, p = 0.05$

Using the Binomial Probability Formula. In Exercises 17–20, assume that a procedure yields a binomial distribution with a trial repeated n times. Use the binomial probability formula to find the probability of x successes given the probability p of success on a single trial.

17. $n = 6, x = 4, p = 0.55$ 18. $n = 6, x = 2, p = 0.45$

19. $n = 8, x = 3, p = 1/4$ 20. $n = 10, x = 8, p = 1/3$

Minitab

```
Binomial with n = 6
and p = 0.167000

    x    |  P(X = x)
  ----------------------
   0.00  |   0.3341
   1.00  |   0.4019
   2.00  |   0.2014
   3.00  |   0.0538
   4.00  |   0.0081
   5.00  |   0.0006
   6.00  |   0.0000
```

Using Computer Results. In Exercises 21–24, refer to the Minitab display in the margin. The probabilities were obtained by entering the values of $n = 6$ and $p = 0.167$. In a clinical test of the drug Lipitor (atorvastatin), 16.7% of the subjects treated with 10 mg of atorvastatin experienced headaches (based on data from Parke-Davis). In each case, assume that 6 subjects are randomly selected and treated with 10 mg of atorvastatin, then find the indicated probability.

21. Find the probability that at least five of the subjects experience headaches. Is it unusual to have at least five of six subjects experience headaches?

22. Find the probability that at most two subjects experience headaches. Is it unusual to have at most two of six subjects experience headaches?

23. Find the probability that more than one subject experiences headaches. Is it unusual to *not* have more than one of six subjects experience headaches?

24. Find the probability that at least one subject experiences headaches. Is it unusual to *not* have at least one of six subjects experience headaches?

25. Drug Reaction In a clinical test of the drug Viagra, it was found that 4% of those in a placebo group experienced headaches.
   a. Assuming that the same 4% rate applies to those taking Viagra, find the probability that among 8 Viagra users, 3 experience headaches.
   b. Assuming that the same 4% rate applies to those taking Viagra, find the probability that among 8 randomly selected users of Viagra, all 8 experienced a headache.
   c. If all 8 Viagra users were to experience a headache, would it appear that the headache rate for Viagra users is different than the 4% rate for those in the placebo group? Explain.

26. Color Blindness Nine percent of men and 0.25% of women cannot distinguish between the colors red and green. This is the type of color blindness that causes problems with traffic signals. If 6 men are randomly selected for a study of traffic signal perceptions, find the probability that exactly 2 of them cannot distinguish between red and green.

27. Acceptance Sampling The Medassist Pharmaceutical Company receives large shipments of aspirin tablets and uses this acceptance sampling plan: Randomly select and test 24 tablets, then accept the whole batch if there is only one or none that doesn't meet the required specifications. If a particular shipment of thousands of aspirin tablets actually has a 4% rate of defects, what is the probability that this whole shipment will be accepted?

28. Affirmative Action Programs A study was conducted to determine whether there were significant differences between medical students admitted through special programs (such as affirmative action) and medical students admitted with regular admissions criteria. It was found that the graduation rate was 94% for the medical students admitted through special programs (based on data from the *Journal of the American Medical Association*).
   a. If 10 of the students from the special programs are randomly selected, find the probability that at least 9 of them graduated.
   b. Would it be unusual to randomly select 10 students from the special programs and get only 7 that graduate? Why or why not?

29. Identifying Gender Discrimination After being rejected for employment, Kim Kelly learns that the Bellevue Seed Company has hired only 2 women among the last 20 new employees. She also learns that the pool of applicants is very large, with an approximately equal number of qualified men and women. Help her address the charge of gender discrimination by finding the probability of getting 2 or fewer women when 20 people are hired, assuming that there is no discrimination based on gender. Does the resulting probability really support such a charge?

30. Testing Effectiveness of Gender Selection Technique The Chapter Problem describes the probability distribution for the number of girls x when 14 newborn babies are randomly selected. Assume that another clinical experiment involves 12 newborn babies. Using the same format as Table 4-1, construct a table for the probability distribution that results from 12 births, then determine whether a gender-selection technique appears to be effective if it results in 9 girls and 3 boys.

*31. Geometric Distribution If a procedure meets all the conditions of a binomial distribution except that the number of trials is not fixed, then the **geometric distribution** can be used. The probability of getting the first success on the xth trial is given by $P(x) = p(1 - p)^{x-1}$ where p is the probability of success on any one trial. Assume that the probability of a defective computer component is 0.2. Find the probability that the first defect is found in the seventh component tested.

*32. Hypergeometric Distribution If we sample from a small finite population without replacement, the binomial distribution should not be used because the events are not independent. If sampling is done without replacement and the outcomes belong to one of two types, we can use the **hypergeometric distribution.** If a population has A objects of one type, while the remaining B objects are of the other type, and if n objects are sampled without replacement, then the probability of getting x objects of type A and $n - x$ objects of type B is

$$P(x) = \frac{A!}{(A - x)!x!} \cdot \frac{B!}{(B - n + x)!(n - x)!} \div \frac{(A + B)!}{(A + B - n)!n!}$$

In Lotto 54, a bettor selects six numbers from 1 to 54 (without repetition), and a winning six-number combination is later randomly selected. Find the probability of getting

a. all six winning numbers.
b. exactly five of the winning numbers.
c. exactly three of the winning numbers.
d. no winning numbers.

*33. Multinomial Distribution The binomial distribution applies only to cases involving two types of outcomes, whereas the **multinomial distribution** involves more than two categories. Suppose we have three types of mutually exclusive outcomes denoted by A, B, and C. Let $P(A) = p_1$, $P(B) = p_2$, and $P(C) = p_3$. In n independent trials, the probability of x_1 outcomes of type A, x_2 outcomes of type B, and x_3 outcomes of type C is given by

$$\frac{n!}{(x_1!)(x_2!)(x_3!)} \cdot p_1^{x_1} \cdot p_2^{x_2} \cdot p_3^{x_3}$$

A genetics experiment involves six mutually exclusive genotypes identified as A, B, C, D, E, and F, and they are all equally likely. If 20 offspring are tested, find the probability of getting exactly five A's, four B's, three C's, two D's, three E's, and three F's by expanding the above expression so that it applies to six types of outcomes instead of only three.

4-4 Mean, Variance, and Standard Deviation for the Binomial Distribution

We saw in Chapter 2 that when investigating collections of real data, these characteristics are usually very important: (1) the measure of center, (2) the measure of variation, (3) the nature of the distribution, (4) the presence of outliers, and (5) a pattern over time. (We used the key letters CVDOT as a device for remembering those characteristics.) A key point of this chapter is that probability distributions describe what will *probably* happen instead of what actually did happen. In Section 4-2, we learned methods for analyzing probability distributions by finding the mean, the standard deviation, and a probability histogram. Because a binomial distribution is a special type of probability distribution, we could use Formulas 4-1, 4-3, and 4-4 (from Section 4-2) for finding the mean, variance, and standard deviation. Fortunately, those formulas can be greatly simplified for binomial distributions, as shown below.

For Any Discrete Probability Distribution	**For Binomial Distributions**
Formula 4-1 $\mu = \Sigma[x \cdot P(x)]$	Formula 4-6 $\mu = np$
Formula 4-3 $\sigma^2 = \Sigma[x^2 \cdot P(x)] - \mu^2$	Formula 4-7 $\sigma^2 = npq$
Formula 4-4 $\sigma = \sqrt{\Sigma[x^2 \cdot P(x)] - \mu^2}$	Formula 4-8 $\sigma = \sqrt{npq}$

> **EXAMPLE Gender of Children** In Section 4-2 we included an example illustrating calculations for μ and σ. We used the example of the random variable x representing the number of girls in 14 births. (See Table 4-3 on page 164 for the calculations that illustrate Formulas 4-1 and 4-4.) Use Formulas 4-6 and 4-8 to find the mean and standard deviation for the numbers of girls in groups of 14 births.
>
> **SOLUTION**   Using the values $n = 14$, $p = 0.5$, and $q = 0.5$, Formulas 4-6 and 4-8 can be applied as follows:
>
> $$\mu = np = (14)(0.5) = 7.0$$
> $$\sigma = \sqrt{npq}$$
> $$= \sqrt{(14)(0.5)(0.5)} = 1.9 \quad \text{(rounded)}$$
>
> If you compare these calculations to those required in Table 4-3, it should be obvious that Formulas 4-6 and 4-8 are substantially easier to use.

Formula 4-6 for the mean makes sense intuitively. Ask most people how many girls are expected in 100 births and the usual response is 50, which can be easily generalized as $\mu = np$. The variance and standard deviation are not so easily justified, and we will omit the complicated algebraic manipulations that lead to Formulas 4-7 and 4-8. Instead, refer again to the preceding example and Table 4-3 to verify that for a binomial distribution, Formulas 4-6, 4-7, and 4-8 will produce the same results as Formulas 4-1, 4-3, and 4-4.

EXAMPLE **Gender Selection** The Chapter Problem involved a preliminary trial with 14 couples who wanted to have baby girls. Although the result of 13 girls in 14 births makes it appear that the MicroSort method of gender selection is effective, we would have much more confidence in that conclusion if the sample size had been considerably larger than 14. Suppose the MicroSort method is used with 100 couples, each of whom will have 1 baby. Assume that the result is 68 girls among the 100 babies.

a. Assuming that the MicroSort gender-selection method has no effect, find the mean and standard deviation for the numbers of girls in groups of 100 randomly selected babies.

b. *Interpret* the values from part (a) to determine whether this result (68 girls among 100 babies) supports a claim that the MicroSort method of gender selection is effective.

SOLUTION

a. Assuming that the MicroSort method has no effect and that girls and boys are equally likely, we have $n = 100$, $p = 0.5$, and $q = 0.5$. We can find the mean and standard deviation by using Formulas 4-6 and 4-8 as follows:

$$\mu = np = (100)(0.5) = 50$$
$$\sigma = \sqrt{npq} = \sqrt{(100)(0.5)(0.5)} = 5$$

For groups of 100 couples who each have a baby, the mean number of girls is 50 and the standard deviation is 5.

b. We must now interpret the results to determine whether 68 girls among 100 babies is a result that could easily occur by chance, or whether that result is so unlikely that the MicroSort method of gender selection seems to be effective. We will use the range rule of thumb as follows:

$$\text{maximum usual value: } \mu + 2\sigma = 50 + 2(5) = 60$$
$$\text{minimum usual value: } \mu - 2\sigma = 50 - 2(5) = 40$$

INTERPRETATION According to our range rule of thumb, values are considered to be usual if they are between 40 and 60, so 68 girls is an unusual result because it is not between 40 and 60. It is very unlikely that we will get 68 girls in 100 births just by chance. If we did get 68 girls in 100 births, we should look for an explanation that is an alternative to chance. If the 100 couples used the MicroSort method of gender selection, it would appear to be effective in increasing the likelihood that a child will be a girl.

You should develop the skills to calculate means and standard deviations using Formulas 4-6 and 4-8, but it is especially important to learn to *interpret* results by using those values. The range rule of thumb, as illustrated in part (b) of the preceding example, suggests that values are unusual if they lie outside of these limits:

$$\text{maximum usual value} = \mu + 2\sigma$$
$$\text{minimum usual value} = \mu - 2\sigma$$

4-4 Exercises

Finding u, σ, and Unusual Values. In Exercises 1–4, assume that a procedure yields a binomial distribution with n trials and the probability of success for one trial is p. Use the given values of n and p to find the mean μ and standard deviation σ. Also, use the range rule of thumb to find the minimum usual value μ − 2σ and the maximum usual value μ + 2σ.

1. $n = 400, p = 0.2$

2. $n = 250, p = 0.45$

3. $n = 1984, p = 3/4$

4. $n = 767, p = 1/6$

5. Guessing Answers Several biology students are unprepared for a surprise true/false test with 10 questions, and all of their answers are guesses.
  a. Find the mean and standard deviation for the number of correct answers for such students.
  b. Would it be unusual for a student to pass by guessing and getting at least 7 correct answers? Why or why not?

6. Guessing Answers Several students are unprepared for a multiple-choice quiz with 10 questions, and all of their answers are guesses. Each question has five possible answers, and only one of them is correct.
  a. Find the mean and standard deviation for the number of correct answers for such students.
  b. Would it be unusual for a student to pass by guessing and getting at least 7 correct answers? Why or why not?

7. Playing Roulette If you bet on any single number in roulette, your probability of winning is 1/38. Assume that you bet on a single number in each of 100 consecutive spins.
  a. Find the mean and standard deviation for the number of wins.
  b. Would it be unusual to not win once in the 100 trials? Why or why not?

8. Left-Handed People Ten percent of American adults are left-handed. A biology class has 25 students in attendance.
  a. Find the mean and standard deviation for the number of left-handed students in such classes of 25 students.
  b. Would it be unusual to survey a class of 25 students and find that 5 of them are left-handed? Why or why not?

9. Analyzing Results of Experiment in Gender Selection An experiment involving a gender-selection method includes a control group of 15 couples who are not given any treatment intended to influence the genders of their babies. Each of the 15 couples has one child.
  a. Construct a table listing the possible values of the random variable *x* (which represents the number of girls among the 15 births) and the corresponding probabilities.
  b. Find the mean and standard deviation for the numbers of girls in such groups of 15.
  c. If the couples have 10 girls and 5 boys, is that result unusual? Why or why not?

10. Cell Phones and Brain Cancer In a study of 420,095 cell phone users in Denmark, it was found that 135 developed cancer of the brain or nervous system. If we assume that such cancer is not affected by cell phones, the probability of a person having such a cancer is 0.000340.

a. Assuming that cell phones have no effect on cancer, find the mean and standard deviation for the numbers of people in groups of 420,095 that can be expected to have cancer of the brain or nervous system.

b. Based on the results from part (a), is it unusual to find that among 420,095 people, there are 135 cases of cancer of the brain or nervous system? Why or why not?

c. What do these results suggest about the publicized concern that cell phones are a health danger because they increase the risk of cancer of the brain or nervous system?

11. Cholesterol-Reducing Drug In a clinical trial of Lipitor (atorvastatin), a common drug used to lower cholesterol, 863 patients were given a treatment of 10 mg tablets. That group consists of 19 patients who experienced flu symptoms (based on data from Pfizer, Inc.). The probability of flu symptoms for a person not receiving any treatment is 0.019.

a. Assuming that Lipitor has no effect on flu symptoms, find the mean and standard deviation for the numbers of people in groups of 863 that can be expected to have flu symptoms.

b. Based on the result from part (a), is it unusual to find that among 863 people, there are 19 who experience flu symptoms? Why or why not?

c. Based on the preceding results, do flu symptoms appear to be an adverse reaction that should be of concern to those who use Lipitor?

12. Opinions About Cloning A recent Gallup poll consisted of 1012 randomly selected adults who were asked whether "cloning of humans should or should not be allowed." Results showed that 89% of those surveyed indicated that cloning should not be allowed.

a. Among the 1012 adults surveyed, how many said that cloning should not be allowed?

b. Assume that people are indifferent so that 50% believe that cloning of humans should not be allowed, and find the mean and standard deviation for the numbers of people in groups of 1012 that can be expected to believe that such cloning should not be allowed.

c. Based on the preceding results, does the 89% result for the Gallup poll appear to be unusually higher than the assumed rate of 50%? Does it appear that an overwhelming majority of adults believe that cloning of humans should not be allowed?

13. Car Crashes For drivers in the 20–24 age bracket, there is a 34% rate of car accidents in one year (based on data from the National Safety Council). An insurance investigator finds that in a group of 500 randomly selected drivers aged 20–24 living in New York City, 42% had accidents in the last year.

a. How many drivers in the New York City group of 500 had accidents in the last year?

b. Assuming that the same 34% rate applies to New York City, find the mean and standard deviation for the numbers of people in groups of 500 that can be expected to have accidents.

c. Based on the preceding results, does the 42% result for the New York City drivers appear to be unusually high when compared to the 34% rate for the general population? Does it appear that higher insurance rates for New York City drivers are justified?

Not At Home

Pollsters cannot simply ignore those who were not at home when they were called the first time. One solution is to make repeated callback attempts until the person can be reached. Alfred Politz and Willard Simmons describe a way to compensate for those missing results without making repeated callbacks. They suggest weighting results based on how often people are not at home. For example, a person at home only two days out of six will have a 2/6 or 1/3 probability of being at home when called the first time. When such a person is reached the first time, his or her results are weighted to count three times as much as someone who is always home. This weighting is a compensation for the other similar people who are home two days out of six and were not at home when called the first time. This clever solution was first presented in 1949.

*14. Using the Empirical Rule and Chebyshev's Theorem An experiment is designed to test the effectiveness of the MicroSort method of gender selection, and 100 couples try to have baby girls using the MicroSort method. In an example included in this section, the range rule of thumb was used to conclude that among 100 births, the number of girls should usually fall between 40 and 60.

a. The empirical rule (see Section 2-5) applies to distributions that are bell-shaped. Is the binomial probability distribution for this experiment (approximately) bell-shaped? How do you know?

b. Assuming that the distribution is bell-shaped, how likely is it that the number of girls will fall between 40 and 60 (according to the empirical rule)?

c. Assuming that the distribution is bell-shaped, how likely is it that the number of girls will fall between 35 and 65 (according to the empirical rule)?

d. Using Chebyshev's theorem, what do we conclude about the likelihood that the number of girls will fall between 40 and 60?

4-5 The Poisson Distribution

The preceding two sections included the binomial distribution, which is one of many probability distributions that can be used for different situations. This section introduces the *Poisson distribution*. It is particularly important because it is often used as a mathematical model for describing the behavior of rare events (with small probabilities). It is used for describing behavior such as radioactive decay, arrivals of people in a line, eagles nesting in a region, patients arriving at an emergency room, and Internet users logging onto a Web site. For example, suppose your local hospital experiences a mean of 2.3 patients arriving at the emergency room on Fridays between 10:00 P.M. and 11:00 P.M. We can find the probability that for a randomly selected Friday between 10:00 P.M. and 11:00 P.M., exactly four patients arrive. We use the Poisson distribution, defined as follows.

Definition

The **Poisson distribution** is a discrete probability distribution that applies to occurrences of some event *over a specified interval*. The random variable x is the number of occurrences of the event in an interval. The interval can be time, distance, area, volume, or some similar unit. The probability of the event occurring x times over an interval is given by Formula 4-9.

Formula 4-9 $$P(x) = \frac{\mu^x \cdot e^{-\mu}}{x!} \quad \text{where } e \approx 2.71828$$

The Poisson distribution has the following requirements:

- The random variable x is the number of occurrences of an event *over some interval*.
- The occurrences must be *random*.
- The occurrences must be *independent* of each other.

- The occurrences must be *uniformly distributed* over the interval being used.

The Poisson distribution has these parameters:

- The mean is μ.
- The standard deviation is $\sigma = \sqrt{\mu}$.

A Poisson distribution differs from a binomial distribution in these fundamental ways:

1. The binomial distribution is affected by the sample size n and the probability p, whereas the Poisson distribution is affected only by the mean μ.

2. In a binomial distribution, the possible values of the random variable x are $0, 1, \ldots, n$, but a Poisson distribution has possible x values of $0, 1, 2, \ldots$, with no upper limit.

EXAMPLE **World War II Bombs** In analyzing hits by V-1 buzz bombs in World War II, South London was subdivided into 576 regions, each with an area of 0.25 km^2. A total of 535 bombs hit the combined area of 576 regions.

a. If a region is randomly selected, find the probability that it was hit exactly twice.

b. Based on the probability found in part (a), how many of the 576 regions are expected to be hit exactly twice?

SOLUTION

a. The Poisson distribution applies because we are dealing with the occurrences of an event (bomb hits) over some interval (a region with area of 0.25 km^2). The mean number of hits per region is

$$\mu = \frac{\text{number of bomb hits}}{\text{number of regions}} = \frac{535}{576} = 0.929$$

Because we want the probability of exactly two hits in a region, we let $x = 2$ and use Formula 4-9 as follows:

$$P(x) = \frac{\mu^x \cdot e^{-\mu}}{x!} = \frac{0.929^2 \cdot 2.71828^{-0.929}}{2!} = \frac{0.863 \cdot 0.395}{2} = 0.170$$

The probability of a particular region being hit exactly twice is $P(2) = 0.170$.

b. Because there is a probability of 0.170 that a region is hit exactly twice, we expect that among the 576 regions, the number that are hit exactly twice is $576 \cdot 0.170 = 97.9$.

In the preceding example, we can also calculate the probabilities and expected values for 0, 1, 3, 4, and 5 hits. (We stop at $x = 5$ because no region was hit more than five times, and the probabilities for $x > 5$ are 0.000 when rounded to three decimal places.) Those probabilities and expected values are listed in Table 4-4. The fourth column of Table 4-4 describes the results that actually occurred during World War II. There were 229 regions that had no hits, 211 regions that were hit

Table 4-4	V-1 Buzz Bomb Hits for 576 Regions in South London		
Number of Bomb Hits	Probability	Expected Number of Regions	Actual Number of Regions
0	0.395	227.5	229
1	0.367	211.4	211
2	0.170	97.9	93
3	0.053	30.5	35
4	0.012	6.9	7
5	0.002	1.2	1

once, and so on. We can now compare the frequencies *predicted* with the Poisson distribution (third column) to the *actual* frequencies (fourth column) to conclude that there is very good agreement. In this case, the Poisson distribution does a good job of predicting the results that actually occurred. (Section 10-2 describes a statistical procedure for determining whether such expected frequencies constitute a good "fit" to the actual frequencies. That procedure does suggest that there is a good fit in this case.)

4-5 Exercises

Using a Poisson Distribution to Find Probability. In Exercises 1–4, assume that the Poisson distribution applies and proceed to use the given mean to find the indicated probability.

1. If $\mu = 2$, find $P(3)$.
2. If $\mu = 0.5$, find $P(2)$.
3. If $\mu = 100$, find $P(99)$.
4. If $\mu = 500$, find $P(512)$.

In Exercises 5–12, use the Poisson distribution to find the indicated probabilities.

5. Radioactive Decay Radioactive atoms are unstable because they have too much energy. When they release their extra energy, they are said to decay. When studying cesium 137, it is found that during the course of decay over 365 days, 1,000,000 radioactive atoms are reduced to 977,287 radioactive atoms.
   a. Find the mean number of radioactive atoms lost through decay in a day.
   b. Find the probability that on a given day, 50 radioactive atoms decayed.

6. Births Currently, 11 babies are born in the village of Westport (population 760) each year (based on data from the U.S. National Center for Health Statistics).
   a. Find the mean number of births per day.
   b. Find the probability that on a given day, there are no births.
   c. Find the probability that on a given day, there is at least one birth.
   d. Based on the preceding results, should medical birthing personnel be on permanent standby, or should they be called in as needed? Does this mean that Westport

mothers might not get the immediate medical attention that they would be likely to get in a more populated area?

7. Deaths from Horse Kicks A classic example of the Poisson distribution involves the number of deaths caused by horse kicks of men in the Prussian Army between 1875 and 1894. Data for 14 corps were combined for the 20-year period, and the 280 corps-years included a total of 196 deaths. After finding the mean number of deaths per corps-year, find the probability that a randomly selected corps-year has the following numbers of deaths.
a. 0   **b.** 1   **c.** 2   **d.** 3   **e.** 4

The actual results consisted of these frequencies: 0 deaths (in 144 corps-years); 1 death (in 91 corps-years); 2 deaths (in 32 corps-years); 3 deaths (in 11 corps-years); 4 deaths (in 2 corps-years). Compare the actual results to those expected from the Poisson probabilities. Does the Poisson distribution serve as a good device for predicting the actual results?

8. Homicide Deaths In one year, there were 116 homicide deaths in Richmond, Virginia (based on "A Classroom Note on the Poisson Distribution: A Model for Homicidal Deaths in Richmond, VA for 1991," *Mathematics and Computer Education,* by Winston A. Richards). For a randomly selected day, find the probability that the number of homicide deaths is
a. 0   **b.** 1   **c.** 2   **d.** 3   **e.** 4

Compare the calculated probabilities to these actual results: 268 days (no homicides); 79 days (1 homicide); 17 days (2 homicides); 1 day (3 homicides); no days with more than 3 homicides.

9. Dandelions Dandelions are studied for their effects on crop production and lawn growth. In one region, the mean number of dandelions per square meter was found to be 7.0 (based on data from Manitoba Agriculture and Food).
a. Find the probability of no dandelions in an area of 1 m^2.
b. Find the probability of at least one dandelion in an area of 1 m^2.
c. Find the probability of at most two dandelions in an area of 1 m^2.

10. Earthquakes For a recent period of 100 years, there were 93 major earthquakes (at least 6.0 on the Richter scale) in the world (based on data from the *World Almanac and Book of Facts*). Assuming that the Poisson distribution is a suitable model, find the mean number of major earthquakes per year, then find the probability that the number of earthquakes in a randomly selected year is
a. 0   **b.** 1   **c.** 2   **d.** 3   **e.** 4   **f.** 5   **g.** 6   **h.** 7

Here are the actual results: 47 years (0 major earthquakes); 31 years (1 major earthquake); 13 years (2 major earthquakes); 5 years (3 major earthquakes); 2 years (4 major earthquakes); 0 years (5 major earthquakes); 1 year (6 major earthquakes); 1 year (7 major earthquakes). After comparing the calculated probabilities to the actual results, is the Poisson distribution a good model?

Review

The concept of a probability distribution is a key element of statistics. A probability distribution describes the probability for each value of a random variable. This chapter includes only discrete probability distributions, but the following chapters will include continuous probability distributions. The following key points were discussed:

- A *random variable* has values that are determined by chance.
- A *probability distribution* consists of all values of a random variable, along with their corresponding probabilities. A probability distribution must satisfy two requirements: $\Sigma P(x) = 1$ and, for each value of x, $0 \le P(x) \le 1$.
- Important characteristics of a *probability distribution* can be explored by constructing a probability histogram and by computing its mean and standard deviation using these formulas:

$$\mu = \Sigma[x \cdot P(x)]$$
$$\sigma = \sqrt{\Sigma[x^2 \cdot P(x)] - \mu^2}$$

- In a *binomial distribution,* there are two categories of outcomes and a fixed number of independent trials with a constant probability. The probability of x successes among n trials can be found by using the binomial probability formula, or Table A-1, or software (such as STATDISK, Minitab, or Excel), or a TI-83/84 Plus calculator.
- In a binomial distribution, the mean and standard deviation can be easily found by calculating the values of $\mu = np$ and $\sigma = \sqrt{npq}$.
- A *Poisson probability distribution* applies to occurrences of some event over a specific interval, and its probabilities can be computed with Formula 4-9.
- *Unusual outcomes:* This chapter stressed the importance of interpreting results by distinguishing between outcomes that are usual and those that are unusual. We used two different criteria. With the range rule of thumb we have

maximum usual value = $\mu + 2\sigma$

minimum usual value = $\mu - 2\sigma$

We can also determine whether outcomes are unusual by using probability values.

Unusually high: x successes among n trials is an *unusually high* number of successes if $P(x$ or more) is very small (such as 0.05 or less).

Unusually low: x successes among n trials is an *unusually low* number of successes if $P(x$ or fewer) is very small (such as 0.05 or less).

Review Exercises

1. **a.** What is a random variable?
   b. What is a probability distribution?
   c. *Consumer Reports* magazine stated that "for condoms, the typical failure rate is about 12%, somewhat worse than birth-control pills (8%), ... " The accompanying table is based on an assumed 12% failure rate for condoms, where x is the number of failures among 5 trials. Does this table describe a probability distribution? Why or why not?
   d. Assuming that the table does describe a probability distribution, find its mean.
   e. Assuming that the table does describe a probability distribution, find its standard deviation.
   f. Is it unusual to randomly select and test 5 condoms and find that they all fail? Why or why not?

2. Employee Drug Testing Among companies doing highway or bridge construction, 80% test employees for substance abuse (based on data from the Construction Finan-

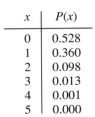

x	$P(x)$
0	0.528
1	0.360
2	0.098
3	0.013
4	0.001
5	0.000

cial Management Association). A study involves the random selection of 10 such companies.

a. Find the probability that 5 of the 10 companies test for substance abuse.

b. Find the probability that at least half of the companies test for substance abuse.

c. For such groups of 10 companies, find the mean and standard deviation for the number (among 10) that test for substance abuse.

d. Would it be unusual to find that 6 of 10 companies test for substance abuse? Why or why not?

3. Reasons for Being Fired "Inability to get along with others" is the reason cited in 17% of worker firings (based on data from Robert Half International, Inc.). Concerned about her company's working conditions, the personnel manager at the Kansas Agriculture Company plans to investigate the five employee firings that occurred over the past year.

a. Assuming that the 17% rate applies, find the probability that among those five employees, the number fired because of an inability to get along with others is at least four.

b. If the personnel manager actually does find that at least four of the firings are due to an inability to get along with others, does this company appear to be very different from other typical companies? Why or why not?

4. Deaths Currently, an average of 7 residents of the village of Westport (population 760) die each year (based on data from the U.S. National Center for Health Statistics).

a. Find the mean number of deaths per day.

b. Find the probability that on a given day, there are no deaths.

c. Find the probability that on a given day, there is one death.

d. Find the probability that on a given day, there is more than one death.

e. Based on the preceding results, should Westport have a contingency plan to handle more than one death per day? Why or why not?

Cumulative Review Exercises

1. Weights: Analysis of Last Digits The accompanying table lists the last digits of weights of the subjects listed in Data Set 1 in Appendix B. The last digits of a data set can sometimes be used to determine whether the data have been measured or simply reported. The presence of disproportionately more 0s and 5s is often a sure indicator that the data have been reported instead of measured.

a. Find the mean and standard deviation of those last digits.

b. Construct the relative frequency table that corresponds to the given frequency table.

c. Construct a table for the probability distribution of randomly selected digits that are all equally likely. List the values of the random variable x (0, 1, 2, ... , 9) along with their corresponding probabilities (0.1, 0.1, 0.1, ... , 0.1), then find the mean and standard deviation of this probability distribution.

d. Recognizing that sample data naturally deviate from the results we theoretically expect, does it seem that the given last digits roughly agree with the distribution we expect with random selection? Or does it seem that there is something about the sample data (such as disproportionately more 0s and 5s) suggesting that the given last digits are not random? (In Chapter 10, we will present a method for answering such questions much more objectively.)

x	f
0	7
1	14
2	5
3	11
4	8
5	4
6	5
7	6
8	12
9	8

2. Determining the Effectiveness of an HIV Training Program The New York State Health Department reports a 10% rate of the HIV virus for the "at-risk" population. In one region, an intensive education program is used in an attempt to lower that 10% rate. After running the program, a follow-up study of 150 at-risk individuals is conducted.

    a. Assuming that the program has no effect, find the mean and standard deviation for the number of HIV cases in groups of 150 at-risk people.

    b. Among the 150 people in the follow-up study, 8% (or 12 people) tested positive for the HIV virus. If the program has no effect, is that rate unusually low? Does this result suggest that the program is effective?

Cooperative Group Activities

1. *In-class activity:* Suppose we want to identify the probability distribution for the number of children born to randomly selected couples. For each student in the class, find the number of brothers and sisters and record the total number of children (including the student) in each family. Construct the relative frequency table for the result obtained. (The values of the random variable x will be 1, 2, 3, ...) What is wrong with using this relative frequency table as an estimate of the probability distribution for the number of children born to randomly selected couples?

2. *Out-of-class activity:* See Cumulative Review Exercise 1, which suggests that an analysis of the last digits of data can sometimes reveal whether the data have been collected through actual measurements or reported by the subjects. Refer to an almanac or the Internet and find a collection of data (such as lengths of rivers in the world), then analyze the distribution of last digits to determine whether the values were obtained through actual measurements.

Technology Project

In mornings and afternoons, the Newport Hospital emergency room can accommodate a maximum of 12 patients in one hour, and the admissions procedure is designed to accommodate a maximum of five patients in one hour. Past experience suggests that for this hospital, an average of 28% of emergency room patients are admitted to the hospital for further care. We can analyze the probability distribution of admissions by considering a binomial distribution with $n = 12$ and $p = 0.28$.

Find the probability that when 12 patients show up in the emergency room in one hour, the admissions procedure is overloaded. That is, find the probability of at least 6 patients requiring admission, assuming that 12 patients show up in the emergency room in one hour. Because of the value of $p = 0.28$, Table A-1 cannot be used, and calculations with the binomial probability formula would be tedious. The best approach is to use statistics software or a TI-83/84 Plus calculator. Use technology to find the relevant probability. Is the probability of an overloaded admissions procedure small enough so that it does not happen very often, or does it seem too high so that changes must be made to make it lower?

from DATA to DECISION

Critical Thinking: Determining criteria for concluding that a gender-selection method is effective

You are responsible for analyzing results from a clinical trial of the effectiveness of a new method of gender selection. Assume that the sample size of $n = 50$ couples has already been established, and each couple will have one child. Further assume that each of the couples will be subjected to a treatment that supposedly increases the likelihood that the child will be a girl.

There is a danger in obtaining results first, then making conclusions about the results. If the results are close to showing the effectiveness of a treatment, it might be tempting to conclude that there is an effect when, in reality, there is no effect. It is better to establish criteria *before* obtaining results. Using the methods of this chapter, identify the criteria that should be used for concluding that the treatment is effective in increasing the likelihood of a girl. Among the 50 births, how many girls would you require in order to conclude that the gender-selection procedure is effective? Explain how you arrived at this result.

Normal Probability Distributions

Ergonomics: Engineering with biology

A relatively new discipline is *ergonomics,* the study of people fitting into their environments. Good ergonomic design results in an environment that is safe, functional, efficient, and comfortable. Ergonomics is used in applications including the design of ambulance stretchers, car dashboards, cycle helmets, screw bottle tops, door handles, manhole covers, keyboards, air traffic control centers, and computer assembly lines. For example, in Vail, Colorado, a gondola carries skiers to the top of a mountain. It bears a plaque stating that the maximum capacity is 12 people or 2004 pounds. Reading that plaque causes many passengers to look around wondering whether they are in danger because there are too many fellow passengers or because there are 12 or fewer passengers who, by chance random variation, are all exceptionally heavy. How likely is it that 12 randomly selected people will have a total weight greater than 2004 pounds?

The ability to comfortably endure a long transcontinental flight is affected by the width of the seat we must occupy. Most U.S. commercial aircraft have seat widths of 17 in. to 18 in., which just barely accommodates the 16 in. required by the average passenger. Business and first-class seats usually have widths between 19 in. and 21 in., so the extra room allows a greater degree of comfort. If American Airlines wants to gain business by increasing passenger comfort, what width should be used in redesigned seats?

When visiting preserved living quarters that are hundreds of years old, many people are struck by the fact that the doorways have openings that are too low for most people living today. When walking through a modern doorway, most of us fit comfortably under the top threshold, which is typically 80 in. high. However, some people are exceptionally tall and must bend to avoid a bump on the head. What percent of people are too tall for the current doorway design standards?

In recent years, the United States Air Force recognized that women make perfectly good pilots of fighter jets. Cockpits of fighter jets were originally designed for men only, so various changes were required to better accommodate the new women pilots. One such change involved a redesign of the ACES-II ejection seats. Because they were originally designed for men who weighed between 140 and 211 1b, the ejection seats provided a greater chance of injury for any women pilots weighing less than 140 1b or more than 211 1b. What weights should be used for the new cockpit design?

In this chapter we address questions such as those posed above.

5-1 Overview

In Chapter 2 we considered important measures of data sets, including measures of center and variation, as well as the distribution of data. In Chapter 3 we discussed basic principles of probability, and in Chapter 4 we presented the following concepts:

- A *random variable* is a variable having a single numerical value, determined by chance, for each outcome of some procedure.
- A *probability distribution* for a discrete random variable describes the probability for each value of the random variable.
- A *discrete* random variable has either a finite number of values or a countable number of values. That is, the *number* of possible values that x can assume is 0, or 1, or 2, and so on.
- A *continuous* random variable has infinitely many values, and those values are often associated with measurements on a continuous scale with no gaps or interruptions.

In Chapter 4 we considered only *discrete* probability distributions, but in this chapter we present *continuous* probability distributions. Although we begin with a uniform distribution, most of the chapter will focus on *normal distributions*. Normal distributions are extremely important because they occur so often in real applications and they play such an important role in methods of inferential statistics. Normal distributions will be used often throughout the remainder of this text.

Definition

If a continuous random variable has a distribution with a graph that is symmetric and bell-shaped, as in Figure 5-1, and it can be described by the equation given as Formula 5-1, we say that it has a **normal distribution.**

Formula 5-1

$$y = \frac{e^{-\frac{1}{2}\left(\frac{x-\mu}{\sigma}\right)^2}}{\sigma\sqrt{2\pi}}$$

The complexity of Formula 5-1 can be intimidating, but it isn't really necessary for us to actually use Formula 5-1. That formula does show, however, that any particular normal distribution is determined by two parameters: the mean, μ,

FIGURE 5-1 The Normal Distribution

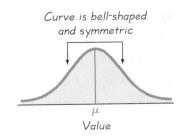

Curve is bell-shaped and symmetric

μ
Value

and standard deviation, σ. Once specific values are selected for μ and σ, we can graph Formula 5-1 as we would graph any equation relating x and y; the result is a continuous probability distribution with a bell shape.

5-2 The Standard Normal Distribution

The focus of this chapter is the concept of a normal probability distribution, but we begin with a *uniform distribution.* The uniform distribution makes it easier for us to see some very important properties, which will be used also with normal distributions.

Uniform Distributions

> ### Definition
>
> A continuous random variable has a **uniform distribution** if its values spread evenly over the range of possibilities. The graph of a uniform distribution results in a rectangular shape.

EXAMPLE Class Length A biology professor plans classes so carefully that the lengths of her classes are uniformly distributed between 50.0 min and 52.0 min. That is, any time between 50.0 min and 52.0 min is possible, and all of the possible values are equally likely. If we randomly select one of her classes and let x be the random variable representing the length of that class, then x has a distribution that can be graphed as in Figure 5-2.

When we discussed *discrete* probability distributions in Section 4-2, we identified two requirements: (1) $\Sigma P(x) = 1$ and (2) $0 \leq P(x) \leq 1$ for all values of x. Also in Section 4-2, we stated that the graph of a discrete probability distribution is called a *probability histogram.* The graph of a continuous probability distribution, such as that of Figure 5-2, is called a *density curve,* and it must satisfy two properties similar to the requirements for discrete probability distributions, as listed in the following definition.

Reliability and Validity

The reliability of data refers to the consistency with which results occur, whereas the validity of data refers to how well the data measure what they are supposed to measure. The reliability of an IQ test can be judged by comparing scores for the test given on one date to scores for the same test given at another time. To test the validity of an IQ test, we might compare the test scores to another indicator of intelligence, such as academic performance. Many critics charge that IQ tests are reliable, but not valid; they provide consistent results, but don't really measure intelligence.

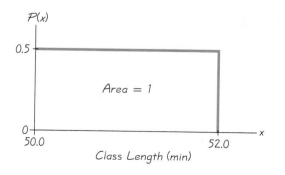

FIGURE 5-2 Uniform Distribution of Class Times

Definition

A **density curve** is a graph of a continuous probability distribution. It must satisfy the following properties:

1. The total area under the curve must equal 1.
2. Every point on the curve must have a vertical height that is 0 or greater. (That is, the curve cannot fall below the *x*-axis.)

By setting the height of the rectangle in Figure 5-2 to be 0.5, we force the enclosed area to be $2 \times 0.5 = 1$, as required. (In general, the area of the rectangle becomes 1 when we make its height equal to the value of 1/range.) This property of having the area under the curve equal to 1 makes it very easy to solve probability problems, so the following statement is important:

Because the total area under the density curve is equal to 1, there is a correspondence between *area* and *probability*.

EXAMPLE **Class Length** Kim, who has developed a habit of living on the edge, has scheduled a job interview immediately following her biology class. If the class runs longer than 51.5 minutes, she will be late for the job interview. Given the uniform distribution illustrated in Figure 5-2, find the probability that a randomly selected class will last longer than 51.5 minutes.

SOLUTION   See Figure 5-3, where we shade the region representing times that are longer than 51.5 minutes. Because the total area under the density curve is equal to 1, there is a correspondence between area and probability. We can therefore find the desired probability by using areas as follows:

$$P(\text{class longer than 51.5 minutes}) = \text{area of shaded region in Figure 5-3}$$
$$= 0.5 \times 0.5$$
$$= 0.25$$

INTERPRETATION   The probability of randomly selecting a class that runs longer than 51.5 minutes is 0.25. Because that probability is so high, Kim should consider making contingency plans that will allow her to get to her job interview on time. Nobody should ever be late for a job interview.

FIGURE 5-3 **Using Area to Find Probability**

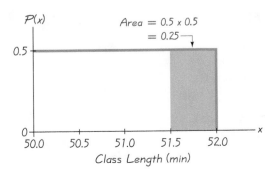

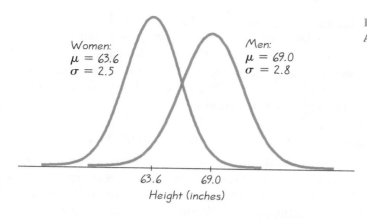

FIGURE 5-4 Heights of
Adult Women and Men

Standard Normal Distribution

The density curve of a uniform distribution is a horizontal line, so it's easy to find the area of any rectangular region: multiply width and height. The density curve of a normal distribution has the more complicated bell shape shown in Figure 5-1, so it's more difficult to find areas, but the basic principle is the same: *There is a correspondence between area and probability.*

Just as there are many different uniform distributions (with different ranges of values), there are also many different normal distributions, with each one depending on two parameters: the population mean, μ, and the population standard deviation, σ. (Recall from Chapter 1 that a *parameter* is a numerical measurement describing some characteristic of a *population*.) Figure 5-4 shows density curves for heights of adult women and men. Because men have a larger mean height, the peak of the density curve for men is farther to the right. Because men's heights have a slightly larger standard deviation, the density curve for men is slightly wider. Figure 5-4 shows two different possible normal distributions. There are infinitely many other possibilities, but one is of special interest.

Definition

The **standard normal distribution** is a normal probability distribution that has a mean of 0 and a standard deviation of 1, and the total area under its density curve is equal to 1. (See Figure 5-5.)

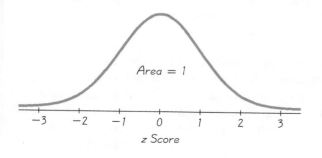

FIGURE 5-5 Standard
Normal Distribution: $\mu = 0$
and $\sigma = 1$

Suppose that we were hired to perform calculations using Formula 5-1. We would quickly see that the easiest values for μ and σ are $\mu = 0$ and $\sigma = 1$. By letting $\mu = 0$ and $\sigma = 1$, mathematicians have calculated many different areas under the curve. As shown in Figure 5-5, the area under the curve is 1, and this allows us to make the correspondence between area and probability, as we did in the preceding example with the uniform distribution.

Finding Probabilities When Given z Scores

Using Table A-2 (in Appendix A and the *Formulas and Tables* insert card), we can find areas (or probabilities) for many different regions. Such areas can be found by using Table A-2, a TI-83/84 Plus calculator, or software such as STATDISK, Minitab, or Excel. The key features of the different methods are summarized in Table 5-1. It is not necessary to know all five methods; you only need to know the method you will be using for your class. Because the following examples and exercises are based on Table A-2, it is essential to understand these points:

1. Table A-2 is designed only for the *standard* normal distribution, which has a mean of 0 and a standard deviation of 1.

2. Table A-2 is on two pages, with one page for *negative z* scores and the other page for *positive z* scores.

3. Each value in the body of the table is a *cumulative area from the left* up to a vertical boundary above a specific z score.

4. When working with a graph, avoid confusion between z scores and areas.

   > **z score:** **Distance along the horizontal scale of the standard normal distribution; refer to the leftmost column and top row of Table A-2**
   >
   > **Area:** **Region under the curve; refer to the values in the body of Table A-2**

5. The part of the z score denoting hundredths is found across the top row of Table A-2.

The following example requires that we find the probability associated with a value less than 1.58. Begin with the z score of 1.58 by locating 1.5 in the left column; next find the value in the adjoining row of probabilities that is directly below 0.08, as shown in this excerpt from Table A-2.

The area (or probability) value of 0.9429 indicates that there is a probability of 0.9429 of randomly selecting a z score less than 1.58. (The following sections will consider cases in which the mean is not 0 or the standard deviation is not 1.)

z	.... 0.08
.	.
.	.
.	.
1.5	.... 0.9429

EXAMPLE **Scientific Thermometers** The Precision Scientific Instrument Company manufactures thermometers that are supposed to give readings of 0°C at the freezing point of water. Tests on a large sample of these instruments reveal that at the freezing point of water, some thermometers give readings below 0°C (denoted by negative numbers) and some give readings above 0°C (denoted by positive numbers). Assume that the mean

Table 5-1	Methods for Finding Normal Distribution Areas

Table A-2

Gives the cumulative area from the left up to a vertical line above a specific value of z.

STATDISK

Gives a few areas, including the cumulative area from the left and the cumulative area from the right.

Minitab

Gives the cumulative area from the left up to a vertical line above a specific value.

Input constant

Excel

Gives the cumulative area from the left up to a vertical line above a specific value.

x

TI-83/84 Plus Calculator

Gives area bounded on the left and bounded on the right by vertical lines above any specific values.

Lower   Upper

reading is 0°C and the standard deviation of the readings is 1.00°C. Also assume that the readings are normally distributed. If one thermometer is randomly selected, find the probability that, at the freezing point of water, the reading is less than 1.58°C.

continued

FIGURE 5-6 **Finding the Area Below $z = 1.58$**

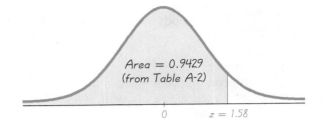

Area = 0.9429
(from Table A-2)

0   z = 1.58

SOLUTION   The probability distribution of readings is a standard normal distribution, because the readings are normally distributed with $\mu = 0$ and $\sigma = 1$. We need to find the area in Figure 5-6 below $z = 1.58$. The *area* below $z = 1.58$ is equal to the *probability* of randomly selecting a thermometer with a reading less than 1.58°C. From Table A-2 we find that this area is 0.9429.

INTERPRETATION   The *probability* of randomly selecting a thermometer with a reading less than 1.58°C (at the freezing point of water) is equal to the *area* of 0.9429 shown as the shaded region in Figure 5-6. Another way to interpret this result is to conclude that 94.29% of the thermometers will have readings below 1.58°C.

EXAMPLE **Scientific Thermometers** Using the thermometers from the preceding example, find the probability of randomly selecting one thermometer that reads (at the freezing point of water) above −1.23°C.

SOLUTION   We again find the desired *probability* by finding a corresponding *area*. We are looking for the area of the region that is shaded in Figure 5-7, but Table A-2 is designed to apply only to cumulative areas from the *left*. Referring to Table A-2 for the page with *negative z* scores, we find that the cumulative area from the left up to $z = -1.23$ is 0.1093 as shown. Knowing that the total area under the curve is 1, we can find the shaded area by subtracting 0.1093 from 1. The result is 0.8907. Even though Table A-2 is designed only for cumulative areas from the left, we can use it to find cumulative areas from the right, as in Figure 5-7.

INTERPRETATION   Because of the correspondence between probability and area, we conclude that the *probability* of randomly selecting a thermometer with a reading above −1.23°C at the freezing point of water is 0.8907 (which is the *area* to the right of $z = -1.23$). In other words, 89.07% of the thermometers have readings above −1.23°C.

FIGURE 5-7 **Finding the Area Above $z = -1.23$**

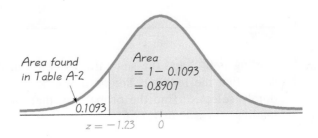

Area found
in Table A-2

Area
= 1 − 0.1093
= 0.8907

0.1093

z = −1.23   0

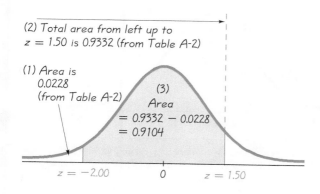

FIGURE 5-8 **Finding the Area Between Two Values**

The preceding example illustrates a way that Table A-2 can be used indirectly to find a cumulative area from the right. The following example illustrates another way that we can find an area by using Table A-2.

EXAMPLE **Scientific Thermometers** Once again, make a random selection from the same sample of thermometers. Find the probability that the chosen thermometer reads (at the freezing point of water) between $-2.00°C$. and $1.50°C$.

SOLUTION   We are again dealing with normally distributed values having a mean of $0°$ and a standard deviation of $1°$. The probability of selecting a thermometer that reads between $-2.00°$ and $1.50°$ corresponds to the shaded area in Figure 5-8. Table A-2 cannot be used to find that area directly, but we can use the table to find that $z = -2.00$ corresponds to the area of 0.0228, and $z = 1.50$ corresponds to the area of 0.9332, as shown in the figure. Refer to Figure 5-8 to see that the shaded area is the difference between 0.9332 and 0.0228. The shaded area is therefore $0.9332 - 0.0228 = 0.9104$.

INTERPRETATION Using the correspondence between probability and area, we conclude that there is a probability of 0.9104 of randomly selecting one of the thermometers with a reading between $-2.00°C$ and $1.50°C$ at the freezing point of water. Another way to interpret this result is to state that if many thermometers are selected and tested at the freezing point of water, then 0.9104 (or 91.04%) of them will read between $-2.00°$ and $1.50°$.

The preceding example can be generalized as a rule stating that the area corresponding to the region between two specific z scores can be found by finding the difference between the two areas found in Table A-2. See Figure 5-9, which shows that the shaded region *B* can be found by finding the *difference* between two areas found from Table A-2: areas *A* and *B* combined (found in Table A-2 as the area corresponding to z_{Right}) and area *A* (found in Table A-2 as the area corresponding to z_{Left}). Recommendation: Don't try to memorize a rule or formula for this case, because it is infinitely better to *understand* the procedure. Understand how Table A-2 works, then draw a graph, shade the desired area, and think of a way to find that area given the condition that Table A-2 provides only cumulative areas from the left.

FIGURE 5-9 **Finding the Area Between Two z Scores**

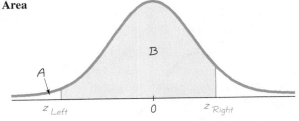

Shaded area B = (areas A and B combined) − (area A)
= (area from Table A-2 using z $_{Right}$) − (area from Table A-2 using z $_{Left}$)

The preceding example concluded with the statement that the probability of a reading between −2.00°C and 1.50°C is 0.9104. Such probabilities can also be expressed with the following notation.

Notation

$P(a < z < b)$ denotes the probability that the z score is between a and b.
$P(z > a)$ denotes the probability that the z score is greater than a.
$P(z < a)$ denotes the probability that the z score is less than a.

Using this notation, we can express the result of the last example as $P(-2.00 < z < 1.50) = 0.9104$, which states in symbols that the probability of a z score falling between −2.00 and 1.50 is 0.9104. With a continuous probability distribution such as the normal distribution, the probability of getting any single *exact* value is 0. That is, $P(z = a) = 0$. For example, there is a 0 probability of randomly selecting someone and getting a person whose height is exactly 68.12345678 in. In the normal distribution, any single point on the horizontal scale is represented not by a region under the curve, but by a vertical line above the point. For $P(z = 1.50)$ we have a vertical line above $z = 1.50$, but that vertical line by itself contains no area, so $P(z = 1.50) = 0$. With any continuous random variable, the probability of any one exact value is 0, and it follows that $P(a \leq z \leq b) = P(a < z < b)$. It also follows that the probability of getting a z score of *at most b* is equal to the probability of getting a z score *less than b*. It is important to correctly interpret key phrases such as *at most, at least, more than, no more than,* and so on.

Finding z Scores from Known Areas

So far, the examples of this section involving the standard normal distribution have all followed the same format: Given z scores, we found areas under the curve. These areas correspond to probabilities. In many other cases, we want a reverse process because we already know the area (or probability), but we need to find the corresponding z score. In such cases, it is very important to avoid confusion between z scores and areas. Remember, z scores are *distances* along the horizontal scale, and they are represented by the numbers in Table A-2 that are in the extreme left column and across the top row. Areas (or probabilities) are regions under the curve, and they are represented by the values in the body of Table A-2. Also, z scores positioned in the left half of the curve are always negative. If we already know a probability and want to determine the corresponding z score, we find it as follows.

Procedure for Finding a z Score from a Known Area

1. Draw a bell-shaped curve and identify the region under the curve that corresponds to the given probability. If that region is not a cumulative region from the left, work instead with a known region that is a cumulative region from the left.

2. Using the cumulative area from the left, locate the closest probability in the *body* of Table A-2 and identify the corresponding z score.

EXAMPLE **Scientific Thermometers** Use the same thermometers as earlier, with temperature readings at the freezing point of water that are normally distributed with a mean of 0°C and a standard deviation of 1°C. Find the temperature corresponding to P_{95}, the 95th percentile. That is, find the temperature separating the bottom 95% from the top 5%. See Figure 5-10.

SOLUTION   Figure 5-10 shows the z score that is the 95th percentile, with 95% of the area (or 0.95) below it. Important: Remember that when referring to Table A-2, the body of the table includes *cumulative areas from the left*. Referring to Table A-2, we search for the area of 0.95 *in the body* of the table and then find the corresponding z score. In Table A-2 we find the areas of 0.9495 and 0.9505, but there's an asterisk with a special note indicating that 0.9500 corresponds to a z score of 1.645. We can now conclude that the z score in Figure 5-10 is 1.645, so the 95th percentile is the temperature reading of 1.645°C.

INTERPRETATION When tested at freezing, 95% of the readings will be less than or equal to 1.645°C, and 5% of them will be greater than or equal to 1.645°C. (The overlap in the preceding statement is not a problem, because there is a probability of 0 that a reading is *equal to* exactly 1.645°C or any other specific value.)

FIGURE 5-10 **Finding the 95th Percentile**

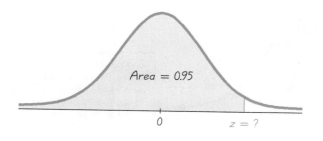

Area = 0.95

0   z = ?

z Score	Cumulative Area from the Left
1.645	0.9500
−1.645	0.0500
2.575	0.9950
−2.575	0.0050

Note that in the preceding solution, Table A-2 led to a z score of 1.645, which is midway between 1.64 and 1.65. When using Table A-2, we can usually avoid interpolation by simply selecting the closest value. There are special cases listed in the accompanying table that are important because they are used so often in a wide variety of applications. (The value of $z = 2.576$ is slightly closer to the area of 0.9950, but $z = 2.575$ has the advantage of being the value midway between $z = 2.57$ and $z = 2.58$.) Except in these special cases, we can select the closest value in the table. (If a desired value is midway between two table values, select the larger value.) Also, for z scores above 3.49, we can use 0.9999 as an approximation of the cumulative area from the left; for z scores below −3.49, we can use 0.0001 as an approximation of the cumulative area from the left.

EXAMPLE **Scientific Thermometers** Using the same thermometers, find the temperatures separating the bottom 2.5% and the top 2.5%.

SOLUTION   Refer to Figure 5-11 where the required z scores are shown. To find the z score located to the left, refer to Table A-2 and search the *body of the table* for an area of 0.025. The result is $z = -1.96$. To find the z score located to the right, refer to Table A-2 and search the *body of the table* for an area of 0.975. (Remember that Table A-2 always gives cumulative areas from the *left*.) The result is $z = 1.96$. The values of $z = -1.96$ and $z = 1.96$ separate the bottom 2.5% and the top 2.5% as shown in Figure 5-11.

INTERPRETATION   When tested at freezing, 2.5% of the thermometer readings will be equal to or less than −1.96°, and 2.5% of the readings will be equal to or greater than 1.96°. Another interpretation is that at the freezing level of water, 95% of all thermometer readings will fall between −1.96° and 1.96°C.

The examples in this section were contrived so that the mean of 0 and the standard deviation of 1 coincided exactly with the parameters of the standard normal distribution. In reality, it is unusual to find such convenient parameters, because typical normal distributions involve means different from 0 and standard deviations different from 1. In the next section we introduce methods for working with such normal distributions, which are much more realistic.

FIGURE 5-11 **Finding z Scores**

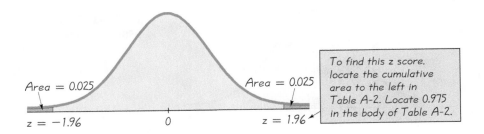

Area = 0.025

Area = 0.025

To find this z score, locate the cumulative area to the left in Table A-2. Locate 0.975 in the body of Table A-2.

$z = -1.96$ 0 $z = 1.96$

5-2 Exercises

Using a Continuous Uniform Distribution. In Exercises 1–4, refer to the continuous uniform distribution depicted in Figure 5-2, assume that a class length between 50.0 min and 52.0 min is randomly selected, and find the probability that the given time is selected.

1. Less than 50.3 min

2. Greater than 51.0 min

3. Between 50.5 min and 50.8 min

4. Between 50.5 min and 51.8 min

Using a Continuous Uniform Distribution. In Exercises 5–8, assume that voltages in a circuit vary between 6 volts and 12 volts, and voltages are spread evenly over the range of possibilities, so that there is a uniform distribution. Find the probability of the given range of voltage levels.

5. Greater than 10 volts

6. Less than 11 volts

7. Between 7 volts and 10 volts

8. Between 6.5 volts and 8 volts

Using the Standard Normal Distribution. In Exercises 9–28, assume that the readings on scientific thermometers are normally distributed with a mean of 0°C and a standard deviation of 1.00°C. A thermometer is randomly selected and tested. In each case, draw a sketch, and find the probability of each reading in degrees Celsius.

9. Less than −0.25

10. Less than −2.75

11. Less than 0.25

12. Less than 2.75

13. Greater than 2.33

14. Greater than 1.96

15. Greater than −2.33

16. Greater than −1.96

17. Between 0.50 and 1.50

18. Between 1.50 and 2.50

19. Between −2.00 and −1.00

20. Between 2.00 and 2.34

21. Between −2.67 and 1.28

22. Between −1.18 and 2.15

23. Between −0.52 and 3.75

24. Between −3.88 and 1.07

25. Greater than 3.57

26. Less than −3.61

27. Greater than 0

28. Less than 0

Basis for Empirical Rule. In Exercises 29–32, find the indicated area under the curve of the standard normal distribution, then convert it to a percentage and fill in the blank. The results form the basis for the empirical rule introduced in Section 2-5.

29. About _____ % of the area is between $z = -1$ and $z = 1$ (or within one standard deviation of the mean).

30. About _____ % of the area is between $z = -2$ and $z = 2$ (or within two standard deviations of the mean).

31. About _____ % of the area is between $z = -3$ and $z = 3$ (or within three standard deviations of the mean).

32. About _____ % of the area is between $z = -3.5$ and $z = 3.5$ (or within 3.5 standard deviations of the mean).

Finding Probability. In Exercises 33–36, assume that the readings on the thermometers are normally distributed with a mean of 0°C and a standard deviation of 1.00C. Find the indicated probability, where z is the reading in degrees.

33. $P(-1.96 < z < 1.96)$

34. $P(z < 1.645)$

35. $P(z > -2.575)$

36. $P(1.96 < z < 2.33)$

Finding Temperature Values. In Exercises 37–40, assume that the readings on the thermometers are normally distributed with a mean of 0°C and a standard deviation of 1.00°C. A thermometer is randomly selected and tested. In each case, draw a sketch, and find the temperature reading corresponding to the given information.

37. Find P_{90}, the 90th percentile. This is the temperature reading separating the bottom 90% from the top 10%.

38. Find P_{20}, the 20th percentile.

39. If 5% of the thermometers are rejected because they have readings that are too low, but all other thermometers are acceptable, find the reading that separates the rejected thermometers from the others.

40. If 3.0% of the thermometers are rejected because they have readings that are too high and another 3.0% are rejected because they have readings that are too low, find the two readings that are cutoff values separating the rejected thermometers from the others.

*41. For a standard normal distribution, find the percentage of data that are
   a. within 1 standard deviation of the mean.
   b. within 1.96 standard deviations of the mean.
   c. between $\mu - 3\sigma$ and $\mu + 3\sigma$.
   d. between 1 standard deviation below the mean and 2 standard deviations above the mean.
   e. more than 2 standard deviations away from the mean.

5-3 Applications of Normal Distributions

Because the examples and exercises in Section 5-2 all involved the *standard* normal distribution (with a mean of 0 and a standard deviation of 1), they were necessarily unrealistic. In this section we include nonstandard normal distributions so that we can work with applications that are real and practical. Fortunately, we can transform values from a nonstandard normal distribution to a standard normal distribution so that we can continue to use the same procedures from Section 5-2.

> **If we convert values to standard scores using Formula 5-2, then procedures for working with all normal distributions are the same as those for the standard normal distribution.**

> Formula 5-2 $z = \dfrac{x - \mu}{\sigma}$ (Round z scores to 2 decimal places.)

Continued use of Table A-2 requires understanding and application of the above principle. (If you use certain calculators or software programs, the conversion to

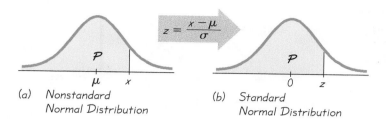

FIGURE 5-12
Converting from Nonstandard to Standard Normal Distribution

z scores is not necessary because probabilities can be found directly.) Regardless of the method used, you need to clearly understand the above basic principle, because it is an important foundation for concepts introduced in the following chapters.

Figure 5-12 illustrates the conversion from a nonstandard to a standard normal distribution. The area in any normal distribution bounded by some score x [as in Figure 5-12(a)] is the same as the area bounded by the equivalent z score in the standard normal distribution [as in Figure 5-12(b)]. This means that when working with a nonstandard normal distribution, you can use Table A-2 the same way it was used in Section 5-2 as long as you first convert the values to z scores. When finding areas with a nonstandard normal distribution, use this procedure:

Procedure for Converting Values in a Nonstandard Normal Distribution to z Scores

1. Sketch a normal curve, label the mean and the specific x values, then *shade* the region representing the desired probability.

2. For each relevant value x that is a boundary for the shaded region, use Formula 5-2 to convert that value to the equivalent z score.

3. Refer to Table A-2 and use the z scores to find the area of the shaded region. This area is the desired probability.

The following example applies these three steps, and it illustrates the relationship between a typical nonstandard normal distribution and the standard normal distribution.

EXAMPLE **Designing Cars** The sitting height (from seat to top of head) of drivers must be considered in the design of a new car model. Men have sitting heights that are normally distributed with a mean of 36.0 in. and a standard deviation of 1.4 in. (based on anthropometric survey data from Gordon, Clauser, et al.). Engineers have provided plans that can accommodate men with sitting heights up to 38.8 in., but taller men cannot fit. If a man is randomly selected, find the probability that he has a sitting height less than 38.8 in. Based on that result, is the current engineering design feasible?

SOLUTION

Step 1:   See Figure 5-13, where we label the mean of 36.0, the maximum sitting height 38.8 in., and we shade the area representing the probability that we want. (We continue to use the same correspondence between *probability* and *area* that was first introduced in Section 5-2.)

continued

Clinical Trial Cut Short

What do you do when you're testing a new treatment and, before your study ends, you find that it is clearly effective? You should cut the study short and inform all participants of the treatment's effectiveness. This happened when hydroxyurea was tested as a treatment for sickle cell anemia. The study was scheduled to last about 40 months, but the effectiveness of the treatment became obvious and the study was stopped after 36 months. (See "Trial Halted as Sickle Cell Treatment Proves Itself" by Charles Marwick, *Journal of the American Medical Association*, Vol. 273, No. 8.)

FIGURE 5-13 **Normal Distribution of Sitting Heights of Men**

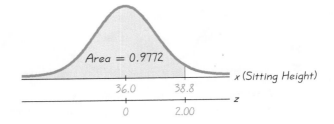

Step 2:   To use Table A-2, we first must use Formula 5-2 to convert the distribution of weights to the standard normal distribution. The height of 38.8 in. is converted to a z score as follows.

$$z = \frac{x - \mu}{\sigma} = \frac{38.8 - 36.0}{1.4} = 2.00$$

This result shows that the sitting height of 38.8 in. is above the mean of 36.0 in. by 2.00 standard deviations.

Step 3:   Referring to Table A-2, we find that $z = 2.00$ corresponds to an area of 0.9772.

INTERPRETATION   There is a probability of 0.9772 of randomly selecting a man with a sitting height less than 38.8 in. This can be expressed in symbols as

$$P(x < 38.8\text{in.}) = P(z < 2.00) = 0.9772$$

Another way to interpret this result is to conclude that 97.72% of men have sitting heights less than 38.8 in. An important consequence of that result is that 2.28% of men will not fit in the car. The manufacturer must now decide whether it can afford to lose 2.28% of all male car drivers.

EXAMPLE **Jet Ejection Seats** In the Chapter Problem, we noted that the Air Force had been using the ACES-II ejection seats designed for men weighing between 140 lb and 211 lb. Given that women's weights are normally distributed with a mean of 143 lb and a standard deviation of 29 lb (based on data from the National Health Survey), what percentage of women have weights that are within those limits?

SOLUTION   See Figure 5-14, which shows the shaded region for women's weights between 140 lb and 211 lb. We can't find that shaded area directly from Table A-2, but we can find it indirectly by using the basic procedures presented in Section 5-2. The way to proceed is first to find the cumulative area from the left up to 140 lb and the cumulative area from the left up to 211 lb, then find the difference between those two areas.

a. *Finding the cumulative area up to 140 lb:*

$$z = \frac{x - \mu}{\sigma} = \frac{140 - 143}{29} = -0.10$$

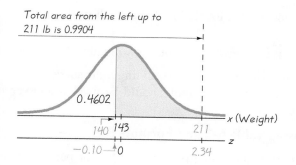

Total area from the left up to 211 lb is 0.9904

0.4602

FIGURE 5-14 **Weights of Women and Ejection Seat Limits**

Using Table A-2, we find that $z = -0.10$ corresponds to an area of 0.4602, as shown in Figure 5-14.

b. *Finding the cumulative area up to 211 lb:*

$$z = \frac{x - \mu}{\sigma} = \frac{211 - 143}{29} = 2.34$$

Using Table A-2, we find that $z = 2.34$ corresponds to an area of 0.9904, as shown in Figure 5-14.

c. *Finding the shaded area between 140 lb and 211 lb:*

$$\text{Shaded area} = 0.9904 - 0.4602 = 0.5302$$

INTERPRETATION The data show that 53.02% of women have weights between the ejection seat limits of 140 lb and 211 lb. This means that 46.98% of women do *not* have weights between the current limits, so far too many women pilots would risk serious injury if ejection became necessary. (We assume in this example that the women who are jet fighter pilots have the same characteristics as women in the general population, but these women probably have different characteristics. These conclusions are likely to be different for these very special women.)

Finding Values from Known Areas

The preceding examples in this section are of the same type: We are given specific limit values and we must find an area (or probability or percentage). In many practical and real cases, the area (or probability or percentage) is known and we must find the relevant value(s). When finding values from known areas, be careful to keep these cautions in mind:

1. *Don't confuse z scores and areas.* Remember, z scores are *distances* along the horizontal scale, but areas are *regions* under the normal curve. Table A-2 lists z scores in the left columns and across the top row, but areas are found in the body of the table.

2. *Choose the correct (right/left) side of the graph.* A value separating the top 10% from the others will be located on the right side of the graph, but a value separating the bottom 10% will be located on the left side of the graph.

3. A z score must be *negative* whenever it is located in the *left* half of the normal distribution.

4. Areas (or probabilities) are positive or zero values, but they are never negative.

Graphs are extremely helpful in visualizing, understanding, and successfully working with normal probability distributions, so they should be used whenever possible.

Procedure for Finding Values Using Table A-2 and Formula 5-2

1. Sketch a normal distribution curve, enter the given probability or percentage in the appropriate region of the graph, and identify the x value(s) being sought.

2. Use Table A-2 to find the z score corresponding to the cumulative left area bounded by x. Refer to the *body* of Table A-2 to find the closest area, then identify the corresponding z score.

3. Using Formula 5-2, enter the values for μ, σ, and the z score found in Step 2, then solve for x. Based on the format of Formula 5-2, we can solve for x as follows:

$$x = \mu + (z \cdot \sigma) \qquad \text{(another form of Formula 5-2)}$$

(If z is located to the left of the mean, be sure that it is a negative number.)

4. Refer to the sketch of the curve to verify that the solution makes sense in the context of the graph and in the context of the problem.

The following example uses this four-step procedure just outlined.

> **EXAMPLE Hip Breadths and Airplane Seats** In designing seats to be installed in commercial aircraft, engineers want to make the seats wide enough to fit 98% of all males. (Accommodating 100% of males would require very wide seats that would be much too expensive.) Men have hip breadths that are normally distributed with a mean of 14.4 in. and a standard deviation of 1.0 in. (based on anthropometric survey data from Gordon, Clauser, et al.). Find P_{98}. That is, find the hip breadth of men that separates the bottom (no pun intended) 98% from the top 2%.

SOLUTION

Step 1:   We begin with the graph shown in Figure 5-15. We have entered the mean of 14.4 in., shaded the area representing the bottom 98%, and identified the desired value as x.

FIGURE 5-15 **Distribution of Hip Breadths of Men**

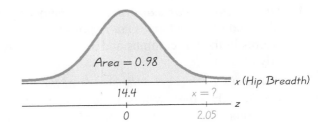

Step 2:   In Table A-2 we search for an area of 0.9800 in the *body* of the table. (The area of 0.98 shown in Figure 5-15 is a cumulative area from the left, and that is exactly the type of area listed in Table A-2.) The area closest to 0.98 is 0.9798, and it corresponds to a *z* score of 2.05.

Step 3:   With $z = 2.05$, $\mu = 14.4$, and $\sigma = 1.0$, we solve for *x* by using Formula 5-2 directly or by using the following version of Formula 5-2:

$$x = \mu + (z \cdot \sigma) = 14.4 + (2.05 \cdot 1.0) = 16.45$$

Step 4:   If we let $x = 16.45$ in Figure 5-15, we see that this solution is reasonable because the 98th percentile should be greater than the mean of 14.4.

INTERPRETATION   The hip breadth of 16.5 in. (rounded to one decimal place, as in μ and σ) separates the lowest 98% from the highest 2%. That is, seats designed for a hip breadth up to 16.5 in. will fit 98% of men. This type of analysis was used to design the seats currently used in commercial aircraft.

EXAMPLE   **Designing Car Dashboards**   When designing the placement of a CD player in a new-model car, engineers must consider the forward grip reach of the driver. If the CD player is placed beyond the forward grip reach, the driver must move his or her body in a way that could be distracting and dangerous. (We wouldn't want anyone injured while trying to hear the best of Barry Manilow.) Design engineers decided that the CD player should be placed so that it is within the forward grip reach of 95% of women. Women have forward grip reaches that are normally distributed with a mean of 27.0 in. and a standard deviation of 1.3 in. (based on anthropometric survey data from Gordon, Churchill, et al.). Find the forward grip reach of women that separates the longest 95% from the others.

SOLUTION

Step 1:   We begin with the graph shown in Figure 5-16. We have entered the mean of 27.0 in., and we have identified the area representing the longest 95% of forward grip reaches. Even though the problem refers to the top 95%, Table A-2 requires that we work with a cumulative *left* area, so we subtract 0.95 from 1 to get 0.05 which is shown as the shaded region.

Step 2:   In Table A-2 we search for an area of 0.05 in the *body* of the table. The areas closest to 0.05 are 0.0505 and 0.0495, but there is an asterisk indicating that an area of 0.05 corresponds to a *z* score of -1.645.

Step 3:   With $z = -1.645$, $\mu = 27.0$, and $\sigma = 1.3$, we solve for *x* by using Formula 5-2 directly or by using the following version of Formula 5-2:

$$x = \mu + (z \cdot \sigma) = 27.0 + (-1.645 \cdot 1.3) = 24.8615$$

continued

FIGURE 5-16 **Finding the Value Separating the Top 95%**

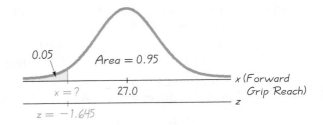

Step 4:   If we let $x = 24.8615$ in Figure 5-16, we see that this solution is reasonable because the forward grip reach separating the top 95% from the bottom 5% should be less than the mean of 27.0 in.

INTERPRETATION   The forward grip reach of 24.9 in. (rounded) separates the top 95% from the others. That is, 95% of women have forward grip reaches that are longer than 24.9 in. and 5% of women have forward grip reaches shorter than 24.9 in.

5-3 Exercises

IQ Scores. In Exercises 1–8, assume that adults have IQ scores that are normally distributed with a mean of 100 and a standard deviation of 15 (as on the Wechsler test). (Hint: Draw a graph in each case.)

1. Find the probability that a randomly selected adult has an IQ that is less than 115.

2. Find the probability that a randomly selected adult has an IQ greater than 131.5 (the requirement for membership in the Mensa organization).

3. Find the probability that a randomly selected adult has an IQ between 90 and 110 (referred to as the *normal* range).

4. Find the probability that a randomly selected adult has an IQ between 110 and 120 (referred to as *bright normal*).

5. Find P_{20}, which is the IQ score separating the bottom 20% from the top 80%.

6. Find P_{80}, which is the IQ score separating the bottom 80% from the top 20%.

7. Find the IQ score separating the top 15% from the others.

8. Find the IQ score separating the top 55% from the others.

9. Body Temperatures Based on the sample results in Data Set 2 of Appendix B, assume that human body temperatures are normally distributed with a mean of 98.20°F and a standard deviation of 0.62°F.
   a. Bellevue Hospital in New York City uses 100.6°F as the lowest temperature considered to be a fever. What percentage of normal and healthy persons would be considered to have a fever? Does this percentage suggest that a cutoff of 100.6°F is appropriate?
   b. Physicians want to select a minimum temperature for requiring further medical tests. What should that temperature be, if we want only 5.0% of healthy people to

exceed it? (Such a result is a *false positive*, meaning that the test result is positive, but the subject is not really sick.)

10. Lengths of Pregnancies The lengths of pregnancies are normally distributed with a mean of 268 days and a standard deviation of 15 days.
    a. One classic use of the normal distribution is inspired by a letter to "Dear Abby" in which a wife claimed to have given birth 308 days after a brief visit from her husband, who was serving in the Navy. Given this information, find the probability of a pregnancy lasting 308 days or longer. What does the result suggest?
    b. If we stipulate that a baby is *premature* if the length of pregnancy is in the lowest 4%, find the length that separates premature babies from those who are not premature. Premature babies often require special care, and this result could be helpful to hospital administrators in planning for that care.

11. Designing Helmets Engineers must consider the breadths of male heads when designing motorcycle helmets. Men have head breadths that are normally distributed with a mean of 6.0 in. and a standard deviation of 1.0 in. (based on anthropometric survey data from Gordon, Churchill, et al.). Due to financial constraints, the helmets will be designed to fit all men except those with head breadths that are in the smallest 2.5% or largest 2.5%. Find the minimum and maximum head breadths that will fit men.

12. CD Player Warranty Replacement times for CD players that fail are normally distributed with a mean of 7.1 years and a standard deviation of 1.4 years (based on data from "Getting Things Fixed," *Consumer Reports*).
    a. Find the probability that a randomly selected CD player will have a replacement time less than 8.0 years.
    b. If you want to provide a warranty so that only 2% of the CD players will fail before the warranty expires, what is the time length of the warranty?

Heights of Women. In Exercises 13–16, assume that heights of women are normally distributed with a mean given by $\mu = 63.6$ in. and a standard deviation given by $\sigma = 2.5$ in. (based on data from the National Health Survey). In each case, draw a graph.

13. Beanstalk Club Height Requirement The Beanstalk Club, a social organization for tall people, has a requirement that women must be at least 70 in. (or 5 ft 10 in.) tall. What percentage of women meet that requirement?

14. Height Requirement for Women Soldiers The U.S. Army requires women's heights to be between 58 in. and 80 in. Find the percentage of women meeting that height requirement. Are many women being denied the opportunity to join the Army because they are too short or too tall?

15. Height Requirement for Rockettes In order to have a precision dance team with a uniform appearance, height restrictions are placed on the famous Rockette dancers at New York's Radio City Music Hall. Because women have grown taller, a more recent change now requires that a Rockette dancer must have a height between 66.5 in. and 71.5 in. If a woman is randomly selected, what is the probability that she meets this new height requirement? What percentage of women meet this new height requirement? Does it appear that Rockettes are generally taller than typical women?

16. Height Requirement for Rockettes Exercise 15 identified specific height requirements for Rockettes. Suppose that those requirements must be changed because too few women now meet them. What are the new minimum and maximum allowable heights if the shortest 20% and the tallest 20% are excluded?

17. Birth Weights Birth weights in the United States are normally distributed with a mean of 3420 g and a standard deviation of 495 g. If a hospital plans to set up special observation conditions for the lightest 2% of babies, what weight is used for the cutoff separating the lightest 2% from the others?

18. Birth Weights Birth weights in Norway are normally distributed with a mean of 3570 g and a standard deviation of 500 g. Repeat Exercise 17 for babies born in Norway. Is the result very different from the result found in Exercise 17?

*19. Units of Measurement Weights of women are normally distributed with a mean of 143 lb and a standard deviation of 29 lb.
  a. If weights of individual women are expressed in units of pounds, what are the units used for the z scores that correspond to individual weights?
  b. If weights of all woman are converted to z scores, what are the mean, standard deviation, and distribution of these z scores?
  c. What are the distribution, mean, and standard deviation of women's weights after they have all been converted to kilograms? (1 lb = 0.4536 kg)

*20. Using Continuity Correction There are many situations in which a normal distribution can be used as a good approximation to a random variable that has only *discrete* values. In such cases, we can use this *continuity correction:* Represent each whole number by the interval extending from 0.5 below the number to 0.5 above it. Assume that IQ scores are all whole numbers having a distribution that is approximately normal with a mean of 100 and a standard deviation of 15.
  a. Without using any correction for continuity, find the probability of randomly selecting someone with an IQ score greater than 105.
  b. Using the correction for continuity, find the probability of randomly selecting someone with an IQ score greater than 105.
  c. Compare the results from parts (a) and (b).

5-4 Sampling Distributions and Estimators

We are beginning to embark on a journey that allows us to learn about populations by obtaining data from samples. Sections 5-5 and 5-6 provide important concepts revealing the behavior of sample means and sample proportions. Before considering those concepts, let's begin by focusing on the behavior of sample statistics in general. The main objective of this section is to learn what we mean by a *sampling distribution of a statistic*, and another important objective is to learn a basic principle about the sampling distribution of sample means and the sampling distribution of sample proportions.

Let's begin with sample means. Instead of getting too abstract, let's consider the *population* consisting of the values 1, 2, 5. Because of extremely rude personnel, the Newport Health Center was open for only three days. On the first day, 1 patient was treated, 2 were treated on the second day, and only 5 were treated on the third day. Because 1, 2, 5 constitute the entire population, it is easy to find the values of the population parameters: $\mu = 2.7$ and $\sigma = 1.7$. When finding the value of the population standard deviation σ, we use

$$\sigma = \sqrt{\frac{\Sigma(x - \mu)^2}{N}} \text{ instead of } s = \sqrt{\frac{\Sigma(x - \bar{x})^2}{n - 1}}$$

It is rare that we know all values in an entire population. The more common case is that there is some really large unknown population that we want to investigate. Because it is not practical to survey every member of the population, we obtain a sample and, based on characteristics of the sample, we make estimates about characteristics of the population. For example, the National Institutes of Health wants to know about various health characteristics, so sample subjects are examined and the results are used to estimate the characteristics of the whole population.

Because the values of 1, 2, 5 constitute an entire population, let's consider samples of size 2. With only three population values, there are only 9 different possible samples of size 2, assuming that we sample with replacement. That is, each selected value is replaced before another selection is made.

Why sample *with* replacement? For small samples of the type that we are discussing, sampling *without replacement* has the very practical advantage of avoiding wasteful duplication whenever the same item is selected more than once. However, throughout this section and in statistics in general, we are particularly interested in sampling *with replacement* for these reasons: (1) When selecting a relatively small sample from a large population, it makes no significant difference whether we sample with replacement or without replacement. (2) Sampling with replacement results in independent events that are unaffected by previous outcomes, and independent events are easier to analyze and they result in simpler formulas. For these reasons, we focus on the behavior of samples that are randomly selected *with replacement*.

When we sample 2 values with replacement from the population of 1, 2, 5, each of the 9 possible samples is equally likely, and they each have probability 1/9. Table 5-2 lists the 9 possible samples of size 2 along with statistics for each sample. That table contains much information, but let's first consider the column of sample means. Because we have all possible values of $\bar{x}$ listed, and because the probability of each is known to be 1/9, we have a probability distribution. (Remember, a probability distribution describes the probability for each value of a random variable, and the random variable in this case is the value of the sample mean $\bar{x}$). Because several important methods of statistics begin with a sample mean that is subsequently used for making inferences about the population mean, it is important to understand the behavior of such sample means. Other important methods of statistics begin with a sample proportion that is subsequently used for making inferences about the population proportion, so it is also important to understand the behavior of such sample proportions. In general, it is often important to understand the behavior of sample statistics. The "behavior" of a statistic can be known by understanding its distribution.

Growth Charts Updated

Pediatricians typically use standardized growth charts to compare their patient's weight and height with a sample of other children. Children are considered to be in the normal range if their weight and height fall between the 5th and 95th percentiles. If they fall outside of that range, they are often given tests to ensure that there are no serious medical problems. Pediatricians became increasingly aware of a major problem with the charts: Because they were based on children living between 1929 and 1975, the growth charts were found to be inaccurate. To rectify this problem, the charts were updated in 2000 to reflect the current measurements of millions of children. The weights and heights of children are good examples of populations that change over time. This is the reason for including changing characteristics of data over time as an important consideration for a population.

Definition

The **sampling distribution of the mean** is the probability distribution of sample means, with all samples having the same sample size *n*. (In general, the sampling distribution of any statistic is the probability distribution of that statistic.)

Table 5-2   Sampling Distributions of Different Statistics (for Samples of Size 2 Drawn with Replacement from the Population 1, 2, 5)

Sample	Mean $\bar{x}$	Median	Range	Variance s^2	Standard Deviation s	Proportion of Odd Numbers	Probability
1, 1	1.0	1.0	0	0.0	0.000	1	1/9
1, 2	1.5	1.5	1	0.5	0.707	0.5	1/9
1, 5	3.0	3.0	4	8.0	2.828	1	1/9
2, 1	1.5	1.5	1	0.5	0.707	0.5	1/9
2, 2	2.0	2.0	0	0.0	0.000	0	1/9
2, 5	3.5	3.5	3	4.5	2.121	0.5	1/9
5, 1	3.0	3.0	4	8.0	2.828	1	1/9
5, 2	3.5	3.5	3	4.5	2.121	0.5	1/9
5, 5	5.0	5.0	0	0.0	0.000	1	1/9
Mean of Statistic Values	2.7	2.7	1.8	2.9	1.3	0.667	
Population Parameter	2.7	2	4	2.9	1.7	0.667	
Does the sample statistic target the population parameter?	Yes	No	No	Yes	No	Yes	

Table 5-3

Sampling Distribution of the Mean

Mean $\bar{x}$	Probability
1.0	1/9
1.5	1/9
3.0	1/9
1.5	1/9
2.0	1/9
3.5	1/9
3.0	1/9
3.5	1/9
5.0	1/9

This table lists the means from the samples in Table 5-2, but it could be condensed by listing 1.0, 1.5, 2.0, 3.0, 3.5, and 5.0 along with their corresponding probabilities of 1/9, 2/9, 1/9, 2/9, 2/9, and 1/9.

EXAMPLE Sampling Distribution of Sample Means A *population* consists of the values 1, 2, 5, and Table 5-2 lists all of the different possible samples of size $n = 2$. The probability of each sample is listed in Table 5-2 as 1/9. For samples of size $n = 2$ randomly selected with replacement from the population 1, 2, 5, identify the specific sampling distribution of the mean. Also, find the mean of this sampling distribution. Do the sample means target the value of the population mean?

SOLUTION   The sampling distribution of the mean is the probability distribution describing the probability for each value of the sample mean, and all of those values are included in Table 5-2. The sampling distribution of the mean can therefore be described by Table 5-3. We could calculate the mean of the sampling distribution using two different approaches: (1) Use $\mu = \Sigma[x \cdot P(x)]$, which is Formula 4-2, or (2) find the mean of those 9 values (because all of the 9 sample means are equally likely). Because the population mean is also 2.7, it appears that the sample means "target" the value of the population mean, instead of systematically underestimating or overestimating the population mean.

From the preceding example we see that the mean of all of the different possible sample means is equal to the mean of the original population, which is $\mu = 2.7$. We can generalize this as a property of sample means: For a fixed sample size, the mean of all possible sample means is equal to the mean of the population. We will revisit this important property in the next section, but let's first make another obvious but important observation: *Sample means vary.* See Table 5-3 and observe how the sample means are different. The first sample mean is 1.0, the second sample mean is 1.5, and so on. This leads to the following definition.

Definition

The value of a statistic, such as the sample mean $\bar{x}$, depends on the particular values included in the sample, and it generally varies from sample to sample. This variability of a statistic is called **sampling variability.**

In Chapter 2 we introduced the important characteristics of a data set: center, variation, distribution, outliers, and pattern over time (summarized with the mnemonic of CVDOT). In examining the samples in Table 5-2, we have already identified a property describing the behavior of sample means: The mean of sample means is equal to the mean of the population. This property addresses the characteristic of center, and we will investigate other characteristics in the next section. We will see that as the sample size increases, the sampling distribution of sample means tends to become a *normal distribution* (not too surprising, given that the title of this chapter is "Normal Probability Distributions"). Consequently, the normal distribution assumes an importance that goes far beyond the applications illustrated in Section 5-3. The normal distribution will be used for many cases in which we want to use a sample mean $\bar{x}$ for the purpose of making some inference about a population mean μ.

Sampling Distribution of Proportions

When making inferences about a population proportion, it is also important to understand the behavior of sample proportions. We define the distribution of sample proportions as follows.

Definition

The **sampling distribution of the proportion** is the probability distribution of sample proportions, with all samples having the same sample size n.

A typical use of inferential statistics is to find some sample proportion and use it to make an inference about the population proportion. Pollsters from the Gallup organization asked 1012 randomly selected adults if they think the cloning of humans should or should not be allowed. Results showed that 901 (or 89%) of those surveyed thought that cloning of humans should not be allowed. That sample result leads to the inference that "89% of all adults think that cloning of humans should not be allowed." The sample proportion of 901/1012 was used to estimate a population proportion p, but we can learn much more by understanding the sampling distribution of such sample proportions.

EXAMPLE Sampling Distribution of Proportions A *population* consists of the values 1, 2, 5, and Table 5-2 lists all of the different possible samples of size $n = 2$ selected with replacement. For each sample, consider the proportion of numbers that are *odd*. Identify the sampling distribution for the proportion of odd numbers, then find its mean. Do sample proportions target the value of the population proportion?

SOLUTION See Table 5-2 where the nine sample proportions are listed as 1, 0.5, 1, 0.5, 0, 0.5, 1, 0.5, 1. Combining those sample proportions with their probabilities of 1/9 in each case, we get the sampling distribution of proportions summarized in Table 5-4. The mean of the *sample* proportions is 0.667. Because the population 1, 2, 5 contains two odd numbers, the *population* proportion of odd numbers is also 2/3 or 0.667. In general, sample proportions tend to target the value of the population proportion, instead of systematically tending to underestimate or overestimate that value.

Table 5-4	
Sampling Distribution of Proportions	
Proportion of Odd Numbers	Probability
0	1/9
0.5	4/9
1	4/9

The preceding example involves a fairly small population, so let's now consider the genders of the senators in the 107th Congress. Because there are only 100 members [13 females (F) and 87 males (M)], we can list the entire population:

```
M F M M F M M M M M M M F M M M M M M M
M M M M M M M M M M M M M F F M M M M M M
M M M F M M M M M F M M M M M M M M M M M
F M M M M M M M M M M M M M M M M M M F F F
M M M F M F M M M M M M M M M M M M M M M
```

The population proportion of female Senators is $p = 13/100 = 0.13$. Usually, we don't know all of the members of the population, so we must estimate it from a sample. For the purpose of studying the behavior of sample proportions, we list a few samples of size $n = 10$:

Sample 1: M F M M F M M M M M → sample proportion is 0.2
Sample 2: M F M M M M M M M M → sample proportion is 0.1
Sample 3: M M M M M M F M M M → sample proportion is 0.1
Sample 4: M M M M M M M M M M → sample proportion is 0
Sample 5: M M M M M M M M F M → sample proportion is 0.1

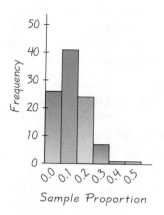

FIGURE 5-17 100 Sample Proportions with $n = 10$ in Each Sample

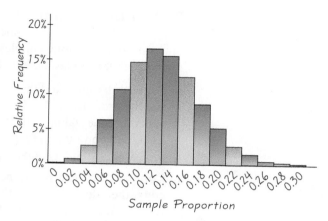

FIGURE 5-18 10,000 Sample Proportions with $n = 50$ in Each Sample

Because there is a very large number of such samples, we cannot list all of them. The authors randomly selected 95 additional samples. Combining these additional 95 samples with the five listed here, we get 100 samples summarized in Table 5-5.

We can see from Table 5-5 that the mean of the 100 sample proportions is 0.119, but if we were to include all other possible samples of size 10, the mean of the sample proportions would equal 0.13, which is the value of the population proportion. Figure 5-17 shows the distribution of the 100 sample proportions summarized in Table 5-5. The shape of that distribution is reasonably close to the shape that would have been obtained with all possible samples of size 10. We can see that the distribution depicted in Figure 5-17 is somewhat skewed to the right, but with a bit of a stretch, it might be approximated very roughly by a normal distribution. In Figure 5-18 we show the results obtained from 10,000 samples of size 50 randomly selected with replacement from the above list of 100 genders. Figure 5-18 very strongly suggests that the distribution is approaching the characteristic bell shape of a normal distribution. The results from Table 5-5 and Figure 5-18 therefore suggest the following.

Table 5-5	
Results from 100 Samples	
Proportion of Female Senators	Frequency
0.0	26
0.1	41
0.2	24
0.3	7
0.4	1
0.5	1
Mean:	0.119
Standard Deviation:	0.100

Properties of the Distribution of Sample Proportions

- **Sample proportions tend to target the value of the population proportion.**

- **Under certain conditions, the distribution of sample proportions approximates a *normal* distribution.**

Which Statistics Make Good Estimators of Parameters?

Chapter 6 will introduce formal methods for using sample statistics to make estimates of the values of population parameters. Some statistics work much better than others, and we can judge their value by examining their sampling distributions, as in the following example.

EXAMPLE **Sampling Distributions** A *population* consists of the values 1, 2, 5. If we randomly select samples of size 2 with replacement, there are nine different possible samples, and they are listed in Table 5-2. Because the nine different samples are equally likely, each sample has a probability of 1/9.

a. For each sample, find the mean, median, range, variance, standard deviation, and the proportion of sample values that are odd. (For each statistic, this will generate nine values which, when associated with nine probabilities of 1/9 each, will combine to form a *sampling distribution* for the statistic.)

b. For each statistic, find the mean of the results from part (a).

c. Compare the means from part (b) with the corresponding population parameters, then determine whether each statistic targets the value of the population parameter. For example, the sample means tend to center about the value of the population mean, which is 8/3 = 2.7, so the sample mean targets the value of the population mean.

SOLUTION

a. See Table 5-2. The individual statistics are listed for each sample.

b. The means of the sample statistics are shown near the bottom of Table 5-2. The mean of the sample means is 2.7, the mean of the sample medians is 3, and so on.

c. The bottom row of Table 5-2 is based on a comparison of the population parameter and results from the sample statistics. For example, the population mean of 1, 2, 5 is $\mu = 2.7$, and the sample means "target" that value of 2.7 in the sense that the mean of the sample means is also 2.7.

INTERPRETATION Based on the results in Table 5-2, we can see that when using a sample statistic to estimate a population parameter, some statistics are good in the sense that they target the population parameter and are therefore likely to yield good results. Such statistics are called *unbiased estimators*. Other statistics are not so good (because they are *biased estimators*). Here is a summary.

- **Statistics that target population parameters:** Mean, Variance, Proportion
- **Statistics that do not target population parameters:** Median, Range, Standard Deviation

Although the sample standard deviation does not target the population standard deviation, the bias is relatively small in large samples, so s is often used to estimate σ. Consequently, means, proportions, variances, and standard deviations will all be considered as major topics in following chapters, but the median and range will be rarely used.

The key point of this section is to introduce the concept of a sampling distribution of a statistic. Consider the goal of trying to find the mean body temperature of all adults. Because that population is so large, it is not practical to measure the temperature of every adult. Instead, we obtain a sample of body temperatures and use it to estimate the population mean. Data Set 2 in Appendix B includes a sample of 106 such body temperatures, and the mean for that sample is $\bar{x} = 98.20°F$. Conclusions that we make about the population mean temperature of all adults require that we understand the behavior of the sampling distribution of all such sample means. Even though it is not practical to obtain every possible sample and we are stuck with just one sample, we can form some very meaningful and important conclusions about the population of all body temperatures. A major goal of the following sections and chapters is to learn how we can effectively use a sample to form conclusions about a population. In Section 5-5 we consider more details about the sampling distribution of sample means, and in Section 5-6 we consider more details about the sampling distribution of sample proportions.

5-4 Exercises

1. Survey of Voters Based on a random sample of $n = 400$ voters, the NBC news division predicts that the Democratic candidate for the presidency will get 49% of the vote, but she actually gets 51%. Should we conclude that the survey was done incorrectly? Why or why not?

2. Sampling Distribution of Cholesterol Levels Data Set 1 in Appendix B includes a sample of measured cholesterol levels (in mg/dL) for 40 randomly selected women. The mean of the 40 cholesterol values is 240.9 mg/dL. The value of 240.9 mg/dL is one value that is part of a sampling distribution. Describe that sampling distribution.

3. Sampling Distribution of Body Temperatures Data Set 2 in Appendix B includes a sample of 106 body temperatures of adults. If we construct a histogram to depict the shape of the distribution of that sample, does that histogram show the shape of a *sampling distribution* of sample means? Why or why not?

4. Sampling Distribution of Survey Results The Gallup organization conducted a poll of 1038 randomly selected adults and found that 52% said that secondhand smoke is "very harmful." (Other options were "somewhat harmful," "not too harmful," "not at all harmful," and some had no opinion.)
   a. Is the 52% (or 0.52) result a statistic or a parameter? Explain.
   b. What is the sampling distribution suggested by the given data?
   c. Would you feel more confident in the results if the sample size were 2000 instead of 1038? Why or why not?

5. Phone Center The Nome Cryogenics Company was in business for only three days. Here are the numbers of phone calls received on each of those days: 10, 6, 5. Assume that samples of size 2 are randomly selected *with replacement* from this population of three values.
   a. List the 9 different possible samples and find the mean of each of them.
   b. Identify the probability of each sample and describe the sampling distribution of sample means. (*Hint*: See Table 5-3.)
   c. Find the mean of the sampling distribution.
   d. Is the mean of the sampling distribution [from part (c)] equal to the mean of the population of the three listed values? Are those means *always* equal?

6. Telemarketing Here are the numbers of sales per day that were made by Kim Ryan, a courteous telemarketer for the Medassist Pharmaceutical Company, when she worked four days before being fired: 1, 11, 9, 3. Assume that samples of size 2 are randomly selected *with replacement* from this population of four values.
   a. List the 16 different possible samples and find the mean of each of them.
   b. Identify the probability of each sample, then describe the sampling distribution of sample means (*Hint*: See Table 5-3.)
   c. Find the mean of the sampling distribution.
   d. Is the mean of the sampling distribution [from part (c)] equal to the mean of the population of the four listed values? Are those means *always* equal?

7. Heights of L.A. Lakers Here are the heights (in inches) of five starting basketball players for the L.A. Lakers: 85, 79, 82, 73, 78. Assume that samples of size 2 are randomly selected *with replacement* from the above population of five heights.
   a. After identifying the 25 different possible samples, find the mean of each of them.
   b. Describe the sampling distribution of means. (*Hint*: See Table 5-3.)
   c. Find the mean of the sampling distribution.
   d. Is the mean of the sampling distribution [from part (c)] equal to the mean of the population of the five heights listed above? Are those means *always* equal?

8. Genetics A genetics experiment involves a population of fruit flies consisting of 1 male named Mike and 3 females named Anna, Barbara, and Chris. Assume that two fruit flies are randomly selected *with replacement*.
   a. After identifying the 16 different possible samples, find the proportion of females in each of them.
   b. Describe the sampling distribution of proportions of females. (*Hint*: See Table 5-5.)
   c. Find the mean of the sampling distribution.
   d. Is the mean of the sampling distribution [from part (c)] equal to the population proportion of females? Does the mean of the sampling distribution of proportions *always* equal the population proportion?

9. Quality Control After constructing a new heart pacemaker, 5 prototype models are produced and it is found that 2 are defective (D) and 3 are acceptable (A). Assume that two pacemakers are randomly selected *with replacement* from this population.
   a. After identifying the 25 different possible samples, find the proportion of defects in each of them.
   b. Describe the sampling distribution of proportions of defects. (*Hint*: See Table 5-5.)
   c. Find the mean of the sampling distribution.
   d. Is the mean of the sampling distribution [from part (c)] equal to the population proportion of defects? Does the mean of the sampling distribution of proportions *always* equal the population proportion?

10. Women Senators Let a population consist of the 10 Democrats and 3 Republicans, all of whom are women in the 107th Congress of the United States.
   a. Develop a procedure for randomly selecting (with replacement) a sample of size 5 from the population of 10 Democrats and 3 Republicans, then select such a sample and list the results.
   b. Find the proportion of Democrats in the sample from part (a).
   c. Is the proportion from part (b) a statistic or a parameter?
   d. Does the sample proportion from part (b) equal the population proportion of Democrats? Can any random sample of size 5 result in a sample proportion that is equal to the population proportion?

e. Assume that all different possible samples of size 5 are listed, and the sample proportion is found for each of them. What can be concluded about the value of the mean of those sample proportions?

*11. Mean Absolute Deviation The population of 1, 2, 5 was used to develop Table 5-2. Identify the sampling distribution of the mean absolute deviation (defined in Section 2-5), then determine whether the mean absolute deviation is a good statistic for estimating a population parameter.

*12. Median as an Estimator In Table 5-2, the sampling distribution of the median has a mean of 2.7. Because the population mean is also 2.7, it might appear that the median is a good statistic for estimating the value of the population mean. Using the population values 1, 2, 5, find the 27 samples of size $n = 3$ that can be selected with replacement, then find the median and mean for each of the 27 samples. After obtaining those results, find the mean of the sampling distribution of the median, and find the mean of the sampling distribution of the mean. Compare the results with the population mean of 2.7. What do you conclude?

5-5 The Central Limit Theorem

This section is *extremely* important because it presents the central limit theorem, which forms the foundation for estimating population parameters and hypothesis testing—topics discussed at length in the following chapters. As you study this section, try to avoid confusion caused by the fact that the central limit theorem involves two different distributions: the distribution of the original population and the distribution of the sample means.

These key terms and concepts were presented in earlier sections:

- A *random variable* is a variable that has a single numerical value, determined by chance, for each outcome of a procedure. (Section 4-2)

- A *probability distribution* is a graph, table, or formula that gives the probability for each value of a random variable. (Section 4-2)

- The *sampling distribution of the mean* is the probability distribution of sample means, with all samples having the same sample size n. (Section 5-4)

For specific examples of these abstract concepts, see the following example.

EXAMPLE **Random Digits** Consider the population of digits 0, 1, 2, 3, 4, 5, 6, 7, 8, 9, which are randomly selected with replacement.

a. *Random variable:* If we conduct trials that consist of randomly selecting a single digit, and if we represent the value of the selected digit by x, then x is a random variable (because its value depends on chance).

b. *Probability distribution:* Assuming that the digits are randomly selected, the probability of each digit is 1/10, which can be expressed as the formula $P(x) = 1/10$. This is a probability distribution (because it describes the probability for each value of the random variable x).

c. *Sampling distribution*: Now suppose that we randomly select all of the different possible samples, each of size $n = 4$. (Remember, we are sampling with

continued

The Fuzzy Central Limit Theorem

In *The Cartoon Guide to Statistics* by Gonick and Smith, the authors describe the Fuzzy Central Limit Theorem as follows: "Data that are influenced by many small and unrelated random effects are approximately normally distributed. This explains why the normal is everywhere: stock market fluctuations, student weights, yearly temperature averages, SAT scores: All are the result of many different effects." People's heights for example, are the results of hereditary factors, environmental factors, nutrition, health care, geographic region, and other influences which, when combined, produce normally distributed values.

replacement, so any particular sample might have the same digit occurring more than once.) In each sample, we calculate the sample mean $\bar{x}$ (which is itself a random variable because its value depends on chance). The probability distribution of the sample means $\bar{x}$ is a sampling distribution.

Part (c) of the above example illustrates a specific sampling distribution of sample means. In Section 5-4 we saw that the mean of sample means is equal to the mean of the population, and as the sample size increases, the corresponding sample means tend to vary less. The *central limit theorem* tells us that if the sample size is large enough, the distribution of sample means can be approximated by a *normal distribution*, even if the original population is not normally distributed. Although we discuss a "theorem," we do not include rigorous proofs. Instead, we focus on the *concepts* and how to apply them. Here are the key points that form such an important foundation for the following chapters:

The Central Limit Theorem and the Sampling Distribution of $\bar{x}$

Given:

1. The random variable x has a distribution (which may or may not be normal) with mean μ and standard deviation σ.
2. Simple random samples all of the same size n are selected from the population. (The samples are selected so that all possible samples of size n have the same chance of being selected.)

Conclusions:

1. The distribution of sample means $\bar{x}$ will, as the sample size increases, approach a *normal* distribution.
2. The mean of all sample means is the population mean μ. (That is, the normal distribution from Conclusion 1 has mean μ.)
3. The standard deviation of all sample means is $\sigma/\sqrt{n}$. (That is, the normal distribution from Conclusion 1 has standard deviation $\sigma/\sqrt{n}$.)

Practical Rules Commonly Used:

1. If the original population is not itself normally distributed, here is a common guideline: For samples of size n greater than 30, the distribution of the sample means can be approximated reasonably well by a normal distribution. (There are exceptions, such as populations with very non-normal distributions requiring sample sizes much larger than 30, but such exceptions are relatively rare.) The approximation gets better as the sample size n becomes larger.
2. If the original population is itself normally distributed, then the sample means will be normally distributed for *any* sample size n (not just the values of n larger than 30).

The central limit theorem involves two different distributions: the distribution of the original population and the distribution of the sample means. As in previous chapters, we use the symbols μ and σ to denote the mean and standard deviation of the original population, but we now need new notation for the mean and standard deviation of the distribution of sample means.

Notation for the Sampling Distribution of $\bar{x}$

If all possible random samples of size n are selected from a population with mean μ and standard deviation σ, the mean of the sample means is denoted by $\mu_{\bar{x}}$, so

$$\mu_{\bar{x}} = \mu$$

Also, the standard deviation of the sample means is denoted by $\sigma_{\bar{x}}$, so

$$\sigma_{\bar{x}} = \frac{\sigma}{\sqrt{n}}$$

$\sigma_{\bar{x}}$ is often called the **standard error of the mean**.

EXAMPLE Random Digits Again consider the population of digits 0, 1, 2, 3, 4, 5, 6, 7, 8, 9, which are randomly selected with replacement. Assume that we randomly select samples of size $n = 4$. In the original population of digits, all of the values are equally likely. Based on the "Practical Rules Commonly Used" (listed in the central limit theorem box), we *cannot* conclude that the sample means are normally distributed, because the original population does not have a normal distribution and the sample size of 4 is not larger than 30. However, we will explore the sampling distribution to see what we can learn.

Table 5-6 was constructed by recording the last four digits of social security numbers from each of 50 different students. The last four digits of social security numbers are random, unlike the beginning digits, which are used to code particular information. If we combine the four digits from each student into one big collection of 200 numbers, we get a mean of $\bar{x} = 4.5$, a standard deviation of $s = 2.8$, and a distribution with the graph shown in Figure 5-19. Now see what happens when we find the 50 sample means, as shown in Table 5-6. (For example, the first student has digits of 1, 8, 6, and 4, and the mean of these four digits is 4.75.) Even though the original collection of data does *not* have a normal distribution, the sample means have a distribution that is approximately *normal*. This can be a confusing concept, so you should stop right here and study this paragraph until its major point becomes clear: The original set of 200 individual numbers does *not* have a normal distribution (because the digits 0–9 occur with approximately equal frequencies), but the 50 sample means *do* have a distribution that is approximately normal. (One of the "Practical Rules Commonly Used" states that samples with $n > 30$ can be approximated with a normal distribution, but smaller samples, such as $n = 4$ in this example, can sometimes have a distribution that is approximately normal.) It's a truly fascinating and intriguing phenomenon in statistics that by sampling from any distribution, we can create a distribution of sample means that is normal or at least approximately normal.

Table 5-6

SSN digits				$\bar{x}$
1	8	6	4	4.75
5	3	3	6	4.25
9	8	8	8	8.25
5	1	2	5	3.25
9	3	3	5	5.00
4	2	6	2	3.50
7	7	1	6	5.25
9	1	5	4	4.75
5	3	3	9	5.00
7	8	4	1	5.00
0	5	6	1	3.00
9	8	2	2	5.25
6	1	5	7	4.75
8	1	3	0	3.00
5	9	6	9	7.25
6	2	3	4	3.75
7	4	0	7	4.50
5	7	5	6	5.75
4	1	5	7	4.25
1	2	0	6	2.25
4	0	2	8	3.50
3	1	2	5	2.75
0	3	4	0	1.75
1	5	1	0	1.75
9	7	4	0	5.00
7	3	1	1	3.00
9	1	1	3	3.50
8	6	5	9	7.00
5	6	4	1	4.00
9	3	9	5	6.50
6	0	7	3	4.00
8	2	9	6	6.25
0	2	8	6	4.00
2	0	9	7	4.50
5	8	9	0	5.50
6	5	4	9	6.00
4	8	7	6	6.25
7	1	2	0	2.50
2	9	5	0	4.00
8	3	2	2	3.75
2	7	1	6	4.00
6	7	7	1	5.25
2	3	3	9	4.25
2	4	7	5	4.50
5	4	3	7	4.75
0	4	3	8	3.75
2	5	8	6	5.25
7	1	3	4	3.75
8	3	7	0	4.50
5	6	6	7	6.00

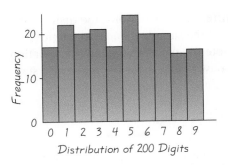

FIGURE 5-19 **Distribution of 200 Digits from Social Security Numbers**

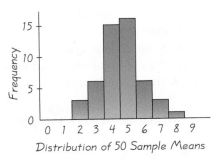

FIGURE 5-20 **Distribution of 50 Sample Means**

Figure 5-20 shows that the distribution of the sample means from the preceding example is approximately normal, even though the original population does not have a normal distribution and the size of $n = 4$ for the individual samples does not exceed 30. If you closely examine Figure 5-20, you can see that it is not an exact normal distribution, but it would become closer to an exact normal distribution as the sample size increases far beyond 4.

As the sample size increases, the sampling distribution of sample means approaches a *normal* distribution.

Applying the Central Limit Theorem

Many important and practical problems can be solved with the central limit theorem. When working on such problems, remember that if the sample size is greater than 30, or if the original population is normally distributed, treat the distribution of sample means as if it were a normal distribution with mean μ and standard deviation $\sigma/\sqrt{n}$.

In the following example, part (a) involves an *individual* value, but part (b) involves the mean for a *sample* of 12 men, so we must use the central limit theorem in working with the random variable $\bar{x}$. Study this example carefully to understand the significant difference between the procedures used in parts (a) and (b). See how this example illustrates the following working procedure:

- **When working with an *individual* value from a normally distributed population, use the methods of Section 5-3. Use $z = \dfrac{x - \mu}{\sigma}$.**

- **When working with a mean for some *sample* (or group), be sure to use the value of $\sigma/\sqrt{n}$ for the standard deviation of the sample means. Use**

$$z = \frac{\bar{x} - \mu}{\sigma/\sqrt{n}}$$

EXAMPLE **Ski Gondola Safety** In the Chapter Problem we noted that a ski gondola in Vail, Colorado, carries skiers to the top of a mountain. It bears a plaque stating that the maximum capacity is 12 people or 2004 pounds. That capacity will be exceeded if 12 people have weights with a mean greater than $2004/12 = 167$ pounds. Because men tend to weigh more than women, a "worst case" scenario involves 12 passengers who are all men. Men have weights that are normally distributed with a mean of 172 lb and a standard deviation of 29 lb (based on data from the National Health Survey).

a. Find the probability that if an individual man is randomly selected, his weight will be greater than 167 lb.

b. Find the probability that 12 randomly selected men will have a mean weight that is greater than 167 lb (so that their total weight is greater than the gondola maximum capacity of 2004 lb).

SOLUTION

a. *Approach: Use the methods presented in Section 5-3 (because we are dealing with an individual value from a normally distributed population).* We seek the area of the green-shaded region in Figure 5-21(a). Before using Table A-2, we convert the weight of 167 to the corresponding z score:

$$z = \frac{x - \mu}{\sigma} = \frac{167 - 172}{29} = -0.17$$

We now refer to Table A-2 using $z = -0.17$ and find that the cumulative area to the left of 167 lb is 0.4325. The green-shaded region is therefore $1 - 0.4325 = 0.5675$. The probability of a randomly selected man weighing more than 167 lb is 0.5675.

b. *Approach: Use the central limit theorem (because we are dealing with the mean for a sample of 12 men, not an individual man).* Even though the sample size is not greater than 30, we use a normal distribution for this reason: The original population of men has a normal distribution, so samples of any size will yield means that are normally distributed. Because we are now dealing with a distribution of sample means, we must use the parameters $\mu_{\bar{x}}$ and $\sigma_{\bar{x}}$, which are evaluated as follows:

$$\mu_{\bar{x}} = \mu = 172$$

$$\sigma_{\bar{x}} = \frac{\sigma}{\sqrt{n}} = \frac{29}{\sqrt{12}} = 8.37158$$

continued

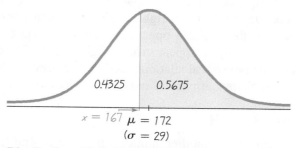

FIGURE 5-21(a) **Distribution of Individual Men's Weights**

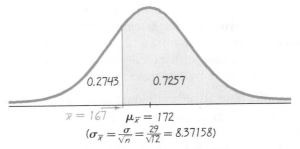

FIGURE 5-21(b) Means of Samples of 12 Men's Weights

Here is a really important point: We must use the computed standard deviation of 8.37158, not the original standard deviation of 29 (because we are working with the distribution of sample *means* for which the standard deviation is 8.37158, not the distribution of *individual* weights for which the standard deviation is 29). We want to find the green-shaded area shown in Figure 5-21(b). Using Table A-2, we find the relevant z score, which is calculated as follows:

$$z = \frac{\bar{x} - \mu_{\bar{x}}}{\sigma_{\bar{x}}} = \frac{167 - 172}{\dfrac{29}{\sqrt{12}}} = \frac{-5}{8.37158} = -0.60$$

Referring to Table A-2, we find that $z = -0.60$ corresponds to a cumulative left area of 0.2743, so the green-shaded region is $1 - 0.2743 = 0.7257$. The probability that the 12 men have a mean weight greater than 167 lb is 0.7257.

INTERPRETATION There is a 0.5675 probability that an individual man will weigh more than 167 lb, and there is a 0.7257 probability that 12 men will have a mean weight of more than 167 lb. Given that the gondola maximum capacity is 2004 lb, it is likely (with probability 0.7257) to be overloaded if it is filled with 12 randomly selected men. However, passenger safety isn't quite so bad because of factors such as these: (1) Male skiers probably have a mean weight less than the mean of 172 lb for the general population of men; (2) women skiers are also likely to be passengers, and they tend to weigh less than men; (3) even though the maximum capacity is listed as 2004 lb, the gondola is designed to operate safely for weights well above that conservative load of 2004 lb. However, the gondola operators would be wise to avoid a load of 12 men, especially if they all appear to have high weights. The calculations used here are exactly the type of calculations used by engineers when they design ski lifts, elevators, escalators, airplanes, and other devices that carry people.

Interpreting Results

The next example illustrates another application of the central limit theorem, but carefully examine the conclusion that is reached. This example shows the type of thinking that is the basis for the important procedure of hypothesis testing (discussed in Chapter 7). This example illustrates the rare event rule for inferential statistics, first presented in Section 3-1.

Rare Event Rule

If, under a given assumption, the probability of a particular observed event is exceptionally small, we conclude that the assumption is probably not correct.

EXAMPLE Body Temperatures Assume that the population of human body temperatures has a mean of 98.6°F, as is commonly believed. Also assume that the population standard deviation is 0.62°F (based on data from University of Maryland researchers). If a sample of size $n = 106$ is randomly selected, find the probability of getting a mean of 98.2°F or lower. (The value of 98.2°F was actually obtained; see the midnight temperatures for Day 2 in Data Set 2 of Appendix B.)

SOLUTION We weren't given the distribution of the population, but because the sample size $n = 106$ exceeds 30, we use the central limit theorem and conclude that the distribution of sample means is a normal distribution with these parameters:

$$\mu_{\bar{x}} = \mu = 98.6$$

$$\sigma_{\bar{x}} = \frac{\sigma}{\sqrt{n}} = \frac{0.62}{\sqrt{106}} = 0.0602197$$

Figure 5-22 shows the shaded area (see the tiny left tail of the graph) corresponding to the probability we seek. Having already found the parameters that apply to the distribution shown in Figure 5-22, we can now find the shaded area by using the same procedures developed in the preceding section. Using Table A-2, we first find the z score:

$$z = \frac{\bar{x} - \mu_{\bar{x}}}{\sigma_{\bar{x}}} = \frac{98.20 - 98.6}{0.0602197} = -6.64$$

continued

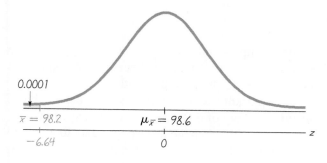

FIGURE 5-22 Distribution of Mean Body Temperatures for Samples of Size $n = 106$

Referring to Table A-2, we find that $z = -6.64$ is off the chart, but for values of z below -3.49, we use an area of 0.0001 for the cumulative left area up to $z = -3.49$. We therefore conclude that the shaded region in Figure 5-22 is 0.0001. (More precise tables or TI-83/84 Plus calculator results show that the area of the shaded region is closer to 0.00000000002, but even those results are only approximations. We can safely report that the probability is quite small, such as less than 0.001.)

INTERPRETATION   The result shows that if the mean of our body temperatures is really 98.6°F, then there is an extremely small probability of getting a sample mean of 98.2°F or lower when 106 subjects are randomly selected. University of Maryland researchers did obtain such a sample mean, and there are two possible explanations: Either the population mean really is 98.6°F and their sample represents a chance event that is extremely rare, or the population mean is actually lower than 98.6°F and so their sample is typical. Because the probability is so low, it seems more reasonable to conclude that the population mean is lower than 98.6°F. This is the type of reasoning used in *hypothesis testing,* to be introduced in Chapter 7. For now, we should focus on the use of the central limit theorem for finding the probability of 0.0001, but we should also observe that this theorem will be used later in developing some very important concepts in statistics.

Correction for a Finite Population

In applying the central limit theorem, our use of $\sigma_{\bar{x}} = \sigma/\sqrt{n}$ assumes that the population has infinitely many members. When we sample with replacement (that is, put back each selected item before making the next selection), the population is effectively infinite. Yet many realistic applications involve sampling without replacement, so successive samples depend on previous outcomes. In manufacturing, quality-control inspectors typically sample items from a finite production run without replacing them. For such a finite population, we may need to adjust $\sigma_{\bar{x}}$. Here is a common rule of thumb:

When sampling without replacement and the sample size n is greater than 5% of the finite population size N (that is, $n > 0.05N$), adjust the standard deviation of sample means $\sigma_{\bar{x}}$ by multiplying it by the *finite population correction factor:*

$$\sqrt{\frac{N - n}{N - 1}}$$

Except for Exercises 15 and 16, the examples and exercises in this section assume that the finite population correction factor does *not* apply, because the population is infinite or the sample size doesn't exceed 5% of the population size.

The central limit theorem is so important because it allows us to use the basic normal distribution methods in a wide variety of different circumstances. In Chapter 6, for example, we will apply the theorem when we use sample data to estimate means of populations. In Chapter 7 we will apply it when we use sample data to test claims made about population means. Such applications of estimating population parameters and testing claims are extremely important uses of statistics, and the central limit theorem makes them possible.

5-5 Exercises

Using the Central Limit Theorem. In Exercises 1–6, assume that men's weights are normally distributed with a mean given by $\mu = 172$ lb and a standard deviation given by $\sigma = 29$ lb (based on data from the National Health Survey).

1. **a.** If 1 man is randomly selected, find the probability that his weight is less than 167 lb.
   b. If 36 men are randomly selected, find the probability that they have a mean weight less than 167 lb.

2. **a.** If 1 man is randomly selected, find the probability that his weight is greater than 180 lb.
   b. If 100 men are randomly selected, find the probability that they have a mean weight greater than 180 lb.

3. **a.** If 1 man is randomly selected, find the probability that his weight is between 170 lb and 175 lb.
   b. If 64 men are randomly selected, find the probability that they have a mean weight between 170 lb and 175 lb.

4. **a.** If 1 man is randomly selected, find the probability that his weight is between 100 lb and 165 lb.
   b. If 81 men are randomly selected, find the probability that they have a mean weight between 100 lb and 165 lb.

5. **a.** If 25 men are randomly selected, find the probability that they have a mean weight greater than 160 lb.
   b. Why can the central limit theorem be used in part (a), even though the sample size does not exceed 30?

6. **a.** If 4 men are randomly selected, find the probability that they have a mean weight between 160 lb and 180 lb.
   b. Why can the central limit theorem be used in part (a), even though the sample size does not exceed 30?

7. Redesign of Ejection Seats In the Chapter Problem, it was noted that engineers were redesigning fighter jet ejection seats to better accommodate women. Before women became fighter jet pilots, the ACES-II ejection seats were designed for men weighing between 140 lb and 211 lb. The population of women has normally distributed weights with a mean of 143 lb and a standard deviation of 29 lb (based on data from the National Health Survey).
   a. If 1 woman is randomly selected, find the probability that her weight is between 140 lb and 211 lb.
   b. If 36 different women are randomly selected, find the probability that their mean weight is between 140 lb and 211 lb.
   c. When redesigning the fighter jet ejection seats to better accommodate women, which probability is more relevant: the result from part (a) or the result from part (b)? Why?

8. Designing Motorcycle Helmets Engineers must consider the breadths of male heads when designing motorcycle helmets. Men have head breadths that are normally distributed with a mean of 6.0 in. and a standard deviation of 1.0 in. (based on anthropometric survey data from Gordon, Churchill, et al.).

   a. If one male is randomly selected, find the probability that his head breadth is less than 6.2 in.

   b. The Safeguard Helmet company plans an initial production run of 100 helmets. Find the probability that 100 randomly selected men have a mean head breadth less than 6.2 in.

   c. The production manager sees the result from part (b) and reasons that all helmets should be made for men with head breadths less than 6.2 in., because they would fit all but a few men. What is wrong with that reasoning?

9. Designing a Roller Coaster The Rock 'n' Roller Coaster at Disney–MGM Studios in Orlando has two seats in each row. When designing that roller coaster, the total width of the two seats in each row had to be determined. In the "worst case" scenario, both seats are occupied by men. Men have hip breadths that are normally distributed with a mean of 14.4 in. and a standard deviation of 1.0 in. (based on anthropometric survey data from Gordon, Churchill, et al.). Assume that two male riders are randomly selected.

   a. Find the probability that their mean hip width is greater than 16.0 in.

   b. If each row of two seats is designed to fit two men only if they have a mean hip breadth of 16.0 in. or less, would too many riders be unable to fit? Does this design appear to be acceptable?

10. Uniform Random-Number Generator The random-number generator on a TI-83/84 Plus calculator and many other calculators and computers yields numbers from a uniform distribution of values between 0 and 1, with a mean of 0.500 and a standard deviation of 0.289. If 100 random numbers are generated, find the probability that their mean is greater than 0.57. Would it be *unusual* to generate 100 such numbers and get a mean greater than 0.57? Why or why not?

11. Blood Pressure For women aged 18–24, systolic blood pressures (in mm Hg) are normally distributed with a mean of 114.8 and a standard deviation of 13.1 (based on data from the National Health Survey). Hypertension is commonly defined as a systolic blood pressure above 140.

   a. If a woman between the ages of 18 and 24 is randomly selected, find the probability that her systolic blood pressure is greater than 140.

   b. If 4 women in that age bracket are randomly selected, find the probability that their mean systolic blood pressure is greater than 140.

   c. Given that part (b) involves a sample size that is not larger than 30, why can the central limit theorem be used?

   d. If a physician is given a report stating that 4 women have a mean systolic blood pressure below 140, can she conclude that none of the women has hypertension (with a blood pressure greater than 140)?

12. Reduced Nicotine in Cigarettes The amounts of nicotine in Dytusoon cigarettes have a mean of 0.941 g and a standard deviation of 0.313 g. The Huntington Tobacco Company, which produces Dytusoon cigarettes, claims that it has now reduced the amount of nicotine. The supporting evidence consists of a sample of 40 cigarettes with a mean nicotine amount of 0.882 g.

   a. Assuming that the given mean and standard deviation have not changed, find the probability of randomly selecting 40 cigarettes with a mean of 0.882 g or less.

b. Based on the result from part (a), is it valid to claim that the amount of nicotine is lower? Why or why not?

13. Elevator Design Women's weights are normally distributed with a mean of 143 lb and a standard deviation of 29 lb, and men's weights are normally distributed with a mean of 172 lb and a standard deviation of 29 lb (based on data from the National Health Survey). You need to design an elevator for the Westport Shopping Center, and it must safely carry 16 people. Assuming a worst case scenario of 16 male passengers, find the maximum total allowable weight if we want a 0.975 probability that this maximum will not be exceeded when 16 males are randomly selected.

14. Seating Design You need to build a bench that will seat 18 male college football players, and you must first determine the length of the bench. Men have hip breadths that are normally distributed with a mean of 14.4 in. and a standard deviation of 1.0 in.
    a. What is the minimum length of the bench if you want a 0.975 probability that it will fit the combined hip breadths of 18 randomly selected men?
    b. What would be wrong with actually using the result from part (a) as the bench length?

*15. Correcting for a Finite Population The Boston Women's Club needs an elevator limited to 8 passengers. The club has 120 women members with weights that approximate a normal distribution with a mean of 143 lb and a standard deviation of 29 lb. (*Hint:* See the discussion of the finite population correction factor.)
    a. If 8 different members are randomly selected, find the probability that their total weight will not exceed the maximum capacity of 1300 lb.
    b. If we want a 0.99 probability that the elevator will not be overloaded whenever 8 members are randomly selected as passengers, what should be the maximum allowable weight?

*16. Population Parameters A *population* consists of these values: 2, 3, 6, 8, 11, 18.
    a. Find μ and σ.
    b. List all samples of size $n = 2$ that can be obtained without replacement.
    c. Find the population mean of all values of $\bar{x}$ by finding the mean of each sample from part (b).
    d. Find the mean $\mu_{\bar{x}}$ and standard deviation $\sigma_{\bar{x}}$ for the population of sample means found in part (c).
    e. Verify that

$$\mu_{\bar{x}} = \mu \quad \text{and} \quad \sigma_{\bar{x}} = \frac{\sigma}{\sqrt{n}} \sqrt{\frac{N - n}{N - 1}}$$

5-6 Normal as Approximation to Binomial

Instead of "Normal as Approximation to Binomial," the proper title for this section should be "Using the Normal Distribution as an Approximation to the Binomial Distribution," but the shorter title has more pizazz. The longer title does a better job of conveying the purpose of this section. Let's begin by reviewing the conditions required for a *binomial probability distribution,* which was introduced in Section 4-3:

1. The procedure must have a *fixed number of trials.*
2. The trials must be *independent.*

3. Each trial must have all outcomes classified into *two categories*.

4. The probabilities must remain *constant* for each trial.

In Section 4-3 we presented three methods for finding binomial probabilities: (1) using the binomial probability formula, (2) using Table A-1, and (3) using software (such as STATDISK, Minitab, or Excel) or a TI-83/84 Plus calculator. In many cases, however, none of those methods is practical, because the calculations require too much time and effort. We now present a new method, which uses a normal distribution as an approximation of a binomial distribution. The following box summarizes the key point of this section.

Normal Distribution as Approximation to Binomial Distribution

When working with a binomial distribution, if $np \geq 5$ and $nq \geq 5$, then the binomial random variable has a probability distribution that can be approximated by a normal distribution with the mean and standard deviation given as

$$\mu = np$$
$$\sigma = \sqrt{npq}$$

To better understand how the normal distribution can be used to approximate a binomial distribution, refer to Figure 5-18 in Section 5-4. That figure is a relative frequency histogram for values of 10,000 sample *proportions*, where each of the 10,000 samples consists of 50 genders randomly selected with replacement from a population in which the proportion of females is 0.13. Those sample proportions can be considered to be binomial probabilities, so Figure 5-18 shows that under suitable conditions, binomial probabilities have a sampling distribution that is approximately normal. The formal justification that allows us to use the normal distribution as an approximation to the binomial distribution results from more advanced mathematics, but Figure 5-18 is a convincing visual argument supporting that approximation.

When solving binomial probability problems, first try to get more exact results by using computer software or a calculator, or Table A-1, or the binomial probability formula. If the binomial probability cannot be found using those more exact procedures, try the technique of using the normal distribution as an approximation to the binomial distribution. This approach involves the following procedure, which is also shown as a flowchart in Figure 5-23.

Procedure for Using a Normal Distribution to Approximate a Binomial Distribution

1. Establish that the normal distribution is a suitable approximation to the binomial distribution by verifying that $np \geq 5$ and $nq \geq 5$. (If these conditions are not both satisfied, then you must use software, or a calculator, or Table A-1, or calculations with the binomial probability formula.)

2. Find the values of the parameters μ and σ by calculating $\mu = np$ and $\sigma = \sqrt{npq}$.

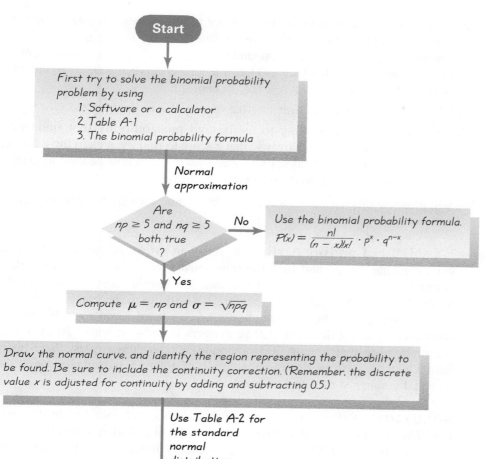

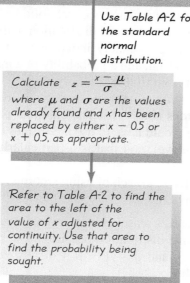

FIGURE 5-23 Using a Normal Approximation to a Binomial Distribution

3. Identify the discrete value x (the number of successes). Change the *discrete* value x by replacing it with the *interval* from $x - 0.5$ to $x + 0.5$. (For further clarification, see the discussion under the subheading "Continuity Corrections" found later in this section.) Draw a normal curve and enter the values of μ, σ, and either $x - 0.5$ or $x + 0.5$, as appropriate.

4. Change x by replacing it with $x - 0.5$ or $x + 0.5$, as appropriate.

5. Find the area corresponding to the desired probability by first finding the z score: $z = (x - \mu)/\sigma$. Now use that z score to find the area to the left of either $x - 0.5$ or $x + 0.5$, as appropriate. That area can now be used to identify the area corresponding to the desired probability.

We will illustrate this normal approximation procedure with the following example.

EXAMPLE **Loading Airliners** When an airliner is loaded with passengers, baggage, cargo, and fuel, the pilot must verify that the gross weight is below the maximum allowable limit, and the weight must be properly distributed so that the balance of the aircraft is within safe acceptable limits. Air America has established a procedure whereby extra cargo must be reduced whenever a plane filled with 200 passengers includes at least 120 men. Find the probability that among 200 randomly selected passengers, there are at least 120 men. Assume that the population of potential passengers consists of an equal number of men and women.

SOLUTION   Refer to Figure 5-23 for the procedure followed in this solution. The given problem does involve a binomial distribution with a fixed number of trials ($n = 200$), which are presumably independent, two categories (man, woman) of outcome for each trial, and a probability of a male ($p = 0.5$) that presumably remains constant from trial to trial.

We will assume that neither software nor a calculator is available. Table A-1 does not apply, because it stops at $n = 15$. The binomial probability formula is not practical, because we would have to use it 81 times (once for each value of x from 120 to 200 inclusive), and that is a particularly unappealing prospect.

Let's proceed with the five-step approach of using a normal distribution to approximate the binomial distribution.

Step 1:   We must first verify that it is reasonable to approximate the binomial distribution by the normal distribution because $np \geq 5$ and $nq \geq 5$. With $n = 200$, $p = 0.5$, and $q = 1 - p = 0.5$, we verify the required conditions as follows.

$$np = 200 \cdot 0.5 = 100 \quad \text{(Therefore } np \geq 5.\text{)}$$
$$nq = 200 \cdot 0.5 = 100 \quad \text{(Therefore } nq \geq 5.\text{)}$$

Step 2:   We now proceed to find the values for μ and σ that are needed for the normal distribution. We get the following:

$$\mu = np = 200 \cdot 0.5 = 100$$
$$\sigma = \sqrt{npq} = \sqrt{200 \cdot 0.5 \cdot 0.5} = 7.0710678$$

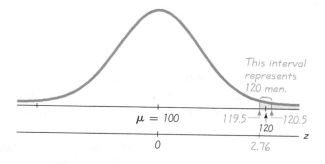

Step 3:   The discrete value of 120 is represented by the vertical strip bounded by 119.5 and 120.5. (See the discussion of continuity corrections, which follows this example.)

Step 4:   Because we want the probability of *at least* 120 men, we want the area representing the discrete number of 120 (the region bounded by 119.5 and 120.5), as well as the area to the right, as shown in Figure 5-24.

Step 5:   We can now proceed to find the shaded area of Figure 5-24 by using the same methods used in Section 5-3. In order to use Table A-2 for the standard normal distribution, we must first convert 119.5 to a z score, then use the table to find the area to the left of 119.5, which is then subtracted from 1. The z score is found as follows:

$$z = \frac{x - \mu}{\sigma} = \frac{119.5 - 100}{7.0710678} = 2.76$$

Using Table A-2, we find that $z = 2.76$ corresponds to an area of 0.9971, so the shaded region is $1 - 0.9971 = 0.0029$.

INTERPRETATION There is a 0.0029 probability of getting at least 120 men among 200 passengers. Because that probability is so small, we can conclude that a roster of 200 passengers will rarely include at least 120 men, so the reduction of extra cargo is not something to be very concerned about.

Continuity Corrections

The procedure for using a normal distribution to approximate a binomial distribution includes a step in which we change a discrete number to an interval that is 0.5 below and 0.5 above the discrete number. See the preceding solution, where we changed 120 to the interval between 119.5 and 120.5. This particular step, called a *continuity correction,* can be somewhat difficult to understand, so we will now consider it in more detail.

Definition

When we use the normal distribution (which is a *continuous* probability distribution) as an approximation to the binomial distribution (which is *discrete*), a **continuity correction** is made to a discrete whole number x in the binomial distribution by representing the single value x by the *interval* from $x - 0.5$ to $x + 0.5$ (that is, adding and subtracting 0.5).

The following practical suggestions should help you use continuity corrections properly.

Procedure for Continuity Corrections

1. When using the normal distribution as an approximation to the binomial distribution, *always* use the continuity correction. (It is required because we are using the *continuous* normal distribution to approximate the *discrete* binomial distribution.)

2. In using the continuity correction, first identify the discrete whole number x that is relevant to the binomial probability problem. For example, if you're trying to find the probability of getting at least 120 men in 200 randomly selected people, the discrete whole number of concern is $x = 120$. First focus on the x value itself, and temporarily ignore whether you want at least x, more than x, fewer than x, or whatever.

3. Draw a normal distribution centered about μ, then draw a *vertical strip area* centered over x. Mark the left side of the strip with the number equal to $x - 0.5$, and mark the right side with the number equal to $x + 0.5$. For $x = 120$, for example, draw a strip from 119.5 to 120.5. *Consider the entire area of the strip to represent the probability of the discrete number x itself.*

4. Now determine whether the value of x itself should be included in the probability you want. (For example, "at least x" does include x itself, but "more than x" does not include x itself.) Next, determine whether you want the probability of at least x, at most x, more than x, fewer than x, or exactly x. Shade the area to the right or left of the strip, as appropriate; also shade the interior of the strip itself *if and only if x itself* is to be included. This total shaded region corresponds to the probability being sought.

To see how this procedure results in continuity corrections, see the common cases illustrated in Figure 5-25. Those cases correspond to the statements in the following list.

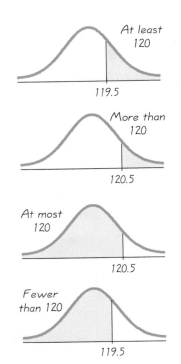

FIGURE 5-25 Using Continuity Corrections

Statement	Area
At least 120 (includes 120 and above)	To the *right* of 119.5
More than 120 (doesn't include 120)	To the *right* of 120.5
At most 120 (includes 120 and below)	To the *left* of 120.5
Fewer than 120 (doesn't include 120)	To the *left* of 119.5
Exactly 120	Between 119.5 and 120.5

EXAMPLE Gender Selection In a test of a gender-selection technique, assume that 100 couples using a particular treatment give birth to 52 girls (and 48 boys). If the technique has no effect, then the probability of a girl is approximately 0.5. If the probability of a girl is 0.5, find the probability that among 100 newborn babies, *exactly* 52 are girls. Based on the result, is there strong evidence supporting a claim that the gender-selection technique increases the likelihood that a baby is a girl? *Caution: This is somewhat of a trick question.*

SOLUTION With $n = 100$ independent trials, and $x = 52$ girls and an assumed population proportion of $p = 0.5$, the conditions satisfy the requirements for a binomial distribution.

For the purposes of this example, we assume that neither computer software nor a TI-83/84 Plus calculator is available. Table A-1 cannot be used, because $n = 100$ exceeds the largest table value of $n = 15$. If we were to use the binomial probability formula, we would need to evaluate an expression that includes 100!, but many calculators and software programs can't handle that. We therefore proceed by using a normal distribution to approximate the binomial distribution.

Step 1:   We begin by checking to determine whether the approximation is suitable:

$$np = 100 \cdot 0.5 = 50 \quad \text{(Therefore } np \geq 5.\text{)}$$
$$nq = 100 \cdot 0.5 = 50 \quad \text{(Therefore } nq \geq 5.\text{)}$$

Step 2:   We now proceed to find the values for μ and σ that are needed for the normal distribution. We get the following:

$$\mu = np = 100 \cdot 0.5 = 50$$
$$\sigma = \sqrt{npq} = \sqrt{100 \cdot 0.5 \cdot 0.5} = 5$$

Step 4:   We draw the normal curve shown in Figure 5-26. The shaded region of the figure represents the probability we want. Use of the continuity correction results in the representation of 52 by the region between 51.5 and 52.5.

Step 5:   Here is the approach used to find the shaded region in Figure 5-26: First find the total area to the left of 52.5, then find the total area to the left of 51.5, then find the *difference* between those two areas. Beginning with the total area to the left of 52.5, we must first find the z score corresponding to 52.5, then refer to Table A-2. We get

$$z = \frac{52.5 - 50}{5} = 0.50$$

continued

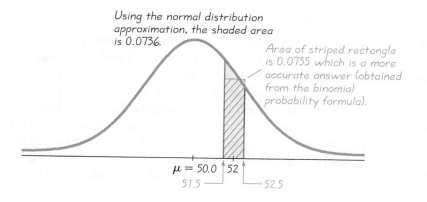

Using the normal distribution approximation, the shaded area is 0.0736.

Area of striped rectangle is 0.0735 which is a more accurate answer (obtained from the binomial probability formula).

$\mu = 50.0$ 52

51.5   52.5

FIGURE 5-26 Using Continuity Correction

We use Table A-2 to find that $z = 0.50$ corresponds to a probability of 0.6915, which is the total area to the left of 52.5. Now we proceed to find the area to the left of 51.5 by first finding the z score corresponding to 51.5:

$$z = \frac{51.5 - 50}{5} = 0.30$$

We use Table A-2 to find that $z = 0.30$ corresponds to a probability of 0.6179, which is the total area to the left of 51.5. The shaded area is $0.6915 - 0.6179 = 0.0736$.

INTERPRETATION The probability of exactly 52 girls (out of 100) is approximately 0.0736. The statement of the problem also asks us to determine whether the sample result of 52 girls among 100 constitutes strong evidence to conclude that the gender technique is effective. However, instead of considering the probability of *exactly* 52 girls, we must consider the probability of *52 or more*. [In Section 4-2, we noted that x successes among n trials is an *unusually high* number of successes if $P(x$ or more$)$ is very small, such as 0.05 or less.] Using the continuity correction, the probability of 52 or more successes is $P(\text{more than } 51.5) = 0.3821$, which is the area to the right of 51.5. Because 0.3821 is not very small (such as 0.05 or less), we do *not* have strong evidence to conclude that the gender technique is effective.

If we find the probability of exactly 52 girls among 100 by using STATDISK, Minitab, or a calculator, we get a result of 0.0735, but the normal approximation method resulted in a value of 0.0736. The discrepancy of 0.0001 occurs because the use of the normal distribution results in an *approximate* value that is the area of the shaded region in Figure 5-26, whereas the exact correct area is a rectangle centered above 52. (Figure 5-26 illustrates this discrepancy.) The area of the rectangle is 0.0735, but the area of the approximating shaded region is 0.0736. This very small discrepancy can be somewhat larger, depending on the sample sizes and probabilities involved.

Interpreting Results

In reality, when we use a normal distribution as an approximation to a binomial distribution, our ultimate goal is not simply to find a probability number. We often need to make some *judgment* based on the probability value, as in the final conclusion of the preceding example. We should understand that low probabilities correspond to events that are very unlikely, whereas large probabilities correspond to likely events. The probability value of 0.05 is often used as a cutoff to distinguish between unlikely events and likely events. The following criterion (from Section 4-2) describes the use of probabilities for distinguishing between results that could easily occur by chance and those results that are highly unusual.

Using Probabilities to Determine When Results Are Unusual

- **Unusually high:** x successes among n trials is an *unusually high* number of successes if $P(x$ or more) is very small (such as 0.05 or less)

- **Unusually low:** x successes among n trials is an *unusually low* number of successes if $P(x$ or fewer) is very small (such as 0.05 or less)

5-6 Exercises

Applying Continuity Correction. In Exercises 1–8, the given values are discrete. Use the continuity correction and describe the region of the normal distribution that corresponds to the indicated probability. For example, the probability of "more than 20 girls" corresponds to the area of the normal curve described with this answer: "the area to the right of 20.5."

1. Probability of more than 15 males with blue eyes

2. Probability of at least 24 students understanding continuity correction

3. Probability of fewer than 100 bald eagles sighted in a week

4. Probability that the number of working vending machines in the United States is exactly 27

5. Probability of no more than 4 absent students in a biostatistics class

6. Probability that the number of Canada geese residing in one pond is between 15 and 20 inclusive

7. Probability that the number of rabbit offspring is between 8 and 10 inclusive

8. Probability of exactly 3 American elm trees with Dutch elm disease

Using Normal Approximation. In Exercises 9–12, do the following. (a) Find the indicated binomial probability by using Table A-1 in Appendix A. (b) If $np \geq 5$ and $nq \geq 5$, also estimate the indicated probability by using the normal distribution as an approximation to the binomial distribution; if $np < 5$ or $nq < 5$, then state that the normal approximation is not suitable.

9. With $n = 14$ and $p = 0.5$, find $P(9)$.

10. With $n = 12$ and $p = 0.8$, find $P(7)$.

11. With $n = 15$ and $p = 0.9$, find $P($at least 14$)$.

12. With $n = 13$ and $p = 0.4$, find $P($fewer than 3$)$.

13. Probability of More Than 55 Girls Estimate the probability of getting more than 55 girls in 100 births. Assume that boys and girls are equally likely. Is it unusual to get more than 55 girls in 100 births?

14. Probability of at Least 65 Girls Estimate the probability of getting at least 65 girls in 100 births. Assume that boys and girls are equally likely. Is it unusual to get at least 65 girls in 100 births?

15. **Probability of at Least Passing** Estimate the probability of passing a true/false test of 100 questions if 60% (or 60 correct answers) is the minimum passing grade and all responses are random guesses. Is that probability high enough to risk passing by using random guesses instead of studying?

16. **Multiple-Choice Test** A multiple-choice test consists of 25 questions with possible answers of a, b, c, d, and e. Estimate the probability that with random guessing, the number of correct answers is between 3 and 10 inclusive.

17. **Mendel's Hybridization Experiment** When Mendel conducted his famous hybridization experiments, he used peas with green pods and yellow pods. One experiment involved crossing peas in such a way that 25% (or 145) of the 580 offspring peas were expected to have yellow pods. Instead of getting 145 peas with yellow pods, he obtained 152. Assuming that Mendel's 25% rate is correct, estimate the probability of getting at least 152 peas with yellow pods among the 580 offspring peas. Is there strong evidence to suggest that Mendel's rate of 25% is wrong?

18. **Cholesterol-Reducing Drug** The probability of flu symptoms for a person not receiving any treatment is 0.019. In a clinical trial of Lipitor (atorvastatin), a common drug used to lower cholesterol, 863 patients were given a treatment of 10 mg atorvastatin tablets, and 19 of those patients experienced flu symptoms (based on data from Pfizer, Inc.). Assuming that Lipitor has no affect on flu symptoms, estimate the probability that at least 19 of 863 people experience flu symptoms. What do these results suggest about flu symptoms as an adverse reaction to Lipitor?

19. **Probability of at Least 50 Color-Blind Men** Nine percent of men and 0.25% of women cannot distinguish between the colors red and green. This is the type of color blindness that causes problems with traffic signals. Researchers need at least 50 men with this type of color blindness, so they randomly select 600 men for a study of traffic-signal perceptions. Estimate the probability that at least 50 of the men cannot distinguish between red and green. Is the result high enough so that the researchers can be very confident of getting at least 50 men with red and green color blindness?

20. **Cell Phones and Brain Cancer** In a study of 420,095 cell phone users in Denmark, it was found that 135 developed cancer of the brain or nervous system. Assuming that cell phones have no effect, there is a 0.000340 probability of a person developing cancer of the brain or nervous system. We therefore expect about 143 cases of such cancer in a group of 420,095 randomly selected people. Estimate the probability of 135 or fewer cases of such cancer in a group of 420,095 people. What do these results suggest about media reports that cell phones cause cancer of the brain or nervous system?

21. **Identifying Gender Discrimination** After being rejected for employment, Kim Kelly learns that the Bellevue Health Care Company has hired only 21 women among its last 62 new employees. She also learns that the pool of applicants is very large, with an equal number of qualified men and women. Help her in her charge of gender discrimination by estimating the probability of getting 21 or fewer women when 62 people are hired, assuming no discrimination based on gender. Does the resulting probability really support such a charge?

22. **Blood Group** Forty-five percent of us have group O blood, according to data provided by the Greater New York Blood Program. Providence Memorial Hospital is conducting a blood drive because its supply of group O blood is low, and it needs 177 donors of group O blood. If 400 volunteers donate blood, estimate the probability that

the number with group O blood is at least 177. Is the pool of 400 volunteers likely to be sufficient?

23. Acceptance Sampling We stated in Section 3-4 that some companies monitor quality by using a method of acceptance sampling, whereby an entire batch of items is rejected if a random sample of a particular size includes more than some specified number of defects. The Medassist Pharmaceutical Company buys aspirin tablets in batches of 5000 and rejects a batch if, when 50 of them are sampled, at least 2 defects are found. (A tablet is defective if the amount of aspirin is not within specifications.) Estimate the probability of rejecting a batch if the supplier is manufacturing aspirin tablets with a defect rate of 10%. Is this monitoring plan likely to identify the unacceptable rate of defects?

24. Car Crashes For drivers in the 20–24 age bracket, there is a 34% rate of car accidents in one year (based on data from the National Safety Council). An insurance investigator finds that in a group of 500 randomly selected drivers aged 20–24 living in New York City, 40% had accidents in the last year. If the 34% rate is correct, estimate the probability that in a group of 500 randomly selected drivers, at least 40% had accidents in the last year. Based on that result, is there strong evidence supporting the claim that the accident rate in New York City is higher than 34%?

25. Cloning Survey A recent Gallup poll consisted of 1012 randomly selected adults who were asked whether "cloning of humans should or should not be allowed." Results showed that 89% of those surveyed indicated that cloning should not be allowed. A news reporter wants to determine whether these survey results constitute strong evidence that the majority (more than 50%) of people are opposed to such cloning. Assuming that 50% of all people are opposed, estimate the probability of getting at least 89% opposed in a survey of 1012 randomly selected people. Based on that result, is there strong evidence supporting the claim that the majority is opposed to such cloning?

5-7 Assessing Normality

The following chapters include some very important statistical methods requiring sample data randomly selected from a population with a *normal* distribution. It therefore becomes necessary to determine whether sample data appear to come from a population that is normally distributed. In this section we introduce the *normal quantile plot* as one tool that can help us determine whether the requirement of a normal distribution appears to be satisfied. (Quantile plots can be used to assess any probability distribution, but the emphasis in this section is on normal probability distributions.)

Definition

A **normal quantile plot** (or **normal probability plot**) is a graph of points (x, y) where each x value is from the original set of sample data, and each y value is the corresponding z score that is a quantile value expected from the standard normal distribution. (See Step 3 in the following procedure for details on finding these z scores.)

Mannequins ≠ Reality

Health magazine compared measurements of mannequins with measurements of women. The following results were reported as "averages," which were presumably means. Height of mannequins: 6 ft; height of women 5 ft 4 in. Waist of mannequins: 23 in.; waist of women: 29 in. Hip size of mannequins: 34 in.; hip size of women: 40 in. Dress size of mannequins: 6; dress size of women: 11. It becomes apparent that when comparing means, mannequins and real women are very different.

Procedure for Determining Whether Data Have a Normal Distribution

1. *Histogram:* Construct a histogram. Reject normality if the histogram departs dramatically from a bell shape.

2. *Outliers:* Identify outliers. Reject normality if there is more than one outlier present. (Just one outlier could be an error or the result of chance variation, but be careful, because even a single outlier can have a dramatic effect on results.)

3. *Normal quantile plot:* If the histogram is basically symmetric and there is at most one outlier, construct a *normal quantile plot*. The following steps describe one relatively simple procedure for constructing a normal quantile plot, but different statistical software packages use various other approaches. (STATDISK and the TI-83/84 Plus calculator use the procedure given here.) This procedure is messy enough so that we usually use software or a calculator to generate the graph.

   a. First sort the data by arranging the values in order from lowest to highest.

   b. With a sample of size n, each value represents a proportion of $1/n$ of the sample. Using the known sample size n, identify the areas of $1/2n$, $3/2n$, $5/2n$, $7/2n$, and so on. These are the cumulative areas to the left of the corresponding sample values.

   c. Use the standard normal distribution (Table A-2) to find the z scores corresponding to the cumulative left areas found in Step (b). (These are the z scores that are expected from a normally distributed sample.)

   d. Match the original sorted data values with their corresponding z scores found in Step (c), then plot the points (x, y), where each x is an original sample value and y is the corresponding z score.

   e. Examine the normal quantile plot using these criteria: If the points do not lie close to a straight line, or if the points exhibit some systematic pattern that is not a straight-line pattern, then the data appear to come from a population that does *not* have a normal distribution. If the pattern of the points is reasonably close to a straight line, then the data appear to come from a population that has a normal distribution. (These criteria can be used loosely for small samples, but they should be used more strictly for large samples.)

Steps 1 and 2 are straightforward, but we illustrate the construction of a normal quantile plot (Step 3) in the following example.

> **EXAMPLE Heights of Men** Data Set 1 in Appendix B includes heights (in inches) of randomly selected men. Let's consider only the first five heights listed for men: 70.8, 66.2, 71.7, 68.7, 67.6. With only five values, a histogram will not be very helpful in revealing the distribution of the data. Instead, construct a normal quantile plot for these five values and determine whether they appear to come from a population that is normally distributed.
>
> **SOLUTION** The following steps correspond to those listed in the above procedure for constructing a normal quantile plot.
>
> a. First, sort the data by arranging them in order. We get 66.2, 67.6, 68.7, 70.8, 71.7.

b. With a sample of size $n = 5$, each value represents a proportion of 1/5 of the sample, so we proceed to identify the cumulative areas to the left of the corresponding sample values. Those cumulative left areas, which are expressed in general as $1/2n$, $3/2n$, $5/2n$, $7/2n$, and so on, become these specific areas for this example with $n = 5$: 1/10, 3/10, 5/10, 7/10, and 9/10. Those same cumulative left areas expressed in decimal form are 0.1, 0.3, 0.5, 0.7, and 0.9.

c. We now search the body of Table A-2 for the cumulative left areas of 0.1000, 0.3000, 0.5000, 0.7000, and 0.9000. We find these corresponding z scores: -1.28, -0.52, 0, 0.52, and 1.28.

d. We now pair the original sorted heights with their corresponding z scores, and we get these (x, y) coordinates which are plotted as in the accompanying STATDISK display: $(66.2, -1.28)$, $(67.6, -0.52)$, $(68.7, 0)$, $(70.8, 0.52)$, and $(71.7, 1.28)$.

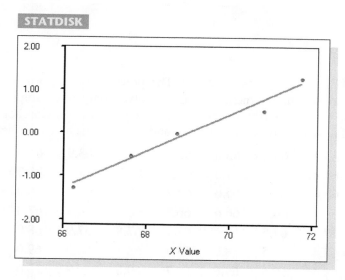

INTERPRETATION We examine the normal quantile plot in the STATDISK display. Because the points appear to lie reasonably close to a straight line, we conclude that the sample of five heights appears to come from a normally distributed population.

The following STATDISK display shows the normal quantile plot for the same data from the preceding example, with one change: The largest value of 71.7 is changed to 717, so it becomes an outlier. Note how the change in that one value affects the graph. This normal quantile plot does *not* result in points that reasonably approximate a straight-line pattern. The following STATDISK display suggests that the values of 66.2, 67.6, 68.7, 70.8, 717 are from a population with a distribution that is *not* a normal distribution.

The following example illustrates the use of a histogram and quantile plot with a larger data set. Such larger data sets typically require the use of software.

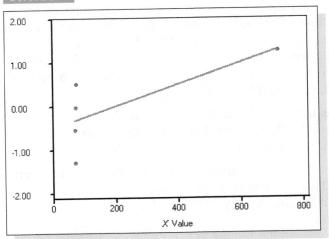

EXAMPLE **Heights of Men** The preceding two normal quantile plots referred to heights of men, but they involved only five sample values. Consider the following 100 heights of men supplied by a researcher who was instructed to randomly select 100 men and measure each of their heights.

63.3	63.4	63.5	63.6	63.7	63.8	63.9	64.0	64.1	64.2
64.3	64.4	64.5	64.6	64.7	64.8	64.9	65.0	65.1	65.2
65.3	65.4	65.5	65.6	65.7	65.8	65.9	66.0	66.1	66.2
66.3	66.4	66.5	66.6	66.7	66.8	66.9	67.0	67.1	67.2
67.3	67.4	67.5	67.6	67.7	67.8	67.9	68.0	68.1	68.2
68.3	68.4	68.5	68.6	68.7	68.8	68.9	69.0	69.1	69.2
69.3	69.4	69.5	69.6	69.7	69.8	69.9	70.0	70.1	70.2
70.3	70.4	70.5	70.6	70.7	70.8	70.9	71.0	71.1	71.2
71.3	71.4	71.5	71.6	71.7	71.8	71.9	72.0	72.1	72.2
72.3	72.4	72.5	72.6	72.7	72.8	72.9	73.0	73.1	73.2

SOLUTION

Step 1:   Construct a histogram. The accompanying STATDISK display shows the histogram of the 100 heights, and that histogram suggests that those heights are *not* normally distributed.

Step 2:   Identify outliers. Examining the list of 100 heights, we find that no values appear to be outliers.

Step 3:   Construct a normal quantile plot. The accompanying STATDISK display shows the normal quantile plot. Examination of the normal quantile plot reveals a systematic pattern that is not a straight-line pattern, suggesting that the data are not from a normally distributed population.

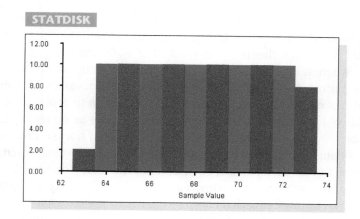

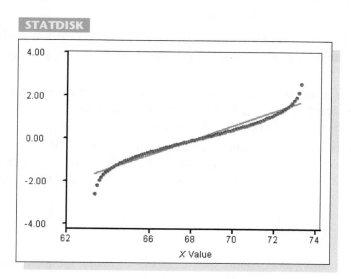

INTERPRETATION Because the histogram does not appear to be bell-shaped, and because the normal quantile plot reveals a pattern that is not a straight-line pattern, we conclude that the heights do not appear to be normally distributed. Some of the statistical procedures in later chapters require that sample data be normally distributed, but that requirement is not satisfied for this data set. We expect that 100 randomly selected heights of men should have a distribution that is approximately normal, so the researcher should be investigated. We could also examine the data more closely. Note that the values, when arranged in order, increase consistently by 0.1. This is further evidence that the heights are not the result of measurements obtained from randomly selected men.

Data Transformations   For many data sets, the distribution is not normal, but we can *transform* the data so that the modified values have a normal distribution. One common transformation is to replace each value of x with $\log (x + 1)$.

[Instead of using log $(x + 1)$, we could use the more direct transformation of replacing each value x with log x, but the use of log $(x + 1)$ has some advantages, including the property that if $x = 0$, then log $(x + 1)$ can be evaluated, whereas log x is undefined.] If the distribution of the log $(x + 1)$ values is a normal distribution, the distribution of the x values is referred to as a **lognormal distribution.** (See Exercise 18.) In addition to replacing each x value with log $(x + 1)$, there are other transformations, such as replacing each x value with $\sqrt{x}$, or $1/x$, or x^2. In addition to getting a required normal distribution when the original data values are not normally distributed, such transformations can be used to correct other deficiencies, such as a requirement (found in later chapters) that different data sets have the same variance. (This property of equal variances is referred to as **homoscedasticity.**)

Here are a few final comments about procedures for determining whether data are from a normally distributed population:

- If the requirement of a normal distribution is not too strict, examination of a histogram and consideration of outliers may be all that you need to assess normality.

- Normal quantile plots can be difficult to construct on your own, but they can be generated with a TI-83/84 Plus calculator or suitable software, such as SPSS, SAS, STATDISK, Minitab, and Excel.

- In addition to the procedures discussed in this section, there are other more advanced procedures for assessing normality, such as the chi-square goodness-of-fit test, the Kolmogorov-Smirnov test, and the Lilliefors test.

5-7 Exercises

Interpreting Normal Quantile Plots. In Exercises 1–4, examine the normal quantile plot and determine whether it depicts data that have a normal distribution.

1.

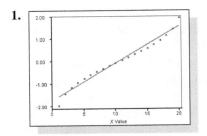

2.

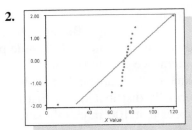

3.

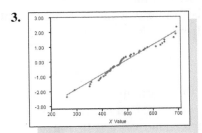

4.

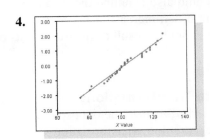

Determining Normality. In Exercises 5–8, refer to the indicated data set and determine whether the requirement of a normal distribution is satisfied. Assume that this requirement is loose in the sense that the population distribution need not be exactly normal, but it must be a distribution that is basically symmetric with only one mode.

5. BMI The measured BMI (body mass index) values of a sample of men, as listed in Data Set 1 in Appendix B.

6. Head Circumferences The head circumferences of males, as listed in Data Set 4 in Appendix B.

7. Water Conductivity The Florida Everglades conductivity levels, as listed below. (The data are from the Garfield Bight hydrology outpost in the Florida Everglades and are provided by Kevin Kotun and the National Park Service.)

57.8	57.8	57.1	57.0	57.3	58.4	59.2	57.7	56.8	56.8
55.2	53.6	52.0	51.9	49.8	49.8	51.7	48.6	44.3	43.2
41.5	40.6	35.9	33.8	32.8	30.5	32.7	32.1	30.3	28.1
29.3	30.2	33.5	40.5	42.4	46.7	46.7	46.5	45.6	47.1
48.1	50.5	51.2	50.4	49.9	49.0	48.5	51.3	52.1	52.4
51.0	52.2	50.3	48.5	49.7	49.9	48.5	48.3	49.0	49.9
51.0									

8. Heights of Poplar Trees The following heights (in meters) of poplar trees, as reported by a field scientist.

2.90	7.07	6.45	2.90	13.01	6.62	2.94	6.83	6.12	4.46
3.62	14.64	6.25	3.25	4.96	6.16	5.49	6.85	6.68	6.27
5.10	6.79	6.90	3.92	1.85	2.86	6.82	6.08	6.88	4.84

Generating Normal Quantile Plots. In Exercises 9–12, use the data from the indicated exercise in this section. Use a TI-83/84 Plus calculator or software (such as SPSS, SAS, STATDISK, Minitab, or Excel) capable of generating normal quantile plots (or normal probability plots). Generate the graph, then determine whether the data appear to come from a normally distributed population.

9. Exercise 5

10. Exercise 6

11. Exercise 7

12. Exercise 8

13. Comparing Data Sets Using the heights of women and the cholesterol levels of women, as listed in Data Set 1 in Appendix B, analyze each of the two data sets and determine whether each appears to come from a normally distributed population. Compare the results and give a possible explanation for any notable differences between the two distributions.

14. Comparing Data Sets Using the systolic blood pressure levels and the elbow breadths of women, as listed in Data Set 1 in Appendix B, analyze each of the two data sets and determine whether each appears to come from a normally distributed population. Compare the results and give a possible explanation for any notable differences between the two distributions.

Constructing Normal Quantile Plots. In Exercises 15 and 16, use the given data values and identify the corresponding z scores that are used for a normal quantile plot, then construct the normal quantile plot and determine whether the data appear to be from a population with a normal distribution.

15. Heights of L.A. Lakers Use this sample of heights (in inches) of players in the starting lineup for the L.A. Lakers professional basketball team: 85, 79, 82, 73, 78.

16. Monitoring Lead in Air On the days immediately following the destruction caused by the terrorist attacks on September 11, 2001, lead amounts (in micrograms per cubic meter) in the air were recorded at Building 5 of the World Trade Center site, and these values were obtained: 5.40, 1.10, 0.42, 0.73, 0.48, 1.10.

*17. Using Standard Scores When constructing a normal quantile plot, suppose that instead of finding z scores using the procedure described in this section, each value in a sample is converted to its corresponding standard score using $z = (x - \bar{x})/s$. If the (x, z) points are plotted in a graph, can this graph be used to determine whether the sample comes from a normally distributed population? Explain.

*18. Lognormal Distribution Test the following phone call times (in seconds) for normality, then replace each value x with log $(x + 1)$ and test the transformed values for normality. What do you conclude?

31.5	75.9	31.8	87.4	54.1	72.2	138.1	47.9	210.6	127.7
160.8	51.9	57.4	130.3	21.3	403.4	75.9	93.7	454.9	55.1

Review

We introduced the concept of probability distributions in Chapter 4, but included only *discrete* distributions. In this chapter we introduced *continuous* probability distributions and focused on the most important category: normal distributions. Normal distributions will be used extensively in the following chapters.

Normal distributions are approximately bell-shaped when graphed. The total area under the density curve of a normal distribution is 1, so there is a convenient correspondence between areas and probabilities. Specific areas can be found using Table A-2 or a TI-83/84 Plus calculator or software. (We do not use Formula 5-1, the equation that is used to define the normal distribution.)

In this chapter we presented important methods for working with normal distributions, including those that use the standard score $z = (x - \mu)/\sigma$ for solving problems such as these:

- Given that gestation times are normally distributed with $\mu = 279$ and $\sigma = 7$, find the probability of randomly selecting a gestation time greater than 290.
- Given that gestation times are normally distributed with $\mu = 279$ and $\sigma = 7$, find the gestation time separating the bottom 85% from the top 15%.

In Section 5-4 we introduced the concept of a sampling distribution. The sampling distribution of the mean is the probability distribution of sample means, with all samples having the same sample size n. The sampling distribution of the proportion is the probability distribution of sample proportions, with all samples having the same sample size n. In general, the sampling distribution of any statistic is the probability distribution of that statistic.

In Section 5-5 we presented the following important points associated with the central limit theorem:

1. The distribution of sample means will, as the sample size n increases, approach a normal distribution.

2. The mean of the sample means is the population mean μ.

3. The standard deviation of the sample means is $\sigma/\sqrt{n}$.

In Section 5-6 we noted that we can sometimes approximate a binomial probability distribution by a normal distribution. If both $np \geq 5$ and $nq \geq 5$, the binomial random variable x is approximately normally distributed with the mean and standard deviation given as $\mu = np$ and $\sigma = \sqrt{npq}$. Because the binomial probability distribution deals with discrete data and the normal distribution deals with continuous data, we apply the continuity correction, which should be used in normal approximations to binomial distributions.

Finally, in Section 5-7 we presented a procedure for determining whether sample data appear to come from a population that has a normal distribution. Some of the statistical methods covered later in this book have a loose requirement of a normally distributed population. In such cases, examination of a histogram and outliers might be all that is needed. In other cases, normal quantile plots might be necessary because of a very strict requirement that the population must have a normal distribution.

Review Exercises

1. High Cholesterol Levels The serum cholesterol levels in men aged 18–24 are normally distributed with a mean of 178.1 and a standard deviation of 40.7. Units are in mg/dL and the data are based on the National Health Survey.
   a. If a man aged 18–24 is randomly selected, find the probability that his serum cholesterol level is greater than 260, a value considered to be "moderately high."
   b. If one man aged 18–24 is randomly selected, find the probability that his serum cholesterol level is between 170 and 200.
   c. If 9 men aged 18–24 are randomly selected, find the probability that their mean serum cholesterol level is between 170 and 200.
   d. The Providence Health Maintenance Organization wants to establish a criterion for recommending dietary changes if cholesterol levels are in the top 3%. What is the cutoff for men aged 18–24?

2. Babies at Risk Weights of newborn babies in the United States are normally distributed with a mean of 3420 g and a standard deviation of 495 g (based on data from "Birth Weight and Perinatal Mortality," by Wilcox et al., *Journal of the American Medical Association,* Vol. 273, No. 9).
   a. A newborn weighing less than 2200 g is considered to be at risk, because the mortality rate for this group is at least 1%. What percentage of newborn babies are in the "at-risk" category? If the Chicago General Hospital has 900 births in a year, how many of those babies are expected to be in the "at-risk" category?
   b. If we redefine a baby to be at risk if his or her birth weight is in the lowest 2%, find the weight that becomes the cutoff separating at-risk babies from those who are not at risk.
   c. If 16 newborn babies are randomly selected, find the probability that their mean weight is greater than 3700 g.
   d. If 49 newborn babies are randomly selected, find the probability that their mean weight is between 3300 g and 3700 g.

3. Blue Genes Some couples have genetic characteristics configured so that one-quarter of all their offspring have blue eyes. A study is conducted of 100 couples believed to have those characteristics, with the result that 19 of their 100 offspring have blue eyes.

continued

Estimate the probability that among 100 offspring, 19 or fewer have blue eyes. Based on that probability, does it seem that the one-quarter rate is wrong? Why or why not?

4. Marine Corps Height Requirements for Men The U.S. Marine Corps requires that men have heights between 64 in. and 78 in. (The National Health Survey shows that heights of men are normally distributed with a mean of 69.0 in. and a standard deviation of 2.8 in.)
   a. Find the percentage of men meeting those height requirements. Are too many men denied the opportunity to join the Marines because they are too short or too tall?
   b. If you are appointed to be the Secretary of Defense and you want to change the requirement so that the shortest 2% and tallest 2% of all men are rejected, what are the new minimum and maximum height requirements?
   c. If 64 men are randomly selected, find the probability that their mean height is greater than 68.0 in.

5. Sampling Distributions
   a. Many different samples of size 100 are randomly selected from the lengths of babies when they are born in the United States. What can be concluded about the shape of the distribution of the means of the different samples?
   b. If the weights of all cars in the United States have a standard deviation of 512 lb, what is the standard deviation of the sample means that are found from many different samples of size 100?
   c. Many different samples of size 1200 are randomly selected from the population of all adults in the United States. In each sample, the proportion of people who voted in the last election is recorded. What can be concluded about the shape of the distribution of those sample proportions?

6. Gender Discrimination When several women are not hired at the Medassist Pharmaceutical Company, they do some research and find that among the many people who applied, 30% were women. However, the 20 people who were hired consist of only 2 women and 18 men. Find the probability of randomly selecting 20 people from a large pool of applicants (30% of whom are women) and getting 2 or fewer women. Based on the result, does it appear that the company is discriminating based on gender?

7. Testing for Normality Refer to the neck sizes of the sample of bears, as listed in Data Set 6 in Appendix B. Do those weights appear to come from a population that has a normal distribution? Explain.

8. Testing for Normality Refer to Data Set 12 in Appendix B and use the first list of systolic blood pressures, which correspond to the pre-exercise phase with no stress. Suppose some more advanced method of statistics has a fairly strict requirement that samples must come from populations having a normal distribution. Do the readings appear to satisfy that requirement? Why or why not?

Cumulative Review Exercises

1. Eye Measurement Statistics The listed sample distances (in millimeters) were obtained by using a pupilometer to measure the distances between the pupils of adults (based on data collected by a student of one of the authors).

   | 67 | 66 | 59 | 62 | 63 | 66 | 66 | 55 |

   a. Find the mean $\bar{x}$ of the distances in this sample.
   b. Find the median of the distances in this sample.

c. Find the mode of the distances in this sample.

d. Find the standard deviation s of this sample.

e. Convert the distance of 59 mm to a z score.

f. Find the actual percentage of these sample values that exceed 59 mm.

g. Assuming a normal distribution, find the percentage of *population* distances that exceed 59 mm. Use the sample values of $\bar{x}$ and s as estimates of μ and σ.

h. What level of measurement (nominal, ordinal, interval, ratio) describes this data set?

i. The listed measurements appear to be rounded to the nearest millimeter, but are the exact unrounded distances discrete data or continuous data?

2. Left-Handedness According to data from the American Medical Association, 10% of us are left-handed.

a. If three people are randomly selected, find the probability that they are all left-handed.

b. If three people are randomly selected, find the probability that at least one of them is left-handed.

c. Why can't we solve the problem in part (b) by using the normal approximation to the binomial distribution?

d. If groups of 50 people are randomly selected, what is the mean number of left-handed people in such groups?

e. If groups of 50 people are randomly selected, what is the standard deviation for the numbers of left-handed people in such groups?

f. Would it be unusual to get 8 left-handed people in a randomly selected group of 50 people? Why or why not?

Cooperative Group Activities

1. *Out-of-class activity:* Divide into groups of three or four students. In each group, develop an original procedure to illustrate the central limit theorem. The main objective is to show that when you randomly select samples from a population, the means of those samples tend to be *normally* distributed, regardless of the nature of the population distribution. In Section 5-5, for example, we used the last four digits of social security numbers as a source of samples from a population of digits that are equally likely; we proceeded to show that even though the original population did not have a normal distribution, the sample means tended to be normally distributed.

2. *In-class activity:* Divide into groups of three or four students. Using a coin to simulate births, each individual group member should simulate 25 births and record the number of simulated births of girls. Combine all the results in the group and record n = total number of births and x = number of girls. Given batches of n births, compute the mean and standard deviation for the number of girls. Is the simulated result usual or unusual? Why?

Technology Project

In Section 5-3 we included an example that involved the design of cars. The solution in that example showed that 97.72% of men have sitting heights less than 38.8 in. That solution involved theoretical calculations based on the assumption that men have sitting heights that are normally distributed with a mean of 36.0 in. and a standard deviation of 1.4 in. (based on anthropometric survey data from Gordon, Clauser, et al.). This project describes a different method of solution that is based on a *simulation* technique: We will use a computer or a TI-83/84 Plus calculator to randomly generate 500 men's sitting heights (from a normally distributed population with $\mu = 36.0$ and $\sigma = 1.4$), then we will find the percentage of those simulated heights that are less than 38.8 in. The SPSS, STATDISK, Minitab, Excel, and TI-83/84 Plus procedures are described as follows. (SAS does not randomly generate values from a normal distribution with a specified mean and standard deviation, as is required for this project.)

SPSS: First enter the value of 1 five hundred times in column 1, then select **Transform** from the main menu bar, then choose the option of **Compute.** In the dialog box, enter "Value" in the box labeled "Target Variable," then enter this expression in the Numeric Expression box: RV.NORMAL(36, 1.4). Click OK. The number of displayed decimal places can be changed by clicking on the **Data View** tab at the bottom. Next, sort the data by clicking on the main menu item of **Sort,** then selecting **Sort Cases.** With this sorted list, it becomes quite easy to count the number of sitting heights less than 38.8 in. Divide that number by 500 to find the percentage of these simulated sitting heights that are less than 38.8 in. Compare the result with the theoretical value of 97.72% that was found in Section 5-3.

STATDISK: Select **Data** from the main menu bar, then choose the option of **Normal Generator.** Proceed to generate 500 values with a mean of 36.0 and a standard deviation of 1.4. (Use one decimal place.) Next, use Copy/Paste to copy the generated data to the data window, then sort the data by clicking on Data Tools. With this sorted list, it becomes quite easy to count the number of sitting heights less than 38.8 in. Divide that number by 500 to find

the percentage of these simulated sitting heights that are less than 38.8 in. Compare the result to the theoretical value of 97.72% that was found in Section 5-3.

Minitab: Select the options of **Calc,** then **Random Data,** then **Normal.** Proceed to enter 500 for the number of rows, C1 for the column in which to store the data, 36.0 for the value of the mean, and 1.4 for the value of the standard deviation. Now select the option of **Manip,** then **Sort,** and proceed to sort column C1, with the sorted column stored in column C1, and with the sorting to be done by column C1. Examine the values in column C1 and determine the number of simulated sitting heights that are less than 38.8, then divide that number by 500 to find the percentage that are less than 38.8 in. Compare the result with the theoretical value of 97.72% that was found in Section 5-3.

Excel: Select **Tools** from the main menu bar, then select **Data Analysis** and **Random Number Generation.** After clicking OK, use the dialog box to enter 1 for the number of variables and 500 for the number of random numbers, then select "normal" for the type of distribution. Enter 36.0 for the mean and 1.4 for the standard deviation. Examine the displayed values and determine the number of simulated sitting heights that are less than 38.8, then divide that number by 500 to find the percentage that are less than 38.8 in. Compare the result with the theoretical value of 97.72% that was found in Section 5-3.

TI-83/84 Plus: Press **MATH,** then select **PRB,** then enter **randNorm** (36.0, 1.4, 500) to generate 500 values from a normally distributed population with $\mu = 36.0$ and $\sigma = 1.4$. Enter **STO → L1** to store the data in list L1. Now press **STAT,** then enter **SortA(L1)** to arrange the data in order. Examine the entries in list L1 to determine the number of simulated sitting heights that are less than 38.8, then divide that number by 500 to find the percentage that are less than 38.8 in. Compare the result with the theoretical value of 97.72% that was found in Section 5-3.

from DATA to DECISION

Critical Thinking: Designing a car seat

If a car seat is too low or too high, it will be uncomfortable and possibly dangerous. Most seats are made to be adjustable, so that men and women with different sitting heights can select a comfortable setting. When designing car seats, the sitting knee heights of men and women become very relevant. Men have sitting knee heights that are normally distributed with a mean of 22.0 in. and a standard deviation of 1.1 in., and women have sitting knee heights that are normally distributed with a mean of 20.3 in. and a standard deviation of 1.0 in. (based on anthropometric survey data from Gordon, Churchill, et al.). Find minimum and maximum sitting knee heights that include at least 95% of all men and at least 95% of all women, but try to find cost-effective limits that are as close together as possible. Find the percentage of men with sitting knee heights between the limits you have determined, then find the percentage of women with sitting knee heights between those same limits. Do your limits favor one gender at the expense of the other? Why is it not practical to simply design car seats so that they accommodate everyone? If you were a design engineer for General Motors, what percentage of the population would you be willing to exclude in your design of car seats? Apart from sitting knee height, what is another important design component that should be considered when determining the adjustment range of car seats?

6

Estimates and Sample Sizes with One Sample

Was Mendel wrong?

When Gregor Mendel conducted his famous genetics experiments with peas, one sample of offspring was obtained by crossing peas with green pods and peas with yellow pods. The offspring consisted of 580 peas. Among the offspring, 428 peas had green pods and 152 peas had yellow pods. Based on his theory of genetics, Mendel expected that 25% of the offspring peas would have yellow pods. However, with 428 peas with green pods and 152 peas with yellow pods, the percentage of peas with yellow pods is 26.2%.

152	peas with yellow pods
428	peas with green pods
580	total number of peas

Percentage of peas with yellow pods =
$$\frac{152}{580} \times 100\% = 26.2\%$$

Mendel expected that 25% of the peas would have yellow pods, but the actual percentage of peas with yellow pods in the experiment is 26.2%. Some key questions now arise. How do we explain that dis-crepancy? Is the discrepancy large enough to suggest that Mendel's 25% rate is incorrect? If we ignore Mendel's theory and use only the sample results from the experiment, what is an estimate of the percentage of peas with yellow pods that would be obtained in similar experiments? After obtaining an estimated percentage of peas with yellow pods based on the sample results from the experiment, what do we know about the accuracy of that estimate? This chapter will present the statistical concepts that we need to answer such questions. We will analyze the results from Mendel's experiment, and in the process, we will learn much about estimating values of population parameters in general. Although the data set we are considering involves the use of a sample proportion to estimate a population proportion, this chapter will also consider the use of a sample mean to estimate a population mean, and we will also see how we can use a sample variance to estimate a population variance.

6-1 Overview

In this chapter we begin working with the true core of inferential statistics as we use sample data to make inferences about population parameters. For example, the Chapter Problem includes results from a genetics experiment consisting of 580 offspring peas, 26.2% of which have yellow pods. Based on the sample statistic of 26.2%, we will estimate the percentage of the population of all such peas with yellow pods.

The two major applications of inferential statistics involve the use of sample data to (1) estimate the value of a population parameter, and (2) test some claim (or hypothesis) about a population. In this chapter we introduce methods for estimating values of population proportions, means, and variances. We also present methods for determining the sample sizes necessary to estimate those parameters. In Chapter 7 we will introduce the basic methods for testing claims (or hypotheses) that have been made about a population parameter.

This chapter and Chapter 7 include important inferential methods involving population proportions, population means, and population variances (or standard deviations). In both chapters we begin with proportions for the following reasons.

1. We all see proportions frequently in the media and in articles in professional journals.

2. Proportions are generally easier to work with than means or variances, so we can better focus on the important principles of estimating parameters and testing hypotheses when those principles are first introduced.

6-2 Estimating a Population Proportion

A Study Strategy: This section contains much information and introduces many concepts. The time devoted to this section will be well spent because the concept of a confidence interval is introduced, and that same general concept will be applied to the following sections of this chapter. We suggest that you use this study strategy: First, read this section with the limited objective of simply trying to understand what confidence intervals are, what they accomplish, and why they are needed. Second, try to develop the ability to construct confidence interval estimates of population proportions. Third, learn how to *interpret* a confidence interval correctly. Fourth, read the section once again and try to understand the underlying theory. You will always enjoy much greater success if you understand what you are doing, instead of blindly applying mechanical steps in order to obtain an answer which may or may not make any sense.

Here is the main objective of this section: Given a sample proportion, estimate the value of the population proportion p. For example, the Chapter Problem includes results from an experiment in which 580 offspring peas were obtained, and 26.2% of them had yellow pods. The sample statistic of 26.2% can be represented as the sample proportion of 0.262. By using the sample size of $n = 580$ total peas and the sample proportion of 0.262 for peas with yellow pods, we will proceed to estimate the proportion p of *all* such peas with yellow pods.

This section will consider only cases in which the normal distribution can be used to approximate the sampling distribution of sample proportions. In

Section 5-6 we noted that in a binomial procedure with n trials and probability p, if $np \geq 5$ and $nq \geq 5$, then the binomial random variable has a probability distribution that can be approximated by a normal distribution. (Remember that $q = 1 - p$.) Those conditions are included among the following requirements for the methods presented in this section.

Requirements for Using a Normal Distribution as an Approximation to a Binomial Distribution

1. The sample is a simple random sample.
2. The conditions for the binomial distribution are satisfied. That is, there is a fixed number of trials, the trials are independent, there are two categories of outcomes, and the probabilities remain constant for each trial. (See Section 4-3.)
3. The normal distribution can be used to approximate the distribution of sample proportions because $np \geq 5$ and $nq \geq 5$ are both satisfied. [Because p and q are unknown, we use the sample proportion to estimate their values. Also, there are other procedures (not included in this book) for dealing with situations in which the normal distribution is not a suitable approximation.]

Recall from Section 1-3 that a simple random sample of n values is obtained if every possible sample of size n has the same chance of being selected. This requirement of random selection means that the methods of this section cannot be used with some other types of sampling, such as stratified, cluster, and convenience sampling. We should be especially clear about this important point:

Data collected in an inappropriate way can be absolutely worthless, even if the sample is quite large.

We know that different samples naturally produce different results. The methods of this section assume that those sample differences are due to chance random fluctuations, not some unsound method of sampling. If you were to conduct a survey of opinions about cloning by surveying members of the American Genetic Association, you should not use the results for making some estimate of the proportion of all adult Americans. The sample of American Genetic Association members is very likely to be a biased sample in the sense that it is not representative of all Americans.

Assuming that we have a simple random sample and the other requirements listed above are also satisfied, we can now proceed with our major objective: using the sample as a basis for estimating the value of the population proportion p. We introduce the new notation $\hat{p}$ (called "p hat") for the sample proportion.

Poll Resistance

Surveys based on relatively small samples can be quite accurate, provided the sample is random or representative of the population. However, increasing survey refusal rates are now making it more difficult to obtain random samples. The Council of American Survey Research Organizations reported that in a recent year, 38% of consumers refused to respond to surveys. The head of one market research company said, "Everyone is fearful of self-selection and worried that generalizations you make are based on cooperators only." Results from the multibillion-dollar market research industry affect the products we buy, the television shows we watch, and many other facets of our lives.

Notation for Proportions

$p =$ proportion of successes in the entire *population*

$\hat{p} = \dfrac{x}{n} =$ *sample* proportion of x *successes* in a sample of size n

$\hat{q} = 1 - \hat{p} =$ sample proportion of *failures* in a sample of size n

Proportion, Probability, and Percent   Although this section focuses on the population proportion p, the procedures discussed here can also be applied to probabilities or percentages, but percentages must be converted to proportions by dropping the percent sign and dividing by 100. For example, 26.2% can be expressed in decimal form as 0.262. The symbol p may therefore represent a proportion, a probability, or the decimal equivalent of a percent. For example, if you survey 80 men and find that 6 of them have red/green colorblindness, then the sample proportion is $\hat{p} = x/n = 6/80 = 0.075$ and $\hat{q} = 0.925$ (calculated from $1 - 0.075$). Instead of computing the value of x/n, the value of $\hat{p}$ is sometimes already known because the sample proportion or percentage is given directly. For example, if it is reported that 580 offspring peas are obtained and 26.2% of them have yellow pods, then $\hat{p} = 0.262$ and $\hat{q} = 0.738$.

If we want to estimate a population proportion with a single value, the best estimate is $\hat{p}$. Because $\hat{p}$ consists of a single value, it is called a point estimate.

> ### Definition
>
> A **point estimate** is a single value (or point) used to approximate a population parameter.

The sample proportion $\hat{p}$ is the best point estimate of the population proportion p.

We use $\hat{p}$ as the point estimate of p because it is unbiased and is the most consistent of the estimators that could be used. It is unbiased in the sense that the distribution of sample proportions tends to center about the value of p; that is, sample proportions $\hat{p}$ do not systematically tend to underestimate p, nor do they systematically tend to overestimate p. (See Section 5-4.) The sample proportion $\hat{p}$ is more efficient than other estimators of p in the sense that the standard deviation of sample proportions tends to be smaller than the standard deviation of other unbiased estimators.

EXAMPLE Genetics Experiment In the Chapter Problem we noted that when peas with green pods were crossed with peas with yellow pods, 580 offspring peas were obtained, and 26.2% of them had yellow pods. Using these experiment results, find the best point estimate of the proportion of all such peas with yellow pods.

SOLUTION   Because the sample proportion is the best point estimate of the population proportion, we conclude that the best point estimate of p is 0.262. When using the sample results to estimate the percentage of all peas with yellow pods, our best estimate is 26.2%.

Why Do We Need Confidence Intervals?

In the preceding example we saw that 0.262 was our *best* point estimate of the population proportion p, but we have no indication of just how *good* our best estimate is. If we had a sample of only 5 peas and 1 of them has yellow pods, our best point estimate would be the sample proportion of $1/5 = 0.2$, but we wouldn't expect this point estimate to be very good because it is based on such a small sample. Because the point estimate has the serious flaw of not revealing anything about how good it is, statisticians have cleverly developed another type of estimate. This estimate, called a *confidence interval* or *interval estimate,* consists of a range (or an interval) of values instead of just a single value.

Definition

A **confidence interval** (or **interval estimate**) is a range (or an interval) of values used to estimate the true value of a population parameter. A confidence interval is sometimes abbreviated as CI.

A confidence interval is associated with a confidence level, such as 0.95 (or 95%). The confidence level gives us the success rate of the procedure used to construct the confidence interval. The confidence level is often expressed as the probability or area $1 - \alpha$ (lowercase Greek alpha). The value of α is the complement of the *confidence level.* For a 0.95 (or 95%) confidence level, $\alpha = 0.05$. For a 0.99 (or 99%) confidence level, $\alpha = 0.01$.

Definition

The **confidence level** is the probability $1 - \alpha$ (often expressed as the equivalent percentage value, such as 95%) that is the proportion of times that the confidence interval actually does contain the population parameter, assuming that the estimation process is repeated a large number of times. (The confidence level is also called the **degree of confidence,** or the **confidence coefficient.**)

The most common choices for the confidence level are 90% (with $\alpha = 0.10$), 95% (with $\alpha = 0.05$), and 99% (with $\alpha = 0.01$). The choice of 95% is most common because it provides a good balance between precision (as reflected in the width of the confidence interval) and reliability (as expressed by the confidence level).

Here's an example of a confidence interval based on the sample data of 580 offspring peas, with 26.2% of them having yellow pods:

The 0.95 (or 95%) confidence interval estimate of the population proportion p is $0.226 < p < 0.298$.

We will now introduce the extremely important concept of critical value(s) as we proceed to describe methods for constructing such confidence intervals, then we will focus on the correct *interpretation* of such confidence intervals.

Critical Values

The methods of this section and many of the other statistical methods found in the following chapters include the use of a standard z score that can be used to distinguish between sample statistics that are likely to occur and those that are unlikely.

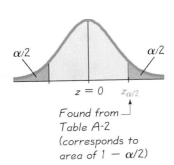

Found from
Table A-2
(corresponds to
area of 1 − α/2)

FIGURE 6-1 Critical Value
$z_{\alpha/2}$ **in the Standard Normal**
Distribution

Such a z score is called a *critical value* (defined below). Critical values are based on the following observations:

1. We know from Section 5-6 that under certain conditions, the sampling distribution of sample proportions can be approximated by a normal distribution, as in Figure 6-1.

2. Sample proportions have a relatively small chance (with probability denoted by α) of falling in one of the red tails of Figure 6-1.

3. Denoting the area of each shaded tail by $\alpha/2$, we see that there is a total probability of α that a sample proportion will fall in either of the two red tails.

4. By the rule of complements (from Chapter 3), there is a probability of $1 - \alpha$ that a sample proportion will fall within the inner green-shaded region of Figure 6-1.

5. The z score separating the right-tail region is commonly denoted by $z_{\alpha/2}$, and is referred to as a *critical value* because it is on the borderline separating sample proportions that are likely to occur from those that are unlikely to occur.

These observations can be formalized with the following notation and definition.

Notation for Critical Value

The critical value $z_{\alpha/2}$ is the positive z value that is at the vertical boundary separating an area of $\alpha/2$ in the right tail of the standard normal distribution. (The value of $-z_{\alpha/2}$ is at the vertical boundary for the area of $\alpha/2$ in the left tail.) The subscript $\alpha/2$ is simply a reminder that the z score separates an area of $\alpha/2$ in the right tail of the standard normal distribution.

Definition

A **critical value** is the number on the borderline separating sample statistics that are likely to occur from those that are unlikely to occur. The number $z_{\alpha/2}$ is a critical value that is a z score with the property that it separates an area of $\alpha/2$ in the right tail of the standard normal distribution. (See Figure 6-1.)

EXAMPLE Finding a Critical Value Find the critical value $z_{\alpha/2}$ corresponding to a 95% confidence level.

SOLUTION *Caution:* To find the critical z value for a 95% confidence level, do *not* look up 0.95 in the body of Table A-2. A 95% confidence level corresponds to $\alpha = 0.05$. See Figure 6-2, where we show that the area in each of the red-shaded tails is $\alpha/2 = 0.025$. We find $z_{\alpha/2} = 1.96$ by noting that all of the area to its left must be $1 - 0.025$, or 0.975. We can refer to Table A-2 and find that the area of 0.9750 (found in the *body* of the table) corresponds exactly to a z score of 1.96. For a 95% confidence level, the critical value is therefore $z_{\alpha/2} = 1.96$. To find the critical z score for a 95% confidence level, look up 0.9750 in the body of Table A-2, not 0.95.

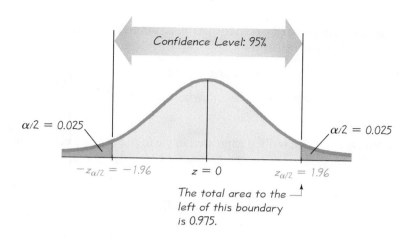

FIGURE 6-2 Finding $z_{\alpha/2}$
for a 95% Confidence Level

The preceding example showed that a 95% confidence level results in a critical value of $z_{\alpha/2} = 1.96$. This is the most common critical value, and it is listed with two other common values in the table that follows.

Confidence Level	α	Critical Value, $z_{\alpha/2}$
90%	0.10	1.645
95%	0.05	1.96
99%	0.01	2.575

Margin of Error

When we collect a set of sample data, such as the experimental pea data given in the Chapter Problem, we can calculate the sample proportion $\hat{p}$ and that sample proportion is typically different from the population proportion p. The difference between the sample proportion and the population proportion can be thought of as an error. We now define the *margin of error E* as follows.

Definition

When data from a simple random sample are used to estimate a population proportion p, the **margin of error,** denoted by **E,** is the maximum likely (with probability $1 - \alpha$) difference between the observed sample proportion $\hat{p}$ and the true value of the population proportion p. The margin of error E is also called the *maximum error of the estimate* and can be found by multiplying the critical value and the standard deviation of sample proportions, as shown in Formula 6-1.

Formula 6-1 $$E = z_{\alpha/2}\sqrt{\frac{\hat{p}\hat{q}}{n}}$$ margin of error for proportions

Given the way that the margin of error E is defined, there is a probability of $1 - \alpha$ that a sample proportion will be in error (different from the population proportion p) by no more than E, and there is a probability of α that the sample proportion will be in error by more than E.

Confidence Interval (or Interval Estimate) for the Population Proportion p

$$\hat{p} - E < p < \hat{p} + E \qquad \text{where} \qquad E = z_{\alpha/2}\sqrt{\frac{\hat{p}\hat{q}}{n}}$$

The confidence interval is often expressed in the following equivalent formats.

$$\hat{p} \pm E$$

or

$$(\hat{p} - E, \hat{p} + E)$$

In Chapter 3, when probabilities were given in decimal form, we rounded to three significant digits. We use that same rounding rule here.

Round-Off Rule for Confidence Interval Estimates of p

Round the confidence interval limits for p to three significant digits.

Based on the preceding results, we can summarize the procedure for constructing a confidence interval estimate of a population proportion p as follows.

Procedure for Constructing a Confidence Interval for p

1. Check that the requirements for this procedure are satisfied. (For this procedure, check that the sample is a simple random sample, the conditions for the binomial distribution are satisfied, and the normal distribution can be used to approximate the distribution of sample proportions because $np \geq 5$ and $nq \geq 5$ are both satisfied.)

2. Refer to Table A-2 and find the critical value $z_{\alpha/2}$ that corresponds to the desired confidence level. (For example, if the confidence level is 95%, the critical value is $z_{\alpha/2} = 1.96$.)

3. Evaluate the margin of error $E = z_{\alpha/2}\sqrt{\hat{p}\hat{q}/n}$

4. Using the value of the calculated margin of error E and the value of the sample proportion $\hat{p}$, find the values of $\hat{p} - E$ and $\hat{p} + E$. Substitute those values in the general format for the confidence interval:

$$\hat{p} - E < p < \hat{p} + E$$

or

$$\hat{p} \pm E$$

or

$$(\hat{p} - E, \hat{p} + E)$$

5. Round the resulting confidence interval limits to three significant digits.

EXAMPLE **Genetics Experiment** In the Chapter Problem we noted that 580 offspring peas were obtained, and 26.2% of them had yellow pods. In a previous example we noted that the best point estimate of the population proportion is 0.262. Use these same results for the following.

a. Find the margin of error E that corresponds to a 95% confidence level.

b. Find the 95% confidence interval estimate of the population proportion p.

c. Based on the results, what can we conclude about Mendel's theory that the percentage of peas with yellow pods should be 25%?

SOLUTION

REQUIREMENT ✓ We should first verify that the necessary requirements are satisfied. (Earlier in this section we listed the requirements for using a normal distribution as an approximation to a binomial distribution.) Given the design of Mendel's experiment, it is reasonable to assume that the sample is a simple random sample. The conditions for a binomial experiment are satisfied, because there is a fixed number of trials (580), the trials are independent (because the color of a pea pod doesn't affect the probability of the color of another pea pod), there are two categories of outcome (yellow, not yellow), and the probability of yellow remains constant. Finally, we use $n = 580$ and $\hat{p} = 0.262$ to find that

$$n\hat{p} = 152 \geq 5$$

and

$$n\hat{q} = 428 \geq 5$$

so the normal distribution can be used to approximate the binomial distribution. The check of requirements has been successfully completed. ✓

a. The margin of error is found by using Formula 6-1 with $z_{\alpha/2} = 1.96$ (as found in the preceding example), $\hat{p} = 0.262$, $\hat{q} = 1 - 0.262 = 0.738$, and $n = 580$. Extra digits are used so that the rounding error will be minimized in the confidence interval limits found in part (b).

$$E = z_{\alpha/2}\sqrt{\frac{\hat{p}\hat{q}}{n}} = 1.96\sqrt{\frac{(0.262)(0.738)}{580}} = 0.035787$$

b. Constructing the confidence interval is quite easy now that we know the values of $\hat{p}$ and E. We simply substitute those values to obtain this result:

$$\hat{p} - E < p < \hat{p} + E$$
$$0.262 - 0.035787 < p < 0.262 + 0.035787$$
$$0.226 < p < 0.298 \text{ (rounded to three significant digits)}$$

This same result could be expressed in the format of 0.262 ± 0.036 or (0.226, 0.298). If we want the 95% confidence interval for the true population *percentage*, we could express the result as $22.6\% < p < 29.8\%$. This confidence interval is often reported with a statement such as this: "It is estimated that 26.2% of the offspring peas will have yellow pods, with a margin

continued

of error of plus or minus 3.6 percentage points." That statement is a verbal expression of this format for the confidence interval: 26.2% ± 3.6%. The level of confidence should also be reported, but it rarely is in the media.

c. Based on the survey results, we are 95% confident that the limits of 22.6% and 29.8% contain the true percentage of offspring peas with yellow pods. The percentage of peas with yellow pods is likely to be any value between 22.6% and 29.8%. That interval does include 25%, so Mendel's expected value of 25% cannot be described as wrong. The results do not appear to provide significant evidence against the 25% rate claimed by Mendel.

6 Interpreting a Confidence Interval

We must be careful to interpret confidence intervals correctly. There is a correct interpretation and many other creative but wrong interpretations of the confidence interval $0.226 < p < 0.298$.

Correct: "We are 95% confident that the interval from 0.226 to 0.298 actually does contain the true value of p." This means that if we were to conduct many different experiments with 580 offspring peas and construct the corresponding confidence intervals, 95% of them would actually contain the value of the population proportion p. (Note that in this correct interpretation, the level of 95% refers to the success rate of the *process* being used to estimate the proportion, and it does not refer to the population proportion itself.)

Wrong: "There is a 95% chance that the true value of p will fall between 0.226 and 0.298."

At any specific point in time, there is a fixed and constant value of p, the proportion of offspring peas that are yellow. If we use sample data to find specific limits, such as 0.226 and 0.298, those limits either enclose the population proportion p or do not, and we cannot determine whether they do or do not without knowing the true value of p. But it's wrong to say that p has a 95% chance of falling within the specific limits of 0.226 and 0.298, because p is a fixed (but unknown) constant, not a random variable. Either p will fall within these limits or it won't; there's no probability involved. This is a confusing concept, so consider the easier example in which we want to find the probability of a baby being born a girl. If the baby has already been born, but the doctor hasn't yet announced the gender, we can't say that there is a 0.5 probability that the baby is a girl, because the baby is already a girl or is not. There is no chance involved, because the gender has been determined. Similarly, a population proportion p is already determined, and the confidence interval limits either contain p or do not, so it's wrong to say that there is a 95% chance that p will fall between 0.226 and 0.298.

A confidence level of 95% tells us that the *process* we are using will, in the long run, result in confidence interval limits that contain the true population proportion 95% of the time. Given the same experimental conditions of peas with green pods crossed with peas with yellow pods, suppose that the true proportion of all offspring peas having yellow pods is $p = 0.250$ as Mendel claimed. Then the confidence interval obtained from the given sample data would contain the

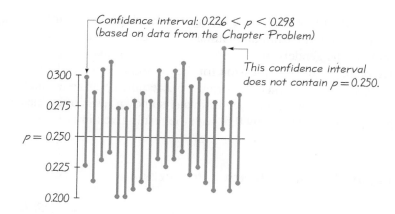

Confidence interval: $0.226 < p < 0.298$
(based on data from the Chapter Problem)

This confidence interval
does not contain $p = 0.250$.

FIGURE 6-3 **Confidence Intervals from 20 Different Samples**

population proportion, because the true population proportion 0.250 is between 0.226 and 0.298. This is illustrated in Figure 6-3. Figure 6-3 shows the first confidence interval for the real experimental data given in the Chapter Problem (with 26.2% of 580 offspring peas having yellow pods), but the other 19 confidence intervals represent hypothetical samples. With 95% confidence, we expect that 19 out of 20 samples should result in confidence intervals that do contain the true value of p, and Figure 6-3 illustrates this with 19 of the confidence intervals containing p, while one confidence interval does not contain p.

Rationale for the Margin of Error   Because the sampling distribution of proportions is approximately normal (because the conditions $np \geq 5$ and $nq \geq 5$ are both satisfied), we can use results from Section 5-6 to conclude that μ and σ are given by $\mu = np$ and $\sigma = \sqrt{npq}$. Both of these parameters pertain to n trials, but we convert them to a per-trial basis by dividing by n as follows:

$$\text{Mean of sample proportions: } \mu = \frac{np}{n} = p$$

$$\text{Standard deviation of sample proportions: } \sigma = \frac{\sqrt{npq}}{n} = \sqrt{\frac{npq}{n^2}} = \sqrt{\frac{pq}{n}}$$

The first result may seem trivial, because we already stipulated that the true population proportion is p. The second result is nontrivial and is useful in describing the margin of error E, but we replace the product pq by $\hat{p}\hat{q}$ because we don't yet know the value of p (it is the value we are trying to estimate). Formula 6-1 for the margin of error reflects the fact that $\hat{p}$ has a probability of $1 - \alpha$ of being within $z_{\alpha/2}\sqrt{pq/n}$ of p. The confidence interval for p, as given previously, reflects the fact that there is a probability of $1 - \alpha$ that $\hat{p}$ differs from p by less than the margin of error $E = z_{\alpha/2}\sqrt{pq/n}$.

Determining Sample Size

Suppose we want to collect sample data with the objective of estimating some population proportion. How do we know *how many* sample items must be obtained? For example, suppose that we want to estimate the proportion of females among children born as twins, triplets, quadruplets, quintuplets, and sextuplets.

How many such children must we observe to develop a reasonable estimate? (The U.S. Department and Health and Human Services tracks the numbers of multiple births, but not the genders of the babies in multiple births.)

If we begin with the expression for the margin of error *E* (Formula 6-1), then solve for *n*, we get Formula 6-2. Formula 6-2 requires $\hat{p}$ as an estimate of the population proportion *p*, but if no such estimate is known (as is often the case), we replace $\hat{p}$ by 0.5 and replace $\hat{q}$ by 0.5, with the result given in Formula 6-3.

Sample Size for Estimating Proportion *p*

When an estimate $\hat{p}$ is known: Formula 6-2 $n = \dfrac{[z_{\alpha/2}]^2 \hat{p}\hat{q}}{E^2}$

When no estimate $\hat{p}$ is known: Formula 6-3 $n = \dfrac{[z_{\alpha/2}]^2 \cdot 0.25}{E^2}$

Round-Off Rule for Determining Sample Size

In order to ensure that the required sample size is at least as large as it should be, if the computed sample size is not a whole number, round it up to the next *higher* whole number.

Use Formula 6-2 when reasonable estimates of $\hat{p}$ can be made by using previous samples, a pilot study, or someone's expert knowledge. When no such estimate can be made, we assign the value of 0.5 to each of $\hat{p}$ and $\hat{q}$ so the resulting sample size will be at least as large as it should be. The underlying reason for the assignment of 0.5 is this: The product $\hat{p} \cdot \hat{q}$ has 0.25 as its largest possible value, which occurs when $\hat{p} = 0.5$ and $\hat{q} = 0.5$. (Try experimenting with different values of $\hat{p}$ to verify that $\hat{p} \cdot \hat{q}$ has 0.25 as the largest possible value.) Note that Formulas 6-2 and 6-3 do not include the population size *N*, so the size of the population is irrelevant. (*Exception:* When sampling without replacement from a relatively small finite population. See Exercise 39.)

EXAMPLE **Sample Size for E-Mail Survey** The ways that we communicate have been dramatically affected by the use of answering machines, fax machines, voice mail, and e-mail. Suppose we want to determine the current percentage of U.S. households using e-mail. How many households must be surveyed in order to be 95% confident that the sample percentage is in error by no more than four percentage points?

a. Use this result from an earlier study: In 1997, 16.9% of U.S. households used e-mail (based on data from *The World Almanac and Book of Facts*).

b. Assume that we have no prior information suggesting a possible value of $\hat{p}$.

SOLUTION

a. The prior study suggests that $\hat{p} = 0.169$, so $\hat{q} = 0.831$ (found from $\hat{q} = 1 - 0.169$). With a 95% level of confidence, we have $\alpha = 0.05$, so $z_{\alpha/2} = 1.96$. Also, the margin of error is $E = 0.04$ (the decimal equivalent of "four percentage points"). Because we have an estimated value of $\hat{p}$, we use Formula 6-2 as follows:

$$n = \frac{[z_{\alpha/2}]^2 \hat{p}\hat{q}}{E^2} = \frac{[1.96]^2(0.169)(0.831)}{0.04^2}$$

$$= 337.194 = 338 \quad \text{(rounded UP)}$$

We must survey at least 338 randomly selected households.

b. As in part (a), we again use $z_{\alpha/2} = 1.96$ and $E = 0.04$, but with no prior knowledge of $\hat{p}$ (or $\hat{q}$), we use Formula 6-3 as follows:

$$n = \frac{[z_{\alpha/2}]^2 \cdot 0.25}{E^2} = \frac{[1.96]^2 \cdot 0.25}{0.04^2}$$

$$= 600.25 = 601 \quad \text{(rounded UP)}$$

INTERPRETATION   To be 95% confident that our sample percentage is within four percentage points of the true percentage for all households, we should randomly select and survey 601 households. By comparing this result to the sample size of 338 found in part (a), we can see that if we have no knowledge of a prior study, a larger sample is required to achieve the same results as when the value of $\hat{p}$ can be estimated. But now let's use some common sense: We know that the use of e-mail is growing so rapidly that the 1997 estimate is too old to be of much use. Today, substantially more than 16.9% of households use e-mail. Realistically, we need a sample larger than 338 households. Assuming that we don't really know the current rate of e-mail usage, we should randomly select 601 households. With 601 households, we will be 95% confident that the sample proportion will be within four percentage points of the true percentage of households using e-mail.

Common Errors   When calculating sample size using Formula 6-2 or 6-3, be sure to substitute the critical z score for $z_{\alpha/2}$. For example, if you are working with 95% confidence, be sure to replace $z_{\alpha/2}$ with 1.96. (Here is the logical sequence: 95% $\Rightarrow \alpha = 0.05 \Rightarrow z_{\alpha/2} = 1.96$ found from Table A-2.) Don't make the mistake of replacing $z_{\alpha/2}$ with 0.95 or 0.05. Also, don't make the mistake of using $E = 4$ as the margin of error corresponding to "four percentage points." When using Formula 6-2 or 6-3, the value of E never exceeds 1. The error of using $E = 4$ instead of $E = 0.04$ causes the sample size to be 1/10,000th of what it should be, so that you might end up with a sample size of only 1 when the answer is rounded up. You really can't estimate a population proportion by surveying only one person (even if the selected person claims to know everything).

Population Size   Part (b) of the preceding example involved application of Formula 6-3, the same formula commonly used by Nielsen, Gallup, and other professional pollsters. Many people incorrectly believe that the sample size should be some percentage of the population, but Formula 6-3 shows that the population size is irrelevant. (In reality, the population size is sometimes used, but only in cases in which we sample without replacement from a relatively small population. See Exercise 39.) Most of the polls featured in newspapers, magazines, and broadcast media involve sample sizes in the range of 1000 to 2000. Even though such polls may involve a very small percentage of the total population, they can provide results that are quite good. When Nielsen surveys 4000 TV households from a population of 104 million households, only 0.004% of the households are surveyed; still, we can be 95% confident that the sample percentage will be within one percentage point of the true population percentage.

Finding the Point Estimate and E from a Confidence Interval   Sometimes we want to better understand a confidence interval that might have been obtained from a journal article, or it might have been generated using software or a calculator. If we already know the confidence interval limits, the sample proportion $\hat{p}$ and the margin of error E can be found as follows:

Point estimate of p:

$$\hat{p} = \frac{(\text{upper confidence limit}) + (\text{lower confidence limit})}{2}$$

Margin of error:

$$E = \frac{(\text{upper confidence limit}) - (\text{lower confidence limit})}{2}$$

EXAMPLE   The article "High-Dose Nicotine Patch Therapy" by Dale, Hurt, et al. (*Journal of the American Medical Association,* Vol. 274, No. 17) includes this statement: "Of the 71 subjects, 70% were abstinent from smoking at 8 weeks (95% confidence interval [CI], 58% to 81%)." Use that statement to find the point estimate $\hat{p}$ and the margin of error E.

SOLUTION   From the given statement, we see that the 95% confidence interval is $0.58 < p < 0.81$. The point estimate $\hat{p}$ is the value midway between the upper and lower confidence interval limits, so we get

$$\hat{p} = \frac{(\text{upper confidence limit}) + (\text{lower confidence limit})}{2}$$

$$= \frac{0.81 + 0.58}{2} = 0.695$$

The margin of error can be found as follows:

$$E = \frac{(\text{upper confidence limit}) - (\text{lower confidence limit})}{2}$$

$$= \frac{0.81 - 0.58}{2} = 0.115$$

6-2 Exercises

Finding Critical Values. In Exercises 1–4, find the critical value $z_{\alpha/2}$ that corresponds to the given confidence level.

1. 99%

2. 90%

3. 98%

4. 92%

5. Express the confidence interval $0.220 < p < 0.280$ in the form of $\hat{p} \pm E$.

6. Express the confidence interval $0.456 < p < 0.496$ in the form of $\hat{p} \pm E$.

7. Express the confidence interval $(0.604, 0.704)$ in the form of $\hat{p} \pm E$.

8. Express the confidence interval 0.742 ± 0.030 in the form of $\hat{p} - E < p < \hat{p} + E$.

Interpreting Confidence Interval Limits. In Exercises 9–12, use the given confidence interval limits to find the point estimate $\hat{p}$ and the margin of error E.

9. $(0.444, 0.484)$

10. $0.278 < p < 0.338$

11. $0.632 < p < 0.678$

12. $0.887 < p < 0.927$

Finding Margin of Error. In Exercises 13–16, assume that a sample is used to estimate a population proportion p. Find the margin of error E that corresponds to the given statistics and confidence level.

13. $n = 800, x = 200, 95\%$

14. $n = 1200, x = 400, 99\%$

15. 99% confidence; the sample size is 1000, of which 45% are successes

16. 95% confidence; the sample size is 500, of which 80% are successes

Constructing Confidence Intervals. In Exercises 17–20, use the sample data and confidence level to construct the confidence interval estimate of the population proportion p.

17. $n = 400, x = 300, 95\%$ confidence

18. $n = 1200, x = 200, 99\%$ confidence

19. $n = 1655, x = 176, 98\%$ confidence

20. $n = 2001, x = 1776, 90\%$ confidence

Determining Sample Size. In Exercises 21–24, use the given data to find the minimum sample size required to estimate a population proportion or percentage.

21. Margin of error: 0.060; confidence level: 99%; $\hat{p}$ and $\hat{q}$ unknown

22. Margin of error: 0.038; confidence level: 95%; $\hat{p}$ and $\hat{q}$ unknown

23. Margin of error: five percentage points; confidence level: 95%; from a prior study, $\hat{p}$ is estimated by the decimal equivalent of 18.5%.

24. Margin of error: three percentage points; confidence level: 90%; from a prior study, $\hat{p}$ is estimated by the decimal equivalent of 8%.

25. Interpreting Calculator Display The Insurance Institute of America wants to estimate the percentage of drivers aged 18–20 who drive a car while impaired because of alcohol consumption. In a large study, 42,772 males aged 18–20 were surveyed, and 5.1% of them said that they drove in the last month while being impaired from alcohol (based on data from "Prevalence of Alcohol-Impaired Driving," by Liu, Siegel, et al., *Journal of the American Medical Association*, Vol. 277, No. 2). Using the sample data and a 95% confidence level, the TI-83/84 Plus calculator display is as shown.

a. Write a statement that interprets the confidence interval.

b. Based on the preceding result, does alcohol-impaired driving appear to be a problem for males aged 18–20? (All states now prohibit the sale of alcohol to persons under the age of 21.)

c. When setting insurance rates for male drivers aged 18–24, what percentage of alcohol-impaired driving would you use if you are working for the insurance company and you want to be conservative by using the likely worst-case scenario?

26. Interpreting Calculator Display In 1920 only 35% of U.S. households had telephones, but that rate is now much higher. A recent survey of 4276 randomly selected households showed that 4019 of them had telephones (based on data from the U.S. Census Bureau). Using those survey results and a 99% confidence level, the TI-83/84 Plus calculator display is as shown.

a. Write a statement that interprets the confidence interval.

b. Based on the preceding result, should pollsters be concerned about results from surveys conducted by telephone?

27. Mendelian Genetics When Mendel conducted his famous genetics experiments with peas, one sample of offspring consisted of 705 peas with red flowers and 224 peas with white flowers.

a. Find a 95% confidence interval estimate of the percentage of peas with red flowers.

b. Based on his theory of genetics, Mendel expected that 75% of the offspring peas would have red flowers. Given that the percentage of offspring yellow peas is not 75%, do the results contradict Mendel's theory? Why or why not?

28. Drug Testing The drug Ziac is used to treat hypertension. In a clinical test, 3.2% of 221 Ziac users experienced dizziness (based on data from Lederle Laboratories).

a. Construct a 99% confidence interval estimate of the percentage of all Ziac users who experience dizziness.

b. In the same clinical test, people in the placebo group didn't take Ziac, but 1.8% of them reported dizziness. Based on the result from part (a), what can we conclude about dizziness as an adverse reaction to Ziac?

29. Smoking and College Education The tobacco industry closely monitors all surveys that involve smoking. One survey showed that among 785 randomly selected subjects who completed four years of college, 18.3% smoke (based on data from the American Medical Association).

a. Construct the 98% confidence interval for the true percentage of smokers among all people who completed four years of college.

b. Based on the result from part (a), does the smoking rate for those with four years of college appear to be substantially different than the 27% rate for the general population?

30. Sample Size for Left-Handed Golfers As a manufacturer of golf equipment, the Spalding Corporation wants to estimate the proportion of golfers who are left-handed. (The company can use this information in planning for the number of right-handed and left-handed sets of golf clubs to make.) How many golfers must be surveyed if we want 99% confidence that the sample proportion has a margin of error of 0.025?

continued

a. Assume that there is no available information that could be used as an estimate of $\hat{p}$.

b. Assume that we have an estimate of $\hat{p}$ found from a previous study suggesting that 15% of golfers are left-handed (based on a *USA Today* report).

c. Assume that instead of using randomly selected golfers, the sample data are obtained by asking TV viewers of the golfing channel to call an "800" phone number to report whether they are left-handed or right-handed. How are the results affected?

31. Sample Size for Plant Growers You have been hired to do market research, and you must estimate the percentage of households in which at least one plant is being grown. How many households must you survey if you want to be 94% confident that your sample percentage has a margin of error of three percentage points?

a. Assume that a previous study suggested that plants are grown in 86% of households.

b. Assume that there is no available information that can be used to estimate the percentage of households in which a plant is being grown.

c. Assume that instead of using randomly selected households, the sample data are obtained by asking readers of the *Washington Post* newspaper to mail in a survey form. How are the results affected?

32. Color Blindness In a study of perception, 80 men are tested and 7 are found to have red/green color blindness (based on data from *USA Today*).

a. Construct a 90% confidence interval estimate of the proportion of all men with this type of color blindness.

b. What sample size would be needed to estimate the proportion of male red/green color blindness if we wanted 96% confidence that the sample proportion is in error by no more than 0.03? Use the sample proportion as a known estimate.

c. Women have a 0.25% rate of red/green color blindness. Based on the result from part (a), can we safely conclude that women have a lower rate of red/green color blindness than men?

33. Cell Phones and Cancer A study of 420,095 Danish cell phone users found that 135 of them developed cancer of the brain or nervous system. Prior to this study of cell phone use, the rate of such cancer was found to be 0.0340% for those not using cell phones. The data are from the *Journal of the National Cancer Institute*.

a. Use the sample data to construct a 95% confidence interval estimate of the percentage of cell phone users who develop cancer of the brain or nervous system.

b. Do cell phone users appear to have a rate of cancer of the brain or nervous system that is different from the rate of such cancer among those not using cell phones? Why or why not?

34. Gender Selection The Genetics and IVF Institute conducted a clinical trial of the XSORT method designed to increase the probability of conceiving a girl. As this book was being written, 325 babies were born to parents using the XSORT method, and 295 of them were girls.

a. Use the sample data to construct a 99% confidence interval estimate of the percentage of girls born to parents using the XSORT method.

b. Based on the result in part (a), does the XSORT method appear to be effective? Why or why not?

35. Gender Selection The Genetics and IVF Institute conducted a clinical trial of the YSORT method designed to increase the probability of conceiving a boy. As this book was being written, 51 babies were born to parents using the YSORT method, and 39 of them were boys.

a. Use the sample data to construct a 99% confidence interval estimate of the percentage of boys born to parents using the YSORT method.

b. Based on the result in part (a), does the YSORT method appear to be effective? Why or why not?

36. Pilot Fatalities Researchers studied crashes of general aviation (noncommercial and nonmilitary) airplanes and found that pilots died in 5.2% of 8411 crash landings (based on data from "Risk Factors for Pilot Fatalities in General Aviation Airplane Crash Landings," by Rostykus, Cummings, and Mueller, *Journal of the American Medical Association*, Vol. 280, No. 11).

a. Construct a 95% confidence interval estimate of the percentage of pilot deaths in all general aviation crashes.

b. Among crashes with an explosion or fire on the ground, the pilot fatality rate is estimated by the 95% confidence interval of (15.5%, 26.9%). Is this result substantially different from the result from part (a)? What can you conclude about an explosion or fire as a risk factor?

c. In planning for the allocation of federal funds to help with medical examinations of deceased pilots, what single percentage should be used? (We want to be reasonably sure that we have enough resources for the worst case scenario.)

37. Wearing Hunter Orange A study of hunting injuries and the wearing of "hunter" orange clothing showed that among 123 hunters injured when mistaken for game, 6 were wearing orange. Among 1115 randomly selected hunters, 811 reported that they routinely wear orange. The data are from the Centers for Disease Control.

a. Construct a 95% confidence interval estimate of the percentage of injured hunters who are wearing orange.

b. Construct a 95% confidence interval estimate of the percentage of hunters who routinely wear orange.

c. Do these results indicate that a hunter who wears orange is less likely to be injured because of being mistaken for game? Why or why not?

38. Effects of Sample Size and Confidence Level Given $n = 1000$, $x = 200$, and a 95% confidence level, we obtain the confidence interval $0.175 < p < 0.225$.

a. What happens to the confidence interval if the values of n and x are changed to 10,000 and 2000, respectively (so that the sample proportion remains at 0.2)? In general, what happens to a confidence interval if the sample size is increased while the sample proportion and confidence level remain the same?

b. What happens to a confidence interval if the confidence level is increased while everything else remains the same?

*39. Using Finite Population Correction Factor This section presented Formulas 6-2 and 6-3, which are used for determining sample size. In both cases we assumed that the population is infinite or very large and that we are sampling with replacement. When we have a relatively small population with size N and sample without replacement, we modify E to include the *finite population correction factor* shown here, and we can solve for n to obtain the result given here. Use this result to repeat part (b) of Exercise 31, assuming that we limit our population to a town with 10,000 households.

$$E = z_{\alpha/2}\sqrt{\frac{\hat{p}\hat{q}}{n}}\sqrt{\frac{N-n}{N-1}} \qquad n = \frac{N\hat{p}\hat{q}[z_{\alpha/2}]^2}{\hat{p}\hat{q}[z_{\alpha/2}]^2 + (N-1)E^2}$$

6-3 Estimating a Population Mean: σ Known

In Section 6-2 we introduced the point estimate and confidence interval as tools for using a sample proportion to estimate a population proportion. We also showed how to determine the minimum sample size required to estimate a population proportion. In this section we again discuss point estimate, confidence interval, and sample size determination, but we now consider the objective of estimating a population mean μ. Estimates of population means are often extremely important. For example, important questions such as these can be addressed using the methods of this section and the following section:

- What is the mean life span of bald eagles in the United States?
- What is the mean weight of elephants in Kenya?
- What is the mean amount of milk obtained from cows in New York State?

The following requirements apply to the methods introduced in this section. (There are other requirements for other such procedures.)

Requirements for Estimating μ when σ is known

1. The sample is a simple random sample. (All samples of the same size have an equal chance of being selected.)
2. The value of the population standard deviation σ is known.
3. Either or both of these conditions is satisfied: The population is normally distributed or $n > 30$. (Because we usually do not know all of the population values, we can check for normality by using tools such as histograms, normal quantile plots, and outliers found from the known *sample* data.)

When using the procedures of this section to estimate an unknown population mean μ, the above requirements indicate that we must know the value of the population standard deviation σ. It would be an unusual set of circumstances that would allow us to know σ without knowing μ. After all, the only way to find the value of σ is to compute it from all of the known population values, so the computation of μ would also be possible and, if we can find the true value of μ, there is no need to estimate it. Although the confidence interval methods of this section are not very realistic, they do reveal the basic concepts of important statistical reasoning, and they form the foundation for sample size determination discussed later in this section.

Requirement of Normality   In this section we use the requirement that we have a simple random sample, the value of σ is known, and either the population is normally distributed or $n > 30$. Technically, the population need not have a distribution that is exactly normal, but it should be approximately normal, meaning that the distribution is somewhat symmetric with one mode and no outliers. Investigate normality by using the sample data to construct a histogram, then determine whether it is approximately bell-shaped. A normal quantile plot (Section 5-7) could be constructed, but the methods of this section are said to be

Estimating Wildlife Population Sizes

The National Forest Management Act protects endangered species, including the northern spotted owl, with the result that the forestry industry was not allowed to cut vast regions of trees in the Pacific Northwest. Biologists and statisticians were asked to analyze the problem, and they concluded that survival rates and population sizes were decreasing for the female owls, known to play an important role in species survival. Biologists and statisticians also studied salmon in the Snake and Columbia Rivers in Washington, and penguins in New Zealand. In the article "Sampling Wildlife Populations" (*Chance*, Vol. 9, No. 2), authors Bryan Manly and Lyman McDonald comment that in such studies, "biologists gain through the use of modeling skills that are the hallmark of good statistics. Statisticians gain by being introduced to the reality of problems by biologists who know what the crucial issues are."

robust, which means that these methods are not strongly affected by departures from normality, provided that those departures are not very extreme. We can usually consider the population to be normally distributed after using the sample data to confirm that there are no outliers and the histogram has a shape that is not too far from a normal distribution. However, it is difficult to see the shape of a distribution if the sample size is small with $n \le 30$.

Sample Size Requirement This section uses the normal distribution as the distribution of sample means. If the original population is itself normally distributed, then the means from samples of any size will be normally distributed. If the original population is not itself normally distributed, then we say that means of samples with size $n > 30$ have a distribution that can be approximated by a normal distribution. The condition that the sample size is $n > 30$ is commonly used as a guideline, but it is not possible to identify a specific minimum sample size that is sufficient for all cases. The minimum sample size actually depends on how much the population distribution departs from a normal distribution. Sample sizes of 15 to 30 are adequate if the population appears to have a distribution that is not far from being normal, but some other populations have distributions that are extremely far from normal and sample sizes of 50 or even 100 or higher might be necessary. We will use the simplified criterion of $n > 30$ as justification for treating the distribution of sample means as a normal distribution.

In Section 6-2 we saw that the sample proportion $\hat{p}$ is the best point estimate of the population proportion p. For similar reasons, the sample mean $\overline{x}$ is the best point estimate of the population mean μ.

The sample mean $\overline{x}$ is the best point estimate of the population mean.

Although we could use another statistic such as the sample median, midrange, or mode as an estimate of the population mean μ, studies have shown that the sample mean $\overline{x}$ usually provides the best estimate, for the following two reasons:

1. For many populations, the distribution of sample means $\overline{x}$ tends to be more consistent (with *less variation*) than the distributions of other sample statistics. (That is, if you use sample means to estimate the population mean μ, those sample means will have a smaller standard deviation than would other sample statistics, such as the median or the mode. The differences between $\overline{x}$ and μ therefore tend to be smaller than the differences obtained with some other statistic, such as the median.)

2. For all populations, the sample mean $\overline{x}$ is an **unbiased estimator** of the population mean μ, meaning that the distribution of sample means tends to center about the value of the population mean μ. (That is, sample means do not systematically tend to overestimate the value of μ, nor do they systematically tend to underestimate μ. Instead, they tend to target the value of μ itself. See Section 5-4 where we illustrated the principle that sample means tend to target the value of the population mean.)

EXAMPLE **Body Temperatures** Data Set 2 in Appendix B includes 106 body temperatures taken at 12 A.M. on day 2. Here are statistics for that sample: $n = 106$, $\bar{x} = 98.20°F$, and $s = 0.62°F$. Use this sample to find the best point estimate of the population mean μ of all body temperatures.

SOLUTION   For the sample data, $\bar{x} = 98.20°F$. Because the sample mean $\bar{x}$ is the best point estimate of the population mean μ, we conclude that the best point estimate of the population mean μ of all body temperatures is $\bar{x} = 98.20°$ F.

Confidence Intervals

We saw in Section 6-2 that although a point estimate is the *best* single value for estimating a population parameter, it does not give us any indication of just how *good* the best estimate is. Statisticians developed the confidence interval (or interval estimate), which consists of a range (or an interval) of values instead of just a single value. The confidence interval is associated with a confidence level, such as 0.95 (or 95%). The confidence level gives us the success rate of the procedure used to construct the confidence interval. As in Section 6-2, the confidence level is often expressed as the probability or area $1 - \alpha$ (lowercase Greek alpha), where α is the complement of the *confidence level*. For a 0.95 (or 95%) confidence level, $\alpha = 0.05$. For a 0.99 (or 99%) confidence level, $\alpha = 0.01$.

Margin of Error   When we collect a set of sample data, such as the set of 106 body temperatures listed for 12 A.M. of day 2 in Data Set 2 in Appendix B, we can calculate the sample mean $\bar{x}$ and that sample mean is typically different from the population mean μ. The difference between the sample mean and the population mean is an error. In Section 5-5 we saw that $\sigma/\sqrt{n}$ is the standard deviation of sample means. Using $\sigma/\sqrt{n}$ and the $z_{\alpha/2}$ notation introduced in Section 6-2, we now use the *margin of error E* expressed as follows:

Formula 6-4 $E = z_{\alpha/2} \cdot \dfrac{\sigma}{\sqrt{n}}$ margin of error for mean (when σ is known)

Formula 6-4 reflects the fact that the sampling distribution of sample means $\bar{x}$ is *exactly* a normal distribution with mean μ and standard deviation $\sigma/\sqrt{n}$ whenever the population has a normal distribution with mean μ and standard deviation σ. If the population is not normally distributed, large samples yield sample means with a distribution that is *approximately* normal.

   Given the way that the margin of error E is defined, there is a probability of $1 - \alpha$ that a sample mean will be in error (different from the population mean μ) by no more than E, and there is a probability of α that the sample mean will be in error by more than E. The calculation of the margin of error E as given in Formula 6-4 requires that you know the population standard deviation σ, but Section 6-4 will present a method for calculating the margin of error E when σ is not known.

   Using the margin of error E, we can now identify the confidence interval for the population mean μ (if the requirements for this section are satisfied). The three commonly used formats for expressing the confidence interval are shown in the following box.

Confidence Interval Estimate of the Population Mean μ (With σ Known)

$$\bar{x} - E < \mu < \bar{x} + E \qquad \text{where} \qquad E = z_{\alpha/2} \cdot \frac{\sigma}{\sqrt{n}}$$

or

$$\bar{x} \pm E$$

or

$$(\bar{x} - E, \bar{x} + E)$$

Definition

The two values $\bar{x} - E$ and $\bar{x} + E$ are called **confidence interval limits.**

Procedure for Constructing a Confidence Interval for μ (with Known σ)

1. Check that the requirements are satisfied. (Requirements: We have a simple random sample, σ is known, and either the population appears to be normally distributed or $n > 30$.)

2. Refer to Table A-2 and find the critical value $z_{\alpha/2}$ that corresponds to the desired confidence level. (For example, if the confidence level is 95%, the critical value is $z_{\alpha/2} = 1.96$.)

3. Evaluate the margin of error $E = z_{\alpha/2}\sigma/\sqrt{n}$.

4. Using the value of the calculated margin of error E and the value of the sample mean $\bar{x}$, find the values of $\bar{x} - E$ and $\bar{x} + E$. Substitute those values in the general format for the confidence interval:

$$\bar{x} - E < \mu < \bar{x} + E$$

or

$$\bar{x} \pm E$$

or

$$(\bar{x} - E, \bar{x} + E)$$

5. Round the resulting values by using the following round-off rule.

Round-Off Rule for Confidence Intervals Used to Estimate μ

1. When using the *original set of data* to construct a confidence interval, round the confidence interval limits to one more decimal place than is used for the original set of data.

2. When the original set of data is unknown and only the *summary statistics* $(n, \bar{x}, s)$ are used, round the confidence interval limits to the same number of decimal places used for the sample mean.

Interpreting a Confidence Interval

As in Section 6-2, we must be careful to interpret confidence intervals correctly. After obtaining a confidence interval estimate of the population mean μ, such as a 95% confidence interval of $98.08 < \mu < 98.32$, there is a correct interpretation and many wrong interpretations.

Correct: "We are 95% confident that the interval from 98.08 to 98.32 actually does contain the true value of μ." This means that if we were to select many different samples of the same size and construct the corresponding confidence intervals, in the long run 95% of them would actually contain the value of μ. (As in Section 6-2, this correct interpretation refers to the success rate of the *process* being used to estimate the population mean.)

Wrong: Because μ is a fixed constant, it would be wrong to say "there is a 95% chance that μ will fall between 98.08 and 98.32." The confidence interval does not describe the behavior of individual sample values, so it would also be wrong to say that "95% of all data values are between 98.08 and 98.32." Also, the confidence interval does not describe the behavior of individual sample means, so it would also be wrong to say that "95% of sample means fall between 98.08 and 98.32."

> **EXAMPLE   Body Temperatures**   For the sample of body temperatures in Data Set 2 in Appendix B (for 12 A.M. on day 2), we have $n = 106$ and $\bar{x} = 98.20°F$. Assume that the sample is a simple random sample and that σ is somehow known to be $0.62°F$. Using a 0.95 confidence level, find both of the following:
>
> **a.** The margin of error E.
>
> **b.** The confidence interval for μ.

SOLUTION

REQUIREMENT ✓ We should first verify that the necessary requirements are satisfied. (Earlier in this section we listed the requirements for using the methods of this section to estimate μ.) From the statement of the problem we are assuming that we have a simple random sample. We are also assuming that σ is known to be $0.62°F$. The third and final requirement is that we have a "normally distributed population or $n > 30$." Because the sample size is $n = 106$, we have $n > 30$, so there is no need to check that the sample comes from a normally distributed population. (However, there are no outliers and a histogram of the 106 body temperatures would show that the sample data have a distribution that is approximately bell-shaped, suggesting that the population of body temperatures is normally distributed.) The check of requirements has been successfully completed and we can proceed with the methods of this section. ✓

a. The 0.95 confidence level implies that $\alpha = 0.05$, so $z_{\alpha/2} = 1.96$ (as was shown in an example in Section 6-2). The margin of error E is calculated by

continued

using Formula 6-4 as follows. Extra decimal places are used to minimize rounding errors in the confidence interval found in part (b).

$$E = z_{\alpha/2} \cdot \frac{\sigma}{\sqrt{n}} = 1.96 \cdot \frac{0.62}{\sqrt{106}} = 0.118031$$

b. With $\bar{x} = 98.20$ and $E = 0.118031$, we construct the confidence interval as follows:

$$\bar{x} - E < \mu < \bar{x} + E$$
$$98.20 - 0.118031 < \mu < 98.20 + 0.118031$$
$$98.08 < \mu < 98.32 \quad \text{(rounded to two decimal places as in } \bar{x}\text{)}$$

INTERPRETATION This result could also be expressed as 98.20 ± 0.12 or as (98.08, 98.32). Based on the sample with $n = 106$, $\bar{x} = 98.20$, and σ assumed to be 0.62, the confidence interval estimate of the population mean μ is $98.08°F < \mu < 98.32°F$ and this interval has a 0.95 confidence level. This means that if we were to select many different samples of size 106 and construct the confidence intervals as we did here, 95% of them would actually contain the value of the population mean μ.

Note that the confidence interval limits of 98.08°F and 98.32°F do not contain 98.6°F, the value generally believed to be the mean body temperature. Based on these results, it seems very unlikely that 98.6°F is the correct mean body temperature of the entire population.

Rationale for the Confidence Interval The basic idea underlying the construction of confidence intervals relates to the central limit theorem, which indicates that if we have a simple random sample from a normally distributed population, or a simple random sample of size $n > 30$ from any population, the distribution of sample means is approximately normal with mean μ and standard deviation $\sigma/\sqrt{n}$. The confidence interval format is really a variation of the equation that was already used with the central limit theorem. In the expression $z = (\bar{x} - \mu_{\bar{x}})/\sigma_{\bar{x}}$, replace $\sigma_{\bar{x}}$ with $\sigma/\sqrt{n}$, then solve for μ to get

$$\mu = \bar{x} - z\frac{\sigma}{\sqrt{n}}$$

Using the positive and negative values for z results in the confidence interval limits we are using.

Let's consider the specific case of a 95% confidence level, so $\alpha = 0.05$ and $z_{\alpha/2} = 1.96$. For this case, there is a probability of 0.05 that a sample mean will be more than 1.96 standard deviations (or $z_{\alpha/2}\sigma/\sqrt{n}$ which we denote by E) away from the population mean μ. Conversely, there is a 0.95 probability that a sample mean will be within 1.96 standard deviations (or $z_{\alpha/2}\sigma/\sqrt{n}$) of μ. (See Figure 6-4.) If the sample mean $\bar{x}$ is within $z_{\alpha/2}\sigma/\sqrt{n}$ of the population mean μ, then μ must be between $\bar{x} - z_{\alpha/2}\sigma/\sqrt{n}$ and $\bar{x} + z_{\alpha/2}\sigma/\sqrt{n}$; this is expressed in the general format of our confidence interval (with $z_{\alpha/2}\sigma/\sqrt{n}$ denoted as E): $\bar{x} - E < \mu < \bar{x} + E$.

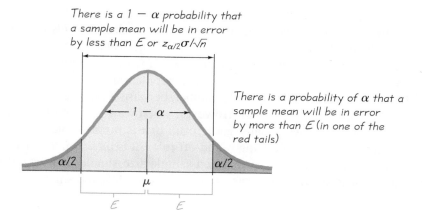

There is a $1 - \alpha$ probability that a sample mean will be in error by less than E or $z_{\alpha/2}\sigma/\sqrt{n}$

There is a probability of α that a sample mean will be in error by more than E (in one of the red tails)

FIGURE 6-4 **Distribution of Sample Means with Known σ**

A key feature of the methods we are using in this section is that we want to estimate an unknown population mean μ, and the population standard deviation σ is known. In the next section we present a method for estimating an unknown population mean μ when the population standard deviation is not known. The conditions of the following section are much more likely to occur in real circumstances. Although the methods of this section are unrealistic because they are based on knowledge of the population standard deviation σ, they do enable us to see the basic method for constructing a confidence interval estimate of μ by using the same normal distribution that has been used often in Chapter 5 and Section 6-2. Also, the methods we have discussed so far in this section lead to a very practical method for determining sample size.

Determining Sample Size Required to Estimate μ

We now want to address this key question: When we plan to collect a simple random sample of data that will be used to estimate a population mean μ, *how many* sample values must be obtained? In other words, we will find the sample size n that is required to estimate the value of a population mean. For example, suppose we want to estimate the mean weight of airline passengers (an important value for reasons of safety). How many passengers must be randomly selected and weighed? Determining the size of a simple random sample is a very important issue, because samples that are needlessly large waste time and money, and samples that are too small may lead to poor results. In many cases we can find the minimum sample size needed to estimate some parameter, such as the population mean μ.

If we begin with the expression for the margin of error E (Formula 6-4) and solve for the sample size n, we get the following:

Sample Size for Estimating Mean μ

Formula 6-5
$$n = \left[\frac{z_{\alpha/2}\sigma}{E} \right]^2$$

where $z_{\alpha/2}$ = critical z score based on the desired confidence level
   E = desired margin of error
   σ = population standard deviation

Formula 6-5 is remarkable because it shows that the sample size does not depend on the size (N) of the population; the sample size depends on the desired confidence level, the desired margin of error, and the value of the standard deviation σ. (See Exercise 31 for dealing with cases in which a relatively large sample is selected without replacement from a finite population.)

The sample size must be a whole number, because it represents the number of sample values that must be found. However, when we use Formula 6-5 to calculate the sample size n, we usually get a result that is not a whole number. When this happens, we use the following round-off rule. (It is based on the principle that when rounding is necessary, the required sample size should be rounded *upward* so that it is at least adequately large as opposed to slightly too small.)

Round-Off Rule for Sample Size *n*

When finding the sample size n, if the use of Formula 6-5 does not result in a whole number, always *increase* the value of n to the next *larger* whole number.

Dealing with Unknown σ When Finding Sample Size   When applying Formula 6-5, there is one very practical dilemma: The formula requires that we substitute some value for the population standard deviation σ, but in reality, it is usually unknown. When determining a required sample size (not constructing a confidence interval), here are some ways that we can work around this problem:

1. Use the range rule of thumb (see Section 2-5) to estimate the standard deviation as follows: $\sigma \approx$ range/4. (With a sample of 87 or more values randomly selected from a normally distributed population, range/4 will yield a value that is greater than or equal to σ at least 95% of the time. See "Using the Sample Range as a Basis for Calculating Sample Size in Power Calculations" by Richard Browne, *The American Statistician*, Vol. 55, No. 4.)

2. Conduct a pilot study by starting the sampling process. Based on the first collection of at least 31 randomly selected sample values, calculate the sample standard deviation s and use it in place of σ. The estimated value of σ can then be improved as more sample data are obtained.

3. Estimate the value of σ by using the results of some other study that was done earlier.

In addition, we can sometimes be creative in our use of other known results. For example, IQ tests are typically designed so that the mean is 100 and the standard deviation is 15. Statistics professors have IQ scores with a mean greater than 100 and a standard deviation less than 15 (because they are a more homogeneous group than people randomly selected from the general population). We do not know the specific value of σ for statistics professors, but we can play it safe by using $\sigma = 15$. Using a value for σ that is larger than the true value will make the sample size larger than necessary, but using a value for σ that is too small would result in a sample size that is inadequate. *When calculating the sample size n, any errors should always be conservative in the sense that they make n too large in-stead of too small.*

EXAMPLE IQ Scores of Statistics Professors Assume that we want to estimate the mean IQ score for the population of statistics professors. How many statistics professors must be randomly selected for IQ tests if we want 95% confidence that the sample mean is within 2 IQ points of the population mean?

SOLUTION   The values required for Formula 6-5 are found as follows:

$z_{\alpha/2} = 1.96$ (This is found by converting the 95% confidence level to $\alpha = 0.05$, then finding the critical z score as described in Section 6-2.)

$E = 2$ (Because we want the sample mean to be within 2 IQ points of μ, the desired margin of error is 2.)

$\sigma = 15$ (See the discussion in the paragraph that immediately precedes this example.)

With $z_{\alpha/2} = 1.96$, $E = 2$, and $\sigma = 15$, we use Formula 6-5 as follows:

$$n = \left[\frac{z_{\alpha/2}\sigma}{E}\right]^2 = \left[\frac{1.96 \cdot 15}{2}\right]^2 = 216.09 = 217 \quad \text{(rounded up)}$$

INTERPRETATION Among the thousands of statistics professors, we need to obtain a simple random sample of at least 217 of them, then we need to get their IQ scores. With a simple random sample of only 217 statistics professors, we will be 95% confident that the sample mean $\bar{x}$ is within 2 IQ points of the true population mean μ.

If we are willing to settle for less accurate results by using a larger margin of error, such as 4, the sample size drops to 54.0225, which is rounded *up* to 55. Doubling the margin of error causes the required sample size to decrease to one-fourth its original value. Conversely, halving the margin of error quadruples the sample size. Consequently, if you want more accurate results, the sample size must be substantially increased. Because large samples generally require more time and money, there is often a need for a trade-off between the sample size and the margin of error E.

6-3 Exercises

Finding Critical Values. In Exercises 1–4, find the critical value $z_{\alpha/2}$ that corresponds to the given confidence level.

1. 98%

2. 95%

3. 96%

4. 99.5%

Verifying Requirements. In Exercises 5–8, determine whether the given conditions justify using the margin of error $E = z_{\alpha/2}\sigma/\sqrt{n}$ when finding a confidence interval estimate of the population mean μ.

5. The sample size is $n = 200$ and $\sigma = 15$.

6. The sample size is $n = 5$ and σ is not known.

7. The sample size is $n = 5$, $\sigma = 12.4$, and the original population is normally distributed.

8. The sample size is $n = 9$, σ is not known, and the original population is normally distributed.

Finding Margin of Error and Confidence Interval. In Exercises 9–12, use the given confidence level and sample data to find (a) the margin of error E and (b) a confidence interval for estimating the population mean μ.

9. Salaries of statistics professors: 95% confidence; $n = 100$, $\bar{x} = \$95,000$ (we wish), and σ is known to be \$12,345.

10. Ages of drivers occupying the passing lane while driving 25 mi/h with the left signal flashing: 99% confidence; $n = 50$, $\bar{x} = 80.5$ years, and σ is known to be 4.6 years.

11. Times between uses of a TV remote control by males during commercials: 90% confidence; $n = 25$, $\bar{x} = 5.24$ sec, the population is normally distributed, and σ is known to be 2.50 sec.

12. Starting salaries of college graduates who have taken a statistics course: 95% confidence; $n = 28$, $\bar{x} = \$45,678$, the population is normally distributed, and σ is known to be \$9900.

Finding Sample Size. In Exercises 13–16, use the given margin of error, confidence level, and population standard deviation σ to find the minimum sample size required to estimate an unknown population mean μ.

13. Margin of error: \$125, confidence level: 95%, $\sigma = \$500$

14. Margin of error: 3 lb, confidence level: 99%, $\sigma = 15$ lb

15. Margin of error: 5 min, confidence level: 90%, $\sigma = 48$ min

16. Margin of error: \$500, confidence level: 94%, $\sigma = \$9877$

Interpreting Results. In Exercises 17–20, refer to the accompanying TI-83/84 Plus calculator display of a 95% confidence interval generated using the methods of this section. The sample display results from using a sample of 80 measured cholesterol levels of randomly selected adults.

TI-83/84 Plus

```
ZInterval
 (262.09,374.11)
 x̄=318.1
 n=80
■
```

17. Identify the value of the point estimate of the population mean μ.

18. Express the confidence interval in the format of $\bar{x} - E < \mu < \bar{x} + E$.

19. Express the confidence interval in the format of $\bar{x} \pm E$.

20. Write a statement that interprets the 95% confidence interval.

21. Everglades Temperatures In order to monitor the ecological health of the Florida Everglades, various measurements are recorded at different times. The bottom temperatures are recorded at the Garfield Bight station and the mean of 30.4°C is obtained for 61 temperatures recorded on 61 different days. Assuming that $\sigma = 1.7$°C, find a 95%

confidence interval estimate of the population mean of all such temperatures. What aspect of this problem is not realistic?

22. Weights of Bears The health of the bear population in Yellowstone National Park is monitored by periodic measurements taken from anesthetized bears. A sample of 54 bears has a mean weight of 182.9 lb. Assuming that σ is known to be 121.8 lb, find a 99% confidence interval estimate of the mean of the population of all such bear weights. What aspect of this problem is not realistic?

23. Cotinine Levels of Smokers When people smoke, the nicotine they absorb is converted to cotinine, which can be measured. A sample of 40 smokers has a mean cotinine level of 172.5. Assuming that σ is known to be 119.5, find a 90% confidence interval estimate of the mean cotinine level of all smokers. (All cotinine amounts are in ng/ml.) What aspect of this problem is not realistic?

24. Head Circumferences In order to help identify baby growth patterns that are unusual, we need to construct a confidence interval estimate of the mean head circumference of all babies that are two months old. A random sample of 100 babies is obtained, and the mean head circumference is found to be 40.6 cm. Assuming that the population standard deviation is known to be 1.6 cm, find a 99% confidence interval estimate of the mean head circumference of all two-month-old babies. What aspect of this problem is not realistic?

25. Sample Size for Mean IQ of Biology Majors The Wechsler IQ test is designed so that the mean is 100 and the standard deviation is 15 for the population of normal adults. Find the sample size necessary to estimate the mean IQ score of college students majoring in biology. We want to be 95% confident that our sample mean is within 2 IQ points of the true mean. The mean for this population is clearly greater than 100. The standard deviation for this population is probably less than 15 because it is a group with less variation than a group randomly selected from the general population; therefore, if we use σ = 15, we are being conservative by using a value that will make the sample size at least as large as necessary. Assume then that σ = 15 and determine the required sample size.

26. Sample Size for Estimating Income An economist wants to estimate the mean income for the first year of work for college graduates who majored in biology. How many such incomes must be found if we want to be 95% confident that the sample mean is within $500 of the true population mean? Assume that a previous study has revealed that for such incomes, σ = $6250.

27. Sample Size Using Range Rule of Thumb Estimate the minimum and maximum ages for typical textbooks currently used in college courses, then use the range rule of thumb to estimate the standard deviation. Next, find the size of the sample required to estimate the mean age (in years) of textbooks currently used in college courses. Use a 90% confidence level and assume that the sample mean will be in error by no more than 0.25 year.

28. Sample Size Using Sample Data You want to estimate the mean pulse rate of adult males. Refer to Data Set 1 in Appendix B and find the maximum and minimum pulse rates for males, then use those values with the range rule of thumb to estimate σ. How many adult males must you randomly select and test if you want to be 95% confident that the sample mean pulse rate is within 2 beats per minute of the true population mean μ? If, instead of using the range rule of thumb, the standard deviation of the male pulse rates in Data Set 1 is used as an estimate of σ, is the required sample size very different? Which sample size is likely to be closer to the correct sample size?

29. Sample Size Using Sample Data You want to estimate the mean diastolic blood pressure level of adult females. Refer to Data Set 1 in Appendix B and find the maximum and minimum diastolic blood pressure level for females, then use those values with the range rule of thumb to estimate σ. How many adult females must you randomly select and test if you want to be 95% confident that the sample mean diastolic blood pressure level is within 3 mm Hg of the true population mean μ? If, instead of using the range rule of thumb, the standard deviation of the female diastolic blood pressure levels in Data Set 1 is used as an estimate of σ, is the required sample size very different? Which sample size is likely to be closer to the correct sample size?

*30. Confidence Interval with Finite Population Correction Factor The standard error of the mean is $\sigma/\sqrt{n}$ provided that the population size is infinite. If the population size is finite and is denoted by N, then the correction factor $\sqrt{(N-n)/(n-1)}$ should be used whenever $n > 0.05N$. This correction factor multiplies the margin of error E given in Formula 6-4, so that the margin of error is as shown below. Find the 95% confidence interval for the mean of 250 IQ scores if a sample of 35 of those scores produces a mean of 110. Assume that $\sigma = 15$.

$$E = z_{\alpha/2}\frac{\sigma}{\sqrt{n}}\sqrt{\frac{N-n}{N-1}}$$

*31. Sample Size with Finite Population Correction Factor In Formula 6-4 for the margin of error E, we assume that the population is infinite, that we are sampling with replacement, or that the population is very large. If we have a relatively small population and sample without replacement, we should modify E to include a *finite population correction factor*, so that the margin of error is as shown in Exercise 30, where N is the population size. That expression for the margin of error can be solved for n to yield

$$n = \frac{N\sigma^2(z_{\alpha/2})^2}{(N-1)E^2 + \sigma^2(z_{\alpha/2})^2}$$

Repeat Exercise 25, assuming that the statistics students are randomly selected without replacement from a population of $N = 200$ biology students.

6-4 Estimating a Population Mean: σ Not Known

In Section 6-3 we presented methods for constructing a confidence interval estimate of an unknown population mean μ, but we considered only cases in which the population standard deviation σ is known. We noted that the requirement of a known σ is not very realistic, because the calculation of σ requires that we know all of the population values, but if we know all of the population values we can easily find the value of the population mean μ, so there is no need to estimate μ. In this section we present a method for constructing confidence interval estimates of μ without the requirement that σ is known. The usual procedure is to collect sample data and find the value of the statistics n, $\bar{x}$, and s. Because the methods of this section are based on those statistics and σ is not required, the methods of this section are very realistic, practical, and they are used often. Note that the following requirements for the methods of this section do *not* include a requirement that σ is known.

Requirements for Estimating μ when σ Is Unknown

1. The sample is a simple random sample.
2. Either the sample is from a normally distributed population or $n > 30$.

As in Section 6-3, the requirement of a normally distributed population is not a strict requirement. We can usually consider the population to be normally distributed after using the sample data to confirm that there are no outliers and the histogram has a shape that is not extremely far from a normal distribution. Also, as in Section 6-3, the requirement that the sample size is $n > 30$ is commonly used as a guideline, but the minimum sample size actually depends on how much the population distribution departs from a normal distribution. We will use the simplified criterion of $n > 30$ as justification for treating the distribution of sample means as a normal distribution. The sampling distribution of sample means $\bar{x}$ is *exactly* a normal distribution with mean μ and standard deviation $\sigma/\sqrt{n}$ whenever the population has a normal distribution with mean μ and standard deviation σ. If the population is not normally distributed, large samples yield sample means with a distribution that is *approximately* normal with mean μ and standard deviation $\sigma/\sqrt{n}$.

As in Section 6-3, the sample mean $\bar{x}$ is the best point estimate (or single-valued estimate) of the population mean μ. As in Section 6-3, the distribution of sample means $\bar{x}$ tends to be more consistent (with *less variation*) than the distributions of other sample statistics, and the sample mean $\bar{x}$ is an *unbiased estimator* that targets the population mean μ.

The sample mean $\bar{x}$ is the best point estimate of the population mean μ.

In Sections 6-2 and 6-3 we noted that there is a serious limitation to the usefulness of a point estimate: The single value of a point estimate does not reveal how good that estimate is. Confidence intervals give us much more meaningful information by providing a range of values associated with a degree of likelihood that the range actually does contain the true value of μ.

Here is the key point of this section: If σ is not known, but the above requirements are satisfied, instead of using the normal distribution, we use the *Student t distribution* developed by William Gosset (1876–1937). Gosset was a Guinness Brewery employee who needed a distribution that could be used with small samples. The Irish brewery where he worked did not allow the publication of research results, so Gosset published under the pseudonym *Student*.

Because we do not know the value of σ, we estimate it with the value of the sample standard deviation s, but this introduces another source of unreliability, especially with small samples. In order to keep a confidence interval at some desired level, such as 95%, we compensate for this additional unreliability by making the confidence interval wider: We use critical values larger than the critical values of $z_{\alpha/2}$ that were used in Section 6-3 where σ was known. Instead of using critical values of $z_{\alpha/2}$, we use the larger critical values of $t_{\alpha/2}$ found from the Student t distribution.

Publication Bias

A concerning trend in modern biomedical research is the so-called "publication bias" in which there is a tendency to publish research with positive results (such as showing that some treatment is effective) more often than negative results (such as showing that some treatment has no effect). This bias is described in the article "Registering Clinical Trials" by Kay Dickersin and Drummond Rennie, *Journal of the American Medical Association*, Vol. 290, No. 4. The authors state that the bias distorts the research evidence, reduces the ability of funding agencies and the public to know what research has been pursued, and is "invariably against the interest of those who offered to participate in trials and of patients in general." They support a process in which all clinical trials are registered in one central system so that the research is known, regardless of the findings and whether they were published.

Student t Distribution

If the distribution of a population is essentially normal (approximately bell-shaped), then the distribution of

$$t = \frac{\bar{x} - \mu}{\frac{s}{\sqrt{n}}}$$

is essentially a **Student t distribution** for all samples of size n. The Student t distribution, often referred to simply as the t **distribution**, is used to find critical values denoted by $t_{\alpha/2}$.

We will soon discuss some of the important properties of the t distribution, but we first present the components needed for the construction of confidence intervals. Let's start with the critical value denoted by $t_{\alpha/2}$. A value of $t_{\alpha/2}$ can be found in Table A-3. To find a critical value $t_{\alpha/2}$, in Table A-3, locate the appropriate number of *degrees of freedom* in the left column and proceed across the corresponding row until reaching the number directly below the appropriate area at the top.

Definition

The number of **degrees of freedom** for a collection of sample data is the number of sample values that can vary after certain restrictions have been imposed on all data values.

For example, if 10 students have quiz scores with a mean of 80, we can freely assign values to the first 9 scores, but the 10th score is then determined. The sum of the 10 scores must be 800, so the 10th score must equal 800 minus the sum of the first 9 scores. Because those first 9 scores can be freely selected to be any values, we say that there are 9 degrees of freedom available. For the applications of this section, the number of degrees of freedom is simply the sample size minus 1.

$$\text{degrees of freedom} = n - 1$$

EXAMPLE **Finding a Critical Value** A sample of size $n = 15$ is a simple random sample selected from a normally distributed population. Find the critical value $t_{\alpha/2}$ corresponding to a 95% confidence level.

SOLUTION   Because $n = 15$, the number of degrees of freedom is given by $n - 1 = 14$. Using Table A-3, we locate the 14th row by referring to the column at the extreme left. As in Section 6-2, a 95% confidence level corresponds to $\alpha = 0.05$, so we find the column listing values for an *area of 0.05 in two tails*. The value corresponding to the row for 14 degrees of freedom and the column for an area of 0.05 in two tails is 2.145, so $t_{\alpha/2} = 2.145$.

We use critical values denoted by $t_{\alpha/2}$ for the margin of error E and the confidence interval.

Margin of Error E for the Estimate of μ [With σ Not Known]

Formula 6-6 $E = t_{\alpha/2}\dfrac{s}{\sqrt{n}}$

where $t_{\alpha/2}$ has $n - 1$ degrees of freedom

Confidence Interval for the Estimate of μ [With σ Not Known]

$$\bar{x} - E < \mu < \bar{x} + E \qquad \text{where} \qquad E = t_{\alpha/2}\dfrac{s}{\sqrt{n}}$$

or

$$\bar{x} \pm E$$

or

$$(\bar{x} - E, \bar{x} + E)$$

The following procedure uses the above margin of error in the construction of confidence interval estimates of μ.

Procedure for Constructing a Confidence Interval for μ (With σ Unknown)

1. Check that the requirements are satisfied. (Requirements: We have a simple random sample, and either the population appears to be normally distributed or $n > 30$.)

2. Using $n - 1$ degrees of freedom, refer to Table A-3 and find the critical value $t_{\alpha/2}$ that corresponds to the desired confidence level.

3. Evaluate the margin of error $E = t_{\alpha/2}s/\sqrt{n}$.

4. Using the value of the calculated margin of error E and the value of the sample mean $\bar{x}$, find the values of $\bar{x} - E$ and $\bar{x} + E$. Substitute those values in the general format for the confidence interval:

$$\bar{x} - E < \mu < \bar{x} + E$$

or

$$\bar{x} \pm E$$

or

$$(\bar{x} - E, \bar{x} + E)$$

5. Round the resulting confidence interval limits. If using the original set of data, round to one more decimal place than is used for the original set of data. If using summary statistics $(n, \bar{x}, s)$, round the confidence interval limits to the same number of decimal places used for the sample mean.

EXAMPLE **Constructing a Confidence Interval** In Section 6-3 we included an example illustrating the construction of a confidence interval to estimate μ. We used the sample of body temperatures in Data Set 2 in Appendix B (for 12 A.M. on day 2), with $n = 106$ and $\bar{x} = 98.20°$F and we also

continued

assumed that the sample is a simple random sample and that σ is "somehow known to be 0.62°F." In reality, σ is not known. Using the statistics $n = 106, \bar{x} = 98.20°F$, and $s = 0.62°F$ (with σ not known) obtained from a simple random sample, find both of the following:

a. The margin of error E.

b. The confidence interval for μ.

SOLUTION

REQUIREMENT ✓ **1.** The first step in our procedure is that we verify that the necessary requirements are satisfied. We must have a simple random sample, which we are assuming in the statement of the problem. Also, we must have either a sample from a normally distributed population or we must have $n > 30$. Because the sample size is $n = 106$, we do have $n > 30$. (Because $n > 30$, verification of normality is not required, but it would be wise to explore the data with graphs.) The check of requirements for the procedure of this section has been successfully completed. We can now continue with the process of constructing a 95% confidence interval by using the t distribution. ✓

2. Next we find the critical value of $t_{\alpha/2} = 1.984$. It is found in Table A-3 as the critical value corresponding to $n - 1 = 105$ degrees of freedom (left column of Table A-3) and an area in two tails of 0.05. (Remember, a 95% confidence level corresponds to $\alpha = 0.05$, which is divided equally between the two tails.) Table A-3 does not include 105 degrees of freedom, so we select the closest number of degrees of freedom, which is 100. The correct value of $t_{\alpha/2}$ for 105 degrees of freedom is 1.983, so using the closest Table A-3 value of 1.984 produces a negligible error here.

3. *Find the margin of error E:* The margin of error $E = 0.11947593$ is computed using Formula 6-6 as shown below, with extra decimal places used to minimize rounding error in the confidence interval found in Step 4.

$$E = t_{\alpha/2}\frac{s}{\sqrt{n}} = 1.984 \cdot \frac{0.62}{\sqrt{106}} = 0.11947593$$

4. *Find the confidence interval:* The confidence interval can now be found by using $\bar{x} = 98.20$ and $E = 0.11947593$ as shown below.

$$\bar{x} - E < \mu < \bar{x} + E$$
$$98.20 - 0.11947593 < \mu < 98.20 + 0.11947593$$
$$98.08052407 < \mu < 98.31947593$$

5. Round the confidence interval limits. Because the sample mean of 98.20 uses two decimal places, round the result to two decimal places to get this result: $98.08 < \mu < 98.32$.

INTERPRETATION This result could also be expressed in the format of 98.20 ± 0.12 or $(98.08, 98.32)$. On the basis of the given sample results, we are 95% confident that the limits of 98.08°F and 98.32°F actually do contain the value of the population mean μ. Note that the confidence interval limits do not contain 98.6°F, the value commonly believed to be the mean body temperature. Based on these results, it appears that the commonly believed value of 98.6°F is *wrong*.

The confidence interval found in the preceding example appears to be the same as the one found in Section 6-3, where we used the normal distribution and the assumption that σ is known to be 0.62°F. Actually, the two confidence intervals are the same only after rounding. Without rounding, the confidence interval in Section 6-3 is (98.08196934, 98.31803066) and the confidence interval found here is (98.08052407, 98.31947593). In some other cases, the differences might be much greater.

We now list the important properties of the *t* distribution that we are using in this section.

Important Properties of the Student *t* Distribution

1. The Student *t* distribution is different for different sample sizes. (See Figure 6-5 for the cases $n = 3$ and $n = 12$.)

2. The Student *t* distribution has the same general symmetric bell shape as the standard normal distribution, but it reflects the greater variability (with wider distributions) that is expected with small samples.

3. The Student *t* distribution has a mean of $t = 0$ (just as the standard normal distribution has a mean of $z = 0$).

4. The standard deviation of the Student *t* distribution varies with the sample size, but it is greater than 1 (unlike the standard normal distribution, which has $\sigma = 1$).

5. As the sample size *n* gets larger, the Student *t* distribution gets closer to the standard normal distribution.

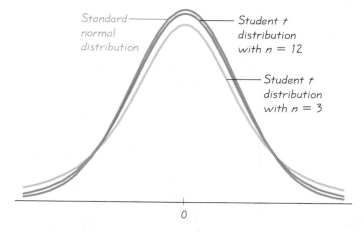

FIGURE 6-5 **Student *t* Distributions for $n = 3$ and $n = 12$**

The Student *t* distribution has the same general shape and symmetry as the standard normal distribution, but it reflects the greater variability that is expected with small samples.

The following is a summary of the conditions indicating use of a t distribution instead of the standard normal distribution. (These same conditions will also apply in Chapter 7.)

Conditions for Using the Student t Distribution

1. σ is unknown; and
2. Either the population has a distribution that is essentially normal, or $n > 30$.

Choosing the Appropriate Distribution

It is sometimes difficult to decide whether to use the standard normal z distribution or the Student t distribution. The flowchart in Figure 6-6 and the accompanying Table 6-1 both summarize the key points to be considered when constructing confidence intervals for estimating μ, the population mean. In Figure 6-6 or Table 6-1, note that if we have a small ($n \leq 30$) sample drawn from a distribution that differs dramatically from a normal distribution, we can't use the methods described in this chapter. One alternative is to use nonparametric methods (see Chapter 12), and another alternative is to use the computer bootstrap method. In both of those approaches, no assumptions are made about the original population. The bootstrap method is described in the Technology Project at the end of this chapter.

The following example focuses on choosing the correct approach by using the methods of this section and Section 6-3.

FIGURE 6-6 Choosing between z and t

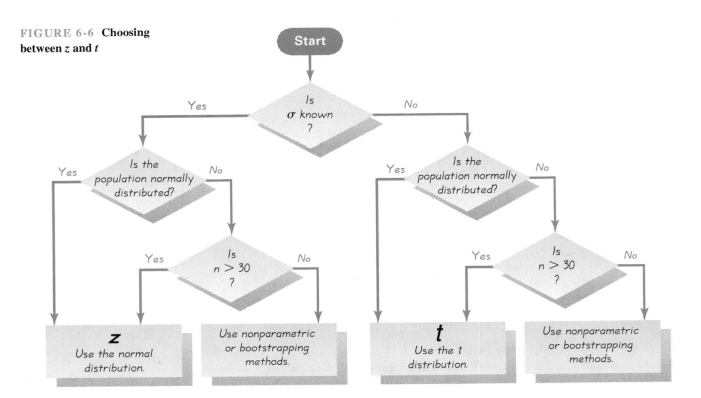

Table 6-1	Choosing between z and t
Method	**Conditions**
Use normal (z) distribution.	σ known and normally distributed population or σ known and $n > 30$
Use t distribution.	σ not known and normally distributed population or σ not known and $n > 30$
Use a nonparametric method or bootstrapping.	Population is not normally distributed and $n \leq 30$

Notes: **1. Criteria for deciding whether the population is normally distributed:** Population need not be exactly normal, but it should appear to be somewhat symmetric with one mode and no outliers.

2. Sample size $n > 30$: This is a commonly used guideline, but sample sizes of 15 to 30 are adequate if the population appears to have a distribution that is not far from being normal and there are no outliers. For some population distributions that are extremely far from normal, the sample size might need to be larger than 50 or even 100.

EXAMPLE **Choosing Distributions** Assuming that you plan to construct a confidence interval for the population mean μ, use the given data to determine whether the margin of error E should be calculated using a critical value of $z_{\alpha/2}$ (from the normal distribution), a critical value of $t_{\alpha/2}$ (from a t distribution), or neither (so that the methods of Sections 6-3 and this section cannot be used).

a. $n = 150, \bar{x} = 100, s = 15$, and the population has a skewed distribution.
b. $n = 8, \bar{x} = 100, s = 15$, and the population has a normal distribution.
c. $n = 8, \bar{x} = 100, s = 15$, and the population has a very skewed distribution.
d. $n = 150, \bar{x} = 100, \sigma = 15$, and the distribution is skewed. (This situation almost never occurs.)
e. $n = 8, \bar{x} = 100, \sigma = 15$, and the distribution is extremely skewed. (This situation almost never occurs.)

SOLUTION   Refer to Figure 6-6 or Table 6-1 to determine the following.

a. Because the population standard deviation σ is not known and the sample is large ($n > 30$), the margin of error is calculated using $t_{\alpha/2}$ in Formula 6-6.

b. Because the population standard deviation σ is not known and the population is normally distributed, the margin of error is calculated using $t_{\alpha/2}$ in Formula 6-6.

continued

c. Because the sample is small and the population does not have a normal distribution, the margin of error E should not be calculated using a critical value of $z_{\alpha/2}$ or $t_{\alpha/2}$. The methods of Section 6-3 and this section do not apply.

d. Because the population standard deviation σ is known and the sample is large ($n > 30$), the margin of error is calculated using $z_{\alpha/2}$ in Formula 6-4.

e. Because the population is not normally distributed and the sample is small ($n \leq 30$), the margin of error E should not be calculated using a critical value of $z_{\alpha/2}$ or $t_{\alpha/2}$. The methods of Section 6-3 and this section do not apply.

EXAMPLE Historical Corn Data In "The Probable Error of a Mean" by William Gosset (*Biometrika*, Vol. VI, No.1) in 1908, the following values were listed for the yields of head corn in pounds per acre. These values resulted from using regular seeds (instead of kiln-dried seeds). Construct a 95% confidence interval estimate of the mean yield.

1903 1935 1910 2496 2108 1961 2060 1444 1612 1316 1511

SOLUTION

REQUIREMENT ✔ **1.** The first step in our procedure is that we verify that the necessary requirements are satisfied. We will assume that we have a simple random sample, which is a reasonable assumption given the way in which the sample data were collected. Also, we must have either a sample from a normally distributed population or we must have $n > 30$. Because the sample size $n = 11$ does not exceed 30, we must verify that the population has a distribution that is approximately normal. The accompanying SPSS screen display shows that the 11 sample values result in a histogram that is roughly bell-shaped, so we can assume that the population has a distribution that is approximately normal. The same SPSS display also shows that $\bar{x} = 1841.5$ and

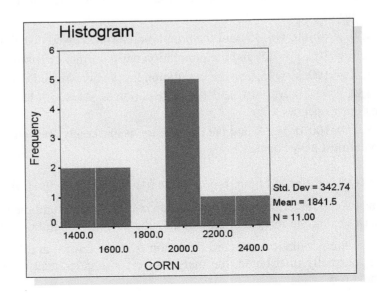

$s = 342.7$ for the sample of 11 values. The check of requirements for the procedure of this section has been successfully completed and we can now continue with the next step in the process of constructing a 95% confidence interval by using the t distribution. ✓

2. Next we find the critical value of $t_{\alpha/2} = 2.228$. It is found in Table A-3 as the critical value corresponding to $n - 1 = 10$ degrees of freedom (left column of Table A-3) and an area in two tails of 0.05. (Remember, a 95% confidence level corresponds to $\alpha = 0.05$, which is divided equally between the two tails.)

3. *Find the margin of error E:* The margin of error $E = 230.215$ is computed using Formula 6-6 as shown below, with extra decimal places used to minimize rounding error in the confidence interval found in Step 4.

$$E = t_{\alpha/2}\frac{s}{\sqrt{n}} = 2.228 \cdot \frac{342.7}{\sqrt{11}} = 230.215$$

4. *Find the confidence interval:* The confidence interval can now be found by using $\bar{x} = 1841.5$ and $E = 230.215$ as shown below.

$$\bar{x} - E < \mu < \bar{x} + E$$
$$1841.5 - 230.215 < \mu < 1841.5 + 230.215$$
$$1611.285 < \mu < 2071.715$$

5. Round the confidence interval limits. Because the original sample data are whole numbers, the result is rounded to one *additional* place to yield this result with one decimal place: $1611.3 < \mu < 2071.7$.

INTERPRETATION   On the basis of the sample data, we are 95% confident that the limits of 1611.3 and 2071.7 actually do contain the value of the population mean.

Finding Point Estimate and E from a Confidence Interval

When software and calculators are used to find a confidence interval, a typical usage requires that you enter a confidence level and sample statistics or raw data, and the display shows the confidence interval limits. The sample mean $\bar{x}$ is the value midway between those limits, and the margin of error E is one-half the difference between those limits (because the upper limit is $\bar{x} + E$ and the lower limit is $\bar{x} - E$, the distance separating them is $2E$).

Point estimate of μ:

$$\bar{x} = \frac{(\text{upper confidence limit}) + (\text{lower confidence limit})}{2}$$

Margin of error:

$$E = \frac{(\text{upper confidence limit}) - (\text{lower confidence limit})}{2}$$

EXAMPLE **Pulse Rates of Women** In analyzing the pulse rates of women listed in Data Set 1 in Appendix B, the SPSS display shown below is obtained. Use the display to find the point estimate $\bar{x}$ and the margin of error E. The sample of pulse rates is randomly selected from a large population of women.

SPSS

Descriptives

			Statistic	Std. Error
Pulse rate (beats/min)	Mean		76.30	1.976
	95% Confidence Interval for Mean	Lower Bound	72.30	
		Upper Bound	80.30	

SOLUTION   In the following calculations, results are rounded to one decimal place, which is one additional decimal place beyond the rounding used for the original list of ages.

$$\bar{x} = \frac{(\text{upper confidence limit}) + (\text{lower confidence limit})}{2}$$

$$= \frac{80.30 + 72.30}{2} = 76.30 \text{ beats per minute}$$

$$E = \frac{(\text{upper confidence limit}) - (\text{lower confidence limit})}{2}$$

$$= \frac{80.30 - 72.30}{2} = 4.00 \text{ beats per minute}$$

Using Confidence Intervals to Describe, Explore, or Compare Data

In some cases, we might use a confidence interval to achieve an ultimate goal of estimating the value of a population parameter. For the body temperature data used in this section, an important goal might be to estimate the mean body temperature of healthy adults, and our results strongly suggest that the commonly used value of 98.6°F is incorrect (because we have 95% confidence that the limits of 98.08°F and 98.32°F contain the true value of the population mean). In other cases, a confidence interval might be one of several different tools used to describe, explore, or compare data sets.

6-4 Exercises

Using Correct Distribution. In Exercises 1–8, do one of the following, as appropriate: (a) Find the critical value $z_{\alpha/2}$, (b) find the critical value $t_{\alpha/2}$, (c) state that neither the normal nor the t distribution applies.

1. 95%; $n = 5$; σ is unknown; population appears to be normally distributed.

2. 95%; $n = 10$; σ is unknown; population appears to be normally distributed.

3. 99%; $n = 15$; σ is known; population appears to be very skewed.

4. 99%; $n = 45$; σ is known; population appears to be very skewed.

5. 90%; $n = 92$; σ is unknown; population appears to be normally distributed.

6. 90%; $n = 9$; $\sigma = 4.2$; population appears to be very skewed.

7. 98%; $n = 7$; $\sigma = 27$; population appears to be normally distributed.

8. 98%; $n = 37$; σ is unknown; population appears to be normally distributed.

Finding Confidence Intervals. In Exercises 9 and 10, use the given confidence level and sample data to find (a) the margin of error and (b) the confidence interval for the population mean μ. Assume that the population has a normal distribution.

9. Math SAT scores for women: 95% confidence; $n = 15, \bar{x} = 496, s = 108$

10. Elbow to fingertip length of men: 99% confidence; $n = 32, \bar{x} = 14.50$ in., $s = 0.70$ in.

Interpreting Calculator Display. In Exercises 11 and 12, use the given data and the corresponding TI-83/84 Plus calculator display to express the confidence interval in the format of $\bar{x} - E < \mu < \bar{x} + E$. Also write a statement that interprets the confidence interval.

11. IQ scores of biology students: 95% confidence; $n = 32, \bar{x} = 117.2, s = 12.1$

12. Heights of NBA players: 99% confidence; $n = 16, \bar{x} = 77.875$ in., $s = 3.50$ in.

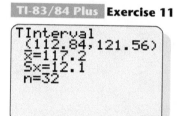

TI-83/84 Plus Exercise 11

```
TInterval
(112.84,121.56)
x̄=117.2
Sx=12.1
n=32
```

Constructing Confidence Intervals. In Exercises 13–24, construct the indicated confidence intervals.

13. Historical Corn Data In "The Probable Error of a Mean" by William Gosset (*Biometrika*, Vol. VI, No.1) published in 1908, the following values were listed for the yields of head corn in pounds per acre. These values resulted from using seeds that were kiln-dried.
   a. Construct a 95% confidence interval estimate of the mean yield.
   b. Compare the result to this confidence interval found from using samples of seed that were not kiln-dried: $1611.3 < \mu < 2071.7$.

   2009 1915 2011 2463 2180 1925 2122 1482 1542 1443 1535

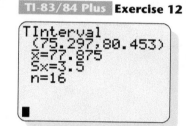

TI-83/84 Plus Exercise 12

```
TInterval
(75.297,80.453)
x̄=77.875
Sx=3.5
n=16
```

14. Crash Hospital Costs A study was conducted to estimate hospital costs for accident victims who wore seat belts. Twenty randomly selected cases have a distribution that appears to be bell-shaped with a mean of $9004 and a standard deviation of $5629 (based on data from the U.S. Department of Transportation).
   a. Construct the 99% confidence interval for the mean of all such costs.
   b. If you are a manager for an insurance company that provides lower rates for drivers who wear seat belts, and you want a conservative estimate for a worst case scenario, what amount should you use as the possible hospital cost for an accident victim who wears seat belts?

15. Heights of Parents Data Set 3 in Appendix B includes the heights of parents of 20 males. If the difference in height for each set of parents is found by subtracting the mother's height from the father's height, the result is a list of 20 values with a mean of 4.4 in. and a standard deviation of 4.2 in. A histogram and a normal quantile plot suggest that the population has a distribution that is not far from normal.

continued

a. Construct a 99% confidence interval estimate of the mean difference between the heights of the mothers and fathers.

b. Does the confidence interval include 0 in.? If a sociologist claims that women tend to marry men who are taller than themselves, does the confidence interval support that claim? Why or why not?

16. Monitoring Lead in Air Listed below are measured amounts of lead (in micrograms per cubic meter or $\mu g/m^3$) in the air. The Environmental Protection Agency has established an air quality standard for lead: 1.5 $\mu g/m^3$. The measurements shown below were recorded at Building 5 of the World Trade Center site on different days immediately following the destruction caused by the terrorist attacks of September 11, 2001. After the collapse of the two World Trade Center buildings, there was considerable concern about the quality of the air. Use the given values to construct a 95% confidence interval estimate of the mean amount of lead in the air. Is there anything about this data set suggesting that the confidence interval might not be very good? Explain.

5.40 1.10 0.42 0.73 0.48 1.10

17. Shoveling Heart Rates Because cardiac deaths appear to increase after heavy snowfalls, an experiment was designed to compare cardiac demands of snow shoveling to those of using an electric snow thrower. Ten subjects cleared tracts of snow using both methods, and their maximum heart rates (beats per minute) were recorded during both activities. The following results were obtained (based on data from "Cardiac Demands of Heavy Snow Shoveling," by Franklin et al., *Journal of the American Medical Association*, Vol. 273, No. 11):

Manual Snow Shoveling Maximum Heart Rates: $n = 10, \bar{x} = 175, s = 15$

Electric Snow Thrower Maximum Heart Rates: $n = 10, \bar{x} = 124, s = 18$

a. Find the 95% confidence interval estimate of the population mean for those people who shovel snow manually.

b. Find the 95% confidence interval estimate of the population mean for those people who use the electric snow thrower.

c. If you are a physician with concerns about cardiac deaths fostered by manual snow shoveling, what single value in the confidence interval from part (a) would be of greatest concern?

d. Compare the confidence intervals from parts (a) and (b) and interpret your findings.

18. Pulse Rates A physician wants to develop criteria for determining whether a patient's pulse rate is atypical, and she wants to determine whether there are significant differences between males and females. Using the sample pulse rates in Data Set 1 in Appendix B, the male pulse rates can be summarized with the statistics $n = 40$, $\bar{x} = 69.4, s = 11.3$. For females, the statistics are $n = 40, \bar{x} = 76.3, s = 12.5$.

a. Construct a 95% confidence interval estimate of the mean pulse rate for males.

b. Compare the 95% confidence intervals for males and females. [The confidence interval for females is given in this section as (72.30, 80.30)]. Can we conclude that the population means for males and females are different? Why or why not?

19. Skull Breadths Maximum breadth of samples of male Egyptian skulls from 4000 B.C. and 150 A.D. (based on data from *Ancient Races of the Thebaid* by Thomson and Randall-Maciver):

4000 B.C. : 131 119 138 125 129 126 131 132 126 128 128 131
150 A.D.:   136 130 126 126 139 141 137 138 133 131 134 129

Changes in head sizes over time suggest interbreeding with people from other regions. Use confidence intervals to determine whether the head sizes appear to have changed from 4000 B.C. to 150 A.D. Explain your result.

20. Head Circumferences In order to correctly diagnose the disorder of hydrocephalus, a pediatrician investigates head circumferences of two-year-old males and females. Use the sample data from Data Set 4 to construct confidence intervals, then determine whether there appears to be a difference between the two genders.

21. Body Mass Index Refer to Data Set 1 in Appendix B and use the sample data.
  a. Construct a 99% confidence interval estimate of the mean body mass index for men.
  b. Construct a 99% confidence interval estimate of the mean body mass index for women.
  c. Compare and interpret the results. We know that men have a mean weight that is greater than the mean weight for women, and the mean height of men is greater than the mean height of women, but do men also have a mean body mass index that is greater than the mean body mass index of women?

22. Petal Lengths of Irises Refer to the petal lengths of irises listed in Data Set 7 in Appendix B. Construct a 95% confidence interval estimate of the mean for each of the three classes. Based on the results, are there any dramatic differences?

23. Sepal Widths of Irises Refer to the sepal widths of irises listed in Data Set 7 in Appendix B. Construct a 95% confidence interval estimate of the mean for each of the three classes. Based on the results, are there any dramatic differences?

24. Hemoglobin Counts Refer to the hemoglobin counts listed in Data Set 10 in Appendix B. Construct a 95% confidence interval estimate of the mean for males and then do the same for females. Are there any notable differences?

25. Yeast Cell Counts Refer to Data Set 11 in Appendix B. Construct a 95% confidence interval estimate of the mean yeast cell count for all such squares. When using yeast in the fermentation process of brewing beer, too few yeast cells result in incomplete fermentation, and too many yeast cells result in beer with a bitter taste. If a brewing process requires that the cell count must be between 4 and 5 cells, do the sample data appear to be acceptable?

26. Effects of Exercise and Stress Refer to Data Set 12 in Appendix B. Construct 95% confidence interval estimates for the mean systolic blood pressure readings corresponding to these two samples: (1) pre-exercise with no stress from the speech test and (2) pre-exercise with stress from the speech test. Compare the results. What do the results suggest about the stress caused by the speech test?

*27. Using the Wrong Distribution Assume that a small simple random sample is selected from a normally distributed population for which σ is unknown. Construction of a confidence interval should use the t distribution, but how are the confidence interval limits affected if the normal distribution is incorrectly used instead?

6-5 Estimating a Population Variance

In this section we consider the same three concepts introduced earlier in this chapter: (1) point estimate, (2) confidence interval, and (3) determining the required sample size. Whereas the preceding sections applied these concepts to estimates

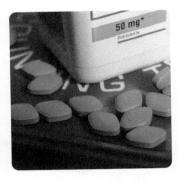

The Value of Drugs

Pfizer, Inc. is a pharmaceutical company that won much profit and fame from at least two very notable drugs: Viagra and Lipitor. It is likely that many more people are familiar with Viagra, even though Lipitor provides Pfizer with much more revenue. In a recent year, Pfizer enjoyed income of $1.75 *billion* from Viagra sales, but Lipitor generated $7.4 *billion* in sales. Lipitor will likely be the first drug to generate more than $10 billion in annual sales. With such income levels and with more than 44 million people taking Lipitor to reduce high cholesterol levels, it has become the most profitable and most used prescription drug ever. Lipitor gained its approval from the Food and Drug Administration through a long process with controlled clinical trials. Results were analyzed with standard statistical methods, including hypothesis testing.

of proportions and means, this section applies them to the population variance σ^2 or standard deviation σ. Here are the main objectives of this section:

1. Given sample values, estimate the population standard deviation σ or the population variance σ^2.

2. Determine the sample size required to estimate a population standard deviation or variance.

Many real situations, such as quality control in a manufacturing process, require that we estimate values of population variances or standard deviations. In addition to making products with measurements yielding a desired mean, the manufacturer must make products of *consistent* quality that do not run the gamut from extremely good to extremely poor. As this consistency can often be measured by the variance or standard deviation, these become vital statistics in maintaining the quality of products and services.

Requirements for Estimating σ or σ^2

1. The sample is a simple random sample.

2. The population must have normally distributed values (even if the sample is large). (The assumption of a normally distributed population was made in earlier sections, but that requirement is more critical here. For the methods of this section, departures from normal distributions can lead to gross errors. Consequently, the requirement of having a normal distribution is much stricter, and we should check the distribution of data by constructing histograms and quantile plots, as described in Section 5-7.)

When we considered estimates of proportions and means, we used the normal and Student t distributions. When developing estimates of variances or standard deviations, we use another distribution, referred to as the chi-square distribution. We will examine important features of that distribution before proceeding with the development of confidence intervals.

Chi-Square Distribution

In a normally distributed population with variance σ^2, we randomly select independent samples of size n and compute the sample variance s^2 (see Formula 2-5) for each sample. The sample statistic $\chi^2 = (n-1)s^2/\sigma^2$ has a distribution called the **chi-square distribution.**

Chi-Square Distribution

Formula 6-7 $$\chi^2 = \frac{(n-1)s^2}{\sigma^2}$$

where n = sample size
s^2 = sample variance
σ^2 = population variance

We denote chi-square by χ^2, pronounced "kigh square." (The specific mathematical equations used to define this distribution are not given here because they are beyond the scope of this text.) To find critical values of the chi-square distribution, refer to Table A-4. The chi-square distribution is determined by the number of degrees of freedom, and in this chapter we use $n - 1$ degrees of freedom.

$$\textbf{degrees of freedom} = n - 1$$

In later chapters we will encounter situations in which the degrees of freedom are not $n - 1$, so we should not make the incorrect generalization that the number of degrees of freedom is always $n - 1$.

Properties of the Distribution of the Chi-Square Statistic

1. The chi-square distribution is not symmetric, unlike the normal and Student t distributions (see Figure 6-7). (As the number of degrees of freedom increases, the distribution becomes more symmetric, as Figure 6-8 illustrates.)

2. The values of chi-square can be zero or positive, but they cannot be negative (see Figure 6-7).

3. The chi-square distribution is different for each number of degrees of freedom (see Figure 6-8), and the number of degrees of freedom is given by $df = n - 1$ in this section. As the number of degrees of freedom increases, the chi-square distribution approaches a normal distribution.

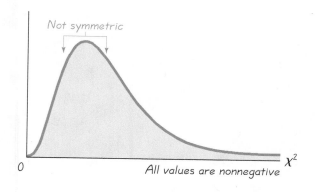

FIGURE 6-7 Chi-Square Distribution

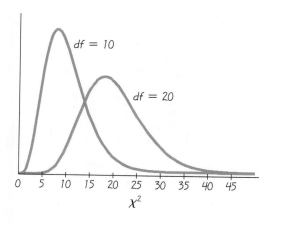

FIGURE 6-8 Chi-Square Distribution for df = 10 and df = 20

Because the chi-square distribution is skewed instead of symmetric, the confidence interval does not fit a format of $s^2 \pm E$ and we must do separate calculations for the upper and lower confidence interval limits. There is a different procedure for finding critical values, illustrated in the following example. Note the following essential feature of Table A-4:

In Table A-4, each critical value of χ^2 corresponds to an area given in the top row of the table, and that area represents the *total region located to the right* of the critical value.

Table A-2 for the standard normal distribution provides cumulative areas from the *left*, but Table A-4 for the chi-square distribution provides cumulative areas from the *right*.

EXAMPLE **Critical Values** Find the critical values of χ^2 that determine critical regions containing an area of 0.025 in each tail. Assume that the relevant sample size is 10 so that the number of degrees of freedom is $10 - 1$, or 9.

SOLUTION See Figure 6-9 and refer to Table A-4. The critical value to the right ($\chi^2 = 19.023$) is obtained in a straightforward manner by locating 9 in the degrees-of-freedom column at the left and 0.025 across the top. The critical value of $\chi^2 = 2.700$ to the left once again corresponds to 9 in the degrees-of-

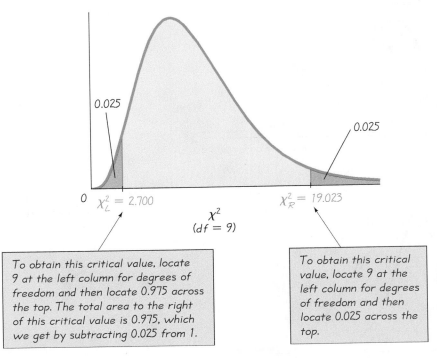

FIGURE 6-9 **Critical Values of the Chi-Square Distribution**

freedom column, but we must locate 0.975 (found by subtracting 0.025 from 1) across the top because the values in the top row are always *areas to the right* of the critical value. Refer to Figure 6-9 and see that the total area to the right of $\chi^2 = 2.700$ is 0.975. Figure 6-9 shows that, for a sample of 10 values taken from a normally distributed population, the chi-square statistic $(n - 1)s^2/\sigma^2$ has a 0.95 probability of falling between the chi-square critical values of 2.700 and 19.023.

When obtaining critical values of χ^2 from Table A-4, note that the numbers of degrees of freedom are consecutive integers from 1 to 30, followed by 40, 50, 60, 70, 80, 90, and 100. When a number of degrees of freedom (such as 52) is not found on the table, you can usually use the closest critical value. For example, if the number of degrees of freedom is 52, refer to Table A-4 and use 50 degrees of freedom. (If the number of degrees of freedom is exactly midway between table values, such as 55, simply find the mean of the two χ^2 values.) For numbers of degrees of freedom greater than 100, use the equation given in Exercise 20, or a more detailed table, or a statistical software package.

Estimators of σ^2

In Section 5-4 we showed that sample variances s^2 (found by using Formula 2-5) tend to target (or center on) the value of the population variance σ^2, so we say that s^2 is an *unbiased estimator* of σ^2. That is, sample variances s^2 do not systematically tend to overestimate the value of σ^2, nor do they systematically tend to underestimate σ^2. Instead, they tend to target the value of σ^2 itself. Also, the values of s^2 tend to produce smaller errors by being closer to σ^2 than do other unbiased measures of variation. For these reasons, s^2 is generally used to estimate σ^2. [However, there are other estimators of σ^2 that could be considered better than s^2. For example, even though $(n - 1)s^2/(n + 1)$ is a biased estimator of σ^2, it has the very desirable property of minimizing the mean of the squares of the errors and therefore has a better chance of being closer to σ^2.]

The sample variance s^2 is used as the best point estimate of the population variance σ^2.

Because s^2 is an unbiased estimator of σ^2, we might expect that s would be an unbiased estimator of σ, but this is not the case. (See Section 5-4.) If the sample size is large, however, the bias is small so that we can use s as a reasonably good estimate of σ. Even though it is a biased estimate, s is often used as a point estimate of σ.

The sample standard deviation s is commonly used as a point estimate of σ (even though it is a biased estimator).

Although s^2 is the best point estimate of σ^2, there is no indication of how good it actually is. To compensate for that deficiency, we develop an interval estimate (or confidence interval) that is more informative.

Confidence Interval (or Interval Estimate) for the Population Variance σ^2

$$\frac{(n-1)s^2}{\chi_R^2} < \sigma^2 < \frac{(n-1)s^2}{\chi_L^2}$$

This expression is used to find a confidence interval for the variance σ^2, but the confidence interval (or interval estimate) for the standard deviation σ is found by taking the square root of each component, as shown below.

$$\sqrt{\frac{(n-1)s^2}{\chi_R^2}} < \sigma < \sqrt{\frac{(n-1)s^2}{\chi_L^2}}$$

The notations χ_R^2 and χ_L^2 in the preceding expressions are described as follows. (Note that some other texts use $\chi_{\alpha/2}^2$ in place of χ_R^2 and they use $\chi_{1-\alpha/2}^2$ in place of χ_L^2.)

Notation

With a total area of α divided equally between the two tails of a chi-square distribution, χ_L^2 denotes the *left*-tailed critical value and χ_R^2 denotes the *right*-tailed critical value (as illustrated in Figure 6-10).

Based on the preceding results, we can summarize the procedure for constructing a confidence interval estimate of σ or σ^2 as follows.

Procedure for Constructing a Confidence Interval for σ or σ^2

1. Check that the requirements for the methods of this section are satisfied. (Requirements: The sample is a simple random sample and a histogram or normal quantile plot suggests that the population has a distribution that is very close to a normal distribution.)

2. Using $n - 1$ degrees of freedom, refer to Table A-4 and find the critical values χ_R^2 and χ_L^2 that correspond to the desired confidence level.

3. Evaluate the upper and lower confidence interval limits using this format of the confidence interval:

$$\frac{(n-1)s^2}{\chi_R^2} < \sigma^2 < \frac{(n-1)s^2}{\chi_L^2}$$

4. If a confidence interval estimate of σ is desired, take the square root of the upper and lower confidence interval limits and change σ^2 to σ.

5. Round the resulting confidence interval limits. If using the original set of data, round to one more decimal place than is used for the original set of data. If using the sample standard deviation or variance, round the confidence interval limits to the same number of decimal places.

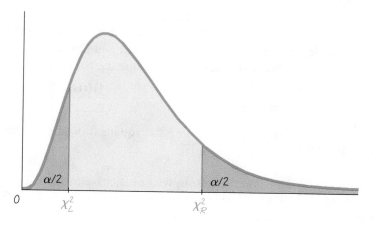

FIGURE 6-10 Chi-Square Distribution with Critical Values χ_L^2 and χ_R^2

The critical values χ_L^2 and χ_R^2 separate the extreme areas corresponding to sample variances that are unlikely (with probability α).

EXAMPLE Body Temperatures Data Set 2 in Appendix B lists 106 body temperatures (at 12:00 A.M. on day 2) obtained by University of Maryland researchers. Assume that the sample is a simple random sample and use the following characteristics of the data set to construct a 95% confidence interval estimate of σ, the standard deviation of the body temperatures of the whole population.

a. As revealed by a histogram and a normal quantile plot, the sample data appear to come from a population with a normal distribution.

b. The sample mean is 98.20°F.

c. The sample standard deviation is $s = 0.62$°F.

d. The sample size is $n = 106$.

e. There are no outliers.

SOLUTION

REQUIREMENT ✓ We should first verify that the necessary requirements are satisfied. (Earlier in this section we listed these requirements: The sample is a simple random sample and the sample appears to come from a normally distributed population.) We are assuming that the sample is a simple random sample. In considering the normality of the population, we note that there are no outliers, and a histogram and normal quantile plot suggest that the population does have a normal distribution. The check of requirements for the procedure of this section has been successfully completed. ✓

We begin by finding the critical values of χ^2. With a sample of 106 values, we have 105 degrees of freedom. This isn't too far away from the 100 degrees of freedom found in Table A-4, so we will go with that. (See Exercise 20 for a method that will yield more accurate critical values.) For a 95% confidence level, we divide $\alpha = 0.05$ equally between the two tails of the chi-square distribution,

continued

and we refer to the values of 0.975 and 0.025 across the top row of Table A-4. The critical values of χ^2 are $\chi^2_L = 74.222$ and $\chi^2_R = 129.561$. Using these critical values, the sample standard deviation of $s = 0.62$, and the sample size of 106, we construct the 95% confidence interval by evaluating the following:

$$\frac{(106 - 1)(0.62)^2}{129.561} < \sigma^2 < \frac{(106 - 1)(0.62)^2}{74.222}$$

This becomes $0.31 < \sigma^2 < 0.54$. Finding the square root of each part (before rounding) yields $0.56°F < \sigma < 0.74°F$.

INTERPRETATION Based on this result, we have 95% confidence that the limits of 0.56°F and 0.74°F contain the true value of σ. We are 95% confident that the standard deviation of body temperatures of all healthy people is between 0.56°F and 0.74°F.

The confidence interval $0.56 < \sigma < 0.74$ can also be expressed as (0.56, 0.74), but the format of $s \pm E$ *cannot* be used because the confidence interval does not have s at its center.

Instead of approximating the critical values by using 100 degrees of freedom, we could use software or the method described in Exercise 20, and the confidence interval becomes $0.55°F < \sigma < 0.72°F$, which is very close to the result obtained here.

Rationale   We now explain why the confidence intervals for σ and σ^2 have the forms just given. If we obtain samples of size n from a population with variance σ^2, the distribution of the $(n - 1)s^2/\sigma^2$ values will be as shown in Figure 6-10. For a simple random sample, there is a probability of $1 - \alpha$ that the statistic $(n - 1)s^2/\sigma^2$ will fall between the critical values of χ^2_L and χ^2_R. In other words (and symbols), there is a $1 - \alpha$ probability that both of the following are true:

$$\frac{(n - 1)s^2}{\sigma^2} < \chi^2_R \qquad \text{and} \qquad \frac{(n - 1)s^2}{\sigma^2} > \chi^2_L$$

If we multiply both of the preceding inequalities by σ^2 and divide each inequality by the appropriate critical value of χ^2, we see that the two inequalities can be expressed in the equivalent forms

$$\frac{(n - 1)s^2}{\chi^2_R} < \sigma^2 \qquad \text{and} \qquad \frac{(n - 1)s^2}{\chi^2_L} > \sigma^2$$

These last two inequalities can be combined into one inequality:

$$\frac{(n - 1)s^2}{\chi^2_R} < \sigma^2 < \frac{(n - 1)s^2}{\chi^2_L}$$

There is a probability of $1 - \alpha$ that these confidence interval limits contain the population variance σ^2. Remember that we should be very careful when interpreting such a confidence interval. It is wrong to say that there is a probability of $1 - \alpha$ that σ^2 will fall between the two confidence interval limits. Instead, we should say that we have $1 - \alpha$ confidence that the limits contain σ^2. Also remember that the requirements are very important. If the sample data were collected in an inappropriate way, the resulting confidence interval may be very wrong.

Determining Sample Size

The procedures for finding the sample size necessary to estimate σ^2 are much more complex than the procedures given earlier for means and proportions. Instead of using very complicated procedures, we will use Table 6-2. STATDISK also provides sample sizes. SPSS, SAS, Minitab, Excel, and the TI-83/84 Plus calculator do not provide such sample sizes.

> **EXAMPLE** We want to estimate σ, the standard deviation of all human body temperatures. We want to be 95% confident that our estimate is within 10% of the true value of σ. How large should the sample be? Assume that the population is normally distributed.
>
> **SOLUTION** From Table 6-2, we can see that 95% confidence and an error of 10% for σ correspond to a sample of size 191. We should randomly select 191 values from the population of body temperatures.

Table 6-2	Sample Size for σ^2		Sample Size for σ	
To be 95% confident that s^2 is within	of the value σ^2, the sample size n should be at least	To be 95% confident that s is within	of the value of σ, the sample size n should be at least	
1%	77,207	1%	19,204	
5%	3,148	5%	767	
10%	805	10%	191	
20%	210	20%	47	
30%	97	30%	20	
40%	56	40%	11	
50%	37	50%	7	
To be 99% confident that s^2 is within	of the value σ^2, the sample size n should be at least	To be 99% confident that s is within	of the value of σ, the sample size n should be at least	
1%	133,448	1%	33,218	
5%	5,457	5%	1,335	
10%	1,401	10%	335	
20%	368	20%	84	
30%	171	30%	37	
40%	100	40%	21	
50%	67	50%	13	

6-5 Exercises

Finding Critical Values. *In Exercises 1–4, find the critical values χ_L^2 and χ_R^2 that correspond to the given confidence level and sample size.*

1. 95%; $n = 16$

2. 95%; $n = 51$

3. 99%; $n = 80$

4. 90%; $n = 40$

Finding Confidence Intervals. *In Exercises 5–8, use the given confidence level and sample data to find a confidence interval for the population standard deviation σ. In each case, assume that a simple random sample has been selected from a population that has a normal distribution.*

5. Salaries of biology professors: 95% confidence; $n = 20$, $\bar{x} = \$95,000$, $s = \$12,345$.

6. Ages of drivers occupying the passing lane while driving 25 mi/h with the left signal flashing: 99% confidence; $n = 27$, $\bar{x} = 80.5$ years, $s = 4.6$ years.

7. Times between uses of a TV remote control by males during commercials: 90% confidence; $n = 30$, $\bar{x} = 5.24$ sec, $s = 2.50$ sec.

8. Starting salaries of college graduates who have taken a biology course: 95% confidence; $n = 51$, $\bar{x} = \$45,678$, $s = \$9900$.

Determining Sample Size. *In Exercises 9–12, assume that each sample is a simple random sample obtained from a normally distributed population.*

9. Find the minimum sample size needed to be 95% confident that the sample standard deviation s is within 10% of σ.

10. Find the minimum sample size needed to be 95% confident that the sample standard deviation s is within 30% of σ.

11. Find the minimum sample size needed to be 99% confident that the sample variance is within 1% of the population variance. Is such a sample size practical in most cases?

12. Find the minimum sample size needed to be 95% confident that the sample variance is within 20% of the population variance.

Finding Confidence Intervals. *In Exercises 13–20, assume that each sample is a simple random sample obtained from a population with a normal distribution.*

13. Historical Corn Data In "The Probable Error of a Mean" by William Gosset (*Biometrika*, Vol. VI, No.1) in 1908, the following values are listed for the yields of head corn in pounds per acre. These values resulted from using regular seeds (instead of kiln-dried seeds). Construct a 95% confidence interval estimate of the standard deviation.

1903 1935 1910 2496 2108 1961 2060 1444 1612 1316 1511

14. Historical Corn Data In "The Probable Error of a Mean" by William Gosset (*Biometrika*, Vol. VI, No.1) in 1908, the following values are listed for the yields of head corn in pounds per acre. These values resulted from using seeds that were kiln-dried.
    a. Construct a 95% confidence interval estimate of the standard deviation.

b. Compare the result to the confidence interval found from using the sample data in Exercise 13.

2009 1915 2011 2463 2180 1925 2122 1482 1542 1443 1535

15. Monitoring Lead in Air Listed below are measured amounts of lead (in micrograms per cubic meter or $\mu g/m^3$) in the air. The Environmental Protection Agency has established an air quality standard for lead: $1.5 \ \mu g/m^3$. The measurements shown below were recorded at Building 5 of the World Trade Center site on different days immediately following the destruction caused by the terrorist attacks of September 11, 2001. After the collapse of the two World Trade buildings, there was considerable concern about the quality of the air. Use the given values to construct a 95% confidence interval estimate of the standard deviation of the amounts of lead in the air. Is there anything about this data set suggesting that the confidence interval might not be very good? Explain.

5.40 1.10 0.42 0.73 0.48 1.10

16. Shoveling Heart Rates Because cardiac deaths appear to increase after heavy snowfalls, an experiment was designed to compare cardiac demands of snow shoveling to those of using an electric snow thrower. Ten subjects cleared tracts of snow using both methods, and their maximum heart rates (beats per minute) were recorded during both activities. The following results were obtained (based on data from "Cardiac Demands of Heavy Snow Shoveling," by Franklin et al., *Journal of the American Medical Association,* Vol. 273, No. 11):

Manual Snow Shoveling Maximum Heart Rates: $n = 10, \bar{x} = 175, s = 15$

Electric Snow Thrower Maximum Heart Rates: $n = 10, \bar{x} = 124, s = 18$

a. Construct a 95% confidence interval estimate of the population standard deviation σ for those who did manual snow shoveling.

b. Construct a 95% confidence interval estimate of the population standard deviation σ for those who used the automated electric snow thrower.

c. Compare and interpret the results. Does the variation appear to be different for the two groups?

17. Pulse Rates A medical researcher wants to determine whether male pulse rates vary more or less than female pulse rates. Using the sample pulse rates in Data Set 1 in Appendix B, the male pulse rates can be summarized with the statistics $n = 40$, $\bar{x} = 69.4, s = 11.3$. For females, the statistics are $n = 40, \bar{x} = 76.3, s = 12.5$.

a. Construct a 95% confidence interval estimate of the population standard deviation σ of pulse rates for males.

b. Construct a 95% confidence interval estimate of the population standard deviation σ of pulse rates for females.

c. Compare the preceding results. Does it appear that the population standard deviations for males and females are different? Why or why not?

18. Patient Waiting Times The Newport Health Clinic experiments with two different configurations for serving patients. In one configuration, patients all enter a single waiting line that feeds three different physicians. In another configuration, patients wait in individual lines at three different physician stations. Waiting times (in minutes) are recorded for 10 patients from each configuration.

a. Construct a 95% confidence interval estimate of the population standard deviation σ of waiting times for the single line.

b. Construct a 95% confidence interval estimate of the population standard deviation σ of waiting times for multiple lines.

c. Compare the preceding results. Does it appear that the population standard deviations are different? Why or why not?

Single line: 65   66   67   68   71   73   74   77   77   77

Multiple lines: 42   54   58   62   67   77   77   85   93   100

19. Body Mass Index Refer to Data Set 1 in Appendix B and use the sample data.
  a. Construct a 99% confidence interval estimate of the standard deviation of body mass indexes for men.
  b. Construct a 99% confidence interval estimate of the standard deviation of body mass indexes for women.
  c. Compare and interpret the results.

*20. Finding Critical Values In constructing confidence intervals for σ or σ^2, we use Table A-4 to find the critical values χ_L^2 and χ_R^2, but that table applies only to cases in which $n \leq 101$, so the number of degrees of freedom is 100 or fewer. For larger numbers of degrees of freedom, we can approximate χ_L^2 and χ_R^2 by using

$$\chi^2 = \frac{1}{2}\left[\pm z_{\alpha/2} + \sqrt{2k - 1}\right]^2$$

where k is the number of degrees of freedom and $z_{\alpha/2}$ is the critical z score first described in Section 6-2. Construct the 95% confidence interval for σ by using the following sample data: The measured heights of 772 men between the ages of 18 and 24 have a standard deviation of 2.8 in. (based on data from the National Health Survey).

Review

The two main activities of inferential statistics are estimating population parameters and testing claims made about population parameters. In this chapter we introduced basic methods for finding *estimates* of population proportions, means, and variances and developed procedures for finding each of the following:

- point estimate
- confidence interval
- required sample size

We discussed point estimate (or single-valued estimate) and formed these conclusions:

- **Proportion:** The best point estimate of p is $\hat{p}$.
- **Mean:** The best point estimate of μ is $\bar{x}$.
- **Variation:** The value of s is commonly used as a point estimate of σ, even though it is a biased estimate. Also, s^2 is the best point estimate of σ^2.

Because the above point estimates consist of single values, they have the serious disadvantage of not revealing how good they are, so confidence intervals (or interval estimates) are commonly used as more revealing and useful estimates. We also considered ways of determining the sample sizes necessary to estimate parameters to within given margins of error. This chapter also introduced the Student t and chi-square distributions. We must be careful to use the correct probability distribution for each set of circumstances. This chapter used the following criteria for selecting the appropriate distribution:

Confidence interval for proportion p: Use the *normal* distribution (assuming that the requirements are satisfied and $np \geq 5$ and $nq \geq 5$ so that the normal distribution can be used to approximate the binomial distribution).

Confidence interval for μ: See Figure 6-6 or Table 6-1 to choose between the *normal* or t distributions (or conclude that neither applies).

Confidence interval for σ or σ^2: Use the *chi-square* distribution (assuming that the requirements are satisfied).

For the confidence interval and sample size procedures in this chapter, it is very important to verify that the listed requirements are satisfied. If they are not, then we cannot use the methods of this chapter and we may need to use other methods, such as the bootstrap method (see the Technology Project at the end of this chapter) or nonparametric methods, such as those discussed in Chapter 12.

Review Exercises

1. Mendel's Genetics Experiments In one of the experiments of Mendel's theory of genetics, peas with long stems were crossed with peas having short stems. (The experiment involved monohybrid crosses.) Results consisted of 787 peas with long stems and 277 peas with short stems.
   a. Find the point estimate of the *percentage* of offspring peas having short stems.
   b. Find a 95% confidence interval estimate of the *percentage* of offspring peas having short stems.
   c. Another characteristic of interest is the constriction of the pea pods. Some offspring peas have inflated pods and others have constricted pods. If you want to estimate the percentage of peas with inflated pods, how many peas must be examined if you want to be 99% confident that your sample percentage is within 2.5 percentage points of the correct population percentage? Assume that we have no prior knowledge about the percentage of peas with inflated pods.

2. Poplar Trees Listed below are heights (in meters) of trees included in an experiment.
   a. Construct a 95% confidence interval estimate of the mean height for the population of control group trees.
   b. Construct a 95% confidence interval estimate of the mean height for the population of trees given the irrigation treatment.
   c. Compare the results from parts (a) and (b).

Control Group:	3.2	1.9	3.6	4.6	4.1	5.1	5.5	4.8	2.9	4.9
	3.9	6.9	6.8	6.0	6.9	6.5	5.5	5.5	4.1	6.3
Irrigation Treatment:	4.1	2.5	4.5	3.6	2.8	5.1	2.9	5.1	1.2	1.6
	6.4	6.8	6.5	6.8	5.8	2.9	5.5	3.3	6.1	6.1

3. Estimating Variation Refer to the same sample data used in Exercise 2.
   a. The trees in the control group were randomly selected from a population. Construct a 95% confidence interval estimate of the standard deviation of heights of trees from this population.

b. Assume that the trees in the irrigation treatment group were randomly selected from a population. Construct a 95% confidence interval estimate of the standard deviation of heights of trees from this population.

c. Compare the results from parts (a) and (b).

4. Estimates from Voter Surveys In a recent presidential election, 611 voters were surveyed and 308 of them said that they voted for the candidate who won (based on data from the ICR Survey Research Group).

   a. Find the point estimate of the *percentage* of voters who said that they voted for the candidate who won.

   b. Find a 98% confidence interval estimate of the *percentage* of voters who said that they voted for the candidate who won.

   c. Of those who voted, 43% actually voted for the candidate who won. Is this result consistent with the survey results? How might a discrepancy be explained?

5. Determining Sample Size You want to estimate the percentage of U.S. biology students who get grades of B or higher. How many such students must you survey if you want 97% confidence that the sample percentage is off by no more than two percentage points?

6. Alcohol Service Policy: Determining Sample Size In a Gallup poll of 1004 adults, 93% indicated that restaurants and bars should refuse service to patrons who have had too much to drink. If you plan to conduct a new poll to confirm that the percentage continues to be correct, how many randomly selected adults must you survey if you want 98% confidence that the margin of error is four percentage points?

7. Investigating Effect of Exercise Refer to Data Set 12 in Appendix B. Construct 95% confidence intervals for the mean systolic blood pressure readings from these two samples: (1) pre-exercise with stress from math test and (2) post-exercise with stress from math test. Based on the results, does the exercise appear to affect systolic blood pressure?

Cumulative Review Exercises

1. Analyzing Weights of Supermodels Supermodels are sometimes criticized on the grounds that their low weights encourage unhealthy eating habits among young women. Listed below are the weights (in pounds) of nine randomly selected supermodels.

125 (Taylor)	119 (Auermann)	128 (Schiffer)	128 (MacPherson)
119 (Turlington)	127 (Hall)	105 (Moss)	123 (Mazza)
115 (Hume)			

Find each of the following:

   a. mean
   b. median
   c. mode
   d. midrange
   e. range
   f. variance
   g. standard deviation
   h. Q_1
   i. Q_2
   j. Q_3
   k. What is the level of measurement of these data (nominal, ordinal, interval, ratio)?
   l. Construct a boxplot for the data.
   m. Construct a 99% confidence interval for the population mean.
   n. Construct a 99% confidence interval for the standard deviation σ.

o. Find the sample size necessary to estimate the mean weight of all supermodels so that there is 99% confidence that the sample mean is in error by no more than 2 lb. Use the sample standard deviation s from part (g) as an estimate of the population standard deviation σ.

p. When women are randomly selected from the general population, their weights are normally distributed with a mean of 143 lb and a standard deviation of 29 lb (based on data from the National Health Survey). Based on the given sample values, do the weights of supermodels appear to be substantially less than the weights of randomly selected women? Explain.

2. X-Linked Recessive Disorders A genetics expert has determined that for certain couples, there is a 0.25 probability that any child will have an X-linked recessive disorder.

a. Find the probability that among 200 such children, at least 65 have the X-linked recessive disorder.

b. A subsequent study of 200 actual births reveals that 65 of the children have the X-linked recessive disorder. Based on these sample results, construct a 95% confidence interval for the proportion of all such children having the disorder.

c. Based on parts (a) and (b), does it appear that the expert's determination of a 0.25 probability is correct? Explain.

3. Analyzing Survey Results In a Gallup poll, adult survey subjects were asked "Do you have a gun in your home?" The respondents consist of 413 people who answered "yes," and 646 other respondents who either answered "no" or had no opinion.

a. What percentage of respondents answered "yes"?

b. Construct a 95% confidence interval estimate of the percentage of all adults who answer "yes" when asked if they have a gun in their home.

c. Can we safely conclude that less than 50% of adults answer "yes" when asked if they have a gun in their home? Why or why not?

d. What is a sensible response to the criticism that the Gallup poll cannot provide good results because the sample size is only 1059 adults selected from a huge population with more than 200 million adults?

Cooperative Group Activities

1. *Out-of-class activity:* Collect sample data, and use the methods of this chapter to construct confidence interval estimates of population parameters. Here are some suggestions for parameters:

- Proportion of students at your college who can raise one eyebrow without raising the other eyebrow. [These sample results are easy to obtain because survey subjects tend to raise one eyebrow (if they can) when they are approached by someone asking questions.]

- Proportion of trees on your campus that are deciduous (or coniferous, or some other relevant characteristic).

- Mean age of cars driven by statistics students and/or the mean age of cars driven by faculty.

- Mean age of biology books and mean age of math books in your college library (based on the copyright dates).

- Mean length of words in *New York Times* editorials and mean length of words in editorials found in your local newspaper.

- Mean length of words in local newspaper and mean length of words in professional journals.

- Proportion of students at your college who can correctly identify the president, and vice president, and the secretary of state.

- Mean age of full-time students at your college.

- Mean pulse rates of male and female full-time students at your college.

- Mean body temperature of full-time students at your college.

2. *In-class activity:* Divide into groups of two. First find the sample size required to estimate the proportion of times that a coin turns up heads when tossed, assuming that you want 80% confidence that the sample proportion is within 0.08 of the true population proportion. Then toss a coin the required number of times and record your results. What percentage of such confidence intervals should actually contain the true value of the population proportion, which we know is $p = 0.5$? Verify this last result by comparing your confidence interval with the confidence intervals found in other groups.

Technology Project

Bootstrap Resampling The *bootstrap method* can be used to construct confidence intervals for situations in which traditional methods cannot (or should not) be used. For example, the following sample of 10 values was randomly selected from a population with a distribution that is very far from normal, so any methods requiring a normal distribution cannot be used.

2.9   564.2   1.4   4.7   67.6   4.8   51.3   3.6   18.0   3.6

If we want to use the above sample data for the construction of a confidence interval estimate of the population mean μ, we note that the sample is small and there is an outlier, so we cannot assume that the sample is from a population with a normal distribution. The bootstrap method, which makes no assumptions about the original population, typically requires a computer to build a bootstrap population by replicating (duplicating) a sample many times. We can draw from the sample with replacement, thereby creating an approximation of the original population. In this way, we pull the sample up "by its own bootstraps" to simulate the original population. Using the sample data given above, construct a 95% confidence interval estimate of the population mean μ by using the bootstrap procedure listed below. Various technologies can be used for this procedure. (If using the STATDISK statistical software program that is on the CD included with this book, enter the 10 sample values in column 1 of the Data Window, then select the main menu item of Analysis, and select the menu item of Bootstrap Resampling.)

a. Create 500 new samples, each of size 10, by selecting 10 values with replacement from the 10 sample values given above.

b. Find the means of the 500 bootstrap samples generated in part (a).

c. Sort the 500 means (arrange them in order).

d. Find the percentiles $P_{2.5}$ and $P_{97.5}$ for the sorted means that result from the preceding step. ($P_{2.5}$ is the mean of the 12th and 13th scores in the sorted list of means; $P_{97.5}$ is the mean of the 487th and 488th scores in the sorted list of means.) Identify the resulting confidence interval by substituting the values for $P_{2.5}$ and $P_{97.5}$ in $P_{2.5} < \mu < P_{97.5}$. Does this confidence interval contain the true value of μ, which is 148?

Now use the bootstrap method to find a 95% confidence interval for the population standard deviation σ. (Use the same steps listed above, but use *standard deviations* instead of means.) Does the bootstrap procedure yield a confidence interval for σ that contains 232.1, verifying that the bootstrap method is effective?

There is a special software package designed specifically for bootstrap resampling methods: Resampling Stats, available from Resampling Stats, Inc., 612 N. Jackson St., Arlington, VA, 22201; telephone number: (703) 522-2713.

from DATA to DECISION

Critical Thinking: For the body mass index (BMI), is there a gender difference?

Exercise 14 in Section 2-4 and Exercise 14 in Section 2-5 use the body mass index (BMI) values from Data Set 1 in Appendix B. Those exercises require comparisons of BMI values for the characteristics of center and variation, but this chapter allows us to use more sophisticated tools. It is well known that men tend to be taller than women and men tend to weigh more than women. The body mass index (BMI) of an individual is computed from height and weight. Refer to Data Set 1 in Appendix B and use the BMI values of men and the BMI values of women. Apply the methods of this chapter to determine whether there appears to be a difference between the two genders. To be thorough, be sure to include verification of any required assumptions, such as any requirements of normal distributions.

7

Hypothesis Testing with One Sample

When is 26.2% more than 25%?

When is 26.2% more than 25%? One simple answer to that question is "always." But, instead of being governed by absolute distinct values, the world of statistics is governed by principles of chance.

Let's again consider the results from an experiment designed to test Mendel's theory of genetics. In the opening problem for Chapter 6, we noted that when monohybrid peas with green pods were crossed with monohybrid peas with yellow pods, there were 580 offspring peas. Among those 580 offspring peas, 152 (or 26.2%) had yellow pods. But, according to Mendel's theory, the predicted percentage of peas with yellow pods is 25%. In the simple world of absolute distinct values, we might say that the experiment results did not support Mendel's theory because we did not get 25% of the peas with yellow pods. But in the world of statistics, we take a much more enlightened, realistic, and practical approach. We recognize that the pea pods were not strictly determined by some rigid rule. The pea pods were affected by an element of chance. Even if Mendel's predicted rate of 25% is correct, we should not expect that every experiment will yield an exact precise value of 25% for the percentage of peas with yellow pods. Some experiments will have a little more than 25%, and other experiments will have a little less than 25%. So how do we know when the result of an experiment does contradict Mendel's predicted value of 25%? The key lies in one word that is at the very core of modern statistics: *significance.* Rejection of Mendel's claimed rate of 25% requires that we obtain results that not only differ from 25%, but they must differ by a *significant* amount. How do we decide when results actually differ from 25% by a significant amount? We use a very standard procedure, which we call *hypothesis testing*. This procedure is extremely important, and is the central focus of this chapter.

7-1 Overview

This chapter describes the statistical procedure for testing hypotheses, which is a very standard procedure that is commonly used by professionals in a wide variety of different disciplines. Professional publications such as the *Journal of the American Medical Association, New England Journal of Medicine, American Journal of Psychiatry*, and *International Journal of Advertising* routinely include the same basic procedures presented in this chapter. Consequently, the work done in studying the methods of this chapter is work that applies to all disciplines, not just statistics.

    The two major activities of inferential statistics are the estimation of population parameters (introduced in Chapter 6) and hypothesis testing (introduced in this chapter). A hypothesis test is a standard procedure for testing some *claim*.

> ### Definition
>
> In statistics, a **hypothesis** is a claim or statement about a property of a population.
>
> A **hypothesis test** (or **test of significance**) is a standard procedure for testing a claim about a property of a population.

    The following statements are typical of the hypotheses (claims) that can be tested by the procedures we develop in this chapter.

- Mendel claims that under certain circumstances, the percentage of offspring peas with yellow pods is 25%.
- Medical researchers claim that the mean body temperature of healthy adults is not equal to 98.6°F.
- When new equipment is used to manufacture heart pacemaker batteries, the new batteries are better because the variation in times to failure is reduced so that the batteries run more consistently.

    Before beginning to study this chapter, you should recall—and understand clearly—this basic rule, first introduced in Section 3-1.

Rare Event Rule for Inferential Statistics

> **If, under a given assumption, the probability of a particular observed event is exceptionally small, we conclude that the assumption is probably not correct.**

    Following this rule, we test a claim by analyzing sample data in an attempt to distinguish between results that can *easily occur by chance* and results that are *highly unlikely to occur by chance*. We can explain the occurrence of highly unlikely results by saying that either a rare event has indeed occurred or that the underlying assumption is not true. Let's apply this reasoning in the following example.

EXAMPLE **Gender Selection** ProCare Industries, Ltd., once provided a product called "Gender Choice," which, according to advertising claims, allowed couples to "increase your chances of having a boy up to 85%, a girl up to 80%." Gender Choice was available in blue packages for couples wanting a baby boy and (you guessed it) pink packages for couples wanting a baby girl. Suppose we conduct an experiment with 100 couples who want to have baby girls, and they all follow the Gender Choice "easy-to-use in-home system" described in the pink package. For the purpose of testing the claim of an increased likelihood for girls, we will assume that Gender Choice has no effect. Using common sense and no formal statistical methods, what should we conclude about the assumption of no effect from Gender Choice if 100 couples using Gender Choice have 100 babies consisting of

a. 52 girls?

b. 97 girls?

SOLUTION

a. We normally expect around 50 girls in 100 births. The result of 52 girls is close to 50, so we should not conclude that the Gender Choice product is effective. If the 100 couples used no special methods of gender selection, the result of 52 girls could easily occur by chance. With 52 girls among 100 babies, there isn't sufficient evidence to say that Gender Choice is effective.

b. The result of 97 girls in 100 births is extremely unlikely to occur by chance. We could explain the occurrence of 97 girls in one of two ways: Either an *extremely* rare event has occurred by chance, or Gender Choice is effective. The extremely low probability of getting 97 girls is strong evidence against the assumption that Gender Choice has no effect. In this case, Gender Choice would appear to be effective.

The key point of the preceding example is that we should conclude that the product is effective only if we get *significantly* more girls than we would normally expect. Although the outcomes of 52 girls and 97 girls are both "above average," the result of 52 girls is not significant, whereas 97 girls is a significant result.

This brief example illustrates the basic approach used in testing hypotheses. The formal method involves a variety of standard terms and conditions incorporated into an organized procedure. We suggest that you begin the study of this chapter by first reading Sections 7-2 and 7-3 casually to obtain a general idea of their concepts and then rereading Section 7-2 more carefully to become familiar with the terminology.

7-2   Basics of Hypothesis Testing

In this section we describe the formal components used in hypothesis testing: null hypothesis, alternative hypothesis, test statistic, critical region, significance level, critical value, *P*-value, type I error, and type II error. *The focus in this section is on the individual components of the hypothesis test, whereas the following sections will bring those components together in comprehensive procedures.* Here are the objectives for this section.

Objectives for This Section

- Given a claim, identify the null hypothesis and the alternative hypothesis, and express them both in symbolic form.
- Given a claim and sample data, calculate the value of the test statistic.
- Given a significance level, identify the critical value(s).
- Given a value of the test statistic, identify the *P*-value.
- State the conclusion of a hypothesis test in simple, nontechnical terms.
- Identify the type I and type II errors that could be made when testing a given claim.

You should study the following example until you thoroughly understand it. Once you do, you will have captured a major concept of statistics.

> **EXAMPLE** Let's again refer to the Gender Choice product that was once distributed by ProCare Industries. In Section 7-1 we noted that the pink packages of Gender Choice were intended to help couples increase the likelihood that when a couple has a baby, the baby will be a girl. ProCare Industries claimed that couples using the pink packages of Gender Choice would have girls at a rate that is greater than 50% or 0.5. Let's again consider an experiment whereby 100 couples use Gender Choice in an attempt to have a baby girl, let's assume that the 100 babies include exactly 52 girls, and let's formalize some of the analysis.
>
> Under normal circumstances the proportion of girls is 0.5, so a claim that Gender Choice is effective can be expressed as $p > 0.5$ (that is, the proportion of girls is greater than 0.5). The outcome of 52 girls supports that claim if the probability of getting *at least 52 girls* is small, such as less than or equal to 0.05. [*Important note:* The probability of getting *exactly* 52 girls or any other specific number of girls is relatively small, but we need the probability of getting a result that is *at least as extreme* as the result of 52 girls. If this point is confusing, review the subsection of "Using Probabilities to Determine When Results Are Unusual" in Section 4-2, where we noted that "*x* successes among *n* trials is an *unusually high* number of successes if $P(x$ or more$)$ is very small (such as 0.05 or less)." Using that criterion, 52 girls in 100 births would be an unusually high number of girls if $P(52$ or more girls$) \leq 0.05$.]
>
> Using a normal distribution as an approximation to the binomial distribution (see Section 5-6), we find $P(52$ or more girls in 100 births$) = 0.3821$. Because we need to determine whether a result of at least 52 girls has a small probability under normal circumstances, we assume that the probability of a girl is 0.5. Figure 7-1 shows that with a probability of 0.5, the outcome of 52 girls in 100 births is not unusual, so we do *not* reject random chance as a reasonable explanation. We conclude that the proportion of girls born to couples using Gender Choice is not significantly greater than the number that we would expect by random chance. Here are the key points:

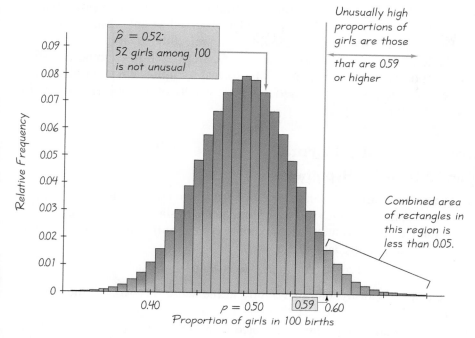

FIGURE 7-1 **Sampling Distribution of Proportions of Girls in 100 Births**

- Claim: For couples using Gender Choice, the proportion of girls is $p > 0.5$.

- Working assumption: The proportion of girls is $p = 0.5$ (with no effect from Gender Choice).

- The sample resulted in 52 girls among 100 births, so the sample proportion is $\hat{p} = 52/100 = 0.52$.

- Assuming that $p = 0.5$, we use a normal distribution as an approximation to the binomial distribution to find that P(at least 52 girls in 100 births) = 0.3821. (Using the methods of Section 5-6 with the normal distribution as an approximation to the binomial distribution, we have $n = 100, p = 0.5$. The observed value of 52 girls is changed to 51.5 as a correction for continuity, and 51.5 converts to $z = 0.30$.)

- There are two possible explanations for the result of 52 girls in 100 births: Either a random chance event (with probability 0.3821) has occurred, or the proportion of girls born to couples using Gender Choice is greater than 0.5. Because the probability of getting at least 52 girls by chance is so high (0.3821), we choose random chance as a reasonable explanation. There isn't sufficient evidence to support a claim that Gender Choice is effective in producing more girls than expected by chance. It was actually this type of analysis that led to the removal of Gender Choice from the market.

The preceding example illustrates well the basic method of reasoning we will use throughout this chapter. Focus on the use of the rare event rule of inferential statistics: **If, under a given assumption, the probability of a particular observed**

event is exceptionally small, we conclude that the assumption is probably not correct. But if the probability of a particular observed sample result is *not* very small, then we do *not* have sufficient evidence to reject the assumption.

In Section 7-3 we will describe the specific steps used in hypothesis testing, but let's first describe the components of a formal **hypothesis test,** or **test of significance.** These terms are often used in a wide variety of disciplines when statistical methods are required.

Components of a Formal Hypothesis Test

Null and Alternative Hypotheses

- The **null hypothesis** (denoted by H_0) is a statement that the value of a population parameter (such as proportion, mean, or standard deviation) is *equal to* some claimed value. Here are some typical null hypotheses of the type considered in this chapter:

$$H_0\text{: } p = 0.5 \qquad H_0\text{: } \mu = 98.6 \qquad H_0\text{: } \sigma = 15$$

  We test the null hypothesis directly in the sense that we assume it is true and reach a conclusion to either reject H_0 or fail to reject H_0.

- The **alternative hypothesis** (denoted by H_1 or H_a) is the statement that the parameter has a value that somehow differs from the null hypothesis. For the methods of this chapter, the symbolic form of the alternative hypothesis must use one of these symbols: $<$ or $>$ or $\neq$. Here are nine different examples of alternative hypotheses involving proportions, means, and standard deviations:

Proportions:	$H_1\text{: } p > 0.5$	$H_1\text{: } p < 0.5$	$H_1\text{: } p \neq 0.5$
Means:	$H_1\text{: } \mu > 98.6$	$H_1\text{: } \mu < 98.6$	$H_1\text{: } \mu \neq 98.6$
Standard Deviations:	$H_1\text{: } \sigma > 15$	$H_1\text{: } \sigma < 15$	$H_1\text{: } \sigma \neq 15$

Note About Always Using the Equal Symbol in H_0: Although a few textbooks use the symbols $\leq$ and $\geq$ in the null hypothesis H_0, most professional journals use only the equal symbol for equality. We conduct the hypothesis test by assuming that the proportion, mean, or standard deviation is *equal to* some specified value so that we can work with a single distribution having a specific value. (Where this textbook uses an expression such as $p = 0.5$ for a null hypothesis, some other textbooks might use $p \leq 0.5$ or $p \geq 0.5$ instead. This textbook uses the working assumption that the parameter is equal to some specified value, but some other textbooks allow other expressions for the null hypothesis.)

Note About Forming Your Own Claims (Hypotheses): If you are conducting a study and want to use a hypothesis test to *support* your claim, the claim must be worded so that it becomes the alternative hypothesis. This requires that your claim must be expressed using only these symbols: $<$ or $>$ or $\neq$. You cannot use a hypothesis test to support a claim that some parameter is *equal to* some specified value.

For example, suppose you have developed a magic potion that raises IQ scores so that the mean becomes greater than 100. If you want to provide evidence of the potion's effectiveness, you must state the claim as $\mu > 100$. (In this context

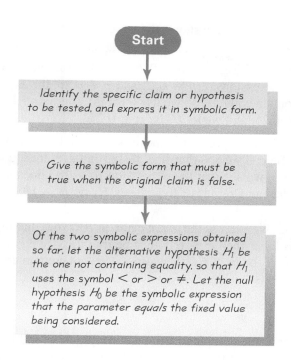

FIGURE 7-2
Identifying H_0 and H_1

of trying to support the goal of the research, the alternative hypothesis is sometimes referred to as the *research hypothesis*. Also in this context, the null hypothesis of $\mu = 100$ is assumed to be true for the purpose of conducting the hypothesis test, but it is hoped that the conclusion will be rejection of the null hypothesis so that the claim of $\mu > 100$ is supported.)

Note About Identifying H_0 and H_1: Figure 7-2 summarizes the procedures for identifying the null and alternative hypotheses. Note that the original statement could become the null hypothesis, it could become the alternative hypothesis, or it might not correspond exactly to either the null hypothesis or the alternative hypothesis.

For example, we sometimes test the validity of someone else's claim, such as the claim of the McKesson Corporation that "the mean amount of rubbing alcohol in its containers is at least 16 fluid ounces." That claim can be expressed in symbols as $\mu \geq 16$. In Figure 7-2 we see that if that original claim is false, then $\mu < 16$. The alternative hypothesis becomes $\mu < 16$, but the null hypothesis is $\mu = 16$. We will be able to address the original claim after determining whether there is sufficient evidence to reject the null hypothesis of $\mu = 16$.

EXAMPLE Identifying the Null and Alternative Hypotheses
Refer to Figure 7-2 and use the given claims to express the corresponding null and alternative hypotheses in symbolic form.

a. The proportion of peas with yellow pods is equal to 0.25.

b. The mean height of an adult male is at most 183 cm.

c. The standard deviation of heights of adult women is greater than 6.0 cm.

continued

SOLUTION   See Figure 7-2, which shows the three-step procedure.

a. In Step 1 of Figure 7-2, we express the given claim as $p = 0.25$. In Step 2 we see that if $p = 0.25$ is false, then $p \neq 0.25$ must be true. In Step 3, we see that with the two expressions of $p = 0.25$ and $p \neq 0.25$, the expression $p \neq 0.25$ does not contain equality, so we use it for the alternative hypothesis. We let H_1 be $p \neq 0.25$, and we let H_0 be $p = 0.25$.

b. In Step 1 of Figure 7-2, we express "the mean is at most 183 cm" in symbols as $\mu \leq 183$. In Step 2 we see that if $\mu \leq 183$ is false, then $\mu > 183$ must be true. In Step 3, we see that the expression $\mu > 183$ does not contain equality, so we let the alternative hypothesis H_1 be $\mu > 183$, and we let H_0 be $\mu = 183$.

c. In Step 1 of Figure 7-2, we express the given claim as $\sigma > 6.0$. In Step 2 we see that if $\sigma > 6.0$ is false, then $\sigma \leq 6.0$ must be true. In Step 3, we let the alternative hypothesis H_1 be $\sigma > 6.0$ (because it does not contain equality), and we let H_0 be $\sigma = 6.0$.

Test Statistic

- The **test statistic** is a value computed from the sample data, and it is used in making the decision about the rejection of the null hypothesis. The test statistic is found by converting the sample statistic (such as the sample proportion $\hat{p}$, or the sample mean $\bar{x}$, or the sample standard deviation s) to a score (such as z, t, or χ^2) with the assumption that the null hypothesis is true. The test statistic can therefore be used for determining whether there is significant evidence against the null hypothesis. In this chapter, we consider hypothesis tests involving proportions, means, and standard deviations (or variances). Based on results from preceding chapters about the sampling distributions of proportions, means, and standard deviations, we use the following test statistics:

Test statistic for proportion
$$z = \frac{\hat{p} - p}{\sqrt{\dfrac{pq}{n}}}$$

Test statistic for mean
$$z = \frac{\bar{x} - \mu}{\dfrac{\sigma}{\sqrt{n}}} \quad \text{or} \quad t = \frac{\bar{x} - \mu}{\dfrac{s}{\sqrt{n}}}$$

Test statistic for standard deviation
$$\chi^2 = \frac{(n - 1)s^2}{\sigma^2}$$

The above test statistic for a proportion is based on the results given in Section 5-6, but it does not include the continuity correction that we usually use when approximating a binomial distribution by a normal distribution. When working with proportions in this chapter, we will work with large samples, so the continuity correction can be ignored because its effect is small. Also, the test statistic for a mean can be based on the normal or Student t distribution, depending on the conditions that are satisfied. When choosing between the normal or Student t distributions, this chapter will use the same criteria described in Section 6-4. (See Figure 6-6 and Table 6-1.)

EXAMPLE **Finding the Test Statistic** A genetics experiment results in 580 offspring peas, and 26.2% (or $\hat{p} = 0.262$) of them have yellow pods. Find the value of the test statistic for Mendel's claim that the proportion of peas with yellow pods is equal to 0.25. (In Section 7-3 we will see that there are requirements that must be verified. For this example, assume that the requirements are satisfied and focus on finding the indicated test statistic.)

SOLUTION The preceding example showed that the given claim results in the following null and alternative hypotheses: $H_0: p = 0.25$ and $H_1: p \neq 0.25$. Because we work under the assumption that the null hypothesis is true with $p = 0.25$, we get the following test statistic:

$$z = \frac{\hat{p} - p}{\sqrt{\frac{pq}{n}}} = \frac{0.262 - 0.25}{\sqrt{\frac{(0.25)(0.75)}{580}}} = 0.67$$

INTERPRETATION We know from previous chapters that a z score of 0.67 is not very large. (A large z score would be one that is greater than 2 or less than -2, indicating that the sample data are more than two standard deviations away from the claimed value.) It appears that in this case, 26.2% is not *significantly* different from 25%. See Figure 7-3 where we show that the sample proportion of 0.262 (from 26.2%) does not fall within the range of values considered to be significant because they are so far away from 0.25 that they are not likely to occur by chance (assuming that the population proportion is $p = 0.25$).

Critical Region, Significance Level, Critical Value, and P-Value

- The **critical region** (or **rejection region**) is the set of all values of the test statistic that cause us to reject the null hypothesis. For example, see the regions of shaded tails in Figure 7-3.

- The **significance level** (denoted by α) is the probability that the test statistic will fall in the critical region when the null hypothesis is actually true. If the test statistic falls in the critical region, we will reject the null hypothesis, so α is the probability of making the mistake of rejecting the null hypothesis when it is true. This is the same α introduced in Section 6-2, where we defined the confidence level for a confidence interval to be the probability $1 - \alpha$. Common choices for α are 0.05, 0.01, and 0.10, with 0.05 being most common.

FIGURE 7-3 Critical Region, Critical Value, Test Statistic

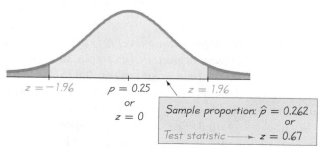

Proportion of Peas with Yellow Pods

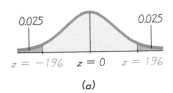

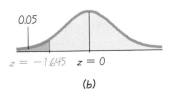

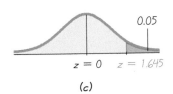

FIGURE 7-4 Finding
Critical Values

- A **critical value** is any value that separates the critical region (where we reject the null hypothesis) from the values of the test statistic that do not lead to rejection of the null hypothesis. The critical values depend on the nature of the null hypothesis, the sampling distribution that applies, and the significance level α. See Figure 7-3 where the critical values of $z = \pm 1.96$ correspond to a significance level of $\alpha = 0.05$. (Critical values were also discussed in Chapter 6.)

EXAMPLE **Finding Critical Values** Using a significance level of $\alpha = 0.05$, find the critical z values for each of the following alternative hypotheses (assuming that the normal distribution can be used to approximate the binomial distribution):

a. $p \neq 0.25$ (so the critical region is in *both* tails of the normal distribution)

b. $p < 0.25$ (so the critical region is in the *left* tail of the normal distribution)

c. $p > 0.25$ (so the critical region is in the *right* tail of the normal distribution)

SOLUTION

a. See Figure 7-4(a). The shaded tails contain a total area of $\alpha = 0.05$, so each tail contains an area of 0.025. Using the methods of Section 5-2, the values of $z = 1.96$ and $z = -1.96$ separate the right and left tail regions. The critical values are therefore $z = 1.96$ and $z = -1.96$.

b. See Figure 7-4(b). With an alternative hypothesis of $p < 0.25$, the critical region is in the left tail. With a left-tail area of 0.05, the critical value is found to be $z = -1.645$ (by using the methods of Section 5-2).

c. See Figure 7-4(c). With an alternative hypothesis of $p > 0.25$, the critical region is in the right tail. With a right-tail area of 0.05, the critical value is found to be $z = 1.645$ (by using the methods of Section 5-2).

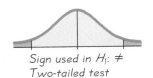

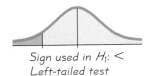

FIGURE 7-5 Two-Tailed,
Left-Tailed, Right-Tailed Tests

Two-Tailed, Left-Tailed, Right-Tailed   The *tails* in a distribution are the extreme regions bounded by critical values. Some hypothesis tests are two-tailed, some are right-tailed, and some are left-tailed.

- **Two-tailed test:** The critical region is in the two extreme regions (tails) under the curve.
- **Left-tailed test:** The critical region is in the extreme left region (tail) under the curve.
- **Right-tailed test:** The critical region is in the extreme right region (tail) under the curve.

In two-tailed tests, the significance level α is divided equally between the two tails that constitute the critical region. For example, in a two-tailed test with a significance level of $\alpha = 0.05$, there is an area of 0.025 in each of the two tails. In tests that are right- or left-tailed, the area of the critical region in one tail is α.

By examining the alternative hypothesis, we can see whether a test is right-tailed, left-tailed, or two-tailed. The tail will correspond to the critical region containing the values that would conflict significantly with the null hypothesis. A use-

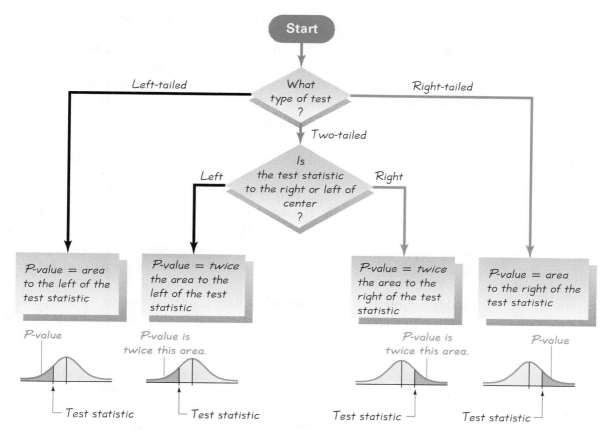

FIGURE 7-6 **Procedure for Finding P-Values**

ful check is summarized in the margin figures (see Figure 7-5), which show that the inequality sign in H_1 points in the direction of the critical region. The symbol $\neq$ is often expressed in programming languages as $< >$, and this reminds us that an alternative hypothesis such as $p \neq 0.5$ corresponds to a two-tailed test.

- The **P-value** (or **p-value** or **probability value**) is the probability of getting a value of the test statistic that is *at least as extreme* as the one representing the sample data, assuming that the null hypothesis is true. The null hypothesis is rejected if the P-value is very small, such as 0.05 or less. P-values can be found by using the procedure summarized in Figure 7-6.

Decisions and Conclusions

We have seen that the original claim sometimes becomes the null hypothesis and at other times becomes the alternative hypothesis. However, our standard procedure of hypothesis testing requires that we always test the null hypothesis, so our initial conclusion will always be one of the following:

1. Reject the null hypothesis.
2. Fail to reject the null hypothesis.

Decision Criterion: The decision to reject or fail to reject the null hypothesis is usually made using either the traditional method (or classical method) of testing hypotheses, the *P*-value method, or the decision is sometimes based on confidence intervals. In recent years, use of the traditional method has been declining, partly because statistical software packages are often designed for the *P*-value method, but not the traditional method.

Traditional method:	*Reject H_0 if the test statistic falls within the critical region.*
	Fail to reject H_0 if the test statistic does not fall within the critical region.
P-value method:	*Reject H_0 if P-value $\leq \alpha$* (where α is the significance level, such as 0.05).
	Fail to reject H_0 if P-value $> \alpha$.
Another option:	Instead of using a significance level such as $\alpha = 0.05$, simply identify the *P* value and leave the decision to the reader.
Confidence intervals:	Because a confidence interval estimate of a population parameter contains the likely values of that parameter, reject a claim that the population parameter has a value that is not included in the confidence interval.

Many statisticians consider it good practice to always select a significance level *before* doing a hypothesis test. This is a particularly good procedure when using the *P*-value method because we may be tempted to adjust the significance level based on the results. For example, with a 0.05 significance level and a *P*-value of 0.06, we should fail to reject the null hypothesis, but it is sometimes tempting to say that a probability of 0.06 is small enough to warrant rejection of the null hypothesis. Other statisticians argue that prior selection of a significance level reduces the usefulness of *P*-values. They contend that no significance level should be specified and that the conclusion should be left to the reader. In this book we use the decision criterion that involves a comparison of a significance level and the *P*-value.

> **EXAMPLE Finding *P*-Values** First determine whether the given conditions result in a two-tailed test, a left-tailed test, or a right-tailed test, then use Figure 7-6 to find the *P*-value, then state a conclusion about the null hypothesis.
>
> **a.** A significance level of $\alpha = 0.05$ is used in testing the claim that $p > 0.25$, and the sample data result in a test statistic of $z = 2.34$.
>
> **b.** A significance level of $\alpha = 0.05$ is used in testing the claim that $p \neq 0.25$, and the sample data result in a test statistic of $z = 0.67$.
>
> **SOLUTION**
>
> **a.** With a claim of $p > 0.25$, the test is right-tailed (see Figure 7-5.) We can find the *P*-value by using Figure 7-6. Because the test is right-tailed, Figure 7-6

shows that the *P*-value is the area to the right of the test statistic $z = 2.34$. Using the methods of Section 5-2, we refer to Table A-2 and find that the area to the *right* of $z = 2.34$ is $1 - 0.9904 = 0.0096$. The *P*-value of 0.0096 is less than or equal to the significance level $\alpha = 0.05$, so we reject the null hypothesis. The *P*-value of 0.0096 is relatively small, indicating that the sample results could *not* easily occur by chance.

b. With a claim of $p \neq 0.25$, the test is two-tailed (see Figure 7-5). We can find the *P*-value by using Figure 7-6. Because the test is two-tailed, and because the test statistic of $z = 0.67$ is to the right of the center, Figure 7-6 shows that the *P*-value is *twice* the area to the right of $z = 0.67$. Using the methods of Section 5-2, we refer to Table A-2 and find that the area to the right of $z = 0.67$ is $1 - 0.7486 = 0.2514$, so the *P*-value $= 2 \times 0.2514 = 0.5028$. The *P*-value of 0.5028 is greater than the significance level, so we fail to reject the null hypothesis. The large *P*-value of 0.5028 shows that the sample results could easily occur by chance.

Wording of the Final Conclusion: The conclusion of rejecting the null hypothesis or failing to reject it is fine for those of us with the wisdom to take a statistics course, but we should use simple, nontechnical terms in stating what the conclusion really means. Figure 7-7 summarizes a procedure for wording of the

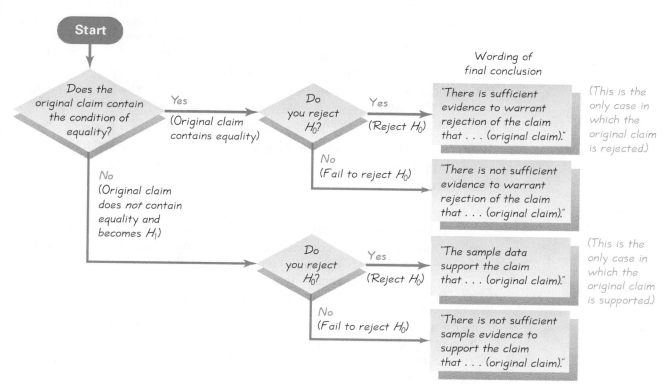

FIGURE 7-7 **Wording of the Final Conclusion**

final conclusion. Note that only one case leads to a statement that the sample data actually *support* the conclusion. If you want to support some claim, state it in such a way that it becomes the alternative hypothesis, and then hope that the null hypothesis gets rejected. For example, to support the claim that the mean body temperature is different from 98.6°F, make the claim that $\mu \neq 98.6$°F. This claim will be an alternative hypothesis that will be supported if you reject the null hypothesis, H_0: $\mu = 98.6$°F. If, on the other hand, you claim that $\mu = 98.6$°F, you will either reject or fail to reject the claim; in either case, you will never *support* the claim that $\mu = 98.6$°F.

Accept/Fail to Reject: Some textbooks say "accept the null hypothesis" instead of "fail to reject the null hypothesis." Whether we use the term *accept* or *fail to reject*, we should recognize that *we are not proving the null hypothesis;* we are merely saying that the sample evidence is not strong enough to warrant rejection of the null hypothesis. It's like a jury's saying that there is not enough evidence to convict a suspect. The term *accept* is somewhat misleading, because it seems to imply incorrectly that the null hypothesis has been proved. (It is misleading to state that "there is sufficient evidence to accept the null hypothesis.") The phrase *fail to reject* says more correctly that the available evidence isn't strong enough to warrant rejection of the null hypothesis. In this book we will use the terminology *fail to reject the null hypothesis,* instead of *accept the null hypothesis.*

Multiple Negatives: When stating the final conclusion in nontechnical terms, it is possible to get correct statements with up to three negative terms. (Example: "There is *not* sufficient evidence to warrant *rejection* of the claim of *no* difference between 0.5 and the population proportion.") Such conclusions with so many negative terms can be confusing, so it would be good to restate them in a way that makes them understandable, but care must be taken to not change the meaning. For example, instead of saying that "there is not sufficient evidence to warrant rejection of the claim of no difference between 0.5 and the population proportion," better statements would be these:

- Fail to reject the claim that the population proportion is equal to 0.5.
- Until stronger evidence is obtained, assume that the population proportion is equal to 0.5.

EXAMPLE **Stating the Final Conclusion** Suppose we test Mendel's claim that the proportion of peas with yellow pods is equal to 25%. This claim of $p = 0.25$ becomes the null hypothesis, while the alternative hypothesis becomes $p \neq 0.25$. Assuming that we fail to reject the null hypothesis of $p = 0.25$, state the conclusion in simple, nontechnical terms.

SOLUTION   Refer to Figure 7-7. The original claim *does* contain the condition of equality, and we fail to reject the null hypothesis. The wording of the final conclusion should therefore be as follows: "There is not sufficient evidence to warrant rejection of the claim that 25% of the peas have yellow pods."

Table 7-1		Type I and Type II Errors	
		True State of Nature	
		The null hypothesis is true.	The null hypothesis is false.
Decision	We decide to reject the null hypothesis.	**Type I error** (rejecting a true null hypothesis) α	Correct decision
	We fail to reject the null hypothesis.	Correct decision	**Type II error** (failing to reject a false null hypothesis) β

Type I and Type II Errors   When testing a null hypothesis, we arrive at a conclusion of rejecting or failing to reject that null hypothesis. Such conclusions are sometimes correct and sometimes wrong (even if we do everything correctly). Table 7-1 summarizes the two different types of errors that can be made, along with the two different types of correct decisions. We distinguish between the two types of errors by calling them type I and type II errors.

- **Type I error:** The mistake of rejecting the null hypothesis when it is actually true. The symbol α (alpha) is used to represent the probability of a type I error.

- **Type II error:** The mistake of failing to reject the null hypothesis when it is actually false. The symbol β (beta) is used to represent the probability of a type II error.

Notation

α (alpha) = probability of a type I error (the probability of rejecting the null hypothesis when it is true)

β (beta) = probability of a type II error (failing to reject a false null hypothesis)

Because students usually find it difficult to remember which error is type I and which is type II, we recommend a mnemonic device, such as "ROUTINE FOR FUN." Using only the consonants from those words (**R**ou**T**i**N**e **F**o**R** **F**u**N**), we can easily remember that a type I error is RTN: reject true null (hypothesis), whereas a type II error is FRFN: failure to reject a false null (hypothesis).

EXAMPLE **Identifying Type I and Type II Errors** Assume that we are conducting a hypothesis test of the claim that $p > 0.25$. Here are the null and alternative hypotheses:

$$H_0: p = 0.25$$
$$H_1: p > 0.25$$

Give statements identifying

a. a type I error.
b. a type II error.

SOLUTION

a. A type I error is the mistake of rejecting a true null hypothesis, so this is a type I error: Conclude that there is sufficient evidence to support $p > 0.25$, when in reality $p = 0.25$.

b. A type II error is the mistake of failing to reject the null hypothesis when it is false, so this is a type II error: Fail to reject $p = 0.25$ (and therefore fail to support $p > 0.25$) when in reality $p > 0.25$.

Controlling Type I and Type II Errors: One step in our standard procedure for testing hypotheses involves the selection of the significance level α, which is the probability of a type I error. However, we don't select β [P(type II error)]. It would be great if we could always have $\alpha = 0$ and $\beta = 0$, but in reality that is not possible, so we must attempt to manage the α and β error probabilities. Mathematically, it can be shown that α, β, and the sample size n are all related, so when you choose or determine any two of them, the third is automatically determined. The usual practice in research and industry is to select the values of α and n, so the value of β is determined. (However, a value of β greater than 0.2 is often considered too high for a hypothesis test to provide significant results.) Depending on the seriousness of a type I error, try to use the largest α that you can tolerate. For type I errors with more serious consequences, select smaller values of α. Then choose a sample size n as large as is reasonable, based on considerations of time, cost, and other relevant factors. (Sample size determinations were discussed in Sections 6-2 and 6-3.) The following practical considerations may be relevant:

1. For any fixed α, an increase in the sample size n will cause a decrease in β. That is, a larger sample will lessen the chance that you make the error of not rejecting the null hypothesis when it's actually false.

2. For any fixed sample size n, a decrease in α will cause an increase in β. Conversely, an increase in α will cause a decrease in β.

3. To decrease both α and β, increase the sample size.

To make sense of these abstract ideas, let's consider M&Ms (produced by Mars, Inc.) and Bufferin brand aspirin tablets (produced by Bristol-Myers Products).

- The mean weight of the M&M candies is supposed to be at least 0.9085 g (in order to conform to the weight printed on the package label).
- The Bufferin tablets are supposed to have a mean weight of 325 mg of aspirin.

Because M&Ms are candies used for enjoyment, whereas Bufferin tablets are drugs used for treatment of health problems, we are dealing with two very different levels of seriousness. If the M&Ms don't have a mean weight of 0.9085 g, the consequences are not very serious, but if the Bufferin tablets don't contain a mean of 325 mg of aspirin, the consequences could be very serious, possibly including consumer lawsuits and actions on the part of the U.S. Food and Drug Administration. Consequently, in testing the claim that $\mu = 0.9085$ g for M&Ms, we might choose $\alpha = 0.05$ and a sample size of $n = 100$; in testing the claim that $\mu = 325$ mg for Bufferin tablets, we might choose $\alpha = 0.01$ and a larger sample size of $n = 500$. (The larger sample size allows us to decrease β while we are also decreasing α.) The smaller significance level α and larger sample size n are chosen because of the more serious consequences associated with testing a commercial drug.

Power of a Test: We use β to denote the probability of failing to reject a false null hypothesis (type II error). It follows that $1 - \beta$ is the probability of rejecting a false null hypothesis. Statisticians refer to this probability as the *power* of a test, and they often use it to gauge the test's effectiveness in recognizing that a null hypothesis is false. A common guideline is to plan an experiment so that the resulting power is at least 0.8 (or 80%), so that the hypothesis test is very effective in rejecting a false null hypothesis.

Definition

The **power** of a hypothesis test is the probability $(1 - \beta)$ of rejecting a false null hypothesis, which is computed by using a particular significance level α, a particular sample size n, a particular assumed value of the population parameter (used in the null hypothesis), and *a particular value of the population parameter that is an alternative to the value assumed in the null hypothesis.*

Note that in the above definition, determination of power requires a specific value that is an alternative to the value assumed in the null hypothesis. Consequently, a hypothesis test can have many different values of power, depending on the particular values chosen as alternatives to the null hypothesis.

EXAMPLE **Power of a Hypothesis Test** Assume that we have the following null and alternative hypotheses, significance level, and sample data.

H_0: $p = 0.5$	Sample size: $n = 100$
H_1: $p \neq 0.5$	Sample proportion: $\hat{p} = 0.57$
Significance level:	$\alpha = 0.05$

Using only the components listed above, we can conduct a complete hypothesis test. (The test statistic is $z = 1.4$, the critical values are $z = \pm 1.96$, the *P*-value is 0.1616, we fail to reject the null hypothesis, and we conclude that there is not sufficient evidence to warrant rejection of the claim that the population proportion is equal to 0.5.) However, determination of power re-

continued

quires another item: a *specific* value of p to be used as an alternative to the value of $p = 0.5$ assumed in the null hypothesis. If we use the above test components along with different alternative values of p, we get the following examples of power values. (The values of power can be found using some software packages, such as Minitab, or values of power can be manually calculated. Because the calculations of power are complicated, in this section only Exercise 43 deals with power, and Exercise 43 includes a procedure for calculating power.)

Specific Alternative Value of p	β	Power of Test $(1 - \beta)$
0.3	0.013	0.987
0.4	0.484	0.516
0.6	0.484	0.516
0.7	0.013	0.987

INTERPRETATION OF POWER: Based on the above list of power values, we see that this hypothesis test has power of 0.987 (or 98.7%) of rejecting H_0: $p = 0.5$ when the population proportion p is actually 0.3. That is, if the true population proportion is actually equal to 0.3, there is a 98.7% chance of making the correct conclusion of rejecting the false null hypothesis that $p = 0.5$. Similarly, there is a 0.516 probability of rejecting $p = 0.5$ when the true value of p is actually 0.4. It makes sense that this test is more effective in rejecting the claim of $p = 0.5$ when the population proportion is actually 0.3 than when the population proportion is actually 0.4. [When identifying animals assumed to be horses, there's a better chance of rejecting an elephant as a horse (because of the greater difference) than rejecting a mule as a horse.] In general, increasing the difference between the assumed parameter value and the actual parameter value results in an increase in power.

Power and the Design of Experiments   If a hypothesis test is conducted with sample data consisting of only a few observations, the power will be low, but the power increases as the sample size increases (and the other components remain the same). In addition to increasing the sample size, there are other ways to increase the power, such as increasing the significance level, using a more extreme value for the population parameter, or decreasing the standard deviation. Just as 0.05 is a common choice for a significance level, a power of at least 0.80 is a common requirement for determining that a hypothesis test is effective. (Some statisticians argue that the power should be higher, such as 0.85 or 0.90.)

When designing an experiment, we might consider how much of a difference between the claimed value of a parameter and its true value is an important amount of difference. If testing the effectiveness of a gender-selection method, a change in the proportion of girls from 0.5 to 0.501 is likely to be unimportant. A change in the proportion of girls from 0.5 to 0.6 might be important. Such magnitudes of differences affect power. When designing an experiment, a goal of having a power value of at least 0.80 can often be used to determine the minimum required sample size. For example, here is a statement similar to a statement taken from an article in the *Journal of the American Medical Association:* "The trial design assumed that with a 0.05 significance level, 153 randomly selected subjects would be needed to achieve 80% power to detect a reduction in the coronary heart disease rate from 0.5 to 0.4." Before conducting the experiment, the researchers determined that in order to achieve power of at least 0.80, they needed at least 153 randomly selected subjects. Due to factors such as dropout rates, the researchers are likely to need somewhat more than 153 subjects.

Comprehensive Hypothesis Test   In this section we described the individual components used in a hypothesis test, but the following sections will combine those components in comprehensive procedures. We can test claims about population parameters by using the traditional method summarized in Figure 7-8, the *P*-value method summarized in Figure 7-9, or we can use a confidence interval (described in Chapter 6). For two-tailed hypothesis tests construct a confidence interval with a confidence level of $1 - \alpha$; but for a one-tailed hypothesis test with significance level α, construct a confidence interval with a confidence level of $1 - 2\alpha$. (See Table 7-2 for common cases.) After constructing the confidence interval, use this criterion:

> **A confidence interval estimate of a population parameter contains the likely values of that parameter. We should therefore reject a claim that the population parameter has a value that is not included in the confidence interval.** *Caution:* **In some cases, a conclusion based on a confidence interval may be different from a conclusion based on a hypothesis test.**

The exercises for this section involve isolated components of hypothesis tests, but the following sections will involve complete and comprehensive hypothesis tests.

7-2 Exercises

Stating Conclusions About Claims. In Exercises 1–4, what do you conclude? (Don't use formal procedures and exact calculations. Use only the rare event rule described in Section 7-1, and make subjective estimates to determine whether events are likely.)

1. Claim: A gender-selection method is effective in helping couples have baby girls, and among 50 babies, 26 are girls.

2. Claim: A gender-selection method is effective in helping couples have baby girls, and among 50 babies, 49 are girls.

3. Claim: The majority of adult Americans are opposed to the cloning of humans, and a Gallup poll of 1012 randomly selected adult Americans shows that 901 of them (or 89%) are opposed to the cloning of humans.

4. Claim: Women born on February 29 have heights that vary less than the general population of women for which $\sigma = 6.5$ cm, and a random sample of 50 women born on February 29 results in heights with $s = 6.4$ cm.

FIGURE 7-8 Traditional Method

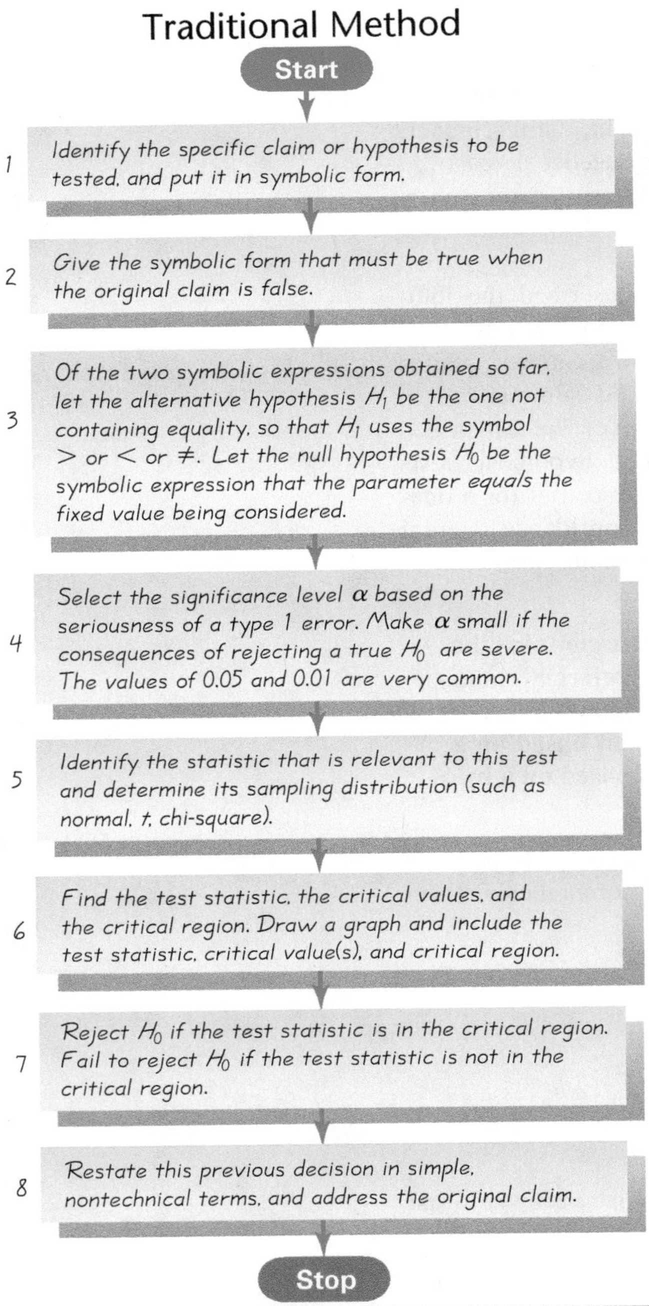

Traditional Method

Start

1 Identify the specific claim or hypothesis to be tested. and put it in symbolic form.

2 Give the symbolic form that must be true when the original claim is false.

3 Of the two symbolic expressions obtained so far. let the alternative hypothesis H_1 be the one not containing equality. so that H_1 uses the symbol $>$ or $<$ or $\neq$. Let the null hypothesis H_0 be the symbolic expression that the parameter *equals* the fixed value being considered.

4 Select the significance level α based on the seriousness of a type 1 error. Make α small if the consequences of rejecting a true H_0 are severe. The values of 0.05 and 0.01 are very common.

5 Identify the statistic that is relevant to this test and determine its sampling distribution (such as normal. t. chi-square).

6 Find the test statistic. the critical values. and the critical region. Draw a graph and include the test statistic. critical value(s), and critical region.

7 Reject H_0 if the test statistic is in the critical region. Fail to reject H_0 if the test statistic is not in the critical region.

8 Restate this previous decision in simple. nontechnical terms. and address the original claim.

Stop

FIGURE 7-9 P-Value Method

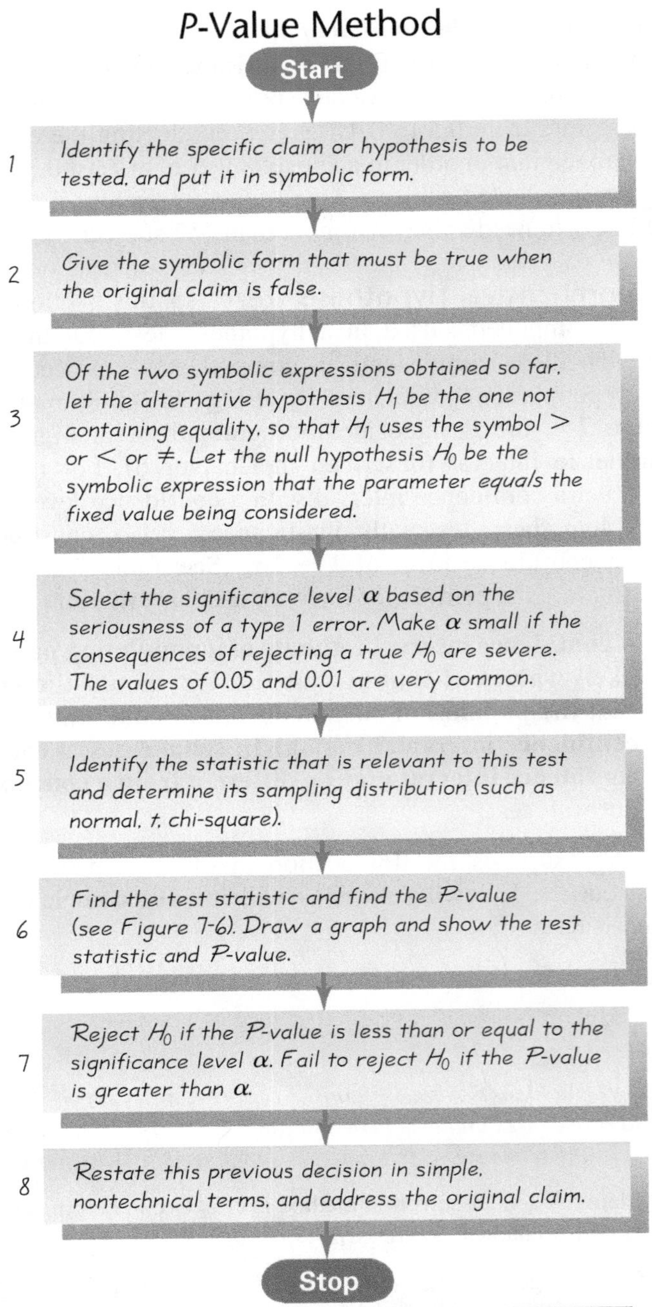

P-Value Method

Start

1 Identify the specific claim or hypothesis to be tested. and put it in symbolic form.

2 Give the symbolic form that must be true when the original claim is false.

3 Of the two symbolic expressions obtained so far. let the alternative hypothesis H_1 be the one not containing equality. so that H_1 uses the symbol $>$ or $<$ or $\neq$. Let the null hypothesis H_0 be the symbolic expression that the parameter *equals* the fixed value being considered.

4 Select the significance level α based on the seriousness of a type 1 error. Make α small if the consequences of rejecting a true H_0 are severe. The values of 0.05 and 0.01 are very common.

5 Identify the statistic that is relevant to this test and determine its sampling distribution (such as normal. t. chi-square).

6 Find the test statistic and find the P-value (see Figure 7-6). Draw a graph and show the test statistic and P-value.

7 Reject H_0 if the P-value is less than or equal to the significance level α. Fail to reject H_0 if the P-value is greater than α.

8 Restate this previous decision in simple. nontechnical terms. and address the original claim.

Stop

Confidence Interval Method

Construct a confidence interval with a confidence level selected as in Table 7-2.
Because a confidence interval estimate of a population parameter contains the likely values of that parameter, reject a claim that the population parameter has a value that is not included in the confidence interval.

Table 7-2 Confidence Level for Confidence Interval

		Two-Tailed Test	One-Tailed Test
Significance	0.01	99%	98%
Level for	0.05	95%	90%
Hypothesis	0.10	90%	80%
Test			

Identifying H_0 and H_1. In Exercises 5–12, examine the given statement, then express the null hypothesis H_0 and alternative hypothesis H_1 in symbolic form. Be sure to use the correct symbol (μ, p, σ) for the indicated parameter.

5. The mean annual income of workers who have studied statistics is greater than $50,000.

6. The mean IQ of statistics students is at least 110.

7. More than one-half of all Internet users make on-line purchases.

8. The percentage of men who watch golf on TV is not 70%, as is claimed by the Madison Advertising Company.

9. Women's heights have a standard deviation less than 2.8 in., which is the standard deviation for men's heights.

10. The percentage of viewers tuned to *60 Minutes* is equal to 24%.

11. The mean amount of rubbing alcohol in containers is at least 12 oz.

12. Salaries among women biologists have a standard deviation greater than $3000.

Finding Critical Values. In Exercises 13–20, find the critical z values. In each case, assume that the normal distribution applies.

13. Two-tailed test; $\alpha = 0.05$.

14. Two-tailed test; $\alpha = 0.01$.

15. Right-tailed test; $\alpha = 0.01$.

16. Left-tailed test; $\alpha = 0.05$.

17. $\alpha = 0.10$; H_1 is $p \neq 0.17$.

18. $\alpha = 0.10$; H_1 is $p > 0.18$.

19. $\alpha = 0.02$; H_1 is $p < 0.19$.

20. $\alpha = 0.005$; H_1 is $p \neq 0.20$.

Finding Test Statistics. In Exercises 21–24, find the value of the test statistic z using

$$z = \frac{\hat{p} - p}{\sqrt{\dfrac{pq}{n}}}$$

21. Gallup Poll The claim is that the proportion of adults who shop using the Internet is less than 0.5 (or 50%), and the sample statistics include $n = 1025$ subjects with 29% saying that they use the Internet for shopping.

22. Genetics Experiment The claim is that the proportion of peas with red flowers is equal to 0.75 (or 75%), and the sample statistics include $n = 929$ peas with 75.9% of them having red flowers.

23. Safety Study The claim is that the proportion of choking deaths of children attributable to balloons is more than 0.25, and the sample statistics include $n = 400$ choking deaths of children with 29.0% of them attributable to balloons.

24. Police Practices The claim is that the proportion of drivers stopped by police in a year is different from the 10.3% rate reported by the Department of Justice. Sample statistics include $n = 800$ randomly selected drivers with 12% of them stopped in the past year.

Finding P-values. In Exercises 25–32, use the given information to find the P-value. (Hint: See Figure 7-6.)

25. The test statistic in a right-tailed test is $z = 0.55$.

26. The test statistic in a left-tailed test is $z = -1.72$.

27. The test statistic in a two-tailed test is $z = 1.95$.

28. The test statistic in a two-tailed test is $z = -1.63$.

29. With $H_1: p > 0.29$, the test statistic is $z = 1.97$.

30. With $H_1: p \neq 0.30$, the test statistic is $z = 2.44$.

31. With $H_1: p \neq 0.31$, the test statistic is $z = 0.77$.

32. With $H_1: p < 0.32$, the test statistic is $z = -1.90$.

Stating Conclusions. In Exercises 33–36, state the final conclusion in simple nontechnical terms. Be sure to address the original claim. (Hint: See Figure 7-7.)

33. Original claim: The proportion of married women is greater than 0.5.
Initial conclusion: Reject the null hypothesis.

34. Original claim: The proportion of college graduates who smoke is less than 0.27.
Initial conclusion: Reject the null hypothesis.

35. Original claim: The proportion of fatal commercial aviation crashes is different from 0.038.
Initial conclusion: Fail to reject the null hypothesis.

36. Original claim: The proportion of M&Ms that are blue is equal to 0.10.
Initial conclusion: Reject the null hypothesis.

Identifying Type I and Type II Errors. In Exercises 37–40, identify the type I error and the type II error that correspond to the given hypothesis.

37. The proportion of married women is greater than 0.5.

38. The proportion of college graduates who smoke is less than 0.27.

39. The proportion of fatal commercial aviation crashes is different from 0.038.

40. The proportion of M&Ms that are blue is equal to 0.10.

*41. Unnecessary Test When testing a claim that the majority of adult Americans are against the death penalty for a person convicted of murder, a random sample of 491 adults is obtained, and 27% of them are against the death penalty (based on data from a Gallup poll). Find the P-value. Why is it not necessary to go through the steps of conducting a formal hypothesis test?

*42. Significance Level If a null hypothesis is rejected with a significance level of 0.05, is it also rejected with a significance level of 0.01? Why or why not?

*43. Power of a Test Assume that you are using a significance level of $\alpha = 0.05$ to test the claim that $p < 0.5$ and that your sample is a simple random sample of size $n = 1998$ with $\hat{p} = 0.48$.   *continued*

a. Find the power of the test, which is the probability of rejecting the null hypothesis, given that the population proportion is actually 0.45. (In the procedure below, we refer to $p = 0.5$ as the "assumed" value, because it is assumed in the null hypothesis; we refer to $p = 0.45$ as the "alternative" value, because it is the value of the population proportion used as an alternative to 0.5.) Use the following procedure.

Step 1:   Using the significance level, find the critical z value(s). (For a right-tailed test, there is a single critical z value that is positive; for a left-tailed test, there is a single critical value of z that is negative; and a two-tailed test will have a critical z value that is negative along with another critical z value that is positive.)

Step 2:   In the expression for the test statistic below, substitute the assumed value of p (used in the null hypothesis). Evaluate $1 - p$ and substitute that result for the entry of q. Also substitute the critical value(s) for z. Then solve for the sample statistic $\hat{p}$. (If the test is two-tailed, substitute the critical value of z that is positive, then solve for the sample statistic $\hat{p}$. Next, substitute the critical value of z that is negative, and solve for the sample statistic $\hat{p}$. A two-tailed test should therefore result in two different values of $\hat{p}$.) The resulting value(s) of $\hat{p}$ separate the region(s) where the null hypothesis is rejected from the region(s) where we fail to reject the null hypothesis.

$$z = \frac{\hat{p} - p}{\sqrt{\dfrac{pq}{n}}}$$

Step 3:   The calculation of power requires a specific value of p that is to be used as an alternative to the value assumed in the null hypothesis. Identify this alternative value of p (not the value used in the null hypothesis), draw a normal curve with this alternative value at the center, and plot the value(s) of $\hat{p}$ found in Step 2.

Step 4:   Refer to the graph in Step 3, and find the area of the new critical region bounded by the value(s) of $\hat{p}$ found in Step 2. (*Caution:* When evaluating $\sqrt{pq/n}$, be sure to use the alternative value of p, not the value of p used for the null hypothesis.) This is the probability of rejecting the null hypothesis, given that the alternative value of p is the true value of the population proportion. Because this is the probability of rejecting H_0 when it is false, the result is the power of the test. What does the value of the power indicate about this hypothesis test?

b. Find β, which is the probability of *failing* to reject the false null hypothesis.

*44. Finding Sample Size   A researcher plans to conduct a hypothesis test using the alternative hypothesis of $H_1: p < 0.4$, and she plans to use a significance level of $\alpha = 0.05$. Find the sample size required to achieve at least 80% power in detecting a reduction in p from 0.4 to 0.3. (*This is a very difficult exercise.* Hint: *See Exercise 43.*)

7-3   Testing a Claim About a Proportion

In Section 7-2 we presented the isolated components of a hypothesis test, but in this section we combine those components in comprehensive hypothesis tests of claims made about population proportions. The proportions can also represent probabilities or the decimal equivalents of percents. The following are examples of the types of claims we will be able to test:

- Fewer than 1/4 of all college graduates smoke.
- Subjects taking the cholesterol-reducing drug Lipitor experience headaches at a rate that is greater than the 7% rate for people who do not take Lipitor.
- Based on a Gallup poll, the majority (more than 50%) of Americans are opposed to the cloning of humans.

The required assumptions, notation, and test statistic are all given below. Basically, claims about a population proportion are usually tested by using a normal distribution as an approximation to the binomial distribution, as we did in Section 5-6. Instead of using the same exact methods of Section 5-6, we use a different but equivalent form of the test statistic shown below, and we don't include the correction for continuity (because its effect tends to be very small with large samples). If the given requirements are not all satisfied, we may be able to use other methods not described in this section. In this section, all examples and exercises involve cases in which the requirements are satisfied, so the sampling distribution of sample proportions can be approximated by the normal distribution.

Testing Claims About a Population Proportion p

Requirements

1. The sample observations are a simple random sample. (Never forget the critical importance of sound sampling methods.)

2. The conditions for a *binomial distribution* are satisfied. (There are a fixed number of independent trials having constant probabilities, and each trial has two outcome categories of "success" and "failure.")

3. The conditions $np \geq 5$ and $nq \geq 5$ are both satisfied, so **the binomial distribution of sample proportions can be approximated by a normal distribution with $\mu = np$ and $\sigma = \sqrt{npq}$** (as described in Section 5-6).

Notation

n = sample size or number of trials

$\hat{p} = \dfrac{x}{n}$ (*sample* proportion)

p = population proportion (used in the null hypothesis)

$q = 1 - p$

Test Statistic for Testing a Claim About a Proportion

$$z = \frac{\hat{p} - p}{\sqrt{\dfrac{pq}{n}}}$$

***P*-values:** Refer to Figure 7-6 and use the standard normal distribution (Table A-2).

Critical values: Use the standard normal distribution (Table A-2).

EXAMPLE **Genetics Experiment** In the Chapter Problem we noted that an experiment involving peas resulted in 580 offspring, with 26.2% of the offspring peas having yellow pods. Mendel claimed that the proportion of peas with yellow pods should equal 25%. Here is a summary of the claim and the sample data:

Claim: The percentage of peas with yellow pods is 25%. That is, $p = 0.25$.

Sample data: $n = 580$ and $\hat{p} = 0.262$

We will illustrate the hypothesis test using the traditional method, the popular *P*-value method, and confidence intervals. Before proceeding, however, we should verify that the necessary requirements are satisfied.

SOLUTION

REQUIREMENT ✓   We should first verify that the necessary requirements are satisfied. (Earlier in this section we listed the requirements for using a normal distribution as an approximation to a binomial distribution.) Given the design of the experiment, it is reasonable to assume that the sample is a simple random sample. The conditions for a binomial experiment are satisfied, because there is a fixed number of trials (580), the trials are independent (because the color of a pea pod doesn't affect the probability of the color of another pea pod), there are two categories of outcome (yellow, not yellow), and the probability of yellow remains constant. Finally, we use $n = 580$ and $p = 0.25$ to find that

$$np = 145 \geq 5$$

and

$$nq = 435 \geq 5$$

so the normal distribution can be used to approximate the binomial distribution. The check of requirements has been successfully completed and we can now proceed to conduct a formal hypothesis test. The traditional method, the *P*-value method, and the use of confidence intervals are illustrated in the following discussion. ✓

The Traditional Method

The traditional method of testing hypotheses is summarized in Figure 7-8. When testing the claim $p = 0.25$ given in the preceding example, the following steps correspond to the procedure in Figure 7-8:

Step 1:   The original claim in symbolic form is $p = 0.25$.

Step 2:   The opposite of the original claim is $p \neq 0.25$.

Step 3:   Of the preceding two symbolic expressions, the expression $p \neq 0.25$ does not contain equality, so it becomes the alternative hypothesis. The null hypothesis is the statement that p equals the fixed value of 0.25. We can therefore express H_0 and H_1 as follows:

$$H_0: p = 0.25$$
$$H_1: p \neq 0.25$$

Step 4:   In the absence of any special circumstances, we will select $\alpha = 0.05$ for the significance level.

Step 5:   Because we are testing a claim about a population proportion p, the sample statistic $\hat{p}$ is relevant to this test, and the sampling distribution of sample proportions $\hat{p}$ is approximated by a *normal distribution*.

Step 6:   The test statistic is evaluated using $n = 580$ and $\hat{p} = 0.262$. In the null hypothesis we are assuming that $p = 0.25$, so $q = 1 - 0.25 = 0.75$. The test statistic is

$$z = \frac{\hat{p} - p}{\sqrt{\dfrac{pq}{n}}} = \frac{0.262 - 0.25}{\sqrt{\dfrac{(0.25)(0.75)}{580}}} = 0.67$$

This is a two-tailed test, so the critical region is an area of $\alpha = 0.05$ divided equally between the two tails. Referring to Table A-2 and applying the methods of Section 5-2, we find the critical values of $z = -1.96$ and $z = 1.96$ at the boundaries of the critical region. See Figure 7-3, which shows the critical region, critical values, and test statistic.

Step 7:   Because the test statistic does not fall within the critical region, we fail to reject the null hypothesis.

Step 8:   We conclude that there is not sufficient sample evidence to warrant rejection of Mendel's claim that 25% of the offspring peas will have yellow pods. (See Figure 7-7 for help with wording this final conclusion.)

The *P*-Value Method

The *P*-value method of testing hypotheses is summarized in Figure 7-9, and it requires the *P*-value that is found using the procedure summarized in Figure 7-6. A comparison of Figures 7-8 and 7-9 shows that the first five steps of the traditional method are the same as the first five steps of the *P*-value method. For the hypothesis test described in the preceding example, the first five steps of the *P*-value method are the same as those shown in the above traditional method, so we now continue with Step 6.

Step 6:   The test statistic is $z = 0.67$ as shown in the preceding traditional method. We now find the *P*-value (instead of the critical value) by using the following procedure, which is shown in Figure 7-6:

Right-tailed test:	*P*-value =	area to right of test statistic z.
Left-tailed test:	*P*-value =	area to left of test statistic z.
Two-tailed test:	*P*-value =	*twice* the area of the extreme region bounded by the test statistic z.

Because the hypothesis test we are considering is two-tailed with a test statistic of $z = 0.67$, the *P*-value is *twice* the area to the right of $z = 0.67$. Referring to Table A-2, we see that for $z = 0.67$, there is an area of 0.7486 for the cumulative area to the *left* of the test statistic. The

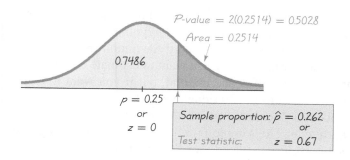

P-value $= 2(0.2514) = 0.5028$

Area $= 0.2514$

0.7486

$p = 0.25$
or
$z = 0$

Sample proportion: $\hat{p} = 0.262$
or
Test statistic: $z = 0.67$

FIGURE 7-10 *P*-Value
Method

area to the right of $z = 0.67$ is therefore $1 - 0.7486 = 0.2514$, which must be doubled to yield a P-value of 0.5028. Figure 7-10 shows the test statistic and P-value for this example.

Step 7:   Because the P-value of 0.5028 is greater than the significance level of $\alpha = 0.05$, we fail to reject the null hypothesis.

Step 8:   As with the traditional method, we conclude that there is not sufficient sample evidence to warrant rejection of Mendel's claim that 25% of the offspring peas have yellow pods. (See Figure 7-7 for help with wording this final conclusion.)

Confidence Interval Method

For two-tailed hypothesis tests construct a confidence interval with a confidence level of $1 - \alpha$; but for a one-tailed hypothesis test with significance level α, construct a confidence interval with a confidence level of $1 - 2\alpha$. (See Table 7-2 for common cases.) For example, the claim of $p > 0.5$ can be tested with a 0.05 significance level by constructing a 90% confidence interval.

Let's now use the confidence interval method to test the claim of $p = 0.25$, with sample data consisting of $n = 580$ and $\hat{p} = 0.262$ (from the example near the beginning of this section). An example in Section 6-2 showed that the 95% confidence interval estimate resulting from these sample data is $0.226 < p < 0.298$. Because we are 95% confident that the true value of p is contained within the limits of 0.226 and 0.298, we do not have sufficient evidence to reject the claim that $p = 0.25$. (The claimed value of 0.25 does fall within the confidence interval, so 0.25 could easily be the true value of the population proportion, and we do not have sufficient evidence to reject 0.25 as the value of the population proportion.)

Caution: When testing claims about a population proportion, the traditional method and the P-value method are equivalent in the sense that they always yield the same results, but the confidence interval method is somewhat different. Both the traditional method and P-value method use the same standard deviation based on the *claimed proportion p*, but the confidence interval uses an estimated standard deviation based on the *sample proportion $\hat{p}$*. Consequently, it is possible that in some cases, the traditional and P-value methods of testing a claim about a proportion might yield a different conclusion than the confidence interval method. (See Exercise 19.) If different conclusions are obtained, realize that the traditional

Test of Touch Therapy

At the age of nine, Emily Rosa entered a school science fair with a project to test touch therapy. Instead of actually touching subjects, touch therapists move their hands a few inches away from the subject's body so that they can improve the human energy field. Emily Rosa tested 21 touch therapists by sitting on one side of a cardboard shield while the therapists placed their hands through the shield. Emily placed her hand above one of the therapist's hands (selected with a coin toss), then the therapist tried to identify the selected hand without seeing Emily's hand. A 50% success rate would be expected with random guesses, but the touch therapists were correct only 44% of the time. Emily Rosa became the youngest author in the *Journal of the American Medical Association* when this article was published: "A Close Look at Therapeutic Touch" by L. Rosa, E. Rosa, L. Sarner, and S. Barrett, Vol. 279, No. 1005.

and *P*-value methods use an *exact* standard deviation based on the assumption that the population proportion has the value given in the null hypothesis. However, the confidence interval is constructed using a standard deviation based on an *estimated* value of the population proportion. If you want to estimate a population proportion, do so by constructing a confidence interval, but if you want to test a hypothesis, use the *P*-value method or the traditional method.

When testing a claim about a population proportion *p,* be careful to identify correctly the sample proportion $\hat{p}$. The sample proportion $\hat{p}$ is sometimes given directly, but in other cases it must be calculated. See the examples below.

Given Statement	Finding $\hat{p}$
10% of the observed patients have blue eyes	$\hat{p}$ is given directly: $\hat{p} = 0.10$
96 surveyed households have cable TV and 54 do not	$\hat{p}$ must be calculated using $\hat{p} = x/n$: $$\hat{p} = \frac{x}{n} = \frac{96}{96 + 54} = 0.64$$

Caution: When a calculator or computer display of $\hat{p}$ results in many decimal places, use all of those decimal places when evaluating the *z* test statistic. Large errors can result from rounding $\hat{p}$ too much.

EXAMPLE **Cell Phones and Cancer** In a Danish study, cell phone users were observed, and it was found that 77 cell phone users developed leukemia while 420,018 cell phone users did not (based on data from the *Journal of the National Cancer Institute,* Vol. 93, No. 3). The logical issue is whether cell phones pose a hazard, so test the claim that cell phone users have leukemia at a rate that is greater than the rate of 0.0190% for people who do not use cell phones. Use a 0.01 significance level.

SOLUTION

REQUIREMENT ✓ We should first verify that the necessary requirements are satisfied. (Earlier in this section we listed the requirements for using a normal distribution as an approximation to a binomial distribution.) Given the way that the data were obtained, it is reasonable to assume that the sample is a simple random sample. The conditions for a binomial experiment are satisfied, because there is a fixed number of trials (77 + 420,018 = 420,095), the trials are independent (because whether a subject has leukemia doesn't affect the probability of another subject having leukemia), there are two categories of outcome (the subject has leukemia or does not), and the probability of leukemia remains constant. Finally, we use $n = 420,095$ and $\hat{p} = 77/420,095$ to find that

$$n\hat{p} = 77 \geq 5$$

and

$$n\hat{q} = 420,018 \geq 5$$

The check of requirements has been successfully completed, and the normal distribution can be used to approximate the binomial distribution. ✓

We now begin with the *P*-value method summarized in Figure 7-9 found in Section 7-2. Note that $n = 77 + 420,018 = 420,095$, $\hat{p} = 77/420,095 = 0.000183$. For the purposes of the test, we assume that $p = 0.000190$, which is the decimal equivalent of 0.0190%. (*Note:* If we think about it, we could stop here without proceeding with the test, because the sample proportion of $\hat{p} = 0.000183$ is *less than* the assumed rate of $p = 0.000190$, so there is no way that the sample proportion could possibly be *significantly greater than* the assumed rate of $p = 0.000190$. However, we will proceed so that the hypothesis testing method can be illustrated.)

Step 1: The original claim is that the leukemia rate for cell phone users is greater than 0.000190. We express this in symbolic form as $p > 0.000190$.

Step 2: The opposite of the original claim is $p \leq 0.000190$.

Step 3: Because $p > 0.000190$ does not contain equality, it becomes H_1. We get

$H_0: p = 0.000190$ (null hypothesis)

$H_1: p > 0.000190$ (alternative hypothesis and original claim)

Step 4: The significance level is $\alpha = 0.01$.

Step 5: Because the claim involves the proportion p, the statistic relevant to this test is the sample proportion $\hat{p}$, and the sampling distribution of sample proportions is approximated by the normal distribution (because the necessary requirements are satisfied).

Step 6: The test statistic of $z = -0.33$ is found as follows:

$$z = \frac{\hat{p} - p}{\sqrt{\dfrac{pq}{n}}} = \frac{0.000183 - 0.000190}{\sqrt{\dfrac{(0.000190)(0.999810)}{420,095}}} = -0.33$$

Refer to Figure 7-6 for the procedure for finding the *P*-value. Figure 7-6 shows that for this right-tailed test, the *P*-value is the area to the *right* of the test statistic. Using Table A-2, $z = -0.33$ has an area of 0.3707 to its left, so the area to its right is $1 - 0.3707 = 0.6293$. The *P*-value is 0.6293.

Step 7: Because the *P*-value of 0.6293 is greater than the significance level of 0.01, we fail to reject the null hypothesis.

INTERPRETATION We failed to reject the null hypothesis, so we consider it to be okay for now. We did not support the alternative hypothesis, which was the original conclusion. Here is the correct final conclusion: There is not sufficient evidence to support the claim that cell phone users have a leukemia rate that is greater than 0.000190, which is the rate for people who do not use cell phones. It appears that the cell phone users do not risk leukemia any more than those who do not use cell phones.

Traditional Method: If we were to repeat the preceding example using the traditional method of testing hypotheses, we would see that in Step 6, the critical value is found to be $z = 2.33$. In Step 7, we would fail to reject the null hypothesis because the test statistic of $z = -0.33$ would not fall within the critical region. We would reach the same conclusion from the P-value method: There is not sufficient evidence to support the claim that cell phone users have a leukemia rate that is greater than 0.000190, which is the rate for people who do not use cell phones.

Confidence Interval Method: If we were to repeat the preceding example using the confidence interval method, we would need to double the significance level of $\alpha = 0.01$, then construct this 98% confidence interval: $0.000135 < p < 0.000232$. The confidence interval limits do contain the value of 0.000190, so we conclude that there is not sufficient evidence to support the claim that the leukemia rate for cell phone users is greater than 0.000190. In this case, the P-value method, traditional method, and confidence interval method all lead to the same conclusion. In some other relatively rare cases, the P-value method and the traditional method might lead to a conclusion that is different from the conclusion reached through the confidence interval method.

Rationale for the Test Statistic: The test statistic used in this section is justified by noting that when using the normal distribution to approximate a binomial distribution, we use $\mu = np$ and $\sigma = \sqrt{npq}$ to get

$$z = \frac{x - \mu}{\sigma} = \frac{x - np}{\sqrt{npq}}$$

We used the above expression in Section 5-6 along with a correction for continuity, but when testing claims about a population proportion, we make two modifications. First, we don't use the correction for continuity because its effect is usually very small for the large samples we are considering. Also, instead of using the above expression to find the test statistic, we use an equivalent expression obtained by dividing the numerator and denominator by n, and we replace x/n by the symbol $\hat{p}$ to get the test statistic we are using. The end result is that the test statistic is simply the same standard score (from Section 2-5) of $z = (x - \mu)/\sigma$, but modified for the binomial notation.

7-3 Exercises

1. Mendel's Hybridization Experiments In one of Mendel's famous hybridization experiments, 8023 offspring peas were obtained, and 24.94% of them had green flowers. The others had white flowers. Consider a hypothesis test that uses a 0.05 significance level to test the claim that green-flowered peas occur at a rate of 25%.
   a. What is the test statistic?
   b. What are the critical values?
   c. What is the P-value?
   d. What is the conclusion?
   e. Can a hypothesis test be used to "prove" that the rate of green-flowered peas is 25%, as claimed?

2. Survey of Drinking In a Gallup survey, 1087 randomly selected adults were asked "Do you have occasion to use alcoholic beverages such as liquor, wine, or beer, or are you a total abstainer?" The result: 62% of the subjects said that they used alcoholic

beverages. Consider a hypothesis test that uses a 0.05 significance level to test the claim that the majority (more than 50%) of adults use alcoholic beverages.
   a. What is the test statistic?
   b. What is the critical value?
   c. What is the *P*-value?
   d. What is the conclusion?
   e. Based on the preceding results, can we conclude that 62% is significantly greater than 50% for all such hypothesis tests? Why or why not?

Testing Claims About Proportions. In Exercises 3–6, assume that a method of gender selection is being tested with couples wanting to have baby girls. Identify the null hypothesis, alternative hypothesis, test statistic, P-value or critical value(s), conclusion about the null hypothesis, and final conclusion that addresses the original claim. Use the P-value method unless your instructor specifies otherwise.

3. Gender Selection Among 100 babies born to 100 couples using a gender-selection method to increase the likelihood of a baby girl, 65 of them are girls. Use a 0.01 significance level to test the claim that for couples using this method, the proportion of girls is greater than 0.5.

4. Gender Selection Among 100 babies born to 100 couples using a gender-selection method, 55 of them are girls. Use a 0.01 significance level to test the claim that for couples using this method, the proportion of girls is not equal to 0.5.

5. Gender Selection Among 50 babies born to 50 couples using a gender-selection method, 15 of them are boys. Use a 0.05 significance level to test the claim that for couples using this method, the proportion of boys is less than 0.5.

6. Gender Selection Among 1000 babies born to 1000 couples using a gender-selection method to increase the likelihood of a baby girl, 540 of them are girls. Use a 0.01 significance level to test the claim that for couples using this method, the proportion of girls is greater than 0.5.

Testing Claims About Proportions. In Exercises 7–18, test the given claim. Identify the null hypothesis, alternative hypothesis, test statistic, P-value or critical value(s), conclusion about the null hypothesis, and final conclusion that addresses the original claim. Use the P-value method unless your instructor specifies otherwise.

7. Cloning Survey In a Gallup poll of 1012 randomly selected adults, 9% said that cloning of humans should be allowed. Use a 0.05 significance level to test the claim that less than 10% of all adults say that cloning of humans should be allowed. Can a newspaper run a headline that "less than 10% of adults say that cloning of humans should be allowed?"

8. Testing Lipitor for Cholesterol Reduction In clinical tests of the drug Lipitor (atorvastatin), 863 patients were treated with 10-mg doses of atorvastatin, and 19 of those patients experienced flu symptoms (based on data from Parke-Davis). Use a 0.01 significance level to test the claim that the percentage of treated patients with flu symptoms is greater than the 1.9% rate for patients not given the treatments. Does it appear that flu symptoms are an adverse reaction of the treatment?

9. Cell Phones and Cancer In a study of 420,095 Danish cell phone users, 135 subjects developed cancer of the brain or nervous system (based on data from the *Journal of the National Cancer Institute*, Vol. 93, No. 3). Test the claim of a once popular belief that such cancers are affected by cell phone use. That is, test the claim that cell phone users develop cancer of the brain or nervous system at a rate that is different from the

rate of 0.0340% for people who do not use cell phones. Because this issue has such great importance, use a 0.005 significance level. Should cell phone users be concerned about cancer of the brain or nervous system?

10. Drug Testing of Job Applicants In 1990, 5.8% of job applicants who were tested for drugs failed the test. At the 0.01 significance level, test the claim that the failure rate is now lower if a simple random sample of 1520 current job applicants results in 58 failures (based on data from the American Management Association). Does the result suggest that fewer job applicants now use drugs?

11. Testing Effectiveness of Nicotine Patches In one study of smokers who tried to quit smoking with nicotine patch therapy, 39 were smoking one year after the treatment, and 32 were not smoking one year after the treatment (based on data from "High-Dose Nicotine Patch Therapy," by Dale et al., *Journal of the American Medical Association,* Vol. 274, No. 17). Use a 0.10 significance level to test the claim that among smokers who try to quit with nicotine patch therapy, the majority are smoking a year after the treatment. Do these results suggest that the nicotine patch therapy is ineffective?

12. Smoking and College Education One survey showed that among 785 randomly selected subjects who completed four years of college, 144 smoke and 641 do not smoke (based on data from the American Medical Association). Use a 0.01 significance level to test the claim that the rate of smoking among those with four years of college is less than the 27% rate for the general population. Why would college graduates smoke at a lower rate than others?

13. Drug Testing for Adverse Reactions The drug Ziac is used to treat hypertension. In a clinical test, 3.2% of 221 Ziac users experienced dizziness (based on data from Lederle Laboratories). Further tests must be conducted for adverse reactions that occur in at least 5% of treated subjects. Using a 0.01 significance level, test the claim that fewer than 5% of all Ziac users experience dizziness.

14. Aviation Fatalities Researchers studied crash landings of general aviation (noncommercial and nonmilitary) airplanes and found that pilots died in 437 of 8411 crash landings (based on data from "Risk Factors for Pilot Fatalities in General Aviation Airplane Crash Landings," by Rostykus, Cummings, and Mueller, *Journal of the American Medical Association,* Vol. 280, No. 11). Use a 0.06 significance level to test the claim that pilots die in more than 5% of such crash landings. From the result, can we conclude that the pilot fatality rate does *not* exceed 5%? Why or why not?

15. Air-Bag Effectiveness In a study of air-bag effectiveness, it was found that in 821 crashes of midsize cars equipped with air-bags, 46 of the crashes resulted in hospitalization of the drivers (based on data from the Highway Loss Data Institute). Use a 0.01 significance level to test the claim that the air-bag hospitalization rate is lower than the 7.8% rate for crashes of midsize cars equipped with automatic safety belts.

16. Birds and Aircraft Crashes Environmental concerns often conflict with modern technology, as is the case with birds that pose a hazard to aircraft during takeoff. An environmental group states that incidents of bird strikes are too rare to justify killing the birds. A pilot's group claims that among aborted takeoffs leading to aircraft going off the end of the runway, 10% are due to bird strikes. Use a 0.05 significance level to test that claim. Sample data consist of 74 aborted takeoffs in which the aircraft overran the runway. Among those 74 cases, 5 were due to bird strikes (based on data from the Air Line Pilots Association and Boeing, as reported in *USA Today*).

17. Females Underrepresented in Textbooks In the article "Sex and Gender Bias in Anatomy and Physical Diagnosis Text Illustrations" (by Meldelsohn, Nieman, Isaacs, Lee, and Levison, *Journal of the American Medical Association*, Vol. 272, No. 16), it is claimed that "males are depicted in a majority of nonreproductive anatomy illustrations." Using the seven most popular anatomy textbooks and ignoring illustrations that were gender neutral, there are 142 nonreproductive illustrations based on females, and there are 554 nonreproductive illustrations based on males. Use a 0.05 significance level to test the claim that males are depicted in a majority of the nonreproductive illustrations.

18. Fruit Flies In an experiment involving fruit flies, parents with normal wings are crossed with parents having vestigial wings. It is claimed that 75% of the offspring would have normal wings, but among 80 offspring, 85% have normal wings. Use a 0.01 significance level to test the claim that the percentage of offspring with normal wings is different from 75%.

*19. Using Confidence Intervals to Test Hypotheses When analyzing the last digits of weights of health exam subjects, it is found that among 1000 randomly selected digits, 119 are zeros. If the digits are randomly selected, the proportion of zeros should be 0.1.
   a. Use the traditional method with a 0.05 significance level to test the claim that the proportion of zeros equals 0.1.
   b. Use the *P*-value method with a 0.05 significance level to test the claim that the proportion of zeros equals 0.1.
   c. Use the sample data to construct a 95% confidence interval estimate of the proportion of zeros. What does the confidence interval suggest about the claim that the proportion of zeros equals 0.1?
   d. Compare the results from the traditional method, the *P*-value method, and the confidence interval method. Do they all lead to the same conclusion?

*20. Proving Claims In the *USA Today* article "Power Lines Not a Cancer Risk for Kids," the first sentence states that "Children who live near high voltage power lines appear to be no more likely to get leukemia than other kids, doctors report today in the most extensive study of one of the most controversial issues ever done." Representing the rate of leukemia for children not living near high voltage power lines by the constant c, write the claim in symbolic form, then identify the null and alternative hypotheses suggested by this statement. Given that we either reject or fail to reject the null hypothesis, what possible conclusions can be made about the original claim? Can the sample data support the claim that children who live near high voltage power lines are no more likely to get leukemia than other children?

*21. Power and Probability of Type II Error For a hypothesis test with a specified significance level α, the probability of a type I error is α, whereas the probability β of a type II error depends on the particular value of p that is used as an alternative to the null hypothesis. Consider a test of the claim that the majority of animated children's movies show the use of alcohol or tobacco or both, with a 0.05 significance level and sample data consisting of 50 such movies (based on data from "Tobacco and Alcohol Use in G-Rated Children's Animated Films" by Goldstein, Sobel, and Newman, *Journal of the American Medical Association*, Vol. 281, No. 12). Assuming that the true value of p is 0.45, find β, the probability of a type II error. Also find the power of the test. What does the value of the power suggest about the effectiveness of the test? [*Hint:* Use the procedure given in Exercise 43 from Section 7-2 and, in Step 4, use the values $p = 0.45$ and $pq/n = (0.45)(0.55)/50$.]

Does Aspirin Help Prevent Heart Attacks?

In a recent study of 22,000 male physicians, half were given regular doses of aspirin while the other half were given placebos. The study ran for six years at a cost of $4.4 million. Among those who took the aspirin, 104 suffered heart attacks. Among those who took the placebos, 189 suffered heart attacks. (The figures are based on data from *Time* and the *New England Journal of Medicine*, Vol. 318, No. 4.) This is a classic experiment involving a treatment group (those who took the aspirin) and a placebo group (those who took pills that looked and tasted like the aspirin pills, but no aspirin was present). We can use methods presented in this chapter to address the issue of whether the results show a statistically significant lower rate of heart attacks among the sample group who took aspirin.

7-4 Testing a Claim About a Mean: σ Known

In this section we consider methods of testing claims made about a population mean μ, and we assume that the population standard deviation σ is known. It would be an unusual set of circumstances that would allow us to know σ without knowing μ, but Section 7-5 deals with cases in which σ is not known. Although this section involves cases that are less realistic than those in Section 7-5, this section is important in describing the same general method used in the following section. Also, there are cases in which the specific value of σ is unknown, but some information about σ could be used. The example presented in this section includes the unrealistic assumption that σ is known to be 0.62°F. The test statistic in that example is found to be $z = -6.64$, which leads to rejection of the common belief that the mean body temperature is equal to 98.6°F. If we analyze the variation of body temperatures, it becomes obvious that σ can't possibly be as high as 2°F, yet using $\sigma = 2$°F would result in a test statistic of $z = -2.05$, which again leads to rejection of the claim that $\mu = 98.6$°F. Because σ must be less than 2°F for body temperatures, the test statistic must be at least as extreme as $z = -2.05$. (See Exercise 13.) This shows that even though we might not know a specific value of σ, there are cases in which the use of very conservative values of σ will allow us to form some meaningful conclusions.

The requirements, test statistic, critical values, and *P*-value are summarized below.

Testing Claims About a Population Mean (with σ Known)

Requirements

1. The sample is a simple random sample. (Remember this very important point made in Chapter 1: *Data carelessly collected may be so completely useless that no amount of statistical torturing can salvage them.*)
2. The value of the population standard deviation σ is known.
3. Either or both of these conditions is satisfied: The population is normally distributed or $n > 30$.

Test Statistic for Testing a Claim About a Mean (with σ Known)

$$z = \frac{\bar{x} - \mu_{\bar{x}}}{\frac{\sigma}{\sqrt{n}}}$$

P-values: Refer to Figure 7-6 and use the standard normal distribution (Table A-2).
Critical values: Use the standard normal distribution (Table A-2).

Before starting the hypothesis testing procedure, we should first *explore* the data set. Using the methods introduced in Chapter 2, investigate center, variation, and distribution by drawing a graph; finding the mean, standard deviation, and

5-number summary; and identifying any outliers. We should verify that the requirements are satisfied. For the sample of 106 body temperatures used in the following example, a histogram shows that the sample data appear to come from a normally distributed population. Also, there are no outliers. The issue of normality is not too important in this example because the sample is so large, but it is important to know that there are no outliers that would dramatically affect results.

EXAMPLE *P*-Value Method Data Set 2 in Appendix B lists a sample of 106 body temperatures having a mean of 98.20°F. Assume that the sample is a simple random sample and that the population standard deviation σ is known to be 0.62°F. Use a 0.05 significance level to test the common belief that the mean body temperature of healthy adults is equal to 98.6°F. Use the *P*-value method by following the procedure outlined in Figure 7-9.

SOLUTION

REQUIREMENT ✓   We should first verify that the necessary requirements are satisfied. From the statement of the problem we are assuming that we have a simple random sample. We are also assuming that σ is known to be 0.62°F. The third and final requirement is that we have a "normally distributed population or $n > 30$." Because the sample size is $n = 106$, we have $n > 30$, so there is no need to check that the sample comes from a normally distributed population. The check of requirements has been successfully completed and we can proceed with the methods of this section. ✓

Refer to Figure 7-9 and follow these steps.

Step 1: The claim that the mean is equal to 98.6 is expressed in symbolic form as $\mu = 98.6$.

Step 2: The alternative (in symbolic form) to the original claim is $\mu \neq 98.6$.

Step 3: Because the statement $\mu \neq 98.6$ does not contain the condition of equality, it becomes the alternative hypothesis. The null hypothesis is the statement that $\mu = 98.6$.

$$H_0: \mu = 98.6 \qquad \text{(original claim)}$$
$$H_1: \mu \neq 98.6$$

Step 4: As specified in the statement of the problem, the significance level is $\alpha = 0.05$.

Step 5: Because the claim is made about the *population mean* μ, the sample statistic most relevant to this test is the *sample mean* $\bar{x} = 98.20$. Because σ is assumed to be known (0.62) and $n > 30$, the central limit theorem indicates that the distribution of sample means can be approximated by a *normal* distribution.

Step 6: The test statistic is calculated as follows:

$$z = \frac{\bar{x} - \mu_{\bar{x}}}{\frac{\sigma}{\sqrt{n}}} = \frac{98.20 - 98.6}{\frac{0.62}{\sqrt{106}}} = -6.64$$

continued

FIGURE 7-11 *P*-Value
Method of Testing
$H_0: \mu = 98.6$

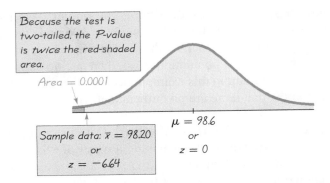

Because the test is two-tailed, the *P*-value is *twice* the red-shaded area.

Area = 0.0001

Sample data: $\bar{x} = 98.20$
or
$z = -6.64$

$\mu = 98.6$
or
$z = 0$

Using the test statistic of $z = -6.64$, we now proceed to find the *P*-value. See Figure 7-6 for the flowchart summarizing the procedure for finding *P*-values. This is a two-tailed test and the test statistic is to the left of the center (because $z = -6.64$ is less than $z = 0$), so the *P*-value is *twice* the area to the left of $z = -6.64$. We now refer to Table A-2 to find that the area to the left of $z = -6.64$ is 0.0001, so the *P*-value is $2(0.0001) = 0.0002$. (More precise results show that the *P*-value is actually much less than 0.0002.) See Figure 7-11.

Step 7: Because the *P*-value of 0.0002 is less than the significance level of $\alpha = 0.05$, we reject the null hypothesis.

INTERPRETATION The *P*-value of 0.0002 is the probability of getting a sample mean as extreme as 98.20°F (with a sample size of $n = 106$) by chance, assuming that $\mu = 98.6$°F and $\sigma = 0.62$°F. Because that probability is so small, we reject random chance as a likely explanation, and we conclude that the assumption of $\mu = 98.6$°F must be wrong. We refer to Figure 7-7 in Section 7-2 for help in correctly stating the final conclusion. We are rejecting the null hypothesis, which is the original claim, so we conclude that there is sufficient evidence to warrant rejection of the claim that the mean body temperature of healthy adults is 98.6°F. There is sufficient evidence to conclude that the mean body temperature of all healthy adults differs from 98.6°F.

Traditional Method If the traditional method of testing hypotheses is used for the preceding example, the first five steps would be the same. In Step 6 we would find the critical values of $z = -1.96$ and $z = 1.96$ instead of finding the *P*-value. We would again reject the null hypothesis because the test statistic of $z = -6.64$ would fall in the critical region. The final conclusion will be the same.

Confidence Interval Method We can use a confidence interval for testing a claim about μ when σ is known. For a two-tailed hypothesis test with a 0.05 significance level, we construct a 95% confidence interval. If we use the sample data in the preceding example ($n = 106$ and $\bar{x} = 98.20$) and assume that $\sigma = 0.62$, we can test the claim that $\mu = 98.6$ by using the methods of Section 6-3 to construct this 95% confidence interval: $98.08 < \mu < 98.32$. Because the claimed value of $\mu = 98.6$ is not contained within the confidence interval, we reject the claim that $\mu = 98.6$. We are 95% confident that the limits of 98.08 and 98.32 contain the true value of μ, so it appears that 98.6 cannot be the true value of μ.

In Section 7-3 we saw that when testing a claim about a population proportion, the traditional method and P-value method are equivalent, but the confidence interval method is somewhat different. When testing a claim about a population mean, there is no such difference, and all three methods are equivalent.

Caution: When testing a claim about μ using a confidence interval, be sure to use the confidence level that is appropriate for a specified significance level. With two-tailed tests, it is easy to see that a 0.05 significance level corresponds to a 95% confidence level, but it gets tricky with one-tailed tests. To test the claim that $\mu < 98.6$ with a 0.05 significance level, construct a 90% confidence interval (as suggested by Table 7-2). To test the claim that $\mu > 98.6$ with a 0.01 significance level, construct a 98% confidence interval (as suggested by Table 7-2).

In the remainder of the text, we will apply methods of hypothesis testing to other circumstances. It is easy to become entangled in a complex web of steps without ever understanding the underlying rationale of hypothesis testing. The key to that understanding lies in the rare event rule for inferential statistics: **If, under a given assumption, there is a small probability of getting sample results at least as extreme as the results that were obtained, we conclude that the assumption is probably not correct.**

7-4 Exercises

Verifying Requirements. In Exercises 1–4, determine whether the given conditions justify using the methods of this section when testing a claim about the population mean μ.

1. The sample size is $n = 25$, $\sigma = 6.44$, and the original population is normally distributed.

2. The sample size is $n = 7$, σ is not known, and the original population is normally distributed.

3. The sample size is $n = 11$, σ is not known, and the original population is normally distributed.

4. The sample size is $n = 47$, $\sigma = 12.6$, and the original population is not normally distributed.

Finding Test Components. In Exercises 5–8, find the test statistic, P-value, critical value(s), and state the final conclusion.

5. Claim: The mean IQ score of statistics professors is greater than 118.
Sample data: $n = 50$, $\bar{x} = 120$. Assume that $\sigma = 12$ and the significance level is $\alpha = 0.05$.

6. Claim: The mean body temperature of healthy adults is less than 98.6°F.
Sample data: $n = 106$, $\bar{x} = 98.20$°F. Assume that $\sigma = 0.62$ and the significance level is $\alpha = 0.01$.

7. Claim: The mean time between uses of a TV remote control by males during commercials equals 5.00 sec.
Sample data: $n = 80$, $\bar{x} = 5.25$ sec. Assume that $\sigma = 2.50$ sec and the significance level is $\alpha = 0.01$.

8. Claim: The mean starting salary for college graduates who have taken a statistics course is equal to $46,000.
Sample data: $n = 65$, $\bar{x} = \$45,678$. Assume that $\sigma = \$9900$ and the significance level is $\alpha = 0.05$.

Testing Hypotheses. In Exercises 9–12, test the given claim. Identify the null hypothesis, alternative hypothesis, test statistic, P-value or critical value(s), conclusion about the null hypothesis, and final conclusion that addresses the original claim. Use the P-value method unless your instructor specifies otherwise.

9. Everglades Temperatures In order to monitor the ecological health of the Florida Everglades, various measurements are recorded at different times. The bottom temperatures are recorded at the Garfield Bight station and the mean of 30.4°C is obtained for 61 temperatures recorded on 61 different days. Assuming that $\sigma = 1.7°C$, test the claim that the population mean is greater than 30.0°C. Use a 0.05 significance level.

10. Weights of Bears The health of the bear population in Yellowstone National Park is monitored by periodic measurements taken from anesthetized bears. A sample of 54 bears has a mean weight of 182.9 lb. Assuming that σ is known to be 121.8 lb, use a 0.10 significance level to test the claim that the population mean of all such bear weights is less than 200 lb.

11. Cotinine Levels of Smokers When people smoke, the nicotine they absorb is converted to cotinine, which can be measured. A sample of 40 smokers has a mean cotinine level of 172.5. Assuming that σ is known to be 119.5, use a 0.01 significance level to test the claim that the mean cotinine level of all smokers is equal to 200.0.

12. Head Circumferences A random sample of 100 two-month-old babies is obtained, and the mean head circumference is found to be 40.6 cm. Assuming that the population standard deviation is known to be 1.6 cm, use a 0.05 significance level to test the claim that the mean head circumference of all two-month-old babies is equal to 40.0 cm.

*13. Testing the Assumed σ In the example included in this section, we rejected H_0: $\mu = 98.6$ and supported H_1: $\mu \neq 98.6$ given the assumption that $\sigma = 0.62$ and the sample data consist of $n = 106$ values with $\bar{x} = 98.20$.
   a. What aspect of this example is not realistic?
   b. Find the largest value of σ that results in the same conclusion that was reached assuming that $\sigma = 0.62$.
   c. Given that the 106 body temperatures have a standard deviation of 0.62, is there any reasonable chance that the true value of σ is greater than the value found in part (b)? What does this imply about the assumption that $\sigma = 0.62$?

*14. Power and Probability of Type II Error For a hypothesis test with a given significance level α, the power of the test is the probability of rejecting a false null hypothesis. The value of power depends on the particular value of μ that is used as an alternative to the null hypothesis. Refer to the body temperature example discussed in this section and assume that $\mu = 98.4$.   *continued*

a. Find the power of the test, given that the population mean is actually 98.4. (In the procedure below, we refer to $\mu = 98.6$ as the "assumed" value, because it is assumed in the null hypothesis; we refer to $\mu = 98.4$ as the "alternative" value, because it is the value of the population mean used as an alternative to 98.6.)

b. Find the value of β, which is the probability of failing to reject the false null hypothesis.

> Step 1:   Using the significance level, find the critical z value(s). (For a right-tailed test, there is a single critical z value that is positive; for a left-tailed test, there is a single critical value of z that is negative; and a two-tailed test will have a critical z value that is negative along with another critical z value that is positive.)
>
> Step 2:   In the expression for the test statistic below, substitute the assumed value of μ (used in the null hypothesis). Substitute the known value of σ and substitute the value of the sample size n. Also substitute the critical value(s) for z. Then solve for the sample statistic $\bar{x}$. (If the test is two-tailed, substitute the critical value of z that is positive, then solve for the sample statistic $\bar{x}$. Next, substitute the critical value of z that is negative, and solve for the sample statistic $\bar{x}$. A two-tailed test should therefore result in two different values of $\bar{x}$.) The resulting value(s) of $\bar{x}$ separate the region(s) where the null hypothesis is rejected from the region(s) where we fail to reject the null hypothesis.

$$z = \frac{\bar{x} - \mu}{\frac{\sigma}{\sqrt{n}}}$$

> Step 3:   The calculation of power requires a specific value of μ that is to be used as an alternative to the value assumed in the null hypothesis. Identify this alternative value of μ (not the value used in the null hypothesis), draw a normal curve with this alternative value at the center, and plot the value(s) of $\bar{x}$ found in Step 2.
>
> Step 4:   Refer to the graph in Step 3, and find the area of the new critical region bounded by the value(s) of $\bar{x}$ found in Step 2. This is the probability of rejecting the null hypothesis, given that the alternative value of μ is the true value of the population mean. Because this is the probability of rejecting H_0 when it is false, the result is the power of the test.

*15. *Power of Test* The *power* of a test, expressed as $1 - \beta$, is the probability of rejecting a false null hypothesis. Assume that in testing the claim that $\mu < 98.6$, the sample data are $n = 106$ and $\bar{x} = 98.20$. Assume that $\sigma = 0.62$ and a 0.05 significance level is used. If the test of the claim $\mu < 98.6$ has a power of 0.8, find the mean μ that is being used as an alternative to the value given in H_0 (see Exercise 14).

7-5 Testing a Claim About a Mean: σ Not Known

One great advantage of learning the methods of hypothesis testing described in the earlier sections of this chapter is that those same methods can be easily modified for use in many other circumstances, such as those discussed in this section. The main objective of this section is to develop the ability to test claims made about population means when the population standard deviation σ is not known. Section 7-4 presented methods for testing claims about μ when σ is known, but it is rare that we do not know the value of μ while we do know the value of σ. The methods of this section are much more practical and realistic because they assume that σ is not known, as is usually the case. The requirements, test statistic, P-value, and critical values are summarized as follows.

Testing Claims About a Population Mean (with σ Not Known)

Requirements

1. The sample is a simple random sample.
2. The value of the population standard deviation σ is *not* known.
3. Either or both of these conditions is satisfied: The population is normally distributed or $n > 30$.

Test Statistic for Testing a Claim About a Mean (with σ Not Known)

$$t = \frac{\bar{x} - \mu_{\bar{x}}}{\frac{s}{\sqrt{n}}}$$

P-values and critical values: Use Table A-3 and use df $= n - 1$ for the number of degrees of freedom. (See Figure 7-6 for P-value procedures.)

The requirement of a normally distributed population is not a strict requirement, and we can usually consider the population to be normally distributed after using the sample data to confirm that there are no outliers and the histogram has a shape that is not very far from a normal distribution. Also, we use the simplified criterion of $n > 30$ as justification for treating the distribution of sample means as a normal distribution, but the minimum sample size actually depends on how much the population distribution departs from a normal distribution. Because we do not know the value of σ, we estimate it with the value of the sample standard deviation s, but this introduces another source of unreliability, especially with small samples. We compensate for this added unreliability by finding P-values and critical values using the t distribution instead of the normal distribution that was used in Section 7-4 where σ was assumed to be known. Here are the important properties of the Student t distribution:

Important Properties of the Student t Distribution

1. The Student t distribution is different for different sample sizes (see Figure 6-5 in Section 6-4).
2. The Student t distribution has the same general bell shape as the standard normal distribution; its wider shape reflects greater variability that is expected when s is used to estimate σ.

3. The Student t distribution has a mean of $t = 0$ (just as the standard normal distribution has a mean of $z = 0$).

4. The standard deviation of the Student t distribution varies with the sample size and is greater than 1 (unlike the standard normal distribution, which has $\sigma = 1$).

5. As the sample size n gets larger, the Student t distribution gets closer to the standard normal distribution.

Choosing the Appropriate Distribution

When testing claims made about population means, sometimes the normal distribution applies, sometimes the Student t distribution applies, and sometimes neither applies, so we must use nonparametric methods or bootstrap resampling techniques. (Nonparametric methods, which do not require a particular distribution, are discussed in Chapter 12; the bootstrap resampling technique is described in the Technology Project at the end of Chapter 6.) See pages 290–291 where Figure 6-6 and Table 6-1 both summarize the decisions to be made in choosing between the normal and Student t distributions. They show that when testing claims about population means, the Student t distribution is used under these conditions:

Use the Student t distribution when σ is not known and either or both of these conditions is satisfied:

The population is normally distributed or $n > 30$.

EXAMPLE **Body Temperatures** A pre-med student in a statistics class is required to do a class project. Intrigued by the body temperatures in Data Set 2 of Appendix B, she plans to collect her own sample data to test the claim that the mean body temperature is less than 98.6°F, as is commonly believed. Because of time constraints imposed by other courses and the desire to maintain a social life that goes beyond talking in her sleep, she finds that she has time to collect data from only 12 people. After carefully planning a procedure for obtaining a simple random sample of 12 healthy adults, she measures their body temperatures and obtains the results listed below. Use a 0.05 significance level to test the claim that these body temperatures come from a population with a mean that is less than 98.6°F.

98.0 97.5 98.6 98.8 98.0 98.5 98.6 99.4 98.4 98.7 98.6 97.6

SOLUTION

REQUIREMENT ✓ The first step in our procedure is checking that the necessary requirements are satisfied. We must have a simple random sample, which we are assuming in the statement of the problem. Also, we must have either a sample from a normally distributed population or we must have $n > 30$. Because the sample size is $n = 12$, we must check for a normal distribution. There are no outliers, and a histogram or normal quantile plot suggest that the sample data do not come from a population with a distribution that is far from

continued

Palm Reading

Some people believe that the length of their palm's lifeline can be used to predict longevity. In a letter published in the *Journal of the American Medical Association*, authors M. E. Wilson and L. E. Mather refuted that belief with a study of cadavers. Ages at death were recorded, along with the length of palm lifelines. The authors concluded that there is no significant correlation between age at death and length of lifeline. Palmistry lost, hands down.

normal. (See the accompanying histogram generated by STATDISK.) We can therefore proceed with the methods of this section. ✓

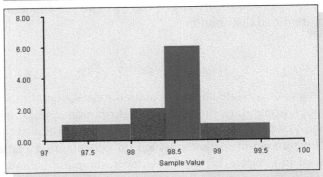

We use the sample data to find these statistics: $n = 12$, $\bar{x} = 98.39$, $s = 0.535$. The sample mean of $\bar{x} = 98.39$ is less than 98.6, but we need to determine whether it is *significantly* less than 98.6. Let's proceed with a formal hypothesis test. We will use the traditional method of hypothesis testing summarized in Figure 7-8.

Step 1: The original claim that "the mean body temperature is less than 98.6°F" can be expressed symbolically as $\mu < 98.6$.

Step 2: The opposite of the original claim is $\mu \geq 98.6$.

Step 3: Of the two symbolic expressions obtained so far, the expression $\mu < 98.6$ does not contain equality, so it becomes the alternative hypothesis H_1. The null hypothesis is the assumption that $\mu = 98.6$.

$$H_0: \mu = 98.6$$
$$H_1: \mu < 98.6 \qquad \text{(original claim)}$$

Step 4: The significance level is $\alpha = 0.05$.

Step 5: In this test of a claim about the population mean, the most relevant statistic is the sample mean. In selecting the correct distribution, we refer to Figure 6-6 or Table 6-1. We select the t distribution because of these conditions: We have a simple random sample, the value of σ is not known, and the sample data appear to come from a population that is normally distributed.

Step 6: The test statistic is

$$t = \frac{\bar{x} - \mu_{\bar{x}}}{\dfrac{s}{\sqrt{n}}} = \frac{98.39 - 98.6}{\dfrac{0.535}{\sqrt{12}}} = -1.360$$

The critical value of $t = -1.796$ is found by referring to Table A-3. First locate $n - 1 = 11$ degrees of freedom in the column at the left. Because this test is left-tailed with $\alpha = 0.05$, refer to the column in-

dicating an area of 0.05 in *one* tail. The test statistic and critical value
are shown in the accompanying STATDISK display.

Step 7:   Because the test statistic of $t = -1.360$ does not fall in the critical
region, we fail to reject H_0.

INTERPRETATION   (Refer to Figure 7-7 for help in wording the final conclusion.) There is not sufficient evidence to support the claim that the sample
comes from a population with a mean less than 98.6°F. This does not "prove"
that the mean is 98.6°F. In fact, μ may well be less than 98.6°F, but the 12 sample values do not provide evidence strong enough to support that claim. If we
use the 106 body temperatures included in Data Set 2 in Appendix B, we
would find that there is sufficient evidence to support the claim that the mean
body temperature is less than 98.6°F, but the 12 sample values included in this
example do not support that claim.

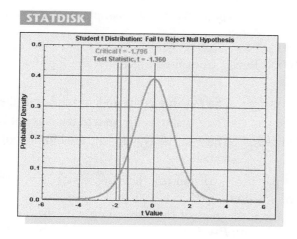

The critical value in the preceding example was $t = -1.796$, but if the normal distribution was being used, the critical value would have been $z = -1.645$.
The Student t critical value is farther to the left, showing that with the Student t
distribution, the sample evidence must be *more extreme* before we consider it to
be significant.

Finding *P*-Values with the Student *t* Distribution

The preceding example followed the traditional approach to hypothesis testing, but
SPSS, SAS, STATDISK, Minitab, the TI-83/84 Plus calculator, and many articles in
professional journals will display *P*-values. For the preceding example, STATDISK
displays a *P*-value of 0.1023, Minitab displays a *P*-value of 0.102, and the TI-83/84
Plus calculator displays a *P*-value of 0.1022565104. (SPSS and SAS display a two-
tailed *P*-value, which must be halved for this one-tailed case; the result is 0.1025 for
SPSS and 0.10225 for SAS.) With a significance level of 0.05 and a *P*-value greater
than 0.05, we fail to reject the null hypothesis, as we did using the traditional

method in the preceding example. If software or a TI-83/84 Plus calculator is not available, we can use Table A-3 to identify a *range of values* containing the *P*-value. We recommend this strategy for finding *P*-values using the *t* distribution:

1. Use software or a TI-83/84 Plus calculator.
2. If the technology is not available, use Table A-3 to identify a range of *P*-values. (See the following example.)

EXAMPLE **Finding *P*-Values** Assuming that neither software nor a TI-83/84 Plus calculator is available, use Table A-3 to find a range of values for the *P*-value corresponding to the given results.

a. In a left-tailed hypothesis test, the sample size is $n = 12$ and the test statistic is $t = -2.007$.

b. In a right-tailed hypothesis test, the sample size is $n = 12$ and the test statistic is $t = 1.222$.

c. In a two-tailed hypothesis test, the sample size is $n = 12$ and the test statistic is $t = -3.456$.

SOLUTION Recall from Figure 7-6 that the *P*-value is the area determined as follows:

Left-tailed test: The *P*-value is the area to the *left* of the test statistic.

Right-tailed test: The *P*-value is the area to the *right* of the test statistic.

Two-tailed test: The *P*-value is *twice* the area in the tail bounded by the test statistic.

In each of parts (a), (b), and (c), the sample size is $n = 12$, so the number of degrees of freedom is df $= n - 1 = 11$. See the accompanying portion of Table A-3 for 11 degrees of freedom along with the boxes describing the procedures for finding *P*-values.

a. The test is left-tailed with test statistic $t = -2.007$, so the *P*-value is the area to the left of -2.007. Because of the symmetry of the *t* distribution, that is the same as the area to the right of $+2.007$. See the accompanying illustration showing that any test statistic between 2.201 and 1.796 has a right-tailed *P*-value that is between 0.025 and 0.05. We conclude that $0.025 <$ *P*-value < 0.05. (The exact *P*-value found using software is 0.0350.)

b. The test is right-tailed with test statistic $t = 1.222$, so the *P*-value is the area to the right of 1.222. See the accompanying illustration showing that any test statistic less than 1.363 has a right-tailed *P*-value that is greater than 0.10. We conclude that the *P*-value is > 0.10. (The exact *P*-value found using software is 0.124.)

c. The test is two-tailed with test statistic $t = -3.456$. The *P*-value is *twice* the area to the left of -3.456, but with the symmetry of the *t* distribution, that is the same as twice the area to the right of $+3.456$. See the accompanying illustration showing that any test statistic greater than 3.456 has a two-tailed *P*-value that is less than 0.01. We conclude that the *P*-value < 0.01. (The exact *P*-value found using software is 0.00537.)

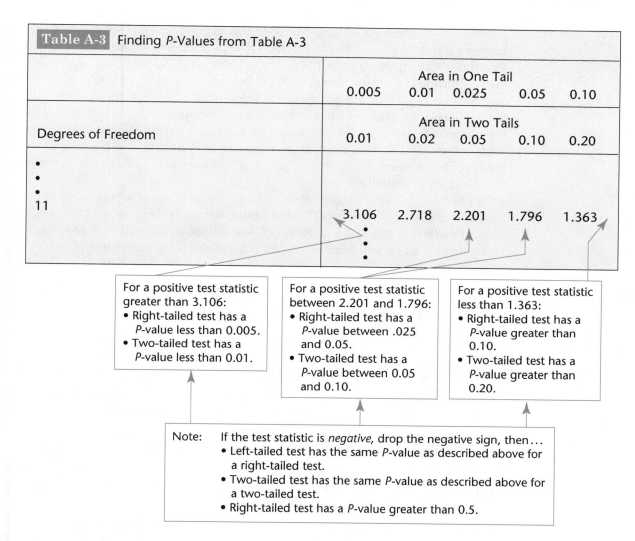

Table A-3   Finding *P*-Values from Table A-3

	Area in One Tail				
	0.005	0.01	0.025	0.05	0.10
	Area in Two Tails				
Degrees of Freedom	0.01	0.02	0.05	0.10	0.20
⋮					
11	3.106	2.718	2.201	1.796	1.363

For a positive test statistic greater than 3.106:
- Right-tailed test has a *P*-value less than 0.005.
- Two-tailed test has a *P*-value less than 0.01.

For a positive test statistic between 2.201 and 1.796:
- Right-tailed test has a *P*-value between .025 and 0.05.
- Two-tailed test has a *P*-value between 0.05 and 0.10.

For a positive test statistic less than 1.363:
- Right-tailed test has a *P*-value greater than 0.10.
- Two-tailed test has a *P*-value greater than 0.20.

Note: If the test statistic is *negative*, drop the negative sign, then...
- Left-tailed test has the same *P*-value as described above for a right-tailed test.
- Two-tailed test has the same *P*-value as described above for a two-tailed test.
- Right-tailed test has a *P*-value greater than 0.5.

Once the format of Table A-3 is understood, it is not difficult to find a range of numbers for *P*-values. Check your results to be sure that they follow the same patterns shown in Table A-3. For example, in part (b), the test statistic of $t = 1.222$ would be positioned to the right of 1.363, so the corresponding right-tailed area would be positioned to the right of 0.10; the pattern of area values is increasing from left to right, so the right-tailed area would be *greater than* 0.10.

Remember, *P*-values can be easily found by using software or a TI-83/84 Plus calculator. Also, the traditional method of testing hypotheses can be used instead of the *P*-value method.

Confidence Interval Method   We can use a confidence interval for testing a claim about μ when σ is not known. For a two-tailed hypothesis test with a 0.05 significance level, we construct a 95% confidence interval, but for a one-tailed hypothesis test with a 0.05 significance level we construct a 90% confidence interval. (See Table 7-2.) Using the sample data from the first example in this section ($n = 12$ and $\bar{x} = 98.39$, $s = 0.535$) with σ not known, and using a 0.05 significance level, we can test the claim that $\mu < 98.6$ by using the confidence interval method. Construct this 90% confidence interval: $98.11 < \mu < 98.67$ (see Section 6-4). Because the assumed value of $\mu = 98.6$ is contained within the confidence interval, we cannot reject the assumption that $\mu = 98.6$. Based on the 12 sample values given in the example, we do not have sufficient evidence to support the claim that the mean body temperature is less than 98.6°F. Based on the confidence interval, the true value of μ is likely to be any value between 98.11 and 98.67, including 98.6.

In Section 7-3 we saw that when testing a claim about a population proportion, the traditional method and P-value method are equivalent, but the confidence interval method is somewhat different. When testing a claim about a population mean, there is no such difference, and all three methods are equivalent.

7-5 Exercises

Using Correct Distribution. In Exercises 1–4, determine whether the hypothesis test involves a sampling distribution of means that is a normal distribution, Student t distribution, or neither. (Hint: See Figure 6-6 and Table 6-1.)

1. Claim: $\mu = 100$. Sample data: $n = 15$, $\bar{x} = 102$, $s = 15.3$. The sample data appear to come from a normally distributed population with unknown μ and σ.

2. Claim: $\mu = 75$. Sample data: $n = 25$, $\bar{x} = 102$, $s = 15.3$. The sample data appear to come from a population with a distribution that is very far from normal, and σ is unknown.

3. Claim: $\mu = 980$. Sample data: $n = 25$, $\bar{x} = 950$, $s = 27$. The sample data appear to come from a normally distributed population with $\sigma = 30$.

4. Claim: $\mu = 2.80$. Sample data: $n = 150$, $\bar{x} = 2.88$, $s = 0.24$. The sample data appear to come from a population with a distribution that is not normal, and σ is unknown.

Finding P-values. In Exercises 5–8, use the given information to find a range of numbers for the P-value. (Hint: See the example and its accompanying display in the subsection of "Finding P-Values with the Student t Distribution.")

5. Right-tailed test with $n = 12$ and test statistic $t = 2.998$

6. Left-tailed test with $n = 12$ and test statistic $t = -0.855$

7. Two-tailed test with $n = 16$ and test statistic $t = 4.629$

8. Two-tailed test with $n = 9$ and test statistic $t = -1.577$

Finding Test Components. In Exercises 9–12 assume that a simple random sample has been selected from a normally distributed population. Find the test statistic, P-value, critical value(s), and state the final conclusion.

9. Claim: The mean IQ score of statistics professors is greater than 118.
   Sample data: $n = 20$, $\bar{x} = 120$, $s = 12$. The significance level is $\alpha = 0.05$.

10. Claim: The mean body temperature of healthy adults is less than 98.6°F.
Sample data: $n = 35, \bar{x} = 98.20°F, s = 0.62$. The significance level is $\alpha = 0.01$.

11. Claim: The mean time between uses of a TV remote control by males during commercials equals 5.00 sec.
Sample data: $n = 81, \bar{x} = 5.25$ sec, $s = 2.50$ sec. The significance level is $\alpha = 0.01$.

12. Claim: The mean starting salary for college graduates who have taken a statistics course is equal to $46,000.
Sample data: $n = 27, \bar{x} = \$45,678, s = \$9900$. The significance level is $\alpha = 0.05$.

Testing Hypotheses. In Exercises 13–24, assume that a simple random sample has been selected from a normally distributed population and test the given claim. Unless specified by your instructor, use either the traditional method or P-value method for testing hypotheses.

13. Effect of Vitamin Supplement on Birth Weight The birth weights (in kilograms) are recorded for a sample of male babies born to mothers taking a special vitamin supplement (based on data from the New York State Department of Health). When testing the claim that the mean birth weight for all male babies of mothers given vitamins is equal to 3.39 kg, which is the mean for the population of all males, SPSS yields the results shown. Based on those results, does the vitamin supplement appear to have an effect on birth weight?

					95% Confidence Interval of the Difference	
	t	df	Sig. (2-tailed)	Mean Difference	Lower	Upper
WEIGHT	1.734	15	.103	.2849994	-.0652608	.6352595

One-Sample Test — Test Value = 3.39

14. Pulse Rates One of the authors, at the peak of an exercise program, claimed that his pulse rate was lower than the mean pulse rate of statistics students. The author's pulse rate was measured to be 60 beats per minute, and the 20 students in his class measured their pulse rates. When testing the claim that statistics students have a mean pulse rate greater than 60 beats per minute, the accompanying TI-83/84 Plus calculator display was obtained. Based on those results, is there sufficient evidence to support the claim that the mean pulse rate of statistics students is greater than 60 beats per minute?

TI-83/84 Plus

```
T-Test
μ>60
t=6.393083002
p=1.9683974E-6
x̄=74.35
Sx=10.03821645
n=20
```

15. Heights of Parents Data Set 3 in Appendix B includes the heights of parents of 20 males. If the difference in height for each set of parents is found by subtracting the mother's height from the father's height, the result is a list of 20 values with a mean of 4.4 in. and a standard deviation of 4.2 in. Use a 0.01 significance level to test the claim that the mean difference is greater than 0. Do the results support a sociologist's claim that women tend to marry men who are taller than themselves?

16. Monitoring Lead in Air Listed below are measured amounts of lead (in micrograms per cubic meter or $\mu g/m^3$) in the air. The Environmental Protection Agency has established an air quality standard for lead: 1.5 $\mu g/m^3$. The measurements shown below were recorded at Building 5 of the World Trade Center site on different days immediately following the destruction caused by the terrorist attacks of September 11, 2001.

continued

After the collapse of the two World Trade buildings, there was considerable concern about the quality of the air. Use a 0.05 significance level to test the claim that the sample is from a population with a mean greater than the EPA standard of 1.5 $\mu g/m^3$. Is there anything about this data set suggesting that the assumption of a normally distributed population might not be valid?

5.40   1.10   0.42   0.73   0.48   1.10

17. Sugar in Cereal A sample of cereal boxes is randomly selected and the sugar contents (grams of sugar per gram of cereal) are recorded. Those amounts are summarized with these statistics: $n = 16$, $\bar{x} = 0.295$ g, $s = 0.168$ g. Use a 0.05 significance level to test the claim of a cereal lobbyist that the mean for all cereals is less than 0.3 g.

18. Treating Chronic Fatigue Syndrome Patients with chronic fatigue syndrome were tested, then retested after being treated with fluorocortisone. Listed below are the changes in fatigue after the treatment (based on data from "The Relationship Between Neurally Mediated Hypotension and the Chronic Fatigue Syndrome" by Bou-Holaigah, Rowe, Kan, and Calkins, Journal of the American Medical Association, Vol. 274, No. 12). A standard scale from -7 to $+7$ was used, with positive values representing improvements. Use a 0.01 significance level to test the claim that the mean change is positive. Does the treatment appear to be effective?

6 5 0 5 6 7 3 3 2 6 5 5 0 6 3 4 3 7 0 4 4

19. Olympic Winners Listed below are the winning times (in seconds) of men in the 100-meter dash for consecutive summer Olympic games, listed in order by row. Assuming that these results are sample data randomly selected from the population of all past and future Olympic games, test the claim that the mean time is less than 10.5 sec. What do you observe about the precision of the numbers? What extremely important characteristic of the data set is not considered in this hypothesis test? Do the results from the hypothesis test suggest that future winning times should be around 10.5 sec, and is such a conclusion valid?

12.0 11.0 11.0 11.2 10.8 10.8 10.8 10.6 10.8 10.3 10.3 10.3
10.4 10.5 10.2 10.0 9.95 10.14 10.06 10.25 9.99 9.92 9.96

20. Nicotine in Cigarettes The Carolina Tobacco Company advertised that its best-selling nonfiltered cigarettes contain at most 40 mg of nicotine, but Consumer Advocate magazine ran tests of 10 randomly selected cigarettes and found the amounts (in mg) shown in the accompanying list. It's a serious matter to charge that the company advertising is wrong, so the magazine editor chooses a significance level of $\alpha = 0.01$ in testing her belief that the mean nicotine content is greater than 40 mg. Using a 0.01 significance level, test the editor's belief that the mean is greater than 40 mg.

47.3 39.3 40.3 38.3 46.3 43.3 42.3 49.3 40.3 46.3

21. Norwegian Babies Weigh More? A random sample of 90 Norwegian babies is obtained, and the mean birth weight is 3570 g with a standard deviation of 498 g (based on data from "Birth Weight and Perinatal Mortality" by Wilcox et al., Journal of the American Medical Association, Vol. 273, No. 9). Use a 0.05 level of significance to test the claim that Norwegian newborn babies have weights with a mean greater than 3420 g, which is the mean for American newborn babies.

22. Conductor Life Span A *New York Times* article noted that the mean life span for 35 male symphony conductors was 73.4 years, in contrast to the mean of 69.5 years for males in the general population. Assuming that the 35 males have life spans with a standard deviation of 8.7 years, use a 0.05 significance level to test the claim that male symphony conductors have a mean life span that is greater than 69.5 years. Does it appear that male symphony conductors live longer than males from the general population? Why doesn't the experience of being a male symphony conductor cause men to live longer? (*Hint*: Are male symphony conductors born, or do they become conductors at a much later age?)

23. Poplar Tree Weights Refer to Data Set 9 in Appendix B and use the weights of the trees from year 1 and site 1 that were treated with fertilizer only. Assume that the population mean for untreated trees from year 1 and site 1 is 0.92 kg and test the claim that the fertilized trees come from a population with a mean less than 0.92 kg. Does the fertilizer appear to have an effect? Does the fertilizer appear to decrease the weights of trees on site 1?

24. Petal Lengths of Irises Refer to Data Set 7 in Appendix B and use the petal lengths of irises in the sertosa class. Using the sample data, test the claim that irises in the sertosa class have a mean petal length that is less than the mean of 4.3 mm for irises from the versicolor class.

*25. Using the Wrong Distribution When testing a claim about a population mean with a simple random sample selected from a normally distributed population with unknown σ, the Student t distribution should be used for finding critical values and/or a P-value. If the standard normal distribution is incorrectly used instead, does that mistake make you more or less likely to reject the null hypothesis, or does it not make a difference? Explain.

*26. Finding Critical t Values When finding critical values, we sometimes need significance levels other than those available in Table A-3. Some computer programs approximate critical t values by calculating

$$t = \sqrt{df \cdot (e^{A^2/df} - 1)}$$

where

$$df = n - 1$$
$$e = 2.718$$
$$A = \frac{z(8 \cdot df + 3)}{(8 \cdot df + 1)}$$

and z is the critical z score. Use this approximation to find the critical t score corresponding to $n = 10$ and a significance level of 0.05 in a right-tailed case. Compare the results to the critical t value found in Table A-3.

7-6 Testing a Claim About a Standard Deviation or Variance

Many service and production organizations share this common goal: Improve quality by reducing variation. Quality-control engineers want to ensure that a product has an acceptable mean, but they also want to produce items of *consistent* quality so that there will be few defects. For example, it is impossible to manufacture Lipitor tablets in such a way that they all include precisely the same desired amount of the key ingredient of atorvastatin. However, quality tablets can be produced by manufacturing them with an acceptable mean level of atorvastatin, and the *variation* among the pills must be low enough to avoid having some pills with unacceptably low levels of atorvastatin while other pills have unacceptably high levels. Consistency is improved by reducing the standard deviation. In the preceding sections of this chapter we described methods for testing claims made about population means and proportions. This section focuses on variation, which is critically important in many applications, including quality control. The main objective of this section is to present methods to test claims made about a population standard deviation σ or variance σ^2. The requirements, test statistic, P-value, and critical values are summarized as follows.

Testing Claims About σ or σ^2

Requirements

1. The sample is a simple random sample.
2. The population has a normal distribution. (This is a much stricter requirement than the requirement of a normal distribution when testing claims about means, as in Sections 7-4 and 7-5.)

Test Statistic for Testing a Claim About σ or σ^2

$$\chi^2 = \frac{(n-1)s^2}{\sigma^2}$$

P-values and Critical values: Use Table A-4 with df $= n - 1$ for the number of degrees of freedom. (Table A-4 is based on *cumulative areas from the right*.)

In Sections 7-4 and 7-5 we saw that the methods of testing claims about means require a normally distributed population, and those methods work reasonably well as long as the population distribution is not very far from being normal. However, tests of claims about standard deviations or variances are not as *robust*, meaning that the results can be very misleading if the population does not have a normal distribution. The condition of a normally distributed population is therefore a much stricter requirement in this section. If the population has a distribution that is far from normal and you use the methods of this section to reject a null hypothesis, you don't really know if the standard deviation is not as assumed or if the rejection is due to the lack of normality.

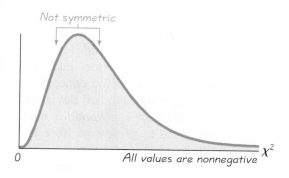

FIGURE 7-12 **Properties of the Chi-Square Distribution**

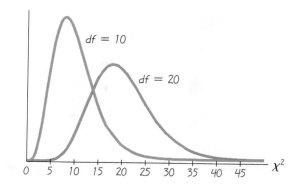

FIGURE 7-13 **Chi-Square Distribution for 10 and 20 Degrees of Freedom**

Don't be confused by reference to both the normal and the chi-square distributions. After verifying that the sample data appear to come from a normally distributed population, we should then shift gears and think in terms of the "chi-square" distribution. The chi-square distribution was introduced in Section 6-5, where we noted the following important properties.

Properties of the Chi-Square Distribution

1. All values of χ^2 are non-negative, and the distribution is not symmetric (see Figure 7-12).

2. There is a different χ^2 distribution for each number of degrees of freedom (see Figure 7-13).

3. The critical values are found in Table A-4 using

$$\text{degrees of freedom} = n - 1$$

Table A-4 is based on cumulative areas from the *right* (unlike the entries in Table A-2, which are cumulative areas from the left). Critical values are found in Table A-4 by first locating the row corresponding to the appropriate number of degrees of freedom (where df $= n - 1$). Next, the significance level α is used to determine the correct column. The following examples are based on a significance level of $\alpha = 0.05$, but any other significance level can be used in a similar manner. Note that in each case, the key area is the region to the *right* of the critical value(s).

Right-tailed test: Because the area to the *right* of the critical value is 0.05, locate 0.05 at the top of Table A-4.

Left-tailed test: With a left-tailed area of 0.05, the area to the *right* of the critical value is 0.95, so locate 0.95 at the top of Table A-4.

Two-tailed test: Divide the significance level of 0.05 between the left and right tails, so the areas to the *right* of the two critical values are 0.975 and 0.025, respectively. Locate 0.975 and 0.025 at the top of Table A-4. (See Figure 6-10 and the example on pages 300–301.)

Delaying Death

University of California sociologist David Phillips has studied the ability of people to postpone their death until after some important event. Analyzing death rates of Jewish men who died near Passover, he found that the death rate dropped dramatically in the week before Passover, but rose the week after. He found a similar phenomenon occurring among Chinese-American women; their death rate dropped the week before their important Harvest Moon Festival, then rose the week after.

EXAMPLE IQ Scores of Statistics Professors IQ scores of adults are normally distributed with a mean of 100 and a standard deviation of 15. A simple random sample of 13 statistics professors yields a standard deviation of $s = 7.2$. A psychologist is quite sure that statistics professors have IQ scores with a mean greater than 100. He doesn't understand the concept of standard deviation very well and does not realize that the standard deviation should be lower than 15 (because statistics professors have less variation than the general population). Instead, he claims that statistics professors have IQ scores with a standard deviation equal to 15, the same standard deviation for the general population. Assume that IQ scores of statistics professors are normally distributed and use a 0.05 significance level to test the claim that $\sigma = 15$. Based on the result, what do you conclude about the standard deviation of IQ scores for statistics professors?

SOLUTION

REQUIREMENT ✔ We should first verify that the necessary requirements are satisfied. (Earlier in this section we listed these requirements: The sample is a simple random sample and the sample appears to come from a normally distributed population.) We are assuming the sample is a simple random sample. Also, the population is known to be normally distributed, so the check of requirements for the procedure of this section has been successfully completed. ✔

We will now proceed by using the traditional method of testing hypotheses as outlined in Figure 7-8.

Step 1: The claim is expressed in symbolic form as $\sigma = 15$.

Step 2: If the original claim is false, then $\sigma \neq 15$.

Step 3: The expression $\sigma \neq 15$ does not contain equality, so it becomes the alternative hypothesis. The null hypothesis is the statement that $\sigma = 15$.

$$H_0: \sigma = 15 \quad \text{(original claim)}$$
$$H_1: \sigma \neq 15$$

Step 4: The significance level is $\alpha = 0.05$.

Step 5: Because the claim is made about σ we use the chi-square distribution.

Step 6: The test statistic is

$$\chi^2 = \frac{(n-1)s^2}{\sigma^2} = \frac{(13-1)(7.2)^2}{15^2} = 2.765$$

The critical values of 4.404 and 23.337 are found in Table A-4, in the 12th row (degrees of freedom $= n - 1 = 12$) in the columns corresponding to 0.975 and 0.025. See the test statistic and critical values shown in Figure 7-14.

Step 7: Because the test statistic is in the critical region, we reject the null hypothesis.

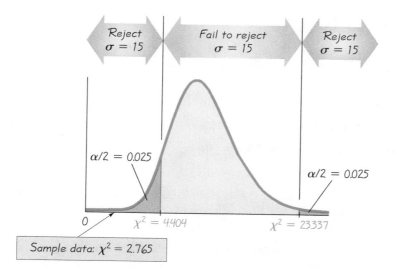

FIGURE 7-14 **Testing the Claim that $\sigma = 15$**

INTERPRETATION There is sufficient evidence to warrant rejection of the claim that the standard deviation is equal to 15. It appears that statistics professors have IQ scores with a standard deviation that is significantly different than the standard deviation of 15 for the general population.

The *P*-Value Method

Instead of using the traditional approach to hypothesis testing for the preceding example, we can also use the *P*-value approach summarized in Figures 7-6 and 7-9. If STATDISK is used for the preceding example, the *P*-value of 0.0060 will be found. If we use Table A-4, we usually cannot find *exact P*-values because that chi-square distribution table includes only selected values of α. (Because of this limitation, testing claims about σ or σ^2 with Table A-4 is easier with the traditional method than the *P*-value method.) If using Table A-4, we can identify limits that contain the *P*-value. The test statistic from the last example is $\chi^2 = 2.765$ and we know that the test is two-tailed with 12 degrees of freedom. Refer to the 12th row of Table A-4 and see that the test statistic of 2.765 is less than every entry in that row, which means that the area to the left of the test statistic is less than 0.005. The *P*-value for a two-tailed test is *twice* the tail area bounded by the test statistic, so we double 0.005 to conclude that the *P*-value is less than 0.01. Because the *P*-value is less than the significance level of $\alpha = 0.05$, we reject the null hypothesis. Again, the traditional method and the *P*-value method are equivalent in the sense that they always lead to the same conclusion.

The Confidence Interval Method

The preceding example can also be solved with the confidence interval method of testing hypotheses. Using the methods described in Section 6-5, we can use the sample data ($n = 13, s = 7.2$) to construct this 95% confidence interval: $5.2 < \sigma < 11.9$. Because the claimed value of $\sigma = 15$ is *not* contained within the confidence interval, we reject the claim that $\sigma = 15$, and we reach the same conclusion from the traditional and *P*-value methods.

7-6 Exercises

Finding Critical Values. *In Exercises 1–4, find the test statistic, then use Table A-4 to find critical value(s) of χ^2 and limits containing the P-value, then determine whether there is sufficient evidence to support the given alternative hypothesis.*

1. $H_1: \sigma \neq 15$, $\alpha = 0.05$, $n = 20$, $s = 10$.

2. $H_1: \sigma > 12$, $\alpha = 0.01$, $n = 5$, $s = 18$.

3. $H_1: \sigma < 50$, $\alpha = 0.01$, $n = 30$, $s = 30$.

4. $H_1: \sigma \neq 4.0$, $\alpha = 0.05$, $n = 81$, $s = 4.7$.

Testing Claims About Variation. *In Exercises 5–12, test the given claim. Assume that a simple random sample is selected from a normally distributed population. Use the traditional method of testing hypotheses unless your instructor indicates otherwise.*

5. Body Temperatures   In Section 7-4, we tested the claim that the mean body temperature is equal to 98.6°F, and we used sample data given in Data Set 2 in Appendix B. The body temperatures taken at 12:00 A.M. on day 2 can be summarized with these statistics: $n = 106$, $\bar{x} = 98.20$°F, $s = 0.62$°F, and a histogram shows that the values have a distribution that is approximately normal. In Section 7-4 we assumed that $\sigma = 0.62$°F, which is an unrealistic assumption. However, the test statistic will cause rejection of $\mu = 98.6$°F as long as the standard deviation is less than 2.11°F. Use the sample statistics and a 0.005 significance level to test the claim that $\sigma < 2.11$°F.

6. Supermodel Weights   Use a 0.01 significance level to test the claim that weights of female supermodels vary less than the weights of women in general. The standard deviation of weights of the population of women is 29 lb. Listed below are the weights (in pounds) of nine randomly selected supermodels.

125 (Taylor)	119 (Auermann)	128 (Schiffer)	128 (MacPherson)
119 (Turlington)	127 (Hall)	105 (Moss)	123 (Mazza)
115 (Hume)			

7. Supermodel Heights   Use a 0.05 significance level to test the claim that heights of female supermodels vary less than the heights of women in general. The standard deviation of heights of the population of women is 2.5 in. Listed below are the heights (in inches) of randomly selected supermodels (Taylor, Harlow, Mulder, Goff, Evangelista, Avermann, Schiffer, MacPherson, Turlington, Hall, Crawford, Campbell, Herzigova, Seymour, Banks, Moss, Mazza, Hume).

71	71	70	69	69.5	70.5	71	72	70
70	69	69.5	69	70	70	66.5	70	71

8. Using Birth Weight Data   Shown below are birth weights (in kilograms) of male babies born to mothers taking a special vitamin supplement (based on data from the New York State Department of Health). Test the claim that this sample comes from a population with a standard deviation equal to 0.470 kg, which is the standard deviation for male birth weights in general. Does the vitamin supplement appear to affect the variation among birth weights?

3.73	4.37	3.73	4.33	3.39	3.68	4.68	3.52
3.02	4.09	2.47	4.13	4.47	3.22	3.43	2.54

9. Is the New Machine Better? The Medassist Pharmaceutical Company uses a machine to pour cold medicine into bottles in such a way that the standard deviation of the weights is 0.15 oz. A new machine is tested on 71 bottles, and the standard deviation for this sample is 0.12 oz. The Dayton Machine Company, which manufactures the new machine, claims that it fills bottles with less variation. At the 0.05 significance level, test the claim made by the Dayton Machine Company. If Dayton's machine is being used on a trial basis, should its purchase be considered?

10. Systolic Blood Pressure for Women Systolic blood pressure results from contraction of the heart. Based on past results from the National Health Survey, it is claimed that women have systolic blood pressures with a mean and standard deviation of 130.7 and 23.4, respectively. Use the systolic blood pressures of women listed in Data Set 1 in Appendix B and test the claim that the sample comes from a population with a standard deviation of 23.4.

11. Weights of Men Anthropometric survey data are used to publish values that can be used in designing products that are suitable for use by adults. According to Gordon, Churchill, et al., men have weights with a mean of 172.0 lb and a standard deviation of 28.7 lb. Using the sample of weights of men listed in Data Set 1 in Appendix B, test the claim that the standard deviation is 28.7 lb. Use a 0.05 significance level. When designing elevators, what would be a consequence of believing that weights of men vary less than they really vary?

12. Heights of Women Anthropometric survey data are used to publish values that can be used in designing products that are suitable for use by adults. According to Gordon, Churchill, et al., women have heights with a mean of 64.1 in. and a standard deviation of 2.52 in. Using the sample of heights of women listed in Data Set 1 in Appendix B, test the claim that the standard deviation is 2.52 in. Use a 0.05 significance level. When designing car seats for women, what would be a consequence of believing that heights of women vary less than they really vary?

*13. Finding Critical Values of χ^2 For large numbers of degrees of freedom, we can approximate critical values of χ^2 as follows:

$$\chi^2 = \frac{1}{2}(z + \sqrt{2k - 1})^2$$

Here k is the number of degrees of freedom and z is the critical value, found in Table A-2. For example, if we want to approximate the two critical values of χ^2 in a two-tailed hypothesis test with $\alpha = 0.05$ and a sample size of 150, we let $k = 149$ with $z = -1.96$ followed by $k = 149$ and $z = 1.96$.

a. Use this approximation to estimate the critical values of χ^2 in a two-tailed hypothesis test with $n = 101$ and $\alpha = 0.05$. Compare the results to those found in Table A-4.

b. Use this approximation to estimate the critical values of χ^2 in a two-tailed hypothesis test with $n = 150$ and $\alpha = 0.05$.

*14. Finding Critical Values of χ^2 Repeat Exercise 13 using this approximation (with k and z as described in Exercise 13):

$$\chi^2 = k\left(1 - \frac{2}{9k} + z\sqrt{\frac{2}{9k}}\right)^3$$

Review

This chapter presented basic methods for testing claims about a population proportion, population mean, or population standard deviation (or variance). The methods of this chapter are used by professionals in a wide variety of disciplines, as illustrated in their many professional journals.

In Section 7-2 we presented the fundamental concepts of a hypothesis test: null hypothesis, alternative hypothesis, test statistic, critical region, significance level, critical value, P-value, type I error, and type II error. We also discussed two-tailed tests, left-tailed tests, right-tailed tests, and the statement of conclusions. We used those components in identifying three different methods for testing hypotheses:

1. The traditional method (summarized in Figure 7-8)
2. The P-value method (summarized in Figure 7-9)
3. Confidence intervals (discussed in Chapter 6)

In Sections 7-3 through 7-6 we discussed specific methods for dealing with different parameters. Because it is so important to be correct in selecting the distribution and test statistic, we provide Table 7-3, which summarizes the hypothesis testing procedures of this chapter.

Table 7-3   Hypothesis Tests

Parameter	Requirements	Distribution and Test Statistic	Critical and P-Values
Proportion	$np \geq 5$ and $nq \geq 5$	Normal: $z = \dfrac{\hat{p} - p}{\sqrt{\dfrac{pq}{n}}}$	Table A-2
Mean	σ known and normally distributed population *or* σ known and $n > 30$	Normal: $z = \dfrac{\bar{x} - \mu_{\bar{x}}}{\dfrac{\sigma}{\sqrt{n}}}$	Table A-2
	σ not known and normally distributed population *or* σ not known and $n > 30$	Student t: $t = \dfrac{\bar{x} - \mu_{\bar{x}}}{\dfrac{s}{\sqrt{n}}}$	Table A-3
	Population not normally distributed and $n \leq 30$	Use a nonparametric method or bootstrapping.	
Standard Deviation or Variance	Strict requirement of a normally distributed population	Chi-Square: $\chi^2 = \dfrac{(n - 1)s^2}{\sigma^2}$	Table A-4

Review Exercises

1. a. You have just collected a very large ($n = 2575$) sample of responses obtained from adult Americans who mailed responses to a questionnaire printed in *Prevention* magazine. A hypothesis test conducted at the 0.01 significance level leads to the conclusion that most (more than 50%) adults are opposed to estate taxes. Can we conclude that most adult Americans are opposed to estate taxes? Why or why not?

b. When testing a diet control drug, a hypothesis test based on 5000 randomly selected subjects shows that the mean weight loss of 0.2 lb is significant at the 0.01 level. Should this drug be used by subjects wanting to lose weight? Why or why not?

c. You have just developed a new cure for the common cold and you plan to conduct a formal test to justify its effectiveness. Which *P*-value would you most prefer: 0.99, 0.05, 0.5, 0.01, or 0.001?

d. In testing the claim that the mean amount of cold medicine in containers is greater than 12 oz, you fail to reject the null hypothesis. State the final conclusion that addresses the original claim.

e. Complete the statement: "A type I error is the mistake of . . ."

2. Identifying Hypotheses and Distributions Based on the given conditions, identify the alternative hypothesis and the sampling distribution (normal, *t*, chi-square) of the test statistic.

a. Claim: The mean annual income of full-time college students is below $10,000. Sample data: For 750 randomly selected college students, the mean is $3662 and the standard deviation is $2996.

b. Claim: With manual assembly of hearing aids, the assembly times vary more than the times for automated assembly, which are known to have a mean of 27.6 sec and a standard deviation of 1.8 sec.

c. Claim: The majority of college students are women. Sample data: Of 500 randomly selected college students, 58% are women.

d. Claim: When a group of adult survey respondents is randomly selected, their mean IQ is equal to 100. Sample data: $n = 150$ and $\bar{x} = 98.8$. It is reasonable to assume that $\sigma = 15$.

3. Effective Treatment The 10% annual death rate of elm trees in one region prompted a research effort intended to prevent future deaths of those trees. When one treatment was applied to 200 trees, it was observed that 16 of the trees died the following year. Use a 0.05 significance level to test the claim that the proportion of tree deaths for treated trees is less than the 10% rate for untreated trees.

4. X-Linked Recessive Disorder When certain couples have baby boys, it is found that 50% of those boys inherit an X-linked recessive disorder. An experimental method is used in an attempt to lower that 50% rate, and a trial results in 40 baby boys with 15 of them inheriting the X-linked recessive disorder. One researcher announces that the sample result of 37.5% (from 15/40) is evidence that the experimental method is effective. He emphasizes that the 37.5% rate is considerably less than the 50% rate. Is the researcher correct in his announcements? Use a 0.05 significance level.

5. Are Consumers Being Cheated? The Orange County Bureau of Weights and Measures received complaints that the Windsor Bottling Company was cheating consumers by putting less than 12 oz of rubbing alcohol in its containers that are labeled as containing 12 oz. When 24 containers are randomly selected and measured, the amounts are found to have a mean of 11.4 oz and a standard deviation of 0.62 oz. The company president, Harry Windsor, claims that the sample is too small to be meaningful. Use the sample data to test the claim that consumers are being cheated. Does Harry Windsor's argument have any validity?

6. Controlling Cholesterol The Medassist Pharmaceutical Company manufactures pills designed to lower cholesterol levels. Listed below are the amounts (in mg) of drug in a random sample of the pills. Use a 0.05 significance level to test the claim that the pills come from a population in which the mean amount of the drug is equal to 25 mg.

    24.1 24.4 24.3 24.9 24.1 26.2 25.1 24.7 24.4 25.0 24.7 25.1 25.3 25.5 25.5

7. Controlling Variation Refer to Exercise 6 and use a 0.05 significance level to test the claim that the production process is out of control because the population standard deviation is greater than 0.5 mg. What are adverse consequences of having a standard deviation that is too large?

8. Effectiveness of New Chicken Feed When a poultry farmer uses her regular feed, the newborn chickens have normally distributed weights with a mean of 62.2 oz. In an experiment with an enriched feed mixture, nine chickens are born with the weights (in ounces) given below. Use a 0.01 significance level to test the claim that the mean weight is higher with the enriched feed.

    61.4   62.2   66.9   63.3   66.2   66.0   63.1   63.7   66.6

9. Effectiveness of Seat Belts A study included 123 children who were wearing seat belts when injured in motor vehicle crashes. The amounts of time spent in an intensive care unit have a mean of 0.83 day and a standard deviation of 0.16 day (based on data from "Morbidity Among Pediatric Motor Vehicle Crash Victims: The Effectiveness of Seat Belts," by Osberg and Di Scala, *American Journal of Public Health*, Vol. 82, No. 3). Using a 0.01 significance level, test the claim that the seat belt sample comes from a population with a mean of less than 1.39 days, which is the mean for the population who were not wearing seat belts when injured in motor vehicle crashes. Do the seat belts seem to help?

Cumulative Review Exercises

1. Monitoring Dioxin Concerns about adverse affects on health often cause officials to research pollution. Listed below are measured amounts of dioxin in the air at the site of the World Trade Center on days immediately following the terrorist attacks of September 11, 2001. Dioxin includes a group of chemicals produced from burning and some types of manufacturing. The listed amounts are in nanograms per cubic meter (ng/m^3) and they are in order with the earliest recorded values at the left. The data are provided by the U.S. Environmental Protection Agency.

    0.161   0.175   0.176   0.032   0.0524   0.044   0.018   0.0281   0.0268

    a. Find the mean of this sample.
    b. Find the median.
    c. Find the standard deviation.
    d. Find the variance.
    e. Find the range.
    f. Construct a 95% confidence interval estimate of the population mean.

g. The EPA uses 0.16 ng/m^3 as its "screening level," which is "set to protect against significantly increased risks of cancer and other adverse health effects." Use a 0.05 significance level to test the claim that this sample comes from a population with a mean less than 0.16 ng/m^3.

h. Is there any important characteristic of the data not addressed by the preceding results? If so, what is it?

2. Weights of Babies The weights of newborn American babies are normally distributed with a mean of 3420 g and a standard deviation of 495 g (based on data from "Birth Weight and Perinatal Mortality" by Wilcox et al., *Journal of the American Medical Association*, Vol. 273, No. 9).

a. If a newborn American baby is randomly selected, find the probability that his or her weight is greater than 3000 g.

b. If five newborn American babies are randomly selected, find the probability that all five have weights greater than 3000 g.

c. If five newborn American babies are randomly selected, find the probability that their mean is above 3000 g.

d. Find P_{90}, the score separating the bottom 90% from the top 10%.

3. Blood Pressure Readings A medical researcher obtains the systolic blood pressure readings (in mm Hg) in the accompanying list from a sample of women aged 18–24 who have a new strain of viral infection. (Healthy women in that age group have readings that are normally distributed with a mean of 114.8 and a standard deviation of 13.1.)

134.9	78.7	108.9	133.0	123.7	96.1	126.9	89.8
132.0	134.7	132.1	121.7	112.3	150.2	158.3	154.4

a. Find the sample mean $\bar{x}$ and standard deviation s.

b. Use a 0.05 significance level to test the claim that the sample comes from a population with a mean blood pressure equal to 114.8.

c. Use the sample data to construct a 95% confidence interval for the population mean μ. Do the confidence interval limits contain the value of 114.8, which is the mean for healthy women aged 18–24?

d. Use a 0.05 significance level to test the claim that the sample comes from a population with a standard deviation equal to 13.1, which is the standard deviation for healthy women aged 18–24.

e. Based on the preceding results, does it seem that the new strain of viral infection affects systolic blood pressure?

Cooperative Group Activities

1. *In-class activity:* Each student should estimate the length of the classroom. The values should be based on visual estimates, with no actual measurements being taken. After the estimates have been collected, measure the length of the room, then test the claim that the sample mean is equal to the actual length of the classroom. Is there a "collective wisdom," whereby the class mean is approximately equal to the actual room length?

2. *In-class activity:* After dividing into groups with sizes between 10 and 20 people, each group member should record the number of their heartbeats in a minute. After calculating $\bar{x}$ and s, each group should proceed to test the claim that the mean is greater than 60, which is the result for one of the authors. (When people exercise, they tend to have lower pulse rates.)

Technology Project

STATDISK, Minitab, Excel, the TI-83/84 Plus calculator, and many other tools can be used to randomly generate data from a normally distributed population with a given mean and standard deviation.

   a. Use such a tool to randomly generate five values from a normally distributed population with a mean of 100 and a standard deviation of 15 (the parameters for typical IQ tests).

   b. Using the five sample values generated in part (a), test the claim that the sample is from a population with a mean equal to 100. Use a 0.10 significance level.

   c. Repeat parts (a) and (b) nine more times, so that a total of 10 different samples have been generated, and 10 different hypothesis tests have been executed.

   d. With a 0.10 significance level, there is 0.10 probability of making a type I error (rejecting the null hypothesis when it is true). Because of the way that the sample data are generated, we know that 100 is the true population mean, so we do make a type I error in this experiment anytime that we reject the null hypothesis of $\mu = 100$. For the 10 trials of this experiment, how often was the null hypothesis actually rejected? When we conduct 10 such trials, how many times do we expect to reject the null hypothesis? Are the actual results consistent with the theoretical results? Explain.

from DATA to DECISION

Critical Thinking: Questioning survey results

Because surveys are now so pervasive throughout our society, each of us should develop the ability to think critically about them. We should question the process of selecting survey subjects, the wording of the question, the significance of the results, the objectivity of the survey sponsor and the group conducting the survey, as well as other issues. In a Gallup poll of 1038 randomly selected adults 18 or older, each subject was asked this question: "In general, how harmful do you feel secondhand smoke is to adults?" The subjects were given these choices: very harmful, somewhat harmful, not too harmful, not at all harmful. Here are the results:

Very harmful:	52%
Somewhat harmful:	33%
Not too harmful:	9%
Not at all harmful:	5%
No opinion:	1%

Analyzing the Results

a. The above results are recent, but a 1994 poll resulted in a 36% response rate for the choice of "very harmful." Use a 0.05 significance level to test the claim that the more recent percentage of "very harmful" responses is greater than the 36% rate from 1994.

b. Can a hypothesis test be used to verify that the 36% rate from 1994 has not changed? Why or why not?

c. If, instead of a random selection of subjects, *Prevention* magazine obtains results by publishing the survey that readers could return by mail, how are the hypothesis test results affected?

d. Does the wording of the question appear to be neutral and without bias? Provide an example of wording that would be somehow biased.

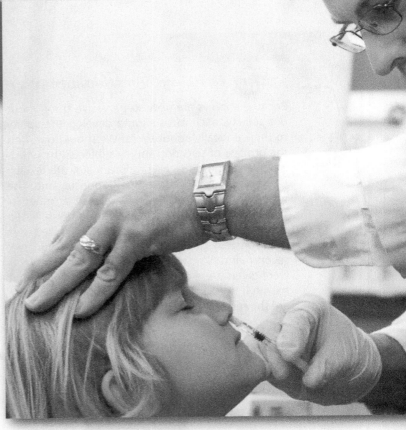

Inferences from Two Samples

How do we conclude that a treatment is effective?

In a *USA Today* article about an experimental nasal spray vaccine for children, the following statement was presented: "In a trial involving 1602 children only 14 (1%) of the 1070 who received the vaccine developed the flu, compared with 95 (18%) of the 532 who got a placebo." The sample data are summarized in Table 8-1.

An informal examination of the results might suggest that the vaccine appears to be effective, because 1% of the vaccinated children developed a flu, compared to the 18% flu rate for subjects given placebos. But it is dangerous to form conclusions based on subjective judgments. Instead, we need a formal and systematic method that can be used to decide whether such treatments are effective. This chapter presents such methods. The experimental results given here will be considered so that we can objectively determine whether the vaccine is actually effective. This procedure is *extremely* important, because it is the same procedure used by professionals, and it is the same procedure included in many articles found in professional journals.

Table 8-1	Testing Effectiveness of Vaccine	
	Developed Flu	Sample Size
Vaccine treatment group	14	1070
Placebo group	95	532

8-1 Overview

Chapter 6 introduced an important activity of inferential statistics: Sample data were used to construct confidence interval *estimates* of population parameters. Chapter 7 introduced a second important activity of inferential statistics: Sample data were used to *test hypotheses* about population parameters. In Chapters 6 and 7, all examples and exercises involved the use of *one* sample to form an inference about *one* population. In reality, there are many important and meaningful situations in which it becomes necessary to compare *two* sets of sample data. The following are examples typical of those found in this chapter, which presents methods for using sample data from two populations so that inferences can be made about those populations.

- When testing the effectiveness of the Salk vaccine in preventing paralytic polio, determine whether the treatment group had a lower incidence of polio than the group given a placebo.
- When testing the effectiveness of a drug designed to lower cholesterol, determine whether the treatment group had more substantial reductions than the group given a placebo.
- When investigating the accuracy of heights reported by people, determine whether there is a significant difference between the heights they report and the actual measured heights.

Chapters 6 and 7 included methods that were applied to proportions, means, and measures of variation (standard deviation and variance), and this chapter will address those same parameters. This chapter extends the same methods introduced in Chapters 6 and 7 to situations involving comparisons of two samples instead of only one. Also, Section 8-5 presents a method for finding confidence interval estimates of odds ratios. (Odds ratios were first introduced in Section 3-6.)

8-2 Inferences About Two Proportions

There are many real and important situations in which it is necessary to use sample data to compare two population proportions. Media reports and articles in professional journals include an abundance of situations involving a comparison of two population proportions. Although this section is based on proportions, we can deal with probabilities or we can deal with percentages by using the corresponding decimal equivalents.

When testing a hypothesis made about two population proportions or when constructing a confidence interval estimate of the difference between two population proportions, the methods presented in this section have the following requirements. (In addition to the methods included in this section, there are other methods for making inferences about two proportions, and they have different requirements.)

Requirements for Inferences About Two Proportions

1. We have proportions from two simple random samples that are *independent*, which means that the sample values selected from one population are not related to or somehow paired or matched with the sample values selected from the other population. (For example, in analyzing differences between headache rates among men and headache rates among women, we should not use a sample in which all of the men are husbands of the corresponding women. Such a sample would involve matched sample values instead of independent samples.)

2. For each of the two samples, the number of successes is at least five, and the number of failures is at least five.

Notation for Two Proportions

For population 1 we let

p_1 = *population* proportion

n_1 = size of the sample taken from population 1

x_1 = number of successes in the sample taken from population 1

$\hat{p}_1 = \dfrac{x_1}{n_1}$ (the *sample* proportion)

$\hat{q}_1 = 1 - \hat{p}_1$

The corresponding meanings are attached to p_2, n_2, x_2, $\hat{p}_2$, and $\hat{q}_2$, which come from population 2. Note that we use p_1 to denote the proportion of the entire *population*, but we use $\hat{p}_1$ to denote the proportion found from the *sample*.

Finding the Numbers of Successes x_1 and x_2: The calculations for hypothesis tests and confidence intervals require that we have specific values for $x_1, n_1, x_2,$ and n_2. Sometimes the available sample data include those specific numbers, but sometimes it is necessary to calculate the values of x_1 and x_2.

For example, consider the statement that "when 734 men were treated with Viagra, 16% of them experienced headaches." From that statement we can see that $n_1 = 734$ and $\hat{p}_1 = 0.16$, but the actual number of successes x_1 is not given. However, from $\hat{p}_1 = x_1/n_1$, we know that

$$x_1 = n_1 \cdot \hat{p}_1$$

so that $x_1 = 734 \cdot 0.16 = 117.44$. But you cannot have 117.44 men who experienced headaches, because each man either experiences a headache or does not, and the number of successes x_1 must therefore be a whole number. We can round 117.44 to 117. We can now use $x_1 = 117$ in the calculations that require its value. It's really quite simple: 16% of 734 means 0.16×734, which results in 117.44, which we round to 117.

Hypothesis Tests

In Section 7-2 we discussed tests of hypotheses made about a single population proportion. We will now consider tests of hypotheses made about two population proportions, but *we will be testing only claims that $p_1 = p_2$*, and we will use the following pooled (or combined) estimate of the value that p_1 and p_2 have in common. (For claims that the difference between p_1 and p_2 is equal to a nonzero constant, see Exercise 23 of this section.) You can see from the form of the pooled estimate $\bar{p}$ that it basically combines the two different samples into one big sample.

Pooled Estimate of p_1 and p_2

The **pooled estimate of p_1 and p_2** is denoted by $\bar{p}$ and is given by

$$\bar{p} = \frac{x_1 + x_2}{n_1 + n_2}$$

We denote the complement of $\bar{p}$ by $\bar{q}$, so $\bar{q} = 1 - \bar{p}$.

Test Statistic for Two Proportions (with $H_0: p_1 = p_2$)

$$z = \frac{(\hat{p}_1 - \hat{p}_2) - (p_1 - p_2)}{\sqrt{\frac{\bar{p}\,\bar{q}}{n_1} + \frac{\bar{p}\,\bar{q}}{n_2}}}$$

where $p_1 - p_2 = 0$ (assumed in the null hypothesis)

$$\hat{p}_1 = \frac{x_1}{n_1} \quad \text{and} \quad \hat{p}_2 = \frac{x_2}{n_2}$$

$$\bar{p} = \frac{x_1 + x_2}{n_1 + n_2}$$
$$\bar{q} = 1 - \bar{p}$$

P-value: Use Table A-2. (Use the computed value of the test statistic z and find the P-value by following the procedure summarized in Figure 7-6.)

Critical values: Use Table A-2. (Based on the significance level α, find critical values by using the procedures introduced in Section 7-2.)

The following example will help clarify the roles of x_1, n_1, $\hat{p}_1$, $\bar{p}$, and so on. In particular, you should recognize that under the assumption of equal proportions, the best estimate of the common proportion is obtained by pooling both samples into one big sample, so that $\bar{p}$ becomes a more obvious estimate of the common population proportion.

EXAMPLE **Testing Effectiveness of a Vaccine** In the Chapter Problem, we noted that a *USA Today* article reported experimental results from a nasal spray vaccine for children. Among 1070 children given the vaccine, 14 developed a flu. Among 532 children given a placebo, 95 developed a flu. The sample data are summarized in Table 8-1. Use a 0.05 significance level to test the claim that the proportion of flu cases among the vaccinated children is less than the proportion of flu cases among children given a placebo.

SOLUTION   For notation purposes, we stipulate that Sample 1 is the treatment (vaccinated) group, and Sample 2 is the placebo group. We can summarize the sample data as follows. (We use extra decimal places for the sample proportions $\hat{p}_1$ and $\hat{p}_2$ because those values will be used in subsequent calculations.)

Vaccinated Children	Children Given Placebo
$n_1 = 1070$	$n_2 = 532$
$x_1 = 14$	$x_2 = 95$
$\hat{p}_1 = \dfrac{x_1}{n_1} = \dfrac{14}{1070} = 0.013084$	$\hat{p}_2 = \dfrac{x_2}{n_2} = \dfrac{95}{532} = 0.178571$

REQUIREMENT ✔   We should first verify that the necessary requirements (listed earlier in the section) are satisfied. Given the design of the experiment, it is reasonable to assume that both samples are simple random samples and they are independent. Also, each of the two samples has at least five successes and at least five failures. (The first sample has 14 successes and 1056 failures. The second sample has 95 successes and 437 failures.) The check of requirements has been successfully completed and we can now proceed to conduct the formal hypothesis test. ✔

We will now use the *P*-value method of hypothesis testing, as summarized in Figure 7-9.

Step 1: The claim of a lower flu rate for vaccinated children can be represented by $p_1 < p_2$.

Step 2: If $p_1 < p_2$ is false, then $p_1 \geq p_2$.

Step 3: Because our claim of $p_1 < p_2$ does not contain equality, it becomes the alternative hypothesis. The null hypothesis is the statement of equality, so we have

$$H_0: p_1 = p_2 \qquad H_1: p_1 < p_2 \text{ (original claim)}$$

Step 4: The significance level is $\alpha = 0.05$.

Step 5: We will use the normal distribution (with the test statistic previously given) as an approximation to the binomial distribution.

Step 6: The test statistic uses the values of $\bar{p}$ and $\bar{q}$, which are as follows:

$$\bar{p} = \frac{x_1 + x_2}{n_1 + n_2} = \frac{14 + 95}{1070 + 532} = 0.068040$$

$$\bar{q} = 1 - \bar{p} = 1 - 0.068040 = 0.931960$$

We can now find the value of the test statistic:

$$z = \frac{(\hat{p}_1 - \hat{p}_2) - (p_1 - p_2)}{\sqrt{\dfrac{\bar{p}\,\bar{q}}{n_1} + \dfrac{\bar{p}\,\bar{q}}{n_2}}}$$

$$= \frac{\left(\dfrac{14}{1070} - \dfrac{95}{532}\right) - 0}{\sqrt{\dfrac{(0.068040)(0.931960)}{1070} + \dfrac{(0.068040)(0.931960)}{532}}} = -12.39$$

continued

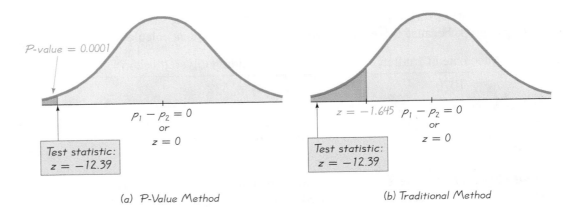

FIGURE 8-1   **Testing Claim that $p_1 < p_2$.**

The P-value of 0.0001 is found as follows: With an alternative hypothesis $H_1: p_1 < p_2$, this is a left-tailed test, so the P-value is the area to the left of the test statistic $z = -12.39$. (See Figure 7-6 which summarizes the procedures for finding P-values in left-tailed tests, right-tailed tests, and two-tailed tests.) Refer to Table A-2 and find that the area to the left of the test statistic $z = -12.39$ is 0.0001, so the P-value is 0.0001. (In Table A-2 we use 0.0001 for the area to the left of any z score of -3.50 or less, but software shows that a more exact P-value is considerably smaller than 0.0001.) The test statistic and P-value are shown in Figure 8-1(a).

Step 7:   Because the P-value of 0.0001 is less than the significance level of $\alpha = 0.05$, we reject the null hypothesis of $p_1 = p_2$.

INTERPRETATION   We must address the original claim that children given the vaccine developed flu at a rate that is less than the rate for those in the placebo group. Because we reject the null hypothesis, we support the alternative hypothesis and conclude that there is sufficient evidence to support the claim of a lower flu rate for the vaccinated children. (See Figure 7-7 for help in wording the final conclusion.)

Traditional Method of Testing Hypotheses

The preceding example illustrates the P-value approach to hypothesis testing, but it would be quite easy to use the traditional approach (or "critical value approach") instead. In Step 6, instead of finding the P-value, we would find the critical value. With a significance level of $\alpha = 0.05$ in a left-tailed test based on the normal distribution, refer to Table A-2 to find that an area of $\alpha = 0.05$ in the left tail corresponds to the critical value of $z = -1.645$. See Figure 8-1(b) where we can see that the test statistic does fall in the critical region bounded by the critical value of $z = -1.645$. We again reject the null hypothesis. Again, we conclude that there is sufficient evidence to support the claim that vaccinated children developed flu at a rate that is less than the flu rate for children from the placebo group.

Confidence Intervals

We can construct a confidence interval estimate of the difference between population proportions $(p_1 - p_2)$ by using the format given below. If a confidence interval estimate of $p_1 - p_2$ does not include zero, we have evidence suggesting that p_1 and p_2 have different values. (If p_1 and p_2 are the same value, then $p_1 = p_2$, which is equivalent to $p_1 - p_2 = 0$.)

Confidence Interval Estimate of $p_1 - p_2$

The confidence interval estimate of the difference $p_1 - p_2$ is:

$$(\hat{p}_1 - \hat{p}_2) - E < (p_1 - p_2) < (\hat{p}_1 - \hat{p}_2) + E$$

where

$$E = z_{\alpha/2} \sqrt{\frac{\hat{p}_1 \hat{q}_1}{n_1} + \frac{\hat{p}_2 \hat{q}_2}{n_2}}$$

EXAMPLE Estimating Vaccine's Effectiveness Use the sample data given in Table 8-1 to construct a 90% confidence interval for the difference between the two population proportions. (The confidence level of 90% is comparable to the significance level of $\alpha = 0.05$ used in the preceding left-tailed hypothesis test. Table 7-2 in Section 7-2 shows that a one-tailed hypothesis test conducted at a 0.05 significance level corresponds to a 90% confidence interval.)

SOLUTION

REQUIREMENT ✓ We should first verify that the necessary requirements (listed earlier in the section) are satisfied. See the preceding example for the verification that the requirements are satisfied. The same verification applies to this example, and we can now proceed to construct the confidence interval. ✓

With a 90% confidence level, $z_{\alpha/2} = 1.645$ (from Table A-2). We first calculate the value of the margin of error E as shown:

$$E = z_{\alpha/2} \sqrt{\frac{\hat{p}_1 \hat{q}_1}{n_1} + \frac{\hat{p}_2 \hat{q}_2}{n_2}} = 1.645 \sqrt{\frac{\left(\frac{14}{1070}\right)\left(\frac{1056}{1070}\right)}{1070} + \frac{\left(\frac{95}{532}\right)\left(\frac{437}{532}\right)}{532}}$$

$$= 0.027906$$

With $\hat{p}_1 = 14/1070 = 0.013084$, $\hat{p}_2 = 95/532 = 0.178571$, and $E = 0.027906$, the confidence interval is evaluated as follows:

$$(\hat{p}_1 - \hat{p}_2) - E < (p_1 - p_2) < (\hat{p}_1 - \hat{p}_2) + E$$

$$(0.013084 - 0.178571) - 0.027906 < (p_1 - p_2) < (0.013084 - 0.178571) + 0.027906$$

$$-0.193 < (p_1 - p_2) < -0.138$$

continued

INTERPRETATION Although the resulting confidence interval is technically correct, its use of negative values for both confidence interval limits causes it to be somewhat unclear. It would be better to say that the vaccine treatment group developed flu at a lower rate than the placebo group, and the difference appears to be between 0.138 and 0.193. Also, the confidence interval limits do not contain zero, suggesting that there *is* a significant difference between the two proportions. (Because the value of 0 is not a likely value of the difference, it is unlikely that the two proportions are equal.)

Rationale: Why Do the Procedures of This Section Work? The test statistic given for hypothesis tests is justified by the following:

1. With $n_1 p_1 \geq 5$ and $n_1 q_1 \geq 5$, the distribution of $\hat{p}_1$ can be approximated by a normal distribution with mean p_1 and standard deviation $\sqrt{p_1 q_1 / n_1}$ and variance $p_1 q_1 / n_1$. These conclusions are based on Sections 5-6 and 6-5, and they also apply to the second sample.

2. Because $\hat{p}_1$ and $\hat{p}_2$ are each approximated by a normal distribution, $\hat{p}_1 - \hat{p}_2$ will also be approximated by a normal distribution with mean $p_1 - p_2$ and variance

$$\sigma^2_{(\hat{p}_1 - \hat{p}_2)} = \sigma^2_{\hat{p}_1} + \sigma^2_{\hat{p}_2} = \frac{p_1 q_1}{n_1} + \frac{p_2 q_2}{n_2}$$

   (The above result is based on this property: The variance of the *differences* between two independent random variables is the *sum* of their individual variances.)

3. Because the values of p_1, q_1, p_2, and q_2 are typically unknown and from the null hypothesis we assume that $p_1 = p_2$, we can pool (or combine) the sample data. The pooled estimate of the common value of p_1 and p_2 is $\bar{p} = (x_1 + x_2)/(n_1 + n_2)$. If we replace p_1 and p_2 by $\bar{p}$ and replace q_1 and q_2 by $\bar{q} = 1 - \bar{p}$, the variance from Step 2 leads to the following standard deviation.

$$\sigma_{(\hat{p}_1 - \hat{p}_2)} = \sqrt{\frac{\bar{p}\,\bar{q}}{n_1} + \frac{\bar{p}\,\bar{q}}{n_2}}$$

4. We now know that the distribution of $p_1 - p_2$ is approximately normal, with mean $p_1 - p_2$ and standard deviation as shown in Step 3, so that the z test statistic has the form given earlier.

The form of the confidence interval requires an expression for the variance different from the one given in Step 3. In Step 3 we are assuming that $p_1 = p_2$, but if we don't make that assumption (as in the construction of a confidence interval), we estimate the variance of $p_1 - p_2$ as

$$\sigma^2_{(\hat{p}_1 - \hat{p}_2)} = \sigma^2_{\hat{p}_1} + \sigma^2_{\hat{p}_2} = \frac{\hat{p}_1 \hat{q}_1}{n_1} + \frac{\hat{p}_2 \hat{q}_2}{n_2}$$

and the standard deviation is estimated by

$$\sqrt{\frac{\hat{p}_1 \hat{q}_1}{n_1} + \frac{\hat{p}_2 \hat{q}_2}{n_2}}$$

In the test statistic

$$z = \frac{(\hat{p}_1 - \hat{p}_2) - (p_1 - p_2)}{\sqrt{\dfrac{\hat{p}_1\hat{q}_1}{n_1} + \dfrac{\hat{p}_2\hat{q}_2}{n_2}}}$$

let z be positive and negative (for two tails) and solve for $p_1 - p_2$. The results are the limits of the confidence interval given earlier.

8-2 Exercises

Finding Number of Successes. In Exercises 1–4, find the number of successes x sug-gested by the given statement.

1. From the Arizona Department of Weights and Measures: Among 37 inspections at NAPA Auto Parts stores, 81% failed.

2. From the *New York Times*: Among 240 vinyl gloves subjected to stress tests, 63% leaked.

3. From *Sociological Methods and Research*: When 294 central-city residents were sur-veyed, 28.9% refused to respond.

4. From a Time/CNN survey: 24% of 205 single women said that they "definitely want to get married."

Calculations for Testing Claims. In Exercises 5 and 6, assume that you plan to use a sig-nificance level of $\alpha = 0.05$ to test the claim that $p_1 = p_2$. Use the given sample sizes and numbers of successes to find (a) the pooled estimate $\bar{p}$, (b) the z test statistic, (c) the criti-cal z values, and (d) the P-value.

5.	Treatment	Placebo
	$n_1 = 436$	$n_2 = 121$
	$x_1 = 192$	$x_2 = 40$

6.	Low Activity	High Activity
	$n_1 = 10{,}239$	$n_2 = 9877$
	$x_1 = 101$	$x_2 = 56$

7. Exercise and Heart Disease In a study of women and heart disease, the following sample results were obtained: Among 10,239 women with a low level of physical ac-tivity (less than 200 kcal/wk), there were 101 cases of heart disease. Among 9877 women with physical activity measured between 200 and 600 kcal/wk, there were 56 cases of heart disease (based on data from "Physical Activity and Coronary Heart Dis-ease in Women" by Lee, Rexrode, et al., *Journal of the American Medical Associa-tion*, Vol. 285, No. 11). Construct a 90% confidence interval estimate for the differ-ence between the two proportions. Does the difference appear to be substantial? Does it appear that physical activity corresponds to a lower rate of heart disease?

8. Exercise and Heart Disease Refer to the sample data in Exercise 7 and use a 0.05 sig-nificance level to test the claim that the rate of heart disease is higher for women with the lower levels of physical activity. What does the conclusion suggest?

9. Effectiveness of Smoking Bans The Joint Commission on Accreditation of Health-care Organizations mandated that hospitals ban smoking by 1994. In a study of the ef-fects of this ban, subjects who smoke were randomly selected from two different pop-ulations. Among 843 smoking employees of hospitals with the smoking ban, 56 quit smoking one year after the ban. Among 703 smoking employees from workplaces

without a smoking ban, 27 quit smoking a year after the ban (based on data from "Hospital Smoking Bans and Employee Smoking Behavior" by Longo, Brownson, et al., *Journal of the American Medical Association*, Vol. 275, No. 16). Is there a significant difference between the two proportions at a 0.05 significance level? Is there a significant difference between the two proportions at a 0.01 significance level? Does it appear that the ban had an effect on the smoking quit rate?

10. Effectiveness of the Salk Vaccine In 1954 an experiment was conducted to test the effectiveness of the Salk vaccine as protection against the devastating effects of polio: 200,745 children were injected with an ineffective salt solution and 201,229 other children were injected with the vaccine. The experiment was "double blind" because the children being injected didn't know whether they were given the real vaccine or the placebo, and the doctors giving the injections and evaluating the results didn't know either. Only 33 of the 200,745 vaccinated children later developed paralytic polio, whereas 115 of the 201,229 injected with the salt solution later developed paralytic polio. Use a 0.01 significance level to test the claim that the vaccine was effective in lowering the incidence of polio.

11. Color Blindness in Men and Women In a study of red/green color blindness, 500 men and 2100 women are randomly selected and tested. Among the men, 45 have red/green color blindness. Among the women, 6 have red/green color blindness (based on data from *USA Today*).
   a. Is there sufficient evidence to support the claim that men have a higher rate of red/green color blindness than women? Use a 0.01 significance level.
   b. Construct the 98% confidence interval for the difference between the color blindness rates of men and women. Does there appear to be a substantial difference?
   c. Why would the sample size for women be so much larger than the sample size for men?

12. Seat Belts and Hospital Time A study was made of 413 children who were hospitalized as a result of motor vehicle crashes. Among 290 children who were not using seat belts, 50 were injured severely. Among 123 children using seat belts, 16 were injured severely (based on data from "Morbidity Among Pediatric Motor Vehicle Crash Victims: The Effectiveness of Seat Belts," by Osberg and Di Scala, *American Journal of Public Health*, Vol. 82, No. 3). Is there sufficient sample evidence to conclude, at the 0.05 significance level, that the rate of severe injuries is lower for children wearing seat belts? Based on these results, what action should be taken?

13. Drinking and Crime Karl Pearson, who developed many important concepts in statistics, collected crime data in 1909. Of those convicted of arson, 50 were drinkers and 43 abstained. Of those convicted of fraud, 63 were drinkers and 144 abstained. Use a 0.01 significance level to test the claim that the proportion of drinkers among convicted arsonists is greater than the proportion of drinkers convicted of fraud. Does it seem reasonable that drinking might have had an effect on the type of crime? Why?

14. Testing Laboratory Gloves The *New York Times* ran an article about a study in which Professor Denise Korniewicz and other Johns Hopkins researchers subjected laboratory gloves to stress. Among 240 vinyl gloves, 63% leaked. Among 240 latex gloves, 7% leaked. At the 0.005 significance level, test the claim that vinyl gloves have a larger leak rate than latex gloves.

15. Written Survey and Computer Survey In a study of 1700 teens aged 15–19, half were given written surveys and half were given surveys using an anonymous computer pro-

gram. Among those given the written surveys, 7.9% say that they carried a gun within the last 30 days. Among those given the computer surveys, 12.4% say that they carried a gun within the last 30 days (based on data from the Urban Institute).

a. The sample percentages of 7.9% and 12.4% are obviously not equal, but is the difference significant? Explain.

b. Construct a 99% confidence interval estimate of the difference between the two population percentages, and interpret the result.

16. Adverse Drug Reactions The drug Viagra has had a substantial economic impact on its producer, Pfizer Pharmaceuticals. In preliminary tests for adverse reactions, it was found that when 734 men were treated with Viagra, 16% of them experienced headaches. Among 725 men in a placebo group, 4% experienced headaches (based on data from Pfizer Pharmaceuticals).

a. Using a 0.01 significance level, is there sufficient evidence to support the claim that among those men who take Viagra, headaches occur at a rate that is greater than the rate for those who do not take Viagra?

b. Construct a 99% confidence interval estimate of the difference between the rate of headaches among Viagra users and the headache rate for those who use a placebo. What does the confidence interval suggest about the two rates?

17. Testing Adverse Drug Reactions In clinical tests of adverse reactions to the drug Viagra, 7% of the 734 subjects in the treatment group experienced dyspepsia (indigestion), but 2% of the 725 subjects in the placebo group experienced dyspepsia (based on data from Pfizer Pharmaceuticals).

a. Is there sufficient evidence to support the claim that dyspepsia occurs at a higher rate among Viagra users than those who do not use Viagra?

b. Construct a 95% confidence interval estimate of the difference between the dyspepsia rate for Viagra users and the dyspepsia rate for those who use a placebo. Do the confidence interval limits contain 0, and what does this suggest about the two dyspepsia rates?

18. Gender-Selection Methods The Genetics and IVF Institute conducted a clinical trial of its methods for gender selection. As this book was written, results included 325 babies born to parents using the XSORT method to increase the probability of conceiving a girl, and 295 of those babies were girls. Also, 51 babies were born to parents using the YSORT method to increase the probability of conceiving a boy, and 39 of those babies were boys.

a. Construct a 95% confidence interval estimate of the difference between the proportion of girls and the proportion of boys.

b. Does there appear to be a difference? Do the XSORT and YSORT methods appear to be effective?

19. Which Treatment Is Better? When treating carpal tunnel syndrome, should a patient be treated with surgery or should a splint be applied? In a study of this issue, 73 patients were treated with surgery and 83 were treated with a splint. After 12 months, success was noted in 67 of the patients treated with surgery. After 12 months, success was noted in 60 of the patients treated with a splint. (Based on data from "Splinting vs Surgery in the Treatment of Carpal Tunnel Syndrome" by Gerritsen et al., *Journal of the American Medical Association,* Vol. 288, No. 10.) Test for a difference in success rates. If there is a difference, which treatment appears to be better?

Gender Gap in Drug Testing

A study of the relationship between heart attacks and doses of aspirin involved 22,000 male physicians. This study, like many others, excluded women. The General Accounting Office recently criticized the National Institutes of Health for not including both sexes in many studies because results of medical tests on males do not necessarily apply to females. For example, women's hearts are different from men's in many important ways. When forming conclusions based on sample results, we should be wary of an inference that extends to a population larger than the one from which the sample was drawn.

20. Health Survey Refer to Data Set 1 in Appendix B and use the sample data to test the claim that the proportion of men over the age of 30 is equal to the proportion of women over the age of 30.

*21. Interpreting Overlap of Confidence Intervals In the article "On Judging the Significance of Differences by Examining the Overlap Between Confidence Intervals" by Schenker and Gentleman (*The American Statistician*, Vol. 55, No. 3), the authors consider sample data in this statement: "Independent simple random samples, each of size 200, have been drawn, and 112 people in the first sample have the attribute, whereas 88 people in the second sample have the attribute."

   a. Use the methods of this section to construct a 95% confidence interval estimate of the difference $p_1 - p_2$. What does the result suggest about the equality of p_1 and p_2?

   b. Use the methods of Section 6-2 to construct individual 95% confidence interval estimates for each of the two population proportions. After comparing the overlap between the two confidence intervals, what do you conclude about the equality of p_1 and p_2?

   c. Use a 0.05 significance level to test the claim that the two population proportions are equal. What do you conclude?

   d. Based on the preceding results, what should you conclude about equality of p_1 and p_2? Which of the three preceding methods is least effective in testing for equality of p_1 and p_2?

*22. Equivalence of Hypothesis Test and Confidence Interval Two different simple random samples are drawn from two different populations. The first sample consists of 20 people with 10 having a common attribute. The second sample consists of 2000 people with 1404 of them having the same common attribute. Compare the results from a hypothesis test of $p_1 = p_2$ (with a 0.05 significance level) and a 95% confidence interval estimate of $p_1 - p_2$.

*23. Testing for Constant Difference To test the null hypothesis that the difference between two population proportions is equal to a nonzero constant c, use the test statistic

$$z = \frac{(\hat{p}_1 - \hat{p}_2) - c}{\sqrt{\dfrac{\hat{p}_1\hat{q}_1}{n_1} + \dfrac{\hat{p}_2\hat{q}_2}{n_2}}}$$

As long as n_1 and n_2 are both large, the sampling distribution of the test statistic z will be approximately the standard normal distribution. Refer to Exercise 16 and use a 0.05 significance level to test the claim that the headache rate of Viagra users is 10 percentage points more than the percentage for those who use a placebo.

8-3 Inferences About Two Means: Independent Samples

In this section we present methods for using sample data from two independent samples to test hypotheses made about two population means or to construct confidence interval estimates of the difference between two population means. We begin by distinguishing between *independent* and *dependent* samples.

Definitions

Two samples are **independent** if the sample values selected from one population are not related to or somehow paired or matched with the sample values selected from the other population.

Two samples are **dependent** if the members of one sample can be used to determine the members of the other sample. [Samples consisting of **matched pairs** (such as husband/wife data) are dependent. In addition to matched pairs of sample data, dependence could also occur with samples related through associations such as family members or littermates.]

EXAMPLE Drug Testing

Independent samples: One group of subjects is treated with the cholesterol-reducing drug Lipitor, while a second and separate group of subjects is given a placebo. These two sample groups are independent because the individuals in the treatment group are in no way paired or matched with corresponding members in the placebo group.

Matched pairs (and also dependent samples): The effectiveness of a diet is tested using weights of subjects measured before and after the diet treatment. Each "before" value is matched with the "after" value because each before/after pair of measurements comes from the same person.

This section considers two independent samples, and the following section addresses matched pairs. When using two independent samples to test a claim about the difference $\mu_1 - \mu_2$, or to construct a confidence interval estimate of $\mu_1 - \mu_2$, use the following.

Requirements

1. The two samples are *independent*.

2. Both samples are *simple random samples*.

3. Either or both of these conditions is satisfied: The two sample sizes are both *large* (with $n_1 > 30$ and $n_2 > 30$) or both samples come from populations having normal distributions. (This test is *robust* against departures from normality. That is, the normality requirement is not very strict, so the procedures perform well as long as there are no outliers and the data are not extremely far from being normally distributed, as verified with histograms and/or normal quantile plots. The robustness increases with larger sample sizes. If normality is very questionable, consider the nonparametric rank sum test in Section 12-4.)

continued

Hypothesis Test Statistic for Two Means: Independent Samples

$$t = \frac{(\bar{x}_1 - \bar{x}_2) - (\mu_1 - \mu_2)}{\sqrt{\dfrac{s_1^2}{n_1} + \dfrac{s_2^2}{n_2}}}$$

Degrees of Freedom: When finding critical values or P-values, use the following for determining the number of degrees of freedom, denoted by df. (Although these two methods typically result in different numbers of degrees of freedom, the conclusion of a hypothesis test is rarely affected by which method is used to compute the number of degrees of freedom.)

1. In this book we use this simple and conservative estimate:
df = smaller of $n_1 - 1$ and $n_2 - 1$.

2. Statistical software packages typically use the more accurate but more difficult estimate given in Formula 8-1. (We will not use Formula 8-1 for the examples and exercises in this book.)

Formula 8-1

$$df = \frac{(A + B)^2}{\dfrac{A^2}{n_1 - 1} + \dfrac{B^2}{n_2 - 1}}$$

where

$$A = \frac{s_1^2}{n_1} \quad \text{and} \quad B = \frac{s_2^2}{n_2}$$

P-values: Refer to Table A-3. Use the procedure summarized in Figure 7-6. (See also the subsection "Finding P-Values with the Student t Distribution" in Section 7-5.)

Critical values: Refer to the t values in Table A-3.

Confidence Interval Estimate of $\mu_1 - \mu_2$: Independent Samples

The confidence interval estimate of the difference $\mu_1 - \mu_2$ is

$$(\bar{x}_1 - \bar{x}_2) - E < (\mu_1 - \mu_2) < (\bar{x}_1 - \bar{x}_2) + E$$

$$\text{where} \quad E = t_{\alpha/2}\sqrt{\frac{s_1^2}{n_1} + \frac{s_2^2}{n_2}}$$

and the number of degrees of freedom (df) is as described above for hypothesis tests. (In this book, we use df = smaller of $n_1 - 1$ and $n_2 - 1$, but software packages typically use Formula 8-1.)

Because the hypothesis test and confidence interval use the same distribution and standard error, they are equivalent in the sense that they result in the same conclusions. Consequently, the null hypothesis of $\mu_1 = \mu_2$ (or $\mu_1 - \mu_2 = 0$) can be tested by determining whether the confidence interval includes 0. For two-tailed hypothesis tests construct a confidence interval with a confidence level of $1 - \alpha$; but for a one-tailed hypothesis test with significance level α, construct a confidence interval with a confidence level of $1 - 2\alpha$. (See Table 7-2 for common cases.) For example, the claim of $\mu_1 > \mu_2$ can be tested with a 0.05 significance level by constructing a 90% confidence interval.

We will discuss the rationale for the above expressions later in this section. For now, note that the listed assumptions do not include the conditions that the population standard deviations σ_1 and σ_2 must be known, nor do we assume that the two populations have the same standard deviation. Alternative methods based on these additional assumptions are discussed later in the section.

Exploring the Data Sets

We should verify the requirements for this section when using two samples to make inferences about two population means. Instead of immediately conducting a hypothesis test or constructing a confidence interval, we should first *explore* the two samples using the methods described in Chapter 2. For each of the two samples, we should investigate center, variation, distribution, outliers, and whether the population appears to be changing over time (CVDOT). It could be very helpful to do the following:

- Find descriptive statistics for both data sets, including n, $\bar{x}$, and s.
- Create boxplots of both data sets, drawn on the same scale so that they can be compared.
- Create histograms of both data sets, so that their distributions can be seen.
- Identify any outliers.

EXAMPLE Hypothesis Test of Treatment for Bipolar Depression In an experiment designed to test the effectiveness of paroxetine for treating bipolar depression, subjects were measured using the Hamilton Depression Scale with the results given below (based on data from "Double-Blind, Placebo-Controlled Comparison of Imipramine and Paroxetine in the Treatment of Bipolar Depression" by Nemeroff, Evans, et al., *American Journal of Psychiatry*, Vol. 158, No. 6). Use a 0.05 significance level to test the claim that the treatment group and placebo group come from populations with the same mean. What does the result of the hypothesis test suggest about paroxetine as a treatment for bipolar depression?

Placebo group:	$n = 43, \bar{x} = 21.57, s = 3.87$
Paroxetine treatment group:	$n = 33, \bar{x} = 20.38, s = 3.91$

SOLUTION

REQUIREMENT ✓ Before beginning the hypothesis test, we should verify that the requirements are satisfied. Based on the design of the experiment, it is reasonable to assume that the two samples are independent simple random samples. The above given data are only known through their summary statistics, so

continued

we cannot construct boxplots, histograms, or check for outliers. However, because the sample sizes are large ($n > 30$), we can proceed with the methods of this section. Assuming that outliers are not a problem, we have satisfied the requirements and we can proceed with a formal hypothesis test to determine whether the difference between the two sample means is really significant. ✓

Because it's a bit tricky to find the P-value in this example, we will use the traditional method of hypothesis testing.

Step 1: The claim of equal means can be expressed symbolically as $\mu_1 = \mu_2$.

Step 2: If the original claim is false, then $\mu_1 \neq \mu_2$.

Step 3: The alternative hypothesis is the expression not containing equality, and the null hypothesis is an expression of equality, so we have

$$H_0: \mu_1 = \mu_2 \text{ (original claim)} \qquad H_1: \mu_1 \neq \mu_2$$

We now proceed with the assumption that $\mu_1 = \mu_2$, or $\mu_1 - \mu_2 = 0$.

Step 4: The significance level is $\alpha = 0.05$.

Step 5: Because we have two independent samples and we are testing a claim about the two population means, we use a t distribution with the test statistic given earlier in this section.

Step 6: The test statistic is calculated as follows:

$$t = \frac{(\bar{x}_1 - \bar{x}_2) - (\mu_1 - \mu_2)}{\sqrt{\dfrac{s_1^2}{n_1} + \dfrac{s_2^2}{n_2}}} = \frac{(21.57 - 20.38) - 0}{\sqrt{\dfrac{3.87^2}{43} + \dfrac{3.91^2}{33}}} = 1.321$$

Because we are using a t distribution, the critical values of $t = \pm 2.037$ are found from Table A-3. [With an area of 0.05 in two tails, we want the t value corresponding to 32 degrees of freedom, which is the smaller of $n_1 - 1$ and $n_2 - 1$ (or the smaller of $43 - 1$ and $33 - 1$).] The test statistic, critical values, and critical region are shown in Figure 8-2.

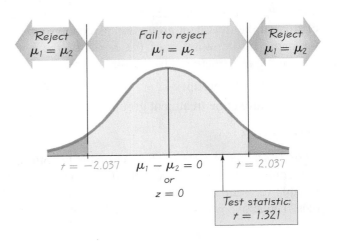

FIGURE 8-2
Distribution of $\bar{x}_1 - \bar{x}_2$ Values

Reject $\mu_1 = \mu_2$ Fail to reject $\mu_1 = \mu_2$ Reject $\mu_1 = \mu_2$

$t = -2.037$ $\mu_1 - \mu_2 = 0$ or $z = 0$ $t = 2.037$

Test statistic: $t = 1.321$

Using software such as STATDISK, Minitab, or a TI-83/84 Plus calculator, we can also find that the P-value is 0.191 and the more accurate critical values are $t = \pm 1.995$ (based on df = 69). (SPSS, SAS, and Excel cannot be used here because they require lists of the original data values.)

Step 7:   Because the test statistic does not fall within the critical region, we fail to reject the null hypothesis $\mu_1 = \mu_2$ (or $\mu_1 - \mu_2 = 0$).

INTERPRETATION   There is not sufficient evidence to warrant rejection of the claim that those given a placebo and those treated with paroxetine have the same mean. Because the means are not significantly different, it appears that the paroxetine treatment does not have a significant effect. Based on these results, paroxetine does not appear to be effective as a treatment for bipolar depression.

EXAMPLE   **Confidence Interval for Treatment of Bipolar Depression** Using the sample data given in the preceding example, construct a 95% confidence interval estimate of the difference between the mean depression score for subjects given a placebo and the mean depression score for subjects treated with paroxetine.

SOLUTION

REQUIREMENT ✔   We should begin by verifying that the necessary requirements are satisfied. See the check of requirements in the preceding example. That same verification applies to this example, so we can proceed with the construction of the confidence interval. ✔

We first find the value of the margin of error E. We use $t_{\alpha/2} = 2.037$, which is found in Table A-3 as the t score corresponding to an area of 0.05 in two tails and df = 32. [As in the preceding example, we want the t score corresponding to 32 degrees of freedom, which is the smaller of $n_1 - 1$ and $n_2 - 1$ (or the smaller of $43 - 1$ and $33 - 1$).]

$$E = t_{\alpha/2}\sqrt{\frac{s_1^2}{n_1} + \frac{s_2^2}{n_2}} = 2.037\sqrt{\frac{3.87^2}{43} + \frac{3.91^2}{33}} = 1.83508$$

We now find the desired confidence interval as follows (with the results rounded to two decimal places, as in the sample statistics):

$$(\bar{x}_1 - \bar{x}_2) - E < (\mu_1 - \mu_2) < (\bar{x}_1 - \bar{x}_2) + E$$
$$(21.57 - 20.38) - 1.83508 < (\mu_1 - \mu_2) < (21.57 - 20.38) + 1.83508$$
$$-0.65 < (\mu_1 - \mu_2) < 3.03$$

Statistics software or the TI-83/84 Plus calculator will use Formula 8-1 for determining the number of degrees of freedom, so more accurate results will be provided. In this example, the use of statistics software or a TI-83/84 Plus calculator will result in the confidence interval of $-0.61 < (\mu_1 - \mu_2) < 2.99$, so we can see that the above confidence interval is quite good.

continued

INTERPRETATION We are 95% confident that the limits of -0.65 and 3.03 actually do contain the difference between the two population means. Because those limits do contain zero, this confidence interval suggests that it is very possible that the two population means are equal. As in the preceding example, it appears that the paroxetine treatment does not have a significant effect and, based on these results, paroxetine does not appear to be effective as a treatment for bipolar depression.

Rationale: Why Do the Test Statistic and Confidence Interval Have the Particular Forms We Have Presented? If the given assumptions are satisfied, the sampling distribution of $\bar{x}_1 - \bar{x}_2$ can be approximated by a t distribution with mean equal to $\mu_1 - \mu_2$ and standard deviation equal to $\sqrt{s_1^2/n_1 + s_2^2/n_2}$. This last expression for the standard deviation is based on the property that the variance of the *differences* between two independent random variables equals the variance of the first random variable *plus* the variance of the second random variable. That is, the variance of sample values $\bar{x}_1 - \bar{x}_2$ will tend to equal $s_1^2/n_1 + s_2^2/n_2$ provided that $\bar{x}_1$ and $\bar{x}_2$ are independent.

Alternative Method: σ_1 and σ_2 are Known. In reality, the population standard deviations σ_1 and σ_2 are almost never known, but if they are known, the test statistic and confidence interval are based on the normal distribution instead of the t distribution. See the following.

Test statistic:
$$z = \frac{(\bar{x}_1 - \bar{x}_2) - (\mu_1 - \mu_2)}{\sqrt{\dfrac{\sigma_1^2}{n_1} + \dfrac{\sigma_2^2}{n_2}}}$$

Confidence interval: $(\bar{x}_1 - \bar{x}_2) - E < (\mu_1 - \mu_2) < (\bar{x}_1 - \bar{x}_2) + E$

where
$$E = z_{\alpha/2}\sqrt{\frac{\sigma_1^2}{n_1} + \frac{\sigma_2^2}{n_2}}$$

Because σ_1 and σ_2 are rarely known in reality, this book will not use this alternative method. See Figure 8-3.

Alternative Method: Assume that $\sigma_1 = \sigma_2$ and *pool* the sample variances. Even when the specific values of σ_1 and σ_2 are not known, if it can be assumed that they have the *same* value, the sample variances s_1^2 and s_2^2 can be *pooled* to obtain an estimate of the common population variance σ^2. The **pooled estimate of σ^2** is denoted by s_p^2 and is a weighted average of s_1^2 and s_2^2, which is included in the following box.

Requirements

1. The two populations have the same standard deviation. That is, $\sigma_1 = \sigma_2$.

2. The two samples are *independent*.

3. Both samples are *simple random samples*.

4. Either or both of these conditions is satisfied: The two sample sizes are both *large* (with $n_1 > 30$ and $n_2 > 30$) or both samples come from populations having normal distributions. (For small samples, the normality requirement is loose in the sense that the procedures perform well as long as there are no

outliers and the samples have distributions that are not extremely far from being normally distributed.)

Hypothesis Test Statistic for Two Means: Independent Samples and $\sigma_1 = \sigma_2$

Test statistic: $t = \dfrac{(\bar{x}_1 - \bar{x}_2) - (\mu_1 - \mu_2)}{\sqrt{\dfrac{s_p^2}{n_1} + \dfrac{s_p^2}{n_2}}}$

where $s_p^2 = \dfrac{(n_1 - 1)s_1^2 + (n_2 - 1)s_2^2}{(n_1 - 1) + (n_2 - 1)}$ (*Pooled* variance)

and the number of degrees of freedom is given by df $= n_1 + n_2 - 2$.

Confidence Interval Estimate of $\mu_1 - \mu_2$: Independent Samples and $\sigma_1 = \sigma_2$

Confidence interval: $(\bar{x}_1 - \bar{x}_2) - E < (\mu_1 - \mu_2) < (\bar{x}_1 - \bar{x}_2) + E$

where $E = t_{\alpha/2}\sqrt{\dfrac{s_p^2}{n_1} + \dfrac{s_p^2}{n_2}}$ and s_p^2 is as given in the above test statistic and

the number of degrees of freedom is given by df $= n_1 + n_2 - 2$.

Inferences About Two Independent Means

FIGURE 8-3 Methods for Inferences About Two Independent Means

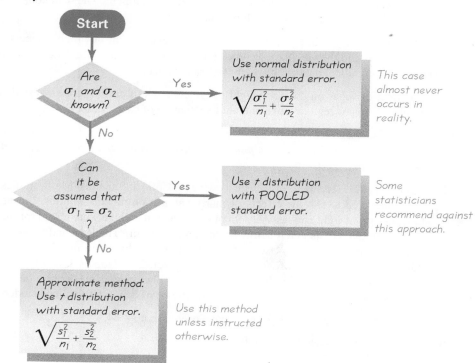

If we want to use this method, how do we determine that $\sigma_1 = \sigma_2$? One approach is to use a hypothesis test of the null hypothesis $\sigma_1 = \sigma_2$, as given in Section 8-5, but that approach is not recommended and, in this book, we will not use the preliminary test of $\sigma_1 = \sigma_2$. In the article "Homogeneity of Variance in the Two-Sample Means Test" (by Moser and Stevens, *The American Statistician*, Vol. 46, No. 1), the authors note that we rarely know that $\sigma_1 = \sigma_2$. They analyze the performance of the difference tests by considering sample sizes and powers of the tests. They conclude that more effort should be spent learning the method given near the beginning of this section, and less emphasis should be placed on the method based on the assumption of $\sigma_1 = \sigma_2$. Unless instructed otherwise, we use this strategy, which is consistent with the recommendations in the article by Moser and Stevens:

> **Assume that σ_1 and σ_2 are unknown, do *not* assume that $\sigma_1 = \sigma_2$, and use the test statistic and confidence interval given near the beginning of this section. See Figure 8-3.**

8-3 Exercises

Independent Samples and Matched Pairs. In Exercises 1–4, determine whether the samples are independent or consist of matched pairs.

1. The effectiveness of Prilosec for treating heartburn is tested by measuring gastric acid secretion in a group of patients treated with Prilosec and another group of patients given a placebo.

2. The effectiveness of Prilosec for treating heartburn is tested by measuring gastric acid secretion in patients before and after the drug treatment. The data consist of the before/after measurements for each patient.

3. The accuracy of verbal responses is tested in an experiment in which subjects report their weights and they are then weighed on a physician's scale. The data consist of the reported weight and measured weight for each subject.

4. The effect of sugar on the net weight of the contents of cans of cola is tested with a sample of cans of regular Coke and another sample of cans of diet Coke.

In Exercises 5–24, assume that the two samples are independent simple random samples selected from normally distributed populations. Do not assume that the population standard deviations are equal.

5. Hypothesis Test for Effect of Marijuana Use on College Students Many studies have been conducted to test the effects of marijuana use on mental abilities. In one such study, groups of light and heavy users of marijuana in college were tested for memory recall, with the results given below (based on data from "The Residual Cognitive Effects of Heavy Marijuana Use in College Students" by Pope and Yurgelun-Todd, *Journal of the American Medical Association*, Vol. 275, No. 7). Use a 0.01 significance level to test the claim that the population of heavy marijuana users has a lower mean than the light users. Should marijuana use be of concern to college students?

   Items sorted correctly by light marijuana users: $n = 64, \bar{x} = 53.3, s = 3.6$
   Items sorted correctly by heavy marijuana users: $n = 65, \bar{x} = 51.3, s = 4.5$

6. Confidence Interval for Effect of Marijuana Use on College Students Refer to the sample data used in Exercise 5 and construct a 98% confidence interval for the difference between the two population means. Does the confidence interval include zero? What does the confidence interval suggest about the equality of the two population means?

7. Confidence Interval for Magnet Treatment of Pain People spend huge sums of money (currently around $5 billion annually) for the purchase of magnets used to treat a wide variety of pains. Researchers conducted a study to determine whether magnets are effective in treating back pain. Pain was measured using the visual analog scale, and the results given below are among the results obtained in the study (based on data from "Bipolar Permanent Magnets for the Treatment of Chronic Lower Back Pain: A Pilot Study" by Collacott, Zimmerman, White, and Rindone, *Journal of the American Medical Association*, Vol. 283, No. 10). Construct a 90% confidence interval estimate of the difference between the mean reduction in pain for those treated with magnets and the mean reduction in pain for those given a sham treatment. Based on the result, does it appear that the magnets are effective in reducing pain?

   Reduction in pain level after magnet treatment: $n = 20, \bar{x} = 0.49, s = 0.96$

   Reduction in pain level after sham treatment: $n = 20, \bar{x} = 0.44, s = 1.4$

8. Hypothesis Test for Magnet Treatment of Pain Refer to the sample data from Exercise 7 and use a 0.05 significance level to test the claim that those treated with magnets have a greater reduction in pain than those given a sham treatment (similar to a placebo). Does it appear that magnets are effective in treating back pain? Is it valid to argue that magnets might appear to be effective if the sample sizes are larger?

9. Confidence Interval for Identifying Psychiatric Disorders Are severe psychiatric disorders related to biological factors that can be physically observed? One study used X-ray computed tomography (CT) to collect data on brain volumes for a group of patients with obsessive-compulsive disorders and a control group of healthy persons. Sample results for volumes (in milliliters) follow for the right cordate (based on data from "Neuroanatomical Abnormalities in Obsessive-Compulsive Disorder Detected with Quantitative X-Ray Computed Tomography," by Luxenberg et al., *American Journal of Psychiatry*, Vol. 145, No. 9). Construct a 99% confidence interval estimate of the difference between the mean brain volume for the healthy control group and the mean brain volume for the obsessive-compulsive group. What does the confidence interval suggest about the difference between the two population means? Based on this result, does it seem that obsessive-compulsive disorders have a biological basis?

   Control group: $n = 10, \bar{x} = 0.45, s = 0.08$

   Obsessive-compulsive patients: $n = 10, \bar{x} = 0.34, s = 0.08$

10. Hypothesis Test for Identifying Psychiatric Disorders Refer to the sample data in Exercise 9 and use a 0.01 significance level to test the claim that there is a difference between the two population means. Based on the result, does it seem that obsessive-compulsive disorders have a biological basis?

11. Confidence Interval for Effects of Alcohol An experiment was conducted to test the effects of alcohol. The errors were recorded in a test of visual and motor skills for a treatment group of people who drank ethanol and another group given a placebo. The results are shown in the accompanying table (based on data from "Effects of Alcohol Intoxication on Risk Taking, Strategy, and Error Rate in Visuomotor Performance," by Streufert et al., *Journal of Applied Psychology*, Vol. 77, No. 4). Construct a 95%

Treatment Group	Placebo Group
$n_1 = 22$	$n_2 = 22$
$\bar{x}_1 = 4.20$	$\bar{x}_2 = 1.71$
$s_1 = 2.20$	$s_2 = 0.72$

confidence interval estimate of the difference between the two population means. Do the results support the common belief that drinking is hazardous for drivers, pilots, ship captains, and so on? Why or why not?

12. **Hypothesis Test for Effects of Alcohol** Refer to the sample data in Exercise 11 and use a 0.05 significance level to test the claim that there is a difference between the treatment group and control group. If there is a significant difference, can we conclude that the treatment causes a decrease in visual and motor skills?

13. **Poplar Trees** Refer to the weights of poplar trees listed in Data Set 9 in Appendix B. Construct a 95% confidence interval estimate of the difference between the mean weights of trees for the two populations with these samples: (1) The sample of trees given no treatment in year 1 at site 1; (2) the sample of trees given fertilizer and irrigation in year 1 at site 1.

14. **Poplar Trees** Refer to the same two samples used in Exercise 13 and test the claim that there is no difference between the two population means.

15. **Petal Lengths of Irises** Using the sample data listed in Data Set 7 in Appendix B, the petal lengths of irises from the Sertosa class have a mean of 1.46 mm and a standard deviation of 0.17 mm. The petal lengths of irises from the Versicolor class have a mean of 4.26 mm and a standard deviation of 0.47 mm. Using a 0.05 significance level, test the claim that the Sertosa and Versicolor irises have the same mean petal length.

16. **Petal Lengths of Irises** Using the sample data from Exercise 15, construct a 95% confidence interval estimate of the difference between the mean petal length of Sertosa irises and the mean petal length of Versicolor irises. What does the result suggest about the equality of those two population means?

17. **BMI of Men and Women** Refer to Data Set 1 in Appendix B and test the claim that the mean body mass index (BMI) of men is equal to the mean BMI of women.

18. **BMI of Men and Women** Refer to Data Set 1 in Appendix B and construct a 95% confidence interval estimate of the difference between the mean body mass index (BMI) of men and the mean BMI of women. What does the result suggest about the equality of those two population means?

19. **Head Circumferences** Refer to Data Set 4 in Appendix B and test the claim that two-month-old boys and two-month-old girls have different mean head circumferences.

20. **Head Circumferences** Refer to Data Set 4 in Appendix B and construct a 95% confidence interval estimate of the difference between the mean head circumference of two-month-old boys and the mean head circumference of two-month-old girls. What does the result suggest about the equality of those two population means?

21. **Seat Belts and Hospital Time** A study of seat belt use involved children who were hospitalized as a result of motor vehicle crashes. For a group of 290 children who were not wearing seat belts, the numbers of days spent in intensive care units (ICUs) have a mean of 1.39 and a standard deviation of 3.06. For a group of 123 children who were wearing seat belts, the numbers of days in the ICU have a mean of 0.83 and a standard deviation of 1.77 (based on data from "Morbidity Among Pediatric Motor Vehicle Crash Victims: The Effectiveness of Seat Belts," by Osberg and Di Scala, *American Journal of Public Health,* Vol. 82, No. 3). Using a 0.01 significance level, is there sufficient evidence to support the claim that the population of children not wearing seat belts has a higher mean number of days spent in the ICU? Based on the result, is there significant evidence in favor of seat belt use among children?

22. Seat Belts and Hospital Time Using the same sample data from Exercise 21, construct a 95% confidence interval estimate of the difference between the two population means. Interpret the result.

23. Effects of Cocaine on Children A study was conducted to assess the effects that occur
when children are exposed to cocaine before birth. Children were tested at age 4 for
object assembly skill, which was described as "a task requiring visual-spatial skills
related to mathematical competence." The 190 children born to cocaine users had a
mean of 7.3 and a standard deviation of 3.0. The 186 children not exposed to cocaine
had a mean score of 8.2 with a standard deviation of 3.0. (The data are based on
"Cognitive Outcomes of Preschool Children with Prenatal Cocaine Exposure" by
Singer et al., *Journal of the American Medical Association*, Vol. 291, No. 20.) Use a
0.05 significance level to test the claim that prenatal cocaine exposure is associated
with lower scores of four-year-old children on the test of object assembly.

24. Birth Weight and IQ When investigating a relationship between birth weight and IQ,
researchers found that 258 subjects with extremely low birth weights (less than 1000
g) had Wechsler IQ scores at age 8 with a mean of 95.5 and a standard deviation of
16.0. For 220 subjects with normal birth weights, the mean at age 8 is 104.9 and the
standard deviation is 14.1. (Based on data from "Neurobehavioral Outcomes of
School-age Children Born Extremely Low Birth Weight or Very Preterm in the
1990s" by Anderson et al., *Journal of the American Medical Association*, Vol. 289,
No. 24.) Does IQ score appear to be affected by birth weight?

*In Exercises 25–28, assume that the two samples are independent simple random samples
selected from normally distributed populations. Also assume that the population standard
deviations are equal ($\sigma_1 = \sigma_2$). so that the standard error of the differences between
means is obtained by pooling the sample variances.*

25. Hypothesis Test with Pooling Do Exercise 5 with the additional assumption that
$\sigma_1 = \sigma_2$. How are the results affected by this additional assumption?

26. Confidence Interval with Pooling Do Exercise 6 with the additional assumption that
$\sigma_1 = \sigma_2$. How are the results affected by this additional assumption?

27. Confidence Interval with Pooling Do Exercise 7 with the additional assumption that
$\sigma_1 = \sigma_2$. How are the results affected by this additional assumption?

28. Hypothesis Test with Pooling Do Exercise 8 with the additional assumption that
$\sigma_1 = \sigma_2$. How are the results affected by this additional assumption?

*29. Verifying a Property of Variances
   a. Find the variance for this *population* of x values: 5, 10, 15. (See Section 2-5 for the
      variance σ^2 of a population.)
   b. Find the variance for this *population* of y values: 1, 2, 3.
   c. List the *population* of all possible differences $x - y$, and find the variance of this
      population.
   d. Use the results from parts (a), (b), and (c) to verify that the variance of the
      differences between two independent random variables is the *sum* of their individ-
      ual variances ($\sigma_{x-y}^2 = \sigma_x^2 + \sigma_y^2$). (This principle is used to derive the test statistic
      and confidence interval given in this section.)
   e. How is the *range* of the differences $x - y$ related to the range of the x values and
      the range of the y values?

Treatment Group	Placebo Group
$n_1 = 22$	$n_2 = 22$
$\bar{x}_1 = 0.049$	$\bar{x}_2 = 0.000$
$s_1 = 0.015$	$s_2 = 0.000$

*30. Effect of No Variation in Sample An experiment was conducted to test the effects of alcohol. The breath alcohol levels were measured for a treatment group of people who drank ethanol and another group given a placebo. The results are given in the accompanying table. Use a 0.05 significance level to test the claim that the two sample groups come from populations with the same mean. The given results are based on data from "Effects of Alcohol Intoxication on Risk Taking, Strategy, and Error Rate in Visuomotor Performance," by Streufert et al., *Journal of Applied Psychology*, Vol. 77, No. 4.

8-4 Inferences from Matched Pairs

In Section 8-3 we defined two samples to be *independent* if the sample values selected from one population are not related to or somehow paired or matched with the sample values selected from the other population. Section 8-3 dealt with inferences about the means of two independent populations, and this section focuses on dependent samples, which we refer to as *matched pairs*. With matched pairs, there is some relationship so that each value in one sample is paired with a corresponding value in the other sample. Here are some typical examples of matched pairs:

- When conducting an experiment to test the effectiveness of a low-fat diet, the weight of each subject is measured once before the diet and once after the diet.

- In a test of the effects of a fertilizer on heights of trees, sample trees are planted in pairs, with one tree given the fertilizer treatment while the other tree is not given the treatment.

When dealing with inferences about the means of matched pairs, summaries of the relevant requirements, notation, hypothesis test statistic, and confidence interval are given below. Because the hypothesis test and confidence interval use the same distribution and standard error, they are equivalent in the sense that they result in the same conclusions. Consequently, we can test the null hypothesis that the mean difference equals zero by determining whether the confidence interval includes zero. [For two-tailed hypothesis tests construct a confidence interval with a confidence level of $1 - \alpha$; but for a one-tailed hypothesis test with significance level α, construct a confidence interval with a confidence level of $1 - 2\alpha$. (See Table 7-2 for common cases.) For example, the claim that the mean difference is greater than 0 can be tested with a 0.05 significance level by constructing a 90% confidence interval.]

Requirements

1. The sample data consist of matched pairs.

2. The samples are simple random samples.

3. Either or both of these conditions is satisfied: The number of matched pairs of sample data is large ($n > 30$) or the pairs of values have differences that

are from a population having a distribution that is approximately normal. (If there is a radical departure from a normal distribution, we should not use the methods given in this section, but we may be able to use nonparametric methods discussed in Chapter 12.)

Notation for Matched Pairs

d = individual difference between the two values in a single matched pair

μ_d = mean value of the differences d for the *population* of all matched pairs

$\overline{d}$ = mean value of the differences d for the paired *sample* data (equal to the mean of the $x - y$ values)

s_d = standard deviation of the differences d for the paired *sample* data

n = number of *pairs* of data

Hypothesis Test Statistic for Matched Pairs

$$t = \frac{\overline{d} - \mu_d}{\frac{s_d}{\sqrt{n}}}$$

where degrees of freedom = $n - 1$.

P-values and Critical values: Table A-3 (t distribution) with $n - 1$ degrees of freedom.

Confidence Intervals for Matched Pairs

$$\overline{d} - E < \mu_d < \overline{d} + E$$

where

$$E = t_{\alpha/2} \frac{s_d}{\sqrt{n}}$$

Critical values of $t_{\alpha/2}$: Use Table A-3 with $n - 1$ degrees of freedom.

Exploring the Data Sets

As always, we should avoid blind and thoughtless rote application of any statistical procedure. We should begin by exploring the data to see what might be learned. We should consider the center, variation, distribution, outliers, and any changes that take place over time. The requirements listed above require that "the number of matched pairs of sample data is large ($n > 30$) or the pairs of values have differences that are from a population having a distribution that is approximately normal." In the following example, we have 11 matched pairs of data, so we should check that the differences have a distribution that is approximately normal. Checking for normality with a histogram might be difficult with only 11 values, but shown below is a Minitab-generated normal probability plot of the 11 differences. This graph serves the same purpose as a normal quantile plot introduced in Section 5-7. Because the pattern of plotted points is reasonably close to a straight line and there is no systematic pattern that is not a straight-line pattern, we conclude that the differences appear to come from a population that has a normal distribution.

Research in Twins

Identical twins occur when a single fertilized egg splits in two, so that both twins share the same genetic makeup. There is now an explosion in research focused on those twins. Speaking for the Center for Study of Multiple Birth, Louis Keith noted that now "we have far more ability to analyze the data on twins using computers with new, built-in statistical packages." A common goal of such studies is to explore the classic issue of "nature versus nurture." For example, Thomas Bouchard, who runs the Minnesota Study of Twins Reared Apart, has found that IQ is 50%–60% inherited, while the remainder is the result of external forces.

Identical twins are matched pairs that provide better results by allowing us to reduce the genetic variation that is inevitable with unrelated pairs of people.

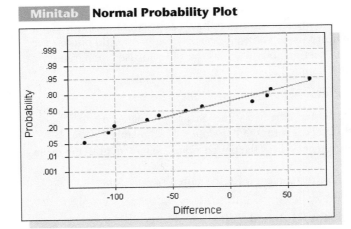

Minitab **Normal Probability Plot**

EXAMPLE **Does the Type of Seed Affect Corn Growth?** In 1908, William Gosset published the article "The Probable Error of a Mean" under the pseudonym of "Student" (*Biometrika*, Vol. 6, No. 1). He included the data listed below for two different types of corn seed (regular and kiln dried) that were used on adjacent plots of land. The listed values in Table 8-2 are the yields of head corn in pounds per acre. Use a 0.05 significance level to test the claim that the type of seed affects the yield.

SOLUTION

REQUIREMENT ✓ The data consist of matched pairs because they are corn yields from *adjacent* plots of land. Based on the design of the original experiment, it is reasonable to assume that the matched pairs constitute a simple random sample. The preceding Minitab display of a normal quantile plot suggests that the differences are from a population with a normal distribution. Note also that the list of differences does not appear to contain an outlier. The requirements are therefore satisfied. ✓

We will follow the same basic method of hypothesis testing that was introduced in Chapter 7, but we will use the above test statistic for matched pairs.

Step 1:   The claim that the type of seed affects the yield is a claim that there is a difference between the yields from regular seed and kiln-dried seed. This difference can be expressed as $\mu_d \neq 0$.

Step 2:   If the original claim is not true, we have $\mu_d = 0$.

Table 8-2	Yields of Corn from Different Seeds										
Regular	1903	1935	1910	2496	2108	1961	2060	1444	1612	1316	1511
Kiln Dried	2009	1915	2011	2463	2180	1925	2122	1482	1542	1443	1535
Difference d	−106	20	−101	33	−72	36	−62	−38	70	−127	−24

Step 3:   The null hypothesis must be a statement of equality and the alternative hypothesis cannot include equality, so we have

$$H_0: \mu_d = 0 \qquad H_1: \mu_d \neq 0 \qquad \text{(original claim)}$$

Step 4:   The significance level is $\alpha = 0.05$.

Step 5:   We use the Student t distribution because the requirements are satisfied. (We are testing a claim about matched pairs of data, we have two simple random samples, and a normal quantile plot of the sample differences shows that they have a distribution that is approximately normal.)

Step 6:   Before finding the value of the test statistic, we must first find the values of $\overline{d}$, and s_d. Refer to Table 8-2 and use the 11 sample differences to find these sample statistics: $\overline{d} = -33.727$ and $s_d = 66.171$. Using these sample statistics and the assumption of the hypothesis test that $\mu_d = 0$, we can now find the value of the test statistic:

$$t = \frac{\overline{d} - \mu_d}{\frac{s_d}{\sqrt{n}}} = \frac{-33.727 - 0}{\frac{66.171}{\sqrt{11}}} = -1.690$$

The critical values of $t = \pm 2.228$ are found from Table A-3 as follows: Use the column for 0.05 (Area in Two Tails), and use the row with degrees of freedom of $n - 1 = 10$. Figure 8-4 shows the test statistic, critical values, and critical region.

Step 7:   Because the test statistic does not fall in the critical region, we fail to reject the null hypothesis.

INTERPRETATION   The sample data in Table 8-2 do not provide sufficient evidence to support the claim that yields from regular seeds and kiln-dried seeds are different. This does *not* prove that the actual yields are equal. Perhaps

continued

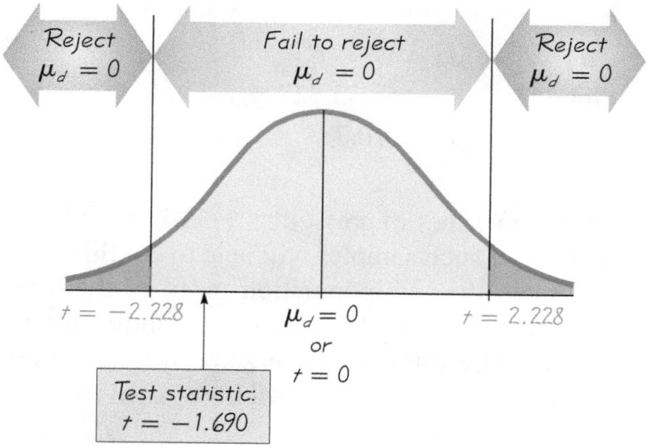

FIGURE 8-4

Distribution of
Differences d between
Values in Matched Pairs

additional sample data might provide the necessary evidence to conclude that the yields from the two types of seed are different. However, based on the available sample data, there does not appear to be a significant difference, so farmers should not treat the seeds by kiln drying them. Kiln drying appears to have no significant effect.

P-Value Method   The preceding example used the traditional method, but the P-value approach could be used by modifying Steps 6 and 7. In Step 6, use the test statistic of $t = -1.690$ and refer to the 10th row of Table A-3 to find that the test statistic (without the negative sign) is between 1.372 and 1.812, indicating that the P-value is between 0.10 and 0.20. Using STATDISK, Excel, Minitab, and a TI-83/84 Plus calculator, the P-value is found to be 0.1218. We again fail to reject the null hypothesis, because the P-value is greater than the significance level of $\alpha = 0.05$.

EXAMPLE Confidence Interval for Differences   Using the same sample matched pairs in Table 8-2, construct a 95% confidence interval estimate of μ_d, which is the mean of the differences between the yields from regular seeds and kiln-dried seeds. Interpret the result.

SOLUTION

REQUIREMENT ✓   See the preceding example which includes the same requirement check that applies here. We can now proceed with the construction of the confidence interval. ✓

We use the values of $\bar{d} = -33.727$, $s_d = 66.171$, $n = 11$, and $t_{\alpha/2} = 2.228$ (found from Table A-3 with $n - 1 = 10$ degrees of freedom and an area of 0.05 in two tails). We first find the value of the margin of error E.

$$E = t_{\alpha/2}\frac{s_d}{\sqrt{n}} = 2.228 \cdot \frac{66.171}{\sqrt{11}} = 44.452$$

The confidence interval can now be found. Note that the final result is rounded using one more decimal place than the original sample differences.

$$\bar{d} - E < \mu_d < \bar{d} + E$$
$$-33.727 - 44.452 < \mu_d < -33.727 + 44.452$$
$$-78.2 < \mu_d < 10.7$$

INTERPRETATION   The result is sometimes expressed as -33.7 ± 44.5 or as $(-78.2, 10.7)$. In the long run, 95% of such samples will lead to confidence interval limits that actually do contain the true population mean of the differences. Note that the confidence interval limits do contain zero, indicating that the true value of μ_d is not significantly different from zero. We cannot conclude that there is a significant difference between the yields of corn from regular seed and kiln-dried seed.

8-4 Exercises

Calculations for Matched Pairs. In Exercises 1 and 2, assume that you want to use a 0.05 significance level to test the claim that the paired sample data come from a population for which the mean difference is $\mu_d = 0$. Find (a) $\overline{d}$, (b) s_d, (c) the t test statistic, and (d) the critical values.

1.

x	1	1	3	5	4
y	0	2	5	8	0

2.

x	5	3	7	9	2	5
y	5	1	2	6	6	4

3. Using the sample paired data in Exercise 1, construct a 95% confidence interval for the population mean of all differences $x - y$.

4. Using the sample paired data in Exercise 2, construct a 99% confidence interval for the population mean of all differences $x - y$.

5. Testing Corn Seeds In 1908, William Gosset published the article "The Probable Error of a Mean" under the pseudonym of "Student" (*Biometrika*, Vol. 6, No. 1). He included the data listed below for yields from two different types of seed (regular and kiln dried) that were used on adjacent plots of land. The listed values are the yields of straw in cwts per acre, where cwt represents 100 pounds.

  a. Using a 0.05 significance level, test the claim that there is no difference between the yields from the two types of seed.

  b. Construct a 95% confidence interval estimate of the mean difference between the yields from the two types of seed.

  c. Does it appear that either type of seed is better?

Regular	19.25	22.75	23	23	22.5	19.75	24.5	15.5	18	14.25	17
Kiln dried	25	24	24	28	22.5	19.5	22.25	16	17.25	15.75	17.25

6. Self-Reported and Measured Female Heights As part of the National Health Survey conducted by the Department of Health and Human Services, self-reported heights and measured heights were obtained for females aged 12–16. Listed below are sample results.

  a. Is there sufficient evidence to support the claim that there is a difference between self-reported heights and measured heights of females aged 12–16? Use a 0.05 significance level.

  b. Construct a 95% confidence interval estimate of the mean difference between reported heights and measured heights. Interpret the resulting confidence interval, and comment on the implications of whether the confidence interval limits contain 0.

Reported height	53	64	61	66	64	65	68	63	64	64	64	67
Measured height	58.1	62.7	61.1	64.8	63.2	66.4	67.6	63.5	66.8	63. 9	62.1	68.5

7. Self-Reported and Measured Male Heights As part of the National Health Survey conducted by the Department of Health and Human Services, self-reported heights and measured heights were obtained for males aged 12–16. Listed on the next page are sample results.

  a. Is there sufficient evidence to support the claim that there is a difference between self-reported heights and measured heights of males aged 12–16? Use a 0.05 significance level.

Crest and Dependent Samples

In the late 1950s, Procter & Gamble introduced Crest toothpaste as the first such product with fluoride. To test the effectiveness of Crest in reducing cavities, researchers conducted experiments with several sets of twins. One of the twins in each set was given Crest with fluoride, while the other twin continued to use ordinary toothpaste without fluoride. It was believed that each pair of twins would have similar eating, brushing, and genetic characteristics. Results showed that the twins who used Crest have significantly fewer cavities than those who did not. This use of twins as dependent samples allowed the researchers to control many of the different variables affecting cavities.

b. Construct a 95% confidence interval estimate of the mean difference between reported heights and measured heights. Interpret the resulting confidence interval, and comment on the implications of whether the confidence interval limits contain 0.

Reported height	68	71	63	70	71	60	65	64	54	63	66	72
Measured height	67.9	69.9	64.9	68.3	70.3	60.6	64.5	67.0	55.6	74.2	65.0	70.8

8. Before/After Treatment Results Captopril is a drug designed to lower systolic blood pressure. When subjects were tested with this drug, their systolic blood pressure readings (in mm Hg) were measured before and after the drug was taken, with the results given in the accompanying table (based on data from "Essential Hypertension: Effect of an Oral Inhibitor of Angiotensin-Converting Enzyme," by MacGregor et al., *British Medical Journal,* Vol. 2).

a. Use the sample data to construct a 99% confidence interval for the mean difference between the before and after readings.

b. Is there sufficient evidence to support the claim that captopril is effective in lowering systolic blood pressure?

Subject	A	B	C	D	E	F	G	H	I	J	K	L
Before	200	174	198	170	179	182	193	209	185	155	169	210
After	191	170	177	167	159	151	176	183	159	145	146	177

9. Effectiveness of Hypnotism in Reducing Pain A study was conducted to investigate the effectiveness of hypnotism in reducing pain. Results for randomly selected subjects are given in the accompanying table (based on "An Analysis of Factors That Contribute to the Efficacy of Hypnotic Analgesia," by Price and Barber, *Journal of Abnormal Psychology,* Vol. 96, No. 1). The values are before and after hypnosis; the measurements are in centimeters on a pain scale.

a. Construct a 95% confidence interval for the mean of the "before–after" differences.

b. Use a 0.05 significance level to test the claim that the sensory measurements are lower after hypnotism.

c. Does hypnotism appear to be effective in reducing pain?

Subject	A	B	C	D	E	F	G	H
Before	6.6	6.5	9.0	10.3	11.3	8.1	6.3	11.6
After	6.8	2.4	7.4	8.5	8.1	6.1	3.4	2.0

10. Parent's Heights Refer to Data Set 3 in Appendix B and use only the data corresponding to male children. Use the paired data consisting of the mother's height and the father's height.

a. Use a 0.01 significance level to test the claim that mothers of male children are shorter than the fathers.

b. Construct a 98% confidence interval estimate of the mean of the differences between the heights of mothers and the heights of fathers.

11. Motion Sickness: Interpreting a Minitab Display The following Minitab display resulted from an experiment in which 10 subjects were tested for motion sickness before and after taking the drug astemizole. The sample data are the differences in the number of head movements that the subjects could endure without becoming nauseous. (The differences were obtained by subtracting the "after" values from the "before" values.)

a. Use a 0.05 significance level to test the claim that astemizole has an effect (for better or worse) on vulnerability to motion sickness. Based on the result, would you use astemizole if you were concerned about motion sickness while on a cruise ship?

b. Instead of testing for some effect (for better or worse), suppose we want to test the claim that astemizole is effective in *preventing* motion sickness? What is the *P*-value, and what do you conclude?

```
95% CI for mean difference: (-48.8, 33.8)
T-Test of mean difference = 0 (vs not = 0):
T-Value = -0.41 P-Value = 0.691
```

12. Dieting: Interpreting SAS Display Researchers obtained weight loss data from a sample of dieters using the New World Athletic Club facilities. The before–after weights are recorded, then the differences (before − after) are computed. The SAS results are shown for a test of the claim that the diet is effective. (*Hint*: If the diet is effective, the before weights should be greater than the corresponding after weights, but the displayed *P*-value is for a two-tailed test.) Is there sufficient evidence to support the claim that the diet is effective? Explain.

SAS

T-Tests			
Variable	DF	t Value	Pr > |t|
diff	6	6.43	0.0007

13. Self-Reported and Measured Weights of Males: Interpreting an Excel Display Refer to the Excel display of the results obtained when testing the claim that there is no difference between self-reported weights and measured weights of males aged 12–16. Is there sufficient evidence to support the claim that there is a difference? The display is based on data from the National Health Survey conducted by the Department of Health and Human Services.

Excel

t-Test: Paired Two Sample for Means		
	Variable 1	Variable 2
Mean	133.75	134.7333333
Variance	291.4772727	280.6151515
Observations	12	12
Pearson Correlation	0.919502265	
Hypothesized Mean Difference	0	
df	11	
t Stat	-0.501440942	
P(T<=t) one-tail	0.312972232	
t Critical one-tail	1.795883691	
P(T<=t) two-tail	0.625944463	
t Critical two-tail	2.200986273	

14. Self-Reported and Measured Heights of Male Statistics Students: Interpreting a STATDISK Display Male statistics students were given a survey that included a question asking them to report their height in inches. They weren't told that their height would be measured, but heights were accurately measured after the survey was completed. Anonymity was maintained, with code numbers used instead of names, so that no personal information would be publicly announced and nobody would be embarrassed by the results. STATDISK results are shown in the margin for a 0.05 significance level. Is there sufficient evidence to support a claim that male statistics students exaggerate their heights?

STATDISK

Sample Size, n	11
Difference Mean, $\bar{x}_d$	0.6727
Difference St Dev, s_d	0.8259
Test Statistic, t	2.7014
Critical t	1.8125
P-Value	0.0111

15. Morning and Night Body Temperatures Refer to Data Set 2 in Appendix B. Use the paired data consisting of body temperatures of women at 8:00 A.M. and at 12:00 A.M. on Day 2.

a. Construct a 95% confidence interval for the mean difference of 8 A.M. temperatures minus 12 A.M. temperatures.

b. Using a 0.05 significance level, test the claim that for those temperatures, the mean difference is 0. Based on the results, do morning and night body temperatures appear to be about the same?

16. Exercise and Stress Refer to the systolic blood pressure readings in Data Set 12 in Appendix B. The math test and speech test were designed to create stress.

    a. Use the matched "pre-exercise with no stress" readings and the "pre-exercise with math stress" readings. Construct a confidence interval. What does the result suggest?

    b. Use the "pre-exercise with math stress" readings and the "post-exercise with math stress" readings. Construct a confidence interval. What does the result suggest?

*17. Using the Correct Procedure

    a. Consider the sample data given below to be matched pairs and use a 0.05 significance level to test the claim that $\mu_d > 0$.

    b. Consider the sample data given below to be two independent samples. Use a 0.05 significance level to test the claim that $\mu_1 > \mu_2$.

    c. Compare the results from parts (a) and (b). Is it critical that the correct method be used? Why or why not?

x	1	3	2	2	1	2	3	3	2	1
y	1	2	1	2	1	2	1	2	1	2

8-5 Odds Ratios

In Section 3-6 we introduced the concept of *odds ratio*, and we presented the following definition.

Definition

In a *retrospective* or *prospective study*, the **odds ratio** (or **OR,** or **relative odds**) is a measure of risk found by evaluating the ratio of the odds for the treatment group (or case group exposed to the risk factor) to the odds for the control group, evaluated as follows:

$$\text{odds ratio} = \frac{\text{odds for treatment (or exposed) group}}{\text{odds for control group}}$$

We noted that if the sample data are in the format of the generalized Table 3-4 reproduced here, then the odds ratio can be calculated as follows:

$$\text{odds ratio} = \frac{ad}{bc}$$

Table 3-4	Generalized Table Summarizing Results of a Prospective Study	
	Disease	No Disease
Treatment (or Exposed)	a	b
Placebo	c	d

Table 8-3	Testing Effectiveness of Vaccine	
	Developed Flu	No Flu
Vaccine treatment group	14	1056
Placebo group	95	437

In the Chapter Problem we included an example involving a clinical trial of a flu vaccine for children. Table 8-3 is a modified form of Table 8-1 given with the Chapter Problem. (Instead of a column listing sample sizes, we list the number of subjects who did not develop a flu.) Using Table 8-3 with the methods of Section 3-6, we can find the following:

- For the vaccine treatment group, the odds in favor of a flu are 14/1056 or 14:1056.
- For the placebo group, the odds in favor of a flu are 95/437 or 95:437.
- Using the two preceding odds results, we find the odds ratio using either of the following two approaches:

$$\text{odds ratio} = \frac{\text{odds for flu in vaccine treatment group}}{\text{odds for flu in placebo group}}$$

$$= \frac{14/1056}{95/437} = 0.0610$$

$$\text{odds ratio} = \frac{ad}{bc} = \frac{(14)(437)}{(1056)(95)} = 0.0610$$

- *Interpretation:* The odds ratio of 0.0610 indicates that the flu rate with the vaccine is only about 6% of the flu rate for those given a placebo. This suggests that the vaccine substantially reduces the flu rate.

Because it is a single value used to estimate the risk of some treatment relative to the risk for a control group, the odds ratio is a *point estimate.* As with all important point estimates, we seek more information about their accuracy. One common way to gain more insight into the accuracy of a point estimate is to construct a *confidence interval estimate* that is associated with some confidence level, such as 95%. We will now do the same as we consider a method for constructing a confidence interval for an odds ratio. One difficulty associated with this method is that the distribution of sample odds ratios is not a normal distribution. If we were to take *logarithms* of sample odds ratios, we would get values with a distribution that is approximately normal. Consequently, logarithms are used for finding the confidence interval limits, but we can transform those limits back to a standard scale by using exponents. The usual method is to use natural logarithms (with base $e = 2.7182818\ldots$). For a particular confidence level, we use the two-tailed areas of $\alpha/2$ to find the critical scores of $z_{\alpha/2}$ and $-z_{\alpha/2}$ as we have done in other sections of this chapter. The variation of natural logarithms of odds ratios can be

Hawthorne and Experimenter Effects

The well-known placebo effect occurs when an untreated subject incorrectly believes that he or she is receiving a real treatment and reports an improvement in symptoms. The Hawthorne effect occurs when treated subjects somehow respond differently, simply because they are part of an experiment. (This phenomenon was called the "Hawthorne effect" because it was first observed in a study of factory workers at Western Electric's Hawthorne plant.) An experimenter effect (sometimes called a Rosenthall effect) occurs when the researcher or experimenter unintentionally influences subjects through such factors as facial expression, tone of voice, or attitude.

described by the expression $\sqrt{\frac{1}{a} + \frac{1}{b} + \frac{1}{c} + \frac{1}{d}}$, which results in the following confidence interval.

Confidence Interval Estimate of Odds Ratio (OR):

$$\frac{ad}{bc} \cdot e^{-z_{\alpha/2}\sqrt{\frac{1}{a}+\frac{1}{b}+\frac{1}{c}+\frac{1}{d}}} < OR < \frac{ad}{bc} \cdot e^{z_{\alpha/2}\sqrt{\frac{1}{a}+\frac{1}{b}+\frac{1}{c}+\frac{1}{d}}}$$

where

$e = 2.7182818\ldots$

a, b, c, d are in the format of the generalized Table 3-4.

Critical values $z_{\alpha/2}$ and $-z_{\alpha/2}$ are found using the normal distribution.

Caution: Because of the way that the confidence interval is constructed, the upper and lower confidence interval limits are not symmetric around the value of the sample odds ratio, so we *cannot* express the confidence interval in the format of "OR ± margin of error." We can, however, use the format of (LCL, UCL) where LCL is the lower confidence interval limit and UCL is the upper confidence interval limit.

Interpreting confidence interval estimates of odds ratios: An odds ratio of 1 indicates that there is no difference between a treatment group and placebo group, so the treatment does not appear to be effective. When interpreting a confidence interval estimate of an odds ratio, it is therefore helpful to consider whether the confidence interval limits include the value of 1.

- **Confidence interval limits include 1: It appears that the treatment has no effect.**
- **Confidence interval limits do not include 1: It appears that the treatment has an effect.**

EXAMPLE **Confidence Interval Estimate of Odds Ratio** Using the vaccine/flu data in Table 8-3, we have already found that the odds ratio is 0.0610 (or 0.06098485 before rounding). Construct a 95% confidence interval estimate of the odds ratio.

SOLUTION Comparing the vaccine/flu data in Table 8-3 to the generalized format of Table 3-4, we see that $a = 14$, $b = 1056$, $c = 95$, and $d = 437$. Also, a 95% confidence level corresponds to $\alpha = 0.05$ which is divided equally between the two tails of the standard normal distribution to yield the critical value of $z_{\alpha/2} = 1.96$.

Let's begin by finding the exponent used for the upper confidence interval limit.

$$z_{\alpha/2}\sqrt{\frac{1}{a} + \frac{1}{b} + \frac{1}{c} + \frac{1}{d}} = 1.96\sqrt{\frac{1}{14} + \frac{1}{1056} + \frac{1}{95} + \frac{1}{437}} = 0.57207222$$

The exponent used in the upper confidence interval limit is 0.57207222, and the exponent used in the lower confidence interval limit is therefore −0.57207222. Using these two exponents and using the unrounded odds ratio of $ad/bc = 0.06098485$, we evaluate the confidence interval as shown below.

$$\frac{ad}{bc} \cdot e^{-z_{\alpha/2}\sqrt{\frac{1}{a}+\frac{1}{b}+\frac{1}{c}+\frac{1}{d}}} < OR < \frac{ad}{bc} \cdot e^{z_{\alpha/2}\sqrt{\frac{1}{a}+\frac{1}{b}+\frac{1}{c}+\frac{1}{d}}}$$

$$0.06098485 \cdot 2.7182818^{-0.57207222} < OR < 0.06098485 \cdot 2.7182818^{0.57207222}$$

$$0.0344 < OR < 0.108$$

Note that the confidence interval limits do not have the odds ratio of 0.06098485 midway between them, so the above confidence interval *cannot* be expressed using a format with $\pm E$, where E is a margin of error. However, the above confidence interval can also be expressed as (0.0344, 0.108).

INTERPRETATION We are 95% confident that the limits of 0.0344 and 0.108 actually do contain the true value of the odds ratio. This means that if we were to repeatedly collect such sample data and construct the confidence interval as we did here, 95% of the confidence intervals will actually contain the true value of the odds ratio.

Also, it is important to note that the confidence interval limits do not include the value of 1. An odds ratio of 1 indicates that there is no difference between the treatment and placebo groups. *Because 1 is not included within the confidence interval, it appears that there is a significant effect from the vaccine.* In fact, Section 8-2 included an example showing that there is a significant difference between the flu rate for the vaccine treatment group and the flu rate for the placebo group.

Relative Risk (or Risk Ratio)   The computation of a confidence interval estimate of the odds ratio can be somewhat difficult, as can be seen in the preceding example. A confidence interval for relative risk can also be computed by using a similar procedure. To obtain a confidence interval estimate of relative risk (RR), use the same basic procedure used with the odds ratio, but apply it to the following confidence interval:

$$\left(\frac{\frac{a}{a+b}}{\frac{c}{c+d}}\right) \cdot e^{-z_{\alpha/2}\sqrt{\frac{b/a}{a+b}+\frac{d/c}{c+d}}} < RR < \left(\frac{\frac{a}{a+b}}{\frac{c}{c+d}}\right) \cdot e^{z_{\alpha/2}\sqrt{\frac{b/a}{a+b}+\frac{d/c}{c+d}}}$$

Fortunately, there are some software packages that automatically calculate such confidence intervals. For example, STATDISK (available on the enclosed CD) can be used to automatically find confidence interval estimates of odds ratio and relative risk.

8-5 Exercises

Headaches and Viagra. In Exercises 1 and 2, use the data in the accompanying table (based on data from Pfizer, Inc.). That table describes results from a clinical trial of the well-known drug Viagra.

	Headache	No Headache
Viagra treatment	117	617
Placebo	29	696

1. Find the odds ratio, then construct a 95% confidence interval estimate of the odds ratio. Determine whether the confidence interval contains the value of 1. Does it appear that the Viagra treatment has an effect on headaches?

2. Construct a 99% confidence interval estimate of the odds ratio. Write a statement that interprets the result.

Facial Injuries and Bicycle Helmets. In Exercises 3 and 4, use the data in the accompanying table (based on data from "A Case-Control Study of the Effectiveness of Bicycle Safety Helmets in Preventing Facial Injury," by Thompson, Rivara, and Wolf, American Journal of Public Health, Vol. 80, No. 12).

	Facial Injuries Received	All Injuries Nonfacial
Helmet worn	30	83
No helmet	182	236

3. Find the odds ratio, then construct a 99% confidence interval estimate of the odds ratio.

4. Construct a 95% confidence interval estimate of the odds ratio and write a statement interpreting the result. Do helmets appear to have an effect on facial injuries?

Smoking and Cancer In Exercises 5 and 6, use the data in the accompanying table.

Retrospective Study of Causes of Death

	Lung Cancer	Other
Smoker	140	532
Nonsmoker	21	1707

5. Find the odds ratio, then construct a 95% confidence interval estimate of the odds ratio.

6. Construct a 99% confidence interval estimate of the odds ratio and write a statement interpreting the result. Does smoking appear to have an effect on whether the cause of death is due to lung cancer?

Clinical Test of Lipitor. In Exercises 7 and 8, use the data in the accompanying table (based on data from Parke-Davis). The cholesterol-reducing drug Lipitor consists of atorvastatin calcium.

	Headache	No Headache
10 mg atorvastatin	15	17
Placebo	65	3

7. Find the odds ratio, then construct a 95% confidence interval estimate of the odds ratio.

8. Construct a 99% confidence interval estimate of the odds ratio and write a statement interpreting the result. Does the treatment appear to affect headaches?

9. Clinical Test of Nicorette Nicorette is a chewing gum designed to help people stop smoking cigarettes. Tests for adverse reactions included 152 subjects treated with Nicorette and 153 subjects given a placebo. Among those in the treatment group, 43 experienced mouth or throat soreness. Among those in the placebo group, 35 experienced mouth or throat soreness. (The results are based on data from Merrell Dow Pharmaceuticals, Inc.)

   a. Use the given information to construct a table in the generalized format of Table 3-4 included in this section.

   b. Find the odds ratio.

   c. Find a 95% confidence interval estimate of the odds ratio and write a statement interpreting the result. Does it appear that mouth or throat soreness should be a concern for those who use Nicorette?

10. Clinical Test of Viagra In clinical tests of the adverse reactions to the drug Viagra, 4.0% of the 734 subjects in the treatment group experienced nasal congestion, but 2.1% of the 725 subjects in the placebo group experienced nasal congestion (based on data from Pfizer Pharmaceuticals).

   a. Use the given information to construct a table in the generalized format of Table 3-4 included in this section.

   b. Find the odds ratio.

   c. Find a 95% confidence interval estimate of the odds ratio and write a statement interpreting the result.

11. Transposing Table Repeat Exercise 7 after reconfiguring the table used for Exercises 7 and 8 as shown below. How are the results affected by this reconfiguration?

	10 mg Atorvastatin	Placebo
Headache	15	65
No headache	17	3

*12. Confidence Interval for Relative Risk The accompanying table summarizes results from a prospective study of polio and the Salk vaccine. In Section 3-6 we found that the relative risk is 0.287. (A more precise value of 0.2876483794 is found by carrying more decimal places in the preliminary calculations.) Using the table, find a 95% confidence interval estimate of the relative risk. Interpret the result.

	Polio	No Polio	Total
Salk vaccine	33	200,712	200,745
Placebo	115	201,114	201,229

*13. Entry of Zero Consider the data in the accompanying table.

   a. What happens when you attempt to calculate the odds ratio?

   b. What happens when you attempt to construct a confidence interval estimate of the odds ratio?

   c. When there is an entry of zero, as in the accompanying table, one way to handle the calculations is to add 0.5 to each entry in the table. Use this approach to construct a 95% confidence interval estimate of the odds ratio. Does the result seem to make sense?

	Disease	No Disease
Treatment	25	75
Placebo	0	100

8-6 Comparing Variation in Two Samples

Because the characteristic of variation among data is extremely important, this section presents a method for using two samples to compare the variances of the two populations from which the samples are drawn. In Section 2-5 we saw that variation in a sample can be measured by the standard deviation, variance, and other measures such as the range and mean absolute deviation. Because standard deviation is a very effective measure of variation, and because it is easier to understand than variance, the early chapters of this book have stressed the use of standard deviation instead of variance. Although the basic procedure of this section is designed for variances, we can also use it for standard deviations. Let's briefly review this relationship between standard deviation and variance: The variance is the square of the standard deviation.

Measures of Variation

s = standard deviation of *sample*

s^2 = variance of *sample* (sample standard deviation squared)

σ = standard deviation of *population*

σ^2 = variance of *population* (population standard deviation squared)

The computations of this section will be greatly simplified if we designate the two samples so that s_1^2 represents the *larger* of the two sample variances. Mathematically, it doesn't really matter which sample is designated as Sample 1, so life will be better if we let s_1^2 represent the larger of the two sample variances, as in the test statistic included in the summary box.

Requirements

1. The two populations are *independent* of each other. (Recall from Section 8-2 that two samples are independent if the sample selected from one population is not related to the sample selected from the other population. The samples are not matched or paired.)

2. The two populations are each *normally distributed*. (*Caution:* This assumption is important because the methods of this section are extremely sensitive to departures from normality.)

Notation for Hypothesis Tests with Two Variances or Standard Deviations

s_1^2 = *larger* of the two sample variances

n_1 = size of the sample with the *larger* variance

σ_1^2 = variance of the population from which the sample with the *larger* variance was drawn

The symbols s_2^2, n_2, and σ_2^2 are used for the other sample and population.

Test Statistic for Hypothesis Tests with Two Variances

$$F = \frac{s_1^2}{s_2^2}$$ (where s_1^2 is the *larger* of the two sample variances)

Critical values: Use Table A-5 to find critical F values that are determined by the following:

1. The significance level α. (Table A-5 has four pages of critical values for $\alpha = 0.025$ and 0.05.)
2. **Numerator degrees of freedom $= n_1 - 1$**
3. **Denominator degrees of freedom $= n_2 - 1$**

For two normally distributed populations with equal variances (that is, $\sigma_1^2 = \sigma_2^2$,) the sampling distribution of the test statistic $F = s_1^2/s_2^2$ is the **F distribution** shown in Figure 8-5 with critical values listed in Table A-5. If you continue to repeat an experiment of randomly selecting samples from two normally distributed populations with equal variances, the distribution of the ratio s_1^2/s_2^2 of the sample variances is the F distribution.

In Figure 8-5, note these properties of the F distribution:

- The F distribution is not symmetric.
- Values of the F distribution cannot be negative.
- The exact shape of the F distribution depends on two different degrees of freedom.

Critical Values: To find a critical value, first refer to Table A-5. The first two pages represent the F distribution with an area of 0.025 in the right tail, so use the first two pages of Table A-5 for either of these two cases: (1) a one-tailed test with a 0.025 significance level, or (2) a two-tailed test with a 0.05 significance level. The

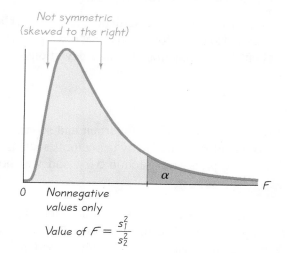

FIGURE 8-5

F Distribution

There is a different F distribution for each different pair of degrees of freedom for the numerator and the denominator.

second two pages of Table A-5 represent the F distribution with an area of 0.05 in the right tail, so use the second two pages for either of these two cases: (1) a one-tailed test with a 0.05 significance level, or (2) a two-tailed test with a 0.10 significance level. Intersect the column representing the degrees of freedom for s_1^2 with the row representing the degrees of freedom for s_2^2. Because we are stipulating that the larger sample variance is s_1^2, all one-tailed tests will be right-tailed and all two-tailed tests will require that we find only the critical value located to the right. Good news: We have no need to find a critical value separating a left-tailed critical region. (Because the F distribution is not symmetric and has only non-negative values, a left-tailed critical value cannot be found by using the negative of the right-tailed critical value; instead, a left-tailed critical value is found by using the reciprocal of the right-tailed value with the numbers of degrees of freedom reversed. See Exercise 10.)

We often have numbers of degrees of freedom that are not included in Table A-5. We could use linear interpolation to approximate missing values, but in most cases that's not necessary because the F test statistic is either less than the lowest possible critical value or greater than the largest possible critical value. For example, Table A-5 shows that for $\alpha = 0.025$ in the right tail, 20 degrees of freedom for the numerator, and 34 degrees of freedom for the denominator, the critical F value is between 2.0677 and 2.1952. Any F test statistic below 2.0677 will result in failure to reject the null hypothesis, any F test statistic above 2.1952 will result in rejection of the null hypothesis, and interpolation is necessary only if the F test statistic happens to fall between 2.0677 and 2.1952. The use of a statistical software package can eliminate this problem by providing critical values or P-values.

Interpreting the F Test Statistic: If the two populations really do have equal variances, then the ratio s_1^2/s_2^2 tends to be close to 1 because s_1^2 and s_2^2 tend to be close in value. But if the two populations have very different variances, s_1^2 and s_2^2 tend to be very different numbers. Denoting the larger of the sample variances by s_1^2, we see that the ratio s_1^2/s_2^2 will be a large number whenever s_1^2 and s_2^2 are far apart in value. Consequently, a value of F near 1 will be evidence in favor of the conclusion that $\sigma_1^2 = \sigma_2^2$, but a large value of F will be evidence against the conclusion of equality of the population variances.

Large values of F are evidence against $\sigma_1^2 = \sigma_2^2$.

Claims About Standard Deviations: The F test statistic applies to a claim made about two variances, but we can also use it for claims about two population standard deviations. Any claim about two population standard deviations can be restated in terms of the corresponding variances.

Exploring the Data

Because the requirement of normal distributions is so important and so strict, we should begin by comparing the two sets of sample data by using tools such as histograms, boxplots, and normal quantile plots (see Section 5-7), and we should search for outliers.

EXAMPLE **Calcium and Blood Pressure** Sample data were collected in a study of calcium supplements and their effects on blood pressure. A placebo group and a calcium group began the study with blood pressure measurements (based on data from "Blood Pressure and Metabolic Effects of Calcium Supplementation in Normotensive White and Black Men," by Lyle et al., *Journal of the American Medical Association,* Vol. 257, No. 13). Sample values are listed below. At the 0.05 significance level, test the claim that the two sample groups come from populations with the same standard deviation. If the experiment requires groups with equal standard deviations, are these two groups acceptable?

Placebo:	124.6	104.8	96.5	116.3	106.1	128.8	107.2	123.1
	118.1	108.5	120.4	122.5	113.6			
Calcium:	129.1	123.4	102.7	118.1	114.7	120.9	104.4	116.3
	109.6	127.7	108.0	124.3	106.6	121.4	113.2	

SOLUTION

REQUIREMENT ✓ We should first verify that the necessary requirements (listed earlier in this section) are satisfied. The two samples are independent instead of being somehow matched or paired. After exploring the two data sets, we obtain these results: There are no outliers, and both data sets result in normal quantile plots suggesting that the sample data come from normally distributed populations. The check of requirements for the procedure of this section has been successfully completed. ✓

Here are the key statistics required for this hypothesis test:

	Sample Size	Standard Deviation
Placebo	$n = 13$	$s_1 = 9.4601$
Calcium	$n = 15$	$s_2 = 8.4689$

Instead of using the sample standard deviations to test the claim of equal standard deviations, we will use the sample variances to test the claim of equal variances. Because we stipulate in this section that the larger variance is denoted by s_1^2, we let $s_1^2 = 9.4601^2$, $n_1 = 13$, $s_2^2 = 8.4689^2$, and $n_2 = 15$. We now proceed to use the traditional method of testing hypotheses as outlined in Figure 7-8.

Step 1:   The claim of equal standard deviations is equivalent to a claim of equal variances, which we express symbolically as $\sigma_1^2 = \sigma_2^2$.

Step 2:   If the original claim is false, then $\sigma_1^2 \neq \sigma_2^2$.

Step 3:   Because the null hypothesis is the statement of equality and because the alternative hypothesis cannot contain equality, we have

$$H_0: \sigma_1^2 = \sigma_2^2 \quad \text{(original claim)} \qquad H_1: \sigma_1^2 \neq \sigma_2^2$$

Step 4:   The significance level is $\alpha = 0.05$.

Step 5:   Because this test involves two population variances, we use the F distribution.

continued

Step 6:   The test statistic is

$$F = \frac{s_1^2}{s_2^2} = \frac{9.4601^2}{8.4689^2} = 1.2478$$

For the critical values, first note that this is a two-tailed test with 0.025 in each tail. As long as we are stipulating that the larger variance is placed in the numerator of the F test statistic, we need to find only the right-tailed critical value. From Table A-5 we see that the critical value of F is 3.0502, which we find by referring to 0.025 in the right tail, with 12 degrees of freedom for the numerator and 14 degrees of freedom for the denominator.

Step 7:   Figure 8-6 shows that the test statistic $F = 1.2478$ does not fall within the critical region, so we fail to reject the null hypothesis of equal variances.

INTERPRETATION   There is not sufficient evidence to warrant rejection of the claim that the two variances are equal. However, we should recognize that the F test is extremely sensitive to populations that are not normally distributed, so a conclusion might make it appear that there is no significant difference between the population variances when there really is a difference that was hidden by non-normal distributions.

We have described the traditional method of testing hypotheses made about two population variances. Exercise 11 deals with the construction of confidence intervals.

FIGURE 8-6   Distribution of s_1^2/s_2^2 for Placebo and Calcium Treatment Samples

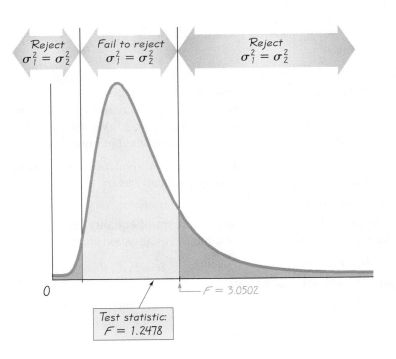

8-6 Exercises

Hypothesis Test of Equal Variances. In Exercises 1 and 2, test the given claim. Use a significance level of $\alpha = 0.05$ *and assume that all populations are normally distributed. Use the traditional method of testing hypotheses outlined in Figure 7-8.*

1. Claim: The treatment population and the placebo population have different variances, so $\sigma_1^2 \neq \sigma_2^2$.

    Treatment group: $n = 25, \bar{x} = 98.6, s = 0.78$
    Placebo group: $n = 30, \bar{x} = 98.2, s = 0.52$

2. Claim: Heights of male statistics students have a larger variance than female statistics students.

    Males: $n = 16, \bar{x} = 68.4, s = 2.54$
    Women: $n = 12, \bar{x} = 63.2, s = 2.39$

3. Hypothesis Test for Magnet Treatment of Pain Researchers conducted a study to determine whether magnets are effective in treating back pain, with results given below (based on data from "Bipolar Permanent Magnets for the Treatment of Chronic Lower Back Pain: A Pilot Study," by Collacott, Zimmerman, White, and Rindone, *Journal of the American Medical Association*, Vol. 283, No. 10). The values represent measurements of pain using the visual analog scale. Use a 0.05 significance level to test the claim that those given a sham treatment (similar to a placebo) have pain reductions that vary more than the pain reductions for those treated with magnets.

    Reduction in pain level after sham treatment: $n = 20, \bar{x} = 0.44, s = 1.4$
    Reduction in pain level after magnet treatment: $n = 20, \bar{x} = 0.49, s = 0.96$

4. Hypothesis Test for Effect of Marijuana Use on College Students In a study of the effects of marijuana use, light and heavy users of marijuana in college were tested for memory recall, with the results given below (based on data from "The Residual Cognitive Effects of Heavy Marijuana Use in College Students," by Pope and Yurgelun-Todd, *Journal of the American Medical Association*, Vol. 275, No. 7). Use a 0.05 significance level to test the claim that the population of heavy marijuana users has a standard deviation different from that of light users.

    Items sorted correctly by light marijuana users: $n = 64, \bar{x} = 53.3, s = 3.6$
    Items sorted correctly by heavy marijuana users: $n = 65, \bar{x} = 51.3, s = 4.5$

5. Cigarette Filters and Nicotine Refer to the sample results listed in the margin for the measured nicotine contents of randomly selected filtered and nonfiltered king-size cigarettes. All measurements are in milligrams, and the data are from the Federal Trade Commission. Use a 0.05 significance level to test the claim that king-size cigarettes with filters have amounts of nicotine that vary more than the amounts of nicotine in nonfiltered king-size cigarettes.

6. Effects of Alcohol An experiment was conducted to test the effects of alcohol. The errors were recorded in a test of visual and motor skills for a treatment group of people who drank ethanol and another group given a placebo. The results are shown in the accompanying table (based on data from "Effects of Alcohol Intoxication on Risk Taking, Strategy, and Error Rate in Visuomotor Performance," by Streufert et al., *Journal of Applied Psychology*, Vol. 77, No. 4). Use a 0.05 significance level to test the claim that the treatment group has scores that vary more than the scores of the placebo group.

Nicotine (mg)	
Filtered Kings	Nonfiltered Kings
$n_1 = 21$	$n_2 = 8$
$\bar{x}_1 = 0.94$	$\bar{x}_2 = 1.65$
$s_1 = 0.31$	$s_2 = 0.16$

Treatment Group	Placebo Group
$n_1 = 22$	$n_2 = 22$
$\bar{x}_1 = 4.20$	$\bar{x}_2 = 1.71$
$s_1 = 2.20$	$s_2 = 0.72$

Zinc Supplement Group	Placebo Group
$n = 294$	$n = 286$
$\bar{x} = 3214$	$\bar{x} = 3088$
$s = 669$	$s = 728$

7. **Testing Effects of Zinc** A study of zinc-deficient mothers was conducted to determine effects of zinc supplementation during pregnancy. Sample data are listed in the margin (based on data from "The Effect of Zinc Supplementation on Pregnancy Outcome," by Goldenberg et al., *Journal of the American Medical Association*, Vol. 274, No. 6). The weights were measured in grams. Using a 0.05 significance level, is there sufficient evidence to support the claim that the variation of birth weights for the placebo population is greater than the variation for the population treated with zinc supplements?

8. **Tobacco and Alcohol Use in Animated Children's Movies** Sample data were collected for the times (in seconds) that animated children's movies show tobacco use and alcohol use. The 50 times of tobacco use have a mean of 57.4 sec and a standard deviation of 104.0 sec. The 50 times of alcohol use have a mean of 32.46 sec and a standard deviation of 66.3 sec (based on data from "Tobacco and Alcohol Use in G-Rated Children's Animated Films," by Goldstein, Sobel, and Newman, *Journal of the American Medical Association*, Vol. 281, No. 12).
   a. Assuming that we want to use the methods of this section to test the claim that the times of tobacco use and the times of alcohol use have different standard deviations, identify the F test statistic, critical value, and conclusion. Use a 0.05 significance level.
   b. Consider the prerequisite of normally distributed populations. Given that the 50 times for tobacco use include 22 times of 0 sec, and the 50 times for alcohol use include 25 times of 0 sec, are the times for tobacco use normally distributed, and are the times for alcohol use normally distributed?
   c. What can be concluded from the results of parts (a) and (b)?

9. **Blanking Out on Tests** Many students have had the unpleasant experience of panicking on a test because the first question was exceptionally difficult. The arrangement of test items was studied for its effect on anxiety. Sample values consisting of measures of "debilitating test anxiety" (which most of us call panic or blanking out) are obtained for a group of subjects with test questions arranged from easy to difficult, and another group with test questions arranged from difficult to easy. The SAS display is shown below (based on data from "Item Arrangement, Cognitive Entry Characteristics, Sex and Test Anxiety as Predictors of Achievement in Examination Performance," by Klimko, *Journal of Experimental Education*, Vol. 52, No. 4). Use a 0.05 significance level to test the claim that the two samples come from populations with the same variance.

SAS

T-Tests					
Variable	Method	Variances	DF	t Value	Pr > |t|
Score	Pooled	Equal	39	-2.40	0.0211
Score	Satterthwaite	Unequal	39	-2.66	0.0114

Equality of Variances				
Variable	Method	Num DF	Den DF	F Value	Pr > F
Score	Folded F	24	15	2.59	0.0599

*10. Finding Lower Critical F Values   In this section, for hypothesis tests that were two-tailed, we found only the upper critical value. Let's denote that value by F_R, where the subscript indicates the critical value for the right tail. The lower critical value F_L (for the left tail) can be found as follows: First interchange the degrees of freedom, and then take the reciprocal of the resulting F value found in Table A-5. (F_R is sometimes denoted by $F_{\alpha/2}$ and F_L is sometimes denoted by $F_{1-\alpha/2}$.) Find the critical values F_L and F_R for two-tailed hypothesis tests based on the following values:

   a. $n_1 = 10, n_2 = 10, \alpha = 0.05$
   b. $n_1 = 10, n_2 = 7, \alpha = 0.05$
   c. $n_1 = 7, n_2 = 10, \alpha = 0.05$

*11. Constructing Confidence Intervals   In addition to testing claims involving σ_1^2 and σ_2^2, we can also construct confidence interval estimates of the ratio σ_1^2/σ_2^2 using the following expression:

$$\left(\frac{s_1^2}{s_2^2} \cdot \frac{1}{F_R}\right) < \frac{\sigma_1^2}{\sigma_2^2} < \left(\frac{s_1^2}{s_2^2} \cdot \frac{1}{F_L}\right)$$

   Here F_L and F_R are as described in the preceding exercise. Refer to the data used in the example from this section and construct a 95% confidence interval estimate for the ratio of the placebo group variance to the calcium-supplement group variance.

Review

In Chapters 6 and 7 we introduced two major concepts of inferential statistics: the estimation of population parameters and the methods of testing hypotheses made about population parameters. Chapters 6 and 7 considered only cases involving a single population, but this chapter considered two samples drawn from two populations.

- Section 8-2 considered inferences made about two population proportions.
- Section 8-3 considered inferences made about the means of two independent populations. Section 8-3 included three different methods, but one method is rarely used because it requires that the two population standard deviations are known. Another method involves pooling the two sample standard deviations to develop an estimate of the standard error, but this method is based on the assumption that the two population standard deviations are known to be equal, and that assumption is often risky. See Figure 8-3 for help in determining which method to apply.
- Section 8-4 considered inferences made about the mean difference for a population consisting of matched pairs.
- Section 8-5 presented a method for constructing a confidence interval estimate of an odds ratio.
- Section 8-6 presented methods for testing claims about the equality of two population standard deviations or variances.

Review Exercises

1. Warmer Surgical Patients Recover Better?   An article published in *USA Today* stated that "in a study of 200 colorectal surgery patients, 104 were kept warm with blankets

and intravenous fluids; 96 were kept cool. The results show: Only 6 of those warmed developed wound infections vs. 18 who were kept cool."

a. Use a 0.05 significance level to test the claim of the article's headline: "Warmer surgical patients recover better." If these results are verified, should surgical patients be routinely warmed?

b. If a confidence interval is to be used for testing the claim in part (a), what confidence level should be used?

c. Using the confidence level from part (b), construct a confidence interval estimate of the difference between the two population proportions.

d. In general, if a confidence interval estimate of the difference between two population proportions is used to test some claim about the proportions, will the conclusion based on the confidence interval always be the same as the conclusion from a standard hypothesis test?

e. Find the odds ratio and construct a 95% confidence interval estimate of the odds ratio. Based on the results, does it appear that "warmer surgical patients recover better?"

2. Brain Volume and Psychiatric Disorders A study used X-ray computed tomography (CT) to collect data on brain volumes for a group of patients with obsessive-compulsive disorders and a control group of healthy persons. Sample results (in mL) are given below for total brain volumes (based on data from "Neuroanatomical Abnormalities in Obsessive-Compulsive Disorder Detected with Quantitative X-Ray Computed Tomography," by Luxenberg et al., *American Journal of Psychiatry,* Vol. 145, No. 9).

a. Construct a 95% confidence interval for the difference between the mean brain volume of obsessive-compulsive patients and the mean brain volume of healthy persons. Assume that the two populations have *unequal* variances.

b. Assuming that the population variances are *unequal,* use a 0.05 significance level to test the claim that there is no difference between the mean for obsessive-compulsive patients and the mean for healthy persons.

c. Based on the results from parts (a) and (b), does it appear that the total brain volume can be used as an indicator of obsessive-compulsive disorders?

Obsessive-compulsive patients: $n = 10, \bar{x} = 1390.03, s = 156.84$

Control group: $n = 10, \bar{x} = 1268.41, s = 137.97$

3. Variation of Brain Volumes Use the same sample data given in Exercise 2 with a 0.05 significance level to test the claim that the populations of total brain volumes for obsessive-compulsive patients and the control group have different amounts of variation.

4. Carbon Monoxide and Cigarettes Refer to the given data for the measured amounts of carbon monoxide (CO) from samples of filtered and nonfiltered king-size cigarettes. All measurements are in milligrams, and the data are from the Federal Trade Commission. Use a 0.05 significance level to test the claim that the mean amount of carbon monoxide in filtered king-size cigarettes is equal to the mean amount of carbon monoxide for nonfiltered king-size cigarettes. Based on this result, are cigarette filters effective in reducing carbon monoxide?

Filtered: 14 12 14 16 15 2 14 16 11 13 13 12 13 12 13 14 14 14 9 17 12

Non-filtered: 14 15 17 17 16 16 14 16

5. Zinc for Mothers A study of zinc-deficient mothers was conducted to determine whether zinc supplementation during pregnancy results in babies with increased weights at birth. Sample data are listed in the margin (based on data from "The Effect of Zinc Supplementation on Pregnancy Outcome," by Goldenberg et al., *Journal of the American Medical Association*, Vol. 274, No. 6). The weights are measured in grams. Using a 0.05 significance level, is there sufficient evidence to support the claim that zinc supplementation does result in increased birth weights?

Zinc Supplement Group	Placebo Group
$n = 294$	$n = 286$
$\bar{x} = 3214$	$\bar{x} = 3088$
$s = 669$	$s = 728$

6. Testing Effects of Physical Training A study was conducted to investigate some effects of physical training. Sample data are listed below, with all weights given in kilograms. (See "Effect of Endurance Training on Possible Determinants of VO_2 During Heavy Exercise," by Casaburi et al., *Journal of Applied Physiology*, Vol. 62, No. 1.)

   a. Is there sufficient evidence to conclude that there is a difference between the pre-training and post-training weights? What do you conclude about the effect of training on weight?

   b. Construct a 95% confidence interval for the mean of the differences between pre-training and post-training weights.

Pretraining:	99	57	62	69	74	77	59	92	70	85
Post-training:	94	57	62	69	66	76	58	88	70	84

7. Drug Solubility Twelve Dozenol tablets are tested for solubility before and after being stored for one year. The indexes of solubility are given in the table below.

Before	472	487	506	512	489	503	511	501	495	504	494	462
After	562	512	523	528	554	513	516	510	524	510	524	508

   a. At the 0.05 significance level, test the claim that the Dozenol tablets are more soluble after the storage period.

   b. Construct a 95% confidence interval estimate of the mean difference (after − before).

Cumulative Review Exercises

1. Gender Difference in Speeding The data in the accompanying table were obtained through a survey of randomly selected subjects (based on data from R. H. Bruskin Associates).

   a. If one of the survey subjects is randomly selected, find the probability of getting someone ticketed for speeding.

   b. If one of the survey subjects is randomly selected, find the probability of getting a man or someone ticketed for speeding.

   c. Find the probability of getting someone ticketed for speeding, given that the selected person is a man.

   d. Find the probability of getting someone ticketed for speeding, given that the selected person is a woman.

   e. Use a 0.05 significance level to test the claim that the percentage of women ticketed for speeding is less than the percentage of men. Can we conclude that men generally speed more than women?

	Ticketed for Speeding Within the Last Year?	
	Yes	No
Men	26	224
Women	27	473

2. Cell Phones and Crashes: Analyzing Newspaper Report In an article from the Associated Press, it was reported that researchers "randomly selected 100 New York motorists who had been in an accident and 100 who had not. Of those in accidents, 13.7 percent owned a cellular phone, while just 10.6 percent of the accident-free drivers had a phone in the car." Identify the most notable feature of these results.

3. Clinical Tests of Viagra In clinical tests of adverse reactions to the drug Viagra, 4.0% of the 734 subjects in the treatment group experienced nasal congestion, but 2.1% of

the 725 subjects in the placebo group experienced nasal congestion (based on data from Pfizer Pharmaceuticals).

a. Construct a 95% confidence interval estimate of the proportion of Viagra users who experience nasal congestion.

b. Construct a 95% confidence interval estimate of the proportion of placebo users who experience nasal congestion.

c. Construct a 95% confidence interval estimate of the difference between the two population proportions.

d. When attempting to determine whether there is a significant difference between the two population proportions, which of the following methods is best?

i. Determine whether the confidence intervals in parts (a) and (b) overlap.

ii. Determine whether the confidence interval in part (c) contains the value of zero.

iii. Conduct a hypothesis test of the null hypothesis $p_1 = p_2$, using a 0.05 significance level.

iv. The methods in parts (i), (ii), and (iii) are all equally good.

Cooperative Group Activities

1. *Out-of-class activity:* Are estimates influenced by anchoring numbers? Refer to the related Chapter 2 Cooperative Group Activity. In Chapter 2 we noted that, according to author John Rubin, when people must estimate a value, their estimate is often "anchored" to (or influenced by) a preceding number. In that Chapter 2 activity, some subjects were asked to quickly estimate the value of $8 \times 7 \times 6 \times 5 \times 4 \times 3 \times 2 \times 1$, and others were asked to quickly estimate the value of $1 \times 2 \times 3 \times 4 \times 5 \times 6 \times 7 \times 8$. In Chapter 2, we could compare the two sets of results by using statistics (such as the mean) and graphs (such as boxplots). The methods of Chapter 8 now allow us to compare the results with a formal hypothesis test. Specifically, collect your own sample data and test the claim that when we begin with larger numbers (as in $8 \times 7 \times 6$), our estimates tend to be larger.

2. *In-class activity:* Divide into groups according to gender, with about 10 or 12 students in each group. Each group member should record his or her pulse rate by counting the number of heartbeats in 1 minute, and the group statistics $(n, \bar{x}, s)$ should be calculated. The groups should test the null hypothesis of no difference between their mean pulse rate and the mean of the pulse rates for the population from which subjects of the same gender were selected for Data Set 1 in Appendix B.

3. *Out-of-class activity:* Randomly select a sample of male students and a sample of female students and ask each selected person whether they modify their diet because of health concerns. Use a formal hypothesis test to determine whether there is a gender gap on this issue. Also, keep a record of the responses according to the gender of the person asking the question. Does the response appear to be influenced by the gender of the interviewer?

Technology Project

STATDISK, Minitab, Excel, the TI-83/84 Plus calculator, and other technologies are capable of generating normally distributed data drawn from a population with a specified mean and standard deviation. Generate two sets of sample data that represent simulated IQ scores, as shown below.

IQ Scores of Treatment Group: Generate 10 sample values from a normally distributed population with mean 100 and standard deviation 15.

IQ Scores of Placebo Group: Generate 12 sample values from a normally distributed population with mean 100 and standard deviation 15.

Here are procedures for some common technologies:

STATDISK: Select **Data,** then select **Normal Generator.**

Minitab: Select **Calc, Random Data, Normal.**

Excel: Select **Tools, Data Analysis, Random Number Generator,** and be sure to select **Normal** for the distribution.

TI-83/84 Plus: Press **MATH,** select **PRB,** then select **6:randNorm(** and proceed to enter the mean, the standard deviation, and the number of scores (such as 100, 15, 10).

You can see from the way the data are generated that both data sets really come from the same population, so there should be no difference between the two sample means.

a. After generating the two data sets, use a 0.10 significance level to test the claim that the two samples come from populations with the same mean.

b. If this experiment is repeated many times, what is the expected percentage of trials leading to the conclusion that the two population means are different? How does this relate to a type I error?

c. If your generated data should lead to the conclusion that the two population means are different, would this conclusion be correct or incorrect in reality? How do you know?

d. If part (a) is repeated 20 times, what is the probability that none of the hypothesis tests leads to rejection of the null hypothesis?

e. Repeat part (a) 20 times. How often was the null hypothesis of equal means rejected? Is this the result you expected?

---

from DATA to DECISION

Critical Thinking: The fear of flying

The lives of many people are affected by a fear that prevents them from flying. Sports announcer John Madden gained notoriety as he crossed the country by rail or motor home, traveling from one football stadium to another. The Marist Institute for Public Opinion conducted a poll of 1014 adults, 48% of whom were men. The results were given in *USA Today* and they show that 12% of the men and 33% of the women fear flying.

Analyzing the Results

1. Is there sufficient evidence to conclude that there is a significant difference between the percentage of men and the percentage of women who fear flying?

2. Construct a 95% confidence interval estimate of the difference between the percentage of men and the percentage of women who fear flying. Do the confidence interval limits contain 0, and what is the significance of whether they do or do not?

3. Construct a 95% confidence interval for the percentage of men who fear flying.

4. Based on the result from Exercise 3, complete the following statement, which is typical of the statement that would be reported in a newspaper or magazine: "Based on the Marist Institute for Public Opinion poll, the percentage of men who fear flying is 22% with a margin of error of _____."

5. Examine the completed statement in Exercise 4. What important piece of information should be included, but is not included?

6. In a separate Gallup poll, 1001 randomly selected adults were asked this question: "If you had to fly on an airplane tomorrow, how would you describe your feelings about flying? Would you be—very afraid, somewhat afraid, not very afraid, or not afraid at all?" Here are the responses: Very afraid (18%), somewhat afraid (26%), not very afraid (17%), not afraid at all (38%), and no opinion (1%). Are these Gallup poll results consistent with those obtained by the survey conducted by the Marist Institute for Public Opinion? Explain. Can discrepancies be explained by the fact that the Gallup survey was conducted after the terrorist attacks of September 11, 2001, whereas the other survey was conducted before that date?

9

Correlation and Regression

Can you use a tape measure to weigh a bear?

Researchers have studied bears by anesthetizing them in order to obtain vital measurements, such as age, gender, length, and weight. (Do not try this at home, because checking the length or gender of a bear not fully anesthetized can be an unpleasant experience.) Because most bears are quite heavy and difficult to lift, researchers and hunters experience considerable difficulty actually weighing a bear in the wild. Can we determine the weight of a bear from other measurements that are easier to obtain?

See the data in Table 9-1 (based on data from Minitab and Gary Alt), which represent eight male bears from Data Set 6 in Appendix B. (For reasons of clarity, we will use this abbreviated data set, but better results could be obtained by using the more complete set of sample values.) Based on these data, does there appear to be an association between the length of a bear and its weight? If so, what is that association? If a researcher anesthetizes a bear and uses a tape measure to find that it is 71.0 in. long, how do we use that length to predict the bear's weight? These questions will be addressed in this chapter.

Table 9-1	Lengths and Weights of Male Bears							
x Length (in.)	53.0	67.5	72.0	72.0	73.5	68.5	73.0	37.0
y Weight (lb)	80	344	416	348	262	360	332	34

9-1 Overview

This chapter introduces important methods for making inferences based on sample data that come in *pairs*. Section 8-4 used matched pairs, but the inferences in Section 8-4 dealt with differences between two population means. This chapter has the objective of determining whether there is an *association* between the two variables and, if such an association exists, we want to describe it with an equation that can be used for predictions.

We begin in Section 9-2 by considering the concept of correlation, which is used to determine whether there is a statistically significant association between two variables. We investigate correlation using the scatterplot (a graph) and the linear correlation coefficient (a measure of the direction and strength of a linear association between two variables). In Section 9-3 we investigate regression analysis; we describe the association between two variables with an equation that relates them and show how to use that equation to predict values of one variable when we know values of the other variable.

In Section 9-4 we analyze the differences between predicted values and actual observed values of a variable. Sections 9-2 through 9-4 involve associations between two variables, but in Section 9-5 we use concepts of multiple regression to describe the association among three or more variables.

9-2 Correlation

The main objective of this section is to analyze a collection of paired sample data (sometimes called **bivariate data**) and determine whether there appears to be an association between the two variables. In statistics, we refer to such an association as a *correlation*. (We will consider only *linear* associations, which means that when graphed, the points approximate a straight-line pattern. Also, we consider only quantitative data.)

This section is partitioned into two parts: (1) Basic Concepts of Correlation; (2) Beyond the Basic Concepts of Correlation. The first part includes core concepts that should be understood well before moving on to the second part.

PART 1   Basic Concepts of Correlation

Definition

A **correlation** exists between two variables when one of them is related to the other in some way.

Table 9-1, for example, consists of paired length/weight data for a sample of eight bears. We will determine whether there is a correlation between the variable x (length) and the variable y (weight).

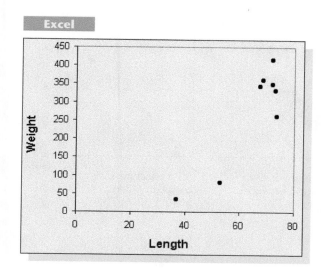

FIGURE 9-1 **Excel-Generated Scatterplot of Length/Weight Bear Data**

Exploring the Data

Before working with the more formal computational methods of this section, we should begin by exploring the data set to see what we can learn. We should always begin an investigation into the association between two variables by constructing a graph called a *scatterplot,* or *scatter diagram.*

Definition

A **scatterplot** (or **scatter diagram**) is a graph in which the paired (x, y) sample data are plotted with a horizontal x-axis and a vertical y-axis. Each individual (x, y) pair is plotted as a single point.

As an example, see Figure 9-1, which shows an Excel display of the 8 pairs of data listed in Table 9-1. When we examine such a scatterplot, we should study the overall pattern of the plotted points. If there is a pattern, we should note its direction. That is, as one variable increases, does the other seem to increase or decrease? We should observe whether there are any outliers, which are points that lie very far away from all of the other points. The Excel-generated scatterplot does appear to reveal a pattern showing that greater lengths appear to be associated with greater weights. This pattern suggests that there is an association between the length of a bear and its weight, but this conclusion is largely subjective because it is based on our perception of whether a pattern is present.

Other examples of scatterplots are shown in Figure 9-2. The graphs in Figure 9-2(a), (b), and (c) depict a pattern of increasing values of y that correspond to increasing values of x. As you proceed from Figure 9-2(a) to (c), the pattern of points becomes closer to a straight line, suggesting that the association between x and y becomes stronger. The scatterplots in Figure 9-2(d), (e), and (f) depict patterns in which the y-values decrease as the x-values increase. Again, as you proceed from Figure 9-2(d) to (f), the association becomes stronger. In contrast to the

Direct Link Between Smoking and Cancer

When we find a statistical correlation between two variables, we must be extremely careful to avoid the mistake of concluding that there is a cause-effect link. The tobacco industry has consistently emphasized that correlation does not imply causality. However, Dr. David Sidransky of Johns Hopkins University now says that "we have such strong molecular proof that we can take an individual cancer and potentially, based on the patterns of genetic change, determine whether cigarette smoking was the cause of that cancer." Based on his findings, he also said that "the smoker had a much higher incidence of the mutation, but the second thing that nailed it was the very distinct pattern of mutations . . . so we had the smoking gun." Although statistical methods cannot prove that smoking *causes* cancer, such proof can be established with physical evidence of the type described by Dr. Sidransky.

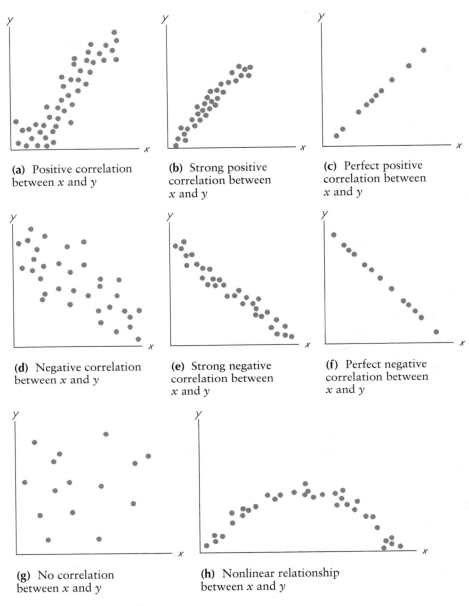

(a) Positive correlation between x and y

(b) Strong positive correlation between x and y

(c) Perfect positive correlation between x and y

(d) Negative correlation between x and y

(e) Strong negative correlation between x and y

(f) Perfect negative correlation between x and y

(g) No correlation between x and y

(h) Nonlinear relationship between x and y

FIGURE 9-2 Scatterplots

first six graphs, the scatterplot of Figure 9-2(g) shows no pattern and suggests that there is no correlation (or association) between x and y. Finally, the scatterplot of Figure 9-2(h) shows a pattern, but it is not a straight-line pattern.

Linear Correlation Coefficient

Because visual examinations of scatterplots are largely subjective, we need more objective measures. We use the linear correlation coefficient r, which is useful for detecting straight-line patterns.

Definition

The **linear correlation coefficient** r measures the strength of the linear association between the paired x- and y-quantitative values in a *sample*. Its value is computed by using Formula 9-1 or any of several other equivalent formulas. (Formula 9-1 is given in the accompanying box, and some equivalent formulas are given later in this section.) [The linear correlation coefficient is sometimes referred to as the **Pearson product moment correlation coefficient** in honor of Karl Pearson (1857–1936), who originally developed it.]

Because the linear correlation coefficient r is calculated using sample data, it is a sample statistic used to measure the strength of the linear correlation between x and y. If we had every pair of population values for x and y, the result of Formula 9-1 would be a population parameter, represented by ρ (Greek rho). The accompanying box includes requirements, notation, and Formula 9-1.

Requirements

Given any collection of sample paired data, the linear correlation coefficient r can always be computed, but the following requirements should be satisfied when testing hypotheses or making other inferences about r.

1. The sample of paired (x, y) data is a *random* sample of quantitative data. (It is important that the sample data have not been collected using some inappropriate method, such as using a voluntary response sample.)

2. Visual examination of the scatterplot must confirm that the points approximate a straight-line pattern.

3. Any outliers must be removed if they are known to be errors. The effects of any other outliers should be considered.

Note: Requirements 2 and 3 above are simplified attempts at checking this formal requirement:

The pairs of (x, y) data must have a **bivariate normal distribution.** (Normal distributions are discussed in Chapter 5, but this assumption basically requires that for any fixed value of x, the corresponding values of y have a distribution that is bell-shaped, and for any fixed value of y, the values of x have a distribution that is bell-shaped.) This assumption is usually difficult to check, so for now, we will use Requirements 2 and 3 as listed above.

Notation for the Linear Correlation Coefficient

n	represents the number of pairs of data present.
Σ	denotes the addition of the items indicated.
Σx	denotes the sum of all x-values.
Σx^2	indicates that each x-value should be squared and then those squares added.

continued

$(\Sigma x)^2$ indicates that the x-values should be added and the total then squared. It is extremely important to avoid confusing Σx^2 and $(\Sigma x)^2$.

Σxy indicates that each x-value should first be multiplied by its corresponding y-value. After obtaining all such products, find their sum.

r represents the linear correlation coefficient for a *sample*.

ρ represents the linear correlation coefficient for a *population*.

Formula 9-1
$$r = \frac{n\Sigma xy - (\Sigma x)(\Sigma y)}{\sqrt{n(\Sigma x^2) - (\Sigma x)^2}\sqrt{n(\Sigma y^2) - (\Sigma y)^2}}$$

Shortcut formula that simplifies manual calculations. The format of this formula makes it easy to use with a spreadsheet or computer program. See other equivalent formulas given later in this section, and also see a rationale for the calculation of r.

Interpreting r Using Table A-6: If the absolute value of the computed value of r exceeds the value in Table A-6, conclude that there is a significant linear correlation. Otherwise, there is not sufficient evidence to support the conclusion of a significant linear correlation.

Interpreting r Using Software: If the computed P-value is less than or equal to the significance level, conclude that there is a significant linear correlation. Otherwise, there is not sufficient evidence to support the conclusion of a significant linear correlation.

Rounding the Linear Correlation Coefficient

Round the linear correlation coefficient r to three decimal places (so that its value can be directly compared to critical values in Table A-6). When calculating r and other statistics in this chapter, rounding in the middle of a calculation often creates substantial errors, so try using your calculator's memory to store intermediate results and round only at the end. Many inexpensive calculators have Formula 9-1 built in so that you can automatically evaluate r after entering the sample data.

EXAMPLE **Calculating r** Using the data given below, find the value of the linear correlation coefficient r.

x	1	1	3	5
y	2	8	6	4

SOLUTION

REQUIREMENT ✓ The accompanying SPSS-generated scatterplot shows a pattern of points that does not appear to be a straight-line pattern. Although we can always compute the value of the linear correlation coefficient, it loses effectiveness in such cases where the pattern of points is not a straight-line pattern. However, in this case, we will proceed for the purpose of illustrating the calculations involved. ✓

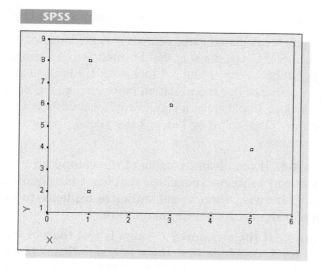

For the given sample of paired data, $n = 4$ because there are four pairs of data. The other components required in Formula 9-1 are found from the calculations in Table 9-2. Note how this vertical format makes the calculations easier.

Using the calculated values and Formula 9-1, we can now evaluate r as follows:

$$r = \frac{n\Sigma xy - (\Sigma x)(\Sigma y)}{\sqrt{n(\Sigma x^2) - (\Sigma x)^2}\sqrt{n(\Sigma y^2) - (\Sigma y)^2}}$$

$$= \frac{4(48) - (10)(20)}{\sqrt{4(36) - (10)^2}\sqrt{4(120) - (20)^2}}$$

$$= \frac{-8}{\sqrt{44}\sqrt{80}} = -0.135$$

These calculations get quite messy with large data sets, so it's fortunate that the linear correlation coefficient can be found automatically with many different calculators and computer programs.

Table 9-2	Finding Statistics Used to Calculate r			
x	y	$x \cdot y$	x^2	y^2
1	2	2	1	4
1	8	8	1	64
3	6	18	9	36
5	4	20	25	16
Total 10	20	48	36	120
↑ Σx	↑ Σy	↑ Σxy	↑ Σx^2	↑ Σy^2

Interpreting the Linear Correlation Coefficient

We need to interpret a calculated value of r, such as the value of -0.135 found in the preceding example. Given the way that Formula 9-1 is constructed, the value of r must always fall between -1 and $+1$ inclusive. If r is close to 0, we conclude that there is no significant linear correlation between x and y, but if r is close to -1 or $+1$ we conclude that there is a significant linear correlation between x and y. Interpretations of "close to" 0 or 1 or -1 are vague, so we use the following specific decision criteria:

> *Using Table A-6:* **If the absolute value of the computed value of r exceeds the value in Table A-6, conclude that there is a significant linear correlation. Otherwise, there is not sufficient evidence to support the conclusion of a significant linear correlation.**

> *Using Software:* **If the computed P-value is less than or equal to the significance level, conclude that there is a significant linear correlation. Otherwise, there is not sufficient evidence to support the conclusion of a significant linear correlation.**

When there really is no linear correlation between x and y, Table A-6 lists values that are "critical" in this sense: They separate *usual* values of r from those that are *unusual*. For example, Table A-6 shows us that with $n = 4$ pairs of sample data, the critical values are 0.950 (for $\alpha = 0.05$) and 0.999 (for $\alpha = 0.01$). Critical values and the role of α are carefully described in Chapters 6 and 7. Here's how we interpret those numbers: With four pairs of data and no linear correlation between x and y, there is a 5% chance that the absolute value of the computed linear correlation coefficient r will exceed 0.950. With $n = 4$ and no linear correlation, there is a 1% chance that $|r|$ will exceed 0.999.

EXAMPLE Lengths and Weights of Bears Given the sample data in Table 9-1, find the value of the linear correlation coefficient r, then refer to Table A-6 to determine whether there is a significant linear correlation between the lengths of bears and their weights. In Table A-6, use a significance level of $\alpha = 0.05$. (With $\alpha = 0.05$ we conclude that there is a significant linear correlation only if the sample is unlikely in this sense: If there is no linear correlation between the two variables, such a value of r occurs 5% of the time or less.)

SOLUTION

REQUIREMENT ✓ Assuming that the sample data have been randomly selected, we examine the Excel-generated scatterplot shown earlier in this section. The pattern of the points is roughly a straight-line pattern and there do not appear to be any outliers. Having completed this simplified check of requirements, we proceed with our analysis. ✓

Using the same procedure illustrated in the preceding example, or using technology, we can find that the 8 pairs of length/weight data in Table 9-1 result in $r = 0.897$. Here is the SPSS display:

SPSS

Correlations

		LENGTH	WEIGHT
LENGTH	Pearson Correlation	1	.897**
	Sig. (2-tailed)	.	.002
	N	8	8
WEIGHT	Pearson Correlation	.897**	1
	Sig. (2-tailed)	.002	.
	N	8	8

**. Correlation is significant at the 0.01 level (2-tailed).

Referring to Table A-6, we locate the row for which $n = 8$ (because there are 8 pairs of data). That row contains the critical values of 0.707 (for $\alpha = 0.05$) and 0.834 (for $\alpha = 0.01$). Using the critical value for $\alpha = 0.05$, we see that there is less than a 5% chance that with no linear correlation, the absolute value of the computed r will exceed 0.707. Because $r = 0.897$, its absolute value does exceed 0.707, so we conclude that there is a significant linear correlation between bear lengths and weights. (Also, SPSS displays the P-value of 0.002. Because the P-value of 0.002 is less than the significance level of 0.05, we again conclude that there is a significant linear correlation between bear lengths and weights.)

We have already noted that the format of Formula 9-1 requires that the calculated value of r always falls between -1 and $+1$ inclusive. We list that property along with other important properties.

Properties of the Linear Correlation Coefficient r

1. The value of r is always between -1 and $+1$ inclusive. That is,

$$-1 \le r \le +1$$

2. *The value of r does not change if all values of either variable are converted to a different scale.*

3. *The value of r is not affected by the choice of x or y.* Interchange all x- and y-values and the value of r will not change.

4. *r measures the strength of a linear association.* It is not designed to measure the strength of an association that is not linear.

Interpreting r: Explained Variation

If we conclude that there is a significant linear correlation between x and y, we can find a linear equation that expresses y in terms of x, and that equation can be used to predict values of y for given values of x. In Section 9-3 we will describe a procedure for finding such equations and show how to predict values of y when given values of x. But a predicted value of y will not necessarily be the exact result, because in addition to x, there are other factors affecting y, such as random variation and other characteristics not included in the study. In Section 9-4 we will present a rationale and more details about this important principle:

> **The value of r^2 is the proportion of the variation in y that is explained by the linear association between x and y.**

EXAMPLE **Lengths and Weights of Bears** Using the length/weight data in Table 9-1, we have found that the linear correlation coefficient is $r = 0.897$. What proportion of the variation in weights of bears can be explained by the linear association between the weights and lengths of bears?

SOLUTION   With $r = 0.897$, we get $r^2 = 0.805$.

INTERPRETATION We conclude that 0.805 (or about 81%) of the variation in weights of bears can be explained by the linear association between the weights of bears and the lengths of bears. This implies that about 19% of the variation in such bear weights can be explained by factors other than their lengths.

Common Errors Involving Correlation

We now identify three of the most common sources of errors made in interpreting results involving correlation:

1. *A common source of error involves concluding that correlation implies causality.* Correlation does not imply causality. Using the sample data in Table 9-1, we can conclude that there is a correlation between bear lengths and bear weights, but we cannot conclude that greater lengths *cause* more weight. The weights of bears may be affected by some other variable lurking in the background. (A **lurking variable** is one that affects the variables being studied, but is not included in the study.) For example, the abundance of the food supply may well affect the weights of bears. The abundance of the food supply can then be a lurking variable.

2. *Another source of error arises with data based on averages.* Averages suppress individual variation and may inflate the correlation coefficient. One study produced a 0.4 linear correlation coefficient for paired data relating income and education among individuals, but the linear correlation coefficient became 0.7 when regional averages were used.

3. *A third source of error involves the property of linearity.* An association may exist between x and y even when there is no significant linear correlation. The data depicted in Figure 9-3 result in a value of $r = 0$, which is an indication of no *linear* correlation between the two variables. However, we can easily see from the figure that there is a pattern reflecting a very strong *nonlinear* association. (Figure 9-3 is a scatterplot that depicts the association between distance above ground and time elapsed for an object thrown upward.)

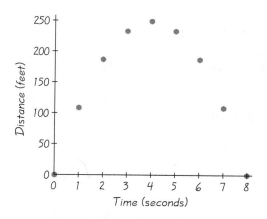

FIGURE 9-3
**Correlation That Is
Nonlinear**

PART 2   Beyond the Basic Concepts of Correlation

Formal Hypothesis Test
(Requires Coverage of Chapter 7)

We present two methods (summarized in the accompanying box and in Figure 9-4) for using a formal hypothesis test to determine whether there is a significant linear correlation between two variables. Some instructors prefer Method 1 because it reinforces concepts introduced in earlier chapters. Others prefer Method 2 because it involves easier calculations. Method 1 uses the Student t distribution with a test statistic having the form $t = r/s_r$, where s_r denotes the sample standard deviation of r values. The test statistic given in the box (for Method 1) reflects the fact that the standard deviation of r values can be expressed as $\sqrt{(1 - r^2)/(n - 2)}$.

Figure 9-4 shows that the decision criterion is to reject the null hypothesis of $\rho = 0$ if the absolute value of the test statistic exceeds the critical values; rejection of $\rho = 0$ means that there is sufficient evidence to support a claim of a linear correlation between the two variables. If the absolute value of the test statistic does not exceed the critical values, then we fail to reject $\rho = 0$; that is, there is not sufficient evidence to conclude that there is a linear correlation between the two variables.

Hypothesis Test for Correlation (See Figure 9-4.)

H_0: $\rho = 0$ (There is no significant linear correlation.)
H_1: $\rho \neq 0$ (There is a significant linear correlation.)

Method 1: Test Statistic Is t

Test statistic: $t = \dfrac{r}{\sqrt{\dfrac{1 - r^2}{n - 2}}}$

Critical values: Use Table A-3 with $n - 2$ degrees of freedom.

continued

P-value: Use Table A-3 with $n - 2$ degrees of freedom.
Conclusion:
- If $|t| >$ critical value from Table A-3, reject H_0 and conclude that there is a significant linear correlation.
- If $|t| \leq$ critical value, fail to reject H_0; there is not sufficient evidence to conclude that there is a significant linear correlation.

Method 2: Test Statistic Is *r*

Test statistic: *r*
Critical values: Refer to Table A-6.
Conclusion:
- If $|r| >$ critical value from Table A-6, reject H_0 and conclude that there is a significant linear correlation.
- If $|r| \leq$ critical value, fail to reject H_0; there is not sufficient evidence to conclude that there is a significant linear correlation.

EXAMPLE **Lengths and Weights of Bears** Using the sample data in Table 9-1, test the claim that there is a linear correlation between the lengths of bears and the weights of bears. For the test statistic, use both (a) Method 1 and (b) Method 2.

SOLUTION

REQUIREMENT ✓ A preceding example already includes verification that the requirements are satisfied. (Assuming that the sample data have been randomly selected, we examine the Excel-generated scatterplot shown earlier in this section. The pattern of the points is roughly a straight-line pattern and there do not appear to be any outliers.) Having completed this simplified check of requirements, we proceed with our analysis. ✓

Refer to Figure 9-4. To claim that there is a significant linear correlation is to claim that the population linear correlation coefficient *r* is different from 0. We therefore have the following hypotheses.

$$H_0: \rho = 0 \quad \text{(There is no significant linear correlation.)}$$
$$H_1: \rho \neq 0 \quad \text{(There is a significant linear correlation.)}$$

No significance level was specified, so use $\alpha = 0.05$.

In a preceding example we already found that $r = 0.897$. With that value, we now find the test statistic and critical value, using each of the two methods just described.

a. *Method 1:* The test statistic is

$$t = \frac{r}{\sqrt{\dfrac{1 - r^2}{n - 2}}} = \frac{0.897}{\sqrt{\dfrac{1 - 0.897^2}{8 - 2}}} = 4.971$$

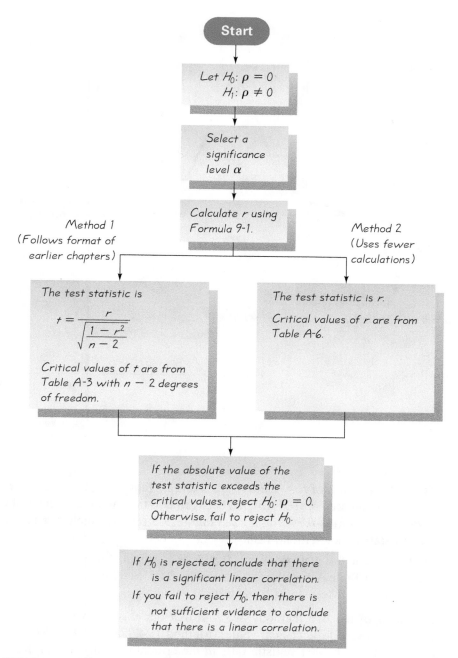

FIGURE 9-4 **Hypothesis Test for a Linear Correlation**

FIGURE 9-5 **Testing**
$H_0: \rho = 0$ **with Method 1**

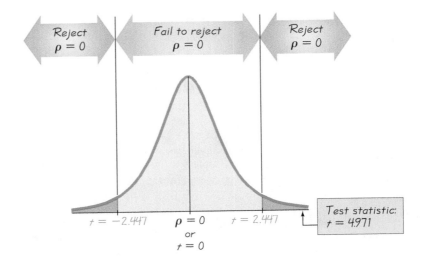

FIGURE 9-6 **Testing**
$H_0: \rho = 0$ **with Method 2**

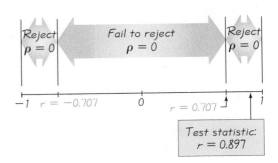

The critical values of $t = \pm 2.447$ are found in Table A-3, where 2.447 corresponds to an area of 0.05 divided between two tails and the number of degrees of freedom is $n - 2 = 6$. See Figure 9-5 for the graph that includes the test statistic and critical values.

b. *Method 2:* The test statistic is $r = 0.897$. The critical values of $r = \pm 0.707$ are found in Table A-6 with $n = 8$ and $\alpha = 0.05$. See Figure 9-6 for a graph that includes this test statistic and critical values.

Using either of the two methods, we find that the absolute value of the test statistic does exceed the critical value. (Method 1: $4.971 > 2.447$; Method 2: $0.897 > 0.707$.) The test statistic falls in the critical region. We therefore reject $H_0: \rho = 0$. There is sufficient evidence to support the claim of a linear correlation between the lengths of bears and their weights.

One-tailed Tests: The preceding example and Figures 9-5 and 9-6 illustrate a two-tailed hypothesis test. The examples and exercises in this section will generally involve two-tailed tests, but one-tailed tests can occur with a claim of a positive linear correlation or a claim of a negative linear correlation. In such cases, the hypotheses will be as shown here.

Claim of *Negative* Correlation (Left-tailed test)	Claim of *Positive* Correlation (Right-tailed test)
$H_0: \rho = 0$ $H_1: \rho < 0$	$H_0: \rho = 0$ $H_1: \rho > 0$

For these one-tailed tests, Method 1 can be handled as in earlier chapters. For Method 2, either calculate the critical value as described in Exercise 28 or modify Table A-6 by replacing the column headings of $\alpha = 0.05$ and $\alpha = 0.01$ by the one-sided critical values of $\alpha = 0.025$ and $\alpha = 0.005$, respectively.

Rationale: We have presented Formula 9-1 for calculating r and have illustrated its use. Formula 9-1 is given below along with some other formulas that are "equivalent" in the sense that they all produce the same values. Different textbook authors prefer different expressions for different reasons, and the formulas given below are the most commonly used expressions for r. These formulas are simply different versions of the same expression. (Algebra enthusiasts are welcome to have fun proving the equivalence of the formulas; others can trust the authors that the various formulas for r will produce the same results.) The format of Formula 9-1 simplifies manual calculations, it simplifies calculations using spreadsheets, and it is easy to include as part of an original computer program.

$$r = \frac{n\Sigma xy - (\Sigma x)(\Sigma y)}{\sqrt{n(\Sigma x^2) - (\Sigma x)^2}\sqrt{n(\Sigma y^2) - (\Sigma y)^2}} \qquad r = \frac{\Sigma\left[\dfrac{(x - \bar{x})}{s_x}\dfrac{(y - \bar{y})}{s_y}\right]}{n - 1}$$

$$r = \frac{\Sigma(x - \bar{x})(y - \bar{y})}{(n - 1)s_x s_y} \qquad r = \frac{s_{xy}}{\sqrt{s_{xx}}\sqrt{s_{yy}}}$$

Although Formula 9-1 simplifies manual calculations, other formulas for r are better for helping us to *understand* how r works. In attempting to understand the reasoning that underlies the development of the linear correlation coefficient, we will use this version of the linear correlation coefficient:

$$r = \frac{\Sigma\left[\dfrac{(x - \bar{x})}{s_x}\dfrac{(y - \bar{y})}{s_y}\right]}{n - 1}$$

We will consider the following paired data, which are depicted in the scatterplot shown in Figure 9-7.

x	1	1	2	4	7
y	4	5	8	15	23

Figure 9-7 includes the point $(\bar{x}, \bar{y}) = (3, 11)$, which is called the *centroid* of the sample points.

Definition

Given a collection of paired (x, y) data, the point $(\bar{x}, \bar{y})$ is called the **centroid.**

FIGURE 9-7 **Scatterplot Partitioned into Quadrants**

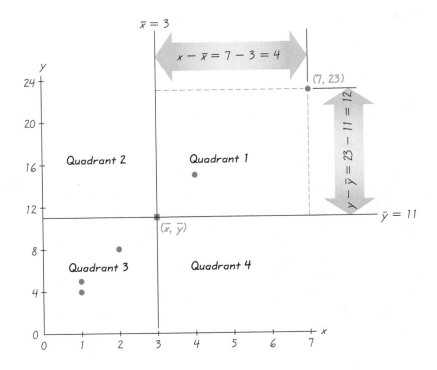

The statistic *r*, sometimes called the *Pearson product moment*, was first developed by Karl Pearson. It is based on the sum of the products $(x - \bar{x})(y - \bar{y})$. The statistic $\sum(x - \bar{x})(y - \bar{y})$ is the basic expression forming the foundation for *r*, and we now explain why that statistic is key.

In any scatterplot, vertical and horizontal lines through the centroid $(\bar{x}, \bar{y})$ divide the diagram into four quadrants, as in Figure 9-7. If the points of the scatterplot tend to approximate an uphill line (as in the figure), individual values of the product $(x - \bar{x})(y - \bar{y})$ tend to be positive because most of the points are found in the first and third quadrants, where the products of $(x - \bar{x})$ and $(y - \bar{y})$ are positive. If the points of the scatterplot approximate a downhill line, most of the points are in the second and fourth quadrants, where $(x - \bar{x})$ and $(y - \bar{y})$ are opposite in sign, so $\sum(x - \bar{x})(y - \bar{y})$ is negative. Points that follow no linear pattern tend to be scattered among the four quadrants, so the value of $\sum(x - \bar{x})(y - \bar{y})$ tends to be close to 0. We can therefore use $\sum(x - \bar{x})(y - \bar{y})$ as a measure of how the points are arranged. A large positive sum suggests that the points are predominantly in the first and third quadrants (corresponding to a positive linear correlation), a large negative sum suggests that the points are predominantly in the second and fourth quadrants (corresponding to a negative linear correlation), and a sum near zero suggests that the points are scattered among the four quadrants (with no linear correlation).

Unfortunately, the sum $\sum(x - \bar{x})(y - \bar{y})$ depends on the magnitude of the numbers used. For example, if you change *x* from inches to feet, that sum will change. To make *r* independent of the particular scale used, we include standardization. In Section 2-6 we saw that we could "standardize" values by converting them to z scores, as in $z = (x - \bar{x})/s$. (Here we use s_x to denote the standard deviation of the sample x values, and we use s_y to denote the standard deviation of the sample y values.) We use a similar technique as we standardize each deviation $(x - \bar{x})$ by dividing it by s_x. We also make the deviations $(y - \bar{y})$ independent of the magnitudes of the numbers by dividing by s_y. We now have the statistic

$$\sum \left[\frac{(x - \bar{x})}{s_x} \frac{(y - \bar{y})}{s_y} \right]$$

which we further modify by introducing the divisor of $n - 1$, which gives us a type of average instead of a sum that grows simply because we have more data. (The reasons for dividing by $n - 1$ instead of n are essentially the same reasons that relate to the standard deviation. See Section 2-5.) The end result is the expression

$$r = \frac{\sum \left[\frac{(x - \bar{x})}{s_x} \frac{(y - \bar{y})}{s_y} \right]}{n - 1}$$

This expression can be algebraically manipulated into the equivalent form of Formula 9-1 or any of the other expressions for r.

Confidence Intervals   In preceding chapters we discussed methods of inferential statistics by addressing methods of hypothesis testing and methods for constructing confidence interval estimates. A similar procedure may be used to find confidence intervals for ρ. However, because the construction of confidence intervals involves somewhat complicated transformations, that process is presented in Exercise 30.

We can use the linear correlation coefficient to determine whether there is a linear association between two variables. Using the data in Table 9-1, we have concluded that there is a linear correlation between the lengths of bears and the weights of bears. Having concluded that an association exists, we would like to determine what that association is so that we can predict the weight of a bear given its length. This next important stage of analysis is addressed in the following section.

9-2 Exercises

In Exercises 1–4, use a significance level of $\alpha = 0.05$.

1. Chest Sizes and Weights of Bears When eight bears were anesthetized, researchers measured the distances (in inches) around the bears' chests and weighed the bears (in pounds). SPSS was used to find that the value of the linear correlation coefficient is $r = 0.993$.
   a. Is there a significant linear correlation between chest size and weight? Explain.
   b. What proportion of the variation in weight can be explained by the linear association between weight and chest size?

2. Guns and Murder Rate Using data collected from the FBI and the Bureau of Alcohol, Tobacco, and Firearms, the number of registered automatic weapons and the murder rate (in murders per 100,000 people) was obtained for each of eight randomly selected states. STATDISK was used to find that the value of the linear correlation coefficient is $r = 0.885$.

  a. Is there a significant linear correlation between the number of registered automatic weapons and the murder rate? Explain.

  b. What proportion of the variation in the murder rate can be explained by the linear association between the murder rate and the number of registered automatic weapons?

3. BMI and Weights of Females Data Set 1 in Appendix B includes data from a national Health Exam of randomly selected subjects. Using the paired BMI/weight data for the females, Excel was used to find that the value of the linear correlation coefficient is $r = 0.936$.

  a. For females, is there a significant linear correlation between BMI and weight? Explain.

  b. What proportion of the variation in weight can be explained by the variation in BMI?

4. Pulse Rates and Weights of Females Data Set 1 in Appendix B includes data from a national Health Exam of randomly selected subjects. Using the paired pulse/weight data for the females, a TI-83/84 Plus calculator was used to find that the value of the linear correlation coefficient is $r = 0.143$.

  a. For females, is there a significant linear correlation between pulse rates and weights? Explain.

  b. What proportion of the variation in weight can be explained by the variation in pulse rate?

Testing for a Linear Correlation. *In Exercises 5 and 6, use a scatterplot and the linear correlation coefficient r to determine whether there is a correlation between the two variables.*

5.

x	0	1	2	3	4
y	4	1	0	1	4

6.

x	1	2	2	5	6
y	2	5	4	15	15

7. Effects of an Outlier Refer to the accompanying Minitab-generated scatterplot.

  a. Examine the pattern of all 10 points and subjectively determine whether there appears to be a correlation between x and y.

  b. After identifying the 10 pairs of coordinates corresponding to the 10 points, find the value of the correlation coefficient r and determine whether there is a significant linear correlation.

  c. Now remove the point with coordinates (10, 10) and repeat parts (a) and (b).

  d. What do you conclude about the possible effect from a single pair of values?

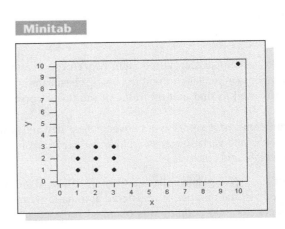

Testing for a Linear Correlation. *In Exercises 8–15, construct a scatterplot, find the value of the linear correlation coefficient r and use a significance level of α = 0.05 to determine whether there is a significant linear correlation between the two variables. Save your work because the same data sets will be used in the next section.*

8. Heights of Fathers and Sons Listed below are the heights of fathers and sons from the first seven cases listed in Data Set 3 in Appendix B. Using only the data listed below, is there a correlation? Do the data suggest that taller fathers tend to have taller sons?

Height of father	70	69	64	71	68	66	74
Height of son	62.5	64.6	69.1	73.9	67.1	64.4	71.1

9. Supermodel Heights and Weights Listed below are heights (in inches) and weights (in pounds) for supermodels Niki Taylor, Nadia Avermann, Claudia Schiffer, Elle MacPherson, Christy Turlington, Bridget Hall, Kate Moss, Valerie Mazza, and Kristy Hume. Is there a correlation between height and weight? If there is a correlation, does it mean that there is a correlation between height and weight of all adult women?

Height (in.)	71	70.5	71	72	70	70	66.5	70	71
Weight (lb)	125	119	128	128	119	127	105	123	115

10. Systolic and Diastolic Blood Pressure Data Set 1 in Appendix B includes data from a national Health Exam of randomly selected subjects. The paired systolic/diastolic blood pressure measurements for the first eight males are listed below. Using only the data listed below, is there a correlation? If a male has a higher systolic level, is he more likely to have a higher diastolic level?

Systolic	125	107	126	110	110	107	113	126
Diastolic	78	54	81	68	66	83	71	72

11. Blood Pressure Measurements Fourteen different second-year medical students took blood pressure measurements of the same patient and the results are listed below. Is there a correlation between systolic and diastolic values? Apart from correlation, is there some other method that might be used to address an important issue suggested by the data?

Systolic	138	130	135	140	120	125	120	130	130	144	143	140	130	150
Diastolic	82	91	100	100	80	90	80	80	80	98	105	85	70	100

12. Temperatures and Marathons In "The Effects of Temperature on Marathon Runner's Performance" by David Martin and John Buoncristiani (*Chance*, Vol. 12, No. 4), high temperatures and times (in minutes) were given for women who won the New York City marathon in recent years. Results are listed below. Is there a correlation between temperature and winning time? Does it appear that winning times are affected by temperature?

x (temperature)	55	61	49	62	70	73	51	57
y (time)	145.283	148.717	148.300	148.100	147.617	146.400	144.667	147.533

13. Smoking and Nicotine When nicotine is absorbed by the body, cotinine is produced. A measurement of cotinine in the body is therefore a good indicator of how much a person smokes. Listed below are the reported numbers of cigarettes smoked per day and the measured amounts of cotinine (in ng/ml). (The values are from randomly selected subjects in the National Health Survey.) Is there a significant linear correlation? Explain the result.

x (cigarettes per day)	60	10	4	15	10	1	20	8	7	10	10	20
y (cotinine)	179	283	75.6	174	209	9.51	350	1.85	43.4	25.1	408	344

14. Tree Circumference and Height Listed below are the circumferences (in feet) and the heights (in feet) of trees in Marshall, Minnesota (based on data from "Tree Measurements" by Stanley Rice, *American Biology Teacher*, Vol. 61, No. 9). Is there a correlation? Why should there be a correlation?

x (circ.)	1.8	1.9	1.8	2.4	5.1	3.1	5.5	5.1	8.3	13.7	5.3	4.9	3.7	3.8
y (ht)	21.0	33.5	24.6	40.7	73.2	24.9	40.4	45.3	53.5	93.8	64.0	62.7	47.2	44.3

15. Testing Duragesic for Pain Relief Duragesic, manufactured by ALZA Corporation, is a drug used for treating chronic pain. An advertisement for this drug included a scatterplot of (x, y) points, where each point corresponds to one study subject, x is the "pain intensity on pre-study analgesic program" and y is the "pain intensity after 1 month on Duragesic." The pain measurements are visual analog scores that are often used to measure pain. The values listed below are based on the published scatterplot, and the y values correspond to the respective x values. Is there a significant linear correlation? Would a significant linear correlation suggest that the drug is effective in treating pain?

x: 1.2 1.3 1.5 1.6 8.0 3.4 3.5 2.8 2.6 2.2 3.0 7.1 2.3 2.1 3.4 6.4 5.0 4.2 2.8
   3.9 5.2 6.9 6.9 5.0 5.5 6.0 5.5 8.6 9.4 10.0 7.6

y: 0.4 1.4 1.8 2.9 6.0 1.4 0.7 3.9 0.9 1.8 0.9 9.3 8.0 6.8 2.3 0.4 0.7 1.2 4.5
   2.0 1.6 2.0 2.0 6.8 6.6 4.1 4.6 2.9 5.4 4.8 4.1

Testing for a Linear Correlation. *In Exercises 16–21, use the data from Appendix B to construct a scatterplot, find the value of the linear correlation coefficient r, and use a significance level of α = 0.05 to determine whether there is a significant linear correlation between the two variables. Save your work because the same data sets will be used in the next section.*

16. Cholesterol and Body Mass Index Refer to Data Set 1 in Appendix B and use the cholesterol levels and body mass index values of the 40 women. Is there a correlation between cholesterol level and body mass index?

17. Cholesterol and Weight Refer to Data Set 1 in Appendix B and use the cholesterol levels and weights of the 40 women. Is there a correlation between cholesterol level and weight?

18. Chest Sizes and Weights of Bears Refer to Data Set 6 in Appendix B and use the chest sizes and weights of all listed bears. Is there a correlation between chest size and weight?

19. Heights of Mothers and Daughters Refer to Data Set 3 in Appendix B and use the heights of the mothers and their female children. Is there a correlation between heights of mothers and daughters?

20. Iris Petal Lengths and Widths Refer to Data Set 7 in Appendix B and use only the petal lengths and petal widths from irises in the Setosa class. Is there a correlation?

21. Iris Petal Lengths and Widths Refer to Data Set 7 in Appendix B and use only the petal lengths and petal widths from irises in the Versicolor class. Is there a correlation? Are the results approximately the same as those obtained in Exercise 20?

Identifying Correlation Errors. In Exercises 22–25, describe the error in the stated conclusion. (See the list of common sources of errors included in this section.)

22. *Given:* The paired sample data of the ages of subjects and their scores on a test of reasoning result in a linear correlation coefficient very close to 0.
    Conclusion: Younger people tend to get higher scores.

23. *Given:* There is a significant linear correlation between personal income and years of education.
    Conclusion: More education causes a person's income to rise.

24. *Given:* Subjects take a test of verbal skills and a test of manual dexterity, and those pairs of scores result in a linear correlation coefficient very close to 0.
    Conclusion: Scores on the two tests are not related in any way.

25. *Given:* There is a significant linear correlation between state average tax burdens and state average incomes.
    Conclusion: There is a significant linear correlation between individual tax burdens and individual incomes.

*26. Using Data from Scatterplot Sometimes, instead of having numerical data, we have only graphical data. The accompanying Excel scatterplot is similar to one that was included in "The Prevalence of Nosocomial Infection in Intensive Care Units in Europe," by Jean-Louis Vincent et al., *Journal of the American Medical Association,* Vol. 274, No. 8. Each point represents a different European country. Estimate the value of the linear correlation coefficient, and determine whether there is a significant linear correlation between the mortality rate and the rate of infections acquired in intensive care units.

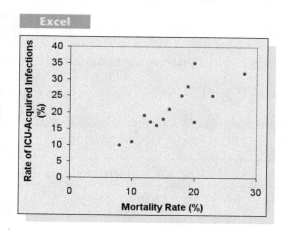

*27. Correlations with Transformed Data In addition to testing for a linear correlation between x and y, we can often use *transformations* of data to explore for other associations. For example, we might replace each x value by x^2 and use the methods of this section to determine whether there is a linear correlation between y and x^2. Given the paired data in the accompanying table, construct the scatterplot and then test for a linear correlation between y and each of the following. Which case results in the largest value of r?

a. x **b.** x^2 **c.** $\log x$ **d.** $\sqrt{x}$ **e.** $1/x$

x	1.3	2.4	2.6	2.8	2.4	3.0	4.1
y	0.11	0.38	0.41	0.45	0.39	0.48	0.61

*28. Finding Critical r-Values The critical values of r in Table A-6 are found by solving

$$t = \frac{r}{\sqrt{\dfrac{1 - r^2}{n - 2}}}$$

for r to get

$$r = \frac{t}{\sqrt{t^2 + n - 2}}$$

where the t value is found from Table A-3 by assuming a two-tailed case with $n - 2$ degrees of freedom. Table A-6 lists the results for selected values of n and α. Use the formula for r given here and Table A-3 (with $n - 2$ degrees of freedom) to find the critical values of r for the given cases.

a. $H_1: \rho \neq 0, n = 50, \alpha = 0.05$
b. $H_1: \rho \neq 0, n = 75, \alpha = 0.10$
c. $H_1: \rho < 0, n = 20, \alpha = 0.05$
d. $H_1: \rho > 0, n = 10, \alpha = 0.05$
e. $H_1: \rho > 0, n = 12, \alpha = 0.01$

*29. Including Categorical Data in a Scatterplot It sometimes becomes important to include categorical data in a scatterplot. Consider the sample data listed below, where weight is in pounds and the "remote" values consist of the number of times the subject used the television remote control during a period of one hour. Minitab was used to generate the scatterplot with the characters F (for females) and M (for males) used to identify gender.

a. Before doing any calculations, examine the Minitab-generated scatterplot. What do you conclude about the correlation between weight and remote control use?
b. Using all 16 pairs of data, is there a correlation between weight and use of the remote?
c. Using only the eight females, is there a correlation between weight and use of the remote?
d. Using only the eight males, is there a correlation between weight and use of the remote?
e. Based on the preceding results, what do you conclude?

Sex	F	F	F	F	F	F	F	F	M	M	M	M	M	M	M	M
Weight	120	126	129	130	131	132	134	140	160	166	168	170	172	174	176	180
Remote	5	3	6	4	2	7	4	3	23	20	16	24	18	21	17	22

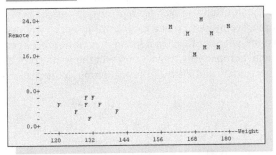

*30. Constructing Confidence Intervals for ρ *Given n pairs of data from which the linear correlation coefficient r can be found, use the following procedure to construct a confidence interval about the population parameter* ρ.

Step a. Use Table A-2 to find $z_{\alpha/2}$ that corresponds to the desired degree of confidence.
Step b. Evaluate the interval limits w_L and w_R:

$$w_L = \frac{1}{2}\ln\left(\frac{1+r}{1-r}\right) - z_{\alpha/2} \cdot \frac{1}{\sqrt{n-3}}$$

$$w_R = \frac{1}{2}\ln\left(\frac{1+r}{1-r}\right) + z_{\alpha/2} \cdot \frac{1}{\sqrt{n-3}}$$

Step c. Now evaluate the confidence interval limits in the expression below.

$$\frac{e^{2w_L}-1}{e^{2w_L}+1} < \rho < \frac{e^{2w_R}-1}{e^{2w_R}+1}$$

Use this procedure to construct a 95% confidence interval for ρ, given 50 pairs of data for which $r = 0.600$.

9-3 Regression

In Section 9-2 we analyzed paired data with the goal of determining whether there is a significant linear correlation between two variables. The main objective of this section is to describe the association between two variables by finding the graph and equation of the straight line that represents the association. This straight line is called the *regression line,* and its equation is called the *regression equation.* Sir Francis Galton (1822–1911) studied the phenomenon of heredity and showed that when tall or short couples have children, the heights of those children tend to *regress,* or revert to the more typical mean height for people of the same gender. We continue to use Galton's "regression" terminology, even though our data do not involve the same height phenomena studied by Galton.

As in Section 9-2, this section is partitioned into two parts: (1) Basic Concepts of Regression; (2) Beyond the Basics of Regression. The first part includes core concepts that should be understood well before moving on to the second part.

PART 1   Basic Concepts of Regression

The accompanying box includes the definition of regression equation and regression line, as well as the requirement, notation, and formulas we are using. The regression equation expresses an association between x (called the **independent variable,** or **predictor variable,** or **explanatory variable**) and $\hat{y}$ (called the **dependent variable,** or **response variable**). The typical equation of a straight line $y = mx + b$ is expressed in the form $\hat{y} = b_0 + b_1 x$, where b_0 is the y-intercept and b_1 is the slope. The given notation shows that b_0 and b_1 are sample statistics used to estimate the population parameters β_0 and β_1. We will use paired sample data to estimate the regression equation. Using only sample data, we can't find the exact values of the population parameters β_0 and β_1, but we can use the sample data to estimate them with b_0 and b_1, which are found by using Formulas 9-2 and 9-3.

Cell Phones and Crashes

Because some countries have banded the use of cell phones in cars while other countries are considering such a ban, researchers studied the issue of whether the use of cell phones while driving increases the chance of a crash. A sample of 699 drivers was obtained. Members of the sample group used cell phones and were involved in crashes. Subjects completed questionnaires and their telephone records were checked. Telephone usage was compared to the time interval immediately preceding a crash to a comparable time period the day before. Conclusion: Use of a cell phone was associated with a crash risk that was about four times as high as the risk when a cell phone was not used. (See "Association between Cellular-Telephone Calls and Motor Vehicle Collisions," by Redelmeier and Tibshirani, *New England Journal of Medicine*, Vol. 336, No. 7.)

Requirements

Given any collection of sample paired data, the equation of the regression line can always be found, but the following requirements should be satisfied when making inferences about the regression line:

1. The sample of paired (x, y) data is a *random* sample of quantitative data. (It is important that the sample data have not been collected using some inappropriate method, such as using a voluntary response sample.)

2. Visual examination of the scatterplot shows that the points approximate a straight-line pattern.

3. Any outliers must be removed if they are known to be errors. Consider the effects of any outliers that are not known errors.

Note: Requirements 2 and 3 above are simplified attempts at checking these formal requirements for regression analysis:

- For each fixed value of x, the corresponding values of y have a distribution that is bell-shaped.
- For the different fixed values of x, the distributions of the corresponding y-values all have the same variance. (This is violated if part of the scatterplot shows points very close to the regression line while another portion of the scatterplot shows points that are much farther away from the regression line. See the discussion of residual plots near the end of this section.)
- For the different fixed values of x, the distributions of the corresponding y-values have means that lie along the same straight line.
- The y-values are independent.

Results are not seriously affected if departures from normal distributions and equal variances are not too extreme.

Definitions

Given a collection of paired sample data, the **regression equation**

$$\hat{y} = b_0 + b_1 x$$

algebraically describes the relationship between the two variables. The graph of the regression equation is called the **regression line** (or *line of best fit*, or *least-squares line*).

Notation for Regression Equation

	Population Parameter	Sample Statistic
y-intercept of regression equation	β_0	b_0
Slope of regression equation	β_1	b_1
Equation of the regression line	$y = \beta_0 + \beta_1 x$	$\hat{y} = b_0 + b_1 x$

Finding the slope b_1 and y-intercept b_0 in the regression equation $\hat{y} = b_0 + b_1 x$

Formula 9-2   **Slope:** $b_1 = \dfrac{n(\Sigma xy) - (\Sigma x)(\Sigma y)}{n(\Sigma x^2) - (\Sigma x)^2}$

Formula 9-3   **y-intercept:** $b_0 = \bar{y} - b_1 \bar{x}$

The y-intercept b_0 can also be found using the formula shown below, but it is much easier to use Formula 9-3 instead.

$$b_0 = \frac{(\Sigma y)(\Sigma x^2) - (\Sigma x)(\Sigma xy)}{n(\Sigma x^2) - (\Sigma x)^2}$$

Formulas 9-2 and 9-3 might look intimidating, but they are programmed into many calculators and computer programs, so the values of b_0 and b_1 can be easily found.

Once we have evaluated b_1 and b_0, we can identify the estimated regression equation, which has the following special property: *The regression line fits the sample points best.* (The specific criterion used to determine which line fits "best" is the least-squares property, which will be described later.) We will now briefly discuss rounding and then illustrate the procedure for finding and applying the regression equation.

Rounding the Slope b_1 and the y-Intercept b_0

It's difficult to provide a simple universal rule for rounding values of b_1 and b_0, but we usually try to round each of these values to *three significant digits* or use the values provided by computer software or a TI-83/84 Plus calculator. Because these values are very sensitive to rounding at intermediate steps of calculations, try to carry at least six significant digits (or use exact values) in the intermediate steps. Depending on how you round, this book's answers to examples and exercises may be slightly different from your answers.

EXAMPLE **Finding the Regression Equation** In Section 9-2 we used the values listed below to find that the linear correlation coefficient of $r = -0.135$ indicates that there is not a significant linear correlation between x and y. Use the given sample data to find the regression equation.

x	1	1	3	5
y	2	8	6	4

SOLUTION

REQUIREMENT ✓   The scatterplot of these points shows a pattern that does not appear to be a straight-line pattern. It appears that the requirements are not all satisfied. However, for the purposes of illustrating the calculations, we will proceed to find the equation of the regression line. ✓

We will find the regression equation by using Formulas 9-2 and 9-3 and these values already found in Table 9-2 in Section 9-2:

$$n = 4 \qquad \Sigma x = 10 \qquad \Sigma y = 20$$
$$\Sigma x^2 = 36 \qquad \Sigma y^2 = 120 \qquad \Sigma xy = 48$$

First find the slope b_1 by using Formula 9-2:

$$b_1 = \frac{n(\Sigma xy) - (\Sigma x)(\Sigma y)}{n(\Sigma x^2) - (\Sigma x)^2}$$

$$= \frac{4(48) - (10)(20)}{4(36) - (10)^2} = \frac{-8}{44} = -0.181818 = -0.182$$

continued

Next, find the y-intercept b_0 by using Formula 9-3 (with $\bar{y} = 20/4 = 5$ and $\bar{x} = 10/4 = 2.5$):

$$b_0 = \bar{y} - b_1\bar{x}$$
$$= 5 - (-0.181818)(2.5) = 5.45$$

Knowing the slope b_1 and y-intercept b_0, we can now express the estimated equation of the regression line as

$$\hat{y} = 5.45 - 0.182x$$

We should realize that this equation is an *estimate* of the true regression equation $y = \beta_0 + \beta_1 x$. This estimate is based on one particular set of sample data, but another sample drawn from the same population would probably lead to a somewhat different equation. We should also realize that because the scatterplot shows a pattern which does not appear to be a straight-line pattern, the resulting regression equation should not be used for inferences or for making predictions.

EXAMPLE **Lengths and Weights of Bears** Using the length/weight data in Table 9-1, we have found that the linear correlation coefficient is $r = 0.897$. Using the same sample data, find the equation of the regression line.

SOLUTION

REQUIREMENT ✓ The scatterplot of these points shows a pattern that is approximately a straight-line pattern. There are no outliers, and we assume that the sample data have been randomly selected. We now proceed to find the equation of the regression line. ✓

Using the same procedure illustrated in the preceding example, or using technology, we can find that the 8 pairs of length/weight data in Table 9-1 result in $b_0 = -352$ and $b_1 = 9.66$. From the SPSS results shown below, we see that the constant is $b_0 = -352$ (rounded to three significant digits) and the coefficient of x is $b_1 = 9.66$ (rounded to three significant digits). Substituting the computed values for b_0 and b_1, we express the regression equation as $\hat{y} = -352 + 9.66x$. Also shown on the following page is the SPSS-generated scatterplot with the graph of the regression line included. We can see that the regression line fits the data well.

SPSS

Coefficients[a]

Model		Unstandardized Coefficients		Standardized Coefficients	t	Sig.
		B	Std. Error	Beta		
1	(Constant)	-351.660	127.410		-2.760	.033
	LENGTH	9.660	1.939	.897	4.981	.002

a. Dependent Variable: WEIGHT

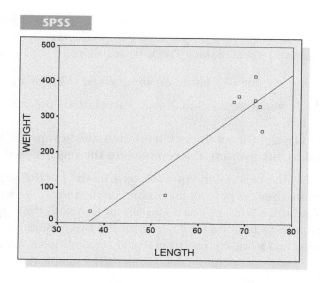

Using the Regression Equation for Predictions

Regression equations are often useful for *predicting* the value of one variable, given some particular value of the other variable. If the regression line fits the data quite well, then it makes sense to use its equation for predictions, provided that we don't go beyond the scope of the available values. That is, do not base predictions on values that are far beyond the boundaries of the known sample data. For example, using the sample data in Table 9-1, a scatterplot suggests that the lengths and weights of bears fit a straight-line pattern reasonably well, so if we measure a bear and find that it is 71.0 in. long, we can predict its weight by substituting $x = 71.0$ into the regression equation of $\hat{y} = -352 + 9.66x$. The following result shows that if a bear is 71.0 in. long, its best predicted weight is 334 lb.

$$\hat{y} = -352 + 9.66x$$
$$= -352 + 9.66(71.0) = 334$$

However, *we should use the equation of the regression line only if the regression equation is a good model for the data.* The suitability of the regression equation can be judged by testing the significance of the linear correlation coefficient *r*. See the following procedure for using the regression equation for making predictions. See also the following example that formalizes the procedure for predicting the weight of a bear that is 71.0 in. long.

PART 2   Beyond the Basics of Regression

Using the Regression Equation for Predictions (*Continued*)

We noted that we should use the equation of the regression line for predictions only if the regression equation is a good model for the data. To be more precise, *we should use the regression equation for predictions only if there is a significant linear correlation. In the absence of a significant linear correlation, we should not use the regression equation for projecting or predicting; instead, our best estimate of the second variable is simply its sample mean.*

In predicting a value of *y* based on some given value of *x* . . .

1. **If there is *not* a significant linear correlation, the best predicted *y*-value is $\bar{y}$.**
2. **If there is a significant linear correlation, the best predicted *y*-value is found by substituting the *x*-value into the regression equation.**

How good is the regression equation as a model for the population data? Figure 9-8 summarizes the process for making predictions, and that process is easier to understand if we think of *r* as a measure of how well the regression line fits the sample data. In addition to thinking about the two conditions of a significant linear correlation and a linear correlation that is not significant, we might also consider this: Regression equations obtained from paired sample data with *r* very close to -1 or $+1$ (because the points in the scatterplot are very close to the regression line) are likely to be much better models for the population data than regression equations from data sets with values of *r* that are not so close to -1 or $+1$ (even if *r* is found to be significant). With some very large samples of paired data, we might find that *r* is significant, even though it is a relatively small value, such as 0.200. In this case, the regression equation might be an acceptable model, but predictions might not be very accurate because *r* is not very close to -1 or $+1$. If *r* indicates that there is not a significant linear correlation, then the regression line fits the data poorly, and the regression equation should not be used for predictions.

FIGURE 9-8 Procedure for Predicting

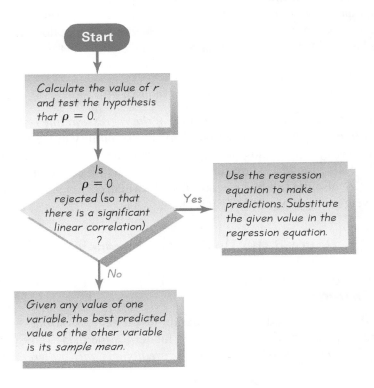

EXAMPLE **Predicting Weight of a Bear** Using the sample data in Table 9-1, we found that there is a significant linear correlation between the lengths of bears and the weights of bears, and we also found that the regression equation is $\hat{y} = -352 + 9.66x$. If a bear is measured and is found to be 71.0 in. long, predict its weight.

SOLUTION

REQUIREMENT ✓ There's a strong temptation to jump in and substitute 71.0 for x in the regression equation, but we should first consider whether there is a significant linear correlation that justifies the use of that equation. In this example, the scatterplot shows that the points approximate a straight-line pattern, and we do have a significant linear correlation (with $r = 0.897$), so our predicted value is found as shown below. ✓

Because there is a significant linear correlation, our predicted value can be found by substitution in the regression equation, as follows:

$$\hat{y} = -352 + 9.66x$$
$$= -352 + 9.66(71.0) = 334$$

The predicted weight of a 71.0 in. long bear is 334 lb. (If there had *not* been a significant linear correlation, our best predicted weight would have been $\bar{y} = 272$ lb.)

EXAMPLE **Hat Size and IQ** There is obviously no significant linear correlation between hat sizes and IQ scores of adults. Given that an individual has a hat size of 7, find the best predicted value of this person's IQ score.

SOLUTION

REQUIREMENT ✓ Because there is no significant linear correlation, we do not use the equation of the regression line for making predictions. ✓

There is no need to collect paired sample data consisting of hat size and IQ score for a sample of randomly selected adults. Instead, the best predicted IQ score is simply the mean IQ of all adults, which is 100.

Carefully compare the solutions to the preceding two examples and note that we used the regression equation when there was a significant linear correlation, but in the absence of such a correlation, the best predicted value of y is simply the value of the sample mean $\bar{y}$. A common error is to use the regression equation for making a prediction when there is no significant linear correlation. That error violates the first of the following guidelines.

Guidelines for Using the Regression Equation

1. *If there is no significant linear correlation, don't use the regression equation to make predictions.*

2. *When using the regression equation for predictions, stay within the scope of the available sample data.* If you find a regression equation that relates women's heights and shoe sizes, it's absurd to predict the shoe size of a woman who is 10 ft tall.

3. *A regression equation based on old data is not necessarily valid now.* The regression equation relating used-car prices and ages of cars is no longer usable if it's based on data from the 1970s.

4. *Don't make predictions about a population that is different from the population from which the sample data were drawn.* If we collect sample data from men and develop a regression equation relating age and TV remote-control usage, the results don't necessarily apply to women. If we use *averages* to develop a regression equation relating SAT math scores and SAT verbal scores, the results don't necessarily apply to *individuals*.

Interpreting the Regression Equation: Marginal Change

We can use the regression equation to see the effect on one variable when the other variable changes by some specific amount.

> ### Definition
>
> In working with two variables related by a regression equation, the **marginal change** in a variable is the amount that it changes when the other variable changes by exactly one unit. The slope b_1 in the regression equation represents the marginal change in y that occurs when x changes by one unit.

For the length/weight data of Table 9-1, the regression line has a slope of 9.66, so if we increase x (the length of a bear) by 1 in., the predicted weight will increase by 9.66 lb.

Outliers and Influential Points

A correlation/regression analysis of bivariate (paired) data should include an investigation of *outliers* and *influential points,* defined as follows.

> ### Definitions
>
> In a scatterplot, an **outlier** is a point lying far away from the other data points.
>
> Paired sample data may include one or more **influential points,** which are points that strongly affect the graph of the regression line.

An outlier is easy to identify: Examine the scatterplot and identify a point that is far away from the others. Here's how to determine whether a point is an influential point: Graph the regression line resulting from the data with the point included, then graph the regression line resulting from the data with the point excluded. If the graph changes by a considerable amount, the point is influential. Influential points are often found by identifying those outliers that are *horizontally* far away from the other points.

For example, refer to the preceding SPSS display (see page 453). If we include an additional bear that is 35 in. long and weighs 400 lb, we have an outlier, because that data point would fall in the upper-left corner of the graph and would be far away from the other points. This additional point would not be an influential point because the regression line would not change its position very much. If this new bear were 1 in. long and weighed 2000 lb (kind of scary, isn't it?), it would become an influential point because the graph of the regression line would change considerably, as shown in the following Excel display. Compare this regression line to the one shown in the previous SPSS graph, and you will see clearly that the addition of that one pair of values has a very dramatic effect on the regression line.

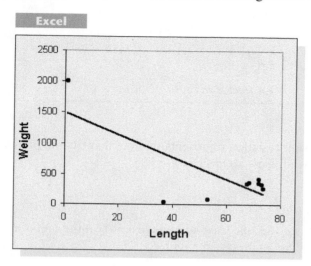

Residuals and the Least-Squares Property

We have stated that the regression equation represents the straight line that fits the data "best," and we will now describe the criterion used in determining the line that is better than all others. This criterion is based on the vertical distances between the original data points and the regression line. Such distances are called *residuals*.

Definition

For a sample of paired (x, y) data, a **residual** is the difference $(y - \hat{y})$ between an observed sample y-value and the value $\hat{y}$, which is the value of y that is predicted by using the regression equation. That is,

$$\text{residual} = (\text{observed } y) - (\text{predicted } y) = y - \hat{y}$$

This definition might seem as clear as tax-form instructions, but you can easily understand residuals by referring to Figure 9-9, which corresponds to the paired sample data in the margin. In Figure 9-9, the residuals are represented by the dashed lines. For a specific example, see the residual indicated as 7, which is directly above $x = 5$. If we substitute $x = 5$ into the regression equation $\hat{y} = 5 + 4x$, we get a predicted value of $\hat{y} = 25$. When $x = 5$, the *predicted* value of y is $\hat{y} = 25$, but the actual *observed* sample value is $y = 32$. The difference $y - \hat{y} = 32 - 25 = 7$ is a residual.

x	1	2	4	5
y	4	24	8	32

FIGURE 9-9 **Residuals and Squares of Residuals**

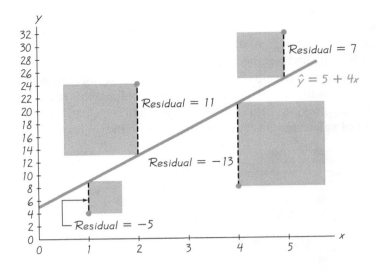

The regression equation represents the line that fits the points "best" according to the following *least-squares property.*

From Figure 9-9, we see that the residuals are -5, 11, -13, and 7, so the sum of their squares is

$$(-5)^2 + 11^2 + (-13)^2 + 7^2 = 364$$

We can visualize the least-squares property by referring to Figure 9-9, where the squares of the residuals are represented by the red square areas. The sum of the red square areas is 364, which is the smallest sum possible. Use any other straight line, and the red squares will combine to produce an area larger than the combined red area of 364.

Fortunately, we need not deal directly with the least-squares property when we want to find the equation of the regression line. Calculus has been used to build the least-squares property into Formulas 9-2 and 9-3. Because the derivations of these formulas require calculus, we don't include them in this text.

Residual Plots

In this section and the preceding section we listed simplified requirements for the effective analyses of correlation and regression results. We noted that we should always begin with a scatterplot, and we should verify that the pattern of points is approximately a straight-line pattern. We should also consider outliers. A *residual plot* can be another helpful device for analyzing correlation and regression results and for checking the requirements necessary for making inferences about correlation and regression.

Definition

A **residual plot** is a scatterplot of the (x, y) values after each of the y-coordinate values has been replaced by the residual value $y - \hat{y}$. That is, a residual plot is a graph of the points $(x, y - \hat{y})$. (Recall that $\hat{y}$ denotes the predicted value of y.)

Because the manual construction of residual plots can be what mathematicians refer to as "tedious," it is recommended that software be used. When examining a residual plot, look for a pattern in the way the points are configured, and use these criteria:

If a residual plot does not reveal any pattern, the regression equation is a good representation of the association between the two variables.

If a residual plot reveals some systematic pattern, the regression equation is not a good representation of the association between the two variables.

Consider the following examples of Minitab displays. With Case 1, all is well. The points are close to the regression line, so the regression equation is a good model for describing the association between the two variables. The corresponding residual plot does not reveal any distinct pattern.

Case 2 results in a scatterplot showing an association between the two variables, but that association is not linear. The corresponding residual plot shows a distinct pattern, confirming that the linear model is not a good model in this case.

Case 3 has a scatterplot in which the points are getting farther away from the regression line, and the residual plot does reveal a pattern of increasing variation. In this case, the regression equation is probably not a good model.

After acquiring the ability to obtain and analyze residual plots along with scatterplots, we are better prepared to check the requirements necessary to ensure the validity of inferences made from using the correlation and regression procedures.

Case 1
Minitab

x	0	1	2	3	4	5	7	8	9	10
y	1	4	8	18	19	24	36	43	42	47

The regression line fits the data well.   The residual plot reveals no pattern.

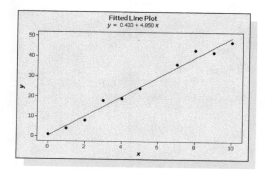
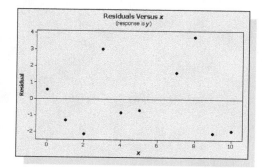

Case 2

Minitab

x	0	1	2	3	4	5	7	8	9	10
y	1	0	2	5	10	20	15	10	7	3

The scatterplot shows that the association is not linear.

The residual plot reveals a distinct pattern.

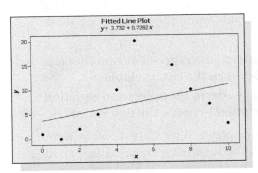

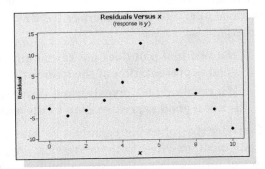

Case 3

Minitab

x	0	1	2	3	4	5	7	8	9	10
y	0	6	9	15	10	35	15	60	75	20

The scatterplot shows increasing variation of points away from the regression line.

The residual plot reveals this pattern: Going from left to right, the points show more spread. (This is contrary to the requirement that for the different values of x, the distributions of y values have the same variance.)

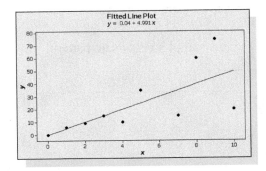

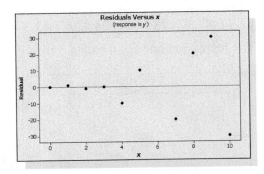

9-3 Exercises

Making Predictions. In Exercises 1–4, use the given data to find the best predicted value of the dependent variable. Be sure to follow the prediction procedure described in this section.

1. In each of the following cases, find the best predicted value of y given that $x = 3.00$. The given statistics are summarized from paired sample data.
   a. $r = 0.987$, $\bar{y} = 5.00$, $n = 20$, and the equation of the regression line is $\hat{y} = 6.00 + 4.00x$.
   b. $r = 0.052$, $\bar{y} = 5.00$, $n = 20$, and the equation of the regression line is $\hat{y} = 6.00 + 4.00x$.

2. In each of the following cases, find the best predicted value of y given that $x = 2.00$. The given statistics are summarized from paired sample data.
   a. $r = -0.123$, $\bar{y} = 8.00$, $n = 30$, and the equation of the regression line is $\hat{y} = 7.00 - 2.00x$.
   b. $r = -0.567$, $\bar{y} = 8.00$, $n = 30$, and the equation of the regression line is $\hat{y} = 7.00 - 2.00x$.

3. Chest Sizes and Weights of Bears When eight bears were anesthetized, researchers measured the distances (in inches) around the bears' chests and weighed the bears (in pounds). Minitab was used to find that the value of the linear correlation coefficient is $r = 0.993$ and the equation of the regression line is $\hat{y} = -187 + 11.3x$, where x represents chest size. Also, the mean weight of the eight bears is 234.5 lb. What is the best predicted weight of a bear with a chest size of 52 in.?

4. Systolic and Diastolic Measurements of Females Data Set 1 in Appendix B includes systolic and diastolic blood pressure measurements of 40 females. Excel was used to find that the value of the linear correlation coefficient is $r = 0.785$ and the regression equation is $\hat{y} = 8.31 + 0.534x$, where x is the systolic reading. Also, the mean of the diastolic readings is 67.4. What is the best predicted value for diastolic pressure given that a woman has a systolic level of 100?

Finding the Equation of the Regression Line. In Exercises 5 and 6, use the given data to find the equation of the regression line.

5.
x	0	1	2	3	4
y	4	1	0	1	4

6.
x	1	2	2	5	6
y	2	5	4	15	15

7. Effects of an Outlier Refer to the Minitab-generated scatterplot given in Exercise 7 of Section 9-2.
   a. Using the pairs of values for all 10 points, find the equation of the regression line.
   b. After removing the point with coordinates (10, 10), use the pairs of values for the remaining nine points and find the equation of the regression line.
   c. Compare the results from parts (a) and (b).

Finding the Equation of the Regression Line and Making Predictions. Exercises 8–21 use the same data sets as the exercises in Section 9-2. In each case, find the regression equation, letting the first variable be the independent (x) variable. Find the indicated predicted values. Caution: When finding predicted values, be sure to follow the prediction procedure described in this section.

8. Heights of Fathers and Sons Find the best predicted height of a son, if it is known that his father is 72.0 in. tall.

Height of father	70	69	64	71	68	66	74
Height of son	62.5	64.6	69.1	73.9	67.1	64.4	71.1

9. Supermodel Heights and Weights Find the best predicted weight of a supermodel who is 69 in. tall.

Height (in.)	71	70.5	71	72	70	70	66.5	70	71
Weight (lb)	125	119	128	128	119	127	105	123	115

10. Systolic and Diastolic Blood Pressure Find the best predicted diastolic blood pressure for a person with a systolic reading of 122.

Systolic	125	107	126	110	110	107	113	126
Diastolic	78	54	81	68	66	83	71	72

11. Blood Pressure Measurements Find the best predicted diastolic blood pressure for a person with a systolic reading of 122.

Systolic	138	130	135	140	120	125	120	130	130	144	143	140	130	150
Diastolic	82	91	100	100	80	90	80	80	80	98	105	85	70	100

12. Temperatures and Marathons Find the best predicted winning time for the 1990 marathon given that the temperature was 73 degrees. How does the predicted value compare to the actual winning time of 150.750 min?

x (temperature)	55	61	49	62	70	73	51	57
y (time)	145.283	148.717	148.300	148.100	147.617	146.400	144.667	147.533

13. Smoking and Nicotine Find the best predicted level of cotinine for a person who smokes 40 cigarettes per day.

x (cigarettes per day)	60	10	4	15	10	1	20	8	7	10	10	20
y (cotinine)	179	283	75.6	174	209	9.51	350	1.85	43.4	25.1	408	344

14. Tree Circumference and Height Find the best predicted height of a tree that has a circumference of 4.0 ft. What is an advantage of being able to determine the height of a tree from its circumference?

x (circ.)	1.8	1.9	1.8	2.4	5.1	3.1	5.5	5.1	8.3	13.7	5.3	4.9	3.7	3.8
y (ht)	21.0	33.5	24.6	40.7	73.2	24.9	40.4	45.3	53.5	93.8	64.0	62.7	47.2	44.3

15. Testing Duragesic for Pain Relief Duragesic, manufactured by ALZA Corporation, is a drug used for treating chronic pain. An advertisement for this drug included a scatterplot of (x, y) points, where each point corresponds to one study subject, x is the "pain intensity on pre-study analgesic program" and y is the "pain intensity after

1 month on Duragesic." The pain measurements are visual analog scores that are often used to measure pain. The values listed below are based on the published scatterplot, and the y values correspond to the respective x values. In addition to finding the equation of the regression line, identify the equation of the line corresponding to a drug that has absolutely no effect.

x: 1.2 1.3 1.5 1.6 8.0 3.4 3.5 2.8 2.6 2.2 3.0 7.1 2.3 2.1 3.4 6.4 5.0 4.2
2.8 3.9 5.2 6.9 6.9 5.0 5.5 6.0 5.5 8.6 9.4 10.0 7.6

y: 0.4 1.4 1.8 2.9 6.0 1.4 0.7 3.9 0.9 1.8 0.9 9.3 8.0 6.8 2.3 0.4 0.7 1.2
4.5 2.0 1.6 2.0 2.0 6.8 6.6 4.1 4.6 2.9 5.4 4.8 4.1

16. **Cholesterol and Body Mass Index** Refer to Data Set 1 in Appendix B and use the cholesterol levels (x) and body mass index values (y) of the 40 women. What is the best predicted value for the body mass index of a woman having a cholesterol level of 500?

17. **Cholesterol and Weight** Refer to Data Set 1 in Appendix B and use the cholesterol levels (x) and weights (y) of the 40 women. What is the best predicted value for the weight of a woman with a cholesterol level of 200?

18. **Chest Sizes and Weights of Bears** Refer to Data Set 6 in Appendix B and use the chest sizes (x) and weights (y) of all listed bears. What is the best predicted value for the weight of a bear with a chest size of 50 in.?

19. **Heights of Mothers and Daughters** Refer to Data Set 3 in Appendix B and use the heights (x) of the mothers and their female children (y). What is the best predicted value for the height of a daughter born to a mother who is 64 in. tall?

20. **Iris Petal Lengths and Widths** Refer to Data Set 7 in Appendix B and use only the petal lengths and petal widths from irises in the Setosa class. What is the best predicted petal width of an iris with a petal length of 1.5 mm?

21. **Iris Petal Lengths and Widths** Refer to Data Set 7 in Appendix B and use only the petal lengths and petal widths from irises in the Versicolor class. What is the best predicted petal width of an iris with a petal length of 1.5 mm? Is the result roughly the same as that obtained in Exercise 20? When predicting the petal length of an iris, is it important to consider the class (Setosa, Versicolor, Virginica)?

22. **Identifying Outliers and Influential Points** Refer to the sample data listed in Table 9-1. If we include a ninth bear that is 120 in. long and weighs 800 lb, is the new point an outlier? Is it an influential point?

23. **Identifying Outliers and Influential Points** Refer to the sample data listed in Table 9-1. If we include a ninth bear that is 120 in. long and weighs 50 lb (affectionately called "Slim"), is the new point an outlier? Is it an influential point?

*24. **Testing Least-Squares Property** According to the least-squares property, the regression line minimizes the sum of the squares of the residuals. We noted that with the paired data in the margin, the regression equation is $\hat{y} = 5 + 4x$ and the sum of the squares of the residuals is 364. Show that the equation $\hat{y} = 8 + 3x$ results in a sum of squares of residuals that is greater than 364.

x	1	2	4	5
y	4	24	8	32

*25. **Using Logarithms to Transform Data** If a scatterplot reveals a nonlinear (not a straight-line) pattern that you recognize as another type of curve, you may be able to apply the methods of this section. For the data given in the margin, find the linear equation ($y = b_0 + b_1x$) that best fits the sample data, and find the logarithmic equation ($y = a + b \ln x$) that best fits the sample data. (*Hint:* Begin by replacing each x-value with $\ln x$.) Which of these two equations fits the data better? Why?

x	2.0	2.5	4.2	10.0
y	12.0	18.7	53.0	225.0

*26. Equivalent Hypothesis Tests Explain why a test of the null hypothesis $H_0: \rho = 0$ is equivalent to a test of the null hypothesis $H_0: \beta_1 = 0$ where ρ is the linear correlation coefficient for a population of paired data, and β_1 is the slope of the regression line for that same population.

*27. Residual Plot A scatterplot is a plot of the paired (x, y) sample data. A residual plot is a graph of the points $(x, y - \hat{y})$. Construct a residual plot for the data in Table 9-1. Are there any noticeable patterns?

9-4 Variation and Prediction Intervals

So far, we have used paired sample data to test for a linear correlation between x and y, and to identify the regression equation. In this section we continue to analyze paired $(x, y,)$ data as we proceed to consider different types of variation that can be used for two major applications:

1. To determine the proportion of the variation in y that can be explained by the linear association between x and y.

2. To construct confidence interval estimates of predicted y-values. Such confidence intervals are called *prediction intervals,* which are formally defined later in this section.

Explained and Unexplained Variation

In Section 9-2 we introduced the concept of correlation and used the linear correlation coefficient r in determining whether there is a significant linear correlation between two variables, denoted by x and y. In addition to serving as a measure of the linear correlation between two variables, the value of r can also provide us with additional information about the variation of sample points about the regression line. We begin with a sample case, which leads to an important definition (coefficient of determination).

Suppose we have a large collection of paired data, which yields the following results:

- There is a significant linear correlation.
- The equation of the regression line is $\hat{y} = 3 + 2x$.
- The mean of the y-values is given by $\bar{y} = 9$.
- One of the pairs of sample data is $x = 5$ and $y = 19$.
- The point $(5, 13)$ is one of the points on the regression line, because substituting $x = 5$ into the regression equation yields $\hat{y} = 13$.

$$\hat{y} = 3 + 2x = 3 + 2(5) = 13$$

Figure 9-10 shows that the point $(5, 13)$ lies on the regression line, but the point $(5, 19)$ is from the original data set and does not lie on the regression line because it does not satisfy the regression equation. Take time to examine Figure 9-10 carefully and note the differences defined as follows.

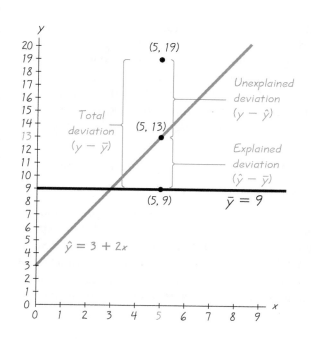

FIGURE 9-10
**Unexplained, Explained,
and Total Deviation**

*Do Air Bags
Save Lives?*

The National Highway Transportation Safety Administration
reported that for a recent year,
3,448 lives were saved because of
air bags. It was reported that for
car drivers involved in frontal
crashes, the fatality rate was reduced 31%; for passengers, there
was a 27% reduction. It was noted
that "calculating lives saved is
done with a mathematical analysis
of the real-world fatality experience of vehicles with air bags
compared with vehicles without
air bags. These are called double-
pair comparison studies, and are
widely accepted methods of statistical analysis."

Definitions

Assume that we have a collection of paired data containing the sample point (x, y), that $\hat{y}$ is the predicted value of y (obtained by using the regression equation), and that the mean of the sample y-values is $\bar{y}$.

The **total deviation** (from the mean) of the particular point (x, y) is the vertical distance $y - \bar{y}$, which is the distance between the point (x, y) and the horizontal line passing through the sample mean $\bar{y}$.

The **explained deviation** is the vertical distance $\hat{y} - \bar{y}$, which is the distance between the predicted y-value and the horizontal line passing through the sample mean $\bar{y}$.

The **unexplained deviation** is the vertical distance $y - \hat{y}$, which is the vertical distance between the point (x, y) and the regression line. (The distance $y - \hat{y}$ is also called a *residual*, as defined in Section 9-3.)

For the specific data under consideration, we get these results:

Total deviation of $(5, 19) = y - \bar{y} = 19 - 9 = 10$

Explained deviation of $(5, 19) = \hat{y} - \bar{y} = 13 - 9 = 4$

Unexplained deviation of $(5, 19) = y - \hat{y} = 19 - 13 = 6$

If we were totally ignorant of correlation and regression concepts and wanted to predict a value of y given a value of x and a collection of paired (x, y) data, our best guess would be $\bar{y}$. But we are not totally ignorant of correlation and regression concepts: We know that in this case (with a significant linear correlation) the way to predict the value of y when $x = 5$ is to use the regression equation, which yields $\hat{y} = 13$, as calculated earlier. We can explain the discrepancy between $\bar{y} = 9$ and $\hat{y} = 13$ by simply noting that there is a significant linear correlation best described by the regression line. Consequently, when $x = 5$, y *should be* 13, not the mean value of 9. But whereas y should be 13, *it is* 19. The discrepancy between 13 and 19 cannot be explained by the regression line, and it is called an *unexplained deviation,* or a *residual.* The specific case illustrated in Figure 9-10 can be generalized as follows:

(total deviation) = (explained deviation) + (unexplained deviation)

or $(y - \bar{y})$ = $(\hat{y} - \bar{y})$ + $(y - \hat{y})$

This last expression applies to a particular point (x, y), and the same association applies to the sums of squares shown in Formula 9-4, even though this last expression is not algebraically equivalent to Formula 9-4. In Formula 9-4, the **total variation** is expressed as the sum of the squares of the total deviation values, the **explained variation** is the sum of the squares of the explained deviation values, and the **unexplained variation** is the sum of the squares of the unexplained deviation values.

Formula 9-4

(total variation) = (explained variation) + (unexplained variation)

or $\Sigma(y - \bar{y})^2$ = $\Sigma(\hat{y} - \bar{y})^2$ + $\Sigma(y - \hat{y})^2$

Coefficient of Determination

The components of Formula 9-4 are used in the following important definition:

Definition

The **coefficient of determination** is the amount of the variation in y that is explained by the regression line. It is computed as

$$r^2 = \frac{\text{explained variation}}{\text{total variation}}$$

We can compute r^2 by using the definition just given with Formula 9-4, or we can simply square the linear correlation coefficient r, which is found by using the methods described in Section 9-2. For example, in Section 9-2 we noted that if $r = 0.897$, then $r^2 = 0.805$, which means that *80.5% of the total variation in y can be explained by the linear association between x and y (as described by the regression equation). It follows that 19.5% of the total variation in y remains unexplained.*

EXAMPLE **Cholesterol and Body Mass** In Exercise 16 in Section 9-2, we find that for the paired data consisting of cholesterol levels and BMI (body mass index) levels of 40 randomly selected women, the linear correlation coefficient is given by $r = 0.482$. Find the percentage of the variation in y (BMI) that can be explained by the linear association between the cholesterol level and body mass index.

SOLUTION   The coefficient of determination is $r^2 = 0.482^2 = 0.232$, indicating that the ratio of explained variation in y to total variation in y is 0.232. We can now state that 23.2% of the total variation in y can be explained by the regression equation. We interpret this to mean that 23.2% of the total variation in BMI can be explained by the variation in their cholesterol levels; the other 76.8% is attributable to other factors, such as weight and height and genetic factors.

Prediction Intervals

In Section 9-3 we used the Table 9-1 sample data to find the regression equation $\hat{y} = -352 + 9.66x$, where $\hat{y}$ represents the predicted bear weight (in pounds) and x represents the bear length (in inches). We then used that equation to predict the y-value, given that $x = 71.0$. We found that the best predicted weight of a 71.0 in. bear is 334 lb. Because 334 is a single value, it is referred to as a *point estimate*. In Chapter 6 we saw that point estimates have the serious disadvantage of not giving us any information about how accurate they might be. Here, we know that 334 is the best predicted value, but we don't know how accurate that value is. In Chapter 6 we developed confidence interval estimates to overcome that disadvantage, and in this section we follow that precedent. We will use a **prediction interval,** which is a confidence interval estimate of a predicted value of y.

The development of a prediction interval requires a measure of the spread of sample points about the regression line. Recall that the unexplained deviation (or residual) is the vertical distance between a sample point and the regression line, as illustrated in Figure 9-10. The *standard error of estimate* is a collective measure of the spread of the sample points about the regression line, and it is formally defined as follows.

Definition

The **standard error of estimate,** denoted by s_e, is a measure of the differences (or distances) between the observed sample y-values and the predicted values $\hat{y}$ that are obtained using the regression equation. It is given as

$$s_e = \sqrt{\frac{\Sigma(y - \hat{y})^2}{n - 2}} \qquad \text{(where } \hat{y} \text{ is the predicted } y\text{-value)}$$

or as the following equivalent formula:

Formula 9-5 $$s_e = \sqrt{\frac{\Sigma y^2 - b_0\Sigma y - b_1\Sigma xy}{n - 2}}$$

SPSS, SAS, STATDISK, Minitab, Excel, and the TI-83/84 Plus calculator are all designed to automatically compute the value of s_e.

The development of the standard error of estimate s_e closely parallels that of the ordinary standard deviation introduced in Chapter 2. Just as the standard deviation is a measure of how values deviate from their mean, the standard error of estimate s_e is a measure of how sample data points deviate from their regression line. (Try to avoid confusion between this "standard error of estimate s_e" and "standard error of the mean," which is $\sigma/\sqrt{n}$.) The reasoning behind dividing by $n - 2$ is similar to the reasoning that led to division by $n - 1$ for the ordinary standard deviation. It is important to note that relatively smaller values of s_e reflect points that stay close to the regression line, and relatively larger values occur with points farther away from the regression line.

Formula 9-5 is algebraically equivalent to the other expression in the definition, but Formula 9-5 is generally easier to work with because it doesn't require that we compute each of the predicted values $\hat{y}$ by substitution in the regression equation. However, Formula 9-5 does require that we find the y-intercept b_0 and the slope b_1 of the estimated regression line.

> **EXAMPLE** Use Formula 9-5 to find the standard error of estimate s_e for the length/weight sample data listed in Table 9-1.
>
> **SOLUTION**   Using the sample data in Table 9-1, we find these values:
>
> $$n = 8 \qquad \Sigma y^2 = 728{,}520 \qquad \Sigma y = 2176 \qquad \Sigma xy = 151{,}879$$
>
> In Section 9-3 we used the Table 9-1 sample data to find the y-intercept and the slope of the regression line. Those values are given here with extra decimal places for greater precision.
>
> $$b_0 = -351.660 \qquad b_1 = 9.65979$$
>
> We can now use these values in Formula 9-5 to find the standard error of estimate s_e.
>
> $$s_e = \sqrt{\frac{\Sigma y^2 - b_0 \Sigma y - b_1 \Sigma xy}{n - 2}}$$
>
> $$= \sqrt{\frac{728{,}520 - (-351.660)(2176) - (9.65979)(151{,}879)}{8 - 2}}$$
>
> $$= 66.5994 = 66.6 \qquad \text{(rounded)}$$
>
> We can measure the spread of the sample points about the regression line with the standard error of estimate $s_e = 66.6$.

We can use the standard error of estimate s_e to construct interval estimates that will help us see how dependable our point estimates of y really are. Assume that for each fixed value of x, the corresponding sample values of y are normally distributed about the regression line, and those normal distributions have the same variance. The following interval estimate applies to an *individual* y-value. (For a confidence interval used to predict the *mean* of all y-values for some given x-value, see Exercise 22.)

Prediction Interval for an Individual y

Given the fixed value x_0, the prediction interval for an individual y is

$$\hat{y} - E < y < \hat{y} + E$$

where the margin of error E is

$$E = t_{\alpha/2}s_e\sqrt{1 + \frac{1}{n} + \frac{n(x_0 - \bar{x})^2}{n(\Sigma x^2) - (\Sigma x)^2}}$$

and x_0 represents the given value of x, $t_{\alpha/2}$ has $n - 2$ degrees of freedom, and s_e is found from Formula 9-5.

EXAMPLE **Lengths and Weights of Bears** For the paired length/weight data in Table 9-1, we have found that when $x = 71.0$ (for length in inches), the best predicted weight of a bear is 334 lb. Construct a 95% prediction interval for the weight of a bear, given that it is 71.0 in. long (so that $x = 71.0$). This will provide a sense of how accurate the predicted value of 334 lb really is.

SOLUTION   In previous sections we have shown that there is a significant linear correlation (at the 0.05 significance level), and the regression equation is $\hat{y} = -352 + 9.66x$. In the preceding example we found that $s_e = 66.5994$, and the following statistics are obtained from the Table 9-1 sample data:

$$n = 8 \qquad \bar{x} = 64.5625 \qquad \Sigma x = 516.5 \qquad \Sigma x^2 = 34{,}525.75$$

From Table A-3 we find $t_{\alpha/2} = 2.447$. (We used $8 - 2 = 6$ degrees of freedom with $\alpha = 0.05$ in two tails.) We first calculate the margin of error E by letting $x_0 = 71.0$, because we want the prediction interval of the weight y for $x = 71.0$.

$$E = t_{\alpha/2}s_e\sqrt{1 + \frac{1}{n} + \frac{n(x_0 - \bar{x})^2}{n(\Sigma x^2) - (\Sigma x)^2}}$$

$$= (2.447)(66.5994)\sqrt{1 + \frac{1}{8} + \frac{8(71.0 - 64.5625)^2}{8(34{,}525.75) - 516.5^2}}$$

$$= (2.447)(66.5994)(1.07710) = 176$$

With $\hat{y} = 334$ and $E = 176$, we get the prediction interval as follows.

$$\hat{y} - E < y < \hat{y} + E$$
$$334 - 176 < y < 334 + 176$$
$$158 < y < 510$$

That is, for a 71.0 in. bear, we have 95% confidence that its true weight is between 158 lb and 510 lb. That's a relatively large range. One factor contributing to the large range is the small sample size of $n = 8$.

continued

In addition to knowing that for a 71.0 in. bear, the predicted weight is 334 lb, we now have a sense of how reliable that estimate really is. The 95% prediction interval found in this example shows that the predicted value of 334 lb can vary substantially.

9-4 Exercises

Interpreting the Coefficient of Determination. In Exercises 1–4, use the value of the linear correlation coefficient r to find the coefficient of determination and the percentage of the total variation that can be explained by the linear association between the two variables.

1. $r = 0.8$

2. $r = -0.6$

3. $r = -0.503$

4. $r = 0.636$

Interpreting a Computer Display. In Exercises 5–8, refer to the Minitab display that was obtained by using the paired data consisting of neck size (in inches) and weight (in pounds) for the sample of bears listed in Data Set 6 in Appendix B. Along with the paired sample data, Minitab was also given a neck size of 25.0 in. to be used for predicting the weight.

5. Testing for Correlation Using the information provided in the display determine the value of the linear correlation coefficient. Given that there are 54 pairs of data, is there a significant linear correlation between bear neck sizes and bear weights?

6. Identifying Total Variation What percentage of the total variation in weight can be explained by the linear association between neck size and weight?

7. Predicting Weight If a bear has a neck size of 25.0 in., what is the single value that is the best predicted weight? (Assume that there is a significant linear correlation between neck size and weight.)

8. Finding Prediction Interval For a given neck size of 25.0 in., identify the 95% prediction interval estimate of the weight, and write a statement interpreting that interval.

```
Minitab

The regression equation is
WEIGHT = - 232 + 20.2 NECK

Predictor       Coef   SE Coef        T       P
Constant     -231.70     22.78   -10.17   0.000
NECK          20.169      1.069    18.86   0.000

S = 43.9131    R-Sq = 87.2%    R-Sq(adj) = 87.0%

Predicted Values for New Observations

New Obs     Fit  SE Fit        95% CI              95% PI
1        272.53    7.64   (257.21, 287.85)   (183.09, 361.97)
```

Finding Measures of Variation. In Exercises 9–12, find the (a) explained variation, (b) unexplained variation, (c) total variation, (d) coefficient of determination, and (e) standard error of estimate s_e. In each case, there is a significant linear correlation so that it is reasonable to use the regression equation when making predictions.

9. Supermodel Heights and Weights Listed below are heights (in inches) and weights (in pounds) for supermodels Niki Taylor, Nadia Avermann, Claudia Schiffer, Elle MacPherson, Christy Turlington, Bridget Hall, Kate Moss, Valerie Mazza, and Kristy Hume.

Height (in.)	71	70.5	71	72	70	70	66.5	70	71
Weight (lb)	125	119	128	128	119	127	105	123	115

10. Blood Pressure Measurements Fourteen different second-year medical students took blood pressure measurements of the same patient and the results are listed below.

Systolic	138	130	135	140	120	125	120	130	130	144	143	140	130	150
Diastolic	82	91	100	100	80	90	80	80	80	98	105	85	70	100

11. Tree Circumference and Height Listed below are the circumferences (in feet) and the heights (in feet) of trees in Marshall, Minnesota (based on data from "Tree Measurements" by Stanley Rice, *American Biology Teacher*, Vol. 61, No. 9).

x (circ.)	1.8	1.9	1.8	2.4	5.1	3.1	5.5	5.1	8.3	13.7	5.3	4.9	3.7	3.8
y (ht)	21.0	33.5	24.6	40.7	73.2	24.9	40.4	45.3	53.5	93.8	64.0	62.7	47.2	44.3

12. Cholesterol and BMI Refer to the paired cholesterol/BMI data for women as listed in Data Set 1 in Appendix B. (Let x represent the cholesterol levels.)

13. Effect of Variation on Prediction Interval Refer to the data given in Exercise 9 and assume that the necessary conditions of normality and variance are met.
   a. Find the predicted weight of a supermodel who is 69 in. tall.
   b. Find a 95% prediction interval estimate of the weight of a supermodel who is 69 in. tall.

14. Finding Predicted Value and Prediction Interval Refer to Exercise 10 and assume that the necessary conditions of normality and variance are met.
   a. Find the predicted diastolic reading given that the systolic reading is 120.
   b. Find a 95% prediction interval estimate of the diastolic reading given that the systolic reading is 120.

15. Finding Predicted Value and Prediction Interval Refer to the data given in Exercise 11 and assume that the necessary conditions of normality and variance are met.
   a. Find the predicted height of a tree that has a circumference of 4.0 ft.
   b. Find a 99% prediction interval estimate of the height of a tree that has a circumference of 4.0 ft.

16. Finding Predicted Value and Prediction Interval Refer to the data described in Exercise 12 and assume that the necessary conditions of normality and variance are met.
   a. Find the predicted BMI level for a woman with a cholesterol level of 200.
   b. Find a 99% prediction interval estimate of the BMI level for a woman with a cholesterol level of 200.

Finding a Prediction Interval. In Exercises 17–20, refer to the Table 9-1 sample data. Let x represent the length of a bear (in inches) and let y represent the weight of a bear (in pounds). Use the given length of a bear and the given confidence level to construct a prediction interval estimate of the weight. (See the example in this section.)

17. $x = 50.0$ in.; 95% confidence

18. $x = 65.0$ in.; 90% confidence

19. $x = 49.7$ in.; 90% confidence

20. $x = 68.0$ in.; 99% confidence

*21. Confidence Intervals for β_0 and β_1 Confidence intervals for the y-intercept β_0 and slope β_1 for a regression line ($y = \beta_0 + \beta_1 x$) can be found by evaluating the limits in the intervals below.

$$b_0 - E < \beta_0 < b_0 + E$$

where
$$E = t_{\alpha/2} s_e \sqrt{\frac{1}{n} + \frac{\bar{x}^2}{\Sigma x^2 - \frac{(\Sigma x)^2}{n}}}$$

$$b_1 - E < \beta_1 < b_1 + E$$

where
$$E = t_{\alpha/2} \cdot \frac{s_e}{\sqrt{\Sigma x^2 - \frac{(\Sigma x)^2}{n}}}$$

In these expressions, the y-intercept b_0 and the slope b_1 are found from the sample data and $t_{\alpha/2}$ is found from Table A-3 by using $n - 2$ degrees of freedom. Using the length/weight data in Table 9-1, find the 95% confidence interval estimates of β_0 and β_1.

*22. Finding Confidence Interval for Mean Predicted Value   From the expression given in this section for the margin of error corresponding to a prediction interval for y, we can get the expression

$$s_{\hat{y}} = s_e \sqrt{1 + \frac{1}{n} + \frac{n(x_0 - \bar{x})^2}{n(\Sigma x^2) - (\Sigma x)^2}}$$

which is the *standard error of the prediction* when predicting for a *single* y, given that $x = x_0$. When predicting for the *mean* of all values of y for which $x = x_0$, the point estimate $\hat{y}$ is the same, but $s_{\hat{y}}$ is as follows:

$$s_{\hat{y}} = s_e \sqrt{\frac{1}{n} + \frac{n(x_0 - \bar{x})^2}{n(\Sigma x^2) - (\Sigma x)^2}}$$

Use the data from Table 9-1 and extend the last example of this section to find a point estimate and a 95% confidence interval estimate of the mean weight of all bears that are 71.0 in. long.

9-5   Multiple Regression

So far, we have used methods of correlation and regression to investigate associations between exactly *two* variables, but some circumstances require more than two variables. In predicting the weight of a bear, for example, we might consider variables such as length, chest size, head size, and neck size, so that a total of five variables are involved. This section presents a method for analyzing such associations involving *more than two* variables. We will focus on three key elements: (1) the multiple regression equation, (2) the value of adjusted R^2, and (3) the *P*-value. As in the previous sections of this chapter, we will work with *linear* associations only. We begin with the *multiple regression equation*.

Multiple Regression Equation

Definition

A **multiple regression equation** expresses a linear association between a dependent variable y and two or more independent variables $(x_1, x_2, \ldots, x_k)$. The general form of a multiple regression equation is

$$\hat{y} = b_0 + b_1 x_1 + b_2 x_2 + \cdots + b_k x_k$$

We will use the following notation, which follows naturally from the notation used in Section 9-3.

Notation

$\hat{y} = b_0 + b_1 x_1 + b_2 x_2 + \cdots + b_k x_k$ (General form of the estimated multiple regression equation)

n = sample size

k = number of *independent* variables. (The independent variables are also called **predictor variables** or x variables.)

$\hat{y}$ = predicted value of the dependent variable y (computed by using the multiple regression equation)

$x_1, x_2, \ldots, x_k$ are the independent variables

β_0 = the y-intercept, or the value of y when all of the predictor variables are 0. (This value is a population parameter.)

b_0 = estimate of β_0 based on the sample data. (b_0 is a sample statistic.)

$\beta_1, \beta_2, \ldots, \beta_k$ are the coefficients of the independent variables $x_1, x_2, \ldots, x_k$

$b_1, b_2, \ldots, b_k$ are the sample estimates of the coefficients $\beta_1, \beta_2, \ldots, \beta_k$

Requirements

When attempting to find a multiple regression equation involving k independent (x) variables, there must be at least $k + 2$ sample values for each of the independent and dependent variables. This requirement is usually not an issue. If this requirement is not satisfied, the software will either provide an error message (such as "not enough data") or provide results that are missing some key elements, such as the *P*-value. Given sufficient sample data for each variable, a multiple regression equation can always be calculated, but the usefulness of the multiple regression is determined through analyses discussed later in this section.

Table 9-3	Data from Anesthetized Male Bears								
Variable	Name	Sample Data							
y	WEIGHT	80	344	416	348	262	360	332	34
x_2	AGE	19	55	81	115	56	51	68	8
x_3	HEADLEN	11.0	16.5	15.5	17.0	15.0	13.5	16.0	9.0
x_4	HEADWDTH	5.5	9.0	8.0	10.0	7.5	8.0	9.0	4.5
x_5	NECK	16.0	28.0	31.0	31.5	26.5	27.0	29.0	13.0
x_6	LENGTH	53.0	67.5	72.0	72.0	73.5	68.5	73.0	37.0
x_7	CHEST	26	45	54	49	41	49	44	19

The computations required for multiple regression are so complicated that a statistical software package *must* be used, so we will focus on *interpreting* computer displays.

EXAMPLE **Bears** For reasons of safety, a study of bears involved the collection of various measurements that were taken after the bears were anesthetized. When obtaining measurements from an anesthetized bear in the wild, it is relatively easy to use a tape measure for finding values such as the chest size, neck size, and overall length, but it is difficult to find the weight because the bear must be lifted. Instead of actually weighing a bear, can we predict its weight based on other measurements that are easier to find? Data Set 6 in Appendix B has measurements taken from 54 bears, but we will consider the data from only eight of those bears, as listed in Table 9-3. Using the data in Table 9-3, find the multiple regression equation in which the dependent (y) variable is weight and the independent variables are head length (HEADLEN) and total overall length (LENGTH).

SOLUTION

REQUIREMENT ✓   As we noted in the requirements for this section, a multiple regression equation can always be found (provided that there are enough sample values). The SPSS result shown below does yield all of the key elements, so the multiple regression equation can be identified. ✓

Using SPSS, we obtain the results shown in the accompanying display.

SPSS

Coefficients[a]						
		Unstandardized Coefficients		Standardized Coefficients		
Model		B	Std. Error	Beta	t	Sig.
1	(Constant)	-374.303	134.093		-2.791	.038
	HEADLEN	18.820	23.148	.383	.813	.453
	LENGTH	5.875	5.065	.546	1.160	.299

a. Dependent Variable: WEIGHT

Using our notation presented earlier in this section, we could write this equation as

$$\hat{y} = -374 + 18.8 \text{ HEADLEN} + 5.88 \text{ LENGTH}$$

or $$\hat{y} = -374 + 18.8x_3 + 5.88x_6$$

If a multiple regression equation fits the sample data well, it can be used for predictions. For example, if we determine that the equation is suitable for predictions, and if we have a bear with a 14.0-in. head length and a 71.0-in. overall length, we can predict its weight by substituting those values into the regression equation to get a predicted weight of 307 lb. Also, the coefficients $b_3 = 18.8$ and $b_6 = 5.88$ can be used to determine marginal change, as described in Section 9-3. For example, the coefficient $b_3 = 18.8$ shows that when the overall length of a bear remains constant, the predicted weight increases by 18.8 lb for each 1-in. increase in the head length.

Adjusted R^2

R^2 denotes the **multiple coefficient of determination,** which is a measure of how well the multiple regression equation fits the sample data. A perfect fit would result in $R^2 = 1$, and a very good fit results in a value near 1. A very poor fit results in a value of R^2 close to 0. The following display is included along with the preceding SPSS display.

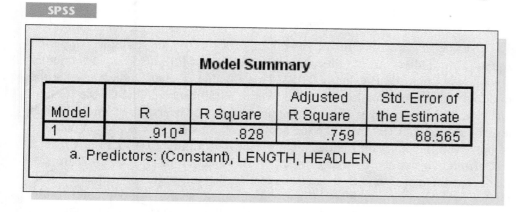

SPSS

Model Summary

Model	R	R Square	Adjusted R Square	Std. Error of the Estimate
1	.910[a]	.828	.759	68.565

a. Predictors: (Constant), LENGTH, HEADLEN

The value of $R^2 = 0.828$ in the SPSS display indicates that 82.8% of the variation in bear weight can be explained by the head length x_3 and the overall length x_6. However, the multiple coefficient of determination R^2 has a serious flaw: As more variables are included, R^2 increases. (R^2 could remain the same, but it usually increases.) The largest R^2 is obtained by simply including all of the available variables, but the best multiple regression equation does not necessarily use all of the available variables. Because of that flaw, comparison of different multiple regression equations is better accomplished with the adjusted coefficient of determination, which is R^2 adjusted for the number of variables and the sample size.

Definition

The **adjusted coefficient of determination** is the multiple coefficient of determination R^2 modified to account for the number of variables and the sample size. It is calculated by using Formula 9-6.

Formula 9-6 $$\text{adjusted } R^2 = 1 - \frac{(n-1)}{[n-(k+1)]}(1 - R^2)$$

where n = sample size

k = number of independent (x) variables

The preceding SPSS display for the data in Table 9-3 shows the adjusted coefficient of determination is 0.759. If we use Formula 9-6 with the R^2 value of 0.828, $n = 8$ and $k = 2$, we find that the adjusted R^2 value is 0.759, confirming the value displayed by SPSS. For the weight, head length, and length data in Table 9-3, the R^2 value of 0.828 indicates that 82.8% of the variation in weight can be explained by the head length x_3 and overall length x_6, but when we compare this multiple regression equation to others, it is better to use the adjusted R^2 of 0.759.

P-Value

The *P*-value is a measure of the overall significance of the multiple regression equation. In addition to the preceding two displays, SPSS also displays another box showing that the *P*-value is 0.012. This *P*-value is small, indicating that the multiple regression equation has good overall significance and is usable for predictions. That is, it makes sense to predict weights of bears based on their head lengths and overall lengths. Like the adjusted R^2, this *P*-value is a good measure of how well the equation fits the sample data. The value of 0.012 results from a test of the null hypothesis that $\beta_3 = \beta_6 = 0$. Rejection of $\beta_3 = \beta_6 = 0$ implies that at least one of β_3 and β_6 is not 0, indicating that this regression equation is effective in determining weights of bears. A complete analysis of computer results might include other important elements, such as the significance of the individual coefficients, but we will limit our discussion to the three key components—multiple regression equation, adjusted R^2, and *P*-value. Keep in mind that we are not necessarily finished when we have found a good multiple regression equation. Based on the nature of the multiple regression equation, adjusted R^2, and *P*-value, we might correctly conclude that we have a usable result. However, the particular multiple regression equation is not necessarily the *best* choice, due to such factors as the inclusion of some independent variables that don't really add predictive value, or independent variables that might be redundant because they are highly correlated with other independent variables. Let us therefore consider the objective of finding the *best* multiple regression equation.

Finding the Best Multiple Regression Equation

Table 9-3 includes seven different variables of measurement for eight different bears. The preceding SPSS displays are based on the selection of weight as the dependent variable and the selection of head length and overall length as the independent variables. But if we want to predict the weight of a bear, is there some other combination of variables that might be better than head length and overall length? Table 9-4 lists a few of the combinations of variables, and we are now confronted with the important objective of finding the *best* multiple regression equation. Because determination of the best multiple regression requires a good dose of judgment, there is no exact and automatic procedure that can be used. *Determination of the best multiple regression equation is often quite difficult and beyond the scope of this book,* but the following guidelines should provide some help.

Guidelines for Finding the Best Multiple Regression Equation

1. *Use common sense and practical considerations to include or exclude variables.* For example, we might exclude the variable of age because inexperienced researchers might not know how to determine the age of a bear and, when questioned, bears are reluctant to reveal their ages. Unlike the other independent variables, the age of a bear cannot be easily obtained with a tape measure. It therefore makes sense to exclude age as an independent variable.

2. *Consider the P-value.* Select an equation having overall significance, as determined by the P-value found in the computer display. For example, see the values of overall significance in Table 9-4. The use of all six independent variables results in an overall significance of 0.046, which is just barely significant at the $\alpha = 0.05$ level; we're better off with the single variable CHEST, which has overall significance of 0.000.

3. *Consider equations with high values of adjusted R^2, and try to include only a few variables.* Instead of including almost every available variable, try to include relatively few independent (x) variables. Use these guidelines:

   - Select an equation having a value of adjusted R^2 with this property: If an additional independent variable is included, the value of adjusted R^2 does

	LENGTH	CHEST	HEADLEN/ LENGTH	AGE/NECK/ LENGTH/CHEST	AGE/HEADLEN/HEADWDTH/ NECK/LENGTH/CHEST
R^2	0.805	0.983	0.828	0.999	0.999
Adjusted R^2	0.773	0.980	0.759	0.997	0.996
Overall significance	0.002	0.000	0.012	0.000	0.046

Table 9-4   Searching for the Best Multiple Regression Equation

not increase by a substantial amount. For example, Table 9-4 shows that if we use only the independent variable CHEST, the adjusted R^2 is 0.980, but when we include all six variables, the adjusted R^2 increases to 0.996. Using six variables instead of only one is too high a price to pay for such a small increase in the adjusted R^2. We're better off using the single independent variable CHEST than using all six independent variables.

- For a given number of independent (x) variables, select the equation with the largest value of adjusted R^2.

- In weeding out independent variables that don't have much of an effect on the dependent variable, it might be helpful to find the linear correlation coefficient r for each pair of variables being considered. For example, using the data in Table 9-3, we will find that there is a 0.955 linear correlation for the paired NECK/HEADLEN data. Because there is such a high correlation between neck size and head length, there is no need to include both of those variables. (Also, it appears that NECK and HEADLEN are not independent variables, which violates one of the requirements for multiple regression.) In choosing between NECK and HEADLEN, we should select NECK for this reason: NECK is a better predictor of WEIGHT because the NECK/WEIGHT paired data have a linear correlation coefficient of $r = 0.971$, which is higher than $r = 0.884$ for the paired HEADLEN/WEIGHT data.

Using these guidelines in an attempt to find the best equation for predicting weights of bears, we find that for the data of Table 9-3, the best regression equation uses the single independent variable of chest size (CHEST). The best regression equation appears to be

$$\text{WEIGHT} = -195 + 11.4\,\text{CHEST}$$

or

$$\hat{y} = -195 + 11.4x_7$$

Some statistical software packages include a program for performing **stepwise regression,** whereby computations are performed with different combinations of independent variables, but there are some problems associated with it, including these: Stepwise regression will not necessarily yield the best model if some predictor variables are highly correlated (indicating that the variables are not independent as required); it yields inflated values of R^2; and it allows us to not *think* about the problem. As always, we should be careful to use computer results as a tool that helps us make intelligent decisions; we should not let the computer become the decision-maker. Instead of relying *solely* on the result of a computer stepwise regression program, consider the preceding factors when trying to identify the best multiple regression equation.

If we eliminate the variable AGE (as in Guideline 1) and then run Minitab's stepwise regression program, we will get a display suggesting that the best regression equation is the one in which CHEST is the only independent variable. (If we include all six independent variables, Minitab selects a regression equation with the independent variables AGE, NECK, LENGTH, and CHEST, with an adjusted R^2 value of 0.997 and overall significance of 0.000.) It appears that we can estimate the weight of a bear based on its chest size, and the regression equation leads to this rule: The weight of a bear (in pounds) is estimated to be 11.4 times the chest size (in inches) minus 195.

Dummy Variables and Logistic Regression

In this section, all variables have been continuous in nature. The weight of a bear can be any value over a continuous range of weights, so weight is a good example of a continuous variable. However, the biological and health sciences involve many cases involving a **dichotomous variable**, which has only *two* possible discrete values (such as male/female or dead/alive or cured/not cured). A common procedure is to represent the two possible discrete values by 0 and 1, where 0 represents a "failure" (such as death) and 1 represents a success. A dichotomous variable with the two possible values of 0 and 1 is called a **dummy variable**.

Procedures of analysis differ dramatically, depending on whether the dummy variable is an independent (x) variable or the dependent (y) variable. If we include a dummy variable as another *independent* (x) variable, we can use the methods of this section.

EXAMPLE **Using a Dummy Variable** Use the height, weight, waist, and pulse rates of the combined data set of 80 women and men as listed in Data Set 1 in Appendix B. Let the dependent y variable represent height and, for the first independent variable, use the dummy variable of *gender* (coded as $0 =$ female, $1 =$ male). Given a weight of 150 lb, a waist size of 80 cm, and a pulse rate of 75 beats per minute, find the multiple regression equation and use it to predict the height of (a) a female and (b) a male.

SOLUTION   Using the methods of this section with a software package, we get this regression equation:

$$HT = 64.4 + 3.47(GENDER) + 0.118(WT) \\ - 0.222(WAIST) + 0.00602(PULSE)$$

To find the predicted height of a female, we substitute 0 for the gender variable. Also substituting 150 for weight, 80 for waist, and 75 for pulse results in a predicted height of 64.8 in. (or 5 ft 5 in.) for a female.

To find the predicted height of a male, we substitute 1 for the gender variable. Also substituting the other values results in a predicted height of 68.3 in. (or 5 ft 8 in.) for a male. Note that when all other variables are the same, a male will have a predicted height that is about 3 in. more than the height of a female.

In the preceding example, we could use the methods of this section because the dummy variable of gender is an *independent* variable. If the dummy variable is the dependent (y) variable, we cannot use the methods of this section, and we should use a method known as **logistic regression**. Suppose, for example, that we use height, weight, waist, and pulse rates of women and men as listed in Data

Set 1 in Appendix B. Let the dependent y variable represent gender (0 = female, 1 = male). Using the 80 values of y (with female coded as 0 and male coded as 1) and the combined list of corresponding heights, weights, waist sizes, and pulse rates, we can use logistic regression to obtain this model:

$$\ln\left(\frac{p}{1-p}\right) = -41.8193 + 0.679195(\text{HT}) - 0.0106791(\text{WT})$$
$$+ 0.0375373(\text{WAIST}) - 0.0606805(\text{PULSE})$$

In the above expression, p represents a probability. A value of $p = 0$ indicates that the person is a female and $p = 1$ indicates a male. A value of $p = 0.2$ indicates that there is a 0.2 chance of the person being a male, so it follows that there is a 0.8 chance that the person is a female. If we use the above model and substitute a height of 72 in., a weight of 200 lb, a waist circumference of 90 cm, and a pulse rate of 85 beats per minute, we can solve for p to get $p = 0.960$, indicating that such a large person is very likely to be a male. In contrast, a small person with a height of 60 in., a weight of 90 lb, a waist size of 68 cm, and a pulse rate of 85 beats per minute results in a value of $p = 0.00962$, indicating that such a small person is unlikely to be a male and is very likely to be a female. This book does not include detailed procedures for using logistic regression, but several books are devoted to this topic, and several other textbooks include detailed information about this method.

When we discussed regression in Section 9-3, we listed four common errors that should be avoided when using regression equations to make predictions. These same errors should be avoided when using multiple regression equations. Be especially careful about concluding that a cause-effect association exists.

9-5 Exercises

Interpreting a Computer Display. *In Exercises 1–4, refer to the SPSS display that follows and answer the given questions or identify the indicated items. The SPSS display is based on the sample of 54 bears listed in Data Set 6 in Appendix B.*

1. Bear Measurements Identify the multiple regression equation that expresses weight in terms of head length, length, and chest size.

2. Bear Measurements Identify the following:
   a. The P-value corresponding to the overall significance of the multiple regression equation
   b. The value of the multiple coefficient of determination R^2
   c. The adjusted value of R^2

3. Bear Measurements Is the multiple regression equation usable for predicting a bear's weight based on its head length, length, and chest size? Why or why not?

4. Bear Measurements A bear is found to have a head length of 14.0 in., a length of 70.0 in., and a chest size of 50.0 in.
   a. Find the predicted weight of the bear.
   b. The bear in question actually weighed 320 lb. How accurate is the predicted weight from part (a)?

SPSS

Model Summary

Model	R	R Square	Adjusted R Square	Std. Error of the Estimate
1	.963[a]	.928	.924	33.656

a. Predictors: (Constant), Distance around the chest, Length of Head, Length of body

ANOVA[b]

Model		Sum of Squares	df	Mean Square	F	Sig.
1	Regression	729645.5	3	243215.153	214.711	.000[a]
	Residual	56637.875	50	1132.758		
	Total	786283.3	53			

a. Predictors: (Constant), Distance around the chest, Length of Head, Length of body
b. Dependent Variable: Measured Weight

Coefficients[a]

Model		Unstandardized Coefficients B	Std. Error	Standardized Coefficients Beta	t	Sig.
1	(Constant)	-271.711	31.617		-8.594	.000
	Length of Head	-.870	5.676	-.015	-.153	.879
	Length of body	.554	1.259	.049	.440	.662
	Distance around the chest	12.153	1.116	.933	10.891	.000

a. Dependent Variable: Measured Weight

Health Data: Finding the Best Multiple Regression Equation. In Exercises 5–8, refer to the accompanying table, which was obtained by using the data for males in Data Set 1 in Appendix B. The dependent variable is weight (in pounds), and the independent variables are HT (height in inches), WAIST (waist circumference in cm), and CHOL (cholesterol in mg).

Independent Variables	P-value	R^2	Adjusted R^2	Regression Equation
HT, WAIST, CHOL	0.000	0.880	0.870	$\hat{y} = -199 + 2.55HT + 2.18WAIST - 0.00534CHOL$
HT, WAIST	0.000	0.877	0.870	$\hat{y} = -206 + 2.66HT + 2.15WAIST$
HT, CHOL	0.002	0.277	0.238	$\hat{y} = -148 + 4.65 HT + 0.00589CHOL$
WAIST, CHOL	0.000	0.804	0.793	$\hat{y} = -42.8 + 2.41WAIST - 0.0106CHOL$
HT	0.001	0.273	0.254	$\hat{y} = -139 + 4.55 HT$
WAIST	0.000	0.790	0.785	$\hat{y} = -44.1 + 2.37WAIST$
CHOL	0.874	0.001	0.000	$\hat{y} = 173 - 0.00233CHOL$

5. If only one independent variable is used to predict weight, which single variable is best? Why?

6. If exactly two independent variables are to be used to predict weight, which two variables should be chosen? Why?

7. Which regression equation is best for predicting the weight? Why?

8. If a male has a height of 72 in., a waist circumference of 105 cm, and a cholesterol level of 250 mg, what is the best predicted value of his weight? Is that predicted value likely to be a good estimate? Is that predicted value likely to be very accurate?

9. Heights of Parents and Children   Refer to Data Set 3 in Appendix B.
   a. Find the regression equation that expresses the dependent variable of a child's height in terms of the independent variable of the height of the mother.
   b. Find the regression equation that expresses the dependent variable of a child's height in terms of the independent variable of the height of the father.
   c. Find the regression equation that expresses the dependent variable of a child's height in terms of the independent variables of the height of the mother and the height of the father.
   d. For the regression equations found in parts (a), (b), and (c), which is the best equation for predicting the height of a child? Why?
   e. Is the best regression equation identified in part (d) a *good* equation for predicting a child's height? Why or why not?

10. Petal Lengths of Irises   Refer to Data Set 7 in Appendix B and use only the measurements from irises in the Setosa class.
   a. Find the regression equation that expresses the dependent variable of petal length in terms of the independent variable of the petal width.
   b. Find the regression equation that expresses the dependent variable of a petal length in terms of the independent variable of sepal length.
   c. Find the regression equation that expresses the dependent variable of a petal length in terms of the independent variables of the petal width and sepal length.
   d. For the regression equations found in parts (a), (b), and (c), which is the best equation for predicting the petal length? Why?
   e. Is the best regression equation identified in part (d) a *good* equation for predicting petal length? Why or why not?

11. Using a Dummy Variable   Refer to Data Set 6 in Appendix B and use the sex, age, and weight of the bears. For sex, let 0 represent female and let 1 represent male. (In Data Set 6, males are already represented by 1, but for females change the sex values of 2 to 0.) Letting the dependent variable be weight, use the variable of age and the dummy variable of sex to find the multiple regression equation, then use it to find the predicted weight of a bear with the characteristics given below. Does sex appear to have much of an effect on the weight of a bear?
   a. Female bear that is 20 years of age
   b. Male bear that is 20 years of age

Review

This chapter presents basic methods for investigating associations or correlations between two or more variables.

- Section 9-2 used scatter diagrams and the linear correlation coefficient to decide whether there is a linear correlation between two variables.
- Section 9-3 presented methods for finding the equation of the regression line that (by the least-squares criterion) best fits the paired data. When there is a significant linear correlation, the regression equation can be used to predict the value of a variable, given some value of the other variable.
- Section 9-4 introduced the concept of total variation, with components of explained and unexplained variation. We defined the coefficient of determination r^2 to be the quotient obtained by dividing explained variation by total variation. We also developed methods for constructing prediction intervals, which are helpful in judging the accuracy of predicted values.
- In Section 9-5 we considered multiple regression, which allows us to investigate associations among several variables. We discussed procedures for obtaining a multiple regression equation, as well as the value of the multiple coefficient of determination R^2, the adjusted R^2, and a P-value for the overall significance of the equation.

Review Exercises

1. Crickets and Temperature One classic application of correlation and regression involves the association between the temperature and the number of times a cricket chirps in a minute. Listed below are the numbers of chirps in 1 minute and the corresponding temperature in degrees Fahrenheit (based on data from *The Song of Insects* by George W. Pierce, Harvard University Press).
   a. Is there sufficient evidence to conclude that there is an association between the number of chirps in a minute and the temperature?
   b. Find the equation of the straight line that best fits the sample data.
   c. If a cricket chirps 1234 times in 1 minute, what is the best predicted value for the temperature?
   d. What percentage of the variation in temperature can be explained by the variation in the cricket chirp rate?
   e. Can the correlation/regression results be used to show that changes in temperature cause changes in the cricket chirp rate?

Chirps in one minute	882	1188	1104	864	1200	1032	960	900
Temperature (°F)	69.7	93.3	84.3	76.3	88.6	82.6	71.6	79.6

2. DWI and Jail A study was conducted to investigate the association between age (in years) and BAC (blood alcohol concentration) measured when convicted DWI (driving while intoxicated) jail inmates were first arrested. Sample data are given below for randomly selected subjects (based on data from the Dutchess County STOP-DWI Program). Based on the result, does the BAC level seem to be related to the age of the person tested?

Age	17.2	43.5	30.7	53.1	37.2	21.0	27.6	46.3
BAC	0.19	0.20	0.26	0.16	0.24	0.20	0.18	0.23

Iowa Corn Data. In Exercises 3–6, use the sample data in the accompanying table. The data were collected from Iowa during a 10-year period. The amounts of precipitation are the annual totals (in inches). The average temperatures are annual averages (in degrees Fahrenheit). The corn values are the amounts of corn produced (in millions of bushels). The values of acres harvested are in thousands of acres.

Year	1	2	3	4	5	6	7	8	9	10
Precipitation	32.4	41.8	36.5	37.5	31.6	40.3	33.3	21.6	24.7	39.4
Average temperature	49.80	46.93	48.47	48.28	46.69	49.31	51.87	49.34	47.06	49.88
Corn	1731	1578	744	1445	1707	1627	1320	899	1446	1562
Acres harvested	13,850	13,150	8,550	12,900	13,550	12,050	10,150	10,700	12,250	12,400

3. a. Use a 0.05 significance level to test for a significant linear correlation between the annual precipitation amounts and corn production amounts.

   b. Using the precipitation amounts and the corn production amounts, find the equation of the regression line. Let the precipitation amount be the independent variable.

   c. What is the best predicted amount of corn production for a year in which the precipitation amount is 29.3 in.?

4. a. Use a 0.05 significance level to test for a significant linear correlation between average annual temperature and corn production.

   b. Using the average annual temperatures and the corn production amounts, find the equation of the regression line. Let the average annual temperature be the independent variable.

   c. What is the best predicted amount of corn production for a year in which the average annual temperature is 48.86°F?

5. a. Use a 0.05 significance level to test for a significant linear correlation between the number of acres harvested and corn production amounts.

   b. Using the numbers of acres harvested and the corn production amounts, find the equation of the regression line. Let the acres harvested be the independent variable.

   c. What is the best predicted amount of corn production for a year in which 13,300,000 acres are harvested? (Remember, the table values of acres harvested are in thousands of acres.)

6. Iowa Weather/Corn Let y = corn production, x_1 = annual precipitation, x_2 = average annual temperature, and x_3 = acres harvested (in thousands of acres). Use software to find the multiple regression equation of the form $\hat{y} = b_0 + b_1x_1 + b_2x_2 + b_3x_3$. Also identify the value of the multiple coefficient of determination R^2, the adjusted R^2, and the P-value representing the overall significance of the multiple regression equation. Based on the results, should the multiple regression equation be used for making predictions? Why or why not?

7. Exercise and Stress Refer to the systolic blood pressure readings in Data Set 12 in Appendix B, and use only the "pre-exercise with no stress" and "pre-exercise with math stress" readings for the first six subjects (females and black).

   a. Test for a linear correlation.

b. Find the equation of the regression line.

c. Based on the preceding results, can we conclude that the stress from the math test has an effect on systolic blood pressure? Why or why not?

Cumulative Review Exercises

1. Effects of Heredity and Environment on IQ In studying the effects of heredity and environment on intelligence, it has been helpful to analyze the IQs of identical twins who were separated soon after birth. Identical twins share identical genes. By studying identical twins raised apart, we can eliminate the variable of heredity and better isolate the effects of the environment. The accompanying table shows the IQs of pairs of identical twins (older twins are x) raised apart (based on data from "IQs of Identical Twins Reared Apart," by Arthur Jensen, *Behavioral Genetics*). The sample data given here are typical of those obtained from other studies.

a. Find the mean and standard deviation of the sample of older twins.

b. Find the mean and standard deviation of the sample of younger twins.

c. Based on the results from parts (a) and (b), does there appear to be a difference between the means of the two populations? In exploring the association between IQs of twins, is such a comparison of the two sample means the best approach? Why or why not?

d. Combine all of the sample IQ scores, then use a 0.05 significance level to test the claim that the mean IQ score of twins reared apart is different from the mean IQ of 100.

e. Is there an association between IQs of twins who were separated soon after birth? What method did you use? Write a summary statement about the effect of heredity and environment on intelligence, and note that your conclusions will be based on this relatively small sample of 12 pairs of identical twins.

x	107	96	103	90	96	113	86	99	109	105	96	89
y	111	97	116	107	99	111	85	108	102	105	100	93

2. Measuring Lung Volumes In a study of techniques used to measure lung volumes, physiological data were collected for 10 subjects. The values given in the accompanying table are in liters, representing the measured forced vital capacities of the 10 subjects in a sitting position and in a supine (lying) position. The issue we want to investigate is whether the position (sitting or supine) has an effect on the measured values.

a. If we test for a correlation between the sitting values and the supine values, will the result allow us to determine whether the position (sitting or supine) has an effect on the measured values? Why or why not?

b. Use an appropriate test for the claim that the position has no effect, so the mean difference is zero.

Subject	A	B	C	D	E	F	G	H	I	J
Sitting	4.66	5.70	5.37	3.34	3.77	7.43	4.15	6.21	5.90	5.77
Supine	4.63	6.34	5.72	3.23	3.60	6.96	3.66	5.81	5.61	5.33

Based on data from "Validation of Esophageal Balloon Technique at Different Lung Volumes and Postures," by Baydur et al., *Journal of Applied Physiology*, Vol. 62, No. 1.

Cooperative Group Activities

1. *In-class activity:* Divide into groups of 8 to 12 people. For each group member, measure the person's height and also measure his or her navel height, which is the height from the floor to the navel. Is there a correlation between height and navel height? If so, find the regression equation with height expressed in terms of navel height. According to an old theory, the average person's ratio of height to navel height is the golden ratio: $(1 + \sqrt{5})/2 \approx 1.6$. Does this theory appear to be reasonably accurate?

2. *In-class activity:* Divide into groups of 8 to 12 people. For each group member, *measure* height and arm span. For the arm span, the subject should stand with arms extended, like the wings on an airplane. It's easy to mark the height and arm span on a chalkboard, then measure the distances there. Using the paired sample data, is there a correlation between height and arm span? If so, find the regression equation with height expressed in terms of arm span. Can arm span be used as a reasonably good predictor of height?

3. *In-class activity:* Divide into groups of 8 to 12 people. For each group member, use a string and ruler to measure head circumference and forearm length. Is there an association between these two variables? If so, what is it?

Technology Project

In Exercise 1 of the Cumulative Review Exercises, we noted that when studying the effects of heredity and environment on intelligence, it has been helpful to analyze the IQs of identical twins who were separated soon after birth. In this project, we will simulate 100 sets of twin births, but we will generate their IQ scores in a way that has no common genetic or environmental influences. Using the same procedure described in the Technology Project at the end of Chapter 5, generate a list of 100 simulated IQ scores randomly selected from a normally distributed population having a mean of 100 and a standard deviation of 15. Now use the same procedure to generate a second list of 100 simulated IQ scores that are also randomly selected from a normally distributed population with a mean of 100 and a standard deviation of 15. Even though the two lists were independently generated, treat them as paired data (so that the first score from each list represents the first set of twins, the second score from each list repre-

sents the second set of twins, and so on). Before doing any calculations, first estimate a value of the linear correlation coefficient that you would expect. Now use the methods of Section 9-2 with a 0.05 significance level to test for a significant linear correlation and state your results.

Consider the preceding procedure to be one trial. Given the way that the sample data were generated, what proportion of such trials should lead to the incorrect conclusion that there is a significant linear correlation? By repeating the trials, we can verify that the proportion is approximately correct. Either repeat the trial or combine your results with others to verify that the proportion is approximately correct. Note that a type I error is the mistake of rejecting a true null hypothesis which, in this case, means that we conclude that there is a significant linear correlation when there really is no such linear correlation.

from DATA to DECISION

Critical Thinking: Should boating restrictions be imposed to reduce the killing of manatees?

Manatees, also called "sea cows," are large mammals that live underwater, often in or near waterways with considerable boat traffic. In Florida, manatee deaths from powerboat encounters have been at the center of considerable controversy between environmentalists and boat operators. Andrew Revkin, author of the *New York Times* article "How Endangered a Species?", wrote that "the (manatee) deaths from boating have continued despite the creation of a network of refuges and slow-speed boating zones, and a result is one of the country's most intense debates over an endangered species." The article included two side-by-side graphs similar to those shown below. (The *New York Times* graphs included data from 1976 through 2000.) The graphs below reflect the data in the accompanying table (based on data from the Florida Department of Highway Safety and Motor Vehicles and the Florida Marine Research Institute).

Analysis

Use the methods of this chapter to address these key issues:

- In comparison to the graphs shown below, is there another graph that better illustrates the association between the number of registered boats and the number of manatee deaths from boats?

- Determine whether there is an association between the numbers of registered boats and the numbers of manatees killed by boats.

- If there is an association between the numbers of registered boats and the numbers of manatees killed by boats, describe it with an equation.

- In addition to the numbers of registered boats, are there other important variables that affect the numbers of manatees killed by boats?

- Should regulations be imposed to reduce the killing of manatees?

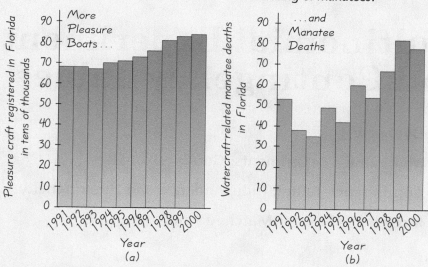

Registered Florida Pleasure Craft (in tens of thousands) and Watercraft-Related Manatee Deaths

Year	1991	1992	1993	1994	1995	1996	1997	1998	1999	2000
x: Boats	68	68	67	70	71	73	76	81	83	84
y: Manatee Deaths	53	38	35	49	42	60	54	67	82	78

10

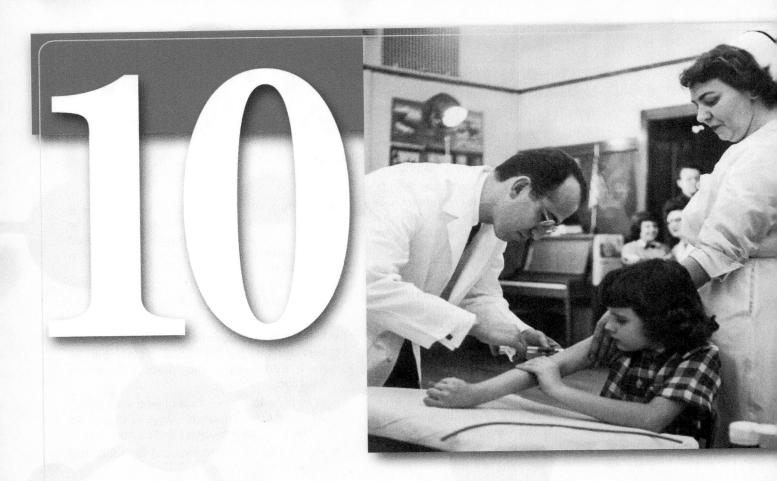

Multinomial Experiments and Contingency Tables

Determining whether a treatment is effective

When Paul Meier of the University of Chicago wrote about the 1954 clinical trial of the Salk poliomyelitis vaccine, he stated that it was "the biggest public health experiment ever." In that experiment, some children were given the Salk poliomyelitis vaccine and others were given a placebo. The experiment results are summarized in Table 10-1. Note that the data in Table 10-1 are categorical data structured in the format of a table with two rows and two columns. There is a row variable: whether the subject was given the Salk vaccine or a placebo. There is also a column variable: whether the subject developed polio.

If we informally examine the results in Table 10-1, it might appear that the Salk vaccine is effective, because the sizes of the treatment group and placebo group are approximately the same, but only 33 vaccinated children developed paralytic polio, compared to 115 children from the placebo group. However, instead of relying on such a subjective judgment, it would be much better if we could apply objective criteria. This chapter will present such objective methods.

Table 10-1	Salk Vaccine Experiment		
		Developed Paralytic Polio?	
		Yes	No
Salk vaccine treatment group		33	200,712
Placebo group		115	201,114

10-1 Overview

In this chapter we continue to apply inferential methods to different configurations of data. Recall from Chapter 1 that categorical (or qualitative, or attribute) data are data that can be separated into different categories (often called **cells**) that are distinguished by some characteristic not based on numerical measures. For example, we might separate a sample of M&Ms into the color categories of red, orange, yellow, brown, blue, and green. After finding the frequency count for each category, we might proceed to test the claim that the frequencies fit (or agree with) the color distribution claimed by the manufacturer (Mars, Inc.). The main objective of this chapter is to test claims about categorical data consisting of frequency counts for different categories. In Section 10-2 we will consider multinomial experiments, which consist of observed frequency counts arranged in a single row or column (called a one-way frequency table), and we will test the claim that the observed frequency counts agree with some claimed distribution. In Section 10-3 we will consider contingency tables (or two-way frequency tables), which consist of frequency counts arranged in a table with at least two rows and two columns. We will use contingency tables for two types of very similar tests: (1) tests of independence, which test the claim that the row and column variables are independent; and (2) tests of homogeneity, which test the claim that different populations have the same proportion of some specified characteristic. In Section 10-4 we consider tables consisting of frequency counts that result from *matched pairs*.

We will see that the methods of this chapter use the same χ^2 (chi-squared) distribution that was first introduced in Section 6-5. Here are important properties of the chi-square distribution:

1. Unlike the normal and Student t distributions, the chi-square distribution is not symmetric. (See Figure 10-1.)

2. The values of the chi-square distribution can be 0 or positive, but they cannot be negative. (See Figure 10-1.)

3. The chi-square distribution is different for each number of degrees of freedom. (See Figure 10-2.)

Critical values of the chi-square distribution are found in Table A-4.

FIGURE 10-1 **The Chi-Square Distribution**

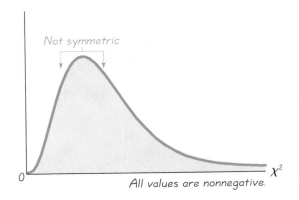

Not symmetric

0

All values are nonnegative.

χ^2

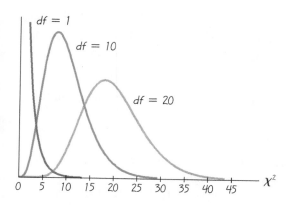

Safest Airplane Seats

Many of us believe that the rear seats are safest in an airplane crash. Safety experts do not agree that any particular part of an airplane is safer than others. Some planes crash nose first when they come down, but others crash tail first on takeoff. Matt McCormick, a survival expert for the National Transportation Safety Board, told *Travel* magazine that "there is no one safe place to sit." Goodness-of-fit tests can be used with a null hypothesis that all sections of an airplane are equally safe. Crashed airplanes could be divided into the front, middle, and rear sections. The observed frequencies of fatalities could then be compared to the frequencies that would be expected with a uniform distribution of fatalities. The χ^2 test statistic reflects the size of the discrepancies between observed and expected frequencies, and it would reveal whether some sections are safer than others.

10-2 Multinomial Experiments: Goodness-of-Fit

Each data set in this section consists of data that have been separated into different categories. The main objective is to determine whether the distribution agrees with or "fits" some claimed distribution. We define a *multinomial experiment* the same way we defined a binomial experiment (Section 4-3), except that a multinomial experiment has more than two categories (unlike a binomial experiment, which has exactly two categories).

Definition

A **multinomial experiment** is an experiment that meets the following conditions:

1. The number of trials is fixed.
2. The trials are independent.
3. All outcomes of each trial must be classified into exactly one of several different categories.
4. The probabilities for the different categories remain constant for each trial.

EXAMPLE Last-Digit Analysis of Weights Thousands of subjects are routinely studied as part of a national health examination. The examination procedures are quite exact. For example, when obtaining weights of subjects, it is extremely important to actually weigh the individuals instead of asking them to report their weights. When asked, people have been known to provide weights that are somewhat lower than their actual weights. So how can researchers verify that weights were obtained through actual measurements instead of asking subjects? One method is to analyze the *last digits* of the weights. When people report weights, they tend to round down—sometimes *way* down. Such reported weights tend to have last digits with disproportionately more 0s and 5s than the last digits of weights obtained through a measurement process. In contrast, if people are actually weighed, the weights tend to have last digits that are uniformly distributed, with 0, 1, 2, . . . , 9 all occurring with roughly the same frequencies.

continued

Table 10-2

Last Digits of Weights

Last Digit	Frequency
0	35
1	0
2	2
3	1
4	4
5	24
6	1
7	4
8	7
9	2

One of the authors obtained weights from 80 randomly selected students, and those weights had last digits summarized in Table 10-2. Later, we will analyze the data, but for now, simply verify that the four conditions of a multinomial experiment are satisfied.

SOLUTION   Here is the verification that the four conditions of a multinomial experiment are all satisfied:

1. The number of trials (last digits) is the fixed number 80.
2. The trials are independent, because the last digit of any individual weight does not affect the last digit of any other weight.
3. Each outcome (last digit) is classified into exactly 1 of 10 different categories. The categories are identified as 0, 1, 2, . . . , 9.
4. In testing the claim that the 10 digits are equally likely, each possible digit has a probability of 1/10, and by assumption, that probability remains constant for each subject.

In this section we are presenting a method for testing a claim that in a multinomial experiment, the frequencies observed in the different categories fit some given distribution. Because we test for how well an observed frequency distribution fits some specified theoretical distribution, this method is often called a *goodness-of-fit test*.

Definition

A **goodness-of-fit test** is used to test the hypothesis that an observed frequency distribution fits (or conforms to) some claimed distribution.

For example, using the data in Table 10-2, we can test the hypothesis that the data fit a uniform distribution, with all of the digits being equally likely. Our goodness-of-fit tests will incorporate the following notation.

Notation

O	represents the *observed frequency* of an outcome.
E	represents the *expected frequency* of an outcome.
k	represents the *number of different categories* or outcomes.
n	represents the total *number of trials*.

Finding Expected Frequencies

In Table 10-2 we see that the observed frequencies O are 35, 0, 2, 1, 4, 24, 1, 4, 7, and 2. The sum of the observed frequencies is 80, so $n = 80$. If we assume that the 80 digits were obtained from a population in which all digits are equally likely, then we *expect* that each digit should occur in $1/10$ of the 80 trials, so each of the 10 expected frequencies is given by $E = 8$. If we generalize this result, we get an easy procedure for finding expected frequencies whenever we are assuming that all of the expected frequencies are equal: Simply divide the total number of observations by the number of different categories ($E = n/k$). In other cases where the expected frequencies are not all equal, we can often find the expected frequency for each category by multiplying the sum of all observed frequencies and the probability p for the category, so $E = np$. We summarize these two procedures here.

- If all expected frequencies are equal, then each expected frequency is the sum of all observed frequencies divided by the number of categories, so that $E = n/k$.

- If the expected frequencies are not all equal, then each expected frequency is found by multiplying the sum of all observed frequencies by the probability for the category, so $E = np$ for each category.

As good as these two formulas for E might be, it would be better to use an informal approach based on an understanding of the circumstances. Just ask, "How can the observed frequencies be split up among the different categories so that there is perfect agreement with the claimed distribution?" Also, recognize that the *observed* frequencies must all be whole numbers because they represent actual counts, but *expected* frequencies need not be whole numbers. For example, when rolling a single die 33 times, the expected frequency for each possible outcome is $33/6 = 5.5$. The expected frequency for the number of 3s occurring is 5.5, even though it is impossible to have the outcome of 3 occur exactly 5.5 times.

We know that sample frequencies typically deviate somewhat from the values we theoretically expect, so we now present the key question: Are the differences between the actual *observed* values O and the theoretically *expected* values E statistically significant? We need a measure of the discrepancy between the O and E values, so we use the test statistic that is given with the requirements and critical values. (Later, we will explain how this test statistic was developed, but you can see that it has differences of $O - E$ as a key component.)

Requirements

1. The data have been randomly selected.

2. The sample data consist of frequency counts for each of the different categories.

3. The sample data come from a multinomial experiment. That is, there is a fixed number of trials, all of the trials are independent, each outcome belongs to exactly one of several different categories, and the probabilities remain constant for each trial.

4. For each category, the *expected* frequency is at least 5. (The expected frequency for a category is the frequency that would occur if the data actually have the distribution that is being claimed. There is no requirement that the *observed* frequency for each category must be at least 5.)

continued

**Test Statistic for Goodness-of-Fit Tests in
Multinomial Experiments**

$$\chi^2 = \sum \frac{(O - E)^2}{E}$$

Critical Values

1. Critical values are found in Table A-4 by using $k - 1$ degrees of freedom, where $k =$ number of categories.

2. Goodness-of-fit hypothesis tests are always *right-tailed*.

The form of the χ^2 test statistic is such that *close agreement* between observed and expected values will lead to a *small* value of χ^2 as well as a *large P*-value. A large discrepancy between observed and expected values will lead to a *large* value of χ^2 and a *small P*-value. The hypothesis tests of this section are therefore always right-tailed, because the critical value and critical region are located at the extreme right of the distribution. These relationships are summarized and illustrated in Figure 10-3.

FIGURE 10-3 Relationships Among the χ^2 Test Statistic, *P*-Value, and Goodness-of-Fit

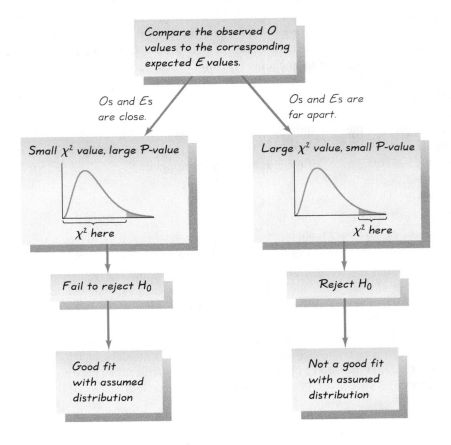

Once we know how to find the value of the test statistic and the critical value, we can test hypotheses by using the procedure introduced in Chapter 7 and summarized in Figure 7-8.

EXAMPLE **Last Digit Analysis of Weights: Equal Expected Frequencies** Let's again refer to Table 10-2 for the last digits of 80 weights. The values of 0 and 5 do seem to occur considerably more often, but is that really significant? Test the claim that the digits do *not* occur with the same frequency. Based on the results, what can we conclude about the procedure used to obtain the weights?

SOLUTION

REQUIREMENT ✓ We require that the sample data are randomly selected, they consist of frequency counts, the data come from a multinomial experiment, and each expected frequency must be at least 5. We have noted earlier that the data come from randomly selected students. The data do consist of frequency counts. The preceding example established that the conditions for a multinomial experiment are satisfied. The preceding discussion of expected values included the result that each expected frequency is 8, so each expected frequency does satisfy the requirement of being a value of at least 5. All of the requirements are satisfied and we can proceed with the hypothesis test. ✓

The claim that the digits do not occur with the same frequency is equivalent to the claim that the relative frequencies or probabilities of the 10 cells $(p_0, p_1, \ldots, p_9)$ are not all equal. We will apply our standard procedure for testing hypotheses.

Step 1:   The original claim is that the digits do not occur with the same frequency. That is, at least one of the probabilities $p_0, p_1, \ldots, p_9$ is different from the others.

Step 2:   If the original claim is false, then all of the probabilities are the same. That is, $p_0 = p_1 = \ldots = p_9$.

Step 3:   The null hypothesis must contain the condition of equality, so we have

$$H_0: p_0 = p_1 = p_2 = p_3 = p_4 = p_5 = p_6 = p_7 = p_8 = p_9$$
$$H_1: \text{At least one of the probabilities is different from the others.}$$

Step 4:   No significance level was specified, so we select $\alpha = 0.05$, a very common choice.

Step 5:   Because we are testing a claim about the distribution of the last digits being a uniform distribution, we use the goodness-of-fit test described in this section. The χ^2 distribution is used with the test statistic given earlier.

Step 6:   The observed frequencies O are listed in Table 10-2. Each corresponding expected frequency E is equal to 8 (because the 80 digits would be

continued

uniformly distributed through the 10 categories). Table 10-3 shows the computation of the χ^2 test statistic. The test statistic is $\chi^2 = 156.500$. The critical value is $\chi^2 = 16.919$ (found in Table A-4 with $\alpha = 0.05$ in the right tail and degrees of freedom equal to $k - 1 = 9$). The test statistic and critical value are shown in Figure 10-4.

Step 7:   Because the test statistic falls within the critical region, there is sufficient evidence to reject the null hypothesis.

Step 8:   There is sufficient evidence to support the claim that the last digits do not occur with the same relative frequency. We now have very strong evidence suggesting that the weights were not actually measured. It is reasonable to speculate that they were reported values instead of actual measurements.

The techniques in this section can be used to test whether an observed frequency distribution is a good fit with some theoretical frequency distribution. The preceding example tested for goodness-of-fit with a uniform distribution. Because many statistical analyses require a normally distributed population, we can use the chi-square test in this section to help determine whether given samples are drawn from normally distributed populations (see Exercise 19).

The preceding example dealt with the null hypothesis that the probabilities for the different categories are all the same. The methods of this section can also be used when the hypothesized probabilities (or frequencies) are different, as shown in the next example.

Table 10-3	Calculating the χ^2 Test Statistic for the Last Digits of Weights				
Last Digit	Observed Frequency O	Expected Frequency E	$O - E$	$(O - E)^2$	$\dfrac{(O - E)^2}{E}$
0	35	8	27	729	91.1250
1	0	8	−8	64	8.0000
2	2	8	−6	36	4.500
3	1	8	−7	49	6.125
4	4	8	−4	16	2.000
5	24	8	16	256	32.000
6	1	8	−7	49	6.125
7	4	8	−4	16	2.000
8	7	8	−1	1	0.125
9	2	8	−6	36	4.500
	80	80		$\chi^2 = \Sigma \dfrac{(O - E)^2}{E} = 156.500$	

(Except for rounding errors, these two totals must agree.)

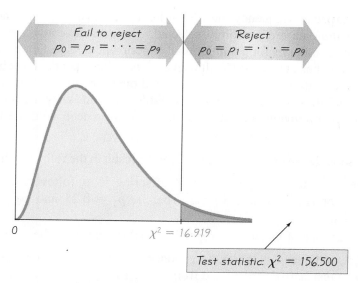

FIGURE 10-4 **Test of** $p_0 = p_1 = p_2 = p_3 = p_4 = p_5 = p_6 = p_7 = p_8 = p_9$

EXAMPLE **Genetics Experiment** Based on the genotypes of parents, offspring are expected to have genotypes distributed in such a way that 25% have genotypes denoted by AA, 50% have genotypes denoted by Aa, and 25% have genotypes denoted by aa. (With a particular disease, AA would represent a healthy normal offspring, Aa would represent a carrier, and aa would represent an offspring afflicted with the disease.) Table 10-4 lists the genotype frequencies for 90 randomly selected offspring. Test the claim that the observed genotype offspring frequencies fit the expected distribution of 25% for AA, 50% for Aa, and 25% for aa. Use a significance level of 0.01.

SOLUTION

REQUIREMENT ✓ We require that the sample data are randomly selected, they consist of frequency counts, the data come from a multinomial experiment,

continued

Genotype	Observed Frequency O	Expected Frequency E	$O - E$	$(O - E)^2$	$\dfrac{(O - E)^2}{E}$
Table 10-4 Genotypes of Offspring					
AA	22	$90 \times 0.25 = 22.5$	-0.5000	0.2500	0.0111
Aa	55	$90 \times 0.50 = 45.0$	10.0000	100.0000	2.2222
aa	13	$90 \times 0.25 = 22.5$	-9.5000	90.2500	4.0111

Total: $\chi^2 = \Sigma \dfrac{(O - E)^2}{E} = 6.244$

and each expected frequency must be at least 5. The statement of the problem notes that the data are randomly selected. The data do consist of frequency counts. The conditions for a multinomial experiment are satisfied because there is a fixed number of trials (90), all of the trials are independent, each offspring belongs to exactly one of the categories, and the probabilities remain constant. The expected frequencies are shown in Table 10-4, and they are all at least 5. All of the requirements are satisfied and we can proceed with the hypothesis test. ✓

In testing the given claim, Steps 1, 2, and 3 result in the following hypotheses:

H_0: The distribution of genotypes of offspring is as follows: 25% for AA, 50% for Aa, and 25% for aa. That is, $p_1 = 0.25$ and $p_2 = 0.50$ and $p_3 = 0.25$.

H_1: At least one of the above proportions is different from the claimed value.

Steps 4, 5, and 6 lead us to use the goodness-of-fit test with a 0.01 significance level and a test statistic calculated from Table 10-4.

The test statistic is $\chi^2 = 6.244$. The critical value of χ^2 is 9.210, and it is found in Table A-4 (using $\alpha = 0.01$ in the right tail with $k - 1 = 2$ degrees of freedom). The test statistic and critical value are shown in Figure 10-5. Because the test statistic does not fall within the critical region, there is not sufficient evidence to warrant rejection of the null hypothesis. There is not sufficient evidence to reject the claim that the distribution of genotypes for offspring is 25% for AA, 50% for Aa, and 25% for aa.

Let's again consider the sample data from the preceding example. In Figure 10-6 we graph the claimed proportions of 0.25, 0.50, and 0.25 along with the observed proportions of $22/90 = 0.244$, $55/90 = 0.611$, and $13/90 = 0.144$, so that we can visualize the discrepancy between the distribution that was claimed

FIGURE 10-5 **Testing for Agreement Between Observed and Expected Frequencies**

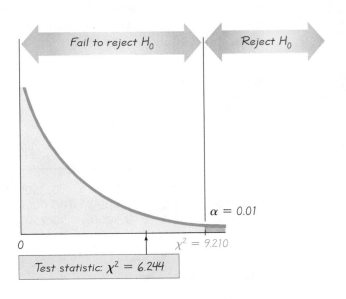

Fail to reject H_0

Reject H_0

$\alpha = 0.01$

0

$\chi^2 = 9.210$

Test statistic: $\chi^2 = 6.244$

and the frequencies that were observed. The points along the dashed line represent the claimed proportions, and the points along the solid line represent the observed proportions. The corresponding pairs of points are not very far apart, showing that the expected frequencies are not very different from the corresponding observed frequencies. However, those same proportions might have been significantly different if the sample sizes had been much larger. But graphs such as Figure 10-6 are helpful in visually comparing expected frequencies and observed frequencies, as well as suggesting which categories result in major discrepancies.

Rationale for the Test Statistic: The preceding examples should be helpful in developing a sense for the role of the χ^2 test statistic. It should be clear that we want to measure the amount of disagreement between observed and expected frequencies. Simply summing the differences between observed and expected values does not result in an effective measure because that sum is always 0, as shown below:

$$\Sigma(O - E) = \Sigma O - \Sigma E = n - n = 0$$

Squaring the $O - E$ values provides a better statistic, which reflects the differences between observed and expected frequencies. (The reasons for squaring the $O - E$ values are essentially the same as the reasons for squaring the $x - \bar{x}$ values in the formula for standard deviation.) The value of $\Sigma(O - E)^2$ measures only the magnitude of the differences, but we need to find the magnitude of the differences relative to what was expected. This relative magnitude is found through division by the expected frequencies, as in the test statistic.

The theoretical distribution of $\Sigma(O - E)^2/E$ is a discrete distribution because the number of possible values is limited to a finite number. The distribution can be approximated by a chi-square distribution, which is continuous. This approximation is generally considered acceptable, provided that all expected values E are at least 5. We included this requirement with the requirements that apply to this section. In Section 5-6 we saw that the continuous normal probability distribution can reasonably approximate the discrete binomial probability distribution, provided that np and nq are both at least 5. We now see that the continuous chi-square distribution

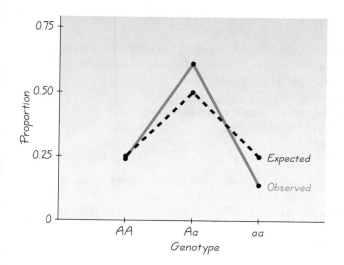

FIGURE 10-6

Comparison of Observed and Expected Frequencies

can reasonably approximate the discrete distribution of $\Sigma(O - E)^2/E$ provided that all values of E are at least 5. (There are ways of circumventing the problem of an expected frequency that is less than 5, such as combining categories so that all expected frequencies are at least 5. Also, another test might be used, such as one that calculates exact probabilities.)

The number of degrees of freedom reflects the fact that we can freely assign frequencies to $k - 1$ categories before the frequency for every category is determined. (Yet, although we say that we can "freely" assign frequencies to $k - 1$ categories, we cannot have negative frequencies nor can we have frequencies so large that their sum exceeds the total of the observed frequencies for all categories combined.)

P-Values

The examples in this section used the traditional approach to hypothesis testing, but the P-value approach can also be used. P-values are automatically provided by SPSS, STATDISK, or the TI-83/84 Plus calculator, or they can be obtained by using the methods described in Chapter 7. For instance, the preceding example resulted in a test statistic of $\chi^2 = 6.244$. That example had $k = 3$ categories, so there were $k - 1 = 2$ degrees of freedom. Referring to Table A-4, we see that for the row with 2 degrees of freedom, the test statistic of 6.244 falls between the row values of 5.991 and 7.378, so the test statistic of $\chi^2 = 6.244$ corresponds to a P-value between 0.05 and 0.025. If the calculations for the preceding example are run on SPSS, STATDISK, or a TI-83/84 Plus calculator, the display will include a P-value of 0.044. Using a 0.01 significance level (as in the preceding example), the P-value suggests that the null hypothesis should *not* be rejected. (Remember, we reject the null hypothesis when the P-value is equal to or less than the significance level.) With the traditional method of testing hypotheses, we failed to reject the claim that the observed genotype frequencies fit the claimed distribution, but the P-value of 0.044 indicates that the probability of getting genotypes like those that were obtained is 0.044.

10-2 Exercises

1. Testing Equally Likely Categories Here are the observed frequencies from four categories: 5, 6, 8, 13. Assume that we want to use a 0.05 significance level to test the claim that the four categories are all equally likely.
   a. What is the null hypothesis?
   b. What is the expected frequency for each of the four categories?
   c. What is the value of the test statistic?
   d. What is the critical value?
   e. What do you conclude about the given claim?

2. Testing Categories with Different Proportions Here are the observed frequencies from five categories: 9, 8, 13, 14, 6. Assume that we want to use a 0.05 significance level to test the claim that the five categories have proportions of 0.2, 0.2, 0.2, 0.3, and 0.1, respectively.
   a. What is the null hypothesis?
   b. What are the expected frequencies for the five categories?

c. What is the value of the test statistic?

d. What is the critical value?

e. What do you conclude about the given claim?

3. Testing Fairness of Roulette Wheel One of the authors observed 500 spins of a roulette wheel at the Mirage Resort and Casino. For each spin, the ball can land in any one of 38 different slots that are supposed to be equally likely. When SPSS was used to test the claim that the slots are in fact equally likely, the test statistic $\chi^2 = 38.232$ was obtained.

a. Find the critical value assuming that the significance level is 0.10.

b. SPSS displayed a P-value of 0.413, but what do you know about the P-value if you must use only Table A-4 along with the given test statistic of 38.232, which results from the 38 spins?

c. Write a conclusion about the claim that the 38 results are equally likely.

4. Testing a Slot Machine One of the authors purchased a slot machine (Bally Model 809), and tested it by playing it 1197 times. When testing the claim that the observed outcomes agree with the expected frequencies, a test statistic of $\chi^2 = 8.185$ was obtained. There are 10 different categories of outcome, including no win, win jackpot, win with three bells, and so on.

a. Find the critical value assuming that the significance level is 0.05.

b. What can you conclude about the P-value from Table A-4 if you know that the test statistic is $\chi^2 = 8.185$ and there are 10 categories?

c. State a conclusion about the claim that the observed outcomes agree with the expected frequencies. Does the author's slot machine appear to be working correctly?

5. Last Digits of Weights A marine biologist obtained weights of salmon caught in Juneau, Alaska. The last digits of those weights were found to have frequencies of 20, 18, 19, 17, 21, 22, 14, 18, 20, 21. Test the claim that the digits do *not* occur with the same frequency. Based on the results, what can we conclude about the procedure used to obtain the weights?

6. Flat Tire and Missed Class A classic tale involves four car-pooling students who missed a test and gave as an excuse a flat tire. On the makeup test, the instructor asked the students to identify the particular tire that went flat. If they really didn't have a flat tire, would they be able to identify the same tire? One of the authors asked 41 other students to identify the tire they would select. The results are listed in the following table (except for one student who selected the spare). Use a 0.05 significance level to test the author's claim that the results fit a uniform distribution. What does the result suggest about the ability of the four students to select the same tire when they really didn't have a flat?

Tire	Left front	Right front	Left rear	Right rear
Number selected	11	15	8	6

7. Do Car Crashes Occur on Different Days with the Same Frequency? It is a common belief that more fatal car crashes occur on certain days of the week, such as Friday or Saturday. A sample of motor vehicle deaths for a year in Montana is randomly selected. The numbers of fatalities for the different days of the week are listed in the accompanying table. At the 0.05 significance level, test the claim that accidents occur with equal frequency on the different days.

Day	Sun	Mon	Tues	Wed	Thurs	Fri	Sat
Number of fatalities	31	20	20	22	22	29	36

Based on data from the Insurance Institute for Highway Safety.

8. **Are DWI Fatalities the Result of Weekend Drinking?** Many people believe that fatal DWI crashes occur because of casual drinkers who tend to binge on Friday and Saturday nights, whereas others believe that fatal DWI crashes are caused by people who drink every day of the week. In a study of fatal car crashes, 216 cases are randomly selected from the pool in which the driver was found to have a blood alcohol content over 0.10. These cases are broken down according to the day of the week, with the results listed in the accompanying table. At the 0.05 significance level, test the claim that such fatal crashes occur on the different days of the week with equal frequency. Does the evidence support the theory that fatal DWI car crashes are due to casual drinkers or that they are caused by those who drink daily?

Day	Sun	Mon	Tues	Wed	Thurs	Fri	Sat
Number	40	24	25	28	29	32	38

Based on data from the Dutchess County STOP-DWI Program.

9. **Measuring Pulse Rates** An example in this section was based on the principle that when certain quantities are measured, the last digits tend to be uniformly distributed, but if they are estimated or reported, the last digits tend to have disproportionately more 0s or 5s. Refer to Data Set 1 in Appendix B and use the last digits of the pulse rates of the 80 men and women. Those pulse rates were obtained as part of the National Health Survey. Test the claim that the last digits of 0, 1, 2, . . . , 9 occur with the same frequency. Based on the observed digits, what can be inferred about the procedure used to obtain the pulse rates?

10. **Measuring Cholesterol Levels** Refer to Data Set 1 in Appendix B and use the last digits of the cholesterol levels of the 80 men and women. Those values were obtained as part of the National Health Survey. Test the claim that the last digits of 0, 1, 2, . . . , 9 occur with the same frequency. Based on the observed digits, what can be inferred about the procedure used to obtain the cholesterol levels?

11. **Genetics Experiment** Based on the genotypes of parents, offspring are expected to have genotypes distributed in such a way that 25% have genotypes denoted by AA, 50% have genotypes denoted by Aa, and 25% have genotypes denoted by aa. When 145 offspring are obtained, it is found that 20 of them have AA genotypes, 90 have Aa genotypes, and 35 have aa genotypes. Test the claim that the observed genotype offspring frequencies fit the expected distribution of 25% for AA, 50% for Aa, and 25% for aa. Use a significance level of 0.05.

12. **Car Crashes and Age Brackets** Among drivers who have had a car crash in the last year, 88 are randomly selected and categorized by age, with the results listed in the accompanying table. If all ages have the same crash rate, we would expect (because of the age distribution of licensed drivers) the given categories to have 16%, 44%, 27%, and 13% of the subjects, respectively. At the 0.05 significance level, test the claim that the distribution of crashes conforms to the distribution of ages. Does any age group appear to have a disproportionate number of crashes?

Age	Under 25	25–44	45–64	Over 64
Drivers	36	21	12	19

Based on data from the Insurance Information Institute.

13. **Eye Color Experiment** A researcher has developed a theoretical model for predicting eye color. After examining a random sample of parents, she predicts the eye color of the first child. The table below lists the eye colors of offspring. Based on her theory,

she predicted that 87% of the offspring would have brown eyes, 8% would have blue eyes, and 5% would have green eyes. Use a 0.05 significance level to test the claim that the actual frequencies correspond to her predicted distribution.

	Brown Eyes	Blue Eyes	Green Eyes
Frequency	132	17	0

14. Time of Birth The times of birth were recorded for 44 babies born in Brisbane, Australia, and the table below summarizes the results based on times recorded using military time (based on data from the Brisbane newspaper *The Sunday Mail* as reported in the *Journal of Statistics Education*). If birth times are uniformly distributed throughout the day, the relative frequencies in the table should be 0.250, 0.583, and 0.167 (because the classes correspond to time intervals of 6 hours, 14 hours, and 4 hours, respectively). Use a 0.05 significance level to test the claim that the births are uniformly distributed throughout the day. Does it appear that babies tend to be born during hours corresponding to regular working hours for physicians?

	24:00–05:59	06:00–19:59	20:00–23:59
Frequency	9	27	8

15. Concussions Among College Football Players Is there an association between concussions among college football players and whether they experienced previous concussions? A study included 122 concussions among college football players with no previous concussions, 41 players with one previous concussion, and 25 players with two or more concussions. Here are corresponding estimated numbers of exposures: 185,060, 42,850, and 12,583 (based on data from "Cumulative Effects Associated with Recurrent Concussion in Collegiate Football Players" by Guskiewicz et al., *Journal of the American Medical Association*, Vol. 290, No. 19). Does the distribution of concussions appear to fit the distribution of exposures? Do previous concussions appear to be associated with a greater likelihood of a concussion?

16. Participation in Clinical Trials by Race A study was conducted to investigate racial disparity in clinical trials of cancer. Among the randomly selected participants, 644 were white, 23 were Hispanic, 69 were black, 14 were Asian/Pacific Islander, and 2 were American Indian/Alaskan Native. The proportions of the U.S. population of the same groups are 0.757, 0.091, 0.108, 0.038, and 0.007, respectively. (Based on data from "Participation in Clinical Trials" by Murthy, Krumholz, and Gross, *Journal of the American Medical Association*, Vol. 291, No. 22.) Use a 0.05 significance level to test the claim that the participants fit the same distribution as the U.S. population. Why is it important to have proportionate representation in such clinical trials?

*17. Detecting Altered Experimental Data When Gregor Mendel conducted his famous hybridization experiments with peas, it appears that his gardening assistant knew the results that Mendel expected, and he altered the results to fit Mendel's expectations. Subsequent analysis of the results led to the conclusion that there is a probability of only 0.00004 that the expected results and reported results would agree so closely. How could the methods of this section be used to detect such results that are just too perfect to be realistic?

*18. Testing Goodness-of-Fit with a Binomial Distribution An observed frequency distribution is as follows:

Number of successes	0	1	2	3
Frequency	89	133	52	26

continued

a. Assuming a binomial distribution with $n = 3$ and $p = 1/3$, use the binomial probability formula to find the probability corresponding to each category of the table.

b. Using the probabilities found in part (a), find the expected frequency for each category.

c. Use a 0.05 significance level to test the claim that the observed frequencies fit a binomial distribution for which $n = 3$ and $p = 1/3$.

*19. Testing Goodness-of-Fit with a Normal Distribution An observed frequency distribution of sample IQ scores is as follows:

IQ score	Less than 80	80–95	96–110	111–120	More than 120
Frequency	20	20	80	40	40

a. Assuming a normal distribution with $\mu = 100$ and $\sigma = 15$, use the methods given in Chapter 5 to find the probability of a randomly selected subject belonging to each class. (Use class boundaries of 79.5, 95.5, 110.5, and 120.5.)

b. Using the probabilities found in part (a), find the expected frequency for each category.

c. Use a 0.01 significance level to test the claim that the IQ scores were randomly selected from a normally distributed population with $\mu = 100$ and $\sigma = 15$.

10-3   Contingency Tables: Independence and Homogeneity

In Section 10-2 we considered categorical data summarized with frequency counts listed in a single row or column. Because the cells of the single row or column correspond to categories of a single variable (such as color), the tables in Section 10-2 are sometimes called *one-way frequency tables*. In this section we again consider categorical data summarized with frequency counts, but the cells correspond to two different variables. The tables we consider in this section are called *contingency tables*, or two-way frequency tables.

Definitions

A **contingency table** (or **two-way frequency table**) is a table in which frequencies correspond to two variables. (One variable is used to categorize rows, and a second variable is used to categorize columns.)

Table 10-1 from the Chapter Problem is reproduced here, and it has two variables: a row variable, which indicates whether the subject was given the Salk vaccine or a placebo; and a column variable, which indicates whether the subject developed paralytic polio.

Contingency tables are especially important because they are often used to analyze survey results. For example, we might ask subjects one question in which they identify their gender (male/female), and we might ask another question in which they describe the frequency of their use of TV remote controls (often/

Table 10-1	Salk Vaccine Experiment		
		Developed Paralytic Polio?	
		Yes	No
Salk vaccine treatment group		33	200,712
Placebo group		115	201,114

sometimes/never). The methods of this section can then be used to determine whether the use of TV remote controls is independent of gender. (We probably already know the answer to that one.) Applications of this type are very numerous, so the methods presented in this section are among those most often used.

This section presents two types of hypothesis testing based on contingency tables. We first consider tests of independence, used to determine whether a contingency table's row variable is *independent* of its column variable. We then consider tests of homogeneity, used to determine whether different populations have the same proportions of some characteristic. Good news: Both types of hypothesis testing use the *same* basic methods. We begin with tests of independence.

Test of Independence

One of the two tests included in this section is a *test of independence* between the row variable and column variable.

> ### Definition
>
> A **test of independence** tests the null hypothesis that there is no association between the row variable and the column variable in a contingency table. (For the null hypothesis, we will use the statement that "the row and column variables are independent.")

It is very important to recognize that in this context, the word *contingency* refers to dependence, but this is only a statistical dependence, and it cannot be used to establish a direct cause-and-effect link between the two variables in question. For example, after analyzing the data in Table 10-1, we might conclude that whether a person develops paralytic polio is dependent on whether that person was given the Salk vaccine or a placebo, but that doesn't necessarily mean that the vaccine/placebo category has some direct causative effect on developing paralytic polio.

When testing the null hypothesis of independence between the row and column variables in a contingency table, the requirements, test statistic, and critical values are described in the accompanying box.

Requirements

1. The sample data consist of frequency counts obtained from independent observations, and each data value can be described by two different variables (or categories). (That is, the frequency counts from independent data can be arranged in a table consisting of at least two rows and at least two columns.)

2. The sample data are randomly selected.

3. The null hypothesis H_0 is the statement that the row and column variables are *independent;* the alternative hypothesis H_1 is the statement that the row and column variables are dependent.

4. For every cell in the contingency table, the *expected* frequency E is at least 5. (There is no requirement that every *observed* frequency must be at least 5. Also, there is no requirement that the population must have a normal distribution or any other specific distribution.)

Test Statistic for a Test of Independence

$$\chi^2 = \sum \frac{(O - E)^2}{E}$$

Critical Values

1. The critical values are found in Table A-4 by using

degrees of freedom $= (r - 1)(c - 1)$

where r is the number of rows and c is the number of columns.

2. In a test of independence with a contingency table, the critical region is located in the *right tail only.*

The test statistic allows us to measure the degree of disagreement between the frequencies actually observed and those that we would theoretically expect when the two variables are independent. Small values of the χ^2 test statistic result from close agreement between frequencies observed and frequencies expected with independent row and column variables. Large values of the χ^2 test statistic are in the rightmost region of the chi-square distribution, and they reflect significant differences between observed and expected frequencies. In repeated large samplings, the distribution of the test statistic χ^2 can be approximated by the chi-square distribution, provided that all expected frequencies are at least 5. The number of degrees of freedom $(r - 1)(c - 1)$ reflects the fact that because we know the total of all frequencies in a contingency table, we can freely assign frequencies to only $r - 1$ rows and $c - 1$ columns before the frequency for every cell is determined. [However, we cannot have negative frequencies or frequencies so large that any row (or column) sum exceeds the total of the observed frequencies for that row (or column).]

In the preceding section we knew the corresponding probabilities and could easily determine the expected values, but the typical contingency table does not come with the relevant probabilities. For each cell in the frequency table, the expected frequency E can be calculated by applying the multiplication rule of probability for independent events. Assuming that the row and column variables are independent (which is assumed in the null hypothesis), the probability of a value being in a particular cell is the probability of being in the row containing the cell (namely, the row total divided by the sum of all frequencies) multiplied by the probability of being in the column containing the cell (namely, the column total divided by the sum of all frequencies) multiplied by the sum of all frequencies. Sound too complicated? The expected frequency for a cell can be simplified to the following equation.

Expected Frequency for a Contingency Table

$$E = \frac{(\text{row total})(\text{column total})}{(\text{grand total})} \qquad \text{expected frequency for a cell}$$

Here *grand total* refers to the total of all observed frequencies in the table. For example, the expected frequency for the upper left cell of Table 10-5 (a duplicate of Table 10-1 with expected frequencies inserted in parentheses) is 73.911, which is found by noting that the total of all frequencies for the first row is 200,745, the total of the column frequencies is 148, and the sum of all frequencies in the table is 401,974, so we get an expected frequency of

$$E = \frac{(\text{row total})(\text{column total})}{(\text{grand total})} = \frac{(200,745)(148)}{401,974} = 73.911$$

EXAMPLE **Finding Expected Frequency** The expected frequency for the upper left cell of Table 10-5 is 73.911. Find the expected frequency for the lower left cell, assuming independence between the row variable (whether the person was given the Salk vaccine or a placebo) and the column variable (whether the person developed paralytic polio).

SOLUTION   The lower left cell lies in the second row (with total 201,229) and the first column (with total 148). The expected frequency is

$$E = \frac{(\text{row total})(\text{column total})}{(\text{grand total})} = \frac{(201,229)(148)}{401,974} = 74.089$$

continued

Table 10-5   Observed and Expected Frequencies

	Developed Paralytic Polio?	
	Yes	No
Salk vaccine treatment group	33 (73.911)	200,712 (200,671.089)
Placebo group	115 (74.089)	201,114 (201,154.911)

INTERPRETATION To interpret this result for the lower left cell, we can say that although 115 subjects given a placebo actually developed paralytic polio, we would have expected that number to be 74.089 subjects, assuming that the vaccine/placebo group is independent of developing paralytic polio. There is a discrepancy between $O = 115$ and $E = 74.089$, and such discrepancies are key components of the test statistic.

To better understand the rationale for finding expected frequencies with this procedure, let's pretend that we know only the row and column totals and that we must fill in the cell expected frequencies by assuming independence (or no relationship) between the two variables involved—that is, we pretend that we know only the row and column totals shown in Table 10-5. Let's begin with the cell in the upper left corner. Because 200,745 of the subjects were given the Salk vaccine, we have $P(\text{vaccine}) = 200{,}745/401{,}974$. Similarly, 148 of the subjects developed paralytic polio, so $P(\text{polio}) = 148/401{,}974$. Because we are assuming independence between the vaccine/treatment category and the column polio category of yes/no, we can use the multiplication rule of probability to get

$$P(\text{vaccine and polio}) = P(\text{vaccine}) \cdot P(\text{polio}) = \frac{200{,}745}{401{,}974} \cdot \frac{148}{401{,}974}$$

This equation is an application of the multiplication rule for independent events, which is expressed in general as follows: $P(A \text{ and } B) = P(A) \cdot P(B)$. Knowing the probability of being in the upper left cell, we can now find the *expected value* for that cell, which we get by multiplying the probability for that cell by the total number of subjects, as shown in the following expression:

$$E = n \cdot p = 401{,}974 \left[\frac{200{,}745}{401{,}974} \cdot \frac{148}{401{,}974} \right] = 73.911$$

The form of this product suggests a general way to obtain the expected frequency of a cell:

$$\text{Expected frequency } E = (\text{grand total}) \cdot \frac{(\text{row total})}{(\text{grand total})} \cdot \frac{(\text{column total})}{(\text{grand total})}$$

This expression can be simplified to

$$E = \frac{(\text{row total}) \cdot (\text{column total})}{(\text{grand total})}$$

We can now proceed to use contingency table data for testing hypotheses, as in the following example.

EXAMPLE **Testing Effectiveness of Salk Vaccine** Refer to the experiment results in Table 10-5. Using a 0.05 significance level, test the claim that whether someone is given the Salk vaccine or a placebo is independent of whether the person develops paralytic polio.

SOLUTION

REQUIREMENT ✓   As required, the data do consist of independent frequency counts, each observation can be categorized according to two variables, we are testing the null hypothesis that the variables are independent, and the expected frequencies (shown in parentheses in Table 10-5) are all at least 5. [The two variables are (1) whether the subject was given the Salk vaccine or a placebo, and (2) whether the subject developed polio or did not.] Because all of the requirements are satisfied, we can proceed with the hypothesis test. ✓

The null hypothesis and alternative hypothesis are as follows:

H_0: Whether someone is given the Salk vaccine or a placebo is independent of whether the person develops paralytic polio.

H_1: Being given the Salk vaccine or a placebo and developing polio are dependent. The significance level is $\alpha = 0.05$.

Because the data are in the form of a contingency table, we use the χ^2 distribution with this test statistic:

$$\chi^2 = \sum \frac{(O - E)^2}{E}$$

$$= \frac{(33 - 73.911)^2}{73.911} + \frac{(200{,}712 - 200{,}671.089)^2}{200{,}671.089}$$

$$+ \frac{(115 - 74.089)^2}{74.089} + \frac{(201{,}114 - 201{,}154.911)^2}{201{,}154.911}$$

$$= 22.645 + 0.008 + 22.591 + 0.008$$

$$= 45.252$$

The critical value is $\chi^2 = 3.841$ and it is found from Table A-4 by noting that $\alpha = 0.05$ in the right tail and the number of degrees of freedom is given by $(r - 1)(c - 1) = (2 - 1)(2 - 1) = 1$.

INTERPRETATION The test statistic and critical value are shown in Figure 10-7. Because the test statistic falls within the critical region, we reject the null hypothesis that whether a person gets the vaccine or placebo is independent of whether the person develops polio. It appears that whether a person gets the vaccine or placebo and whether that person develops paralytic polio are dependent variables. However, the real issue here is whether the Salk vaccine is effective. Note that the polio rate for the Salk vaccine treatment group is $33/200{,}745 = 0.000164$ (or 0.016%), whereas the polio rate for the placebo group is $115/201{,}229 = 0.000571$ (or 0.057%). Both rates are very small, but the polio rate for the Salk vaccine treatment group is roughly $1/4$ the rate for the placebo group. Some might argue that although the difference is statistically significant, the polio rates for both groups are so small that the vaccine is not really practical, but the seriousness of paralytic polio makes the vaccine extremely practical. Also, the two sample proportions of $33/200{,}745$ and $115/201{,}229$ can be

continued

FIGURE 10-7 **Test of Independence for the Salk Vaccine Experiment**

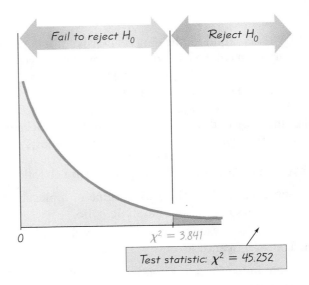

compared by conducting a hypothesis test with the alternative hypothesis of $p_1 < p_2$. Such a test can be conducted using the methods of Section 8-2. (See Exercise 10 in Section 8-2.) We might also note that some statisticians use *Yates's correction for continuity* in cells with an expected frequency of less than 10 or in all cells of a contingency table with two rows and two columns, as in Table 10-5. (See Exercise 17.)

P-Values

The preceding example used the traditional approach to hypothesis testing, but we can easily use the *P*-value approach. SPSS, SAS, STATDISK, Minitab, Excel, and the TI-83/84 Plus calculator all provide *P*-values for tests of independence in contingency tables. If a suitable calculator or statistical software package is not available, *P*-values can be estimated from Table A-4 in Appendix A. Locate the appropriate number of degrees of freedom to isolate a particular row in that table. Find where the test statistic falls in that row, and you can identify a range of possible *P*-values by referring to the areas given at the top of each column. In the preceding example, there is 1 degree of freedom, so go to the first row of Table A-4. Now use the test statistic of $\chi^2 = 45.252$ to see that this test statistic is greater than (and farther to the right of) every critical value of χ^2 that is in the first row, so the *P*-value is less than 0.005. On the basis of this small *P*-value, we again reject the null hypothesis and conclude that there is sufficient sample evidence to warrant rejection of the null hypothesis of independence.

As in Section 10-2, if observed and expected frequencies are close, the χ^2 test statistic will be small and the *P*-value will be large. If observed and expected frequencies are far apart, the χ^2 test statistic will be large and the *P*-value will be small. These relationships are summarized and illustrated in Figure 10-8.

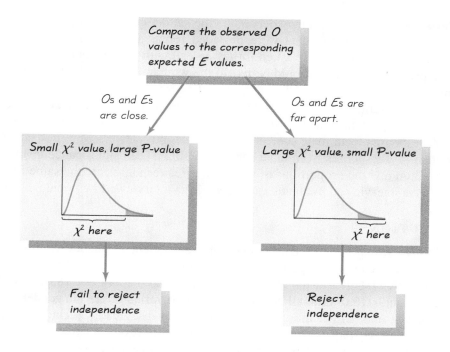

FIGURE 10-8 **Relationships Among Key Components in Test of Independence**

Test of Homogeneity

In the preceding example, we illustrated a test of independence by using a sample of 401,974 people who were included in an experiment to test the effectiveness of the Salk vaccine. We treated those 401,974 people as a random sample drawn from *one* population of all people who would find themselves in similar circumstances. However, some other samples are drawn from *different* populations, and we want to determine whether those populations have the same proportions of the characteristics being considered. The *test of homogeneity* can be used in such cases. (The word *homogeneous* means "having the same quality," and in this context, we are testing to determine whether the proportions are the same.)

> ### Definition
>
> In a **test of homogeneity,** we test the claim that *different populations* have the same proportions of some characteristics.

In conducting a test of homogeneity, we can use the same procedures already presented in this section, as illustrated in the following example.

EXAMPLE **Influence of Gender** Does a pollster's gender have an effect on poll responses by men? A *U.S. News & World Report* article about polls stated: "On sensitive issues, people tend to give 'acceptable' rather than honest

continued

responses; their answers may depend on the gender or race of the interviewer." To support that claim, data were provided for an Eagleton Institute poll in which surveyed men were asked if they agreed with this statement: "Abortion is a private matter that should be left to the woman to decide without government intervention." We will analyze the effect of gender on male survey subjects only. Table 10-6 is based on the responses of surveyed men. Assume that the survey was designed so that male interviewers were instructed to obtain 800 responses from male subjects, and female interviewers were instructed to obtain 400 responses from male subjects. Using a 0.05 significance level, test the claim that the proportions of agree/disagree responses are the same for the subjects interviewed by males and the subjects interviewed by females.

SOLUTION

REQUIREMENT ✔ The data consist of independent frequency counts, each observation can be categorized according to two variables, and the expected frequencies (shown in the accompanying Minitab display as 578.67, 289.33, 221.33, and 110.67) are all at least 5. [The two variables are (1) gender of interviewer, and (2) whether the subject agreed or disagreed.] Because this is a test of homogeneity, we test the claim that the proportions of agree/disagree responses are the same for the subjects interviewed by males and the subjects interviewed by females. All of the requirements are satisfied, so we can proceed with the hypothesis test. ✔

Because we have two separate populations (subjects interviewed by males and subjects interviewed by females), we test for homogeneity with these hypotheses:

H_0: The proportions of agree/disagree responses are the same for the subjects interviewed by males and the subjects interviewed by females.

H_1: The proportions are different.

The significance level is $\alpha = 0.05$. We use the same χ^2 test statistic described earlier, and it is calculated by using the same procedure. Instead of listing the details of that calculation, we provide the Minitab display that results from the data in Table 10-6.

Table 10-6 Gender and Survey Responses		
	Gender of Interviewer	
	Male	Female
Men who agree	560	308
Men who disagree	240	92

Minitab

```
Expected counts are printed below observed counts

              C1          C2       Total
   1          560         308       868
            578.67      289.33

   2          240          92       332
            221.33      110.67

   Total      800         400      1200

   Chi-Sq = 0.6062 + 1.204 +
            1.574 + 3.149 = 6.529

   DF = 1, P-Value = 0.011
```

The Minitab display shows the expected frequencies of 578.67, 289.33, 221.33, and 110.67. The display also includes the test statistic of $\chi^2 = 6.529$ and the P-value of 0.011. Using the P-value approach to hypothesis testing, we reject the null hypothesis of equal (homogeneous) proportions. There is sufficient evidence to warrant rejection of the claim that the proportions are the same. It appears that response and the gender of the interviewer are dependent. Although this statistical analysis cannot be used to justify any statement about causality, it does appear that men are influenced by the gender of the interviewer.

Fisher Exact Test

For the analysis of 2 × 2 tables, we have included the requirement that every cell must have an expected frequency of 5 or greater. This requirement is necessary for the χ^2 distribution to be a suitable approximation to the exact distribution of the test statistic $\sum \frac{(O - E)^2}{E}$. Consequently, if a 2 × 2 table has a cell with an expected frequency less than 5, the preceding procedures should not be used, because the χ^2 distribution is not a suitable approximation. The *Fisher exact test* is often used for such a 2 × 2 table, because it provides an *exact* P-value and does not require an approximation technique.

Consider the data in Table 10-7, with expected frequencies shown in parentheses below the observed frequencies. The first cell has an expected frequency less than 5, so the preceding methods should not be used. With the Fisher exact test, we calculate the probability of getting the observed results by chance (assuming that wearing a helmet and receiving facial injuries are independent), and we also calculate the probability of any result that is *more extreme*. This use of "more extreme" results can be a somewhat confusing concept, so let's again consider a situation first discussed in Section 4-2. (See Section 4-2 for the subsection of "Using Probabilities to Determine When Results Are Unusual.") If you are flipping a coin 1000 times to determine whether it favors heads, a result of 501 heads is not

Table 10-7	Helmets and Facial Injuries in Bicycle Accidents (Expected frequencies are in parentheses)	
	Helmet Worn	No Helmet
Facial injuries received	2	13
	(3)	(12)
All injuries nonfacial	6	19
	(5)	(20)

evidence that the coin favors heads, because it is very easy to get a result like 501 heads in 1000 tosses just by chance. Yet, the probability of getting *exactly* 501 heads in 1000 tosses is actually quite small: 0.0252. This low probability reflects the fact that with 1000 tosses, any *specific* number of heads will have a very low probability. However, we do not consider 501 heads among 1000 tosses to be *unusual*, because the probability of getting *at least* 501 heads is high: 0.487. We generalized this principle as follows:

- **Unusually high:** *x* successes among *n* trials is an *unusually high* number of successes if $P(x$ or more$)$ is very small (such as 0.05 or less)

- **Unusually low:** *x* successes among *n* trials is an *unusually low* number of successes if $P(x$ or fewer$)$ is very small (such as 0.05 or less)

When testing the null hypothesis of independence between wearing a helmet and receiving a facial injury, the frequencies of 2, 13, 6, 19 can be replaced by 1, 14, 7, 18, respectively, to obtain *more extreme* results with the same row and column totals. (The Fisher exact test is sometimes criticized because the use of fixed row and column totals is often unrealistic.) The Fisher exact test requires that we find the probabilities for the observed frequencies and each set of more extreme frequencies. Those probabilities are then added to provide an exact *P*-value. [The individual probabilities are computed using this hypergeometric probability formula: $P = \dfrac{(a+b)!(c+d)!(a+c)!(b+d)!}{a!b!c!d!(a+b+c+d)!}$, where *a, b, c, d* are the cell frequencies.] Because the calculations are typically quite complex, it's a good idea to use software. For the data in Table 10-7, STATDISK, SPSS, SAS, and Minitab use Fisher's exact test to obtain an exact *P*-value of 0.686. Because this exact *P*-value is not small (such as less than 0.05), we fail to reject the null hypothesis that wearing a helmet and receiving facial injuries are independent. (See Exercise 2.)

Matched Pairs In addition to the requirement that each cell must have an expected frequency of at least 5, the methods of this section also require that the individual observations must be *independent*. If a 2 × 2 table consists of frequency counts that result from *matched pairs*, we do not have the required independence. For such cases, we can use McNemar's test, introduced in the following section.

10-3 Exercises

1. Is There Racial Profiling? *Racial profiling* is the controversial practice of targeting someone for criminal behavior on the basis of the person's race, national origin, or ethnicity. The accompanying table summarizes results for randomly selected drivers stopped by police in a recent year (based on data from the U.S. Department of Justice, Bureau of Justice Statistics). Using the data in this table results in the Minitab display. Use a 0.05 significance level to test the claim that being stopped is independent of race and ethnicity. Based on the available evidence, can we conclude that racial profiling is being used?

Race and Ethnicity

	Black and Non-Hispanic	White and Non-Hispanic
Stopped by police	24	147
Not stopped by police	176	1253

Minitab

```
Chi-Sq = 0.322 + 0.046 +
               0.039 + 0.006 = 0.413
DF = 1, P-Value = 0.521
```

2. Testing Effectiveness of Bicycle Helmets A study was conducted of 531 persons injured in bicycle crashes, and randomly selected sample results are summarized in the accompanying table. The SPSS results also are shown; see the results corresponding to the Pearson chi-square test. At the 0.05 significance level, test the claim that wearing a helmet has no effect on whether facial injuries are received. Based on these results, does a helmet seem to be effective in helping to prevent facial injuries in a crash?

	Helmet Worn	No Helmet
Facial injuries received	30	182
All injuries nonfacial	83	236

Based on data from "A Case-Control Study of the Effectiveness of Bicycle Safety Helmets in Preventing Facial Injury," by Thompson, Thompson, Rivara, and Wolf, *American Journal of Public Health,* Vol. 80, No. 12.

SPSS

Chi-Square Tests

	Value	df	Asymp. Sig. (2-sided)	Exact Sig. (2-sided)	Exact Sig. (1-sided)
Pearson Chi-Square	10.708[b]	1	.001		
Continuity Correction[a]	10.011	1	.002		
Likelihood Ratio	11.147	1	.001		
Fisher's Exact Test				.001	.001
N of Valid Cases	531				

a. Computed only for a 2x2 table

b. 0 cells (.0%) have expected count less than 5. The minimum expected count is 45.11.

3. Accuracy of Polygraph Tests The data in the accompanying table summarize results from tests of the accuracy of polygraphs (based on data from the Office of Technology Assessment). Use a 0.05 significance level to test the claim that whether the subject lies is independent of the polygraph indication. What do the results suggest about the effectiveness of polygraphs?

	Polygraph Indicated Truth	Polygraph Indicated Lie
Subject actually told the truth	65	15
Subject actually told a lie	3	17

4. **Testing Influence of Gender** Table 10-6 summarizes data for male survey subjects, but the accompanying table summarizes data for a sample of women. Using a 0.01 significance level, and assuming that the sample sizes of 800 men and 400 women are predetermined, test the claim that the proportions of agree/disagree responses are the same for the subjects interviewed by males and the subjects interviewed by females.

	Gender of Interviewer	
	Male	Female
Women who agree	512	336
Women who disagree	288	64

Based on data from the Eagleton Institute.

5. **Fear of Flying Gender Gap** The Marist Institute for Public Opinion conducted a poll of 1014 adults, 48% of whom were men. The poll results show that 12% of the men and 33% of the women fear flying. After constructing a contingency table that summarizes the data in the form of frequency counts, use a 0.05 significance level to test the claim that gender is independent of the fear of flying.

6. **No Smoking** The accompanying table summarizes successes and failures when subjects used different methods in trying to stop smoking. The determination of smoking or not smoking was made five months after the treatment was begun, and the data are based on results from the Centers for Disease Control and Prevention. Use a 0.05 significance level to test the claim that success is independent of the method used. If someone wants to stop smoking, does the choice of the method make a difference?

	Nicotine Gum	Nicotine Patch
Smoking	191	263
Not smoking	59	57

7. **No Smoking** Repeat Exercise 6 after including the additional data shown in the table.

	Nicotine Gum	Nicotine Patch	Nicotine Inhaler
Smoking	191	263	95
Not smoking	59	57	27

8. **Smoking in China** The table below summarizes results from a survey of males aged 15 or older living in the Minhang District of China (based on data from "Cigarette Smoking in China" by Gong, Koplan, Feng, et al., *Journal of the American Medical Association*, Vol. 274, No. 15). Using a 0.05 significance level, test the claim that

smoking is independent of education level. What do you conclude about the relationship between smoking and education in China?

	Primary School	Middle School	College
Smoker	606	1234	100
Never smoked	205	505	137

9. Occupational Hazards Use the data in the table to test the claim that occupation is independent of whether the cause of death was homicide. The table is based on data from the U.S. Department of Labor, Bureau of Labor Statistics. Does any particular occupation appear to be most prone to homicides? If so, which one?

	Police	Cashiers	Taxi Drivers	Guards
Homicide	82	107	70	59
Cause of death other than homicide	92	9	29	42

10. Survey Refusals and Age Bracket A study of people who refused to answer survey questions provided the randomly selected sample data shown in the table. At the 0.01 significance level, test the claim that the cooperation of the subject (response or refusal) is independent of the age category. Does any particular age group appear to be particularly uncooperative?

	Age					
	18–21	22–29	30–39	40–49	50–59	60 and over
Responded	73	255	245	136	138	202
Refused	11	20	33	16	27	49

Based on data from "I Hear You Knocking But You Can't Come In," by Fitzgerald and Fuller, *Sociological Methods and Research*, Vol. 11, No. 1.

11. Firearm Training and Safety Does firearm training result in safer practices by gun owners? In one study, randomly selected subjects were surveyed with the results given in the accompanying table. Use a 0.05 significance level to test the claim that formal firearm training is independent of how firearms are stored. Does the formal training appear to have a positive effect?

	Guns Stored Loaded and Unlocked?	
	Yes	No
Had formal firearm training	122	329
Had no formal firearm training	49	299

Based on data from "Firearm Training and Storage," by Hemenway, Solnick, Azrael, *Journal of the American Medical Association*, Vol. 273, No. 1.

12. Is Seat Belt Use Independent of Cigarette Smoking? A study of seat belt users and nonusers yielded the randomly selected sample data summarized in the given table. Test the claim that the amount of smoking is independent of seat belt use. A plausible theory is that people who smoke more are less concerned about their health and safety

and are therefore less inclined to wear seat belts. Is this theory supported by the sample data?

	Number of Cigarettes Smoked per Day			
	0	1–14	15–34	35 and over
Wear seat belts	175	20	42	6
Don't wear seat belts	149	17	41	9

Based on data from "What Kinds of People Do Not Use Seat Belts?" by Helsing and Comstock, *American Journal of Public Health*, Vol. 67, No. 11.

13. Clinical Test of Lipitor The cholesterol-reducing drug Lipitor consists of atorvastatin calcium, and results summarizing headaches as an adverse reaction in clinical tests are given in the table (based on data from Parke-Davis). Using a 0.05 significance level, test the claim that getting a headache is independent of the amount of atorvastatin used as a treatment. (*Hint*: Because not all of the expected values are 5 or greater, combine the results for the treatments consisting of 20 mg of atorvastatin and 40 mg of atorvastatin.) Do headaches appear to be a concern for those who are treated with Lipitor?

	Placebo	10 mg Atorvastatin	20 mg Atorvastatin	40 mg Atorvastatin	80 mg Atorvastatin
Headache	19	47	6	2	6
No headache	251	816	30	77	88

14. Exercise and Smoking A study of the effects of exercise by women included results summarized in the table (based on data from "Physical Activity and Coronary Heart Disease in Women" by Lee, Rexrode, Cook, Manson, and Buring, *Journal of the American Medical Association*, Vol. 285, No. 11). Exercise values are in kilocalories of physical activity per week. Use a 0.05 significance level to test the claim that the level of smoking is independent of the level of exercise.

	Exercise Level (kilocalories per week)			
	Below 200	200–599	600–1499	1500 or greater
Never smoked	4997	5205	5784	4155
Smoke less than 15 cigarettes per day	604	484	447	359
Smoke 15 or more cigarettes per day	1403	830	644	350

15. Exercise and Smoking Nicorette is a chewing gum designed to help people stop smoking cigarettes. Tests for adverse reactions yielded the results given in the accompanying table. At the 0.05 significance level, test the claim that the treatment (drug or placebo) is independent of the reaction (whether or not mouth or throat soreness was experienced). If you are thinking about using Nicorette as an aid to stop smoking, should you be concerned about mouth or throat soreness?

	Drug	Placebo
Mouth or throat soreness	43	35
No mouth or throat soreness	109	118

Based on data from Merrell Dow Pharmaceuticals, Inc.

16. *Titanic* The table below summarizes the fate of the passengers and crew when the *Titanic* sank on Monday, April 15, 1912. Use a 0.05 significance level to test the claim that surviving is independent of the demographic category (men, women, boys, girls). Does it appear that the rule of "women and children first" was followed when the *Titanic* sank?

	Men	Women	Boys	Girls	Total
Survived	332	318	29	27	706
Died	1360	104	35	18	1517
Total	1692	422	64	45	2223

*17. Using Yates's Correction for Continuity The chi-square distribution is continuous, whereas the test statistic used in this section is discrete. Some statisticians use *Yates's correction for continuity* in cells with an expected frequency of less than 10 or in all cells of a contingency table with two rows and two columns. With Yates's correction, we replace

$$\sum \frac{(O-E)^2}{E} \quad \text{with} \quad \sum \frac{(|O-E|-0.5)^2}{E}$$

Given the contingency table in Exercise 1, find the value of the χ^2 test statistic with and without Yates's correction. In general, what effect does Yates's correction have on the value of the test statistic?

*18. Equivalent Tests Assume that a contingency table has two rows and two columns with frequencies of a and b in the first row and frequencies of c and d in the second row.

a. Verify that the test statistic can be expressed as

$$\chi^2 = \frac{(a+b+c+d)(ad-bc)^2}{(a+b)(c+d)(b+d)(a+c)}$$

b. Let $\hat{p} = a/(a+c)$ and let $\hat{p}_2 = b/(b+d)$. Show that the test statistic

$$z = \frac{(\hat{p}_1 - \hat{p}_2) - 0}{\sqrt{\dfrac{\overline{p}\overline{q}}{n_1} + \dfrac{\overline{p}\overline{q}}{n_2}}}$$

where

$$\overline{p} = \frac{a+b}{a+b+c+d}$$

and

$$\overline{q} = 1 - \overline{p}$$

is such that $z^2 = \chi^2$ [the same result as in part (a)]. This result shows that the chi-square test involving a 2×2 table is equivalent to the test for the difference between two proportions, as described in Section 8-2.

10-4 McNemar's Test for Matched Pairs

The contingency table procedures in Section 10-3 are based on *independent* data. For 2×2 tables consisting of frequency counts that result from *matched pairs*, we do not have independence. For example, assume that each of several test subjects is afflicted with tinea pedis (athlete's foot) on each foot, and each subject is

Table 10-8	2 × 2 Table with Frequency Counts from Matched Pairs		
		Treatment X	
		Cured	Not Cured
Treatment Y	Cured	a	b
	Not cured	c	d

Expensive Diet Pill

There are many past examples in which ineffective treatments were marketed for substantial profits. Capsules of "Fat Trapper" and "Exercise in a Bottle," manufactured by the Enforma Natural Products company, were advertised as being effective treatments for weight reduction. Advertisements claimed that after taking the capsules, fat would be blocked and calories would be burned, even without exercise. Because the Federal Trade Commission identified claims that appeared to be unsubstantiated, the company was fined $10 million for deceptive advertising.

The effectiveness of such treatments can be determined with experiments in which one group of randomly selected subjects is given the treatment, while another group of randomly selected subjects is given a placebo. The resulting weight losses can be compared using statistical methods, such as those described in this section.

given a treatment X on one foot and a treatment Y on the other foot. Table 10-8 is a general table summarizing the frequency counts that result from the matched pairs of feet given the two different treatments. If $a = 12$ in Table 10-8, then 12 subjects enjoyed a cure on each foot. If $b = 8$ in Table 10-8, then each of 8 subjects had one foot not cured by treatment X while their other foot was cured by treatment Y.

Because the frequency counts in Table 10-8 result from *matched* pairs of feet, the data are not independent and we cannot use the contingency table procedures from Section 10-3. Instead, we use McNemar's test.

Definition

McNemar's test uses frequency counts from *matched pairs* of nominal data from two categories to test the null hypothesis that for a table such as Table 10-8, the frequencies b and c occur in the same proportion.

Requirements

1. The sample data have been randomly selected.
2. The sample data consist of *matched pairs* of frequency counts.
3. The data are at the nominal level of measurement, and each observation can be classified two ways: (1) According to the category distinguishing values with each matched pair (such as left foot and right foot), and (2) according to another category with two possible values (such as cured/not cured).
4. For tables such as Table 10-8, the frequencies are such that $b + c \geq 10$.

Test statistic (for testing the null hypothesis that in tables, such as Table 10-8, the frequencies b and c occur in the same proportion):

$$\chi^2 = \frac{(|b - c| - 1)^2}{b + c}$$

where the frequencies of b and c are obtained from the 2 × 2 table with a format similar to Table 10-8. (The frequencies b and c must come from "discordant" pairs, as described later in this section.)

Critical Values

1. The critical region is located in the *right tail only*.
2. The critical values are found in Table A-4 by using **degrees of freedom = 1**.

EXAMPLE **Comparing Treatments** Two different creams are used to treat tinea pedis (athlete's foot). Each subject with this fungal infection on both feet is given a treatment of Pedacream on one foot while their other foot is treated with Fungacream. The sample results are summarized in Table 10-9. Using a 0.05 significance level, apply McNemar's test to test the null hypothesis that the following two proportions are the same:

- The proportion of subjects with no cure on the Pedacream-treated foot and a cure on the Fungacream-treated foot.
- The proportion of subjects with a cure on the Pedacream-treated foot and no cure on the Fungacream-treated foot.

Based on the results, does there appear to be a difference between the two treatments? Does one of the treatments appear to be better than the other?

SOLUTION

REQUIREMENT ✓ The data consist of matched pairs of frequency counts from randomly selected subjects, and each observation can be categorized according to two variables. (One variable has values of "Pedacream" and "Fungacream," and the other variable has values of "cured" and "not cured.") Also, for tables such as Table 10-8, the frequencies must be such that $b + c \geq 10$. For Table 10-9, $b = 8$ and $c = 40$, so that $b + c = 48$, which is at least 10. All of the requirements are therefore satisfied. Although Table 10-9 might appear to be a 2 × 2 contingency table, we cannot use the procedures of Section 10-3 because the data come from *matched pairs* (instead of being independent). Instead, we use McNemar's test. ✓

continued

Table 10-9	Clinical Trials of Treatments for Athlete's Foot		
	Treatment with Pedacream		80 subjects treated on 160 feet:
	Cured	Not cured	12 had both feet cured.
			20 had neither foot cured.
Treatment with Fungacream — Cured	12	8	8 had cures with Fungacream, but not Pedacream.
Treatment with Fungacream — Not cured	40	20	40 had cures with Pedacream, but not Fungacream.

After comparing the frequency counts in Table 10-8 to those given in Table 10-9, we see that $b = 8$ and $c = 40$, so the test statistic can be calculated as follows:

$$\chi^2 = \frac{(|b - c| - 1)^2}{b + c} = \frac{(|8 - 40| - 1)^2}{8 + 40} = 20.021$$

With a 0.05 significance level and degrees of freedom given by df = 1, we refer to Table A-4 to find the critical value of $\chi^2 = 3.841$ for this right-tailed test. The test statistic of $\chi^2 = 20.021$ exceeds the critical value of $\chi^2 = 3.841$, so we reject the null hypothesis. It appears that the two creams produce different results. Analyzing the frequencies of 8 and 40, we see that many more feet were cured with Pedacream than Fungacream, so the Pedacream treatment appears to be more effective.

Note that in the calculation of the test statistic in the preceding example, we did not use the 12 subjects with both feet cured (one foot from each cream) and we did not use the 20 subjects with neither foot cured. Instead of including the cure/cure results and the no cure/no cure results, we used only the cure/no cure results and the no cure/cure results. That is, we are using only the results from the categories that are *different*. Such different categories are referred to as *discordant pairs*.

> **Definition**
>
> **Discordant pairs** of results come from pairs of categories in which the two categories are different (as in cure/no cure or no cure/cure).

When trying to determine whether there is a significant difference between the two cream treatments in Table 10-9, we are not helped by the subjects with both feet cured, and we are not helped by those subjects with neither foot cured. The differences are reflected in the discordant results from the subjects with one foot cured while the other foot was not cured. Consequently, the test statistic includes only the two frequencies that result from the two discordant (or different) pairs of categories.

Caution: When applying McNemar's test, be careful to use only the frequencies from the pairs of categories that are different. Do not blindly use the frequencies in the upper right and lower left corners, because they do not necessarily represent the discordant pairs. If Table 10-9 were reconfigured as shown below, it would be inconsistent in its format, but it would be technically correct in summarizing the same results as the preceding table; however, blind use of the frequencies of 20 and 12 would result in the *wrong* test statistic.

		Treatment with Pedacream	
		Cured	Not cured
Treatment with Fungacream	Not cured	40	20
	Cured	12	8

In this reconfigured table, the discordant pairs of frequencies are these:

cured/not cured: 40

not cured/cured: 8

With this reconfigured table, we should again use the frequencies of 40 and 8, not 20 and 12. In a more perfect world, all such 2 × 2 tables would be configured with a consistent format, and we would be much less likely to use the wrong frequencies.

In addition to comparing treatments given to matched pairs (as in the preceding example), McNemar's test is often used to test a null hypothesis of no change in before/after types of experiments. (See Exercises 1–8.)

10-4 Exercises

In Exercises 1–8, refer to the following table. The table summarizes results from an experiment in which subjects were first classified as smokers or nonsmokers, then they were given a treatment, then later they were again classified as smokers or nonsmokers.

		Before treatment	
		Smoke	Don't smoke
After treatment	Smoke	50	6
	Don't smoke	8	80

1. **Sample Size** How many subjects are included in the experiment?

2. **Treatment Effectiveness** How many subjects changed their smoking status after the treatment?

3. **Treatment Ineffectiveness** How many subjects appear to be unaffected by the treatment one way or the other?

4. **Why Not *t* Test?** Section 8-3 presented procedures for dealing with data consisting of matched pairs. Why can't we use the procedures of Section 8-3 for the analysis of the results summarized in the table?

5. **Discordant Pairs** Which of the following pairs of before/after results are *discordant*?
   a. smoke/smoke
   b. smoke/don't smoke
   c. don't smoke/smoke
   d. don't smoke/don't smoke

6. **Test Statistic** Using the appropriate frequencies, find the value of the test statistic.

7. **Critical Value** Using a 0.01 significance level, find the critical value.

8. **Conclusion** Based on the preceding results, what do you conclude? How does the conclusion make sense in terms of the original sample results?

9. **Treating Athlete's Foot** As in the example of this section, assume that subjects are inflicted with athlete's foot on each of their feet. Also assume that for each subject, one foot is treated with a fungicide solution while the other foot is given a placebo. The

results are given in the accompanying table. Using a 0.05 significance level, test the effectiveness of the treatment.

		Fungicide Treatment	
		Cure	No Cure
Placebo	Cure	5	12
	No cure	22	55

10. Treating Athlete's Foot Repeat Exercise 9 after changing the frequency of 22 to 66.

11. PET/CT Compared to MRI In the article "Whole-Body Dual-Modality PET/CT and Whole Body MRI for Tumor Staging in Oncology" (by Antoch et al., *Journal of the American Medical Association*, Vol. 290, No. 24), the authors cite the importance of accurately identifying the stage of a tumor. Accurate staging is critical for determining appropriate therapy. The article discusses a study involving the accuracy of positron emission tomography (PET) and computed tomography (CT) compared to magnetic resonance imaging (MRI). Using the data in the given table for 50 cancers analyzed with both technologies, does there appear to be a difference in accuracy? Does either technology appear to be better?

		PET/CT	
		Correct	Incorrect
MRI	Correct	36	1
	Incorrect	11	2

12. Testing a Treatment In the article "Eradication of Small Intestinal Bacterial Overgrowth Reduces Symptoms of Irritable Bowel Syndrome" (by Pimentel, Chow, Lin, *American Journal of Gastroenterology*, Vol. 95, No. 12), the authors include a discussion of whether antibiotic treatment of bacteria overgrowth reduces intestinal complaints. McNemar's test was used to analyze results for those subjects with eradication of bacterial overgrowth. Using the data in the given table, does the treatment appear to be effective against abdominal pain?

		Abdominal Pain Before Treatment?	
		Yes	No
Abdominal pain after treatment?	Yes	11	1
	No	14	3

*13. Correction for Continuity The test statistic given in this section includes a correction for continuity. The test statistic given below does not include the correction for continuity, and it is sometimes used as the test statistic for McNemar's test. Refer to the example in this section, find the value of the test statistic using the expression given below, and compare the result to the one found in the example.

$$\chi^2 = \frac{(b - c)^2}{b + c}$$

*14. Using Common Sense Consider the table given below, and use a 0.05 significance level.
   a. What does McNemar's test suggest about the effectiveness of the treatment?
   b. The values of a and d are not used in the calculations, but what does common sense suggest if $a = 5000$ and $d = 4000$?

		Before treatment	
		Smoke	Don't Smoke
After treatment	Smoke	a	5
	Don't smoke	20	d

*15. Small Sample Case The assumptions for McNemar's test include the requirement that $b + c \geq 10$ so that the distribution of the test statistic can be approximated by the chi-square distribution. Refer to the example in this section and replace the table data with the values given below. McNemar's test should not be used because the condition of $b + c \geq 10$ is not satisfied with $b = 2$ and $c = 6$. Instead, use the binomial distribution to find the probability that among 8 equally likely outcomes, the results consist of 6 items in one category and 2 in the other category, or the results are more extreme. That is, use a probability of 0.5 to find the probability that among $n = 8$ trials, the number of successes x is 6 or 7 or 8. Double that probability to find the P-value for this test. Compare the result to the P-value of 0.289 that results from using the chi-square approximation, even though the condition of $b + c \geq 10$ is violated. What do you conclude about the two treatments?

		Treatment with Pedacream	
		Cured	Not Cured
Treatment with Fungacream	Cured	12	2
	Not cured	6	20

Review

In this chapter we worked with data summarized as frequency counts for different categories. In Section 10-2 we described methods for testing goodness-of-fit in a multinomial experiment, which is similar to a binomial experiment except that there are more than two categories of outcomes. Multinomial experiments result in frequency counts arranged in a single row or column, and we tested to determine whether the observed sample frequencies agree with (or "fit") some claimed distribution.

In Section 10-3 we described methods for testing claims involving contingency tables (or two-way frequency tables), which have at least two rows and two columns. Contingency tables incorporate two variables: One variable is used for determining the row that describes a sample value, and the second variable is used for determining the column that describes a sample value. Section 10-3 included two types of hypothesis test: (1) a test of independence between the row and column variables; (2) a test of homogeneity to decide whether different populations have the same proportions of some characteristics.

In Section 10-4 we described McNemar's test for analyzing 2×2 tables of data consisting of frequency counts from matched pairs. When the frequency counts for a 2×2 table consist of matched pairs instead of being independent, we cannot use the methods of Section 10-3, and we must use the methods of Section 10-4 instead.

The following are some key components of the methods discussed in this chapter.

Section 10-2 (Test for goodness-of-fit):

- Test statistic is $\chi^2 = \sum \dfrac{(O - E)^2}{E}$

- Test is right-tailed with $k - 1$ degrees of freedom. All expected frequencies must be at least 5.

Section 10-3 (Contingency table test of independence or homogeneity):

- Test statistic is $\chi^2 = \sum \dfrac{(O - E)^2}{E}$

- Test is right-tailed with $(r - 1)(c - 1)$ degrees of freedom. All expected frequencies must be at least 5.

Section 10-4 (McNemar's Test Using Matched Pairs):

- Test statistic: $\chi^2 = \dfrac{(|b - c| - 1)^2}{b + c}$

- Test is right-tailed with 1 degree of freedom.

Review Exercises

1. Do Gunfire Deaths Occur More Often on Weekends? When *Time* magazine tracked U.S. deaths by gunfire during a one-week period, the results shown in the accompanying table were obtained. At the 0.05 significance level, test the claim that gunfire death rates are the same for the different days of the week. Is there any support for the theory that more gunfire deaths occur on weekends when more people are at home?

Weekday	Mon	Tues	Wed	Thurs	Fri	Sat	Sun
Number of deaths by gunfire	74	60	66	71	51	66	76

2. Is Drinking Independent of Type of Crime? The accompanying table lists sample data that statistician Karl Pearson used in 1909. Does the type of crime appear to be related to whether the criminal drinks or abstains? Are there any crimes that appear to be associated with drinking?

	Arson	Rape	Violence	Stealing	Coining (Counterfeiting)	Fraud
Drinker	50	88	155	379	18	63
Abstainer	43	62	110	300	14	144

3. Testing for Independence Between Early Discharge and Rehospitalization of Newborn Is it safe to discharge newborns from the hospital early after their births? The accompanying table shows results from a study of this issue. Use a 0.05 significance level to test the claim that whether the newborn was discharged early or late is independent of whether the newborn was rehospitalized within a week of discharge. Does the conclusion change if the significance level is changed to 0.01?

	Rehospitalized Within 1 Week of Discharge?	
	Yes	No
Early discharge (less than 30 hours)	622	3997
Late discharge (30–78 hours)	631	4660

Based on data from "The Safety of Newborn Early Discharge," by Liu and others, *Journal of the American Medical Association,* Vol. 278, No. 4.

4. Comparing Treatments Two different creams are used to treat subjects with poison ivy irritation on both hands. Each subject is given a treatment of Ivy Ease on one hand while their other hand is treated with a placebo. The sample results are summarized in the table below. Use a 0.05 significance level to test the null hypothesis that the following two proportions are the same:

- The proportion of subjects with relief on the hand treated with Ivy Ease and no relief on the hand treated with a placebo.

- The proportion of subjects with no relief on the hand treated with Ivy Ease and relief on the hand treated with a placebo.

Does the Ivy Ease treatment appear to be effective?

		Treatment with Ivy Ease	
		Relief	No Relief
Placebo	Relief	12	8
	No relief	32	19

Cumulative Review Exercises

1. Finding Statistics Assume that in Table 10-10, the row and column titles have no meaning so that the table contains test scores for eight randomly selected prisoners who were convicted of removing labels from pillows. Find the mean, median, range, variance, standard deviation, and 5-number summary.

2. Finding Probability Assume that in Table 10-10, the letters A, B, C, and D represent the choices on the first question of a multiple-choice quiz. Also assume that x represents men and y represents women and that the table entries are frequency counts, so 66 men chose answer A, 77 women chose answer A, 80 men chose answer B, and so on.

a. If one response is randomly selected, find the probability that it is response C.

b. If one response is randomly selected, find the probability that it was made by a man.

c. If one response is randomly selected, find the probability that it is response C or was made by a man.

d. If two different responses are randomly selected, find the probability that they were both made by a woman.

3. Testing for Equal Proportions Using the same assumptions as in Exercise 2, test the claim that men and women choose the different answers in the same proportions.

Table 10-10				
	A	B	C	D
x	66	80	82	75
y	77	89	94	84

4. Testing for a Relationship Assume that Table 10-10 lists test scores for four people, where the x-score is from a test of memory and the y-score is from a test of reasoning. Test the claim that there is a relationship between the x- and y-scores.

5. Testing for Effectiveness of Training Assume that Table 10-10 lists test scores for four people, where the x-score is from a pretest taken before a training session on memory improvement and the y-score is from a post-test taken after the training. Test the claim that the training session is effective in raising scores.

6. Testing for Equality of Means Assume that in Table 10-10, the letters A, B, C, and D represent different versions of the same test of reasoning. The x-scores were obtained by four randomly selected men and the y-scores were obtained by four randomly selected women. Test the claim that men and women have the same mean score.

Cooperative Group Activities

1. *Out-of-class activity:* Divide into groups of four or five students. Each group member should survey at least 15 male students and 15 female students at the same college by asking two questions: (1) Does the subject exercise on a regular basis? (2) If the subject were to make up an absence excuse of a flat tire, which tire would he or she say went flat if the instructor asked? (See Exercise 6 in Section 10-2.) Ask the subject to write the two responses on an index card, and also record the gender of the subject and whether the subject wrote with the right or left hand. Use the methods of this chapter to analyze the data collected. Include these tests:

● Exercise is independent of the gender of the subject.
● The tire identified as being flat is independent of the gender of the subject.
● Exercise is independent of whether the subject is right- or left-handed.
● The tire identified as being flat is independent of whether the subject is right- or left-handed.
● Gender is independent of whether the subject is right- or left-handed.
● Exercise is independent of the tire identified as being flat.

2. *Out-of-class activity:* Divide into groups of four or five students. Each group member should select about 15 other students and first ask them to "randomly" select four digits each. After the four digits have been recorded, ask each subject to write the last four digits of his or her social security number. Take the "random" sample results and mix them into one big sample, then mix the social security digits into a second big sample. Using the "random" sample set, test the claim that students select digits randomly. Then use the social security digits to test the claim that they come from a population of random digits. Compare the results. Does it appear that students can randomly select digits? Are they likely to select any digits more often than others? Are they likely to select any digits less often than others? Do the last digits of social security numbers appear to be randomly selected?

3. *Out-of-class activity:* Divide into groups of two or three students. Some examples and exercises of this chapter were based on the analysis of last digits of values. It was noted that the analysis of last digits can sometimes reveal whether values are the results of actual measurements or whether they are reported estimates. Collect data different from the data used in the examples and exercises in this chapter, then conduct an analysis of the last digits.

Technology Project

Use SPSS, SAS, STATDISK, or Minitab, or Excel, or a TI-83/84 Plus calculator, or any other software package or calculator capable of generating equally likely random digits between 0 and 9 inclusive. Generate 500 digits and record the results in the accompanying table. Use a 0.05 significance level to test the claim that the sample digits come from a population with a uniform distribution (so that all digits are equally likely). Does the random number generator appear to be working as it should?

Digit	0	1	2	3	4	5	6	7	8	9
Frequency										

from DATA to DECISION

Critical Thinking: Should the treatment be distributed?

The table below summarizes results from an experiment involving randomly selected subjects who experience some degree of sleeplessness. Some of the subjects were treated with a drug, while others were given a placebo.

	Experienced Improvement	Did Not Experience Any Improvement
Treatment group	10,222	49,875
Placebo group	10,587	53,252

Analyze the Results

1. Use the methods of this chapter to test for independence between the treatment/placebo group and whether the subject experienced improvement in the degree of sleeplessness. What do you conclude? Does it appear that the treatment is effective? What does the result suggest about the decision to distribute the treatment?

2. Use the methods of Section 8-2 to test the claim that the treatment results in a greater rate of improvement than the rate achieved with a placebo. What do you conclude? Does it appear that the treatment is effective? What does the result suggest about the decision to distribute the treatment?

3. If you are the chief executive officer of a pharmaceutical company and you are given the results in the accompanying table, what should you do? Should you market the treatment? Why or why not?

Analysis of Variance

Do different treatments affect the weights of poplar trees?

Data Set 9 in Appendix B includes weights (in kilograms) of poplar trees given different treatments at different sites. Suppose we know only those weights given for Year 1 at Site 1. Site 1 has rich and moist soil, and it is located near a creek. The weights we will consider are summarized in Table 11-1.

In the spirit of exploring data by investigating center, variation, distribution, outliers, and changing patterns over time (CVDOT), we begin by obtaining the sample statistics listed at the bottom of Table 11-1. Examining the sample means, we see that they appear to vary considerably, going from a low of 0.164 to a high of 1.334. Also, the sample standard deviations appear to vary considerably, going from a low of 0.126 to a high of 0.859. It's difficult to analyze distributions because each sample consists of only 5 values, but normal quantile plots suggest that three of the samples are from populations having distributions that are approximately normal. However, analysis of the weights of poplar trees given a fertilizer treatment suggests that the weight

of 1.34 kg is an outlier. With only one outlier present, we will proceed under the assumption that the samples come from populations with distributions that are approximately normal. We could do additional analyses later to determine whether the weight of 1.34 kg has much of an effect on our results. (See Exercise 1 in Section 11-2.)

It appears that the differences among the sample means indicate that the samples come from populations with different means, but instead of considering only the sample means, we should also consider the amounts of variation, the sample sizes, and the nature of the distribution of sample means. One way of taking all of these relevant factors into account is to conduct a formal hypothesis test that automatically includes them. Such a test will be introduced in this chapter, and we will use it to determine whether there is sufficient evidence to conclude that the means are not all equal. We will then know whether the different treatments have an effect.

Table 11-1	Weights (kg) of Poplar Trees			
		Treatment		
	None	Fertilizer	Irrigation	Fertilizer and Irrigation
	0.15	1.34	0.23	2.03
	0.02	0.14	0.04	0.27
	0.16	0.02	0.34	0.92
	0.37	0.08	0.16	1.07
	0.22	0.08	0.05	2.38
n	5	5	5	5
$\bar{x}$	0.184	0.332	0.164	1.334
s	0.127	0.565	0.126	0.859

11-1 Overview

The Placebo Effect

It has long been believed that placebos actually help some patients. In fact, some formal studies have shown that when given a placebo (a treatment with no medicinal value), many test subjects show some improvement. Estimates of improvement rates have typically ranged between one-third and two-thirds of the patients. However, a more recent study suggests that placebos have no real effect. An article in the *New England Journal of Medicine* (Vol. 334, No. 21) was based on research of 114 medical studies over 50 years. The authors of the article concluded that placebos appear to have some effect only for relieving pain, but not for other physical conditions. They concluded that apart from clinical trials, the use of placebos "cannot be recommended."

Instead of "Analysis of Variance," a better title for this chapter might be "Testing for Equality of Three or More Population Means." Although it is not very catchy, the latter title does a better job of describing the objective of this chapter. We want to introduce a procedure for testing the hypothesis that three or more population means are equal, so a typical null hypothesis will be $H_0: \mu_1 = \mu_2 = \mu_3 = \mu_4$ and the alternative hypothesis will be the statement that at least one mean is different from the others. In Section 8-3 we already presented procedures for testing the hypothesis that *two* population means are equal, but the methods of that section do not apply when three or more means are involved. Instead of referring to the main objective of testing for equal means, the term *analysis of variance* refers to the *method* we use, which is based on an analysis of sample variances.

> **Definition**
>
> **Analysis of variance (ANOVA)** is a method of testing the equality of three or more population means by analyzing sample variances.

ANOVA is used in applications such as the following:

- If we treat one group with two aspirin tablets each day and a second group with one aspirin tablet each day, while a third group is given a placebo each day, we can test to determine if there is sufficient evidence to support the claim that the three groups have different mean blood pressure levels.
- The claim has been made that supermarkets place high-sugar cereals on shelves that are at eye-level for children, so we can test the claim that the cereals on the shelves have the same mean sugar content.

Why Can't We Just Test Two Samples at a Time? Why do we need a new procedure when we can test for equality of two means by using the methods presented in Chapter 8? For example, if we want to use the sample data from Table 11-1 to test the claim that the four populations have the same mean, why not simply pair them off and do two at a time by testing the six null hypotheses listed below?

$$H_0: \mu_1 = \mu_2 \qquad H_0: \mu_1 = \mu_3 \qquad H_0: \mu_1 = \mu_4$$
$$H_0: \mu_2 = \mu_3 \qquad H_0: \mu_2 = \mu_4 \qquad H_0: \mu_3 = \mu_4$$

With four populations, this approach (doing two at a time) requires six different hypothesis tests, so the level of confidence could be as low as 0.95^6 (or 0.735). In general, as we increase the number of individual tests of significance, we increase the likelihood of finding a difference by chance alone (instead of a real difference in the means). The risk of a type I error—finding a difference in one of the pairs when no such difference actually exists—is far too high. The method of analysis of variance helps us avoid that particular pitfall (rejecting a true null hypothesis) by using one test for equality of several means.

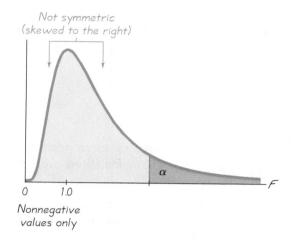

Not symmetric (skewed to the right)

0   1.0

Nonnegative values only

α

F

F Distribution

The ANOVA methods of this chapter require the *F* distribution, which was first introduced in Section 8-6. In Section 8-6 we noted that the *F* distribution has the following important properties (see Figure 11-1):

1. The *F* distribution is not symmetric; it is skewed to the right.
2. The values of *F* can be 0 or positive, but they cannot be negative.
3. There is a different *F* distribution for each pair of degrees of freedom for the numerator and denominator.

Critical values of *F* are given in Table A-5.

Analysis of variance (ANOVA) is based on a comparison of two different estimates of the variance common to the different populations. Those estimates (the *variance between samples* and the *variance within samples*) will be described in Section 11-2. The term *one-way* is used because the sample data are separated into groups according to one characteristic, or factor. For example, the weights summarized in Table 11-1 are separated into four different groups according to the one characteristic (or factor) of treatment (none, fertilizer, irrigation, fertilizer and irrigation). In Section 11-3 we will introduce two-way analysis of variance, which allows us to compare populations separated into categories using two characteristics (or factors). For example, we might separate weights of poplar trees using the following two factors: (1) site location (moist or dry) and (2) treatment (none, fertilizer, irrigation, fertilizer and irrigation). (In this section, we will illustrate methods of one-way analysis of variance by using only the data in Table 11-1, which are taken from Data Set 9 in Appendix B. However, the methods of two-way analysis of variance presented in Section 11-3 would be more appropriate if the more complete Data Set 9 is used.)

Suggested Study Strategy: Because the procedures used in this chapter require complicated calculations, we will emphasize the use and interpretation of computer software, such as SPSS, SAS, STATDISK, Minitab, Excel, or a TI-83/84 Plus calculator. We suggest that you begin Section 11-2 by focusing on this key concept: We are using a procedure to test a claim that three or more means are equal. Although the details of the calculations are complicated, our

procedure will be easy because it is based on a *P*-value. If the *P*-value is small, such as 0.05 or lower, reject equality of means. Otherwise, fail to reject equality of means. After understanding that basic and simple procedure, proceed to understand the underlying rationale.

11-2 One-Way ANOVA

In this section we consider tests of hypotheses that three or more population means are all equal, as in H_0: $\mu_1 = \mu_2 = \mu_3 = \mu_4$. The calculations are very complicated, so we recommend the following approach:

1. Understand that a small *P*-value (such as 0.05 or less) leads to rejection of the null hypothesis of equal means. With a large *P*-value (such as greater than 0.05), fail to reject the null hypothesis of equal means.

2. Develop an understanding of the underlying rationale by studying the example in this section.

3. Become acquainted with the nature of the SS (sum of squares) and MS (mean square) values and their role in determining the *F* test statistic, but use statistical software packages or a calculator for finding those values.

The method we use is called **one-way analysis of variance** (or **single-factor analysis of variance**) because we use a single property, or characteristic, for categorizing the populations. This characteristic is sometimes referred to as a *treatment*, or *factor*.

> **Definition**
>
> A **treatment** (or **factor**) is a property, or characteristic, that allows us to distinguish the different populations from one another.

For example, the weights summarized in Table 11-1 are distinguished according to the category of treatment (or factor). The term *treatment* is used because early applications of analysis of variance involved agricultural experiments in which different plots of farmland were treated with different fertilizers, seed types, insecticides, and so on. The accompanying box includes the requirements and the procedure we will use.

> **Requirements**
>
> 1. The populations have distributions that are approximately normal. (This is a loose requirement, because the method works well unless a population has a distribution that is very far from normal. If a population does have a distribution that is far from normal, use the Kruskal-Wallis test described in Section 12-5.)
>
> 2. The populations have the same variance σ^2 (or standard deviation σ). [This method is *robust* against variances that are not equal, which means that the method works well unless the population variances differ by large amounts.

University of Wisconsin statistician George E. P. Box showed that as long as the sample sizes are equal (or nearly equal), the variances can differ by amounts that make the largest up to nine times the smallest and the results of ANOVA will continue to be essentially reliable.]

3. The samples are simple random samples of quantitative data. (That is, samples of the same size have the same probability of being selected, and the sample values are at the interval or ratio levels of measurement.)

4. The samples are independent of each other. (The samples are not matched or paired in any way.)

5. The different samples are from populations that are categorized in only one way. (This is the basis for the name of the method: *one-way* analysis of variance.)

Procedure for Testing $H_0: \mu_1 = \mu_2 = \mu_3 = \ldots$

1. **Use a technology such as SPSS, SAS, STATDISK, Minitab, Excel, or a TI-83/84 Plus calculator to obtain results.**

2. **Identify the *P*-value from the display.**

3. **Form a conclusion based on these criteria:**

   - **If the *P*-value $\leq \alpha$, reject the null hypothesis of equal means and conclude that at least one of the population means is different from the others.**

   - **If the *P*-value $> \alpha$, fail to reject the null hypothesis of equal means.**

Caution when interpreting results: When we conclude that there is sufficient evidence to reject the claim of equal population means, we cannot conclude from ANOVA that any *particular* mean is different from the others. (There are several other tests that can be used to identify the specific means that are different, and those procedures are called **multiple comparison procedures.** Comparison of confidence intervals, the Scheffé test, the extended Tukey test, and the Bonferroni test are common multiple comparison procedures.)

EXAMPLE **Weights of Poplar Trees** Given the weights of poplar trees listed in Table 11-1 and a significance level of $\alpha = 0.05$, use SPSS, SAS, STATDISK, Minitab, Excel, or a TI-83/84 PLUS calculator to test the claim that the four samples come from populations with means that are not all the same.

SOLUTION

REQUIREMENT ✓ When investigating the normality requirement for the four different populations, the only questionable sample is the second sample of the fertilizer treatment group. For that sample alone, the normal quantile plot and the histogram suggest that normality might not be satisfied, and the problem is the value of 1.34, which appears to be an outlier, so we should be careful here. It would be wise to do the analysis with and without that value included. See Exercise 1 where we see that in this case, the outlier of 1.34 does not have a dramatic effect on the results. The variances are very different, but the largest

continued

ANOVA

WEIGHT

	Sum of Squares	df	Mean Square	F	Sig.
Between Groups	4.682	3	1.561	5.731	.007
Within Groups	4.357	16	.272		
Total	9.040	19			

Source	DF	Sum of Squares	Mean Square	F Value	Pr > F
Model	3	4.68241500	1.56080500	5.73	0.0073
Error	16	4.35724000	0.27232750		
Corrected Total	19	9.03965500			

Source:	DF:	SS:	MS:	Test Stat, F:	Critical F:	P-Value:
Treatment:	3	4.682415	1.560805	5.731353	3.238868	0.0073483
Error:	16	4.35724	0.2723275			
Total:	19	9.039655	0.4757713			

Reject the Null Hypothesis
Reject equality of means

One-way ANOVA

Source:	DF:	SS:	MS:	F:	P
Factor:	3	4.682	1.561	5.73	0.007
Error:	16	4.357	0.272		
Total:	19	9.040			

```
One-way ANOVA
 F=5.731352875
 P=.0073482944
 Factor
  df=3
  SS=4.682415
↓ MS=1.560805
```

```
One-way ANOVA
↑ MS=1.560805
 Error
  df=16
  SS=4.35724
  MS=.2723275
 Sxp=.521850074
■
```

Anova: Single Factor

SUMMARY

Groups	Count	Sum	Average	Variance
Column 1	5	0.92	0.184	0.01613
Column 2	5	1.66	0.332	0.31932
Column 3	5	0.82	0.164	0.01593
Column 4	5	6.67	1.334	0.73793

ANOVA

Source of Variation	SS	df	MS	F	P-value	F crit
Between Groups	4.682415	3	1.560805	5.731353	0.007348	3.238872
Within Groups	4.35724	16	0.272328			
Total	9.039655	19				

is no more than nine times the smallest, so the loose requirement of equal variances is satisfied. We are assuming that we have simple random samples. We know that the samples are independent, and each value belongs to exactly one group. The requirements are therefore satisfied and we can proceed with the hypothesis test. ✓

The null hypothesis is $H_0: \mu_1 = \mu_2 = \mu_3 = \mu_4$, and the alternative hypothesis is the claim that at least one of the means is different from the others.

Step 1:   Use technology to obtain ANOVA results, such as one of those shown on the facing page.

Step 2:   The displays all show that the P-value is 0.007 when rounded.

Step 3:   Because the P-value of 0.007 is less than the significance level of $\alpha = 0.05$, we reject the null hypothesis of equal means.

INTERPRETATION   There is sufficient evidence to support the claim that the four population means are not all the same. Based on the sample of weights listed in Table 11-1, we conclude that those weights come from populations having means that are not all the same. On the basis of this ANOVA test, we cannot conclude that any particular mean is different from the others.

Rationale

The method of analysis of variance is based on this fundamental concept: With the assumption that the populations all have the same variance σ^2, we estimate the common value of σ^2 using two different approaches. The F test statistic is the ratio of those estimates, so that a significantly *large* F test statistic (located far to the right in the F distribution graph) is evidence against equal population means. Figure 11-2 shows the relationship between the F test statistic and the P-value.

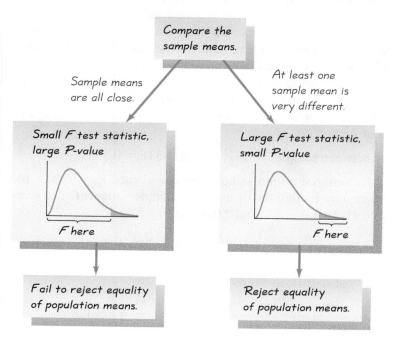

FIGURE 11-2 **Relationship Between the F Test Statistic and P-Value**

The two approaches for estimating the common value of σ^2 are as follows:

1. The **variance between samples** (also called **variation due to treatment**) is an estimate of the common population variance σ^2 that is based on the variation among the sample *means*.

2. The **variance within samples** (also called **variation due to error**) is an estimate of the common population variance σ^2 based on the sample *variances*.

Test Statistic for One-Way ANOVA

$$F = \frac{\text{variance between samples}}{\text{variance within samples}}$$

The numerator of the test statistic F measures variation between sample means. The estimate of variance in the denominator depends only on the sample variances and is not affected by differences among the sample means. Consequently, sample means that are close in value result in an F test statistic that is close to 1, and we conclude that there are no significant differences among the sample means. But if the value of F is excessively *large,* then we reject the claim of equal means. (The vague terms "close to 1" and "excessively large" are made objective by the corresponding P-value, which tells us whether the F test statistic is or is not in the critical region.) Because excessively large values of F reflect unequal means, the test is right-tailed.

Calculations with Equal Sample Sizes n

Refer to Data Set A in Table 11-2. If the data sets all have the same sample size (as in $n = 4$ for Data Set A in Table 11-2), the required calculations aren't overwhelmingly difficult. First, find the variance between samples by evaluating $ns_{\bar{x}}^2$, where $s_{\bar{x}}^2$ is the variance of the sample means and n is the size of each of the samples. That is, consider the sample means to be an ordinary set of values and calculate the variance. (From the central limit theorem, $\sigma_{\bar{x}} = \sigma/\sqrt{n}$ can be solved for σ to get $\sigma = \sqrt{n} \cdot \sigma_{\bar{x}}$, so that we can estimate σ^2 with $ns_{\bar{x}}^2$.) For example, the sample means for Data Set A in Table 11-2 are 5.5, 6.0, and 6.0. Those three values have a variance of $s_{\bar{x}}^2 = 0.0833$, so that

$$\text{variance between samples} = ns_{\bar{x}}^2 = 4(0.0833) = 0.3332$$

Next, estimate the variance within samples by calculating s_p^2, which is the pooled variance obtained by finding the mean of the sample variances. The sample variances in Table 11-2 are 3.0, 2.0, and 2.0, so that

$$\text{variance within samples} = s_p^2 = \frac{3.0 + 2.0 + 2.0}{3} = 2.3333$$

Finally, evaluate the F test statistic as follows:

$$F = \frac{\text{variance between samples}}{\text{variance within samples}} = \frac{ns_{\bar{x}}^2}{s_p^2} = \frac{0.3332}{2.3333} = 0.1428$$

Table 11-2   Effect of a Mean on the F Test Statistic

	A	add 10			B	
Sample 1	Sample 2	Sample 3		Sample 1	Sample 2	Sample 3
7	6	4		17	6	4
3	5	7		13	5	7
6	5	6		16	5	6
6	8	7		16	8	7

	A				B	
$n_1 = 4$	$n_2 = 4$	$n_3 = 4$		$n_1 = 4$	$n_2 = 4$	$n_3 = 4$
$\bar{x}_1 = 5.5$	$\bar{x}_2 = 6.0$	$\bar{x}_3 = 6.0$		$\bar{x}_1 = 15.5$	$\bar{x}_2 = 6.0$	$\bar{x}_3 = 6.0$
$s_1^2 = 3.0$	$s_2^2 = 2.0$	$s_3^2 = 2.0$		$s_1^2 = 3.0$	$s_2^2 = 2.0$	$s_3^2 = 2.0$

Variance between samples	$ns_{\bar{x}}^2 = 4\,(0.0833) = 0.3332$	$ns_{\bar{x}}^2 = 4\,(30.0833) = 120.3332$
Variance within samples	$s_p^2 = \dfrac{3.0 + 2.0 + 2.0}{3} = 2.3333$	$s_p^2 = \dfrac{3.0 + 2.0 + 2.0}{3} = 2.3333$
F test statistic	$F = \dfrac{ns_{\bar{x}}^2}{s_p^2} = \dfrac{0.3332}{2.3333} = 0.1428$	$F = \dfrac{ns_{\bar{x}}^2}{s_p^2} = \dfrac{120.3332}{2.3333} = 51.5721$
P-value (found from Excel)	P-value = 0.8688	P-value = 0.0000118

The critical value of F is found by assuming a right-tailed test, because large values of F correspond to significant differences among means. With k samples each having n values, the numbers of degrees of freedom are computed as follows.

Degrees of Freedom:
(k = number of samples and n = sample size)

numerator degrees of freedom $= k - 1$
denominator degrees of freedom $= k(n - 1)$

For Data Set A in Table 11-2, $k = 3$ and $n = 4$, so the degrees of freedom are 2 for the numerator and $3(4 - 1) = 9$ for the denominator. With $\alpha = 0.05$, 2 degrees of freedom for the numerator, and 9 degrees of freedom for the denominator, the critical F value from Table A-5 is 4.2565. If we were to use the traditional method of hypothesis testing with Data Set A in Table 11-2, we would see that this right-tailed test has a test statistic of $F = 0.1428$ and a critical value of $F = 4.2565$, so the test statistic is not in the critical region and we therefore fail to reject the null hypothesis of equal means.

To really see how the F test statistic works, consider both collections of sample data in Table 11-2. Note that the three samples in part A are identical to the three samples in part B, except that in part B we have added 10 to each value of Sample 1 from part A. The three sample means in part A are very close, but there are substantial differences in part B. The three sample variances in part A are identical to those in part B.

Adding 10 to each data value in the first sample of Table 11-2 has a dramatic effect on the test statistic, with F changing from 0.1428 to 51.5721. Adding 10 to each data value in the first sample also has a dramatic effect on the P-value, which changes from 0.8688 (not significant) to 0.0000118 (significant). Note that the variance between samples in part A is 0.3332, but for part B it is 120.3332 (indicating that the sample means in part B are farther apart). Note also that the variance within samples is 2.3333 in both parts, because the variance within a sample isn't affected when we add a constant to every sample value. *The change in the F test statistic and the P-value is attributable only to the change in* $\bar{x}_1$. This illustrates that the F test statistic is very sensitive to sample *means*, even though it is obtained through two different estimates of the common population *variance*.

Here is the key point of Table 11-2: Data Sets A and B are identical except that in Data Set B, 10 is added to each value of the first sample. Adding 10 to each value of the first sample causes the three sample means to grow farther apart, with the result that the F test statistic increases and the P-value decreases.

Calculations with Unequal Sample Sizes

While the calculations required for cases with equal sample sizes are reasonable, they become really complicated when the sample sizes are not all the same. The same basic reasoning applies because we calculate an F test statistic that is the ratio of two different estimates of the common population variance σ^2, but those estimates involve *weighted* measures that take the sample sizes into account, as shown below.

$$F = \frac{\text{variance between samples}}{\text{variance within samples}} = \frac{\left[\dfrac{\Sigma n_i(\bar{x}_i - \bar{\bar{x}})^2}{k - 1}\right]}{\left[\dfrac{\Sigma(n_i - 1)s_i^2}{\Sigma(n_i - 1)}\right]}$$

where $\bar{\bar{x}}$ = mean of all sample values combined

         k = number of population means being compared

         n_i = number of values in the ith sample

         $\bar{x}_i$ = mean of values in the ith sample

         s_i^2 = variance of values in the ith sample

The factor of n_i is included so that larger samples carry more weight. The denominator of the test statistic is simply the mean of the sample variances, but it is a weighted mean with the weights based on the sample sizes.

Because calculating this test statistic can lead to large rounding errors, the various software packages typically use a different (but equivalent) expression that involves SS (for sum of squares) and MS (for mean square) notation. Although the following notation and components are complicated and involved, the basic idea is the same: The test statistic F is a ratio with a numerator reflecting variation *between* the means of the samples and a denominator reflecting variation *within* the samples. If the populations have equal means, the F ratio tends to be close to 1, but if the population means are not equal, the F ratio tends to be significantly larger than 1. Key components in our ANOVA method are described as follows.

SS(total), or total sum of squares, is a measure of the total variation (around $\bar{\bar{x}}$) in all of the sample data combined.

Formula 11-1 $$\text{SS total} = \Sigma(x - \bar{\bar{x}})^2$$

SS(total) can be broken down into the components of SS(treatment) and SS(error), described as follows:

SS(treatment), also referred to as **SS(factor)** or **SS(between groups)** or **(SS between samples)**, is a measure of the variation *between* the sample means.

Formula 11-2
$$\text{SS(treatment)} = n_1(\bar{x}_1 - \bar{\bar{x}})^2 + n_2(\bar{x}_2 - \bar{\bar{x}})^2 + \cdots + n_k(\bar{x}_k - \bar{\bar{x}})^2$$
$$= \Sigma n_i(\bar{x}_i - \bar{\bar{x}})^2$$

If the population means $(\mu_1, \mu_2, \ldots, \mu_k)$ are equal, then the sample means $\bar{x}_1, \bar{x}_2, \ldots, \bar{x}_k$ will all tend to be close together and also close to $\bar{\bar{x}}$. The result will be a relatively small value of SS(treatment). If the population means are not all equal, however, then at least one of $\bar{x}_1, \bar{x}_2, \ldots, \bar{x}_k$ will tend to be far apart from the others and also far apart from $\bar{\bar{x}}$. The result will be a relatively large value of SS(treatment).

SS(error), also referred to as **SS(within groups)** or **SS(within samples)**, is a sum of squares representing the variation that is assumed to be common to all the populations being considered.

Formula 11-3
$$\text{SS(error)} = (n_1 - 1)s_1^2 + (n_2 - 1)s_2^2 + \cdots + (n_k - 1)s_k^2$$
$$= \Sigma(n_i - 1)s_i^2$$

Given the preceding expressions for SS(total), SS(treatment), and SS(error), the following relationship will always hold.

Formula 11-4   SS(total) = SS(treatment) + SS(error)

SS(treatment) and SS(error) are both sums of squares, and if we divide each by its corresponding number of degrees of freedom, we get *mean* squares. Some of the following expressions for mean squares include the notation N:

N = **total number of values in all samples combined**

MS(treatment) is a mean square for treatment, obtained as follows:

Formula 11-5 $MS(treatment) = \dfrac{SS(treatment)}{k - 1}$

MS(error) is a mean square for error, obtained as follows:

Formula 11-6 $MS(error) = \dfrac{SS(error)}{N - k}$

MS(total) is a mean square for the total variation, obtained as follows:

Formula 11-7 $MS(total) = \dfrac{SS(total)}{N - 1}$

Test Statistic for ANOVA with Unequal Sample Sizes

In testing the null hypothesis $H_0: \mu_1 = \mu_2 = \cdots = \mu_k$ against the alternative hypothesis that these means are not all equal, the test statistic

Formula 11-8 $F = \dfrac{MS(treatment)}{MS(error)}$

has an F distribution (when the null hypothesis H_0 is true) with degrees of freedom given by

numerator degrees of freedom = $k - 1$
denominator degrees of freedom = $N - k$

This test statistic is essentially the same as the one given earlier, and its interpretation is also the same as described earlier. The denominator depends only on the sample variances that measure variation within the treatments and is not affected by the differences among the sample means. In contrast, the numerator does depend on differences among the sample means. If the differences among the sample means are extreme, they will cause the numerator to be excessively large, so F will also be excessively large. Consequently, very large values of F suggest unequal means, and the ANOVA test is therefore right-tailed.

Tables are a convenient format for summarizing key results in ANOVA calculations, and Table 11-3 has a format often used in computer displays. (See the preceding SPSS, SAS, STATDISK, Minitab, and Excel displays.) The entries in Table 11-3 result from the weights of the poplar trees in Table 11-1.

Table 11-3	ANOVA Table for Weights of Poplar Trees			
Source of Variation	Sum of Squares (SS)	Degrees of Freedom	Mean Square (MS)	F Test Statistic
Treatments	4.6824	3	1.5608	5.7314
Error	4.3572	16	0.2723	
Total	9.0397	19		

Designing the Experiment   With one-way (or single-factor) analysis of variance, we use one factor as the basis for partitioning the data into different categories. If we conclude that the differences among the means are significant, we can't be absolutely sure that the differences can be explained by the factor being used. It is possible that the variation of some other unknown factor is responsible. One way to reduce the effect of the extraneous factors is to design the experiment so that it has a **completely randomized design,** in which each element is given the same chance of belonging to the different categories, or treatments. For example, you might assign subjects to a treatment group, placebo group, and control group through a process of random selection equivalent to picking slips from a bowl. Another way to reduce the effect of extraneous factors is to use a **rigorously controlled design,** in which elements are carefully chosen so that all other factors have no variability. In general, good results require that the experiment be carefully designed and executed.

Identifying Means That Are Different

After conducting a test of analysis of variance, we might conclude that there is sufficient evidence to reject a claim of equal population means, but we cannot conclude from ANOVA that any *particular* mean is different from the others. There are several other formal and informal procedures that can be used to identify the specific means that are different. One informal approach is to use the same scale for constructing boxplots of the data sets to see if one or more of the data sets is very different from the others. Another approach is to construct confidence interval estimates of the means from the data sets, then compare those confidence intervals to see if one or more of them does not overlap with the others.

We noted earlier that it is unwise to pair off the samples and conduct individual hypothesis tests using the procedure described in Section 8-3. With four populations, this approach (doing two at a time) requires six different hypothesis tests, so if each test is conducted with a 0.05 significance level, the overall level of confidence for the six tests could be as low as 0.95^6 (or 0.735), so the significance level could be as high as $1 - 0.735 = 0.265$. This high significance level indicates that the risk of a type I error—finding a difference in one of the pairs when no such difference actually exists—is far too high.

There are several procedures for identifying which means differ from the others. Some of the tests, called **range tests**, allow us to identify subsets of means that are not significantly different from each other. Other tests, called **multiple comparison tests**, use pairs of means, but they make adjustments to overcome the problem of having a significance level that increases as the number of individual tests increases. There is no consensus on which test is best, but some of the more common tests are the Duncan test, Student-Newman-Keuls test (or SNK test), Tukey test (or Tukey honestly significant difference test), Scheffé test, Dunnett test, least significant difference test, and the Bonferroni test. Let's consider the Bonferroni test. Here is the procedure:

Bonferroni Multiple Comparison Test

1. Do a separate t test for each pair of samples, but make the adjustments described in the following steps.

2. For an estimate of the variance σ^2 that is common to all of the involved populations, use the value of MS(error), which uses all of the available sample data. The value of MS(error) is typically obtained when conducting the analysis of variance test. Using the value of MS(error), calculate the value of the test statistic t, as shown below. This particular test statistic is based on the choice of Sample 1 and Sample 4; change the subscripts and use another pair of samples until all of the different possible pairs of samples have been included.

$$t = \frac{\bar{x}_1 - \bar{x}_4}{\sqrt{MS(error) \cdot \left(\frac{1}{n_1} + \frac{1}{n_4} \right)}}$$

3. After calculating the value of the test statistic t for a particular pair of samples, find either the critical t value or the P-value, but make the following adjustment so that the overall significance level does not run amok.

   P-value: Use the test statistic t with df $= N - k$, where N is the total number of sample values and k is the number of samples and find the P-value the usual way, but adjust the P-value by multiplying it by the number of different possible pairings of two samples. (For example, with four samples, there are six different possible pairings, so adjust the P-value by multiplying it by 6.)

   Critical value: When finding the critical value, adjust the significance level α by dividing it by the number of different possible pairings of two samples. (For example, with four samples, there are six different possible pairings, so adjust the significance level by dividing it by 6.)

Note that in Step 3 of the preceding Bonferroni procedure, an individual test is conducted with a much lower significance level, or the P-value is greatly increased. For example, if we have four samples, there are six different possible pairings of two samples. If we are using a 0.05 significance level, each of the six individual paired comparison tests uses a significance level of $0.05/6 = 0.00833$. Rejection of equality of means therefore requires differences that are much farther apart. The degree of confidence for all six tests can be as low as $(1 - 0.00833)^6 = 0.95$, so the chance of a type 1 error in any of the six tests is around 0.05. This adjustment in Step 3 compensates for the fact that we are doing several tests instead of only one test.

EXAMPLE **Using the Bonferroni Test** A previous example in this section used analysis of variance with the sample data in Table 11-1. We concluded that there is sufficient evidence to warrant rejection of the claim of equal means. Use the Bonferroni test with a 0.05 significance level to identify which mean is different from the others.

SOLUTION   The Bonferroni test requires a separate t test for each different possible pair of samples. Here are the null hypotheses to be tested:

$$H_0: \mu_1 = \mu_2 \quad H_0: \mu_1 = \mu_3 \quad H_0: \mu_1 = \mu_4$$
$$H_0: \mu_2 = \mu_3 \quad H_0: \mu_2 = \mu_4 \quad H_0: \mu_3 = \mu_4$$

We begin with $H_0: \mu_1 = \mu_4$. From Table 11-1 we see that $\bar{x}_1 = 0.184$, $n_1 = 5, \bar{x}_4 = 1.334$, and $n_4 = 5$. From the preceding results of analysis of variance, we also know that for the data in Table 11-3, MS(error) $= 0.2723275$. We can now evaluate the test statistic:

$$t = \frac{\bar{x}_1 - \bar{x}_4}{\sqrt{\text{MS(error)} \cdot \left(\frac{1}{n_1} + \frac{1}{n_4}\right)}} = \frac{0.184 - 1.334}{\sqrt{0.2723275 \cdot \left(\frac{1}{5} + \frac{1}{5}\right)}} = -3.484$$

The number of degrees of freedom is df $= N - k = 20 - 4 = 16$. With a test statistic of $t = -3.484$ and with df $= 16$, the two-tailed P-value is 0.003065, but we adjust this P-value by multiplying it by 6 (the number of different possible pairs of samples) to get a final P-value of 0.01839. Because this P-value is small (less than 0.05), we reject the null hypothesis. It appears that Samples 1 and 4 have significantly different means. [If we want to find the critical value, use df $= 16$ and use an adjusted significance level of $0.05/6 = 0.00833$. (This adjusted significance level isn't too far from the area of 0.01, which is included in Table A-3. The critical t values for the two-tailed test are therefore close to -2.921 and 2.921.) The actual critical values are -3.008 and 3.008. The test statistic of -3.484 is in the critical region, so again reject the null hypothesis that $\mu_1 = \mu_4$.)

Instead of continuing with separate hypothesis tests for the other five pairings, see the SPSS display showing all of the Bonferroni test results. (The third row of numerical results corresponds to the results found here.) The display shows that the mean for Sample 4 is significantly different from each of the other three sample means. Based on the Bonferroni test, it appears that the weights of the poplar trees treated with both fertilizer and irrigation have a mean that is significantly different from each of the other three treatment categories.

SPSS   **Bonferroni Results**

Multiple Comparisons

Dependent Variable: WEIGHT
Bonferroni

(I) SAMPLE	(J) SAMPLE	Mean Difference (I-J)	Std. Error	Sig.	95% Confidence Interval Lower Bound	95% Confidence Interval Upper Bound
1.00	2.00	-.1480	.33005	1.000	-1.1409	.8449
	3.00	.0200	.33005	1.000	-.9729	1.0129
	4.00	-1.1500*	.33005	.018	-2.1429	-.1571
2.00	1.00	.1480	.33005	1.000	-.8449	1.1409
	3.00	.1680	.33005	1.000	-.8249	1.1609
	4.00	-1.0020*	.33005	.047	-1.9949	-.0091
3.00	1.00	-.0200	.33005	1.000	-1.0129	.9729
	2.00	-.1680	.33005	1.000	-1.1609	.8249
	4.00	-1.1700*	.33005	.016	-2.1629	-.1771
4.00	1.00	1.1500*	.33005	.018	.1571	2.1429
	2.00	1.0020*	.33005	.047	.0091	1.9949
	3.00	1.1700*	.33005	.016	.1771	2.1629

*. The mean difference is significant at the .05 level.

This example illustrates that a suitable software package is an invaluable tool for conducting a multiple comparison or range test.

11-2 Exercises

1. Weights of Poplar Trees We noted in the Chapter Problem that the weight of 1.34 kg appears to be an outlier. If we delete that value, the SPSS results are as shown below.

SPSS

WEIGHT

	Sum of Squares	df	Mean Square	F	Sig.
Between Groups	5.216	3	1.739	8.448	.002
Within Groups	3.087	15	.206		
Total	8.303	18			

a. What is the null hypothesis?
b. What is the alternative hypothesis?
c. Identify the value of the test statistic.
d. Find the critical value for a 0.05 significance level.
e. Identify the P-value.
f. Based on the preceding results, what do you conclude about equality of the population means?

g. Compare these results to those obtained with the weight of 1.34 kg included. Does the apparent outlier of 1.34 kg have much of an effect on the results? Does the conclusion change?

2. Weights of Poplar Trees The Chapter Problem uses the weights of poplar trees for Year 1 and Site 1. (See Data Set 9 in Appendix B.) If we use the weights for Year 2 and Site 1, the analysis of variance results from SAS are as shown in the accompanying display. Assume that we want to use a 0.05 significance level in testing the null hypothesis that the four treatments result in weights with the same population mean.
   a. What is the null hypothesis?
   b. What is the alternative hypothesis?
   c. Identify the value of the test statistic.
   d. Find the critical value for a 0.05 significance level.
   e. Identify the P-value.
   f. Based on the preceding results, what do you conclude about equality of the population means?

SAS

Source	DF	Sum of Squares	Mean Square	F Value	Pr > F
Model	3	3.34645500	1.11548500	6.14	0.0056
Error	16	2.90620000	0.18163750		
Corrected Total	19	6.25265500			

3. Marathon Times A random sample of males who finished the New York Marathon is partitioned into three categories with ages of 21–29, 30–39, and 40 or over. The times (in seconds) are obtained from a random sample of those who finished. The analysis of variance results obtained from Excel are shown below.
   a. What is the null hypothesis?
   b. What is the alternative hypothesis?
   c. Identify the value of the test statistic.
   d. Find the critical value for a 0.05 significance level.
   e. Identify the P-value.
   f. Is there sufficient evidence to support the claim that men in the different age categories have different mean times?

Excel

Source of Variation	SS	df	MS	F	P-value	F crit
Between Groups	3532063.284	2	1766031.642	0.188679406	0.828324293	3.080387501
Within Groups	1010875649	108	9359959.71			
Total	1014407712	110				

4. Systolic Blood Pressure in Different Age Groups A random sample of 40 women is partitioned into three categories with ages of below 20, 20 through 40, and over 40. The systolic blood pressure levels are obtained from Data Set 1 in Appendix B. The analysis of variance results obtained from Minitab are shown on the next page.

a. What is the null hypothesis?
b. What is the alternative hypothesis?
c. Identify the value of the test statistic.
d. Identify the P-value.
e. Is there sufficient evidence to support the claim that women in the different age categories have different mean blood pressure levels?

Minitab

Source	DF	SS	MS	F	P
Factor	2	938	469	1.65	0.205
Error	37	10484	283		
Total	39	11422			

5. Weights of Poplar Trees Refer to Data Set 9 in Appendix B and use the weights of poplar trees in Year 1 and Site 2. Use a 0.05 significance level to test the claim that those weights are from populations with the same mean. Site 2 has dry and sandy soil. Does it appear that the weights are different for the various treatments at this drier site?

6. Weights of Poplar Trees Refer to Data Set 9 in Appendix B and use the weights of poplar trees in Year 2 and Site 2. Use a 0.05 significance level to test the claim that those weights are from populations with the same mean. Site 2 has dry and sandy soil. Does it appear that the second-year weights are different for the various treatments at this drier site?

In Exercises 7 and 8, use the listed sample data from car crash experiments conducted by the National Transportation Safety Administration. New cars were purchased and crashed into a fixed barrier at 35 mi/h, and the listed measurements were recorded for the dummy in the driver's seat. The subcompact cars are the Ford Escort, Honda Civic, Hyundai Accent, Nissan Sentra, and Saturn SL4. The compact cars are Chevrolet Cavalier, Dodge Neon, Mazda 626 DX, Pontiac Sunfire, and Subaru Legacy. The midsize cars are Chevrolet Camaro, Dodge Intrepid, Ford Mustang, Honda Accord, and Volvo S70. The full-size cars are Audi A8, Cadillac Deville, Ford Crown Victoria, Oldsmobile Aurora, and Pontiac Bonneville.

7. Head Injury in a Car Crash The head injury data (in units of a standard head injury condition, or hic) are given below. Use a 0.05 significance level to test the null hypothesis that the different weight categories have the same mean. Do the data suggest that larger cars are safer?

Subcompact:	681	428	917	898	420
Compact:	643	655	442	514	525
Midsize:	469	727	525	454	259
Full-size:	384	656	602	687	360

8. Chest Deceleration in a Car Crash The chest deceleration data (g) are given below. Use a 0.05 significance level to test the null hypothesis that the different weight categories have the same mean. Do the data suggest that larger cars are safer?

Subcompact:	55	47	59	49	42
Compact:	57	57	46	54	51
Midsize:	45	53	49	51	46
Full-size:	44	45	39	58	44

9. Archeology: Skull Breadths from Different Epochs The values in the following table are measured maximum breadths of male Egyptian skulls from different epochs (based on data from *Ancient Races of the Thebaid,* by Thomson and Randall-Maciver). Changes in head shape over time suggest that interbreeding occurred with immigrant populations. Use a 0.05 significance level to test the claim that the different epochs do not all have the same mean.

4000 B.C.	1850 B.C.	150 A.D.
131	129	128
138	134	138
125	136	136
129	137	139
132	137	141
135	129	142
132	136	137
134	138	145
138	134	137

10. Secondhand Smoke in Different Groups Refer to Data Set 5 in Appendix B. Use a 0.05 significance level to test the claim that the mean cotinine level is different for these three groups: nonsmokers who are not exposed to environmental tobacco smoke, nonsmokers who are exposed to tobacco smoke, and people who smoke. What do the results suggest about secondhand smoke?

11. Iris Sepal Lengths Refer to Data Set 7 in Appendix B. Use a 0.05 significance level to test the claim that the three different species come from populations having the same mean sepal length.

12. Iris Petal Widths Refer to Data Set 7 in Appendix B. Use a 0.05 significance level to test the claim that the three different species come from populations having the same mean petal width.

13. Exercise and Stress Refer to Data Set 12 in Appendix B. Using only the systolic blood pressure readings from the "pre-exercise with no stress" list, test the claim that the four different combinations of gender and race (female and black, male and black, female and white, male and white) have the same mean. What does the result suggest about the suitability of the sample?

14. Exercise and Stress Refer to Data Set 12 in Appendix B. Using only the systolic blood pressure readings from the "pre-exercise with math stress" list, test the claim that the four different combinations of gender and race (female and black, male and black, female and white, male and white) have the same mean. Assuming that the four strata of gender/race combinations have baseline characteristics that are essentially the same, what does the result suggest about the effect of the stress caused by the math test?

*15. **Using the Tukey Test** This section included a display of the Bonferroni test results from Table 11-1. Shown here is the SPSS-generated display of results from the Tukey test. Compare the Tukey test results to those from the Bonferroni test.

Multiple Comparisons

Dependent Variable: WEIGHT
Tukey HSD

(I) SAMPLE	(J) SAMPLE	Mean Difference (I-J)	Std. Error	Sig.	95% Confidence Interval Lower Bound	95% Confidence Interval Upper Bound
1.00	2.00	-.1480	.33005	.969	-1.0923	.7963
	3.00	.0200	.33005	1.000	-.9243	.9643
	4.00	-1.1500*	.33005	.015	-2.0943	-.2057
2.00	1.00	.1480	.33005	.969	-.7963	1.0923
	3.00	.1680	.33005	.956	-.7763	1.1123
	4.00	-1.0020*	.33005	.036	-1.9463	-.0577
3.00	1.00	-.0200	.33005	1.000	-.9643	.9243
	2.00	-.1680	.33005	.956	-1.1123	.7763
	4.00	-1.1700*	.33005	.013	-2.1143	-.2257
4.00	1.00	1.1500*	.33005	.015	.2057	2.0943
	2.00	1.0020*	.33005	.036	.0577	1.9463
	3.00	1.1700*	.33005	.013	.2257	2.1143

*. The mean difference is significant at the .05 level.

*16. **Using the Bonferroni Test** Shown below are partial results from using the Bonferroni test with the sample data given in Exercise 9. Assume that a 0.05 significance level is being used.
   a. What do the displayed results tell us?
   b. Use the Bonferroni test procedure to find the significance value for the test of a significant difference between the mean skull breadth from 1850 B.C. and 150 A.D. Identify the test statistic and either the P-value or critical values. What do the results indicate?

(I) SAMPLE	(J) SAMPLE	Mean Difference (I-J)	Std. Error	Sig.	95% Confidence Interval Lower Bound	95% Confidence Interval Upper Bound
1.00	2.00	-1.7778	1.95105	1.000	-6.7991	3.2435
	3.00	-5.4444*	1.95105	.030	-10.4657	-.4231

11-3 Two-Way ANOVA

In Section 11-2 we used analysis of variance to decide whether three or more populations have the same mean. That section used procedures referred to as *one*-way analysis of variance (or single-factor analysis of variance) because the data are categorized into groups according to a *single* factor (or treatment). Recall that a factor, or treatment, is a property that is the basis for categorizing the different groups of data. See Table 11-4 which lists weights (in kilograms) of poplar trees. The listed weights are from Data Set 9 in Appendix B and they are partitioned into eight categories according to two variables: (1) the row variable of site and (2) the column variable of treatment. **Two-way analysis of variance** involves *two* factors, such as site and treatment in Table 11-4. The eight subcategories in Table 11-4 are called *cells,* so Table 11-4 has eight cells containing five values each.

In analyzing the sample data in Table 11-4, we have already discussed the one-way analysis of variance for a single factor, so it might seem reasonable to simply proceed with one-way ANOVA for the factor of site and another one-way ANOVA for the factor of treatment. Unfortunately, conducting two separate one-way ANOVA tests wastes information and totally ignores a very important feature: the effect of any *interaction* between the two factors.

Definition

There is an **interaction** between two factors if the effect of one of the factors changes for different categories of the other factor.

As an example of an *interaction* between two factors, consider the pairings of food and wine at a quality restaurant. It is known that certain foods and wines interact well to produce an enjoyable taste, while others interact poorly to produce an unpleasant taste. There is a good interaction between Chablis wine and oysters; the limestone in the soil where Chablis is made leaves a residue in the wine that

Data Mining

The term *data mining* is commonly used to describe the now popular practice of analyzing an existing large set of data for the purpose of finding relationships, patterns, or any interesting results that were not found in the original studies of the data set. Some statisticians express concern about ad hoc inference—a practice in which a researcher goes on a fishing expedition through old data, finds something significant, and then identifies an important question that has already been answered. Robert Gentleman, a column editor for *Chance* magazine, writes that "there are some interesting and fundamental statistical issues that data mining can address. We simply hope that its current success and hype don't do our discipline (statistics) too much damage before its limitations are discussed."

Table 11-4	Weights (kg) of Poplar Trees in Year 1			
			Treatment	
	None	Fertilizer	Irrigation	Fertilizer and Irrigation
Site 1 (rich, moist)	0.15	1.34	0.23	2.03
	0.02	0.14	0.04	0.27
	0.16	0.02	0.34	0.92
	0.37	0.08	0.16	1.07
	0.22	0.08	0.05	2.38
Site 2 (sandy, dry)	0.60	1.16	0.65	0.22
	1.11	0.93	0.08	2.13
	0.07	0.30	0.62	2.33
	0.07	0.59	0.01	1.74
	0.44	0.17	0.03	0.12

Table 11-5	Means (kg) of Cells from Table 11-4			
	Treatment			
	None	Fertilizer	Irrigation	Fertilizer and Irrigation
Site 1 (rich, moist)	0.184	0.332	0.164	1.334
Site 2 (sandy, dry)	0.458	0.630	0.278	1.308

interacts well with oysters. Peanut butter and jelly also interact well. In contrast, ketchup and ice cream interact in a way that results in a bad taste. Physicians must be careful to avoid prescribing drugs with interactions that produce adverse effects. It was found that the antifungal drug Nizoral (ketoconazole) interacted with the antihistamine drug Seldane (terfenadine) in such a way that Seldane was not metabolized properly, causing abnormal heart rhythms in some patients. Seldane was subsequently removed from the market.

Let's explore the data in Table 11-4 by calculating the mean for each cell and by constructing the graph shown in Figure 11-3. The individual cell means are shown in Table 11-5, and Figure 11-3 includes graphs of those means. A visual comparison of the means in Table 11-5 shows that they vary from a low of 0.164 kg to a high of 1.334 kg, so they appear to vary considerably. The graph in Figure 11-3 shows that the Site 2 means appear to be greater than the Site 1 means for three of the four treatment categories. Because the Site 2 line segments appear to be approximately *parallel* to the corresponding Site 1 line segments, it appears that the site weights behave the same for the different treatment categories, so there does not appear to be an interaction effect. In general, if a graph such as Fig-

FIGURE 11-3 **Means (kg) of Cells in Table 11-4**

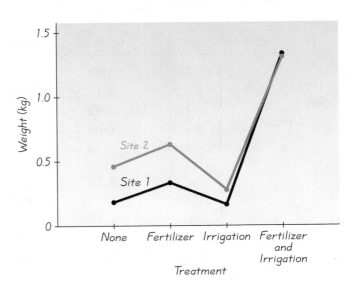

ure 11-3 results in line segments that are approximately *parallel*, we have evidence that there is *not an interaction* between the row and column variables. If the Site 2 line segments were far from parallel to the Site 1 line segments, we would have evidence of an interaction between site and treatment. These observations based on Table 11-5 and Figure 11-3 are largely subjective, so we now proceed with the more objective method of two-way analysis of variance.

In using two-way ANOVA for the data of Table 11-4, we consider three possible effects on the weights of the poplar trees: (1) the effects of an interaction between site and treatment; (2) the effects of site; (3) the effects of treatment. The calculations are quite involved, so *we will assume that a software package or a TI-83/84 Plus calculator is being used.* The Minitab and SPSS displays for the data in Table 11-4 are shown here.

Minitab

```
Analysis of Variance for Weight

Source        DF        SS        MS        F        P
Site           1     0.272     0.272     0.81     0.374
Treatment      3     7.547     2.516     7.50     0.001
Interaction    3     0.172     0.057     0.17     0.915
Error         32    10.727     0.335
Total         39    18.718
```

SPSS

Dependent Variable: WEIGHT

Source	Type III Sum of Squares	df	Mean Square	F	Sig.
Model	21.727[a]	8	2.716	8.102	.000
TREAT	7.547	3	2.516	7.505	.001
SITE	.272	1	.272	.812	.374
TREAT * SITE	.172	3	5.721E-02	.171	.915
Error	10.727	32	.335		
Total	32.453	40			

a. R Squared = .669 (Adjusted R Squared = .587)

The Minitab and SPSS displays include SS (sum of squares) components similar to those described in Section 11-2. Because the circumstances of Section 11-2 involved only a single factor, we used SS(treatment) as a measure of the variation due to the different treatment categories, and we used SS(error) as a measure of the variation due to sampling error. Here we use SS(site) as a measure of variation among the sites. We use SS(treatment) as a measure of variation among the treatments. We continue to use SS(error) as a measure of variation due to sampling error. Similarly, we use MS(site) and MS(treatment) for the two different mean squares and continue to use MS(error) as before. Also, we use df(site) and df(treatment) for the two different degrees of freedom.

Here are the requirements and basic procedure for two-way analysis of variance. The procedure is also summarized in Figure 11-4.

FIGURE 11-4 **Procedure for
Two-Way ANOVA**

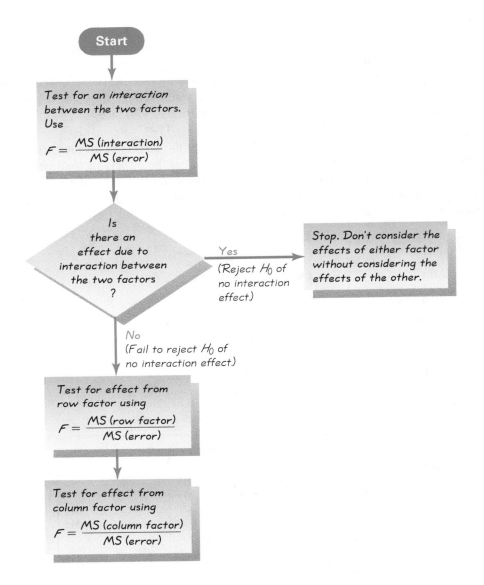

Requirements

1. For each cell, the sample values come from a population with a distribution that is normal. (This is a somewhat loose requirement in the sense that we can assume that normality is satisfied as long as no cell has a distribution that is dramatically different from a normal distribution.)

2. For each cell, the sample values come from populations having the same variance σ^2 (or standard deviation σ). (The method is robust against unequal variances. That is, the method works well with variances that are not dramatically different. Consequently, the requirement of equal variances is not a strict requirement.)

3. The samples are simple random samples. (That is, samples of the same size have the same probability of being selected.)

4. The samples are independent of each other. (The samples are not matched or paired in any way.)

5. The sample values are categorized two ways. (This is the basis for the name of the method: *two-way* analysis of variance.)

6. All of the cells have the same number of sample values. (This requirement, called a *balanced* design, is used for the methods of this section, but it is not technically required for general methods of ANOVA.)

We can use the following procedure when the above requirements are satisfied. We will illustrate the procedure with the data in Table 11-4.

Procedure for Two-Way ANOVA

Step 1:   *Interaction Effect:* In two-way analysis of variance, begin by testing the null hypothesis that there is no interaction between the two factors. Using either the Minitab or the SPSS display for the data in Table 11-4, we get the following test statistic (where SPSS denotes the interaction value as TREAT*SITE, and the MS value of 5.721E-02 is expressed in standard form as 0.05721).

$$F = \frac{MS(\text{interaction})}{MS(\text{error})} = \frac{0.057}{0.335} = 0.17$$

Interpretation: The corresponding P-value is shown as 0.915, so we fail to reject the null hypothesis of no interaction between the two factors. That is, there does not appear to be an interaction effect. It does not appear that the weights of the poplar trees are affected by an interaction between site and treatment. (This is supported by the graph shown in Figure 11-3.)

Step 2:   *Row/Column Effects:* If we do reject the null hypothesis of no interaction between factors, then we should stop now; we should not proceed with the two additional tests. (If there is an interaction between factors, we shouldn't consider the effects of either factor without considering those of the other.)

  If we fail to reject the null hypothesis of no interaction between factors, then we should proceed to test the following two hypotheses:

  H_0:   There are no effects from the row factor (that is, the row means are equal).

  H_0:   There are no effects from the column factor (that is, the column means are equal).

In Step 1, we failed to reject the null hypothesis of no interaction between factors, so we proceed with the next two hypothesis tests identified in Step 2.

  For the row factor of site we get

$$F = \frac{MS(\text{site})}{MS(\text{error})} = \frac{0.272}{0.335} = 0.81$$

Interpretation: This value is not significant because the corresponding P-value is shown in the Minitab and SPSS displays as 0.374. We fail to reject the null hypothesis of no effects from site. That is, the site of the poplar tree does not appear to have an effect on weight.

For the column factor of treatment we get

$$F = \frac{MS(treatment)}{MS(error)} = \frac{2.516}{0.335} = 7.50$$

Interpretation: This value is significant because the corresponding *P*-value is shown as 0.001. We therefore reject the null hypothesis of no effects from treatment. The treatment does appear to have an effect on weight of a poplar tree. Based on the sample data in Table 11-4, we conclude that weights do appear to have unequal means for the different treatment categories, but the weights appear to have equal means for sites. (See Figure 11-3 and note that the line segments appear to change dramatically for the different treatment categories, but the Site 1 and Site 2 line segments are not dramatically different from each other.)

Special Case: One Observation per Cell and No Interaction   Table 11-4 contains 5 observations per cell. If our sample data consist of only one observation per cell, we lose MS(interaction), SS(interaction), and df(interaction) because those values are based on sample variances computed for each individual cell. If there is only one observation per cell, there is no variation within individual cells and those sample variances cannot be calculated. Here's how we proceed when there is one observation per cell: *If it seems reasonable to assume (based on knowledge about the circumstances) that there is no interaction between the two factors, make that assumption and then proceed as before to test the following two hypotheses separately:*

H_0: There are no effects from the row factor.

H_0: There are no effects from the column factor.

As an example, if we know only the first value in each cell of Table 11-4, we have the data shown in Table 11-6. In Table 11-6, the two row means are 0.9375 and 0.6575. Is that difference significant, suggesting that there is an effect due to site? In Table 11-6, the four column means are 0.3750, 1.2500, 0.4400, and 1.1250. Are those differences significant, suggesting that there is an effect due to treatment? Assuming that weights are not affected by some interaction between site and treatment, the Minitab display is as shown below. (If we believe there is an interaction, the method described here does not apply.)

Minitab

```
Analysis of Variance for Weight

Source      DF      SS        MS        F      P
Site         1   0.15680   0.156800   0.28   0.634
Treatment    3   1.23665   0.412217   0.73   0.598
Error        3   1.68690   0.562300
Total        7   3.08035
```

We first use the results from the Minitab display to test the null hypothesis of no effects from the row factor of site.

$$F = \frac{MS(site)}{MS(error)} = \frac{0.157}{0.562} = 0.28$$

Table 11-6	Weights (kg) of Poplar Trees in Year 1			
	Treatment			
	None	Fertilizer	Irrigation	Fertilizer and Irrigation
Site 1 (rich, moist)	0.15	1.34	0.23	2.03
Site 2 (sandy, dry)	0.60	1.16	0.65	0.22

This test statistic is not significant, because the corresponding P-value in the Minitab display is 0.634. We fail to reject the null hypothesis; it appears that the poplar tree weights are not affected by the site.

We now use the Minitab display to test the null hypothesis of no effect from the column factor of treatment category. The test statistic is

$$F = \frac{MS(\text{treatment})}{MS(\text{error})} = \frac{0.412}{0.562} = 0.73$$

This test statistic is not significant because the corresponding P-value is given in the Minitab display as 0.598. We fail to reject the null hypothesis, so it appears that the poplar tree weight is not affected by the treatment. Using the data in Table 11-6, we conclude that the poplar tree weights do not appear to be affected by either site or treatment, but when we used 5 values from each cell (Table 11-4), we concluded that the weights appear to be affected by treatment. Such is the power of larger samples.

In this section we have briefly discussed an important branch of statistics. We have emphasized the interpretation of computer displays while omitting the manual calculations and formulas, which are quite formidable.

11-3 Exercises

Interpreting a Computer Display. *Some of Exercises 1–7 require the given Minitab display, which results from the amounts of the pesticide DDT measured in falcons in three different age categories (young, middle-aged, old) at three different locations (United States, Canada, Arctic region). The data set is included with the Minitab software as the file FALCON.MTW.*

Minitab

```
Analysis of Variance for DDT
Source        DF        SS        MS         F         P
Site           2   17785.41   8892.70   2581.75     0.000
Age            2    1721.19    860.59    249.85     0.000
Interaction    4      17.70      4.43      1.28     0.313
Error         18      62.00      3.44
Total         26   19586.30
```

1. **Meaning of Two-Way ANOVA** The method of this section is referred to as two-way analysis of variance, or two-way ANOVA. Why is the term *two-way* used? Why is the term *analysis of variance* used?

2. **Why Not One-Way ANOVA?** The given Minitab display results from measured amounts of DDT in falcons partitioned into nine cells according to one factor of location and another factor of age of the falcon. Each cell includes three DDT measurements. Why can't we conduct a thorough analysis of the data by simply executing two separate tests using one-way ANOVA (described in Section 11-2), where one test addresses the differences in sites and the other test addresses the differences in age? That is, why is two-way ANOVA required instead of two separate applications of one-way ANOVA?

3. **Interaction Effect** Assume that two-way analysis of variance reveals that there is a significant effect from an interaction between two factors. Why should we *not* proceed to test the effect from the row factor?

4. **Why Not Use Two-Way ANOVA?** Why can't we use the method of two-way analysis of variance with two-way tables described in Section 10-3?

5. **Interaction Effect** Refer to the Minitab display and test the null hypothesis that the amounts of DDT are not affected by an interaction between site and age. What do you conclude?

6. **Effect of Site** Refer to the Minitab display and assume that the amounts of DDT in the falcons are not affected by an interaction between site and age. Is there sufficient evidence to support the claim that the site has an effect on the amount of DDT?

7. **Effect of Age** Refer to the Minitab display and assume that the amounts of DDT in the falcons are not affected by an interaction between site and age. Is there sufficient evidence to support the claim that age has an effect on the amount of DDT?

Interpreting a Computer Display. In Exercises 8–10, use the SPSS display that results from the New York Marathon running times (in seconds) for randomly selected runners who finished.

Times (in seconds) for New York Marathon Runners

	Age		
	21–29	**30–39**	**40 and over**
Male	13,615	14,677	14,528
	18,784	16,090	17,034
	14,256	14,086	14,935
	10,905	16,461	14,996
	12,077	20,808	22,146
Female	16,401	15,357	17,260
	14,216	16,771	25,399
	15,402	15,036	18,647
	15,326	16,297	15,077
	12,047	17,636	25,898

SPSS

Dependent Variable: TIME

Source	Type III Sum of Squares	df	Mean Square	F	Sig.
Model	8202665802ª	6	1367110967	151.422	.000
GENDER	15225412.8	1	15225412.80	1.686	.206
AGE	92086979.4	2	46043489.70	5.100	.014
GENDER * AGE	21042068.6	2	10521034.30	1.165	.329
Error	216683456	24	9028477.350		
Total	8419349258	30			

a. R Squared = .974 (Adjusted R Squared = .968)

8. Interaction Effect Test the null hypothesis that the times are not affected by an interaction between gender and age category. What do you conclude?

9. Effect of Gender Assume that the marathon running times are not affected by an interaction between gender and age category. Is there sufficient evidence to support the claim that gender has an effect on running times?

10. Effect of Age Category Assume that running times are not affected by an interaction between gender and age category. Is there sufficient evidence to support the claim that the age category has an effect on time?

Interpreting a Computer Display. *In Exercises 11 and 12, refer to the given Minitab display. This display results from a study in which 24 subjects were given hearing tests using four different lists of words. The 24 subjects had normal hearing and the tests were conducted with no background noise. The main objective was to determine whether the four lists are equally difficult to understand. In the original table of hearing test scores, each cell has one entry. The original data are from* A Study of the Interlist Equivalency of the CID W-22 Word List Presented in Quiet and in Noise, *by Faith Loven, University of Iowa. The original data are available on the Internet through DASL (Data and Story Library).*

Minitab

```
Analysis of Variance for Hearing
Source      DF        SS        MS        F        P
Subject     23     3231.6     140.5     3.87    0.000
List         3      920.5     306.8     8.45    0.000
Error       69     2506.5      36.3
Total       95     6658.6
```

11. Hearing Tests: Effect of Subject Assuming that there is no effect on hearing test scores from an interaction between subject and list, is there sufficient evidence to support the claim that the choice of subject has an effect on the hearing test score? Interpret the result by explaining why it makes practical sense.

12. Hearing Tests: Effect of Word List Assuming that there is no effect on hearing test scores from an interaction between subject and list, is there sufficient evidence to support the claim that the choice of word list has an effect on the hearing test score?

13. Pulse Rates The following table lists pulse rates from Data Set 1 in Appendix B. Are pulse rates affected by an interaction between gender and age? Are pulse rates affected by gender? Are pulse rates affected by age?

	Age		
	Under 20	**20–40**	**Over 40**
Male	96 64 68 60	64 88 72 64	68 72 60 88
Female	76 64 76 68	72 88 72 68	60 68 72 64

Weights of Poplar Trees. In Exercises 14–18, use the poplar tree weights from Year 2, as listed in Data Set 9 in Appendix B. (The example in this section used the data from Year 1.)

14. Interaction Effect   Use a 0.05 significance level to test for an interaction between the site and treatment. What do you conclude?

15. Effect of Site   Assuming that there is no interaction between site and treatment, use a 0.05 significance level to test the claim that the poplar tree weights are affected by the site. What do you conclude?

16. Effect of Treatment   Assuming that there is no interaction between site and treatment, use a 0.05 significance level to test the claim that the poplar tree weights are affected by the treatment. What do you conclude?

17. Effect of Site   Use only the first weight from each cell and assume that there is no interaction between site and treatment. Does there appear to be an effect from site?

18. Effect of Treatment   Use only the first weight from each cell and assume that there is no interaction between site and treatment. Does there appear to be an effect from treatment?

Review

In Section 8-3 we presented a procedure for testing equality between *two* population means, but in Section 11-2 we used analysis of variance (or ANOVA) to test for equality of three or more population means. This method requires (1) normally distributed populations, (2) populations with the same standard deviation (or variance), and (3) simple random samples that are independent of each other. The methods of one-way analysis of variance are used when we have three or more samples taken from populations that are characterized according to a single factor. The following are key features of one-way analysis of variance:

- The *F* test statistic is based on the ratio of two different estimates of the common population variance σ^2, as shown below.

$$F = \frac{\text{variance between samples}}{\text{variance within samples}} = \frac{\text{MS(treatment)}}{\text{MS(error)}}$$

- Critical values of *F* can be found in Table A-5, but we focused on the interpretation of *P*-values that are included as part of a computer display.

In Section 11-3 we considered two-way analysis of variance, with data categorized according to two different factors. One factor is used to arrange the sample data in different rows, while the other factor is used for different columns. The procedure for two-way analysis of variance is summarized in Figure 11-4, and it requires that we first test for an interaction between the two factors. If there is no significant interaction, we then proceed to conduct individual tests for effects from each of the two factors. We also considered the use of two-way analysis of variance for the special case in which there is only one observation per cell.

Because of the nature of the calculations required throughout this chapter, we emphasized the interpretation of computer displays.

Review Exercises

1. **Drinking and Driving** The Associated Insurance Institute sponsors studies of the effects of drinking on driving. In one such study, three groups of adult men were randomly selected for an experiment designed to measure their blood alcohol levels after consuming five drinks. Members of group A were tested after one hour, members of group B were tested after two hours, and members of group C were tested after four hours. The results are given in the accompanying table; the SPSS display for these data is also shown. At the 0.05 significance level, test the claim that the three groups have the same mean level.

A	B	C
0.11	0.08	0.04
0.10	0.09	0.04
0.09	0.07	0.05
0.09	0.07	0.05
0.10	0.06	0.06
		0.04
		0.05

SPSS

A

	Sum of Squares	df	Mean Square	F	Sig.
Between Groups	.008	2	.004	46.900	.000
Within Groups	.001	14	.000		
Total	.009	16			

2. **Iris Petal Lengths** Using the petal lengths of irises listed in Data Set 7 in Appendix B, the analysis of variance results are obtained from SPSS, and the results are shown below. Using a 0.05 significance level, test the claim that the three species of irises have the same mean petal lengths.

SPSS

PL

	Sum of Squares	df	Mean Square	F	Sig.
Between Groups	436.644	2	218.322	1179.034	.000
Within Groups	27.220	147	.185		
Total	463.864	149			

3. **Iris Sepal Widths** Use the sepal widths of irises listed in Data Set 7 in Appendix B. Using a 0.05 significance level, test the claim that the three species of irises have the same mean sepal widths.

Interpreting a Computer Display. In Exercises 4–6, use the Minitab display that results from the values listed in the accompanying table. The sample data are student estimates (in feet) of the length of their classroom. The actual length of the classroom is 24 ft 7.5 in.

	Major		
	Math	Business	Liberal Arts
Female	28 25 30	35 25 20	40 21 30
Male	25 30 20	30 24 25	25 20 32

Minitab

```
Analysis of Variance for LENGTH
Source        DF        SS        MS        F        P
GENDER         1       29.4      29.4     0.78     0.395
MAJOR          2       10.1       5.1     0.13     0.876
Interaction    2       14.1       7.1     0.19     0.832
Error         12      453.3      37.8
Total         17      506.9
```

4. Interaction Effect Test the null hypothesis that the estimated lengths are not affected by an interaction between gender and major.

5. Effect of Gender Assume that estimated lengths are not affected by an interaction between gender and major. Is there sufficient evidence to support the claim that estimated length is affected by gender?

6. Effect of Major Assume that estimated lengths are not affected by an interaction between gender and major. Is there sufficient evidence to support the claim that estimated length is affected by major?

Cumulative Review Exercises

1. Boston Rainfall Statistics When attempting to study the relationship between weather and plant growth, a researcher collects the Monday rainfall amounts for Boston during a recent year, and those amounts (in inches) are listed below.
   a. Find the mean.
   b. Find the standard deviation.
   c. Find the 5-number summary.
   d. Identify any outliers.
   e. Construct a histogram.
   f. Assume that you want to test the null hypothesis that the mean amount of rainfall is the same for the seven days of the week, and that the additional sample data are available. Can you use one-way ANOVA? Why or why not?
   g. Based on the sample data, estimate the probability that precipitation will fall on a randomly selected Monday in Boston.

0.00	0.00	0.00	0.00	0.05	0.00	0.01	0.00	0.00	0.12
0.00	0.00	1.41	0.00	0.00	0.00	0.47	0.00	0.00	0.92
0.01	0.01	0.00	0.00	0.00	0.00	0.03	0.00	0.01	0.00
0.11	0.01	0.49	0.00	0.01	0.00	0.00	0.00	0.12	0.00
0.00	0.00	0.59	0.00	0.01	0.00	0.00	0.41	0.00	0.00
0.00	0.43								

2. M&M as a Treatment The table below lists 60 SAT scores separated into categories according to the color of the M&M candy used as a treatment. The SAT scores are based on data from the College Board, and the M&M color element is based on author whimsy.

   a. Find the mean of the 20 SAT scores in each of the three categories. Do the three means appear to be approximately equal?

   b. Find the median of the 20 SAT scores in each of the three categories. Do the three medians appear to be approximately equal?

   c. Find the standard deviation of the 20 SAT scores in each of the three categories. Do the three standard deviations appear to be approximately equal?

   d. Test the null hypothesis that there is no difference between the mean SAT score of subjects treated with red M&Ms and the mean SAT score of subjects treated with green M&Ms.

   e. Construct a 95% confidence interval estimate of the mean SAT score for the population of subjects receiving the red M&M treatment.

   f. Test the null hypothesis that the three populations (red, green, and blue M&M treatments) have the same mean SAT score.

Red	1130	621	813	996	1030	1257	898	743	921	1179
	1092	855	896	858	1095	1133	896	1190	908	699
Green	996	630	583	828	1121	993	1025	907	1111	1147
	780	916	793	1188	499	1180	1229	1450	1071	1153
Blue	706	1068	1013	892	1370	1611	939	1004	821	915
	866	848	1408	793	1097	1244	996	1131	1039	1159

3. Weights of Babies: Finding Probabilities In the United States, weights of newborn babies are normally distributed with a mean of 7.54 lb and a standard deviation of 1.09 lb (based on data from "Birth Weight and Perinatal Mortality," by Wilcox, Skjaerven, Buekens, and Kiely, *Journal of the American Medical Association,* Vol. 273, No. 9).

   a. If a newborn baby is randomly selected, what is the probability that he or she weighs more than 8.00 lb?

   b. If 16 newborn babies are randomly selected, what is the probability that their mean weight is more than 8.00 lb?

   c. What is the probability that each of the next three babies will have a birth weight greater than 7.54 lb?

Cooperative Group Activities

1. *Out-of-class activity:* The *World Almanac and Book of Facts* includes a section called "Noted Personalities," with subsections comprised of architects, artists, business leaders, cartoonists, social scientists, military leaders, philosophers, political leaders, scientists, writers, composers, entertainers, and others. Design and conduct an observational study that begins with the selection of samples from select groups, to be followed by a comparison of life spans of people from the different categories. Do any particular groups appear to have life spans that are different from the other groups? Can you explain such differences?

2. *In-class activity:* Begin by asking each student in the class to estimate the length of the classroom. Specify that the length is the distance between the chalkboard and the opposite wall. (See Review Exercises 4–6.) Each student should also write his or her gender (male/female) and major. Then divide into groups of three or four, and use the data from the entire class to address these questions:

   ● Is there a significant difference between the mean estimate for males and the mean estimate for females?

   ● Is there sufficient evidence to reject equality of the mean estimates for different majors? Describe how the majors were categorized.

- Does an interaction between gender and major have an effect on the estimated length?
- Does gender appear to have an effect on estimated length?
- Does major appear to have an effect on estimated length?

3. *Out-of-class activity:* Divide into groups of three or four students. Each group should survey other students at the same college by asking them to identify their major and gender. You might include other factors, such as employment (none, part-time, full-time) and age (under 21, 21–30, over 30). For each surveyed subject, determine the accuracy of the time on his or her wristwatch. First set your own watch to the correct time using an accurate and reliable source ("At the tone, the time is . . ."). For watches that are ahead of the correct time, record positive times. For watches that are behind the correct time, record negative times. Use the sample data to address questions such as these:

- Does gender appear to have an effect on the accuracy of the wristwatch?
- Does major have an effect on wristwatch accuracy?
- Does an interaction between gender and major have an effect on wristwatch accuracy?

Technology Project

Using the pulse rates of females listed in Data Set 1 in Appendix B, we find that $n = 40$, $\bar{x} = 76.3$, and $s = 12.5$. We will assume that these statistics are reasonably good estimates of the parameters for the population of all women. Conduct an experiment with the following steps:

1. Using STATDISK, Minitab, Excel, a TI-83/84 Plus calculator, or some other suitable technology, randomly generate 100 pulse rates from a normally distributed population having a mean of 76.3 and a standard deviation of 12.5.

2. Repeat Step 1 and generate a second sample of 100 values from a normally distributed population having a mean of 76.3 and a standard deviation of 12.5.

3. Generate a third sample of 100 values from a normally distributed population with a standard deviation of 12.5, but use a mean of 81.3 for this third sample.

4. Use the methods of this chapter to test the null hypothesis that the three samples come from populations having the same mean.

5. What do you conclude from Step 4? Given the way that the three samples were generated, we know that they come from populations with means that are not all equal. If Step 4 resulted in failure to reject the null hypothesis, what type of error was made (type I or type II)? Explain how correct application of a sound statistical procedure can result in such an error.

from DATA to DECISION

Critical Thinking: Should you approve this drug?

Drugs must undergo thorough testing before being approved for general use. In addition to testing for adverse reactions, they must also be tested for their effectiveness, and the analysis of such test results typically involves methods of statistics. Consider the development of Xynamine—a new drug designed to lower pulse rates. In order to obtain more consistent results that do not have a confounding variable of gender, the drug is tested using males only. Given below are pulse rates for a placebo group, a group of men treated with Xynamine in 10 mg doses, and a group of men treated with 20 mg doses of Xynamine. The project manager for the drug conducts research and finds that for adult males, pulse rates are normally distributed with a mean around 70 beats per minute and a standard deviation of approximately 11 beats per minute. His summary report states that the drug is effective, based on this evidence: The placebo group has a mean pulse rate of 68.9, which is close to the value of 70 beats per minute

for adult males in general, but the group treated with 10 mg doses of Xynamine has a lower mean pulse rate of 66.2, and the group treated with 20 mg doses of Xynamine has the lowest mean pulse rate of 65.2.

Analyze the Results

Analyze the data using the methods of this chapter. Based on the results, does it appear that there is sufficient evidence to support the claim that the drug lowers pulse rates? Are there any serious problems with the design of the experiment? Given that only males were involved in the experiment, do the results also apply to females? The project manager compared the post-treatment pulse rates to the mean pulse rate for adult males. Is there a better way to measure the drug's effectiveness in lowering pulse rates? How would you characterize the overall validity of the experiment? Based on the available results, should the drug be approved? Write a brief report summarizing your findings.

Placebo Group	10 mg Treatment Group	20 mg Treatment Group
77	67	72
61	48	94
66	79	57
63	67	63
81	57	69
75	71	59
66	66	64
79	85	82
66	75	34
75	77	76
48	57	59
70	45	53

Nonparametric Statistics

CHAPTER PROBLEM

Do different treatments affect the weights of poplar trees?

The Chapter Problem for Chapter 11 included the same data listed in Table 12-1, which lists weights (in kilograms) of poplar trees given different treatments at different sites. These data, taken from Data Set 9 in Appendix B, are the weights given for year 1 at Site 1, which has rich and moist soil and is located near a creek. In Chapter 11 we noted that the methods of analysis of variance require that the samples come from populations having distributions that are approximately normal. We also noted that the second sample (trees treated with fertilizer) included a weight of 1.34 kg, which appears to be an outlier. In Exercise 1 from Section 11-2, the ANOVA results are shown for the data with the outlier of 1.34 kg excluded, and we can see that the results do not change dramatically. When the outlier is excluded, the F test statistic changes from 5.73 to 8.45, and the P-value changes from 0.007 to 0.002.

But what if there are more outliers, or analysis of the samples strongly suggests that the samples come from populations having distributions that are very far from normal? The methods of analysis of variance presented in Chapter 11 could not be used with such a violation of a requirement that the distributions are approximately normal. In this chapter we introduce alternative methods that do not require that populations have distributions that are normal (or any other particular distribution). We will apply one of those methods to the sample weights in Table 12-1, and we will determine whether the different treatments appear to have an effect on the tree weights.

Table 12-1	Weights (kg) of Poplar Trees			
		Treatment		
	None	Fertilizer	Irrigation	Fertilizer and Irrigation
	0.15	1.34	0.23	2.03
	0.02	0.14	0.04	0.27
	0.16	0.02	0.34	0.92
	0.37	0.08	0.16	1.07
	0.22	0.08	0.05	2.38
n	5	5	5	5
$\bar{x}$	0.184	0.332	0.164	1.334
s	0.127	0.565	0.126	0.859

12-1 Overview

The methods of inferential statistics presented in Chapters 6, 7, 8, 9, and 11 are called *parametric methods* because they are based on sampling from a population with specific parameters, such as the mean μ, standard deviation σ, or proportion p. Those parametric methods usually must conform to some fairly strict conditions, such as a requirement that the sample data come from a normally distributed population. This chapter introduces nonparametric methods, which do not have such strict requirements.

Definitions

Parametric tests require assumptions about the nature or shape of the populations involved; **nonparametric tests** do not require that populations must have particular distributions. Consequently, nonparametric tests of hypotheses are often called **distribution-free tests.**

Although the term *nonparametric* suggests that the test is not based on a parameter, there are some nonparametric tests that do depend on a parameter such as the median. The nonparametric tests do not, however, require a particular distribution, so they are sometimes referred to as *distribution-free* tests. Although *distribution-free* is a more accurate description, the term *nonparametric* is more commonly used. The following are major advantages and disadvantages of nonparametric methods.

Advantages of Nonparametric Methods

1. Nonparametric methods can be applied to a wide variety of situations because they do not have the more rigid requirements of the corresponding parametric methods. In particular, nonparametric methods do not require normally distributed populations.

2. Unlike many parametric methods, nonparametric methods can often be applied to categorical data, such as the genders of survey respondents.

3. Nonparametric methods usually involve simpler computations than the corresponding parametric methods and are therefore easier to understand and apply.

4. If a data set includes an outlier that is actually an incorrect observation, the effect of the outlier is much less than its effect in a corresponding parametric test.

Disadvantages of Nonparametric Methods

1. Nonparametric methods tend to waste information because exact numerical data are often reduced to a qualitative form. For example, in the nonparametric sign test (described in Section 12-2), weight losses by dieters are recorded simply as negative signs; the actual magnitudes of the weight losses are ignored.

2. Nonparametric tests are not as efficient as parametric tests, so with a nonparametric test we generally need stronger evidence (such as a larger sample or

greater differences) in order to reject a null hypothesis. (However, some non-parametric tests are almost as efficient as their parametric counterparts when the populations are normally distributed and, for cases involving non-normal populations, nonparametric tests may be more efficient than the corresponding parametric tests.)

3. If outliers are present in the data, and they are not errors, but they are the result of contaminating factors that are changing the data in dramatic ways, then the use of nonparametric methods may result in an underestimation of the effects of those contaminating factors.

When the requirements of population distributions are satisfied, nonparametric tests are generally less efficient than their parametric counterparts, but the reduced efficiency can be compensated for by an increased sample size. For example, Section 12-6 will present a concept called *rank correlation,* which has an efficiency rating of 0.91 when compared to the linear correlation presented in Chapter 9. This means that with all other things being equal, nonparametric rank correlation requires 100 sample observations to achieve the same results as 91 sample observations analyzed through parametric linear correlation, assuming the stricter requirements for using the parametric method are met. Table 12-2 lists the nonparametric methods covered in this chapter, along with the corresponding parametric approach and **efficiency** rating. Table 12-2 shows that several nonparametric tests have efficiency ratings above 0.90, so the lower efficiency might not be a critical factor in choosing between parametric and nonparametric methods. However, because parametric tests do have higher efficiency ratings than their nonparametric counterparts, it's generally better to use the parametric tests when their required assumptions are satisfied.

Ranks

Sections 12-3 through 12-6 use methods based on ranks, which we now describe.

Table 12-2	Efficiency: Comparison of Parametric and Nonparametric Tests		
Application	Parametric Test	Nonparametric Test	Efficiency Rating of Nonparametric Test with Normal Population
Matched pairs of sample data	t test or z test	Sign test	0.63
		Wilcoxon signed-ranks test	0.95
Two independent samples	t test or z test	Wilcoxon rank-sum test	0.95
Several independent samples	Analysis of variance (F test)	Kruskal-Wallis test	0.95
Correlation	Linear correlation	Rank correlation test	0.91

Lie Detectors

Why not require all criminal suspects to take lie detector tests and dispense with trials by jury? The Council of Scientific Affairs of the American Medical Association states, "It is established that classification of guilty can be made with 75% to 97% accuracy, but the rate of false positives is often sufficiently high to preclude use of this (polygraph) test as the sole arbiter of guilt or innocence." A "false positive" is an indication of guilt when the subject is actually innocent. Even with accuracy as high as 97%, the percentage of false positive results can be 50%, so half of the innocent subjects incorrectly appear to be guilty.

Definition

Data are *sorted* when they are arranged according to some criterion, such as smallest to largest or best to worst. A **rank** is a number assigned to an individual sample item according to its order in the sorted list. The first item is assigned a rank of 1, the second item is assigned a rank of 2, and so on.

EXAMPLE The numbers 5, 3, 40, 10, and 12 can be sorted (arranged from lowest to highest) as 3, 5, 10, 12, and 40, and these numbers have ranks of 1, 2, 3, 4, and 5, respectively:

5	3	40	10	12	Original values
3	5	10	12	40	Values sorted (arranged in order)
↑	↑	↑	↑	↑	
1	2	3	4	5	Ranks

Handling ties in ranks: If a tie in ranks occurs, the usual procedure is to find the mean of the ranks involved and then assign this mean rank to each of the tied items, as in the following example.

EXAMPLE The numbers 3, 5, 5, 10, and 12 are given ranks of 1, 2.5, 2.5, 4, and 5, respectively. In this case, ranks 2 and 3 were tied, so we found the mean of 2 and 3 (which is 2.5) and assigned it to the values that created the tie:

3	5	5	10	12	Original values
↑	↑	↑	↑	↑	
1	2.5	2.5	4	5	Ranks

2 and 3 are tied

12-2 Sign Test

The main objective of this section is to understand the *sign test* procedure, which is among the easiest for nonparametric tests.

Definition

The **sign test** is a nonparametric (distribution-free) test that uses plus and minus signs to test different claims, including:

1. Claims involving matched pairs of sample data
2. Claims involving nominal data
3. Claims about the median of a single population

Basic Concept of the Sign Test   The basic idea underlying the sign test is to analyze the frequencies of the plus and minus signs to determine whether they are significantly different. For example, suppose that we test a treatment designed to lower blood pressure. If 100 subjects are treated and 51 of them experience lower blood pressure while the other 49 have increased blood pressure, common sense suggests that there is not sufficient evidence to say that the drug is effective, because 51 decreases out of 100 is not significant. But what about 52 decreases and 48 increases? Or 90 decreases and 10 increases? The sign test allows us to determine when such results are significant.

For consistency and simplicity, we will use a test statistic based on the number of times that the *less frequent* sign occurs. The relevant requirements, notation, test statistic, and critical values are summarized in the accompanying box. Figure 12-1 summarizes the sign test procedure, which will be illustrated with examples that follow.

Sign Test

Requirements

1. The sample data have been randomly selected.

2. There is *no* requirement that the sample data come from a population with a particular distribution, such as a normal distribution.

Notation

x = the number of times the *less frequent* sign occurs

n = the total number of positive and negative signs combined

Test Statistic

If $n \leq 25$, the test statistic is x (the number of times the less frequent sign occurs).

If $n > 25$, the test statistic is $z = \dfrac{(x + 0.5) - \left(\dfrac{n}{2}\right)}{\dfrac{\sqrt{n}}{2}}$

Critical values

1. For $n \leq 25$, critical x values are found in Table A-7.

2. For $n > 25$, critical z values are found in Table A-2.

Caution: When applying the sign test in a one-tailed test, we need to be very careful to avoid making the wrong conclusion when one sign occurs significantly more often than the other, but the sample data contradict the alternative hypothesis. For example, suppose we are testing the claim that a gender-selection technique favors boys, but we get a sample of 10 boys and 90 girls. With a sample proportion of boys equal to 0.10, the data contradict the alternative hypothesis H_1: $p > 0.5$. There is no way we can support a claim of $p > 0.5$ with any sample

FIGURE 12-1 Sign Test Procedure

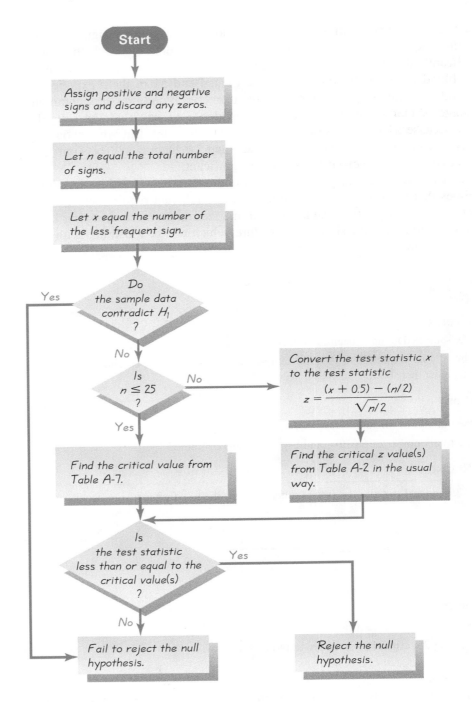

proportion less than 0.5, so we immediately fail to reject the null hypothesis and don't proceed with the sign test. Figure 12-1 summarizes the procedure for the sign test and includes this check: Do the sample data contradict H_1? If the sample data are in the opposite direction of H_1, fail to reject the null hypothesis. *It is always important to think about the data and to avoid relying on blind calculations or computer results.*

Claims Involving Matched Pairs

When using the sign test with data that are matched by pairs, we convert the raw data to plus and minus signs as follows:

1. We subtract each value of the second variable from the corresponding value of the first variable.

2. We record only the *sign* of the difference found in Step 1. We *exclude ties:* that is, we exclude any matched pairs in which both values are equal.

The key concept underlying this use of the sign test is this:

If the two sets of data have equal medians, the number of positive signs should be approximately equal to the number of negative signs.

EXAMPLE **Does the Type of Seed Affect Corn Growth?** In 1908, William Gosset published the article "The Probable Error of a Mean" under the pseudonym of "Student" (*Biometrika*, Vol. 6, No. 1). He included the data listed in Table 12-3 for two different types of corn seed (regular and kiln dried) that were used on *adjacent* plots of land. The listed values are the yields of head corn in pounds per acre. Use the sign test with a 0.05 significance level to test the claim that there is no difference between the yields from the regular and kiln-dried seed.

SOLUTION

REQUIREMENT ✓ The only requirement is that the sample data are randomly selected. There is no requirement about the distribution of the population, such as a common requirement that the sample data come from a normally distributed population. Based on the design of this experiment, we assume that the sample data are random. ✓

Here's the basic idea: If there is no difference between the yields from regular seeds and the yields from kiln-dried seeds, the numbers of positive and negative signs should be approximately equal. In Table 12-3 we have 7 negative and 4 positive signs. Are the numbers of positive and negative signs approximately equal, or are they significantly different? We follow the same basic steps for testing hypotheses as outlined in Figure 7-8, and we apply the sign test procedure summarized in Figure 12-1.

continued

Table 12-3	Yields of Corn from Different Seeds									
Regular	1903	1935	1910	2496	2108	1961	2060	1444	1612	1316	1511
Kiln dried	2009	1915	2011	2463	2180	1925	2122	1482	1542	1443	1535
Sign of difference	−	+	−	+	−	+	−	−	+	−	−

Steps   The null hypothesis is the claim of no difference between the yields
1, 2, 3:   from the regular seed and the yields from the kiln-dried seed, and the
           alternative hypothesis is the claim that there is a difference.

> H_0:   There is no difference. (The median of the differences is
>   equal to 0.)
>
> H_1:   There is a difference. (The median of the differences is not
>   equal to 0.)

Step 4:   The significance level is $\alpha = 0.05$.

Step 5:   We are using the nonparametric sign test.

Step 6:   The test statistic x is the number of times the less frequent sign oc-
           curs. Table 12-3 includes differences with 7 negative signs and 4 pos-
           itive signs. (If there had been any differences of 0, we would have
           discarded them.) We let x equal the smaller of 7 and 4, so $x = 4$.
           Also, $n = 11$ (the total number of positive and negative signs com-
           bined). Our test is two-tailed with $\alpha = 0.05$. We refer to Table A-7
           where the critical value of 1 is found for $n = 11$ and $\alpha = 0.05$ in two
           tails. (See Figure 12-1.)

Step 7:   With a test statistic of $x = 4$ and a critical value of 1, we fail to reject
           the null hypothesis of no difference. [See Note 2 included with Table
           A-7: "Reject the null hypothesis if the number of the less frequent
           sign (x) is less than or equal to the value in the table." Because $x = 4$
           is *not* less than or equal to the critical value of 1, we fail to reject the
           null hypothesis.]

Step 8:   There is not sufficient evidence to warrant rejection of the claim that
           the median of the differences is equal to 0. That is, there is not suffi-
           cient evidence to warrant rejection of the claim of no difference be-
           tween the yields from the regular seed and the yields from the kiln-
           dried seed. This is the same conclusion that was reached using the
           parametric t test with matched pairs in Section 8-4, but sign test re-
           sults do not always agree with parametric test results.

Claims Involving Nominal Data

Recall that nominal data consist of names, labels, or categories only. Although
such a nominal data set limits the calculations that are possible, we can identify
the *proportion* of the sample data that belong to a particular category, and we can
test claims about the corresponding population proportion p. The following example uses nominal data consisting of genders (boy/girl). The sign test is used by
representing boys with positive $(+)$ signs and girls with negative $(-)$ signs.
(Those signs are chosen arbitrarily, honest.) Also note the procedure for handling
cases in which $n > 25$.

EXAMPLE **Testing a Method of Gender Selection** Suppose that
a test of a method of gender selection involves 100 couples who are trying to
have baby girls, and the results include 30 boys and 70 girls. Do these results
provide enough evidence to conclude that the gender-selection technique has

an effect? Use the sign test and a 0.05 significance level to test the null hypothesis that the method of gender selection has no effect.

SOLUTION

REQUIREMENT ✓   The only requirement is that the sample data are randomly selected. Based on the design of this experiment, we assume that the sample data are random. We can now proceed with the sign test. ✓

Let p denote the population proportion of girls born to parents using the method of gender selection. If the method has no effect, then the proportion of girls will be approximately 0.5. The null and alternative hypotheses can therefore be stated as follows:

$$H_0: p = 0.5 \quad \text{(the proportion of girls is equal to 0.5)}$$
$$H_1: p \neq 0.5$$

Denoting boys by plus $(+)$ and girls by minus $(-)$, we have 30 positive signs and 70 negative signs. Refer now to the sign test procedure summarized in Figure 12-1. The test statistic x is the smaller of 30 and 70, so $x = 30$. This test involves two tails because a disproportionately low number of either gender will cause us to reject the claim of equality. The sample data do not contradict the alternative hypothesis because 30 and 70 are not precisely equal. Continuing with the procedure in Figure 12-1, we note that the value of $n = 100$ is above 25, so the test statistic x is converted (using a correction for continuity) to the test statistic z as follows:

$$z = \frac{(x + 0.5) - \left(\dfrac{n}{2}\right)}{\dfrac{\sqrt{n}}{2}}$$

$$= \frac{(30 + 0.5) - \left(\dfrac{100}{2}\right)}{\dfrac{\sqrt{100}}{2}} = -3.90$$

With $\alpha = 0.05$ in a two-tailed test, the critical values are $z = \pm 1.96$ The test statistic $z = -3.90$ is less than -1.96 (see Figure 12-2), so we reject the null hypothesis that the proportion of girls is equal to 0.5. There is sufficient sample evidence to warrant rejection of the null hypothesis that the method of gender selection has no effect. It appears that babies born to parents using this method of gender selection are born in such a way that the proportion of girls is different from 0.5.

Claims About the Median of a Single Population

The next example illustrates the procedure for using the sign test in testing a claim about the median of a single population. See how the negative and positive signs are based on the claimed value of the median.

FIGURE 12-2 **Testing the Null Hypothesis of No Effect**

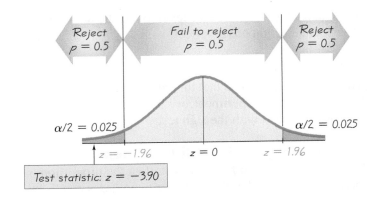

EXAMPLE Body Temperatures Data Set 2 in Appendix B includes measured body temperatures of adults. Use the 106 temperatures listed for 12 A.M. on day 2 with the sign test to test the claim that the median is less than 98.6°F. The data set has 106 subjects—68 subjects with temperatures below 98.6°F, 23 subjects with temperatures above 98.6°F, and 15 subjects with temperatures equal to 98.6°F.

SOLUTION

REQUIREMENT ✓ The only requirement is that the sample data are randomly selected and, based on the design of this experiment, we assume that the sample data are random. We can now proceed with the sign test. ✓

The claim that the median is less than 98.6°F is the alternative hypothesis, while the null hypothesis is the claim that the median is equal to 98.6°F.

H_0: Median is equal to 98.6°F.   (median = 98.6°F)
H_1: Median is less than 98.6°F.   (median < 98.6°F)

Following the procedure outlined in Figure 12-1, we discard the 15 zeros, we use the negative sign (−) to denote each temperature that is below 98.6°F, and we use the positive sign (+) to denote each temperature that is above 98.6°F. We therefore have 68 negative signs and 23 positive signs, so $n = 91$ and $x = 23$ (the number of the less frequent sign). The sample data do not contradict the alternative hypothesis, because most of the 91 temperatures are below 98.6°F. (If the sample data did conflict with the alternative hypothesis, we could immediately terminate the test by concluding that we fail to reject the null hypothesis.) The value of n exceeds 25, so we convert the test statistic x to the test statistic z.

$$z = \frac{(x + 0.5) - \left(\dfrac{n}{2}\right)}{\dfrac{\sqrt{n}}{2}}$$

$$= \frac{(23 + 0.5) - \left(\dfrac{91}{2}\right)}{\dfrac{\sqrt{91}}{2}} = -4.61$$

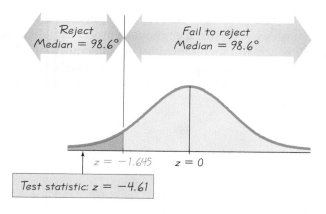

FIGURE 12-3 **Testing the Claim That the Median Is Less Than 98.6°F**

In this one-tailed test with $\alpha = 0.05$, we use Table A-2 to get the critical z value of -1.645. From Figure 12-3 we can see that the test statistic of $z = -4.61$ does fall within the critical region. We therefore reject the null hypothesis. On the basis of the available sample evidence, we support the claim that the median body temperature of healthy adults is less than 98.6°F.

In this sign test of the claim that the median is below 98.6°F, we get a test statistic of $z = -4.61$ with a P-value of 0.00000202, but a parametric test of the claim that $\mu < 98.6°F$ results in a test statistic of $t = -6.611$ with a P-value of 0.000000000813. The P-value from the sign test is not as low as the P-value from the parametric test, illustrating the fact that the sign test isn't as sensitive as the parametric test. Both tests lead to rejection of the null hypothesis, but the sign test doesn't consider the sample data to be as extreme, partly because the sign test uses only information about the *direction* of the data, ignoring the *magnitudes* of the data values. The next section introduces the Wilcoxon signed-ranks test, which largely overcomes that disadvantage.

Rationale for the test statistic used when $n > 25$: When finding critical values for the sign test, we use Table A-7 only for n up to 25. When $n > 25$, the test statistic z is based on a normal approximation to the binomial probability distribution with $p = q = 1/2$. Recall from Section 5-6 that the normal approximation to the binomial distribution is acceptable when both $np \geq 5$ and $nq \geq 5$. Recall from Section 4-4 that $\mu = np$ and $\sigma = \sqrt{npq}$ for binomial probability distributions. Because this sign test assumes that $p = q = 1/2$, we meet the $np \geq 5$ and $nq \geq 5$ prerequisites whenever $n \geq 10$. (However, we use Table A-7 for values of n up to 25 because the table is easier to use and it provides more accurate results.) Also, with the assumption that $p = q = 1/2$, we get $\mu = np = n/2$ and $\sqrt{npq} = \sqrt{n/4} = \sqrt{n}/2$, so

$$z = \frac{x - \mu}{\sigma}$$

becomes

$$z = \frac{x - \left(\dfrac{n}{2}\right)}{\dfrac{\sqrt{n}}{2}}$$

Finally, we replace x by $x + 0.5$ as a correction for continuity. That is, the values of x are discrete, but because we are using a continuous probability distribution, a discrete value such as 10 is actually represented by the interval from 9.5 to 10.5. Because x represents the less frequent sign, we act conservatively by concerning ourselves only with $x + 0.5$; we thus get the test statistic z, as given in the equation and in Figure 12-1.

12-2 Exercises

In Exercises 1–4, assume that matched pairs of data result in the given number of signs when the value of the second variable is subtracted from the corresponding value of the first variable. Use the sign test with a 0.05 significance level to test the null hypothesis of no difference.

1. Positive signs: 10; negative signs: 5; ties: 3

2. Positive signs: 6; negative signs: 16; ties: 2

3. Positive signs: 50; negative signs: 40; ties: 5

4. Positive signs: 10; negative signs: 30; ties: 3

In Exercises 5–14, use the sign test.

5. Testing Corn Seeds In 1908, William Gosset published the article "The Probable Error of a Mean" under the pseudonym of "Student" (*Biometrika*, Vol. 6, No. 1). He included the data listed below for yields from two different types of seed (regular and kiln dried) that were used on adjacent plots of land. The listed values are the yields of straw in cwts per acre, where cwt represents 100 pounds. Using a 0.05 significance level, test the claim that there is no difference between the yields from the two types of seed. Does it appear that either type of seed is better?

Regular	19.25	22.75	23	23	22.5	19.75	24.5	15.5	18	14.25	17
Kiln dried	25	24	24	28	22.5	19.5	22.25	16	17.25	15.75	17.25

6. Before/After Treatment Results Captopril is a drug designed to lower systolic blood pressure. When subjects were tested with this drug, their systolic blood pressure readings (in mm Hg) were measured before and after the drug was taken, with the results given in the accompanying table (based on data from "Essential Hypertension: Effect of an Oral Inhibitor of Angiotensin-Converting Enzyme," by MacGregor et al., *British Medical Journal*, Vol. 2). Is there sufficient evidence to support the claim that captopril is effective in lowering systolic blood pressure?

Subject	A	B	C	D	E	F	G	H	I	J	K	L
Before	200	174	198	170	179	182	193	209	185	155	169	210
After	191	170	177	167	159	151	176	183	159	145	146	177

7. Testing for a Difference Between Reported and Measured Male Heights As part of the National Health and Nutrition Examination Survey conducted by the Department of Health and Human Services, self-reported heights and measured heights were obtained for males aged 12–16. Listed below are sample results. Is there sufficient evidence to support the claim that there is a difference between self-reported heights and measured heights of males aged 12–16? Use a 0.05 significance level.

Reported height	68	71	63	70	71	60	65	64	54	63	66	72
Measured height	67.9	69.9	64.9	68.3	70.3	60.6	64.5	67.0	55.6	74.2	65.0	70.8

8. Testing for a Difference Between Reported and Measured Male Heights The table below lists matched pairs of measured heights of 12 male statistics students. Use a 0.05 significance level to test the claim that there is no difference between reported height and measured height.

Reported height	68	74	82.25	66.5	69	68	71	70	70	67	68	70
Measured height	66.8	73.9	74.3	66.1	67.2	67.9	69.4	69.9	68.6	67.9	67.6	68.8

9. Testing for Difference Between Prone and Supine Measurements In a study of techniques used to measure lung volumes, physiological data were collected for 10 subjects. The values given in the table are in liters and represent the measured functional residual capacities of the 10 subjects both in a prone position and in a supine (lying) position. At the 0.05 significance level, test the claim that there is no significant difference between the measurements taken in the two positions.

Prone	2.96	4.65	3.27	2.50	2.59	5.97	1.74	3.51	4.37	4.02
Supine	1.97	3.05	2.29	1.68	1.58	4.43	1.53	2.81	2.70	2.70

Based on "Validation of Esophageal Balloon Technique at Different Lung Volumes and Postures," by Baydur, Cha, and Sassoon, *Journal of Applied Physiology,* Vol. 62, No. 1.

10. Testing for a Median Body Temperature of 98.6°F A pre-med student in a statistics class is required to do a class project. Intrigued by the body temperatures in Data Set 2 in Appendix B, she plans to collect her own sample data to test the claim that the median body temperature is less than 98.6°F. Because of time constraints, she finds that she has time to collect data from only 12 people. After carefully planning a procedure for obtaining a simple random sample of 12 healthy adults, she measures their body temperatures and obtains the results listed below. Use a 0.05 significance level to test the claim that these body temperatures come from a population with a median that is less than 98.6°F.

97.6   97.5   98.6   98.2   98.0   99.0   98.5   98.1   98.4   97.9   97.9   97.7

11. Testing for Median Underweight The Prince Biomass Company supplies bottles of plant food labeled 12 oz. When the Prince County Department of Weights and Measures tests a random sample of bottles, the amounts listed below are obtained. Using a 0.05 significance level, is there sufficient evidence to file a charge that the company is cheating consumers by giving amounts with a median less than 12 oz?

11.4 11.8 11.7 11.0 11.9 11.9 11.5 12.0 12.1 11.9 10.9 11.3 11.5 11.5 11.6

12. Nominal Data: Success of Treatment In a clinical experiment involving 1002 treated patients, 701 had measurable improvements in symptoms. Is there sufficient evidence to support the claim that among all treated patients, the majority have measurable improvements in symptoms?

13. Nominal Data: Smoking and Nicotine Patches In one study of 71 smokers who tried to quit smoking with nicotine patch therapy, 41 were smoking one year after the treatment (based on data from "High-Dose Nicotine Patch Therapy," by Dale et al., *Journal of the American Medical Association,* Vol. 274, No. 17). Use a 0.05 significance level to test the claim that among smokers who try to quit with nicotine patch therapy, the majority are smoking a year after the treatment.

14. Parents' Heights Refer to Data Set 3 in Appendix B and use only the data corresponding to male children. Use the paired data consisting of the mother's height and the father's height. Use a 0.01 significance level to test the claim that the median height of mothers of male children is less than the median height of the fathers.

*15. Procedures for Handling Ties In the sign test procedure described in this section, we excluded ties (represented by 0 instead of a sign of + or −). A second approach is to treat half of the 0s as positive signs and half as negative signs. (If the number of 0s is odd, exclude one so that they can be divided equally.) With a third approach, in two-tailed tests make half of the 0s positive and half negative; in one-tailed tests make all 0s either positive or negative, whichever supports the null hypothesis. Assume that in using the sign test on a claim that the median value is less than 100, we get 60 values below 100, 40 values above 100, and 21 values equal to 100. Identify the test statistic and conclusion for the three different ways of handling differences of 0. Assume a 0.05 significance level in all three cases.

*16. Finding Critical Values Table A-7 lists critical values for limited choices of α. Use Table A-1 to add a new column in Table A-7 (down to $n = 15$) that represents a significance level of 0.03 in one tail or 0.06 in two tails. For any particular n, use $p = 0.5$, because the sign test requires the assumption that $P(\text{positive sign}) = P(\text{negative sign}) = 0.5$. The probability of x or fewer like signs is the sum of the probabilities for values up to and including x.

*17. Normal Approximation Error The Biosim.com company has hired 18 women among the last 54 new employees. Job applicants are about half men and half women, all of whom are qualified. Using a 0.01 significance level with the sign test, is there sufficient evidence to charge bias? Does the conclusion change if the binomial distribution is used instead of the approximating normal distribution?

12-3 Wilcoxon Signed-Ranks Test for Matched Pairs

In Section 12-2 we used the sign test to analyze three different types of data, including sample data consisting of matched pairs. The sign test used only the signs of the differences and did not use their actual magnitudes (how large the numbers are). This section introduces the *Wilcoxon signed-ranks test,* which is also used with sample paired data. By using ranks, this test takes the magnitudes of the differences into account. (See Section 12-1 for a description of ranks.) Because the Wilcoxon signed-ranks test incorporates and uses more information than the sign test, it tends to yield conclusions that better reflect the true nature of the data.

Definition

The **Wilcoxon signed-ranks test** is a nonparametric test that uses ranks of sample data consisting of matched pairs. It is used to test the null hypothesis that the population of differences has a median of zero, so the null and alternative hypotheses are as follows:

H_0:   The matched pairs have differences that come from a population with a median equal to zero.

H_1:   The matched pairs have differences that come from a population with a nonzero median.

(The Wilcoxon signed-ranks test can also be used to test the claim that a sample comes from a population with a specified median. See Exercise 9 for this application.)

Wilcoxon Signed-Ranks Procedure

Step 1:   For each pair of data, find the difference d by subtracting the second value from the first value. Keep the signs, but discard any pairs for which $d = 0$

Step 2:   *Ignore the signs of the differences,* then sort the differences from lowest to highest and replace the differences by the corresponding rank value (as described in Section 12-1). When differences have the same numerical value, assign to them the mean of the ranks involved in the tie.

Step 3:   Attach to each rank the sign of the difference from which it came. That is, insert those signs that were ignored in Step 2.

Step 4:   Find the sum of the absolute values of the negative ranks. Also find the sum of the positive ranks.

Step 5:   Let T be the *smaller* of the two sums found in Step 4. Either sum could be used, but for a simplified procedure we arbitrarily select the smaller of the two sums. (See the notation for T in the accompanying box.)

Step 6:   Let n be the number of pairs of data for which the difference d is not 0.

Step 7:   Determine the test statistic and critical values based on the sample size, as shown in the accompanying box.

Step 8:   When forming the conclusion, reject the null hypothesis if the sample data lead to a test statistic that is in the critical region—that is, the test statistic is less than or equal to the critical value(s). Otherwise, fail to reject the null hypothesis.

Wilcoxon Signed-Ranks Test

Requirements

1. The data consist of matched pairs that have been randomly selected.

2. The population of differences (found from the pairs of data) has a distribution that is approximately *symmetric* about the median. (There is *no* requirement that the data have a normal distribution.)

continued

Notation

See the accompanying procedure steps for finding the rank sum T.

$T = $ the smaller of the following two sums:

1. The sum of the absolute values of the negative ranks of the nonzero differences d
2. The sum of the positive ranks of the nonzero differences d

Test Statistic

If $n \le 30$, the test statistic is T.

If $n > 30$, the test statistic is $z = \dfrac{T - \dfrac{n(n + 1)}{4}}{\sqrt{\dfrac{n(n + 1)(2n + 1)}{24}}}$

Critical values

1. If $n \le 30$, the critical T value is found in Table A-8.
2. If $n > 30$, the critical z values are found in Table A-2.

EXAMPLE **Does the Type of Seed Affect Corn Growth?** In 1908, William Gosset published the article "The Probable Error of a Mean" under the pseudonym of "Student" (*Biometrika*, Vol. 6, No. 1). He included the data listed in Table 12-4 for two different types of corn seed (regular and kiln dried) that were used on *adjacent* plots of land. The listed values are the yields of head corn in pounds per acre. Use the Wilcoxon signed-ranks test with a 0.05 significance level to test the claim that there is no difference between the yields from the regular and kiln-dried seed.

SOLUTION

REQUIREMENT ✔ We must have matched pairs of randomly selected data. The data are matched and, given the design of this experiment, it is reasonable to assume that the matched pairs have been randomly selected. The accompanying Minitab-generated histogram shows that the distribution of differences isn't very symmetric about the median of 0, but the departure from symmetry isn't too extreme with these small samples. ✔

Table 12-4	Yields of Corn from Different Seeds										
Regular	1903	1935	1910	2496	2108	1961	2060	1444	1612	1316	1511
Kiln dried	2009	1915	2011	2463	2180	1925	2122	1482	1542	1443	1535
Differences d	−106	20	−101	33	−72	36	−62	−38	70	−127	−24
Ranks of |differences|	10	1	9	3	8	4	6	5	7	11	2
Signed ranks	−10	1	−9	3	−8	4	−6	−5	7	−11	−2

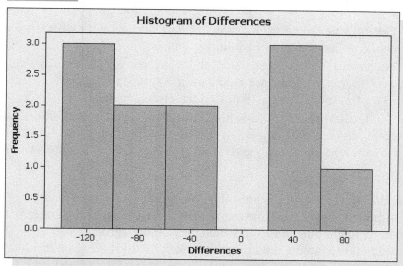

The null and alternative hypotheses are as follows:

H_0:   The yields from regular seed and kiln-dried seed are such that the median of the population of differences is equal to zero.

H_1:   The median of the population of differences is nonzero.

The significance level is $\alpha = 0.05$. We are using the Wilcoxon signed-ranks test procedure, so the test statistic is calculated by using the eight-step procedure presented earlier in this section.

Step 1:   In Table 12-4, the row of differences is obtained by computing this difference for each pair of data:

d = yield from regular seed − yield from kiln dried seed.

Step 2:   Ignoring their signs, we rank the absolute differences from lowest to highest. (If there had been any ties in ranks, they would have been handled by assigning the mean of the involved ranks to each of the tied values. Also, any differences of 0 would have been discarded.)

Step 3:   The bottom row of Table 12-4 is created by attaching to each rank the sign of the corresponding difference. If there really is no difference between the times of the first trial and the times of the second trial (as in the null hypothesis), we expect the sum of the positive ranks to be approximately equal to the sum of the absolute values of the negative ranks.

Step 4:   We now find the sum of the absolute values of the negative ranks, and we also find the sum of the positive ranks.

Sum of absolute values of negative ranks: 51
Sum of positive ranks: 15

continued

Step 5:   Letting T be the smaller of the two sums found in Step 4, we find that $T = 15$.

Step 6:   Letting n be the number of pairs of data for which the difference d is not 0, we have $n = 11$.

Step 7:   Because $n = 11$, we have $n \leq 30$, so we use a test statistic of $T = 15$ (and we do not calculate a z test statistic). Also, because $n \leq 30$, we use Table A-8 to find the critical value of 11.

Step 8:   The test statistic $T = 15$ is not less than or equal to the critical value of 11, so we fail to reject the null hypothesis. It appears that there is no difference between yields from regular seed and kiln-dried seed.

If we use the sign test with the preceding example, we will arrive at the same conclusion. Although the sign test and the Wilcoxon signed-ranks test agree in this particular case, there are other cases in which they do not agree.

Rationale: In this example the unsigned ranks of 1 through 11 have a total of 66, so if there are no significant differences, each of the two signed-rank totals should be around $66 \div 2$, or 33. That is, the negative ranks and positive ranks should split up as 33–33 or something close, such as 31–35. The table of critical values shows that at the 0.05 significance level with 11 pairs of data, a split of 11–55 represents a significant departure from the null hypothesis, and any split that is farther apart (such as 10–56 or 2–64) will also represent a significant departure from the null hypothesis. Conversely, splits like 12–52 do not represent significant departures from a 33–33 split, and they would not justify rejecting the null hypothesis. The Wilcoxon signed-ranks test is based on the lower rank total, so instead of analyzing both numbers constituting the split, we consider only the lower number.

The sum $1 + 2 + 3 + \cdots + n$ of all the ranks is equal to $n(n + 1)/2$, and if this is a rank sum to be divided equally between two categories (positive and negative), each of the two totals should be near $n(n + 1)/4$, which is half of $n(n + 1)/2$. Recognition of this principle helps us understand the test statistic used when $n > 30$. The denominator in that expression represents a standard deviation of T and is based on the principle that

$$1^2 + 2^2 + 3^3 + \cdots + n^2 = \frac{n(n + 1)(2n + 1)}{6}$$

The Wilcoxon signed-ranks test can be used for matched pairs of data only. The next section will describe a rank-sum test that can be applied to two sets of independent data that are not matched in pairs.

12-3 Exercises

Using the Wilcoxon Signed-Ranks Test. In Exercises 1 and 2, refer to the given paired sample data and use the Wilcoxon signed-ranks test to test the claim that the samples come from populations with differences having a median of zero. Use a 0.05 significance level.

1.

x	12	14	17	19	20	27
y	12	15	15	14	12	18	19	20

~ Test for Two Independent Samples

2.

x	8	6	9	12	22	31	34	35	37
y	8	8	12	17	29	39	24	47	49

Using the Wilcoxon Signed-Ranks Test. In Exercises 3–8, refer to the sample data for the given exercises in Section 12-2. Instead of the sign test, use the Wilcoxon signed-ranks test to test the claim that both samples come from populations having differences with a median equal to zero.

3. Exercise 5   **4.** Exercise 6   **5.** Exercise 7

6. Exercise 8   **7.** Exercise 9   **8.** Exercise 14

*9. Using the Wilcoxon Signed-Ranks Test for Claims About a Median The Wilcoxon signed-ranks test can be used to test the claim that a sample comes from a population with a specified median. The procedure used is the same as the one described in this section, except that the differences (Step 1) are obtained by subtracting the value of the hypothesized median from each value. Use the sample data consisting of the 106 body temperatures listed for 12 A.M. on day 2 in Data Set 2 in Appendix B. At the 0.05 significance level, test the claim that healthy adults have a median body temperature that is equal to 98.6°F.

12-4 Wilcoxon Rank-Sum Test for Two Independent Samples

This section introduces the *Wilcoxon rank-sum test,* which is a nonparametric test that two independent sets of sample data come from populations with equal medians. Two samples are independent if the sample values selected from one population are not related or somehow matched or paired with the sample values from the other population. (To avoid confusion between the Wilcoxon rank-sum for independent samples and the Wilcoxon signed-ranks test for matched pairs, consider using the Internal Revenue Service for the mnemonic of IRS to remind us of "independent: rank sum.")

Definition

The **Wilcoxon rank-sum test** is a nonparametric test that uses ranks of sample data from two independent populations. It is used to test the null hypothesis that the two independent samples come from populations with equal medians. The alternative hypothesis is the claim that the two population distributions have different medians.

H_0: The two samples come from populations with equal medians.
H_1: The two samples come from populations with different medians.

The Wilcoxon rank-sum test is equivalent to the **Mann-** ....est (see Exercise 11), which is included in some other textbooks and ...are packages (such as Minitab). The key idea underlying the Wilcoxon rank-sum test is this: If two samples are drawn from populations with equal medians and the individual values are all ranked as one combined collection of values, then the high and low ranks should fall evenly between the two samples. If the low ranks are found predominantly in one sample and the high ranks are found predominantly in the other sample, we have evidence that the populations have different medians. This key idea is reflected in the following procedure for finding the value of the test statistic.

Procedure for Finding the Value of the Test Statistic

1. Temporarily combine the two samples into one big sample, then replace each sample value with its rank. (The lowest value gets a rank of 1, the next lowest value gets a rank of 2, and so on. If values are tied, assign to them the mean of the ranks involved in the tie. See Section 12-1 for a description of ranks and the procedure for handling ties.)

2. Find the sum of the ranks for either one of the two samples.

3. Calculate the value of the z test statistic as shown in the accompanying box, where either sample can be used as "Sample 1." (If testing the null hypothesis of equal medians and if both sample sizes are greater than 10, then the sampling distribution of R is approximately normal with mean μ_R and standard deviation σ_R, and the test statistic is as shown in the box.)

Wilcoxon Rank-Sum Test

Requirements

1. There are two independent samples that were randomly selected.

2. Each of the two samples has more than 10 values. (For samples with 10 or fewer values, special tables are available in reference books, such as *CRC Standard Probability and Statistics Tables and Formulae*, published by CRC Press.)

3. There is *no* requirement that the two populations have a normal distribution or any other particular distribution.

Notation

n_1 = size of Sample 1

n_2 = size of Sample 2

R_1 = sum of ranks for Sample 1

R_2 = sum of ranks for Sample 2

R = same as R_1 (sum of ranks for Sample 1)

μ_R = mean of the sample R values that is expected when the two populations have equal medians

σ_R = standard deviation of the sample R values that is expected when the two populations have equal medians

---

Meta-Analysis

The term *meta-analysis* refers to a technique of doing a study that essentially combines results of other studies. It has the advantage that separate smaller samples can be combined into one big sample, making the collective results more meaningful. It also has the advantage of using work that has already been done. Meta-analysis has the disadvantage of being only as good as the studies that are used. If the previous studies are flawed, the "garbage in, garbage out" phenomenon can occur. The use of meta-analysis is currently popular in medical research and psychological research. As an example, a study of migraine headache treatments was based on data from 46 other studies. (See "Meta-Analysis of Migraine Headache Treatments: Combining Information from Heterogeneous Designs" by Dominici et al., *Journal of the American Statistical Association*, Vol. 94, No. 445.)

Test Statistic

$$z = \frac{R - \mu_R}{\sigma_R}$$

where

$$\mu_R = \frac{n_1(n_1 + n_2 + 1)}{2}$$

$$\sigma_R = \sqrt{\frac{n_1 n_2 (n_1 + n_2 + 1)}{12}}$$

n_1 = size of the sample from which the rank sum R is found

n_2 = size of the other sample

R = sum of ranks of the sample with size n_1

Critical values: Critical values can be found in Table A-2 (because the test statistic is based on the normal distribution).

Note that unlike the corresponding hypothesis tests in Section 8-3, the Wilcoxon rank-sum test does *not* require normally distributed populations. Also, the Wilcoxon rank-sum test can be used with some data at the ordinal level of measurement, such as data consisting of ranks. In contrast, the parametric methods of Section 8-3 cannot be used with data at the ordinal level of measurement. In Table 12-1 we noted that with normally distributed populations, the Wilcoxon rank-sum test has a 0.95 efficiency rating when compared with the parametric t test or z test. This test has a high efficiency rating and involves easier calculations, so it might be preferred over the parametric tests presented in Section 8-3, even when the requirement of normality is satisfied.

The expression for μ_R is based on the following result of mathematical induction: The sum of the first n positive integers is given by $1 + 2 + 3 + \cdots + n = n(n + 1)/2$. The expression for σ_R is based on a result stating that the integers $1, 2, 3, \ldots, n$ have standard deviation $\sqrt{(n^2 - 1)/12}$.

EXAMPLE   BMI of Men and Women Refer to Data Set 1 in Appendix B and use only the first 13 sample values of the body mass index (BMI) of men and the first 12 sample values of the BMI of women. The sample BMI values are listed in Table 12-5. (Only parts of the available sample values are used so that the calculations in this example are easier to follow.) Use a 0.05 significance level to test the claim that the median BMI of men is equal to the median BMI of women.

SOLUTION

REQUIREMENT ✓ The Wilcoxon rank-sum test requires two independent and random samples, each with more than 10 values. The sample data are independent and random, and the sample sizes are 13 and 12. The requirements are satisfied and we can proceed with the test. ✓

continued

Table 12-5

BMI Measurements

Men		Women	
23.8	(11.5)	19.6	(2.5)
23.2	(9)	23.8	(11.5)
24.6	(14)	19.6	(2.5)
26.2	(17)	29.1	(22)
23.5	(10)	25.2	(15.5)
24.5	(13)	21.4	(5)
21.5	(6)	22.0	(7)
31.4	(24)	27.5	(19)
26.4	(18)	33.5	(25)
22.7	(8)	20.6	(4)
27.8	(20)	29.9	(23)
28.1	(21)	17.7	(1)
25.2	(15.5)		
$n_1 = 13$		$n_2 = 12$	
$R_1 = 187$		$R_2 = 138$	

The null and alternative hypotheses are as follows:

H_0:   Men and women have BMI values with equal medians.

H_1:   Men and women have BMI values with medians that are not equal.

Rank all 25 BMI measurements combined, beginning with a rank of 1 (assigned to the lowest value of 17.7). Ties in ranks are handled as described in Section 12-1: Find the mean of the ranks involved and assign this mean rank to each of the tied values. (The 2nd and 3rd values are both 19.6, so assign a rank of 2.5 to each of those values. The 11th and 12th values are both 23.8, so assign the rank of 11.5 to each of those values. The 15th and 16th values are both 25.2, so assign a rank of 15.5 to each of those values.) The ranks corresponding to the individual sample values are shown in parentheses in Table 12-5. R denotes the sum of the ranks for the sample we choose as Sample 1. If we choose the BMI values for men, we get

$$R = 11.5 + 9 + 14 + \cdots + 15.5 = 187$$

Because there are 13 values for men, we have $n_1 = 13$. Also, $n_2 = 12$ because there are 12 values for women. We can now find the values of μ_R, σ_R, and the test statistic z:

$$\mu_R = \frac{n_1(n_1 + n_2 + 1)}{2} = \frac{13(13 + 12 + 1)}{2} = 169$$

$$\sigma_R = \sqrt{\frac{n_1 n_2(n_1 + n_2 + 1)}{12}} = \sqrt{\frac{(13)(12)(13 + 12 + 1)}{12}} = 18.385$$

$$z = \frac{R - \mu_R}{\sigma_R} = \frac{187 - 169}{18.385} = 0.98$$

The test is two-tailed because a large positive value of z would indicate that the higher ranks are found disproportionately in the first sample, and a large negative value of z would indicate that the first sample had a disproportionate share of lower ranks. In either case, we would have strong evidence against the claim that the two samples come from populations with equal medians.

The significance of the test statistic z can be treated in the same manner as in previous chapters. We are now testing (with $\alpha = 0.05$) the hypothesis that the two populations have equal medians, so we have a two-tailed test with critical z values of 1.96 and -1.96. The test statistic of $z = 0.98$ does *not* fall within the critical region, so we fail to reject the null hypothesis that men and women have BMI values with equal medians. It appears that BMI values of men and women are basically the same.

We can verify that if we interchange the two sets of sample values and consider the sample of BMI values of women to be first, $R = 138$, $\mu_R = 156$, $\sigma_R = 18.385$, and $z = -0.98$, so the conclusion is exactly the same.

EXAMPLE **BMI of Men and Women** The preceding example used only 13 of the 40 sample BMI values for men listed in Data Set 1 in Appendix B, and it used only 12 of the 40 BMI values for women. Do the results change if we use all 40 sample values for men and all 40 sample values for women? Use the Wilcoxon rank-sum test.

SOLUTION

REQUIREMENT ✓   As in the preceding example, the sample data are independent and random. Both sample sizes are greater than 10. The requirements are satisfied and we can proceed with the test. ✓

The null and alternative hypotheses are the same as in the preceding example. Instead of manually calculating the rank sums, we refer to the Minitab display shown here. In that Minitab display, "ETA1" and "ETA2" denote the median of the first sample and the median of the second sample, respectively. Here are the key components of the Minitab display: The rank sum for BMI values of men is $W = 1727.5$, the P-value is 0.3032 (or 0.3031 after an adjustment for ties), and the conclusion is that we cannot reject (the null hypothesis) with a significance level of 0.05. Bottom line: The BMI values of men and women appear to be the same.

```
Minitab

BMI (Male     N = 40     Median = 26.200
BMI (Female   N = 40     Median = 23.900

Point estimate for ETA1-ETA2 is 1.250
95.1 Percent CI for ETA1-ETA2 is (-1.300,3.200)

W = 1727.5
Test of ETA1 = ETA2 vs ETA1 not = ETA2 is significant at 0.3032
The test is significant at 0.3031 (adjusted for ties)
Cannot reject at alpha = 0.05
```

12-4 Exercises

Identifying Rank Sums. In Exercises 1 and 2, use a 0.05 significance level with the methods of this section to identify the rank sums R_1 and R_2, μ_R, σ_R, the test statistic z, the critical z values, and then state the conclusion.

1. Sample 1 values: 1   3   4   6   8   12   15   16   17   22   26

   Sample 2 values: 2   5   7   9   11   13   14   18   19   20   25   26

2. Sample 1 values: 1   3   4   6   8   12   15   16   17   22   26

   Sample 2 values: 22   25   28   33   34   35   37   39   41   43   45

Using the Wilcoxon Rank-Sum Test. In Exercises 3–10, use the Wilcoxon rank-sum test.

3. Are Severe Psychiatric Disorders Related to Biological Factors? A study used X-ray computed tomography (CT) to collect data on brain volumes for a group of patients with obsessive-compulsive disorders and a control group of healthy persons. The accompanying list shows sample results (in milliliters) for volumes of the right cordate (based on data from "Neuroanatomical Abnormalities in Obsessive-Compulsive Disorder Detected with Quantitative X-Ray Computed Tomography," by Luxenberg et al., *American Journal of Psychiatry*, Vol. 145, No. 9). Use a 0.01 significance level

to test the claim that obsessive-compulsive patients and healthy persons have the same brain volumes. Based on this result, can we conclude that obsessive-compulsive disorders have a biological basis?

Obsessive-Compulsive Patients				Control Group			
0.308	0.210	0.304	0.344	0.519	0.476	0.413	0.429
0.407	0.455	0.287	0.288	0.501	0.402	0.349	0.594
0.463	0.334	0.340	0.305	0.334	0.483	0.460	0.445

4. Pulse Rates Refer to Data Set 1 in Appendix B for the pulse rates of men and women. Use only the first 13 pulse rates of men and use only the first 12 pulse rates of women to test the claim that the two samples of pulse rates come from populations with the same median. Use a 0.05 significance level.

5. Cholesterol Levels Refer to Data Set 1 in Appendix B for the cholesterol levels of men and women. Use only the first 13 cholesterol levels of men and use only the first 12 cholesterol levels of women to test the claim that the two samples of cholesterol levels come from populations with the same median. Use a 0.05 significance level.

6. Systolic Blood Pressure Refer to Data Set 1 in Appendix B for the systolic blood pressure levels of men and women. Use only the first 15 values of men and use only the first 15 values of women to test the claim that the two samples of systolic blood pressure levels come from populations with the same median. Use a 0.05 significance level.

7. Pulse Rates Repeat Exercise 4 using all 40 pulse rates of men and all 40 pulse rates of women.

8. Cholesterol Levels Repeat Exercise 5 using all 40 cholesterol levels of men and all 40 cholesterol levels of women.

9. Systolic Blood Pressure Repeat Exercise 6 using all 40 systolic blood pressure levels of men and all 40 systolic blood pressure levels of women.

10. Head Circumferences Refer to Data Set 4 in Appendix B and test the claim that two-month-old boys and two-month-old girls have head circumferences with the same median.

*11. Using the Mann-Whitney U Test The Mann-Whitney U test is equivalent to the Wilcoxon rank-sum test for independent samples in the sense that they both apply to the same situations and always lead to the same conclusions. In the Mann-Whitney U test we calculate

$$z = \frac{U - \dfrac{n_1 n_2}{2}}{\sqrt{\dfrac{n_1 n_2 (n_1 + n_2 + 1)}{12}}}$$

where

$$U = n_1 n_2 + \frac{n_1 (n_1 + 1)}{2} - R$$

Using the values listed in Table 12-5 in this section, find the z test statistic for the Mann-Whitney U test and compare it to the z test statistic of 0.98 that was found using the Wilcoxon rank-sum test.

12-5 Kruskal-Wallis Test

This section introduces the *Kruskal-Wallis test,* which is used to test the null hypothesis that three or more independent samples come from populations with equal medians. In Section 11-2 we used one-way analysis of variance (ANOVA) to test the null hypothesis that three or more populations have the same mean, but ANOVA requires that all of the involved populations have normal distributions. The Kruskal-Wallis test does not require normal distributions.

Definition

The **Kruskal-Wallis Test** (also called the **H test**) is a nonparametric test that uses ranks of sample data from three or more independent populations. It is used to test the null hypothesis that the independent samples come from populations with equal medians; the alternative hypothesis is the claim that at least one of the medians is different from the others.

H_0:   The samples come from populations with equal medians.

H_1:   The samples come from populations with medians that are not all equal.

In applying the Kruskal-Wallis test, we compute the *test statistic H, which has a distribution that can be approximated by the chi-square distribution as long as each sample has at least five observations.* When we use the chi-square distribution in this context, the number of degrees of freedom is $k - 1$, where k is the number of samples. (For a quick review of the key features of the chi-square distribution, see Section 6-5.)

Procedure for Finding the Value of the Test Statistic H

1. Temporarily combine all samples into one big sample and assign a rank to each sample value. (Sort the values from lowest to highest, and in cases of ties, assign to each observation the mean of the ranks involved.)

2. For each sample, find the sum of the ranks and find the sample size.

3. Calculate *H* by using the results of Step 2 and the notation and test statistic given in the accompanying box.

Kruskal-Wallis Test

Requirements

1. We have at least three independent samples, all of which are randomly selected.

2. Each sample has at least five observations. (If samples have fewer than five observations, refer to special tables of critical values, such as *CRC Standard Probability and Statistics Tables and Formulae,* published by CRC Press.)

3. There is *no* requirement that the populations have a normal distribution or any other particular distribution.

continued

Notation

N = total number of observations in all samples combined

k = number of samples

R_1 = sum of ranks for Sample 1

n_1 = number of observations in Sample 1

For Sample 2, the sum of ranks is R_2 and the number of observations is n_2, and similar notation is used for the other samples.

Test Statistic

$$H = \frac{12}{N(N+1)}\left(\frac{R_1^2}{n_1} + \frac{R_2^2}{n_2} + \cdots + \frac{R_k^2}{n_k}\right) - 3(N+1)$$

Critical values

1. The test is *right-tailed*.
2. df $= k - 1$. (Because the test statistic H can be approximated by a chi-square distribution, use Table A-4 with $k - 1$ degrees of freedom, where k is the number of different samples.)

The test statistic H is basically a measure of the variance of the rank sums $R_1, R_2, \ldots, R_k$. If the ranks are distributed evenly among the sample groups, then H should be a relatively small number. If the samples are very different, then the ranks will be excessively low in some groups and high in others, with the net effect that H will be large. Consequently, only large values of H lead to rejection of the null hypothesis that the samples come from identical populations. *The Kruskal-Wallis test is therefore a right-tailed test.*

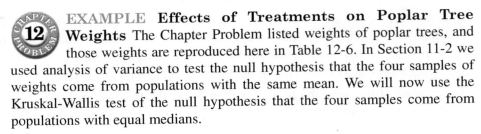

EXAMPLE   Effects of Treatments on Poplar Tree Weights The Chapter Problem listed weights of poplar trees, and those weights are reproduced here in Table 12-6. In Section 11-2 we used analysis of variance to test the null hypothesis that the four samples of weights come from populations with the same mean. We will now use the Kruskal-Wallis test of the null hypothesis that the four samples come from populations with equal medians.

SOLUTION

REQUIREMENT ✓ The Kruskal-Wallis test requires three or more independent and random samples, each with at least 5 values. Each of the four samples is independent and random, and each sample size is 5. Having satisfied all of the requirements, we can proceed with the test. ✓

The null and alternative hypotheses are as follows:

H_0:   The populations of poplar tree weights from the four treatments have equal medians.

H_1:   The four population medians are not all equal.

Table 12-6	Weights (kg) of Poplar Trees		
		Treatment	
None	Fertilizer	Irrigation	Fertilizer and Irrigation
0.15 (8)	1.34 (18)	0.23 (12)	2.03 (19)
0.02 (1.5)	0.14 (7)	0.04 (3)	0.27 (13)
0.16 (9.5)	0.02 (1.5)	0.34 (14)	0.92 (16)
0.37 (15)	0.08 (5.5)	0.16 (9.5)	1.07 (17)
0.22 (11)	0.08 (5.5)	0.05 (4)	2.38 (20)
$n_1 = 5$	$n_2 = 5$	$n_3 = 5$	$n_4 = 5$
$R_1 = 45$	$R_2 = 37.5$	$R_3 = 42.5$	$R_4 = 85$

In determining the value of the test statistic H, we must first rank all of the data. We begin with the lowest values of 0.02 and 0.02. Because there is a tie between the values corresponding to ranks 1 and 2, we assign the mean rank of 1.5 to each of those tied items. In Table 12-6, ranks are shown in parentheses next to the original tree weights. Next we find the sample size, n, and sum of ranks, R, for each sample, and those values are listed at the bottom of Table 12-6. Because the total number of observations is 20, we have $N = 20$. We can now evaluate the test statistic as follows:

$$H = \frac{12}{N(N+1)}\left(\frac{R_1^2}{n_1} + \frac{R_2^2}{n_2} + \cdots + \frac{R_k^2}{n_k}\right) - 3(N+1)$$

$$= \frac{12}{20(20+1)}\left(\frac{45^2}{5} + \frac{37.5^2}{5} + \frac{42.5^2}{5} + \frac{85^2}{5}\right) - 3(20+1)$$

$$= 8.214$$

Because each sample has at least five observations, the distribution of H is approximately a chi-square distribution with $k - 1$ degrees of freedom. The number of samples is $k = 4$, so we have $4 - 1 = 3$ degrees of freedom. Refer to Table A-4 to find the critical value of 7.815, which corresponds to 3 degrees of freedom and a 0.05 significance level (with an area of 0.05 in the right tail).

The test statistic $H = 8.214$ is in the critical region bounded by 7.815, so we reject the null hypothesis of equal medians. (In Section 11-2, we rejected the null hypothesis of equal means.)

INTERPRETATION There is sufficient evidence to reject the claim that the populations of poplar tree weights from the four treatments have equal medians. At least one of the medians appears to be different from the others.

Rationale: The test statistic H, as presented earlier, is the rank version of the test statistic F used in the analysis of variance discussed in Chapter 11. When we deal with ranks R instead of original values x, many components are predetermined. For example, the sum of all ranks can be expressed as $N(N + 1)/2$, where N is the total number of values in all samples combined. The expression

$$H = \frac{12}{N(N + 1)}\Sigma n_i(\overline{R} - \overline{\overline{R}})^2$$

where

$$\overline{R}_i = \frac{R_i}{n_i} \qquad \overline{\overline{R}} = \frac{\Sigma R_i}{\Sigma n_i}$$

combines weighted variances of ranks to produce the test statistic H given here. This expression for H is algebraically equivalent to the expression for H given earlier as the test statistic. The earlier form of H (not the one given here) is easier to work with. In comparing the procedures of the parametric F test for analysis of variance and the nonparametric Kruskal-Wallis test, we see that in the absence of computer software, the Kruskal-Wallis test is much simpler to apply. We need not compute the sample variances and sample means. We do not require normal population distributions. Life becomes so much easier. However, the Kruskal-Wallis test is not as efficient as the F test, so it might require more dramatic differences for the null hypothesis to be rejected.

12-5 Exercises

Interpreting Kruskal-Wallis Test Results. In Exercises 1 and 2, interpret the given Kruskal-Wallis test results and address the given question.

1. Weights of Poplar Trees The Chapter Problem and the example in this section use the weights of poplar trees for Year 1 and Site 1. (See Data Set 9 in Appendix B.) If we use the weights for Year 2 and Site 1, the Kruskal-Wallis test results from Minitab are as shown in the accompanying display. Assume that we want to use a 0.05 significance level in testing the null hypothesis that the four treatments result in weights with equal medians.
   a. What is the null hypothesis?
   b. What is the alternative hypothesis?
   c. Identify the value of the test statistic.
   d. Find the critical value for a 0.05 significance level.
   e. Identify the P-value.
   f. Based on the preceding results, what do you conclude about the distributions of weights from the different treatments?

```
Minitab

Treatment    N   Median   Ave Rank      Z
Fert         5   0.8700        9.7   -0.35
Fert/Irrig   5   1.6400       16.8    2.75
Irrig        5   0.6600        6.5   -1.75
None         5   0.5700        9.0   -0.65
Overall     20               10.5

H = 8.37   DF = 3   P = 0.039
H = 8.37   DF = 3   P = 0.039   (adjusted for ties)
```

2. Marathon Times A random sample of males who finished the New York Marathon is partitioned into three categories with ages of 21–29, 30–39, and 40 or over. The times (in seconds) are obtained, and the Kruskal-Wallis test results from STATDISK are shown below. We want to test the claim that the different age categories have median times that are not all equal.
   a. What is the null hypothesis?
   b. What is the alternative hypothesis?
   c. Identify the value of the test statistic.
   d. Find the critical value for a 0.05 significance level.
   e. Identify the P-value.
   f. Is there sufficient evidence to support the claim that the different age categories have different distributions of times?

STATDISK

Total Num Values:	111
Rank Sum 1:	731.0000
Rank Sum 2:	2393.0000
Rank Sum 3:	3092.0000
Test Statistic, H:	0.5843
Critical H:	5.991471
P-value:	0.7466603

3. Weights of Poplar Trees Refer to Data Set 9 in Appendix B and use the weights of poplar trees in Year 1 and Site 2. Use the Kruskal-Wallis test with a 0.05 significance level to test the claim that those weights are from populations with medians that are not all equal. Site 2 has dry and sandy soil. Does it appear that the weights are different for the various treatments at this drier site?

4. Weights of Poplar Trees Refer to Data Set 9 in Appendix B and use the weights of poplar trees in Year 2 and Site 2. Use the Kruskal-Wallis test with a 0.05 significance level to test the claim that those weights are from populations with medians that are not all equal. Site 2 has dry and sandy soil. Does it appear that the second-year weights are different for the various treatments at this drier site?

5. Skull Breadths from Different Epochs The values in the table in the margin are measured maximum breadths of male Egyptian skulls from different epochs (based on data from *Ancient Races of the Thebaid*, by Thomson and Randall-Maciver). Changes in head shape over time suggest that interbreeding occurred with immigrant populations. Use a 0.05 significance level to test the claim that the different epochs do not all have the same median. Is interbreeding of cultures suggested by the data?

4000 B.C.	1850 B.C.	150 A.D.
131	129	128
138	134	138
125	136	136
129	137	139
132	137	141
135	129	142
132	136	137
134	138	145
138	134	137

6. Systolic Blood Pressure in Different Age Groups Refer to Data Set 1 in Appendix B, and use the age categories of below 20, 20 through 40, and over 40. Use only the systolic blood pressure levels for women, and use the Kruskal-Wallis test to test the claim that the different age categories have median systolic blood pressure levels that are not all equal.

In Exercises 7 and 8, use the listed sample data from car crash experiments conducted by the National Transportation Safety Administration. New cars were purchased and crashed into a fixed barrier at 35 mi/h, and the listed measurements were recorded for the dummy in the driver's seat. The subcompact cars are the Ford Escort, Honda Civic, Hyundai Accent, Nissan Sentra, and Saturn SL4. The compact cars are Chevrolet Cavalier, Dodge Neon, Mazda 626 DX, Pontiac Sunfire, and Subaru Legacy. The midsize cars are Chevrolet Camaro, Dodge Intrepid, Ford Mustang, Honda Accord, and Volvo S70. The full-size cars are Audi A8, Cadillac Deville, Ford Crown Victoria, Oldsmobile Aurora, and Pontiac Bonneville.

7. Head Injury in a Car Crash The head injury data (in units of a standard head injury condition, or hic) are given below. Use the Kruskal-Wallis test with a 0.05 significance level to test the null hypothesis that the different weight categories have different medians. Do the data suggest that larger cars are safer?

Subcompact:	681	428	917	898	420
Compact:	643	655	442	514	525
Midsize:	469	727	525	454	259
Full-size:	384	656	602	687	360

8. Chest Deceleration in a Car Crash The chest deceleration data (g) are given below. Use the Kruskal-Wallis test with a 0.05 significance level to test the null hypothesis that the different weight categories have different medians. Do the data suggest that larger cars are safer?

Subcompact:	55	47	59	49	42
Compact:	57	57	46	54	51
Midsize:	45	53	49	51	46
Full-size:	44	45	39	58	44

9. Secondhand Smoke in Different Groups Refer to Data Set 5 in Appendix B. Use the Kruskal-Wallis test with a 0.05 significance level to test the claim that the median cotinine levels are not all equal for these three groups: nonsmokers who are not exposed to environmental tobacco smoke, nonsmokers who are exposed to tobacco smoke, and people who smoke. What do the results suggest about secondhand smoke?

10. Iris Sepal Lengths Refer to Data Set 7 in Appendix B. Use the Kruskal-Wallis test with a 0.05 significance level to test the claim that the three different species come from populations having median sepal lengths that are not all equal.

11. Iris Petal Widths Refer to Data Set 7 in Appendix B. Use the Kruskal-Wallis test with a 0.05 significance level to test the claim that the three different species come from populations having median petal widths that are not all equal.

***12.** Correcting the *H* Test Statistic for Ties In using the Kruskal-Wallis test, there is a correction factor that should be applied whenever there are many ties: Divide *H* by

$$1 - \frac{\Sigma T}{N^3 - N}$$

For each group of tied observations, calculate $T = t^3 - t$, where t is the number of observations that are tied within the individual group. Find t for each group of tied values, then compute the value of T for each group, then add the T values to get ΣT. The total number of observations in all samples combined is N. Use this procedure to find the corrected value of H for Exercise 3. Does the corrected value of H differ substantially from the value found in Exercise 3?

12-6 Rank Correlation

In this section we describe how the nonparametric method of rank correlation is used with paired data to test for an association between two variables. In Chapter 9 we used paired sample data to compute values for the linear correlation coefficient *r*, but in this section we use *ranks* as the basis for measuring the strength of the correlation between two variables.

Definition

The **rank correlation test** (or **Spearman's rank correlation test**) is a nonparametric test that uses ranks of sample data consisting of matched pairs. It is used to test for an association between two variables, so the null and alternative hypotheses are as follows (where ρ_s denotes the rank correlation coefficient for the entire population):

H_0: $\rho_s = 0$ (There is *no* correlation between the two variables.)
H_1: $\rho_s \neq 0$ (There is a correlation between the two variables.)

Advantages: Rank correlation has some distinct advantages over the parametric methods discussed in Chapter 9:

1. The nonparametric method of rank correlation can be used in a wider variety of circumstances than the parametric method of linear correlation. With rank correlation, we can analyze paired data that are ranks or can be converted to ranks. For example, if two judges rank 30 different gymnasts, we can use rank correlation, but not linear correlation. Unlike the parametric method of Chapter 9, the method of rank correlation does *not* require a normal distribution for any population.

2. Rank correlation can be used to detect some (not all) relationships that are not linear. (See the third example in this section.)

Disadvantage: A disadvantage of rank correlation is its efficiency rating of 0.91, as described in Section 12-1. This efficiency rating shows that with all other circumstances being equal, the nonparametric approach of rank correlation requires 100 pairs of sample data to achieve the same results as only 91 pairs of sample observations analyzed through the parametric approach, assuming the stricter requirements of the parametric approach are met.

The requirements, notation, test statistic, and critical values are summarized in the accompanying box. We use the notation r_s for the rank correlation coefficient so that we don't confuse it with the linear correlation coefficient *r*. The subscript *s* has nothing to do with standard deviation; it is used in honor of Charles Spearman (1863–1945), who originated the rank correlation approach. In fact, r_s is often called **Spearman's rank correlation coefficient.** The rank correlation procedure is summarized in Figure 12-4.

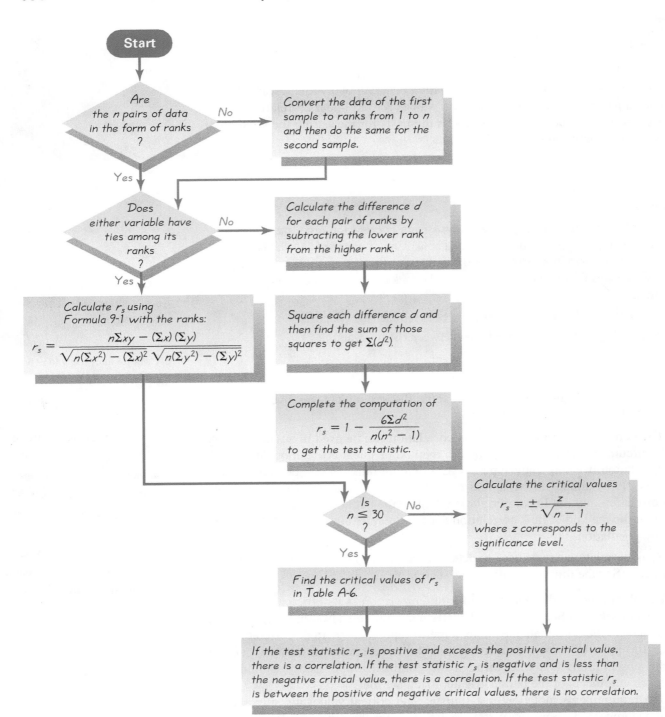

FIGURE 12-4 Rank Correlation Procedure for Testing $H_0: \rho_s = 0$

Rank Correlation

Requirements

1. The sample data consist of independent pairs that have been randomly selected.
2. Unlike the parametric methods of Section 9-2, there is *no* requirement that the pairs of sample data have a bivariate normal distribution (as described in Section 9-2). There is *no* requirement of a normal distribution for any population.

Notation

r_s = rank correlation coefficient for sample paired data (r_s is a sample statistic)

ρ_s = rank correlation coefficient for all the population data (ρ_s is a population parameter)

n = number of pairs of sample data

d = difference between ranks for the two values within a pair

Test Statistic

No ties: After converting the data in each sample to ranks, if there are no ties among ranks for the first variable and there are no ties among ranks for the second variable, the exact value of the test statistic can be calculated using this formula:

$$r_s = 1 - \frac{6\Sigma d^2}{n(n^2 - 1)}$$

Ties: After converting the data in each sample to ranks, if either variable has ties among its ranks, the exact value of the test statistic r_s can be found by using Formula 9-1 with the ranks:

$$r_s = \frac{n\Sigma xy - (\Sigma x)(\Sigma y)}{\sqrt{n(\Sigma x^2) - (\Sigma x)^2}\sqrt{n(\Sigma y^2) - (\Sigma y)^2}}$$

Critical values

1. If $n \le 30$, critical values are found in Table A-9.
2. If $n > 30$, critical values of r_s are found by using Formula 12-1.

Formula 12-1 $r_s = \dfrac{\pm z}{\sqrt{n - 1}}$ (critical values when $n > 30$)

where the value of z corresponds to the significance level, and the value of z is found in Table A-2.

EXAMPLE Correlation Between Salary and Stress Table 12-7 lists salary rankings and stress rankings for randomly selected jobs (based on data from *The Jobs Rated Almanac*). The salaries are ranked with 1 corresponding to the highest salary, and the stress levels are ranked with 1 corresponding to the most stress. Does it appear that salary increases as stress increases?

continued

SOLUTION

REQUIREMENT ✓ The sample data consist of pairs that have been randomly selected. The requirements are satisfied and we can proceed with the test. ✓

The linear correlation coefficient r (Section 9-2) should not be used because it requires normal distributions, but the data consist of ranks, which are not normally distributed. Instead, we use the rank correlation coefficient to test for a relationship between the salary and stress ranks.

The null and alternative hypotheses are as follows:

$$H_0: \rho_s = 0$$

$$H_1: \rho_s \neq 0$$

Following the procedure of Figure 12-4, the data are in the form of ranks and neither of the two variables (salary rank/stress rank) has ties among ranks, so the exact value of the test statistic can be calculated as shown below. We use $n = 10$ (for 10 pairs of data) and $\Sigma d^2 = 24$ (as shown in Table 12-7) to get

$$r_s = 1 - \frac{6\Sigma d^2}{n(n^2 - 1)} = 1 - \frac{6(24)}{10(10^2 - 1)}$$

$$= 1 - \frac{144}{990} = 0.855$$

Now we refer to Table A-9 to determine that the critical values are ± 0.648 (based on $\alpha = 0.05$ and $n = 10$). Because the test statistic $r_s = 0.855$ does exceed the critical value of 0.648, we reject the null hypothesis. There is sufficient evidence to support a claim of a correlation between the salary rankings and the stress rankings. It appears that increases in stress appear to correlate with increases in salary.

Table 12-7	Job Salary and Stress Rankings			
Job	Salary Rank	Stress Rank	Difference d	d^2
Stockbroker	2	2	0	0
Zoologist	6	7	1	1
Electrical engineer	3	6	3	9
School principal	5	4	1	1
Hotel manager	7	5	2	4
Bank officer	10	8	2	4
Occupational safety inspector	9	9	0	0
Home economist	8	10	2	4
Psychologist	4	3	1	1
Airline pilot	1	1	0	0
				Total = 24

EXAMPLE **Large Sample Case** Assume that the preceding example is
expanded by including a total of 40 jobs and that the test statistic r_s is found to
be 0.567. If the significance level is $\alpha = 0.05$, what do you conclude about the
correlation?

SOLUTION

REQUIREMENT ✓   The sample data consist of pairs that have been randomly
selected. The requirements are satisfied and we can proceed with the test. ✓

Because there are 40 pairs of data, we have $n = 40$. Because n exceeds 30,
we find the critical values from Formula 12-1 instead of Table A-9. With
$\alpha = 0.05$ in two tails, we let $z = 1.96$ to get

$$r_s = \frac{\pm 1.96}{\sqrt{40 - 1}} = \pm 0.314$$

The test statistic of $r_s = 0.567$ does exceed the critical value of 0.314, so we
reject the null hypothesis. There is sufficient evidence to support the claim of a
correlation between salary and stress.

The next example is intended to illustrate the principle that rank correlation
can sometimes be used to detect relationships that are not linear.

EXAMPLE **Detecting a Nonlinear Pattern** A *Raiders of the Lost
Ark* pinball machine (model L-7) is used to measure learning that results from
repeating manual functions. Subjects were selected so that they are similar in
important characteristics of age, gender, intelligence, education, and so on.
Table 12-8 lists the numbers of games played and the last scores (in millions)
for subjects randomly selected from the group with similar characteristics. We
expect that there should be an association between the number of games
played and the pinball score. Is there sufficient evidence to support the claim
that there is such an association?

SOLUTION

REQUIREMENT ✓   The sample data consist of pairs that have been randomly
selected. The requirements are satisfied and we can proceed with the test. ✓

continued

Table 12-8   Pinball Scores (Ranks in parentheses)

Number of Games Played	9 (2)	13 (4)	21 (5)	6 (1)	52 (7)	78 (8)	33 (6)	11 (3)	120 (9)
Score	22 (2)	62 (4)	70 (6)	10 (1)	68 (5)	73 (8)	72 (7)	58 (3)	75 (9)
d	0	0	1	0	2	0	1	0	0
d^2	0	0	1	0	4	0	1	0	0

We will test the null hypothesis of no rank correlation ($\rho_s = 0$).

$$H_0: \rho_s = 0 \quad \text{(no correlation)}$$
$$H_1: \rho_s \neq 0 \quad \text{(correlation)}$$

Refer to Figure 12-4, which we follow in this solution. The original scores are not ranks, so we converted them to ranks and entered the results in parentheses in Table 12-8. (Section 12-1 describes the procedure for converting scores into ranks.)

After expressing all data as ranks, we calculate the differences, d, and then square them. The sum of the d^2 values is 6. We now calculate

$$r_s = 1 - \frac{6\Sigma d^2}{n(n^2 - 1)} = 1 - \frac{6(6)}{9(9^2 - 1)}$$

$$= 1 - \frac{36}{720} = 0.950$$

Proceeding with Figure 12-4, we have $n = 9$, so we answer yes when asked if $n \leq 30$. We use Table A-9 to get the critical values of ± 0.700. Finally, the sample statistic of 0.950 exceeds 0.700, so we conclude that there is significant correlation. Higher numbers of games played appear to be associated with higher scores. Subjects appeared to better learn the game by playing more.

In the preceding example, if we compute the linear correlation coefficient r (using Formula 9-1) for the original data, we get $r = 0.586$, which leads to the conclusion that there is not enough evidence to support the claim of a significant linear correlation at the 0.05 significance level. If we examine the Excel scatter diagram, we can see that the pattern of points is not a straight-line pattern. This last example illustrates an advantage of the nonparametric approach over the parametric approach: With rank correlation, we can sometimes detect relationships that are not linear.

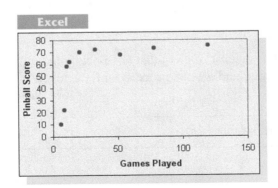

12-6 Exercises

Finding the Test Statistic. In Exercises 1–3, sketch a scatter diagram, use it to estimate the value of r_s, then calculate the value of r_s, and state whether there appears to be correlation between x and y.

1.

x	1	3	5	4	2
y	1	3	5	4	2

2.

x	1	2	3	4	5
y	5	4	3	2	1

3.

x	1	2	3	4	5
y	2	5	3	1	4

Finding Critical Values. In Exercises 4–8, find the critical value(s) for r_s by using either Table A-9 or Formula 12-1, as appropriate. Assume two-tailed cases, where α represents the significance level and n represents the number of pairs of data.

4. $n = 20, \alpha = 0.05$ **5.** $n = 50, \alpha = 0.05$

6. $n = 40, \alpha = 0.02$ **7.** $n = 15, \alpha = 0.01$

8. $n = 82, \alpha = 0.04$

Testing for Rank Correlation. In Exercises 9–16, use the rank correlation coefficient to test for a correlation between the two variables. Use a significance level of $\alpha = 0.05$.

9. Correlation Between Salary and Physical Demand An example in this section included paired salary and stress level ranks for 10 randomly selected jobs. The physical demands of the jobs were also ranked; the salary and physical demand ranks are given below (based on data from *The Jobs Rated Almanac*). Does there appear to be a relationship between the salary of a job and its physical demands?

Salary	2	6	3	5	7	10	9	8	4	1
Physical demand	5	2	3	8	10	9	1	7	6	4

10. Business School Rankings *Business Week* magazine ranked business schools two different ways. Corporate rankings were based on surveys of corporate recruiters, and graduate rankings were based on surveys of MBA graduates. The table below is based on the results for 10 schools. Is there a correlation between the corporate rankings and the graduate rankings? Use a significance level of $\alpha = 0.05$

School	PA	NW	Chi	Sfd	Hvd	MI	IN	Clb	UCLA	MIT
Corporate ranking	1	2	4	5	3	6	8	7	10	9
Graduate ranking	3	5	4	1	10	7	6	8	2	9

11. Correlation Between Heights and Weights of Supermodels Listed below are heights (in inches) and weights (in pounds) for supermodels Niki Taylor, Nadia Avermann, Claudia Schiffer, Elle MacPherson, Christy Turlington, Bridget Hall, Kate Moss, Valerie Mazza, and Kristy Hume.

Height	71	70.5	71	72	70	70	66.5	70	71
Weight	125	119	128	128	119	127	105	123	115

12. Heights of Sons and Fathers Listed below are the heights of sons and fathers from the first seven cases listed in Data Set 3 in Appendix B. Using only the data listed below, is there a correlation? Do the data suggest that taller fathers tend to have taller sons?

Height of son	62.5	64.6	69.1	73.9	67.1	64.4	71.1
Height of father	70	69	64	71	68	66	74

13. Systolic and Diastolic Blood Pressure Data Set 1 in Appendix B includes data from the National Health Examination Survey of randomly selected subjects. The paired systolic/diastolic blood pressure measurements for the first eight males are listed below. Using only the data listed below, is there a correlation? If a male has a higher systolic level, is he more likely to have a higher diastolic level?

Systolic	125	107	126	110	110	107	113	126
Diastolic	78	54	81	68	66	83	71	72

14. Blood Pressure Measurements Fourteen different second-year medical students took blood pressure measurements of the same patient and the results are listed below. Is there a correlation between systolic and diastolic values? Apart from correlation, is there some other method that might be used to address an important issue suggested by the data?

Systolic	138	130	135	140	120	125	120	130	130	144	143	140	130	150
Diastolic	82	91	100	100	80	90	80	80	80	98	105	85	70	100

15. Smoking and Nicotine When nicotine is absorbed by the body, cotinine is produced. A measurement of cotinine in the body is therefore a good indicator of how much a person smokes. Listed below are the reported numbers of cigarettes smoked per day and the measured amounts of cotinine (in ng/ml). (The values are from randomly selected subjects in the National Health Examination Survey.) Is there a significant linear correlation? Explain the result.

x (cigarettes per day)	60	10	4	15	10	1	20	8	7	10	10	20
y (cotinine)	179	283	75.6	174	209	9.51	350	1.85	43.4	25.1	408	344

16. Tree Circumference and Height Listed below are the circumferences (in feet) and the heights (in feet) of trees in Marshall, Minnesota (based on data from "Tree Measurements" by Stanley Rice, American Biology Teacher, Vol. 61, No. 9). Is there a correlation?

x (circ.)	1.8	1.9	1.8	2.4	5.1	3.1	5.5	5.1	8.3	13.7	5.3	4.9	3.7	3.8
y (ht)	21.0	33.5	24.6	40.7	73.2	24.9	40.4	45.3	53.5	93.8	64.0	62.7	47.2	44.3

Testing for a Linear Correlation. *In Exercises 17–20, use the data from Appendix B to construct a scatterplot, find the value of the rank correlation coefficient r_s, and use a significance level of $\alpha = 0.05$ to determine whether there is a significant correlation between the two variables.*

17. Cholesterol and Body Mass Index Refer to Data Set 1 in Appendix B and use the cholesterol levels and body mass index values of the 40 women. Is there a correlation between cholesterol level and body mass index?

18. Cholesterol and Weight Refer to Data Set 1 in Appendix B and use the cholesterol levels and weights of the 40 women. Is there a correlation between cholesterol level and weight?

19. Chest Sizes and Weights of Bears Refer to Data Set 6 in Appendix B and use the chest sizes and weights of all listed bears. Is there a correlation between chest size and weight?

20. Heights of Daughters and Mothers Refer to Data Set 3 in Appendix B and use the heights of the female children and their mothers. Is there a correlation between heights of daughters and mothers?

*21. Finding Critical Values One alternative to using Table A-9 to find critical values is to compute them using this approximation:

$$r_s = \pm \sqrt{\frac{t^2}{t^2 + n - 2}}$$

Here t is the t score from Table A-3 corresponding to the significance level and $n - 2$ degrees of freedom. Apply this approximation to find critical values of r_s for the following cases.
   a. $n = 8, \alpha = 0.05$
   b. $n = 15, \alpha = 0.05$
   c. $n = 30, \alpha = 0.05$
   d. $n = 30, \alpha = 0.01$
   e. $n = 8, \alpha = 0.01$

Review

In this chapter we examined five different nonparametric tests for analyzing sample data. Nonparametric tests are also called distribution-free tests because they do not require that the populations have a particular distribution, such as a normal distribution. However, nonparametric tests are not as efficient as parametric tests, so we generally need stronger evidence before we reject a null hypothesis.

Table 12-9 lists the nonparametric tests presented in this chapter, along with their functions. The table also lists the corresponding parametric tests.

Table 12-9	Summary of Nonparametric Tests	
Nonparametric Test	Function	Parametric Test
Sign test (Section 12-2)	Test for claimed value of median with one sample	z test or t test (Sections 7-4, 7-5)
	Test for differences between matched pairs	t test (Section 8-4)
	Test for claimed value of a proportion	z test (Section 7-3)
Wilcoxon signed-ranks test (Section 12-3)	Test for differences between matched pairs	t test (Section 8-4)
Wilcoxon rank-sum test (Section 12-4)	Test for difference between two independent samples	t test or z test (Section 8-3)
Kruskal-Wallis test (Section 12-5)	Test that more than two independent samples have equal medians	Analysis of variance (Section 11-2)
Rank correlation (Section 12-6)	Test for relationship between two variables	Linear correlation (Section 9-2)

Review Exercises

Using Nonparametric Tests. In Exercises 1–6, use a 0.05 significance level with the indicated test. If no particular test is specified, use the appropriate nonparametric test from this chapter.

1. **Temperatures and Marathons** In "The Effects of Temperature on Marathon Runner's Performance" by David Martin and John Buoncristiani (*Chance*, Vol. 12, No. 4), high temperatures and times (in minutes) were given for women who won the New York City Marathon in recent years. Results are listed below. Is there a correlation between temperature and winning time? Does it appear that winning times are affected by temperature?

x (temperature)	55	61	49	62	70	73	51	57
y (time)	145.283	148.717	148.300	148.100	147.617	146.400	144.667	147.533

2. **Testing Gender Selection** In a test of a method of gender selection, 66 offspring are obtained, and 1/3 of them are females. Use a 0.01 significance level to test the claim that the method has an effect on the gender of the offspring.

3. **Do Beer Drinkers and Liquor Drinkers Have Different BAC Levels?** The sample data in the following list show BAC (blood alcohol concentration) levels at arrest of

randomly selected jail inmates who were convicted of DWI or DUI offenses. The data are categorized by the type of drink consumed (based on data from the U.S. Department of Justice). Test the claim that beer drinkers and liquor drinkers have the same median BAC levels. Based on these results, do both groups seem equally dangerous, or is one group more dangerous than the other?

Beer

0.129	0.146	0.148	0.152
0.154	0.155	0.187	0.212
0.203	0.190	0.164	0.165

Liquor

0.220	0.225	0.185	0.182
0.253	0.241	0.227	0.205
0.247	0.224	0.226	0.234
0.190	0.257		

4. Does Weight of a Car Affect Leg Injuries in a Crash? Data were obtained from car crash experiments conducted by the National Transportation Safety Administration. New cars were purchased and crashed into a fixed barrier at 35 mi/h, and measurements were recorded for the dummy in the driver's seat. Use the sample data listed below to test for differences in left femur load measurements (in lb) among the four weight categories. Is there sufficient evidence to conclude that leg injury measurements for the four car weight categories are not all the same? Do the data suggest that heavier cars are safer in a crash?

Subcompact:	595	1063	885	519	422
Compact:	1051	1193	946	984	584
Midsize:	629	1686	880	181	645
Full-size:	1085	971	996	804	1376

5. Testing Effectiveness of Treatment An experiment involves treating horses with a vitamin supplement, and their weights are measured before and after the treatment, with the results given in the table below. Using a 0.05 significance level, test the claim that the treatment has an effect on weights. Use the sign test.

Subject	A	B	C	D	E	F	G	H	I	J
Weight before treatment	700	840	830	860	840	690	830	1180	930	1070
Weight after treatment	720	840	820	900	870	700	800	1200	950	1080

6. Testing Effectiveness of Treatment Do Exercise 5 using the Wilcoxon signed-ranks test.

Cumulative Review Exercises

Heights of Presidential Winners and Losers. *For Exercises 1–5, refer to the accompanying table, which shows the heights of presidents matched with the heights of the candidates they beat. All heights are in inches, and only the winners and second-place candidates are included. Use a 0.05 significance level for the following tests.*

Winner	76	66	70	70	74	71.5	73	74
Runner-up	64	71	72	72	68	71	69.5	74

1. Use the linear correlation coefficient r to test for a significant linear correlation between the heights of the winners and the heights of the candidates they beat. (See Section 9-2.) Does there appear to be a correlation?

2. Use the rank correlation coefficient r_s to test for a significant correlation between the heights of the winners and the heights of the candidates they beat. (See Section 12-6.) Does there appear to be a correlation?

3. Use the sign test to test the claim that there is a difference between the heights of the winning candidates and the heights of the losing candidates.

4. Use the Wilcoxon signed-ranks test to test the claim that differences between the heights of the winning candidates and heights of the corresponding losing candidates are differences from a population with a median equal to zero.

5. Use the parametric t test (see Section 8-4) to test the claim that there is a difference between the heights of the winning candidates and the heights of the corresponding losing candidates.

Cooperative Group Activities

1. *In-class activity:* Divide into groups of 8 to 12 people. For each group member, *measure* his or her height and *measure* his or her arm span. For the arm span, the subject should stand with arms extended, like the wings on an airplane. It's easy to mark the height and arm span on a chalkboard, then measure the distances there. Divide the following tasks among subgroups of three or four people.
   a. Use rank correlation with the paired sample data to determine whether there is a correlation between height and arm span.
   b. Use the sign test to test for a difference between the two variables.
   c. Use the Wilcoxon signed-ranks test to test for a difference between the two variables.

2. *In-class activity:* Do activity 1 using pulse rate instead of arm span. Measure pulse rates by counting the number of heartbeats in 1 minute.

3. *Out-of-class activity:* Divide into groups of three or four students. Investigate the relationship between two vari-

ables by collecting your own paired sample data and using the methods of Section 12-6 to determine whether there is a significant correlation. Suggested topics:

   ● Is there a relationship between taste and cost of different brands of chocolate chip cookies (or colas)? (Taste can be measured on some number scale, such as 1 to 10.)
   ● Is there a relationship between the lengths of men's (or women's) feet and their heights?
   ● Is there a relationship between student grade-point averages and the amount of television watched?
   ● Is there a relationship between heights of fathers (or mothers) and heights of their first sons (or daughters)?

4. *Out-of-class activity:* Divide into groups of three or four. Survey students by asking them to identify their major and gender. For each surveyed subject, determine the accuracy of the time on his or her wristwatch. First set your own watch to the correct time using an accurate and reliable source ("At the tone, the time is . . ."). For

watches that are ahead of the correct time, record positive times. For watches that are behind the correct time, record negative times. Use the sample data to address these questions:

- Do the errors appear to be the same for both genders?
- Do the errors appear to be the same for the different majors?

5. *In-class activity:* Divide into groups of 8 to 12 people. For each group member, measure the person's height and also measure his or her navel height, which is the height from the floor to the navel. Use the rank correlation coefficient to determine whether there is a correlation between height and navel height.

Technology Project

Methods of gender selection can be tested in experiments that involve the treatment of parents and the subsequent analysis of generated offspring. Instead of actually generating offspring, we will use technology to simulate their generation. Use STATDISK, Minitab, Excel, a TI-83/84 Plus calculator, or any other technology to randomly generate 1000 values, each of which is 0 or 1. Representing boys by 0

and girls by 1, we then have a simulated sample of 1000 offspring. We can then use the sign test to determine whether boys and girls are equally likely. In this case, we are actually testing the technology for a bias in favor of either 0s or 1s. Generate the 1000 values, then use the sign test to analyze the results. Does the technology appear to have some bias? Could you detect a bias using larger sample sizes?

from DATA *to* DECISION

Critical Thinking: Investigating the relationship between numbers of registered boats and the numbers of manatees killed by boats

The Chapter 9 "Data to Decision" project involved manatee deaths from powerboat encounters. The sample data are reproduced in the accompanying table. Is there a correlation between the numbers of registered boats and the numbers of manatees killed by boats?

Analysis

- Identify the requirements for linear correlation introduced in Section 9-2 and also identify the requirements for rank correlation introduced in Section 12-6. Compare those requirements.
- When using linear correlation introduced in Section 9-2, it is difficult to test its requirement

of a bivariate normal distribution, but a partial check can be made by determining whether the values of both variables have distributions that are approximately normal. Use normal quantile plots and determine whether both variables have distributions that are approximately normal.

- Using rank correlation, determine whether there is a relationship between the numbers of registered boats and the numbers of manatees killed by boats.
- When using the given sample data to test for a correlation, which method appears to be better: linear correlation or rank correlation? Why?

Registered Florida Pleasure Craft (in tens of thousands) and Watercraft-Related Manatee Deaths

Year	1991	1992	1993	1994	1995	1996	1997	1998	1999	2000
x: Boats	68	68	67	70	71	73	76	81	83	84
y: Manatee Deaths	53	38	35	49	42	60	54	67	82	78

13

Life Tables

Is the number of deaths unusually high?

In one region around Orange County, there are 5000 people who reach their 16th birthday. Suppose that 25 of them die before their 17th birthday. Parents, expressing understandable concern, claim that this number of deaths is unusually high and the causes of death should be investigated as part of a comprehensive study. Others suggest that the number of deaths of 16-year-olds will naturally vary from year to year, and some years have fewer deaths than average, while other years have disproportionately greater numbers of deaths, so that the 25 deaths is no cause for concern.

How do we objectively address this issue? One essential piece of information is the death rate for people in this age group. That rate can be extracted from a *life table*, which can be very helpful in situations such as the one considered here. We will consider the different components of a life table, and we will use a specific and real life table to address the questions raised above.

Is Parachuting Safe?

About 30 people die each year as more than 100,000 people make about 2.25 million parachute jumps. In comparison, a typical year includes about 200 scuba diving fatalities, 7000 drownings, 900 bicycle deaths, 800 lightning deaths, and 1150 deaths from bee stings. Of course, these figures don't necessarily mean that parachuting is safer than bike riding or swimming. A fair comparison should involve fatality rates, not just the total number of deaths.

One of the authors, with much trepidation, made two parachute jumps but quit after missing the spacious drop zone both times. He has also flown in a hang glider, hot air balloon, and Goodyear blimp.

13-1 Overview

In this chapter we introduce *life tables*. In Section 13-2 we consider one particular life table, and we describe its various components. In Section 13-3 we consider some important applications based on life tables.

Life tables are routinely developed by government agencies, and their use is critical to a variety of different fields. Studies of wildlife populations often involve life tables. Insurance companies could not provide life insurance policies without the information contained in life tables. A term life insurance policy is essentially a bet that a person will live or die within a certain time period, and the amount of the bet (cost of the policy or the premium amount) must reflect the probability that the subject will live or die within that time period. Many professionals monitor the health of the United States, and their analyses require some basis for comparisons. Life tables are commonly used as the basis for such comparisons. Other policy planners need to know how long people are expected to live, so that they can satisfy their health and financial needs. For example, employees in the United States typically have funds contributed to a large Social Security account maintained by the government. It is intended that these funds can be used to provide financial support when employees age and retire. However, as people tend to live longer, the payments continue for longer periods of time and careful planning is required to ensure that future generations will continue to enjoy the benefits of the Social Security system. Such careful planning requires accurate information about changes in longevity—information that is available in life tables. Life tables therefore become key tools in many different applications.

13-2 Elements of a Life Table

Life tables include information about death rates and longevity, but there are different types of life tables. A **cohort** (or **generation**) life table is a record of the actual observed mortality experience for a particular group, such as everyone born in 1980. Because cohort life tables require complete data collected over periods spanning decades of time, they aren't very practical and are not commonly used. The main focus of this chapter will be *period life tables*. In subsequent references to *life tables*, it is assumed that we are referring to period life tables, defined as follows.

Definition

A **period** (or **current**) life table describes mortality and longevity data for a *hypothetical* (or "synthetic") cohort, with the data computed with this assumption: The conditions affecting mortality in a particular year (such as 2000) remain the same throughout the lives of everyone in the hypothetical cohort. A period life table shows the long-term results of mortality conditions that were present during the particular basis year.

Table 13-1 is an example of a period life table. Because Table 13-1 is a period life table for the United States for the year 2000, we know that it was constructed with this important assumption: The death rates for the various age groups that were in effect in the year 2000 continue to remain in effect during the entire lives of the 100,000 hypothetical people assumed to be present at age 0. That is, we pretend that a population of 100,000 people is born, and they each live their entire lives in a world with the same constant death rates that were present in the year 2000.

Table 13-1 was provided by the National Center for Health Statistics, and it is based on actual data from the U.S. Census Bureau (for population estimates), the Medicare program (a government program providing financial support for health care), and death certificates issued in different states. Actual mortality data were collected, but they were then used to describe the experience of 100,000 *hypothetical* people.

Table 13-1 describes 100,000 hypothetical people, where the 100,000 people are representative of the *total population* of the United States, including males and females of various races. Because mortality experiences can be very different for various gender and race groups, it is common to have tables for specific subgroups. As of this writing, period life tables were available for males, females, whites, white males, white females, blacks, black males, and black females. We do not provide these other tables, but they should be used in place of Table 13-1 when appropriate. For example, if calculating the cost of a term life insurance policy for a 27-year-old white female, more accurate information could be obtained from the period life table for white females than the information in Table 13-1, which includes males, females, whites, and blacks.

Components of a Period Life Table

We now proceed to describe the different columns in Table 13-1.

Column 1: The first column of Table 13-1 lists age categories. The first category of 0–1 represents the age interval from birth to the first birthday. Note that some of the other columns are based on the age *at the beginning* of the age interval. For example, the last column lists expected ages, which are the expected remaining lifetime amounts measured from the *beginning* of the age interval. For the age interval of 0–1, the expected remaining lifetime is 76.9, which means that a newborn has an expected lifetime of 76.9 years. Also note that the age intervals overlap with a common value used for the upper limit of one class as well as the lower limit for the following class. When we discussed frequency tables in Chapter 2, we noted that the class limits should not overlap, but the overlapping is not an important issue with period life tables. It would be a very rare event to have someone die on the exact instant of the beginning of their birth date, so we need not make adjustments for this event which will never occur.

Column 2: The second column lists values of the probability of dying *during the age interval* listed in the first column. The first row of values shows that there is a 0.006930 probability of someone dying between birth and their first birthday. The second row shows that there is a 0.000517 probability of someone dying between their first birthday and their second birthday. The values in this column reflect the death rates for the various age groups that were observed in the year 2000.

Sensitive Surveys

Survey respondents are sometimes reluctant to honestly answer questions on a sensitive topic, such as employee theft or sex. Stanley Warner (York University, Ontario) devised a scheme that leads to more accurate results in such cases. As an example, ask employees if they stole within the past year and also ask them to flip a coin. The employees are instructed to answer no if they didn't steal and the coin turns up heads. Otherwise, they should answer yes. The employees are more likely to be honest because the coin flip helps protect their privacy. Probability theory can then be used to analyze responses so that more accurate results can be obtained.

Table 13-1 Life Table for Total Population: United States, 2000

Age Interval	Probability of Dying During the Interval	Number Surviving to the Beginning of the Interval	Number of Deaths During the Interval	Person-Years Lived (Total time lived *during the interval* by those alive at the beginning of the interval)	Total Person-Years Lived (in *this and all subsequent* intervals)	Expected Remaining Lifetime (from the beginning of the interval)
0–1	0.006930	100,000	693	99,392	7,686,810	76.9
1–2	0.000517	99,307	51	99,281	7,587,418	76.4
2–3	0.000347	99,256	34	99,238	7,488,137	75.4
3–4	0.000243	99,221	24	99,209	7,388,898	74.5
4–5	0.000202	99,197	20	99,187	7,289,689	73.5
5–6	0.000189	99,177	19	99,168	7,190,502	72.5
6–7	0.000177	99,158	18	99,150	7,091,334	71.5
7–8	0.000167	99,141	17	99,132	6,992,185	70.5
8–9	0.000154	99,124	15	99,117	6,893,052	69.5
9–10	0.000137	99,109	14	99,102	6,793,936	68.6
10–11	0.000125	99,095	12	99,089	6,694,833	67.6
11–12	0.000130	99,083	13	99,077	6,595,744	66.6
12–13	0.000170	99,070	17	99,062	6,496,668	65.6
13–14	0.000253	99,053	25	99,041	6,397,606	64.6
14–15	0.000366	99,028	36	99,010	6,298,565	63.6
15–16	0.000491	98,992	49	98,968	6,199,555	62.6
16–17	0.000607	98,943	60	98,913	6,100,587	61.7
17–18	0.000706	98,883	70	98,848	6,001,674	60.7
18–19	0.000780	98,814	77	98,775	5,902,826	59.7
19–20	0.000833	98,736	82	98,695	5,804,051	58.8
20–21	0.000888	98,654	88	98,610	5,705,355	57.8
21–22	0.000945	98,567	93	98,520	5,606,745	56.9
22–23	0.000983	98,474	97	98,425	5,508,225	55.9
23–24	0.000996	98,377	98	98,328	5,409,800	55.0
24–25	0.000991	98,279	97	98,230	5,311,472	54.0
25–26	0.000981	98,181	96	98,133	5,213,242	53.1
26–27	0.000977	98,085	96	98,037	5,115,109	52.1
27–28	0.000979	97,989	96	97,941	5,017,072	51.2
28–29	0.000993	97,893	97	97,845	4,919,130	50.2
29–30	0.001019	97,796	100	97,746	4,821,286	49.3
30–31	0.001050	97,696	103	97,645	4,723,539	48.3
31–32	0.001087	97,594	106	97,541	4,625,894	47.4
32–33	0.001141	97,488	111	97,432	4,528,353	46.5
33–34	0.001215	97,376	118	97,317	4,430,921	45.5
34–35	0.001302	97,258	127	97,195	4,333,604	44.6
35–36	0.001395	97,132	135	97,064	4,236,409	43.6
36–37	0.001492	96,996	145	96,924	4,139,345	42.7
37–38	0.001602	96,851	155	96,774	4,042,422	41.7
38–39	0.001728	96,696	167	96,613	3,945,648	40.8
39–40	0.001870	96,529	180	96,439	3,849,035	39.9
40–41	0.002021	96,349	195	96,251	3,752,596	38.9
41–42	0.002181	96,154	210	96,049	3,656,345	38.0
42–43	0.002355	95,944	226	95,831	3,560,296	37.1
43–44	0.002550	95,718	244	95,596	3,464,465	36.2
44–45	0.002768	95,474	264	95,342	3,368,869	35.3
45–46	0.003014	95,210	287	95,066	3,273,527	34.4
46–47	0.003284	94,923	312	94,767	3,178,460	33.5
47–48	0.003567	94,611	337	94,443	3,083,693	32.6
48–49	0.003851	94,274	363	94,092	2,989,250	31.7
49–50	0.004138	93,911	389	93,717	2,895,158	30.8
50–51	0.004443	93,522	415	93,314	2,801,442	30.0

Age Interval	Probability of Dying During the Interval	Number Surviving to the Beginning of the Interval	Number of Deaths During the Interval	Person-Years Lived (Total time lived *during the interval* by those alive at the beginning of the interval)	Total Person-Years Lived (in *this and all subsequent* intervals)	Expected Remaining Lifetime (from the beginning of the interval)
51–52	0.004780	93,107	445	92,884	2,708,127	29.1
52–53	0.005152	92,662	477	92,423	2,615,243	28.2
53–54	0.005579	92,184	514	91,927	2,522,820	27.4
54–55	0.006075	91,670	557	91,392	2,430,893	26.5
55–56	0.006654	91,113	606	90,810	2,339,501	25.7
56–57	0.007309	90,507	661	90,176	2,248,691	24.8
57–58	0.008023	89,845	721	89,485	2,158,515	24.0
58–59	0.008773	89,124	782	88,733	2,069,030	23.2
59–60	0.009563	88,343	845	87,920	1,980,297	22.4
60–61	0.010446	87,498	914	87,041	1,892,377	21.6
61–62	0.011448	86,584	991	86,088	1,805,336	20.9
62–63	0.012521	85,593	1,072	85,057	1,719,248	20.1
63–64	0.013646	84,521	1,153	83,944	1,634,191	19.3
64–65	0.014828	83,368	1,236	82,749	1,550,247	18.6
65–66	0.016058	82,131	1,319	81,472	1,467,498	17.9
66–67	0.017400	80,812	1,406	80,109	1,386,026	17.2
67–68	0.018933	79,406	1,503	78,655	1,305,916	16.4
68–69	0.020701	77,903	1,613	77,097	1,227,262	15.8
69–70	0.022663	76,290	1,729	75,426	1,150,165	15.1
70–71	0.024673	74,561	1,840	73,641	1,074,739	14.4
71–72	0.026741	72,722	1,945	71,749	1,001,098	13.8
72–73	0.029042	70,777	2,056	69,749	929,349	13.1
73–74	0.031663	68,721	2,176	67,633	859,600	12.5
74–75	0.034588	66,545	2,302	65,395	791,966	11.9
75–76	0.037675	64,244	2,420	63,034	726,571	11.3
76–77	0.040886	61,823	2,528	60,560	663,538	10.7
77–78	0.044437	59,296	2,635	57,978	602,978	10.2
78–79	0.048530	56,661	2,750	55,286	545,000	9.6
79–80	0.053313	53,911	2,874	52,474	489,714	9.1
80–81	0.058841	51,037	3,003	49,535	437,240	8.6
81–82	0.065093	48,034	3,127	46,471	387,705	8.1
82–83	0.072140	44,907	3,240	43,287	341,234	7.6
83–84	0.079850	41,668	3,327	40,004	297,947	7.2
84–85	0.088195	38,340	3,381	36,650	257,943	6.7
85–86	0.096751	34,959	3,382	33,268	221,293	6.3
86–87	0.105884	31,577	3,343	29,905	188,025	6.0
87–88	0.115605	28,233	3,264	26,601	158,121	5.6
88–89	0.125917	24,969	3,144	23,397	131,519	5.3
89–90	0.136824	21,825	2,986	20,332	108,122	5.0
90–91	0.148322	18,839	2,794	17,442	87,790	4.7
91–92	0.160404	16,045	2,574	14,758	70,348	4.4
92–93	0.173058	13,471	2,331	12,305	55,590	4.1
93–94	0.186266	11,140	2,075	10,102	43,284	3.9
94–95	0.200006	9,065	1,813	8,158	33,182	3.7
95–96	0.214248	7,252	1,554	6,475	25,024	3.5
96–97	0.228960	5,698	1,305	5,046	18,549	3.3
97–98	0.244099	4,394	1,072	3,857	13,503	3.1
98–99	0.259622	3,321	862	2,890	9,646	2.9
99–100	0.275475	2,459	677	2,120	6,756	2.7
100 and over	1.000000	1,781	1,781	4,636	4,636	2.6

From E. Arias, United States Life Tables, 2000. *National Vital Statistics Reporter,* Vol. 51, No. 3, National Center For Health Statistics, 2002.

Column 3: The third column lists the number of people alive *at the beginning of the age interval*. Note that the first row lists the size of our hypothetical population: 100,000. This number decreases as some of the hypothetical people have hypothetical deaths. The second row shows that among the 100,000 hypothetical people who were born, 99,307 of them are alive on their first birthday. The values in column 3 can be computed by using the death rates in column 2. For example, the first age interval of 0–1 has a death rate of 0.006930, so we expect that 693 of the 100,000 people alive at birth will die, leaving a population of 100,000 − 693 = 99,307 people alive at the beginning of the second age interval of 1–2.

Column 4: The fourth column shows the number of people who died *during the age interval* in the column at the left. This value can be computed by multiplying the number of people alive at the beginning of the age interval (column 3) by the probability of dying during the age interval (column 2). For example, the first row shows that 100,000 people were present at age 0 and the death rate for those in the age interval of 0–1 is 0.006930, so we expect the number of deaths to be 100,000 × 0.006930 = 693.

Column 5: This column lists the total time (in years) lived *during the age interval* by those who were alive at the beginning of the age interval. For example, the first row has an entry of 99,392 (years) in column 5, and this value indicates that the 100,000 people who were present at age 0 lived a total of 99,392 years. If none of those people had died, this entry would have been 100,000 years, but some of them did die, so the total is less than 100,000 years.

Column 6: The sixth column is somewhat similar to the fifth column in the sense that it lists the total number of years lived by those present at the beginning of the age interval, but column 5 lists the number of years lived *during the age interval*, whereas column 6 lists the number of years lived *during the age interval and all of the following age intervals* as well.

Column 7: The last column lists the expected remaining lifetime (in years), measured from the beginning of the age interval. For example, the first row shows that the expected remaining lifetime is 76.9 years, so the average expected lifetime of the 100,000 newborn people is expected to be 76.9 years. This value can be computed by dividing the person-years lived (column 6) by the number of people who were present at the beginning of the age interval. For example, the age interval of 1–2 has an expected remaining lifetime of 76.4 years, and the total person-years lived (column 6) divided by the number of people present at the beginning of the age interval (column 3) yields 7,587,418 ÷ 99,307 = 76.4 (rounded). Again, we should remember that this assumes that the same mortality conditions present in the year 2000 remain constant throughout the lifetimes of the 100,000 hypothetical people. We should also remember that Table 13-1 describes the total population, and expected remaining lifetime values will be different for different genders and races.

Notation: More formal mathematical notation is often used for the above components. For example, the age intervals can be expressed in general form as x to $x + 1$, the probability of dying during an interval is expressed as q_x, the number alive at the beginning of an interval is expressed as l_x, the number dying in an interval is expressed as d_x, and so on. Such notation makes it possible to develop formulas describing components of the period life table. For example, the formula $d_x = q_x \cdot l_x$

shows that the number of deaths in an interval is equal to the probability of dying during the interval multiplied by the number alive at the beginning of the interval. Also, such notation is commonly used to describe the columns of a period life table. In Table 13-1, for example, the original government report labels the columns with headings such as "probability of dying between ages x to $x + 1$" and "person-years lived between ages x to $x + 1$." The use of this more formal notation is not necessary for the purposes of this chapter, but this notation is commonly used in some fields.

Abridged Life Table

Table 13-1 is sometimes referred to as a *complete* life table, because it lists a separate row of data for each year. An *abridged* life table uses age intervals for time periods longer than 1 year. Age intervals of 5 years or 10 years are common.

> **Definition**
>
> An **abridged life table** is a life table in which the age intervals have been combined, so that the table is more convenient to use.

Table 13-2 is an abridged life table that condenses Table 13-1. Table 13-2 was constructed by using the values found in Table 13-1.

Table 13-2 Abridged Life Table for Total Population: United States, 2000

Age Interval	Probability of Dying During the Interval	Number Surviving to the Beginning of the Interval	Number of Deaths During the Interval	Person-Years Lived (Total time lived *during the interval* by those alive at the beginning of the interval)	Total Person-Years Lived (in *this and all subsequent* intervals)	Expected Remaining Lifetime (from the beginning of the interval)
0–10	0.009050	100,000	905	991,977	7,686,810	76.9
10–20	0.004450	99,095	441	989,478	6,694,833	67.6
20–30	0.009711	98,654	958	981,816	5,705,355	57.8
30–40	0.013788	97,696	1,347	970,943	4,723,539	48.3
40–50	0.029341	96,349	2,827	951,154	3,752,596	38.9
50–60	0.064413	93,522	6,024	909,065	2,801,442	30.0
60–70	0.147855	87,498	12,937	817,638	1,892,377	21.6
70–80	0.315500	74,561	23,524	637,499	1,074,739	14.4
80–90	0.630876	51,037	32,198	349,450	437,240	8.6
90–100	0.905462	18,839	17,058	83,154	87,790	4.7
100 and over	1.000000	1,781	1,781	4,636	4,636	2.6

From E. Arias, United States Life Tables, 2000. *National Vital Statistics Reporter*, Vol. 51, No. 3, National Center For Health Statistics, 2002.

EXAMPLE **Expected Remaining Life** An abridged life table is to be constructed using age intervals of 10 years. Use Table 13-1 to find the expected remaining lifetime for someone reaching their 20th birthday.

SOLUTION From Table 13-1 we see that someone reaching their 20th birthday has an expected remaining lifetime of 57.8 years. In the abridged table, the expected remaining lifetime for someone reaching their 20th birthday will be the same value of 57.8 years, so the age interval of 20–30 will also have 57.8 years listed in the last column.

EXAMPLE **Probability of Dying** Use Table 13-1 to find the probability of a person dying between the ages of 10 and 20.

SOLUTION From Table 13-1 we see that there were 99,095 people alive on their 10th birthday, and there are 98,654 people alive on their 20th birthday. The probability of surviving between the 10th and 20th birthdays is therefore $98,654/99,095 = 0.995550$. Using the rule of complements from Chapter 3, it follows that the probability of dying is $1 - 0.995550 = 0.004450$.

EXAMPLE **Person-Years Lived** Use Table 13-1 to find the number of person-years lived during the age interval of 25–30.

SOLUTION From Table 13-1 we see that people lived a total of 5,213,242 years beyond their 25th birthday. We also see from Table 13-1 that people lived a total of 4,723,539 years beyond their 30th birthday. We can use simple subtraction to find that the number of years lived during the interval from the 25th birthday to the 30th birthday is $5,213,242 - 4,723,539 = 489,703$ years.

In this section we focused on the nature of a life table, and we described the different components that make up a life table such as Table 13-1. In the next section we consider some real applications of life tables.

13-2 Exercises

1. Types of Life Tables What is the difference between a *cohort* life table and a *period* life table?

2. Cohort and Period Life Tables Why are period life tables so much more practical than cohort life tables?

3. Population Size Table 13-1 was constructed with the assumption that the initial population size is 100,000. How are the values in the first row affected if a population size of 50,000 is used instead?

4. Interpretation of Table Assume that someone was born on January 1, 2000, and that person is alive now. If we refer to Table 13-1 for their expected remaining lifetime, what basic assumption about the construction of Table 13-1 might make that value inaccurate?

In Exercises 5–8, refer to the accompanying life table for white females in the United States for the year 2000.

Life Table for White Females: United States, 2000

Age Interval	Probability of Dying During the Interval	Number Surviving to the Beginning of the Interval	Number of Deaths During the Interval	Person-Years Lived (Total time lived *during the interval* by those alive at the beginning of the interval)	Total Person-Years Lived (in *this and all subsequent* intervals)	Expected Remaining Lifetime (from the beginning of the interval)
0–1	0.005127	100,000	513	99,550	7,996,958	80.0
1–2		99,487	41	99,467	7,897,408	79.4
2–3	0.000268	99,446		99,433	7,797,941	78.4

From E. Arias, United States Life Tables, 2000. *National Vital Statistics Reporter,* Vol. 51, No. 3, National Center For Health Statistics, 2002.

5. Probability of Dying Find the missing value in the second column of the table.

6. Finding the Number of Deaths Find the missing value in the fourth column of the table.

7. Finding Survival Rate Find the probability that a white female will live from birth to her second birthday.

8. Constructing an Abridged Table Assume that you want to use the given table for the construction of an abridged table with 0–2 as the first age interval. Find the values in the first row of this abridged table.

9. Probability of Surviving Using Table 13-1, find the probability that someone will *survive* from their 20th birthday to their 21st birthday. Given 250 people who reach their 20th birthday, what is the expected number of people who survive to their 21st birthday?

10. Expected Remaining Lifetime Use Table 13-1 to find the following:
   a. Find the expected remaining lifetime for a person who has just reached their 60th birthday, then find the expected remaining lifetime for a person who has just reached their 61st birthday.
   b. Find the expected age at death of someone who has just reached their 60th birthday, and find the expected age at death for someone who has just reached their 61st birthday.
   c. Why don't the results in part (b) differ by one year?

11. Probability of Surviving Find the probability that a person will survive from their 16th birthday to their 66th birthday.

12. Probability of Surviving Find the probability that a person will survive from their 21st birthday to their 80th birthday.

13. Abridged Table Table 13-2 is an abridged life table based on Table 13-1. Table 13-2 includes age intervals of 10 years. Using the numbers of people alive at the beginning

of age intervals in Table 13-1, find the probability of dying during the given age intervals. Given that these age intervals are so close, why are the results so different?
   a. 0–5
   b. 5–10

14. Abridged Table Table 13-2 is an abridged life table based on Table 13-1. Table 13-2 includes age intervals of 10 years. Using Table 13-1, find the person-years lived during the given age intervals.
   a. 0–5
   b. 5–10

15. Abridged Table Use Table 13-1 and the results from Exercises 13 and 14 to find the values in a row with an age interval of 0–5.

16. Abridged Table Use Table 13-1 and the results from Exercises 13 and 14 to find the values in a row with an age interval of 5–10.

13-3 Applications of Life Tables

Section 13-2 presented an example of a life table (Table 13-1) and described the components of a life table. In this section we consider some applications. We begin with the important issue of Social Security, as in the following example.

EXAMPLE **Social Security** Assume that there are 4,453,000 births in the United States this year (based on data from the U.S. National Center for Health Statistics). If the age for receiving full Social Security payments is 67, how many of those born this year are expected to be alive on their 67th birthday?

SOLUTION From Table 13-1 we see that among 100,000 people born, we expect that 79,406 of them will survive to their 67th birthday. That is, the probability of someone surviving from birth to their 67th birthday is 0.79406. If each of 4,453,000 people has a 0.79406 probability of surviving to their 67th birthday, the expected number of such survivors is $4{,}453{,}000 \times 0.79406 = 3{,}535{,}949$. Such results are critically important for those professionals responsible for effective administration of the Social Security program.

Another application is based on the property that a life table includes *probabilities*, and they can be treated as the same probabilities used in preceding chapters of this book. Consider the following example in which a hypothesis test is conducted with a probability value found in Table 13-1.

EXAMPLE **Hypothesis Test** In the Chapter Problem, we noted that for one region around Orange County, there are 5000 people who reach their 16th birthday. If 25 of them die before their 17th birthday, do we have sufficient evidence to conclude that this number of deaths is unusually high?

SOLUTION From Table 13-1, we find that for the age interval of 16–17, the probability of dying is 0.000607. However, the proportion of actual deaths can be expressed by $\hat{p} = x/n = 25/5000 = 0.005$. We can proceed to test the claim that the Orange County deaths are from a population with a death rate greater than the life table value of 0.000607. Using the methods of Section 7-3, we begin with the following null and alternative hypotheses.

$$H_0: \quad p = 0.000607$$

$$H_1: \quad p > 0.000607 \quad \text{(claim being tested)}$$

No significance level was specified, so we will use the common value of $\alpha = 0.05$. We now proceed to find the test statistic (as in Section 7-3):

$$z = \frac{\hat{p} - p}{\sqrt{\dfrac{pq}{n}}} = \frac{0.005 - 0.000607}{\sqrt{\dfrac{(0.000607)(0.999393)}{5000}}} = 12.61$$

Using the P-value approach, we can find from Table A-2 that the test statistic of $z = 12.61$ corresponds to a P-value of 0.0001. Using software, we would find that the P-value is 0.0000 when rounded to four decimal places. Using the traditional method of testing hypotheses, we can use Table A-2 to find the critical value of $z = 1.645$.

Because the P-value is less than 0.05 (or because the test statistic of $z = 12.61$ is greater than the critical value of $z = 1.645$), we reject the null hypothesis. There is sufficient evidence to support the claim that the proportion of deaths is significantly greater than the proportion that is usually expected for this age interval. We should note that this analysis assumed that $p = 0.000607$, which was found in Table 13-1. However, Table 13-1 was constructed under the assumption that the mortality conditions present in the year 2000 will remain constant throughout the lifetimes of the hypothetical population of 100,000 people.

For yet another application, we use this definition from Section 3-6:

Definition

The **actual odds against** event A occurring are the ratio $P(\overline{A})/P(A)$, usually expressed in the form of $m{:}n$ (or "m to n"), where m and n are integers having no common factors.

In addition to the actual odds against event A, we also present the following definition of the *payoff* odds against event A. This is the definition in common use at casinos and racetracks.

Definition

The **payoff odds** against event A represent the ratio of the net profit (if you win) to the amount of the bet. That is,

$$\text{payoff odds} = (\text{net profit}){:}(\text{amount bet})$$

The following example shows how an insurance company can use the above two definitions with a life table to establish the amount it must charge for a life insurance policy.

EXAMPLE **Life Insurance Rates** Table 13-1 shows that the probability of someone dying in the age interval of 31–32 is 0.001087. Someone who has just reached their 31st birthday wants to purchase a one-year term life insurance policy that would pay $100,000 in the event of death. What is the minimum amount that an insurance company must charge for this policy? Assume that this minimum amount would result in no net gain or loss for the insurance company, if many policies are issued under the same circumstances.

SOLUTION First, we find the actual odds against dying. With $P(\text{dying}) = 0.001087$, it follows that $P(\text{surviving}) = P(\overline{\text{dying}}) = 1 - 0.001087 = 0.998913$. We can now calculate the actual odds against dying as follows:

$$\text{actual odds against dying} = \frac{P(\overline{\text{dying}})}{P(\text{dying})} = \frac{0.998913}{0.001087} = \frac{998,913}{1087}$$

For the purposes of this solution, it is convenient to divide the numerator and denominator of the above result by 998,913 to get 1/0.001088, which can be expressed as odds of 1:0.001088 or "1 to 0.001088."

Next, if we consider these actual odds to be in the format of

$$\text{payoff odds} = (\text{net profit}):(\text{amount bet})$$

we see that for each dollar of net profit, an amount of 0.001088 must be "bet." (In this case, the "bet" is the cost of the policy.) To gain a net profit of $100,000, the amount of the bet (or policy cost) must be $108.80. (This amount is equal to $100,000 \times 0.001088$.) The insurance company would need to charge $108.80 for the term life insurance policy of $100,000, assuming that it will break even with a large number of such policies. Of course, the actual cost of the policy will be somewhat greater than $108.80, so that the insurance company can cover its administrative costs, plan for a margin of profit, and receive some payment for its risk. Also, insurance companies typically base their rates on more detailed life tables and risk factors. A 16-year-old male driver who drinks and smokes can expect to pay more than a 16-year-old female who does not drive, drink, or smoke.

Shortcut: The above solution can be generalized with the following simple formula for the break-even cost of the term life insurance policy:

$$\text{Break-even cost of policy} = (\text{Amount of policy}) \times \frac{P(\text{dying})}{1 - P(\text{dying})}$$

Using this simplified expression, a $100,000 policy for someone with a 0.001087 probability of dying will have a break-even cost of

$$(\text{Amount of policy}) \times = \frac{P(\text{dying})}{1 - P(\text{dying})}$$

$$= \$100,000 \times \frac{0.001087}{1 - 0.001087} = \$108.80$$

[The above result is actually $108.81829, which is rounded to $108.80 for two reasons: (1) to make the answer match the preceding solution ("Consistency is the hobgoblin of small minds?"), and (2) using four significant digits in the result is justified by the fact that the probability of dying used four significant digits.]

The preceding example is very relevant for those planning to enter an insurance profession, but a similar approach can be used by health care professionals in a variety of other circumstances. For example, instead of an insurance policy premium of $100,000, the result of a death might be measured in terms of its cost to society in lost productivity.

Follow-Up Tables

So far in this chapter we have considered only life tables describing the mortality experience of a group of people. Variations of the basic life tables are often used in other applications.

Definition

A **follow-up** (or **modified**) life table is a period life table modified so that, instead of deaths, some other factor is used as the basis for describing the probability of surviving some treatment or exposure to a disease or some other characteristic.

For example, medical researchers might be very interested in a follow-up life table describing the experience of patients implanted with heart pacemakers. Instead of deaths, we might collect data on the lengths of times patients survived with pacemakers before experiencing myocardial infarctions (heart attacks).

As another example, see Table 13-3. A follow-up life table with a similar format was suggested as a device that would be helpful in making projections of substance abuse problems. The information in a table such as Table 13-3 would be very helpful to those professionals attempting to address drug-abuse problems.

Table 13-3	Follow-Up Life Table of Substance Abuse						
Age Interval	Number Alive and Drug-Free at Beginning of Interval	Number Who Began Drug Use During Interval	Number Who Died During Interval	Number Who Discontinued Drug Use During Interval	Percent Beginning Drug Use During Interval	Percent Drug-Free During Interval	Percent Drug-Free at End of Interval
0–5	1000	0	8	0	0.000%	100.0%	100.0%
5–10							
10–15							
15–20							

Based on data from S. Korper and C. Council (editors), *Substance Use by Older Adults: Estimates of Future Impact on the Treatment System* (DHHS Publication No. SMA 03-3763, Analytic Series A-21). Rockville, MD: Substance Abuse and Mental Health Services Administration, Office of Applied Studies.

The applications discussed in this section constitute a small sample of the real and important applications in use and the applications that have potential use.

13-3 Exercises

In Exercises 1–4, assume that there are 4,453,000 births in the United States this year (based on data from the U.S. National Center for Health Statistics). Use Table 13-1 to find the given values.

1. Future Drivers If the minimum age for receiving a driver's license is 16, how many of those born this year are expected to be alive on their 16th birthday? Identify one practical use for this result.

2. Future Mothers In planning for future population growth, one key factor is the number of women of childbearing age. For one component of that number, assume that half of the births this year are girls, and find the number of them that are expected to reach their 21st birthday.

3. Future Health Care In planning for future health care needs, it is important to make accurate predictions of the size of the population of elderly people. How many of those born this year are expected to be alive on their 60th birthday?

4. Care for the Elderly Managed-care facilities and nursing homes are populated mostly by people of age 65 or older. How many of those born this year are expected to be alive on their 65th birthday?

5. Hypothesis Test In one region, a researcher finds that among 12,500 people who survived to their 20th birthday, there are 15 deaths before they reached age 21. Using Table 13-1, test the claim that this is an unusually high number of deaths.

6. Hypothesis Test In one region, a researcher finds that among 6772 people who survived to their 50th birthday, there are 36 deaths before they reached age 51. Using Table 13-1, test the claim that this is an unusually high number of deaths.

7. Hypothesis Test In one region, a researcher finds that among 8774 people who survived to their 16th birthday, there are 147 deaths before they reached age 30. Using Table 13-1, test the claim that this is an unusually high number of deaths.

8. Hypothesis Test In one region, a researcher finds that among 4285 people who survived to their 60th birthday, there are 601 deaths before they reached age 70. Using Table 13-1, test the claim that this is an unusually *low* number of deaths.

9. Life Insurance Rates Someone who has just reached their 49th birthday wants to purchase a one-year term life insurance policy that would pay $100,000 in the event of death before age 50. What is the minimum amount that an insurance company must charge for this policy? Use Table 13-1 and assume that this minimum amount would result in no gain or loss for an insurance company issuing many policies under the same circumstances.

10. Life Insurance Rates Someone who has just reached their 55th birthday wants to purchase a one-year term life insurance policy that would pay $50,000 in the event of death before age 56. What is the minimum amount that an insurance company must charge for this policy? Use Table 13-1 and assume that this minimum amount would result in no gain or loss for an insurance company issuing many policies under the same circumstances.

11. Life Insurance Rates Someone who has just reached their 60th birthday wants to purchase a term life insurance policy that would pay $50,000 in the event of death before age 70. What is the minimum amount that an insurance company must charge for this policy? Use Table 13-1 and assume that this minimum amount would result in no gain or loss for an insurance company issuing many policies under the same circumstances.

12. Life Insurance Rates Someone who has just reached their 25th birthday wants to purchase a term life insurance policy that would pay $20,000 in the event of death before age 30. What is the minimum amount that an insurance company must charge for this policy? Use Table 13-1 and assume that this minimum amount would result in no gain or loss for an insurance company issuing many policies under the same circumstances.

13. Life Insurance Rates Repeat Exercise 9 by assuming that the policyholder is a black female. Using a life table different from Table 13-1, it is found that the probability of a black female dying between her 49th birthday and 50th birthday is 0.005829. Compare this result to the one found in Exercise 9. What does this result suggest about the mortality of a 49-year-old black female?

14. Life Insurance Rates Repeat Exercise 10 by assuming that the policyholder is a white male. Using a life table different from Table 13-1, it is found that the probability of a white male dying between his 55th birthday and 56th birthday is 0.007721. Compare this result to the one found in Exercise 10. What does this result suggest about the mortality of a 55-year-old white male?

Review

In this chapter we introduced the concept of a period (or current) life table, which describes mortality and longevity data for a hypothetical group of people. It is important to know that a period life table is computed with the assumption that the conditions affecting mortality in a particular year remain unchanged throughout the lives of everyone in the hypothetical group. We presented Table 13-1 as an example of a period life table. We also described a cohort (or generation) life table, which is based on records of actual observed mortality experiences for a particular group, but serious practical considerations make the construction of cohort life tables extremely difficult. We also described an abridged life table, which combines age categories. Table 13-2 is an example of an abridged life table.

We considered some applications of life tables, including determination of population sizes, hypothesis testing with proportions, and calculation of insurance premiums.

Review Exercises

In Exercises 1–5, refer to the accompanying life table for black females in the United States for the year 2000.

Life Table for Black Females: United States, 2000

Age Interval	Probability of Dying During the Interval	Number Surviving to the Beginning of the Interval	Number of Deaths During the Interval	Person-Years Lived (Total time lived *during the interval* by those alive at the beginning of the interval)	Total Person-Years Lived (in *this and all subsequent* intervals)	Expected Remaining Lifetime (from the beginning of the interval)
0–1	0.012672	100,000	1,267	98,890	7,493,035	74.9
1–2		98,733	78	98,694	7,394,145	74.9
2–3	0.000537	98,655		98,628	7,295,452	73.9

From E. Arias, United States Life Tables, 2000. *National Vital Statistics Reporter,* Vol. 51, No. 3, National Center For Health Statistics, 2002.

1. **Probability of Dying** Find the missing value in the second column of the table.

2. **Finding the Number of Deaths** Find the missing value in the fourth column of the table.

3. **Finding Survival Rate** Find the probability that a black female will live from birth to her second birthday.

4. **Constructing an Abridged Table** Assume that you want to use the given table for the construction of an abridged table with 0–2 as the first age interval. Find the values in the first row of this abridged table.

5. **Expected Remaining Lifetime** What is the expected remaining lifetime of a black female who was just born? What is the expected remaining lifetime of a black female who is celebrating her first birthday? Compare those results and comment on the relationship between them.

Cumulative Review Exercises

In Exercises 1–4, use the same life table used for Review Exercises 1–5.

1. **Hypothesis Test of Experimental Results** In one region, a dramatic community-wide program is implemented in an attempt to reduce the high rate of mortality for young black females. Results show that among 786 births of black females, 782 survived to their first birthday. Using the data in the life table for black females, test the claim that the program was effective in reducing the mortality rate.

2. **Confidence Interval for Experimental Results** Given the sample results from Exercise 1, construct a 95% confidence interval for the survival rate (from birth to the first birthday) for black females born in the region with the mortality reduction program in

effect. Instead of using three significant digits, express the results with five significant digits. What can be concluded about the fact that the confidence interval limits do (or do not) contain the survival rate for black females, as suggested by the life table?

3. Probability of Surviving If three different newborn black females are randomly selected, find the probability that they all survive to their first birthday.

4. Probability of Surviving If four different newborn black females are randomly selected, find the probability that at least one of them does not survive to her first birthday. Does this probability apply to families in which there are four children, all of whom are black females?

Cooperative Group Activities

1. *In-class activity:* Table 13-1 describes the life and death experience of a hypothetical group of 100,000 people. That table was constructed using actual results obtained from a variety of different sources. Instead of using real sources of vital statistics, assume that the death rate of some population is a constant 0.4 for each year. Construct as much of the life table as possible. How many years will pass before none of the hypothetical population of 100,000 people are alive?

2. *Out-of-class activity:* Repeat Cooperative Group Activity 1, but instead of using a constant death rate of 0.4 each year, use a computer or calculator to randomly generate the death rate for each year. For each year, generate a random number between 0 and 1, and assume that it is the death rate that applies to the particular year.

Technology Project

Using the Internet, find or construct a follow-up (or modified) life table. Instead of using life/death as a basis for constructing the table, use some other factor, such as some treatment or exposure to a disease or some other characteristic. Obtain a printed copy of the table and describe at least one way in which the table can be used in some positive way.

from DATA to DECISION

Critical Thinking: How do we reduce the high mortality rate among young black females?

Compare the first three rows of the life table for white females in the United States to the first three rows of the life table for black females in the United States. (See the table that accompanies Exercises 5–8 in Section 13-2, and see the table that accompanies Review Exercises 1–5.)

Analyze the Results
1. What notable differences are apparent?
2. Identify some reasons for the apparent differences.

3. What are some steps that could be taken to effectively reduce the high rate of mortality among young black females?
4. What is a reasonable response to someone who argues that a program for reducing mortality among young black females should not be implemented, because it discriminates on the basis of race?

Appendix A Tables

TABLE A-1   Binomial Probabilities

									p								
n	x	.01	.05	.10	.20	.30	.40	.50	.60	.70	.80	.90	.95	.99		x
2	0	.980	.902	.810	.640	.490	.360	.250	.160	.090	.040	.010	.002	0+		0
	1	.020	.095	.180	.320	.420	.480	.500	.480	.420	.320	.180	.095	.020		1
	2	0+	.002	.010	.040	.090	.160	.250	.360	.490	.640	.810	.902	.980		2
3	0	.970	.857	.729	.512	.343	.216	.125	.064	.027	.008	.001	0+	0+		0
	1	.029	.135	.243	.384	.441	.432	.375	.288	.189	.096	.027	.007	0+		1
	2	0+	.007	.027	.096	.189	.288	.375	.432	.441	.384	.243	.135	.029		2
	3	0+	0+	.001	.008	.027	.064	.125	.216	.343	.512	.729	.857	.970		3
4	0	.961	.815	.656	.410	.240	.130	.062	.026	.008	.002	0+	0+	0+		0
	1	.039	.171	.292	.410	.412	.346	.250	.154	.076	.026	.004	0+	0+		1
	2	.001	.014	.049	.154	.265	.346	.375	.346	.265	.154	.049	.014	.001		2
	3	0+	0+	.004	.026	.076	.154	.250	.346	.412	.410	.292	.171	.039		3
	4	0+	0+	0+	.002	.008	.026	.062	.130	.240	.410	.656	.815	.961		4
5	0	.951	.774	.590	.328	.168	.078	.031	.010	.002	0+	0+	0+	0+		0
	1	.048	.204	.328	.410	.360	.259	.156	.077	.028	.006	0+	0+	0+		1
	2	.001	.021	.073	.205	.309	.346	.312	.230	.132	.051	.008	.001	0+		2
	3	0+	.001	.008	.051	.132	.230	.312	.346	.309	.205	.073	.021	.001		3
	4	0+	0+	0+	.006	.028	.077	.156	.259	.360	.410	.328	.204	.048		4
	5	0+	0+	0+	0+	.002	.010	.031	.078	.168	.328	.590	.774	.951		5
6	0	.941	.735	.531	.262	.118	.047	.016	.004	.001	0+	0+	0+	0+		0
	1	.057	.232	.354	.393	.303	.187	.094	.037	.010	.002	0+	0+	0+		1
	2	.001	.031	.098	.246	.324	.311	.234	.138	.060	.015	.001	0+	0+		2
	3	0+	.002	.015	.082	.185	.276	.312	.276	.185	.082	.015	.002	0+		3
	4	0+	0+	.001	.015	.060	.138	.234	.311	.324	.246	.098	.031	.001		4
	5	0+	0+	0+	.002	.010	.037	.094	.187	.303	.393	.354	.232	.057		5
	6	0+	0+	0+	0+	.001	.004	.016	.047	.118	.262	.531	.735	.941		6
7	0	.932	.698	.478	.210	.082	.028	.008	.002	0+	0+	0+	0+	0+		0
	1	.066	.257	.372	.367	.247	.131	.055	.017	.004	0+	0+	0+	0+		1
	2	.002	.041	.124	.275	.318	.261	.164	.077	.025	.004	0+	0+	0+		2
	3	0+	.004	.023	.115	.227	.290	.273	.194	.097	.029	.003	0+	0+		3
	4	0+	0+	.003	.029	.097	.194	.273	.290	.227	.115	.023	.004	0+		4
	5	0+	0+	0+	.004	.025	.077	.164	.261	.318	.275	.124	.041	.002		5
	6	0+	0+	0+	0+	.004	.017	.055	.131	.247	.367	.372	.257	.066		6
	7	0+	0+	0+	0+	0+	.002	.008	.028	.082	.210	.478	.698	.932		7
8	0	.923	.663	.430	.168	.058	.017	.004	.001	0+	0+	0+	0+	0+		0
	1	.075	.279	.383	.336	.198	.090	.031	.008	.001	0+	0+	0+	0+		1
	2	.003	.051	.149	.294	.296	.209	.109	.041	.010	.001	0+	0+	0+		2
	3	0+	.005	.033	.147	.254	.279	.219	.124	.047	.009	0+	0+	0+		3
	4	0+	0+	.005	.046	.136	.232	.273	.232	.136	.046	.005	0+	0+		4
	5	0+	0+	0+	.009	.047	.124	.219	.279	.254	.147	.033	.005	0+		5
	6	0+	0+	0+	.001	.010	.041	.109	.209	.296	.294	.149	.051	.003		6
	7	0+	0+	0+	0+	.001	.008	.031	.090	.198	.336	.383	.279	.075		7
	8	0+	0+	0+	0+	0+	.001	.004	.017	.058	.168	.430	.663	.923		8

NOTE: 0+ represents a positive probability less than 0.0005.

(continued)

								p							
n	x	.01	.05	.10	.20	.30	.40	.50	.60	.70	.80	.90	.95	.99	x
9	0	.914	.630	.387	.134	.040	.010	.002	0+	0+	0+	0+	0+	0+	0
	1	.083	.299	.387	.302	.156	.060	.018	.004	0+	0+	0+	0+	0+	1
	2	.003	.063	.172	.302	.267	.161	.070	.021	.004	0+	0+	0+	0+	2
	3	0+	.008	.045	.176	.267	.251	.164	.074	.021	.003	0+	0+	0+	3
	4	0+	.001	.007	.066	.172	.251	.246	.167	.074	.017	.001	0+	0+	4
	5	0+	0+	.001	.017	.074	.167	.246	.251	.172	.066	.007	.001	0+	5
	6	0+	0+	0+	.003	.021	.074	.164	.251	.267	.176	.045	.008	0+	6
	7	0+	0+	0+	0+	.004	.021	.070	.161	.267	.302	.172	.063	.003	7
	8	0+	0+	0+	0+	0+	.004	.018	.060	.156	.302	.387	.299	.083	8
	9	0+	0+	0+	0+	0+	0+	.002	.010	.040	.134	.387	.630	.914	9
10	0	.904	.599	.349	.107	.028	.006	.001	0+	0+	0+	0+	0+	0+	0
	1	.091	.315	.387	.268	.121	.040	.010	.002	0+	0+	0+	0+	0+	1
	2	.004	.075	.194	.302	.233	.121	.044	.011	.001	0+	0+	0+	0+	2
	3	0+	.010	.057	.201	.267	.215	.117	.042	.009	.001	0+	0+	0+	3
	4	0+	.001	.011	.088	.200	.251	.205	.111	.037	.006	0+	0+	0+	4
	5	0+	0+	.001	.026	.103	.201	.246	.201	.103	.026	.001	0+	0+	5
	6	0+	0+	0+	.006	.037	.111	.205	.251	.200	.088	.011	.001	0+	6
	7	0+	0+	0+	.001	.009	.042	.117	.215	.267	.201	.057	.010	0+	7
	8	0+	0+	0+	0+	.001	.011	.044	.121	.233	.302	.194	.075	.004	8
	9	0+	0+	0+	0+	0+	.002	.010	.040	.121	.268	.387	.315	.091	9
	10	0+	0+	0+	0+	0+	0+	.001	.006	.028	.107	.349	.599	.904	10
11	0	.895	.569	.314	.086	.020	.004	0+	0+	0+	0+	0+	0+	0+	0
	1	.099	.329	.384	.236	.093	.027	.005	.001	0+	0+	0+	0+	0+	1
	2	.005	.087	.213	.295	.200	.089	.027	.005	.001	0+	0+	0+	0+	2
	3	0+	.014	.071	.221	.257	.177	.081	.023	.004	0+	0+	0+	0+	3
	4	0+	.001	.016	.111	.220	.236	.161	.070	.017	.002	0+	0+	0+	4
	5	0+	0+	.002	.039	.132	.221	.226	.147	.057	.010	0+	0+	0+	5
	6	0+	0+	0+	.010	.057	.147	.226	.221	.132	.039	.002	0+	0+	6
	7	0+	0+	0+	.002	.017	.070	.161	.236	.220	.111	.016	.001	0+	7
	8	0+	0+	0+	0+	.004	.023	.081	.177	.257	.221	.071	.014	0+	8
	9	0+	0+	0+	0+	.001	.005	.027	.089	.200	.295	.213	.087	.005	9
	10	0+	0+	0+	0+	0+	.001	.005	.027	.093	.236	.384	.329	.099	10
	11	0+	0+	0+	0+	0+	0+	0+	.004	.020	.086	.314	.569	.895	11
12	0	.886	.540	.282	.069	.014	.002	0+	0+	0+	0+	0+	0+	0+	0
	1	.107	.341	.377	.206	.071	.017	.003	0+	0+	0+	0+	0+	0+	1
	2	.006	.099	.230	.283	.168	.064	.016	.002	0+	0+	0+	0+	0+	2
	3	0+	.017	.085	.236	.240	.142	.054	.012	.001	0+	0+	0+	0+	3
	4	0+	.002	.021	.133	.231	.213	.121	.042	.008	.001	0+	0+	0+	4
	5	0+	0+	.004	.053	.158	.227	.193	.101	.029	.003	0+	0+	0+	5
	6	0+	0+	0+	.016	.079	.177	.226	.177	.079	.016	0+	0+	0+	6
	7	0+	0+	0+	.003	.029	.101	.193	.227	.158	.053	.004	0+	0+	7
	8	0+	0+	0+	.001	.008	.042	.121	.213	.231	.133	.021	.002	0+	8
	9	0+	0+	0+	0+	.001	.012	.054	.142	.240	.236	.085	.017	0+	9
	10	0+	0+	0+	0+	0+	.002	.016	.064	.168	.283	.230	.099	.006	10
	11	0+	0+	0+	0+	0+	0+	.003	.017	.071	.206	.377	.341	.107	11
	12	0+	0+	0+	0+	0+	0+	0+	.002	.014	.069	.282	.540	.886	12

NOTE: 0+ represents a positive probability less than 0.0005.

(*continued*)

								p								
n	x	.01	.05	.10	.20	.30	.40	.50	.60	.70	.80	.90	.95	.99	x
13	0	.878	.513	.254	.055	.010	.001	0+	0+	0+	0+	0+	0+	0+	0
	1	.115	.351	.367	.179	.054	.011	.002	0+	0+	0+	0+	0+	0+	1
	2	.007	.111	.245	.268	.139	.045	.010	.001	0+	0+	0+	0+	0+	2
	3	0+	.021	.100	.246	.218	.111	.035	.006	.001	0+	0+	0+	0+	3
	4	0+	.003	.028	.154	.234	.184	.087	.024	.003	0+	0+	0+	0+	4
	5	0+	0+	.006	.069	.180	.221	.157	.066	.014	.001	0+	0+	0+	5
	6	0+	0+	.001	.023	.103	.197	.209	.131	.044	.006	0+	0+	0+	6
	7	0+	0+	0+	.006	.044	.131	.209	.197	.103	.023	.001	0+	0+	7
	8	0+	0+	0+	.001	.014	.066	.157	.221	.180	.069	.006	0+	0+	8
	9	0+	0+	0+	0+	.003	.024	.087	.184	.234	.154	.028	.003	0+	9
	10	0+	0+	0+	0+	.001	.006	.035	.111	.218	.246	.100	.021	0+	10
	11	0+	0+	0+	0+	0+	.001	.010	.045	.139	.268	.245	.111	.007	11
	12	0+	0+	0+	0+	0+	0+	.002	.011	.054	.179	.367	.351	.115	12
	13	0+	0+	0+	0+	0+	0+	0+	.001	.010	.055	.254	.513	.878	13
14	0	.869	.488	.229	.044	.007	.001	0+	0+	0+	0+	0+	0+	0+	0
	1	.123	.359	.356	.154	.041	.007	.001	0+	0+	0+	0+	0+	0+	1
	2	.008	.123	.257	.250	.113	.032	.006	.001	0+	0+	0+	0+	0+	2
	3	0+	.026	.114	.250	.194	.085	.022	.003	0+	0+	0+	0+	0+	3
	4	0+	.004	.035	.172	.229	.155	.061	.014	.001	0+	0+	0+	0+	4
	5	0+	0+	.008	.086	.196	.207	.122	.041	.007	0+	0+	0+	0+	5
	6	0+	0+	.001	.032	.126	.207	.183	.092	.023	.002	0+	0+	0+	6
	7	0+	0+	0+	.009	.062	.157	.209	.157	.062	.009	0+	0+	0+	7
	8	0+	0+	0+	.002	.023	.092	.183	.207	.126	.032	.001	0+	0+	8
	9	0+	0+	0+	0+	.007	.041	.122	.207	.196	.086	.008	0+	0+	9
	10	0+	0+	0+	0+	.001	.014	.061	.155	.229	.172	.035	.004	0+	10
	11	0+	0+	0+	0+	0+	.003	.022	.085	.194	.250	.114	.026	0+	11
	12	0+	0+	0+	0+	0+	.001	.006	.032	.113	.250	.257	.123	.008	12
	13	0+	0+	0+	0+	0+	0+	.001	.007	.041	.154	.356	.359	.123	13
	14	0+	0+	0+	0+	0+	0+	0+	.001	.007	.044	.229	.488	.869	14
15	0	.860	.463	.206	.035	.005	0+	0+	0+	0+	0+	0+	0+	0+	0
	1	.130	.366	.343	.132	.031	.005	0+	0+	0+	0+	0+	0+	0+	1
	2	.009	.135	.267	.231	.092	.022	.003	0+	0+	0+	0+	0+	0+	2
	3	0+	.031	.129	.250	.170	.063	.014	.002	0+	0+	0+	0+	0+	3
	4	0+	.005	.043	.188	.219	.127	.042	.007	.001	0+	0+	0+	0+	4
	5	0+	.001	.010	.103	.206	.186	.092	.024	.003	0+	0+	0+	0+	5
	6	0+	0+	.002	.043	.147	.207	.153	.061	.012	.001	0+	0+	0+	6
	7	0+	0+	0+	.014	.081	.177	.196	.118	.035	.003	0+	0+	0+	7
	8	0+	0+	0+	.003	.035	.118	.196	.177	.081	.014	0+	0+	0+	8
	9	0+	0+	0+	.001	.012	.061	.153	.207	.147	.043	.002	0+	0+	9
	10	0+	0+	0+	0+	.003	.024	.092	.186	.206	.103	.010	.001	0+	10
	11	0+	0+	0+	0+	.001	.007	.042	.127	.219	.188	.043	.005	0+	11
	12	0+	0+	0+	0+	0+	.002	.014	.063	.170	.250	.129	.031	0+	12
	13	0+	0+	0+	0+	0+	0+	.003	.022	.092	.231	.267	.135	.009	13
	14	0+	0+	0+	0+	0+	0+	0+	.005	.031	.132	.343	.366	.130	14
	15	0+	0+	0+	0+	0+	0+	0+	0+	.005	.035	.206	.463	.860	15

NOTE: 0+ represents a positive probability less than 0.0005.

From Frederick C. Mosteller, Robert E. K. Rourke, and George B. Thomas, Jr., *Probability with Statistical Applications*, 2nd ed., © 1970 Addison-Wesley Publishing Co., Reading, MA. Reprinted with permission.

NEGATIVE z Scores

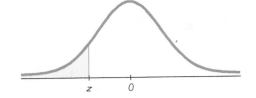

TABLE A-2	Standard Normal (z) Distribution: Cumulative Area from the LEFT								
z	.00	.01	.02	.03	.04	.05	.06	.07	.08	.09
−3.50 and lower	.0001									
−3.4	.0003	.0003	.0003	.0003	.0003	.0003	.0003	.0003	.0003	.0002
−3.3	.0005	.0005	.0005	.0004	.0004	.0004	.0004	.0004	.0004	.0003
−3.2	.0007	.0007	.0006	.0006	.0006	.0006	.0006	.0005	.0005	.0005
−3.1	.0010	.0009	.0009	.0009	.0008	.0008	.0008	.0008	.0007	.0007
−3.0	.0013	.0013	.0013	.0012	.0012	.0011	.0011	.0011	.0010	.0010
−2.9	.0019	.0018	.0018	.0017	.0016	.0016	.0015	.0015	.0014	.0014
−2.8	.0026	.0025	.0024	.0023	.0023	.0022	.0021	.0021	.0020	.0019
−2.7	.0035	.0034	.0033	.0032	.0031	.0030	.0029	.0028	.0027	.0026
−2.6	.0047	.0045	.0044	.0043	.0041	.0040	.0039	.0038	.0037	.0036
−2.5	.0062	.0060	.0059	.0057	.0055	.0054	.0052	.0051	* .0049	.0048
−2.4	.0082	.0080	.0078	.0075	.0073	.0071	.0069	.0068	.0066	.0064
−2.3	.0107	.0104	.0102	.0099	.0096	.0094	.0091	.0089	.0087	.0084
−2.2	.0139	.0136	.0132	.0129	.0125	.0122	.0119	.0116	.0113	.0110
−2.1	.0179	.0174	.0170	.0166	.0162	.0158	.0154	.0150	.0146	.0143
−2.0	.0228	.0222	.0217	.0212	.0207	.0202	.0197	.0192	.0188	.0183
−1.9	.0287	.0281	.0274	.0268	.0262	.0256	.0250	.0244	.0239	.0233
−1.8	.0359	.0351	.0344	.0336	.0329	.0322	.0314	.0307	.0301	.0294
−1.7	.0446	.0436	.0427	.0418	.0409	.0401	.0392	.0384	.0375	.0367
−1.6	.0548	.0537	.0526	.0516	.0505 *	.0495	.0485	.0475	.0465	.0455
−1.5	.0668	.0655	.0643	.0630	.0618	.0606	.0594	.0582	.0571	.0559
−1.4	.0808	.0793	.0778	.0764	.0749	.0735	.0721	.0708	.0694	.0681
−1.3	.0968	.0951	.0934	.0918	.0901	.0885	.0869	.0853	.0838	.0823
−1.2	.1151	.1131	.1112	.1093	.1075	.1056	.1038	.1020	.1003	.0985
−1.1	.1357	.1335	.1314	.1292	.1271	.1251	.1230	.1210	.1190	.1170
−1.0	.1587	.1562	.1539	.1515	.1492	.1469	.1446	.1423	.1401	.1379
−0.9	.1841	.1814	.1788	.1762	.1736	.1711	.1685	.1660	.1635	.1611
−0.8	.2119	.2090	.2061	.2033	.2005	.1977	.1949	.1922	.1894	.1867
−0.7	.2420	.2389	.2358	.2327	.2296	.2266	.2236	.2206	.2177	.2148
−0.6	.2743	.2709	.2676	.2643	.2611	.2578	.2546	.2514	.2483	.2451
−0.5	.3085	.3050	.3015	.2981	.2946	.2912	.2877	.2843	.2810	.2776
−0.4	.3446	.3409	.3372	.3336	.3300	.3264	.3228	.3192	.3156	.3121
−0.3	.3821	.3783	.3745	.3707	.3669	.3632	.3594	.3557	.3520	.3483
−0.2	.4207	.4168	.4129	.4090	.4052	.4013	.3974	.3936	.3897	.3859
−0.1	.4602	.4562	.4522	.4483	.4443	.4404	.4364	.4325	.4286	.4247
−0.0	.5000	.4960	.4920	.4880	.4840	.4801	.4761	.4721	.4681	.4641

NOTE: For values of z below −3.49, use 0.0001 for the area.
*Use these common values that result from interpolation:

z score	Area
−1.645	0.0500
−2.575	0.0050

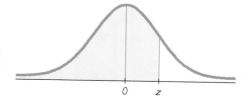

POSITIVE z Scores

TABLE A-2	(continued) Cumulative Area from the LEFT									
z	.00	.01	.02	.03	.04	.05	.06	.07	.08	.09
0.0	.5000	.5040	.5080	.5120	.5160	.5199	.5239	.5279	.5319	.5359
0.1	.5398	.5438	.5478	.5517	.5557	.5596	.5636	.5675	.5714	.5753
0.2	.5793	.5832	.5871	.5910	.5948	.5987	.6026	.6064	.6103	.6141
0.3	.6179	.6217	.6255	.6293	.6331	.6368	.6406	.6443	.6480	.6517
0.4	.6554	.6591	.6628	.6664	.6700	.6736	.6772	.6808	.6844	.6879
0.5	.6915	.6950	.6985	.7019	.7054	.7088	.7123	.7157	.7190	.7224
0.6	.7257	.7291	.7324	.7357	.7389	.7422	.7454	.7486	.7517	.7549
0.7	.7580	.7611	.7642	.7673	.7704	.7734	.7764	.7794	.7823	.7852
0.8	.7881	.7910	.7939	.7967	.7995	.8023	.8051	.8078	.8106	.8133
0.9	.8159	.8186	.8212	.8238	.8264	.8289	.8315	.8340	.8365	.8389
1.0	.8413	.8438	.8461	.8485	.8508	.8531	.8554	.8577	.8599	.8621
1.1	.8643	.8665	.8686	.8708	.8729	.8749	.8770	.8790	.8810	.8830
1.2	.8849	.8869	.8888	.8907	.8925	.8944	.8962	.8980	.8997	.9015
1.3	.9032	.9049	.9066	.9082	.9099	.9115	.9131	.9147	.9162	.9177
1.4	.9192	.9207	.9222	.9236	.9251	.9265	.9279	.9292	.9306	.9319
1.5	.9332	.9345	.9357	.9370	.9382	.9394	.9406	.9418	.9429	.9441
1.6	.9452	.9463	.9474	.9484	.9495 *	.9505	.9515	.9525	.9535	.9545
1.7	.9554	.9564	.9573	.9582	.9591	.9599	.9608	.9616	.9625	.9633
1.8	.9641	.9649	.9656	.9664	.9671	.9678	.9686	.9693	.9699	.9706
1.9	.9713	.9719	.9726	.9732	.9738	.9744	.9750	.9756	.9761	.9767
2.0	.9772	.9778	.9783	.9788	.9793	.9798	.9803	.9808	.9812	.9817
2.1	.9821	.9826	.9830	.9834	.9838	.9842	.9846	.9850	.9854	.9857
2.2	.9861	.9864	.9868	.9871	.9875	.9878	.9881	.9884	.9887	.9890
2.3	.9893	.9896	.9898	.9901	.9904	.9906	.9909	.9911	.9913	.9916
2.4	.9918	.9920	.9922	.9925	.9927	.9929	.9931	.9932	.9934	.9936
2.5	.9938	.9940	.9941	.9943	.9945	.9946	.9948	.9949 *	.9951	.9952
2.6	.9953	.9955	.9956	.9957	.9959	.9960	.9961	.9962	.9963	.9964
2.7	.9965	.9966	.9967	.9968	.9969	.9970	.9971	.9972	.9973	.9974
2.8	.9974	.9975	.9976	.9977	.9977	.9978	.9979	.9979	.9980	.9981
2.9	.9981	.9982	.9982	.9983	.9984	.9984	.9985	.9985	.9986	.9986
3.0	.9987	.9987	.9987	.9988	.9988	.9989	.9989	.9989	.9990	.9990
3.1	.9990	.9991	.9991	.9991	.9992	.9992	.9992	.9992	.9993	.9993
3.2	.9993	.9993	.9994	.9994	.9994	.9994	.9994	.9995	.9995	.9995
3.3	.9995	.9995	.9995	.9996	.9996	.9996	.9996	.9996	.9996	.9997
3.4	.9997	.9997	.9997	.9997	.9997	.9997	.9997	.9997	.9997	.9998
3.50 and up	.9999									

NOTE: For values of z above 3.49, use 0.9999 for the area.
*Use these common values that result from interpolation:

z score	Area
1.645	0.9500
2.575	0.9950

Common Critical Values

Confidence Level	Critical Value
0.90	1.645
0.95	1.96
0.99	2.575

TABLE A-3	t Distribution: Critical t Values				
			Area in One Tail		
	0.005	0.01	0.025	0.05	0.10
Degrees of Freedom			**Area in Two Tails**		
	0.01	0.02	0.05	0.10	0.20
1	63.657	31.821	12.706	6.314	3.078
2	9.925	6.965	4.303	2.920	1.886
3	5.841	4.541	3.182	2.353	1.638
4	4.604	3.747	2.776	2.132	1.533
5	4.032	3.365	2.571	2.015	1.476
6	3.707	3.143	2.447	1.943	1.440
7	3.499	2.998	2.365	1.895	1.415
8	3.355	2.896	2.306	1.860	1.397
9	3.250	2.821	2.262	1.833	1.383
10	3.169	2.764	2.228	1.812	1.372
11	3.106	2.718	2.201	1.796	1.363
12	3.055	2.681	2.179	1.782	1.356
13	3.012	2.650	2.160	1.771	1.350
14	2.977	2.624	2.145	1.761	1.345
15	2.947	2.602	2.131	1.753	1.341
16	2.921	2.583	2.120	1.746	1.337
17	2.898	2.567	2.110	1.740	1.333
18	2.878	2.552	2.101	1.734	1.330
19	2.861	2.539	2.093	1.729	1.328
20	2.845	2.528	2.086	1.725	1.325
21	2.831	2.518	2.080	1.721	1.323
22	2.819	2.508	2.074	1.717	1.321
23	2.807	2.500	2.069	1.714	1.319
24	2.797	2.492	2.064	1.711	1.318
25	2.787	2.485	2.060	1.708	1.316
26	2.779	2.479	2.056	1.706	1.315
27	2.771	2.473	2.052	1.703	1.314
28	2.763	2.467	2.048	1.701	1.313
29	2.756	2.462	2.045	1.699	1.311
30	2.750	2.457	2.042	1.697	1.310
31	2.744	2.453	2.040	1.696	1.309
32	2.738	2.449	2.037	1.694	1.309
34	2.728	2.441	2.032	1.691	1.307
36	2.719	2.434	2.028	1.688	1.306
38	2.712	2.429	2.024	1.686	1.304
40	2.704	2.423	2.021	1.684	1.303
45	2.690	2.412	2.014	1.679	1.301
50	2.678	2.403	2.009	1.676	1.299
55	2.668	2.396	2.004	1.673	1.297
60	2.660	2.390	2.000	1.671	1.296
65	2.654	2.385	1.997	1.669	1.295
70	2.648	2.381	1.994	1.667	1.294
75	2.643	2.377	1.992	1.665	1.293
80	2.639	2.374	1.990	1.664	1.292
90	2.632	2.368	1.987	1.662	1.291
100	2.626	2.364	1.984	1.660	1.290
200	2.601	2.345	1.972	1.653	1.286
300	2.592	2.339	1.968	1.650	1.284
400	2.588	2.336	1.966	1.649	1.284
500	2.586	2.334	1.965	1.648	1.283
750	2.582	2.331	1.963	1.647	1.283
1000	2.581	2.330	1.962	1.646	1.282
2000	2.578	2.328	1.961	1.646	1.282
Large	2.576	2.326	1.960	1.645	1.282

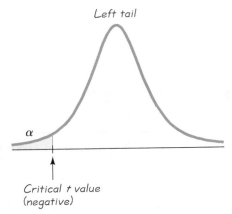

Left tail

α

Critical t value
(negative)

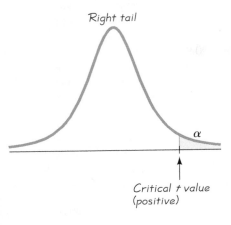

Right tail

α

Critical t value
(positive)

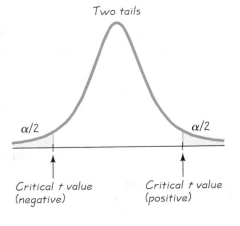

Two tails

$\alpha/2$ $\alpha/2$

Critical t value
(negative)

Critical t value
(positive)

TABLE A-4 | Chi-Square (χ^2) Distribution

Degrees of Freedom	Area to the Right of the Critical Value									
	0.995	0.99	0.975	0.95	0.90	0.10	0.05	0.025	0.01	0.005
1	—	—	0.001	0.004	0.016	2.706	3.841	5.024	6.635	7.879
2	0.010	0.020	0.051	0.103	0.211	4.605	5.991	7.378	9.210	10.597
3	0.072	0.115	0.216	0.352	0.584	6.251	7.815	9.348	11.345	12.838
4	0.207	0.297	0.484	0.711	1.064	7.779	9.488	11.143	13.277	14.860
5	0.412	0.554	0.831	1.145	1.610	9.236	11.071	12.833	15.086	16.750
6	0.676	0.872	1.237	1.635	2.204	10.645	12.592	14.449	16.812	18.548
7	0.989	1.239	1.690	2.167	2.833	12.017	14.067	16.013	18.475	20.278
8	1.344	1.646	2.180	2.733	3.490	13.362	15.507	17.535	20.090	21.955
9	1.735	2.088	2.700	3.325	4.168	14.684	16.919	19.023	21.666	23.589
10	2.156	2.558	3.247	3.940	4.865	15.987	18.307	20.483	23.209	25.188
11	2.603	3.053	3.816	4.575	5.578	17.275	19.675	21.920	24.725	26.757
12	3.074	3.571	4.404	5.226	6.304	18.549	21.026	23.337	26.217	28.299
13	3.565	4.107	5.009	5.892	7.042	19.812	22.362	24.736	27.688	29.819
14	4.075	4.660	5.629	6.571	7.790	21.064	23.685	26.119	29.141	31.319
15	4.601	5.229	6.262	7.261	8.547	22.307	24.996	27.488	30.578	32.801
16	5.142	5.812	6.908	7.962	9.312	23.542	26.296	28.845	32.000	34.267
17	5.697	6.408	7.564	8.672	10.085	24.769	27.587	30.191	33.409	35.718
18	6.265	7.015	8.231	9.390	10.865	25.989	28.869	31.526	34.805	37.156
19	6.844	7.633	8.907	10.117	11.651	27.204	30.144	32.852	36.191	38.582
20	7.434	8.260	9.591	10.851	12.443	28.412	31.410	34.170	37.566	39.997
21	8.034	8.897	10.283	11.591	13.240	29.615	32.671	35.479	38.932	41.401
22	8.643	9.542	10.982	12.338	14.042	30.813	33.924	36.781	40.289	42.796
23	9.260	10.196	11.689	13.091	14.848	32.007	35.172	38.076	41.638	44.181
24	9.886	10.856	12.401	13.848	15.659	33.196	36.415	39.364	42.980	45.559
25	10.520	11.524	13.120	14.611	16.473	34.382	37.652	40.646	44.314	46.928
26	11.160	12.198	13.844	15.379	17.292	35.563	38.885	41.923	45.642	48.290
27	11.808	12.879	14.573	16.151	18.114	36.741	40.113	43.194	46.963	49.645
28	12.461	13.565	15.308	16.928	18.939	37.916	41.337	44.461	48.278	50.993
29	13.121	14.257	16.047	17.708	19.768	39.087	42.557	45.722	49.588	52.336
30	13.787	14.954	16.791	18.493	20.599	40.256	43.773	46.979	50.892	53.672
40	20.707	22.164	24.433	26.509	29.051	51.805	55.758	59.342	63.691	66.766
50	27.991	29.707	32.357	34.764	37.689	63.167	67.505	71.420	76.154	79.490
60	35.534	37.485	40.482	43.188	46.459	74.397	79.082	83.298	88.379	91.952
70	43.275	45.442	48.758	51.739	55.329	85.527	90.531	95.023	100.425	104.215
80	51.172	53.540	57.153	60.391	64.278	96.578	101.879	106.629	112.329	116.321
90	59.196	61.754	65.647	69.126	73.291	107.565	113.145	118.136	124.116	128.299
100	67.328	70.065	74.222	77.929	82.358	118.498	124.342	129.561	135.807	140.169

From Donald B. Owen, *Handbook of Statistical Tables*, © 1962 Addison-Wesley Publishing Co., Reading, MA. Reprinted with permission of the publisher.

Degrees of Freedom

$n - 1$ for confidence intervals or hypothesis tests with a standard deviation or variance
$k - 1$ for multinomial experiments or goodness-of-fit with k categories
$(r - 1)(c - 1)$ for contingency tables with r rows and c columns
$k - 1$ for Kruskal-Wallis test with k samples

0.025

TABLE A-5 F Distribution ($\alpha = 0.025$ in the right tail)

df₂ \ df₁	1	2	3	4	5	6	7	8	9
1	647.79	799.50	864.16	899.58	921.85	937.11	948.22	956.66	963.28
2	38.506	39.000	39.165	39.248	39.298	39.331	39.335	39.373	39.387
3	17.443	16.044	15.439	15.101	14.885	14.735	14.624	14.540	14.473
4	12.218	10.649	9.9792	9.6045	9.3645	9.1973	9.0741	8.9796	8.9047
5	10.007	8.4336	7.7636	7.3879	7.1464	6.9777	6.8531	6.7572	6.6811
6	8.8131	7.2599	6.5988	6.2272	5.9876	5.8198	5.6955	5.5996	5.5234
7	8.0727	6.5415	5.8898	5.5226	5.2852	5.1186	4.9949	4.8993	4.8232
8	7.5709	6.0595	5.4160	5.0526	4.8173	4.6517	4.5286	4.4333	4.3572
9	7.2093	5.7147	5.0781	4.7181	4.4844	4.3197	4.1970	4.1020	4.0260
10	6.9367	5.4564	4.8256	4.4683	4.2361	4.0721	3.9498	3.8549	3.7790
11	6.7241	5.2559	4.6300	4.2751	4.0440	3.8807	3.7586	3.6638	3.5879
12	6.5538	5.0959	4.4742	4.1212	3.8911	3.7283	3.6065	3.5118	3.4358
13	6.4143	4.9653	4.3472	3.9959	3.7667	3.6043	3.4827	3.3880	3.3120
14	6.2979	4.8567	4.2417	3.8919	3.6634	3.5014	3.3799	3.2853	3.2093
15	6.1995	4.7650	4.1528	3.8043	3.5764	3.4147	3.2934	3.1987	3.1227
16	6.1151	4.6867	4.0768	3.7294	3.5021	3.3406	3.2194	3.1248	3.0488
17	6.0420	4.6189	4.0112	3.6648	3.4379	3.2767	3.1556	3.0610	2.9849
18	5.9781	4.5597	3.9539	3.6083	3.3820	3.2209	3.0999	3.0053	2.9291
19	5.9216	4.5075	3.9034	3.5587	3.3327	3.1718	3.0509	2.9563	2.8801
20	5.8715	4.4613	3.8587	3.5147	3.2891	3.1283	3.0074	2.9128	2.8365
21	5.8266	4.4199	3.8188	3.4754	3.2501	3.0895	2.9686	2.8740	2.7977
22	5.7863	4.3828	3.7829	3.4401	3.2151	3.0546	2.9338	2.8392	2.7628
23	5.7498	4.3492	3.7505	3.4083	3.1835	3.0232	2.9023	2.8077	2.7313
24	5.7166	4.3187	3.7211	3.3794	3.1548	2.9946	2.8738	2.7791	2.7027
25	5.6864	4.2909	3.6943	3.3530	3.1287	2.9685	2.8478	2.7531	2.6766
26	5.6586	4.2655	3.6697	3.3289	3.1048	2.9447	2.8240	2.7293	2.6528
27	5.6331	4.2421	3.6472	3.3067	3.0828	2.9228	2.8021	2.7074	2.6309
28	5.6096	4.2205	3.6264	3.2863	3.0626	2.9027	2.7820	2.6872	2.6106
29	5.5878	4.2006	3.6072	3.2674	3.0438	2.8840	2.7633	2.6686	2.5919
30	5.5675	4.1821	3.5894	3.2499	3.0265	2.8667	2.7460	2.6513	2.5746
40	5.4239	4.0510	3.4633	3.1261	2.9037	2.7444	2.6238	2.5289	2.4519
60	5.2856	3.9253	3.3425	3.0077	2.7863	2.6274	2.5068	2.4117	2.3344
120	5.1523	3.8046	3.2269	2.8943	2.6740	2.5154	2.3948	2.2994	2.2217
∞	5.0239	3.6889	3.1161	2.7858	2.5665	2.4082	2.2875	2.1918	2.1136

Numerator degrees of freedom (df₁)

Denominator degrees of freedom (df₂)

TABLE A-5   F Distribution ($\alpha = 0.025$ in the right tail) *(continued)*

df₂	\multicolumn{10}{c}{Numerator degrees of freedom (df₁)}									
Denominator degrees of freedom (df₂)	10	12	15	20	24	30	40	60	120	∞
1	968.63	976.71	984.87	993.10	997.25	1001.4	1005.6	1009.8	1014.0	1018.3
2	39.398	39.415	39.431	39.448	39.456	39.465	39.473	39.481	39.490	39.498
3	14.419	14.337	14.253	14.167	14.124	14.081	14.037	13.992	13.947	13.902
4	8.8439	8.7512	8.6565	8.5599	8.5109	8.4613	8.4111	8.3604	8.3092	8.2573
5	6.6192	6.5245	6.4277	6.3286	6.2780	6.2269	6.1750	6.1225	6.0693	6.0153
6	5.4613	5.3662	5.2687	5.1684	5.1172	5.0652	5.0125	4.9589	4.9044	4.8491
7	4.7611	4.6658	4.5678	4.4667	4.4150	4.3624	4.3089	4.2544	4.1989	4.1423
8	4.2951	4.1997	4.1012	3.9995	3.9472	3.8940	3.8398	3.7844	3.7279	3.6702
9	3.9639	3.8682	3.7694	3.6669	3.6142	3.5604	3.5055	3.4493	3.3918	3.3329
10	3.7168	3.6209	3.5217	3.4185	3.3654	3.3110	3.2554	3.1984	3.1399	3.0798
11	3.5257	3.4296	3.3299	3.2261	3.1725	3.1176	3.0613	3.0035	2.9441	2.8828
12	3.3736	3.2773	3.1772	3.0728	3.0187	2.9633	2.9063	2.8478	2.7874	2.7249
13	3.2497	3.1532	3.0527	2.9477	2.8932	2.8372	2.7797	2.7204	2.6590	2.5955
14	3.1469	3.0502	2.9493	2.8437	2.7888	2.7324	2.6742	2.6142	2.5519	2.4872
15	3.0602	2.9633	2.8621	2.7559	2.7006	2.6437	2.5850	2.5242	2.4611	2.3953
16	2.9862	2.8890	2.7875	2.6808	2.6252	2.5678	2.5085	2.4471	2.3831	2.3163
17	2.9222	2.8249	2.7230	2.6158	2.5598	2.5020	2.4422	2.3801	2.3153	2.2474
18	2.8664	2.7689	2.6667	2.5590	2.5027	2.4445	2.3842	2.3214	2.2558	2.1869
19	2.8172	2.7196	2.6171	2.5089	2.4523	2.3937	2.3329	2.2696	2.2032	2.1333
20	2.7737	2.6758	2.5731	2.4645	2.4076	2.3486	2.2873	2.2234	2.1562	2.0853
21	2.7348	2.6368	2.5338	2.4247	2.3675	2.3082	2.2465	2.1819	2.1141	2.0422
22	2.6998	2.6017	2.4984	2.3890	2.3315	2.2718	2.2097	2.1446	2.0760	2.0032
23	2.6682	2.5699	2.4665	2.3567	2.2989	2.2389	2.1763	2.1107	2.0415	1.9677
24	2.6396	2.5411	2.4374	2.3273	2.2693	2.2090	2.1460	2.0799	2.0099	1.9353
25	2.6135	2.5149	2.4110	2.3005	2.2422	2.1816	2.1183	2.0516	1.9811	1.9055
26	2.5896	2.4908	2.3867	2.2759	2.2174	2.1565	2.0928	2.0257	1.9545	1.8781
27	2.5676	2.4688	2.3644	2.2533	2.1946	2.1334	2.0693	2.0018	1.9299	1.8527
28	2.5473	2.4484	2.3438	2.2324	2.1735	2.1121	2.0477	1.9797	1.9072	1.8291
29	2.5286	2.4295	2.3248	2.2131	2.1540	2.0923	2.0276	1.9591	1.8861	1.8072
30	2.5112	2.4120	2.3072	2.1952	2.1359	2.0739	2.0089	1.9400	1.8664	1.7867
40	2.3882	2.2882	2.1819	2.0677	2.0069	1.9429	1.8752	1.8028	1.7242	1.6371
60	2.2702	2.1692	2.0613	1.9445	1.8817	1.8152	1.7440	1.6668	1.5810	1.4821
120	2.1570	2.0548	1.9450	1.8249	1.7597	1.6899	1.6141	1.5299	1.4327	1.3104
∞	2.0483	1.9447	1.8326	1.7085	1.6402	1.5660	1.4835	1.3883	1.2684	1.0000

From Maxine Merrington and Catherine M. Thompson, "Tables of Percentage Points of the Inverted Beta (F) Distribution," *Biometrika 33* (1943): 80–84. Reproduced with permission of the Biometrika Trustees.

(continued)

TABLE A-5 F Distribution ($\alpha = 0.05$ in the right tail)

	Numerator degrees of freedom (df$_1$)								
Denominator degrees of freedom (df$_2$)	1	2	3	4	5	6	7	8	9
1	161.45	199.50	215.71	224.58	230.16	233.99	236.77	238.88	240.54
2	18.513	19.000	19.164	19.247	19.296	19.330	19.353	19.371	19.385
3	10.128	9.5521	9.2766	9.1172	9.0135	8.9406	8.8867	8.8452	8.8123
4	7.7086	6.9443	6.5914	6.3882	6.2561	6.1631	6.0942	6.0410	6.9988
5	6.6079	5.7861	5.4095	5.1922	5.0503	4.9503	4.8759	4.8183	4.7725
6	5.9874	5.1433	4.7571	4.5337	4.3874	4.2839	4.2067	4.1468	4.0990
7	5.5914	4.7374	4.3468	4.1203	3.9715	3.8660	3.7870	3.7257	3.6767
8	5.3177	4.4590	4.0662	3.8379	3.6875	3.5806	3.5005	3.4381	3.3881
9	5.1174	4.2565	3.8625	3.6331	3.4817	3.3738	3.2927	3.2296	3.1789
10	4.9646	4.1028	3.7083	3.4780	3.3258	3.2172	3.1355	3.0717	3.0204
11	4.8443	3.9823	3.5874	3.3567	3.2039	3.0946	3.0123	2.9480	2.8962
12	4.7472	3.8853	3.4903	3.2592	3.1059	2.9961	2.9134	2.8486	2.7964
13	4.6672	3.8056	3.4105	3.1791	3.0254	2.9153	2.8321	2.7669	2.7144
14	4.6001	3.7389	3.3439	3.1122	2.9582	2.8477	2.7642	2.6987	2.6458
15	4.5431	3.6823	3.2874	3.0556	2.9013	2.7905	2.7066	2.6408	2.5876
16	4.4940	3.6337	3.2389	3.0069	2.8524	2.7413	2.6572	2.5911	2.5377
17	4.4513	3.5915	3.1968	2.9647	2.8100	2.6987	2.6143	2.5480	2.4943
18	4.4139	3.5546	3.1599	2.9277	2.7729	2.6613	2.5767	2.5102	2.4563
19	4.3807	3.5219	3.1274	2.8951	2.7401	2.6283	2.5435	2.4768	2.4227
20	4.3512	3.4928	3.0984	2.8661	2.7109	2.5990	2.5140	2.4471	2.3928
21	4.3248	3.4668	3.0725	2.8401	2.6848	2.5727	2.4876	2.4205	2.3660
22	4.3009	3.4434	3.0491	2.8167	2.6613	2.5491	2.4638	2.3965	2.3419
23	4.2793	3.4221	3.0280	2.7955	2.6400	2.5277	2.4422	2.3748	2.3201
24	4.2597	3.4028	3.0088	2.7763	2.6207	2.5082	2.4226	2.3551	2.3002
25	4.2417	3.3852	2.9912	2.7587	2.6030	2.4904	2.4047	2.3371	2.2821
26	4.2252	3.3690	2.9752	2.7426	2.5868	2.4741	2.3883	2.3205	2.2655
27	4.2100	3.3541	2.9604	2.7278	2.5719	2.4591	2.3732	2.3053	2.2501
28	4.1960	3.3404	2.9467	2.7141	2.5581	2.4453	2.3593	2.2913	2.2360
29	4.1830	3.3277	2.9340	2.7014	2.5454	2.4324	2.3463	2.2783	2.2229
30	4.1709	3.3158	2.9223	2.6896	2.5336	2.4205	2.3343	2.2662	2.2107
40	4.0847	3.2317	2.8387	2.6060	2.4495	2.3359	2.2490	2.1802	2.1240
60	4.0012	3.1504	2.7581	2.5252	2.3683	2.2541	2.1665	2.0970	2.0401
120	3.9201	3.0718	2.6802	2.4472	2.2899	2.1750	2.0868	2.0164	1.9588
∞	3.8415	2.9957	2.6049	2.3719	2.2141	2.0986	2.0096	1.9384	1.8799

(continued)

TABLE A-5 F Distribution ($\alpha = 0.05$ in the right tail) *(continued)*

Numerator degrees of freedom (df_1)

df_2	10	12	15	20	24	30	40	60	120	∞
1	241.88	243.91	245.95	248.01	249.05	250.10	251.14	252.20	253.25	254.31
2	19.396	19.413	19.429	19.446	19.454	19.462	19.471	19.479	19.487	19.496
3	8.7855	8.7446	8.7029	8.6602	8.6385	8.6166	8.5944	8.5720	8.5494	8.5264
4	5.9644	5.9117	5.8578	5.8025	5.7744	5.7459	5.7170	5.6877	5.6581	5.6281
5	4.7351	4.6777	4.6188	4.5581	4.5272	4.4957	4.4638	4.4314	4.3985	4.3650
6	4.0600	3.9999	3.9381	3.8742	3.8415	3.8082	3.7743	3.7398	3.7047	3.6689
7	3.6365	3.5747	3.5107	3.4445	3.4105	3.3758	3.3404	3.3043	3.2674	3.2298
8	3.3472	3.2839	3.2184	3.1503	3.1152	3.0794	3.0428	3.0053	2.9669	2.9276
9	3.1373	3.0729	3.0061	2.9365	2.9005	2.8637	2.8259	2.7872	2.7475	2.7067
10	2.9782	2.9130	2.8450	2.7740	2.7372	2.6996	2.6609	2.6211	2.5801	2.5379
11	2.8536	2.7876	2.7186	2.6464	2.6090	2.5705	2.5309	2.4901	2.4480	2.4045
12	2.7534	2.6866	2.6169	2.5436	2.5055	2.4663	2.4259	2.3842	2.3410	2.2962
13	2.6710	2.6037	2.5331	2.4589	2.4202	2.3803	2.3392	2.2966	2.2524	2.2064
14	2.6022	2.5342	2.4630	2.3879	2.3487	2.3082	2.2664	2.2229	2.1778	2.1307
15	2.5437	2.4753	2.4034	2.3275	2.2878	2.2468	2.2043	2.1601	2.1141	2.0658
16	2.4935	2.4247	2.3522	2.2756	2.2354	2.1938	2.1507	2.1058	2.0589	2.0096
17	2.4499	2.3807	2.3077	2.2304	2.1898	2.1477	2.1040	2.0584	2.0107	1.9604
18	2.4117	2.3421	2.2686	2.1906	2.1497	2.1071	2.0629	2.0166	1.9681	1.9168
19	2.3779	2.3080	2.2341	2.1555	2.1141	2.0712	2.0264	1.9795	1.9302	1.8780
20	2.3479	2.2776	2.2033	2.1242	2.0825	2.0391	1.9938	1.9464	1.8963	1.8432
21	2.3210	2.2504	2.1757	2.0960	2.0540	2.0102	1.9645	1.9165	1.8657	1.8117
22	2.2967	2.2258	2.1508	2.0707	2.0283	1.9842	1.9380	1.8894	1.8380	1.7831
23	2.2747	2.2036	2.1282	2.0476	2.0050	1.9605	1.9139	1.8648	1.8128	1.7570
24	2.2547	2.1834	2.1077	2.0267	1.9838	1.9390	1.8920	1.8424	1.7896	1.7330
25	2.2365	2.1649	2.0889	2.0075	1.9643	1.9192	1.8718	1.8217	1.7684	1.7110
26	2.2197	2.1479	2.0716	1.9898	1.9464	1.9010	1.8533	1.8027	1.7488	1.6906
27	2.2043	2.1323	2.0558	1.9736	1.9299	1.8842	1.8361	1.7851	1.7306	1.6717
28	2.1900	2.1179	2.0411	1.9586	1.9147	1.8687	1.8203	1.7689	1.7138	1.6541
29	2.1768	2.1045	2.0275	1.9446	1.9005	1.8543	1.8055	1.7537	1.6981	1.6376
30	2.1646	2.0921	2.0148	1.9317	1.8874	1.8409	1.7918	1.7396	1.6835	1.6223
40	2.0772	2.0035	1.9245	1.8389	1.7929	1.7444	1.6928	1.6373	1.5766	1.5089
60	1.9926	1.9174	1.8364	1.7480	1.7001	1.6491	1.5943	1.5343	1.4673	1.3893
120	1.9105	1.8337	1.7505	1.6587	1.6084	1.5543	1.4952	1.4290	1.3519	1.2539
∞	1.8307	1.7522	1.6664	1.5705	1.5173	1.4591	1.3940	1.3180	1.2214	1.0000

Denominator degrees of freedom (df_2)

From Maxine Merrington and Catherine M. Thompson, "Tables of Percentage Points of the Inverted Beta (*F*) Distribution," *Biometrika 33* (1943): 80–84. Reproduced with permission of the Biometrika Trustees.

TABLE A-6	Critical Values of the Pearson Correlation Coefficient r	
n	$\alpha = .05$	$\alpha = .01$
4	.950	.999
5	.878	.959
6	.811	.917
7	.754	.875
8	.707	.834
9	.666	.798
10	.632	.765
11	.602	.735
12	.576	.708
13	.553	.684
14	.532	.661
15	.514	.641
16	.497	.623
17	.482	.606
18	.468	.590
19	.456	.575
20	.444	.561
25	.396	.505
30	.361	.463
35	.335	.430
40	.312	.402
45	.294	.378
50	.279	.361
60	.254	.330
70	.236	.305
80	.220	.286
90	.207	.269
100	.196	.256

NOTE: To test $H_0: \rho = 0$ against $H_1: \rho \neq 0$, reject H_0 if the absolute value of r is greater than the critical value in the table.

TABLE A-7   Critical Values for the Sign Test

	α			
n	.005 (one tail) .01 (two tails)	.01 (one tail) .02 (two tails)	.025 (one tail) .05 (two tails)	.05 (one tail) .10 (two tails)
1	*	*	*	*
2	*	*	*	*
3	*	*	*	*
4	*	*	*	*
5	*	*	*	0
6	*	*	0	0
7	*	0	0	0
8	0	0	0	1
9	0	0	1	1
10	0	0	1	1
11	0	1	1	2
12	1	1	2	2
13	1	1	2	3
14	1	2	2	3
15	2	2	3	3
16	2	2	3	4
17	2	3	4	4
18	3	3	4	5
19	3	4	4	5
20	3	4	5	5
21	4	4	5	6
22	4	5	5	6
23	4	5	6	7
24	5	5	6	7
25	5	6	7	7

NOTES:

1. * indicates that it is not possible to get a value in the critical region.

2. Reject the null hypothesis if the number of the less frequent sign (x) is less than or equal to the value in the table.

3. For values of n greater than 25, a normal approximation is used with

$$z = \frac{(x + 0.5) - \left(\frac{n}{2}\right)}{\frac{\sqrt{n}}{2}}$$

TABLE A-8	Critical Values of T for the Wilcoxon Signed-Ranks Test			
	α			
n	.005 (one tail) .01 (two tails)	.01 (one tail) .02 (two tails)	.025 (one tail) .05 (two tails)	.05 (one tail) .10 (two tails)
5	*	*	*	1
6	*	*	1	2
7	*	0	2	4
8	0	2	4	6
9	2	3	6	8
10	3	5	8	11
11	5	7	11	14
12	7	10	14	17
13	10	13	17	21
14	13	16	21	26
15	16	20	25	30
16	19	24	30	36
17	23	28	35	41
18	28	33	40	47
19	32	38	46	54
20	37	43	52	60
21	43	49	59	68
22	49	56	66	75
23	55	62	73	83
24	61	69	81	92
25	68	77	90	101
26	76	85	98	110
27	84	93	107	120
28	92	102	117	130
29	100	111	127	141
30	109	120	137	152

NOTES:

1. * indicates that it is not possible to get a value in the critical region.

2. Reject the null hypothesis if the test statistic T is less than or equal to the critical value found in this table. Fail to reject the null hypothesis if the test statistic T is greater than the critical value found in the table.

From *Some Rapid Approximate Statistical Procedures,* Copyright © 1949, 1964 Lederle Laboratories Division of American Cyanamid Company. Reprinted with the permission of the American Cyanamid Company.

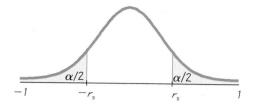

TABLE A-9	Critical Values of Spearman's Rank Correlation Coefficient r_s			
n	$\alpha = 0.10$	$\alpha = 0.05$	$\alpha = 0.02$	$\alpha = 0.01$
5	.900	—	—	—
6	.829	.886	.943	—
7	.714	.786	.893	.929
8	.643	.738	.833	.881
9	.600	.700	.783	.833
10	.564	.648	.745	.794
11	.536	.618	.709	.755
12	.503	.587	.678	.727
13	.484	.560	.648	.703
14	.464	.538	.626	.679
15	.446	.521	.604	.654
16	.429	.503	.582	.635
17	.414	.485	.566	.615
18	.401	.472	.550	.600
19	.391	.460	.535	.584
20	.380	.447	.520	.570
21	.370	.435	.508	.556
22	.361	.425	.496	.544
23	.353	.415	.486	.532
24	.344	.406	.476	.521
25	.337	.398	.466	.511
26	.331	.390	.457	.501
27	.324	.382	.448	.491
28	.317	.375	.440	.483
29	.312	.368	.433	.475
30	.306	.362	.425	.467

NOTES: For $n > 30$, use $r_s = \pm z/\sqrt{n - 1}$, where z corresponds to the level of significance.

For example, if $\alpha = 0.05$, then $z = 1.96$.

If the absolute value of the test statistic r_s exceeds the positive critical value, then reject $H_0: \rho_s = 0$ and conclude that there is a correlation.

Based on data from "*Biostatistical Analysis,* 4th edition," © 1999, by Jerrold Zar, Prentice Hall, Inc., Upper Saddle River, New Jersey.

Appendix B Data Sets

Data Set 1   Health Exam Results
Data Set 2   Body Temperatures of Healthy Adults
Data Set 3   Parent/Child Heights
Data Set 4   Head Circumferences
Data Set 5   Passive and Active Smoke
Data Set 6   Bears
Data Set 7   Iris Measurements
Data Set 8   Cuckoo Egg Lengths
Data Set 9   Poplar Tree Weights
Data Set 10   Blood Measurements
Data Set 11   Counts of Yeast Cells
Data Set 12   Systolic Blood Pressure Measurements

Data Set 1: Health Exam Results

AGE is in years, HT is height (inches), WT is weight (pounds), WAIST is circumference (cm), PULSE is pulse rate (beats per minute), SYS is systolic blood pressure (mmHg), DIAS is diastolic blood pressure (mmHg), CHOL is cholesterol (mg), BMI is body mass index, LEG is upper leg length (cm), ELBOW is elbow breadth (cm), WRIST is wrist breadth (cm), and ARM is arm circumference (cm). Data are from the U.S. Department of Health and Human Services, National Center for Health Statistics, Third National Health and Nutrition Examination Survey.

 Text file names for males are MAGE, MHT, MWT, MWAST, MPULS, MSYS, MDIAS, MCHOL, MBMI, MLEG, MELBW, MWRST, MARM.

Male	Age	HT	WT	Waist	Pulse	SYS	DIAS	CHOL	BMI	Leg	Elbow	Wrist	Arm
	58	70.8	169.1	90.6	68	125	78	522	23.8	42.5	7.7	6.4	31.9
	22	66.2	144.2	78.1	64	107	54	127	23.2	40.2	7.6	6.2	31.0
	32	71.7	179.3	96.5	88	126	81	740	24.6	44.4	7.3	5.8	32.7
	31	68.7	175.8	87.7	72	110	68	49	26.2	42.8	7.5	5.9	33.4
	28	67.6	152.6	87.1	64	110	66	230	23.5	40.0	7.1	6.0	30.1
	46	69.2	166.8	92.4	72	107	83	316	24.5	47.3	7.1	5.8	30.5
	41	66.5	135.0	78.8	60	113	71	590	21.5	43.4	6.5	5.2	27.6
	56	67.2	201.5	103.3	88	126	72	466	31.4	40.1	7.5	5.6	38.0
	20	68.3	175.2	89.1	76	137	85	121	26.4	42.1	7.5	5.5	32.0
	54	65.6	139.0	82.5	60	110	71	578	22.7	36.0	6.9	5.5	29.3
	17	63.0	156.3	86.7	96	109	65	78	27.8	44.2	7.1	5.3	31.7
	73	68.3	186.6	103.3	72	153	87	265	28.1	36.7	8.1	6.7	30.7
	52	73.1	191.1	91.8	56	112	77	250	25.2	48.4	8.0	5.2	34.7
	25	67.6	151.3	75.6	64	119	81	265	23.3	41.0	7.0	5.7	30.6
	29	68.0	209.4	105.5	60	113	82	273	31.9	39.8	6.9	6.0	34.2
	17	71.0	237.1	108.7	64	125	76	272	33.1	45.2	8.3	6.6	41.1
	41	61.3	176.7	104.0	84	131	80	972	33.2	40.2	6.7	5.7	33.1
	52	76.2	220.6	103.0	76	121	75	75	26.7	46.2	7.9	6.0	32.2
	32	66.3	166.1	91.3	84	132	81	138	26.6	39.0	7.5	5.7	31.2
	20	69.7	137.4	75.2	88	112	44	139	19.9	44.8	6.9	5.6	25.9
	20	65.4	164.2	87.7	72	121	65	638	27.1	40.9	7.0	5.6	33.7
	29	70.0	162.4	77.0	56	116	64	613	23.4	43.1	7.5	5.2	30.3
	18	62.9	151.8	85.0	68	95	58	762	27.0	38.0	7.4	5.8	32.8
	26	68.5	144.1	79.6	64	110	70	303	21.6	41.0	6.8	5.7	31.0
	33	68.3	204.6	103.8	60	110	66	690	30.9	46.0	7.4	6.1	36.2
	55	69.4	193.8	103.0	68	125	82	31	28.3	41.4	7.2	6.0	33.6
	53	69.2	172.9	97.1	60	124	79	189	25.5	42.7	6.6	5.9	31.9
	28	68.0	161.9	86.9	60	131	69	957	24.6	40.5	7.3	5.7	32.9
	28	71.9	174.8	88.0	56	109	64	339	23.8	44.2	7.8	6.0	30.9
	37	66.1	169.8	91.5	84	112	79	416	27.4	41.8	7.0	6.1	34.0
	40	72.4	213.3	102.9	72	127	72	120	28.7	47.2	7.5	5.9	34.8
	33	73.0	198.0	93.1	84	132	74	702	26.2	48.2	7.8	6.0	33.6
	26	68.0	173.3	98.9	88	116	81	1252	26.4	42.9	6.7	5.8	31.3
	53	68.7	214.5	107.5	56	125	84	288	32.1	42.8	8.2	5.9	37.6
	36	70.3	137.1	81.6	64	112	77	176	19.6	40.8	7.1	5.3	27.9
	34	63.7	119.5	75.7	56	125	77	277	20.7	42.6	6.6	5.3	26.9
	42	71.1	189.1	95.0	56	120	83	649	26.3	44.9	7.4	6.0	36.9
	18	65.6	164.7	91.1	60	118	68	113	26.9	41.1	7.0	6.1	34.5
	44	68.3	170.1	94.9	64	115	75	656	25.6	44.5	7.3	5.8	32.1
	20	66.3	151.0	79.9	72	115	65	172	24.2	44.0	7.1	5.4	30.7

(continued)

Data Set 1: Health Exam Results (*Continued*)

Text file names for females are FAGE, FHT, FWT, FWAST, FPULS, FSYS, FDIAS, FCHOL, FBMI, FLEG, FELBW, FWRST, FARM.

Female	Age	HT	WT	Waist	Pulse	SYS	DIAS	CHOL	BMI	Leg	Elbow	Wrist	Arm
	17	64.3	114.8	67.2	76	104	61	264	19.6	41.6	6.0	4.6	23.6
	32	66.4	149.3	82.5	72	99	64	181	23.8	42.8	6.7	5.5	26.3
	25	62.3	107.8	66.7	88	102	65	267	19.6	39.0	5.7	4.6	26.3
	55	62.3	160.1	93.0	60	114	76	384	29.1	40.2	6.2	5.0	32.6
	27	59.6	127.1	82.6	72	94	58	98	25.2	36.2	5.5	4.8	29.2
	29	63.6	123.1	75.4	68	101	66	62	21.4	43.2	6.0	4.9	26.4
	25	59.8	111.7	73.6	80	108	61	126	22.0	38.7	5.7	5.1	27.9
	12	63.3	156.3	81.4	64	104	41	89	27.5	41.0	6.8	5.5	33.0
	41	67.9	218.8	99.4	68	123	72	531	33.5	43.8	7.8	5.8	38.6
	32	61.4	110.2	67.7	68	93	61	130	20.6	37.3	6.3	5.0	26.5
	31	66.7	188.3	100.7	80	89	56	175	29.9	42.3	6.6	5.2	34.4
	19	64.8	105.4	72.9	76	112	62	44	17.7	39.1	5.7	4.8	23.7
	19	63.1	136.1	85.0	68	107	48	8	24.0	40.3	6.6	5.1	28.4
	23	66.7	182.4	85.7	72	116	62	112	28.9	48.6	7.2	5.6	34.0
	40	66.8	238.4	126.0	96	181	102	462	37.7	33.2	7.0	5.4	35.2
	23	64.7	108.8	74.5	72	98	61	62	18.3	43.4	6.2	5.2	24.7
	27	65.1	119.0	74.5	68	100	53	98	19.8	41.5	6.3	5.3	27.0
	45	61.9	161.9	94.0	72	127	74	447	29.8	40.0	6.8	5.0	35.0
	41	64.3	174.1	92.8	64	107	67	125	29.7	38.2	6.8	4.7	33.1
	56	63.4	181.2	105.5	80	116	71	318	31.7	38.2	6.9	5.4	39.6
	22	60.7	124.3	75.5	64	97	64	325	23.8	38.2	5.9	5.0	27.0
	57	63.4	255.9	126.5	80	155	85	600	44.9	41.0	8.0	5.6	43.8
	24	62.6	106.7	70.0	76	106	59	237	19.2	38.1	6.1	5.0	23.6
	37	60.6	149.9	98.0	76	110	70	173	28.7	38.0	7.0	5.1	34.3
	59	63.5	163.1	104.7	76	105	69	309	28.5	36.0	6.7	5.1	34.4
	40	58.6	94.3	67.8	80	118	82	94	19.3	32.1	5.4	4.2	23.3
	45	60.2	159.7	99.3	104	133	83	280	31.0	31.1	6.4	5.2	35.6
	52	67.6	162.8	91.1	88	113	75	254	25.1	39.4	7.1	5.3	31.8
	31	63.4	130.0	74.5	60	113	66	123	22.8	40.2	5.9	5.1	27.0
	32	64.1	179.9	95.5	76	107	67	596	30.9	39.2	6.2	5.0	32.8
	23	62.7	147.8	79.5	72	95	59	301	26.5	39.0	6.3	4.9	31.0
	23	61.3	112.9	69.1	72	108	72	223	21.2	36.6	5.9	4.7	27.0
	47	58.2	195.6	105.5	88	114	79	293	40.6	27.0	7.5	5.5	41.2
	36	63.2	124.2	78.8	80	104	73	146	21.9	38.5	5.6	4.7	25.5
	34	60.5	135.0	85.7	60	125	73	149	26.0	39.9	6.4	5.2	30.9
	37	65.0	141.4	92.8	72	124	85	149	23.5	37.5	6.1	4.8	27.9
	18	61.8	123.9	72.7	88	92	46	920	22.8	39.7	5.8	5.0	26.5
	29	68.0	135.5	75.9	88	119	81	271	20.7	39.0	6.3	4.9	27.8
	48	67.0	130.4	68.6	124	93	64	207	20.5	41.6	6.0	5.3	23.0
	16	57.0	100.7	68.7	64	106	64	2	21.9	33.8	5.6	4.6	26.4

Data Set 2: Body Temperatures (in degrees Fahrenheit) of Healthy Adults

Data provided by Dr. Steven Wasserman, Dr. Philip Mackowiak, and Dr. Myron Levine of the University of Maryland.

Text file name for the last column is TMPTR.

Subject	Age	Sex	Smoke	Temperature Day 1		Temperature Day 2	
				8 A.M.	12 A.M.	8 A.M.	12 A.M.
1	22	M	Y	98.0	98.0	98.0	98.6
2	23	M	Y	97.0	97.6	97.4	—
3	22	M	Y	98.6	98.8	97.8	98.6
4	19	M	N	97.4	98.0	97.0	98.0
5	18	M	N	98.2	98.8	97.0	98.0
6	20	M	Y	98.2	98.8	96.6	99.0
7	27	M	Y	98.2	97.6	97.0	98.4
8	19	M	Y	96.6	98.6	96.8	98.4
9	19	M	Y	97.4	98.6	96.6	98.4
10	24	M	N	97.4	98.8	96.6	98.4
11	35	M	Y	98.2	98.0	96.2	98.6
12	25	M	Y	97.4	98.2	97.6	98.6
13	25	M	N	97.8	98.0	98.6	98.8
14	35	M	Y	98.4	98.0	97.0	98.6
15	21	M	N	97.6	97.0	97.4	97.0
16	33	M	N	96.2	97.2	98.0	97.0
17	19	M	Y	98.0	98.2	97.6	98.8
18	24	M	Y	—	—	97.2	97.6
19	18	F	N	—	—	97.0	97.7
20	22	F	Y	—	—	98.0	98.8
21	20	M	Y	—	—	97.0	98.0
22	30	F	Y	—	—	96.4	98.0
23	29	M	N	—	—	96.1	98.3
24	18	M	Y	—	—	98.0	98.5
25	31	M	Y	—	98.1	96.8	97.3
26	28	F	Y	—	98.2	98.2	98.7
27	27	M	Y	—	98.5	97.8	97.4
28	21	M	Y	—	98.5	98.2	98.9
29	30	M	Y	—	99.0	97.8	98.6
30	27	M	N	—	98.0	99.0	99.5
31	32	M	Y	—	97.0	97.4	97.5
32	33	M	Y	—	97.3	97.4	97.3
33	23	M	Y	—	97.3	97.5	97.6
34	29	M	Y	—	98.1	97.8	98.2
35	25	M	Y	—	—	97.9	99.6
36	31	M	N	—	97.8	97.8	98.7
37	25	M	Y	—	99.0	98.3	99.4
38	28	M	N	—	97.6	98.0	98.2
39	30	M	Y	—	97.4	—	98.0
40	33	M	Y	—	98.0	—	98.6
41	28	M	Y	98.0	97.4	—	98.6
42	22	M	Y	98.8	98.0	—	97.2
43	21	F	Y	99.0	—	—	98.4
44	30	M	N	—	98.6	—	98.6
45	22	M	Y	—	98.6	—	98.2
46	22	F	N	98.0	98.4	—	98.0
47	20	M	Y	—	97.0	—	97.8
48	19	M	Y	—	—	—	98.0
49	33	M	N	—	98.4	—	98.4
50	31	M	Y	99.0	99.0	—	98.6
51	26	M	N	—	98.0	—	98.6
52	18	M	N	—	—	—	97.8
53	23	M	N	—	99.4	—	99.0
54	28	M	Y	—	—	—	96.5

(continued)

Data Set 2: Body Temperatures (*Continued*)

Subject	Age	Sex	Smoke	Temperature Day 1 8 A.M.	Temperature Day 1 12 A.M.	Temperature Day 2 8 A.M.	Temperature Day 2 12 A.M.
55	19	M	Y	—	97.8	—	97.6
56	21	M	N	—	—	—	98.0
57	27	M	Y	—	98.2	—	96.9
58	29	M	Y	—	99.2	—	97.6
59	38	M	N	—	99.0	—	97.1
60	29	F	Y	—	97.7	—	97.9
61	22	M	Y	—	98.2	—	98.4
62	22	M	Y	—	98.2	—	97.3
63	26	M	Y	—	98.8	—	98.0
64	32	M	N	—	98.1	—	97.5
65	25	M	Y	—	98.5	—	97.6
66	21	F	N	—	97.2	—	98.2
67	25	M	Y	—	98.5	—	98.5
68	24	M	Y	—	99.2	97.0	98.8
69	25	M	Y	—	98.3	97.6	98.7
70	35	M	Y	—	98.7	97.5	97.8
71	23	F	Y	—	98.8	98.8	98.0
72	31	M	Y	—	98.6	98.4	97.1
73	28	M	Y	—	98.0	98.2	97.4
74	29	M	Y	—	99.1	97.7	99.4
75	26	M	Y	—	97.2	97.3	98.4
76	32	M	N	—	97.6	97.5	98.6
77	32	M	Y	—	97.9	97.1	98.4
78	21	F	Y	—	98.8	98.6	98.5
79	20	M	Y	—	98.6	98.6	98.6
80	24	F	Y	—	98.6	97.8	98.3
81	21	F	Y	—	99.3	98.7	98.7
82	28	M	Y	—	97.8	97.9	98.8
83	27	F	N	98.8	98.7	97.8	99.1
84	28	M	N	99.4	99.3	97.8	98.6
85	29	M	Y	98.8	97.8	97.6	97.9
86	19	M	N	97.7	98.4	96.8	98.8
87	24	M	Y	99.0	97.7	96.0	98.0
88	29	M	N	98.1	98.3	98.0	98.7
89	25	M	Y	98.7	97.7	97.0	98.5
90	27	M	N	97.5	97.1	97.4	98.9
91	25	M	Y	98.9	98.4	97.6	98.4
92	21	M	Y	98.4	98.6	97.6	98.6
93	19	M	Y	97.2	97.4	96.2	97.1
94	27	M	Y	—	—	96.2	97.9
95	32	M	N	98.8	96.7	98.1	98.8
96	24	M	Y	97.3	96.9	97.1	98.7
97	32	M	Y	98.7	98.4	98.2	97.6
98	19	F	Y	98.9	98.2	96.4	98.2
99	18	F	Y	99.2	98.6	96.9	99.2
100	27	M	N	—	97.0	—	97.8
101	34	M	Y	—	97.4	—	98.0
102	25	M	N	—	98.4	—	98.4
103	18	M	N	—	97.4	—	97.8
104	32	M	Y	—	96.8	—	98.4
105	31	M	Y	—	98.2	—	97.4
106	26	M	N	—	97.4	—	98.0
107	23	M	N	—	98.0	—	97.0

Data Set 3: Parent/Child Heights (inches)

Data are from the U.S. Department of Health and Human Services, National Center for Health Statistics, Third National Health and Nutrition Examination Survey.

 Text file names: CHDHT, MOMHT, DADHT.

Gender	Height	Mother's Height	Father's Height
M	62.5	66	70
M	64.6	58	69
M	69.1	66	64
M	73.9	68	71
M	67.1	64	68
M	64.4	62	66
M	71.1	66	74
M	71.0	63	73
M	67.4	64	62
M	69.3	65	69
M	64.9	64	67
M	68.1	64	68
M	66.5	62	72
M	67.5	69	66
M	66.5	62	72
M	70.3	67	68
M	67.5	63	71
M	68.5	66	67
M	71.9	65	71
M	67.8	71	75
F	58.6	63	64
F	64.7	67	65
F	65.3	64	67
F	61.0	60	72
F	65.4	65	72
F	67.4	67	72
F	60.9	59	67
F	63.1	60	71
F	60.0	58	66
F	71.1	72	75
F	62.2	63	69
F	67.2	67	70
F	63.4	62	69
F	68.4	69	62
F	62.2	63	66
F	64.7	64	76
F	59.6	63	69
F	61.0	64	68
F	64.0	60	66
F	65.4	65	68

Data Set 4: Head Circumferences (cm) of Two-Month-Old Babies

Data are from the U.S. Department of Health and Human Services, National Center for Health Statistics, Third National Health and Nutrition Examination Survey.

 Text file names: MHED, FHED.

Male

40.1	39.8	42.3	41.0	42.5	40.9	35.5	35.7	41.1	41.4
42.2	42.3	43.2	42.2	42.4	43.2	39.9	40.9	40.7	41.7
41.7	41.0	40.4	42.0	41.2	39.7	41.9	41.3	40.2	41.0
41.1	40.4	39.2	42.8	41.9	42.8	41.0	40.9	42.0	42.6
41.0	39.6	40.2	40.9	40.2	41.8	41.7	41.7	40.9	42.8

Female

39.3	40.2	41.3	38.1	39.6	40.6	38.6	40.5	40.5	40.3
39.5	40.7	40.2	38.2	40.3	42.6	39.9	40.0	40.7	38.6
41.0	43.7	40.0	40.1	41.0	40.8	41.0	40.3	40.2	39.2
34.4	41.0	39.6	40.9	36.9	43.6	40.2	40.8	37.8	41.2
42.0	38.3	39.6	38.9	36.3	39.9	40.3	40.1	42.0	41.6

Data Set 5: Passive and Active Smoke

All values are measured levels of serum cotinine (in ng/ml), a metabolite
of nicotine. (When nicotine is absorbed by the body, cotinine is produced.)
Data are from the U.S. Department of Health and Human Services,
National Center for Health Statistics, Third National Health and Nutrition
Examination Survey.

 Text file names: NOETS, ETS, SMKR.

Smokers (subjects reported tobacco use)

1	0	131	173	265	210	44	277	32	3
35	112	477	289	227	103	222	149	313	491
130	234	164	198	17	253	87	121	266	290
123	167	250	245	48	86	284	1	208	173

ETS (nonsmokers exposed to environmental tobacco smoke at home or work)

384	0	69	19	1	0	178	2	13	1
4	0	543	17	1	0	51	0	197	3
0	3	1	45	13	3	1	1	1	0
0	551	2	1	1	1	0	74	1	241

NOETS (nonsmokers with no exposure to environmental tobacco smoke at home or work)

0	0	0	0	0	0	0	0	0	0
0	9	0	0	0	0	0	0	244	0
1	0	0	0	90	1	0	309	0	0
0	0	0	0	0	0	0	0	0	0

Data Set 6: Bears (wild bears anesthetized)

AGE is in months; MONTH is the month of measurement (1 = January); SEX is coded with 1 = male and 2 = female; HEADLEN is head length (inches); HEADWTH is width of head (inches); NECK is distance around neck (in inches); LENGTH is length of body (inches); CHEST is distance around chest (inches); and WEIGHT is measured in pounds. Data are from Gary Alt and Minitab, Inc.

Text file names: BAGE, BMNTH, BSEX, BHDLN, BHDWD, BNECK, BLEN, BCHST, BWGHT.

Age	Month	Sex	Headlen	Headwth	Neck	Length	Chest	Weight
19	7	1	11.0	5.5	16.0	53.0	26.0	80
55	7	1	16.5	9.0	28.0	67.5	45.0	344
81	9	1	15.5	8.0	31.0	72.0	54.0	416
115	7	1	17.0	10.0	31.5	72.0	49.0	348
104	8	2	15.5	6.5	22.0	62.0	35.0	166
100	4	2	13.0	7.0	21.0	70.0	41.0	220
56	7	1	15.0	7.5	26.5	73.5	41.0	262
51	4	1	13.5	8.0	27.0	68.5	49.0	360
57	9	2	13.5	7.0	20.0	64.0	38.0	204
53	5	2	12.5	6.0	18.0	58.0	31.0	144
68	8	1	16.0	9.0	29.0	73.0	44.0	332
8	8	1	9.0	4.5	13.0	37.0	19.0	34
44	8	2	12.5	4.5	10.5	63.0	32.0	140
32	8	1	14.0	5.0	21.5	67.0	37.0	180
20	8	2	11.5	5.0	17.5	52.0	29.0	105
32	8	1	13.0	8.0	21.5	59.0	33.0	166
45	9	1	13.5	7.0	24.0	64.0	39.0	204
9	9	2	9.0	4.5	12.0	36.0	19.0	26
21	9	1	13.0	6.0	19.0	59.0	30.0	120
177	9	1	16.0	9.5	30.0	72.0	48.0	436
57	9	2	12.5	5.0	19.0	57.5	32.0	125
81	9	2	13.0	5.0	20.0	61.0	33.0	132
21	9	1	13.0	5.0	17.0	54.0	28.0	90
9	9	1	10.0	4.0	13.0	40.0	23.0	40
45	9	1	16.0	6.0	24.0	63.0	42.0	220
9	9	1	10.0	4.0	13.5	43.0	23.0	46
33	9	1	13.5	6.0	22.0	66.5	34.0	154
57	9	2	13.0	5.5	17.5	60.5	31.0	116
45	9	2	13.0	6.5	21.0	60.0	34.5	182
21	9	1	14.5	5.5	20.0	61.0	34.0	150
10	10	1	9.5	4.5	16.0	40.0	26.0	65
82	10	2	13.5	6.5	28.0	64.0	48.0	356
70	10	2	14.5	6.5	26.0	65.0	48.0	316
10	10	1	11.0	5.0	17.0	49.0	29.0	94
10	10	1	11.5	5.0	17.0	47.0	29.5	86
34	10	1	13.0	7.0	21.0	59.0	35.0	150
34	10	1	16.5	6.5	27.0	72.0	44.5	270
34	10	1	14.0	5.5	24.0	65.0	39.0	202
58	10	2	13.5	6.5	21.5	63.0	40.0	202
58	10	1	15.5	7.0	28.0	70.5	50.0	365
11	11	1	11.5	6.0	16.5	48.0	31.0	79
23	11	1	12.0	6.5	19.0	50.0	38.0	148
70	10	1	15.5	7.0	28.0	76.5	55.0	446
11	11	2	9.0	5.0	15.0	46.0	27.0	62
83	11	2	14.5	7.0	23.0	61.5	44.0	236
35	11	1	13.5	8.5	23.0	63.5	44.0	212
16	4	1	10.0	4.0	15.5	48.0	26.0	60
16	4	1	10.0	5.0	15.0	41.0	26.0	64
17	5	1	11.5	5.0	17.0	53.0	30.5	114
17	5	2	11.5	5.0	15.0	52.5	28.0	76
17	5	2	11.0	4.5	13.0	46.0	23.0	48
8	8	2	10.0	4.5	10.0	43.5	24.0	29
83	11	1	15.5	8.0	30.5	75.0	54.0	514
18	6	1	12.5	8.5	18.0	57.3	32.8	140

Data Set 7: Iris Measurements

From "The Use of Multiple Measurements in Taxonomic Problems" by Ronald A. Fisher, *Annals of Statistics*, Vol. 7. SL denotes sepal length, SW denotes sepal width, PL denotes petal length, and PW denotes petal width. All measurements are in mm.

Text file names are SETSL, SETSW, SETPL, SETPW, VERSL, VERSW, VERPL, VERPW, VIRSL, VIRSW, VIRPL, VIRPW.

Class: Setosa				Class: Versicolor				Class: Virginica			
SL	SW	PL	PW	SL	SW	PL	PW	SL	SW	PL	PW
5.1	3.5	1.4	0.2	7.0	3.2	4.7	1.4	6.3	3.3	6.0	2.5
4.9	3.0	1.4	0.2	6.4	3.2	4.5	1.5	5.8	2.7	5.1	1.9
4.7	3.2	1.3	0.2	6.9	3.1	4.9	1.5	7.1	3.0	5.9	2.1
4.6	3.1	1.5	0.2	5.5	2.3	4.0	1.3	6.3	2.9	5.6	1.8
5.0	3.6	1.4	0.2	6.5	2.8	4.6	1.5	6.5	3.0	5.8	2.2
5.4	3.9	1.7	0.4	5.7	2.8	4.5	1.3	7.6	3.0	6.6	2.1
4.6	3.4	1.4	0.3	6.3	3.3	4.7	1.6	4.9	2.5	4.5	1.7
5.0	3.4	1.5	0.2	4.9	2.4	3.3	1.0	7.3	2.9	6.3	1.8
4.4	2.9	1.4	0.2	6.6	2.9	4.6	1.3	6.7	2.5	5.8	1.8
4.9	3.1	1.5	0.1	5.2	2.7	3.9	1.4	7.2	3.6	6.1	2.5
5.4	3.7	1.5	0.2	5.0	2.0	3.5	1.0	6.5	3.2	5.1	2.0
4.8	3.4	1.6	0.2	5.9	3.0	4.2	1.5	6.4	2.7	5.3	1.9
4.8	3.0	1.4	0.1	6.0	2.2	4.0	1.0	6.8	3.0	5.5	2.1
4.3	3.0	1.1	0.1	6.1	2.9	4.7	1.4	5.7	2.5	5.0	2.0
5.8	4.0	1.2	0.2	5.6	2.9	3.6	1.3	5.8	2.8	5.1	2.4
5.7	4.4	1.5	0.4	6.7	3.1	4.4	1.4	6.4	3.2	5.3	2.3
5.4	3.9	1.3	0.4	5.6	3.0	4.5	1.5	6.5	3.0	5.5	1.8
5.1	3.5	1.4	0.3	5.8	2.7	4.1	1.0	7.7	3.8	6.7	2.2
5.7	3.8	1.7	0.3	6.2	2.2	4.5	1.5	7.7	2.6	6.9	2.3
5.1	3.8	1.5	0.3	5.6	2.5	3.9	1.1	6.0	2.2	5.0	1.5
5.4	3.4	1.7	0.2	5.9	3.2	4.8	1.8	6.9	3.2	5.7	2.3
5.1	3.7	1.5	0.4	6.1	2.8	4.0	1.3	5.6	2.8	4.9	2.0
4.6	3.6	1.0	0.2	6.3	2.5	4.9	1.5	7.7	2.8	6.7	2.0
5.1	3.3	1.7	0.5	6.1	2.8	4.7	1.2	6.3	2.7	4.9	1.8
4.8	3.4	1.9	0.2	6.4	2.9	4.3	1.3	6.7	3.3	5.7	2.1
5.0	3.0	1.6	0.2	6.6	3.0	4.4	1.4	7.2	3.2	6.0	1.8
5.0	3.4	1.6	0.4	6.8	2.8	4.8	1.4	6.2	2.8	4.8	1.8
5.2	3.5	1.5	0.2	6.7	3.0	5.0	1.7	6.1	3.0	4.9	1.8
5.2	3.4	1.4	0.2	6.0	2.9	4.5	1.5	6.4	2.8	5.6	2.1
4.7	3.2	1.6	0.2	5.7	2.6	3.5	1.0	7.2	3.0	5.8	1.6
4.8	3.1	1.6	0.2	5.5	2.4	3.8	1.1	7.4	2.8	6.1	1.9
5.4	3.4	1.5	0.4	5.5	2.4	3.7	1.0	7.9	3.8	6.4	2.0
5.2	4.1	1.5	0.1	5.8	2.7	3.9	1.2	6.4	2.8	5.6	2.2
5.5	4.2	1.4	0.2	6.0	2.7	5.1	1.6	6.3	2.8	5.1	1.5
4.9	3.1	1.5	0.1	5.4	3.0	4.5	1.5	6.1	2.6	5.6	1.4
5.0	3.2	1.2	0.2	6.0	3.4	4.5	1.6	7.7	3.0	6.1	2.3
5.5	3.5	1.3	0.2	6.7	3.1	4.7	1.5	6.3	3.4	5.6	2.4
4.9	3.1	1.5	0.1	6.3	2.3	4.4	1.3	6.4	3.1	5.5	1.8
4.4	3.0	1.3	0.2	5.6	3.0	4.1	1.3	6.0	3.0	4.8	1.8
5.1	3.4	1.5	0.2	5.5	2.5	4.0	1.3	6.9	3.1	5.4	2.1
5.0	3.5	1.3	0.3	5.5	2.6	4.4	1.2	6.7	3.1	5.6	2.4
4.5	2.3	1.3	0.3	6.1	3.0	4.6	1.4	6.9	3.1	5.1	2.3
4.4	3.2	1.3	0.2	5.8	2.6	4.0	1.2	5.8	2.7	5.1	1.9
5.0	3.5	1.6	0.6	5.0	2.3	3.3	1.0	6.8	3.2	5.9	2.3
5.1	3.8	1.9	0.4	5.6	2.7	4.2	1.3	6.7	3.3	5.7	2.5
4.8	3.0	1.4	0.3	5.7	3.0	4.2	1.2	6.7	3.0	5.2	2.3
5.1	3.8	1.6	0.2	5.7	2.9	4.2	1.3	6.3	2.5	5.0	1.9
4.6	3.2	1.4	0.2	6.2	2.9	4.3	1.3	6.5	3.0	5.2	2.0
5.3	3.7	1.5	0.2	5.1	2.5	3.0	1.1	6.2	3.4	5.4	2.3
5.0	3.3	1.4	0.2	5.7	2.8	4.1	1.3	5.9	3.0	5.1	1.8

Data Set 8: Cuckoo Egg Lengths

From *The Methods of Statistics*, 4th Edition, John Wiley and Sons, Inc., by L.H.C. Tippett, as listed in the Data and Story Library (http://lib.stat.cmu.edu/DASL/Datafiles/cuckoodat.html). These data are lengths of cuckoo eggs put in the nests of other birds. All lengths are in mm.

Text file names are MDW, TREE, HEDGE, ROBIN, PIED, WREN.

Meadow Pipit	Tree Pipit	Hedge Sparrow	Robin	Pied Wagtail	Wren
19.65	21.05	20.85	21.05	21.05	19.85
20.05	21.85	21.65	21.85	21.85	20.05
20.65	22.05	22.05	22.05	21.85	20.25
20.85	22.45	22.85	22.05	21.85	20.85
21.65	22.65	23.05	22.05	22.05	20.85
21.65	23.25	23.05	22.25	22.45	20.85
21.65	23.25	23.05	22.45	22.65	21.05
21.85	23.25	23.05	22.45	23.05	21.05
21.85	23.45	23.45	22.65	23.05	21.05
21.85	23.45	23.85	23.05	23.25	21.25
22.05	23.65	23.85	23.05	23.45	21.45
22.05	23.85	23.85	23.05	24.05	22.05
22.05	24.05	24.05	23.05	24.05	22.05
22.05	24.05	25.05	23.05	24.05	22.05
22.05	24.05		23.25	24.85	22.25
22.05			23.85		
22.05					
22.05					
22.05					
22.05					
22.25					
22.25					
22.25					
22.25					
22.25					
22.25					
22.25					
22.25					
22.45					
22.45					
22.45					
22.65					
22.65					
22.85					
22.85					
22.85					
22.85					
23.05					
23.25					
23.25					
23.45					
23.65					
23.85					
24.25					
24.45					

Data Set 9: Poplar Tree Weights (kg)

The data are from a study conducted by researchers at Pennsylvania State University; the data were obtained from Minitab, Inc.

 Text file names are NONE, FERT, IRRIG, FRTIR.

Year 1

	No Treatment	Fertilizer	Irrigation	Fertilizer and Irrigation
Site 1 (rich, moist)	0.15	1.34	0.23	2.03
	0.02	0.14	0.04	0.27
	0.16	0.02	0.34	0.92
	0.37	0.08	0.16	1.07
	0.22	0.08	0.05	2.38
Site 2 (sandy, dry)	0.60	1.16	0.65	0.22
	1.11	0.93	0.08	2.13
	0.07	0.30	0.62	2.33
	0.07	0.59	0.01	1.74
	0.44	0.17	0.03	0.12

Year 2

	No Treatment	Fertilizer	Irrigation	Fertilizer and Irrigation
Site 1 (rich, moist)	1.21	0.94	0.07	0.85
	0.57	0.87	0.66	1.78
	0.56	0.46	0.10	1.47
	0.13	0.58	0.82	2.25
	1.30	1.03	0.94	1.64
Site 2 (sandy, dry)	0.24	0.92	0.96	1.07
	1.69	0.07	1.43	1.63
	1.23	0.56	1.26	1.39
	0.99	1.74	1.57	0.49
	1.80	1.13	0.72	0.95

Data Set 10: Blood Measurements

Data are from the U.S. Department of Health and Human Services, National Center for Health Statistics, Third National Health and Nutrition Examination Survey. BCC denotes blood cell count measured in cells per microliter; hemoglobin is measured in g/dL; platelet count is number per mm³.

Text file names are BLSEX, BLAGE, BLWHT, BLRED, BLHEM, BLPLT.

Sex	Age	White BCC	Red BCC	Hemoglobin	Platelet Count
F	30	8.9	4.39	13.5	224
M	29	5.25	5.18	15.9	264.5
M	21	5.95	4.88	13.9	360
M	33	10.05	5.94	15.85	384.5
F	24	6.5	4.48	13.55	364.5
F	28	9.45	3.58	10.95	468
M	57	5.45	4.4	12.5	171
M	28	5.3	4.88	15.6	328.5
F	48	7.65	4.22	13.05	323.5
F	60	6.4	4.36	12.2	306.5
F	20	5.15	4.59	14.25	264.5
F	19	16.6	3.63	11.5	233
F	53	5.75	4.54	14	254.5
M	34	5.55	4.92	14.05	267
F	26	11.6	3.92	11.4	463
M	18	6.85	5.44	14.65	238
M	26	6.65	5.07	15.4	251
F	25	5.9	4.24	12.8	282.5
F	48	9.3	4.21	14.3	307.5
M	24	6.3	4.8	14.7	321.5
F	48	8.55	5.06	14.9	360.5
M	28	6.4	5.14	15.2	282.5
F	52	10.8	4.71	14.9	315
F	32	4.85	4.23	11.5	284
F	46	4.9	4.68	14.45	259.5
M	20	7.85	5.99	14.05	291.5
F	23	8.75	3.83	12.2	259.5
M	26	7.7	4.82	15.35	164
M	27	5.3	4.57	14.15	199.5
M	46	6.5	4.59	14.5	220
F	37	6.9	3.95	11.1	369
M	27	4.55	4.77	14.85	245
M	25	7.1	5.12	14.6	266
M	55	8	5.37	15.6	369
M	27	4.7	5.17	15.45	210.5
M	53	4.4	4.97	14.6	234
F	50	9.75	3.87	12.75	471
M	18	4.9	4.8	15.2	244.5
M	36	10.75	5.05	15.65	365.5
M	52	11	5.11	15.3	265
F	29	4.05	4.2	12.75	198
F	31	9.05	4.07	13.1	390
F	17	5.05	3.74	11.4	269.5
F	24	6.4	4.72	13.3	344.5
F	37	4.05	4.63	13.7	386.5
F	48	7.6	4.63	15.1	256
F	23	4.95	4.3	13.55	226
M	45	9.6	5.13	15.95	225
F	21	3	4.07	11.9	259
F	22	9.1	4.6	14.2	271.5

Data Set 11: Counts of Yeast Cells

A yeast cell is a living organism. The data are counts of yeast cells using a hemacytometer, and each value is the count over 1 mm^2 divided into 400 squares. The data were collected by William S. Gosset, who developed the Student t distribution. Gosset was an employee of the Guiness Brewery, and his contributions to statistics have origins in the brewing process, which requires yeast used for fermentation. Counts of yeast cells were important because the addition of too little yeast would result in incomplete fermentation, but too much yeast would result in beer with a bitter taste. The data are from *Student's Collected Papers*, edited by E. S. Pearson and John Wishart, Cambridge University Press, London, 1958. (Gosset published with the pseudonym of "A Student.")

 Text file name is YEAST.

```
2   2   4   4   4   5   2   4   7   7   4   7   5   2   8   6   7   4   3   4
3   3   2   4   2   5   4   2   8   6   3   6   6  10   8   3   5   6   4   4
7   9   5   2   7   4   4   2   4   4   4   3   5   6   5   4   1   4   2   6
4   1   4   7   3   2   3   5   8   2   9   5   3   9   5   5   2   4   3   4
4   1   5   9   3   4   4   6   6   5   4   6   5   5   4   3   5   9   6   4
4   4   5  10   4   4   3   8   3   2   1   4   1   5   6   4   2   3   3   3
3   7   4   5   1   8   5   7   9   5   8   9   5   6   6   4   3   7   4   4
7   5   6   3   6   7   4   5   8   6   3   3   4   3   7   4   4   4   5   3
8  10   6   3   3   6   5   2   5   3  11   3   7   4   7   3   5   5   3   4
1   3   7   2   5   5   5   3   3   4   6   5   6   1   6   4   4   4   6   4
4   2   5   4   8   6   3   4   6   5   2   6   6   1   2   2   2   5   2   2
5   9   3   5   6   4   6   5   7   1   3   6   5   4   2   8   9   5   4   3
2   2  11   4   6   6   4   6   2   5   3   5   7   2   6   5   5   1   2   7
5  12   5   8   2   4   2   1   6   4   5   1   2   9   1   3   4   7   3   6
5   6   5   4   4   5   2   7   6   2   7   3   5   4   4   5   4   7   5   4
8   4   6   6   5   3   3   5   7   4   5   5   5   6  10   2   3   8   3   5
6   6   4   2   6   6   7   5   4   5   8   6   7   6   4   2   6   1   1   4
7   2   5   7   4   6   4   5   1   5  10   8   7   5   4   6   4   4   7   5
4   3   1   6   2   5   3   3   3   7   4   3   7   8   4   7   3   1   4   4
7   6   7   2   4   5   1   3  12   4   2   2   8   7   6   7   6   3   5   4
```

Data Set 12: Systolic Blood Pressure Measurements

Data are from "Sympathoadrenergic Mechanisms in Reduced Hemodynamic Stress Responses after Exercise" by Kim Brownley et al., *Medicine and Science in Sports and Exercise*, Vol. 35, No. 6. The blood pressure measurements (mm Hg) are taken before and after a period consisting of 25 minutes of aerobic bicycle exercise. During the pre- and post-exercise periods, subjects were measured during a time of no stress, and a time of stress caused by an arithmetic test, and a time of stress caused by a speech test.

 Text file names are SBSEX, SBRAC, SBBNO, SBBM, SBBS, SBANO, SBAM, SBAS.

SEX	RACE	Pre-Exercise with No Stress	Pre-Exercise with Math Stress	Pre-Exercise with Speech Stress	Post-Exercise with No Stress	Post-Exercise with Math Stress	Post-Exercise with Speech Stress
Female	Black	117.00	131.00	128.50	110.00	122.33	123.50
Female	Black	130.67	142.33	147.50	114.00	129.33	134.50
Female	Black	102.67	109.00	112.00	99.67	108.67	109.50
Female	Black	93.67	103.00	114.00	99.00	101.33	105.00
Female	Black	96.33	106.33	114.00	93.33	105.00	101.50
Female	Black	92.00	112.00	114.50	100.33	107.67	112.50
Male	Black	115.67	129.33	128.00	115.00	135.67	125.50
Male	Black	120.67	147.67	138.50	115.00	129.00	130.00
Male	Black	133.00	159.67	149.00	144.00	154.00	148.50
Male	Black	120.33	131.67	143.00	120.00	128.33	118.50
Male	Black	124.67	160.33	172.50	121.00	145.67	168.50
Male	Black	118.33	136.33	154.50	108.33	128.00	146.00
Female	White	119.67	143.00	141.00	115.33	131.33	124.00
Female	White	106.00	129.33	132.00	113.00	130.00	129.00
Female	White	108.33	148.00	149.50	110.33	139.33	135.00
Female	White	107.33	128.00	135.00	109.33	126.67	127.50
Female	White	117.00	134.67	135.00	125.67	127.67	131.00
Female	White	113.33	121.00	133.00	103.33	109.67	117.00
Male	White	124.33	138.00	144.50	129.00	125.33	134.50
Male	White	111.00	139.00	141.50	109.67	133.33	131.50
Male	White	99.67	123.67	130.00	114.33	124.67	128.50
Male	White	128.33	141.33	142.00	119.00	126.67	130.50
Male	White	102.00	123.00	134.50	108.00	122.00	122.50
Male	White	127.33	156.67	163.50	127.33	154.00	146.50

Appendix C: Answers to Odd-Numbered Exercises (and ALL Review Exercises and Cumulative Review Exercises)

Chapter 1 Answers

Section 1-2

1. Statistic
3. Statistic
5. Discrete
7. Discrete
9. Ratio
11. Interval
13. Nominal
15. Ratio
17. Sample: Captured trout. Population: All trout in the body of water. Not likely to be representative.
19. The 1059 selected adults. Population: All adults. Given the methods used by Gallup, the sample is likely to be representative.

Section 1-3

1. Experiment
3. Observational study
5. Retrospective
7. Cross-sectional
9. Convenience
11. Random
13. Cluster
15. Systematic
17. Stratified
19. Cluster
21. Yes; yes. Each tablet has the same chance of being selected, and all samples of size 50 have the same chance of being selected.
23. Not a random sample, because not all subjects have the same chance of being selected. Not a simple random sample because not all samples of size 36 have the same chance of being selected.
25. Random sample, because each person has the same chance of being selected. Not a simple random sample, because not all samples of the same size have the same chance of being selected. For example, it is impossible to get a sample consisting of people who are all of the same gender.
27. Random sample, because each adult has the same chance of being selected. Not a simple random sample, because not all samples of the same size have the same chance of being selected. For example, it is impossible to get a sample that includes people from 11 different blocks.
29. Prospective: Data are collected in the future from the three groups. Randomized: Subjects are randomly selected and are randomly assigned to one of the three groups. Double-blind: The subjects don't know which treatment or placebo they are given, and those who evaluate the subjects also do not know.
31. Motorcyclists who were killed in crashes.

Chapter 1 Review Exercises

1. No. The sample is a voluntary response sample. Some respondents may have answered multiple times, and those with very strong feelings about the issue are more likely to respond.
2. Answers may vary, but here are some possibilities: For a random procedure, compile a list of all adults and use a computer to randomly select a sample of them. Systematic: Compile a sequential and numbered list of all adults, randomly select one of the first million, then select every millionth adult after the first selection. Convenience: Survey friends and relatives that you encounter. Stratified: Randomly select 500 males and 500 females. Cluster: Randomly select 20 telephone exchanges, then survey all adults with those telephone exchanges.
3. a. Ratio
   b. Ordinal
   c. Nominal
   d. Interval
4. a. Discrete
   b. Ratio
   c. Statistic
   d. It would be a voluntary response sample and would be likely to elicit responses only from those with strong feelings about the issue.
5. a. Systematic; representative
   b. Convenience; not representative
   c. Cluster; not representative
   d. Random; representative
   e. Stratified; not representative
6. a. Blinding occurs when subjects do not know if they are given the treatment or a placebo. Blinding also occurs when those who evaluate the effects do not know whether the subjects were given a treatment or a placebo. Use a placebo consisting of a substance with similar appearance and taste, but which contains no treatment. Use coding so that the results can be evaluated by the researchers.
   b. Blinding is important, because subjects may exhibit real or imagined effects if they believe that they are given the treatment.
   c. The randomly selected subjects are assigned to the treatment group and placebo group through a random process.
   d. Subjects are carefully assigned to the treatment group and placebo group by using a process to ensure that the subjects in the two groups have similar characteristics, at least for those characteristics that might affect the results.
   e. Replication is the repetition of an experiment, and it is used effectively when there are enough subjects to recognize differences between the treatment group and placebo group.

Chapter 1 Cumulative Review Exercises

1. a. 855
   b. 247

2. a. 540
   b. 5%
3. a. 7%
   b. 8
4. 93
5. Removing all of the plaque is equivalent to removing 100% of it. More than 100% cannot be removed.
6. All percentages of success should be multiples of 5. The given percentages cannot be correct.
7. 0.0000000000072744916
8. 4,398,046,500,000
9. 282,429,540,000
10. 0.000000000058207661

Chapter 2 Answers

Section 2-2

1. Class width: 10. Class midpoints: 94.5, 104.5, 114.5, 124.5, 134.5, 144.5, 154.5. Class boundaries: 89.5, 99.5, 109.5, 119.5, 129.5, 139.5, 149.5, 159.5.
3. Class width: 200. Class midpoints: 99.5, 299.5, 499.5, 699.5, 899.5, 1099.5, 1299.5. Class boundaries: −0.5, 199.5, 399.5, 599.5, 799.5, 999.5, 1199.5, 1399.5.

5.
Systolic Blood Pressure of Men	Relative Frequency
90–99	3%
100–109	10%
110–119	43%
120–129	30%
130–139	13%
140–149	0%
150–159	3%

7.
Cholesterol of Men	Relative Frequency
0–199	33%
200–399	28%
400–599	13%
600–799	20%
800–999	5%
1000–1199	0%
1200–1399	3%

9.
Systolic Blood Pressure of Men	Cumulative Frequency
Less than 100	1
Less than 110	5
Less than 120	22
Less than 130	34
Less than 140	39
Less than 150	39
Less than 160	40

11.
Cholesterol of Men	Cumulative Frequency
Less than 200	13
Less than 400	24
Less than 600	29
Less than 800	37
Less than 1000	39
Less than 1200	39
Less than 1400	40

13.
Weight (lb)	Frequency
0–49	6
50–99	10
100–149	10
150–199	7
200–249	8
250–299	2
300–349	4
350–399	3
400–449	3
450–499	0
500–549	1

15. The circumferences for females appear to be slightly lower, but the difference does not appear to be significant.

Circumference (cm)	Males	Females
34.0–35.9	2	1
36.0–37.9	0	3
38.0–39.9	5	14
40.0–41.9	29	27
42.0–43.9	14	5

17. Most common: 4.

Number of Cells	Frequency
1	20
2	43
3	53
4	86
5	70
6	54
7	37
8	18
9	10
10	5
11	2
12	2

Section 2-3

1. 22.46 mm
3. 14% (from 17 out of 120)
5. 40%; 200

7. 183 pounds

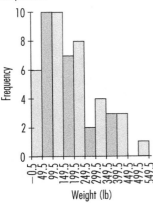

9. The distribution appears to satisfy the requirement of being approximately bell-shaped.

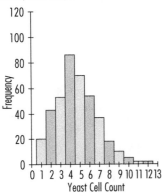

11. Graphs may vary, but a comparison of dotplots suggests no significant difference, so that irrigation does not appear to have an effect.

13. 5012, 5012, 5012, 5055, 5200, 5200, 5200, 5200, 5327, 5327, 5335, 5472

15.

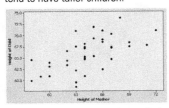

17. There does appear to be a relationship. It appears that taller mothers tend to have taller children.

19. There does not appear to be a relationship between sepal length and petal length.

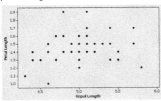

Section 2-4

1. $\bar{x} = 157.8$ sec; median $= 88.0$ sec; mode $= 0$ sec; midrange $= 274.0$ sec
3. $\bar{x} = 25.37$; median $= 24.00$; mode $= 19.6$; midrange $= 27.70$. Yes.
5. $\bar{x} = 26.9$; median $= 26.0$; mode $=$ none; midrange $= 28.0$. Yes.
7. $\bar{x} = 133.9$, median:132.5; mode:130; midrange:135.
   Given that the measurements are taken from the same person, the values appear to vary by large amounts.
9. The single line and multiple lines have these same results:
   $\bar{x} = 71.5$ min; median $= 72.0$ min; mode $= 77$ min; midrange $= 71.0$ min. Although the measures of center are the same for the two sets of data, the single line appears to have values that vary much less than the values for the multiple lines.
11. Control group: $\bar{x} = 4.95$ m; median $= 5.0$ m; mode $= 5.5$ m; midrange $= 4.4$ m. Irrigation group: $\bar{x} = 4.48$ m; median $= 4.8$ m; mode $= 2.9$ m, 5.1 m, 6.1 m, 6.8 m; midrange $= 4.0$ m. Based on the means, the trees treated with irrigation do not have greater heights, so the irrigation does not appear to be effective in increasing height.
13. Males: $\bar{x} = 41.10$ cm; median $= 41.10$ cm.
    Females: $\bar{x} = 40.05$ cm; median $= 40.20$ cm.
    There does appear to be a small difference.
15. Setosa: $\bar{x} = 1.46$ mm; median $= 1.50$ mm.
    Versicolor: $\bar{x} = 4.26$ mm; median $= 4.35$ mm.
    Virginica: $\bar{x} = 5.55$ mm; median $= 5.55$ mm.
    The three classes do not appear to be about the same.
17. The mean of 177.0 differs somewhat from the mean of 172.5 that is found by using the original list of data values.
19. a. Means: 113.722, 133.097, 137.396. Medians: 116.335, 133.17, 136.75. It appears that the math and speech tests correspond to higher systolic blood pressure readings.
    b. Means: 113.499, 126.903, 128.375. Medians: 113.5, 127.835, 128.75. It appears that the exercise has an effect of reducing the response to stress.
21. a. 182.9 lb
    b. 171.0 lb
    c. 159.2 lb
    The results differ by substantial amounts, suggesting that the mean of the original set of weights is strongly affected by extreme values.

Section 2-5

1. range $= 548.0$ sec; $s^2 = 46308.2$ sec^2; $s = 215.2$ sec; vary widely
3. range $= 20.00$; $s^2 = 32.09$; $s = 5.66$; no, because it is within two standard deviations of the mean.
5. range $= 28.0$ years; $s^2 = 75.0$ years^2; $s = 8.7$ years; the variation is likely to be less than the variation for the general population.
7. range $= 30.0$, $s^2 = 81.8$, $s = 9.0$. The values vary widely, suggesting that the measurement process is not very accurate.

9. Single line: range $= 12.0$ min; $s^2 = 22.7$ min^2; $s = 4.8$ min
Multiple lines: range $= 58.0$ min; $s^2 = 331.8$ min^2; $s = 18.2$ min
The waiting times vary much less with a single line than with multiple lines.

11. Control group: range $= 5.00$ m; $s^2 = 2.02$ m^2; $s = 1.42$ m
Irrigation group: range $= 5.60$ m; $s^2 = 3.18$ m^2; $s = 1.78$ m
The amounts of variation do not appear to differ by substantial amounts.

13. Males: 1.50 cm; Females: 1.64 cm; difference does not appear to be very substantial.

15. Setosa: $s = 0.17$ mm; versicolor: $s = 0.47$ mm; virginica: $s = 0.55$ mm. The three classes do not appear to have the same amount of variation.

17. 106.2, which differs from 119.5 by a considerable amount.

19. Answer varies, but 11.5 years is a reasonable estimate (based on a maximum of 70 years of age and a minimum of 24 years of age).

21. a. 68%
    b. 99.7%

23. Heights of men: 3.1%; lengths of cuckoo eggs: 5.1%. The coefficients of variation are not dramatically different.

25. days; days^2

27. a. 6
    b. 6
    c. 3.0
    d. $n - 1$
    e. No. The mean of the sample variances (6) equals the population variance (6), but the mean of the sample standard deviations (1.9) does not equal the mean of the population standard deviation (2.4).

Section 2-6

1. a. 6 cm
   b. 6/7 or 0.86
   c. 0.86
   d. Usual
3. a. 24.9
   b. 2.02
   c. −2.02
   d. Unusual. Exercise often results in lower pulse rates.
5. 2.56; unusual
7. 4.52; yes; patient is ill.
9. Biology test, because $z = -0.50$ is greater than $z = -2.00$.
11. 43
13. 15
15. 46
17. 251.5
19. 121
21. 0
23. a. 165
    b. 169
    c. 279.5
    d. Yes; yes
    e. No; no

Section 2-7

1.

3.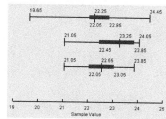

5. The center does not appear to be 98.6.

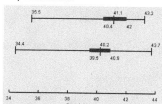

7. The three sets of lengths appear to exhibit some differences, but the differences are not dramatic. (The top boxplot represents meadow pipits, the middle boxplot represents tree pipits, and the bottom boxplot represents robins.)

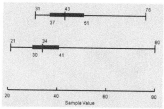

9. The boxplot for males (top) appears to represent greater head circumferences than those of the girls. The difference does not appear to be very substantial.

11. The boxplot for actors (top) appears to represent ages that are considerably higher than the ages of actresses when they won Oscars.

13. The boxplot for the petal lengths of sertosa (top) obscures these values of the 5-number summary: 0.1, 0.2, 0.2, 0.3, 0.6. The boxplot for versicolor is in the middle. The boxplots reveal dramatically different petal lengths, with sertosa having the smallest.

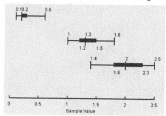

15. The top boxplot represents hemoglobin counts of females. The box-plots appear to reveal dramatic differences between the two data sets.

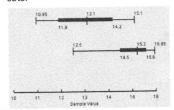

Chapter 2 Review Exercises

Section 2-2

1. a. 4.54
   b. 3.95
   c. 1.8, 3.7, 5.1
   d. 7.75
   e. 11.90
   f. 2.65
   g. 7.03
   h. 3.25
   i. 5.15
   j. 1.85
2. a. 3.46
   b. Yes, because it is more than two standard deviations away from the mean.
   c. There are no other unusual values. (No other values are below −0.76 or above 9.84.)
3.

Tree Circumference (feet)	Frequency
1.0 − 2.9	4
3.0 − 4.9	9
5.0 − 6.9	5
7.0 − 8.9	1
9.0 − 10.9	0
11.0 − 12.9	0
13.0 − 14.9	1

4. Distribution is approximately bell-shaped.

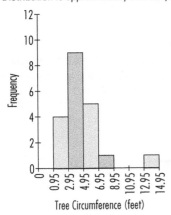

5.

Chapter 2 Cumulative Review Exercises

1. a. Continuous
   b. Ratio
2. a. Mode, because the others are numerical measures that require data at the interval or ratio levels of measurement.
   b. Convenience
   c. Cluster
   d. More consistency can be achieved by reducing the standard deviation. (It is also important to keep the mean at an acceptable value.)

Chapter 3 Answers

Section 3-2

1. a. 1/2 or 0.5
   b. 0.20
   c. 0
3. $-1, 2, 5/3, \sqrt{2}$
5. a. 3/8
   b. 3/8
   c. 1/8
7. 428/580 = 0.738; yes.
9. a. 1/17 or 0.0588
   b. No
11. a. 0.0220 (not 0.0225)
    b. Yes
13. a. 1/365
    b. 31/365
    c. 1
15. 117/734 or 0.159; yes.
17. a. brown-brown, brown-blue, blue-brown, blue-blue
    b. 1/4
    c. 3/4
19. a. brown-blue, brown-blue, brown-blue, brown-blue
    b. 0
    c. 1

Section 3-3

1. a. Not disjoint
   b. Not disjoint
   c. Disjoint
3. a. 0.95
   b. 0.9975
5. 10/14 = 0.714
7. 364/365 or 0.997
9. 0.239
11. 0.341

13. 0.600
15. 0.490
17. 0.140
19. 0.870
21. 0.290
23. No. The example varies.

Section 3-4

1. a. Dependent
   b. Dependent
   c. Dependent (although a strong case could be made for "independent")
3. 1/12
5. 0.702
7. 0.736
9. a. 0.203
   b. 0.2
11. a. 0.001
    b. 0.000742
13. 0.000105
15. a. 1/1024
    b. No, because there are other ways to pass.
17. 1/1024; yes, because the probability of getting 10 girls by chance is so small.
19. 0.739 (or 0.738 assuming dependence); no.
21. 3/4 or 0.75

Section 3-5

1. None of the students has group A blood.
3. At least one of the subjects has the X-linked recessive gene.
5. 1
7. 31/32; yes.
9. 0.5; no.
11. 11/14 or 0.786; get another test.
13. 0.999999; yes, because the likelihood of being awakened increases from 0.99 to 0.999999.
15. 0.271
17. 0.897
19. 0.0805
21. a. 0.8
    b. 0.703
23. a. 0.05
    b. 0.0989
25. a. 0.679
    b. 0.95

Section 3-6

1. Prospective
3. With treatment: 0.159; with placebo: 0.040. The risk of a headache appears to be higher in the treatment group.

5. 9 (rounded up from 8.4)
7. 3.99 or roughly 4. The incidence of headaches among Viagra users is roughly 4 times the incidence of headaches for those who take a placebo.
9. It does appear that headaches occur substantially more among Viagra users than those using a placebo, so there is a basis for concern.
11. 182/418 or 0.435. No, because incidence rates in this retrospective study depend on the design used by the researcher.
13. 0.170
15. 2.13. The risk of facial injuries for those not wearing helmets is 2.13 times the risk for those wearing helmets.
17. Yes, because those not wearing helmets appear to have 2.13 times the risk of facial injuries of those wearing helmets.
19. The odds ratio of 2.13 does not change, but the risk ratio changes from 1.64 to 0.644. In retrospective studies, use odds ratios, but do not use risk ratios because they change according to the design chosen by the researcher.

Section 3-7

1. 4.5
3. 14.2
5. 65.1
7. 3.2
9. 0.0085; the rate uses fewer decimal places and is easier to understand.
11. a. 0.0075
    b. 0.0000563
    c. 0.9850563
13. a. 29.0
    b. 0.358
15. No, the health of the nation is not necessarily declining. The increasing number of deaths each year is probably due to the growing population or better reporting or better data collection.
17. The U.S. has a population distribution of 35.21%, 52.40%, and 12.39% for the three age categories. If the 14,220,632 Florida residents have the same distribution, the three age groups have these numbers of people: 5,007,261; 7,451,231; and 1,762,141 respectively. Using the same mortality rates for the three individual age groups and using the new adjusted population sizes for the three Florida age categories, we get these numbers of deaths: 4110; 41,051; and 80,412, respectively. Using the adjusted numbers of deaths and the adjusted population sizes for the different categories, the crude mortality rate becomes 8.8, which is much closer to the U.S. mortality rate of 8.5 than the mortality rate of 11.8 found for Florida before the adjustments.

Section 3-8

1. 720
3. 600
5. 300

7. 2,598,960

9. 32

11. 1,048,576

13. 1/35,960; it appears that the oldest employees were selected.

15. 1/5005; yes.

17. 1/1,000,000,000

19. a. 11,880

    b. 495

21. a. 1,048,576

    b. 184,756

    c. 0.176

    d. No, although the result is common, it should not happen consistently.

Chapter 3 Review Exercises

1. 0.8

2. 0.32

3. 0.97

4. 0.85

5. 0.460

6. 0.638

7. 15/32 or 0.469

8. 15/80 or 3/16 or 0.188

9. 0.0777

10. 1/4096; yes.

11. a. 1/120

    b. 720

12. 0.979

13. a. 0.0027832

    b. 0.00000775

    c. 0.9944413

Chapter 3 Cumulative Review Exercises

1. a. 4.0

    b. 4.0

    c. 2.2

    d. 4.7

    e. Yes.

    f. 6/7

    g. 0.729

    h. 1/262,144; yes.

2. a. 63.6 in.

    b. 1/4

    c. 3/4

    d. 1/16

    e. 5/16

Chapter 4 Answers

Section 4-2

1. a. Continuous

    b. Discrete

    c. Continuous

    d. Discrete

    e. Discrete

3. $\mu = 1.5, \sigma = 0.9$

5. Not a probability distribution because $\sum P(x) = 0.94 \neq 1$.

7. $\mu = 0.7, \sigma = 0.7$

9. a. 0.122

    b. 0.212

    c. Part b, because we want the probability of a result that is at least as extreme as the one obtained.

    d. No. With $P(9$ or more$) = 0.212$, the result of 9 girls can easily occur by chance.

11. a. $P(11$ or more girls$) = 0.029$

    b. Yes. With $P(11$ or more$) = 0.029$, the result of 11 girls cannot easily occur by chance, so it appears that the result can be better explained by some other factor, such as the effectiveness of the gender-selection technique.

13. $\mu = 2.0, \sigma = 1.0$

x	P(x)
0	0.0625
1	0.25
2	0.375
3	0.25
4	0.0625

Section 4-3

1. Not binomial; more than two outcomes; not a fixed number of trials.

3. Not binomial; more than two outcomes.

5. Binomial

7. Not binomial; more than two outcomes.

9. a. 0.128

    b. WWC, WCW, CWW; 0.128 for each

    c. 0.384

11. 0.980

13. 0.171

15. 0+

17. 0.278

19. 0.208

21. 0.0006; because the probability is so small, the result is unusual.

23. 0.2639; because the probability of not getting more than one who experiences headaches is not small (0.7361), the result is not unusual.

25. a. 0.00292

    b. 0+ (or 0.00000000000655)

    c. Yes, because it is very unlikely (0+) that all 8 would get headaches by chance.

27. 0.751
29. 0.000201; the probability supports the charge in the sense that the event is not likely to occur by chance.
31. 0.0524
33. 0.000535

Section 4-4

1. $\mu = 80.0, \sigma = 8.0$, minimum $= 64.0$, maximum $= 96.0$
3. $\mu = 1488.0, \sigma = 19.3$, minimum $= 1449.4$, maximum $= 1526.6$
5. a. $\mu = 5.0, \sigma = 1.6$
   b. No, because 7 is within two standard deviations of the mean.
7. a. $\mu = 2.6, \sigma = 1.6$
   b. No, because 0 wins is within two standard deviations of the mean.
9. a. The probabilities for 0, 1, 2, 3, . . . , 15 are
   $0+, 0+, 0.003, 0.014, . . . , 0+$ (from Table A-1).
   b. $\mu = 7.5, \sigma = 1.9$
   c. No, because 10 is within two standard deviations of the mean. Also, $P(10$ or more girls$) = 0.151$, showing that it is easy to get 10 or more girls by chance.
11. a. $\mu = 16.4, \sigma = 4.0$
    b. No, because 19 is within two standard deviations of the mean.
    c. No
13. a. 210
    b. $\mu = 170.0, \sigma = 10.6$
    c. Yes, because 210 is more than two standard deviations above the mean. Higher rates are justified.

Section 4-5

1. 0.180
3. 0.0399
5. a. 62.2
   b. 0.0155 (0.0156 using rounded mean)
7. a. 0.497
   b. 0.348
   c. 0.122
   d. 0.0284
   e. 0.00497
   The expected frequencies of 139, 97, 34, 8, and 1.4 compare reasonably well to the actual frequencies, so the Poisson distribution does provide good results.
9. a. 0.000912
   b. 0.999
   c. 0.0296

Chapter 4 Review Exercises

1. a. A random variable is a variable that has a single numerical value (determined by chance) for each outcome of some procedure.
   b. A probability distribution gives the probability for each value of the random variable.

c. Yes, because the sum of the probabilities is 1 and each probability value is between 0 and 1 inclusive.
   d. $\mu = 0.6$
   e. $\sigma = 0.7$
   f. Yes, the result is unusual because its probability is small (less than 0.05), indicating that it is not likely to occur by chance.
2. a. 0.026
   b. 0.992 (or 0.994)
   c. $\mu = 8.0, \sigma = 1.3$
   d. No, because 6 is within two standard deviations of the mean.
3. a. 0.00361
   b. This company appears to be very different because the event of at least four firings is so unlikely with a probability of only 0.00361.
4. a. 7/365
   b. 0.981
   c. 0.0188
   d. 0.0002
   e. No, because the event is so rare.

Chapter 4 Cumulative Review Exercises

1. a. $\mu = 4.4; \sigma = 3.0$
   b. Use the same values of x (0, 1, 2, . . . , 9) but replace the frequencies with these relative frequencies (rounded): 9%, 18%, 6%, 14%, 10%, 5%, 6%, 8%, 15%, 10%.
   c. $\mu = 4.5; \sigma = 2.9$
   d. The last digits appear to agree reasonably well with the distribution we expect with random selection.
2. a. $\mu = 15.0; \sigma = 3.7$
   b. The rate is not unusually low because 12 is within two standard deviations of the mean. The result does not provide strong evidence of the program's effectiveness.

Chapter 5 Answers

Section 5-2

1. 0.15
3. 0.15
5. 1/3
7. 1/2
9. 0.4013
11. 0.5987
13. 0.0099
15. 0.9901
17. 0.2417
19. 0.1359
21. 0.8959
23. 0.6984
25. 0.0001
27. 0.5
29. 68.26%
31. 99.74%

33. 0.9500
35. 0.9950
37. 1.28
39. −1.645
41. a. 68.26%
    b. 95%
    c. 99.74%
    d. 81.85%
    e. 4.56%

Section 5-3

1. 0.8413
3. 0.4972
5. 87.4
7. 115.6
9. a. 0.01%; yes
    b. 99.2°
11. 4.0 in.; 8.0 in.
13. 0.52%
15. 0.1222; 12.22%; yes, they are all well above the mean.
17. 2405 g
19. a. The z scores are real numbers that have no units of measurement.
    b. $\mu = 0$; $\sigma = 1$; distribution is normal.
    c. $\mu = 64.9$ kg, $\sigma = 13.2$ kg, distribution is normal.

Section 5-4

1. No, because of sampling variability, sample proportions will naturally vary from the true population proportion, even if the sampling is done with a perfectly valid procedure.
3. No, the histogram represents the distribution shape of one sample, but a sampling distribution includes all possible samples of the same size, such as all of the means computed from all possible samples of 106 people.
5. a. 10–10; 10–6; 10–5; 6–10; 6–6; 6–5; 5–10; 5–6; 5–5; the means are listed in part b.
    b.

Mean	10.0	8.0	7.5	8.0	6.0	5.5	7.5	5.5	5.0
Probability	1/9	1/9	1/9	1/9	1/9	1/9	1/9	1/9	1/9

    c. 7.0
    d. Yes; yes
7. a. Means: 85.0, 82.0, 83.5, 79.0, 81.5, 82.0, 79.0, 80.5, 76.0, 78.5, 83.5, 80.5, 82.0, 77.5, 80.0, 79.0, 76.0, 77.5, 73.0, 75.5, 81.5, 78.5, 80.0, 75.5, 78.0
    b. The probability of each mean is 1/25. The sampling distribution consists of the 25 sample means paired with the probability of 1/25.
    c. 79.4
    d. Yes; yes

9. a. 1, 1, 0.5, 0.5, 0.5, 1, 1, 0.5, 0.5, 0.5, 0.5, 0.5, 0, 0, 0, 0.5, 0.5, 0, 0, 0, 0.5, 0.5, 0, 0, 0
    b. The sampling distribution consists of the 25 proportions paired with the probability of 1/25.
    c. 0.4
    d. Yes; yes
11. The MAD values are 0, 0.5, 2, 0.5, 0, 1.5, 2, 1.5, 0. They each have probability 1/9. The mean of the sampling distribution is 0.89, but the MAD of the population is 1.6. Because the MAD values do not target the population MAD, a MAD statistic is not good for estimating the population parameter.

Section 5-5

1. a. 0.4325
    b. 0.1515
3. a. 0.0677
    b. 0.5055
5. a. 0.9808
    b. If the original population has a normal distribution, the central limit theorem provides good results for any sample size.
7. a. 0.5302
    b. 0.7323
    c. Part a, because the seats will be occupied by individual women, not groups of women.
9. a. 0.0119
    b. No; yes
11. a. 0.0274
    b. 0.0001
    c. Because the original population is normally distributed, the sampling distribution of sample means will be normally distributed for any sample size.
    d. No, the mean can be less than 140 while individual values are above 140.
13. 2979 lb
15. a. 0.9750
    b. 1329 lb

Section 5-6

1. The area to the right of 15.5
3. The area to the left of 99.5
5. The area to the left of 4.5
7. The area between 7.5 and 10.5
9. Table: 0.122; normal approximation: 0.1218
11. Table: 0.549; normal approximation is not suitable.
13. 0.1357; no
15. 0.0287; no
17. 0.2676; no
19. 0.7389; no, not very confident.
21. 0.0080; yes
23. 0.9505; yes
25. 0.0001; yes

Section 5-7

1. Not normal
3. Normal
5. Normal
7. Not normal
9. Normal
11. Not normal
13. Heights appear to be normal, but cholesterol levels do not appear to be normal. Cholesterol levels are strongly affected by diet, and diets might vary in dramatically different ways that do not yield normally distributed results.
15. $-1.28, -0.52, 0, 0.52, 1.28$; normal

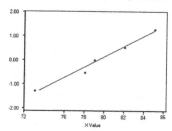

17. No, the transformation to z scores involves subtracting a constant and dividing by a constant, so the plot of the (x, z) points will always be a straight line, regardless of the nature of the distribution.

Chapter 5 Review Exercises

1. a. 0.0222
   b. 0.2847
   c. 0.6720
   d. 254.6
2. a. 0.69% of 900 = 6.21 babies
   b. 2405 g
   c. 0.0119
   d. 0.9553
3. 0.1020; no, assuming that the correct rate is 25%, there is a high probability (0.1020) that 19 or fewer offspring will have blue eyes. Because the observed event could easily occur by chance, there is not strong evidence against the 25% rate.
4. a. 0.9626
   b. 63.3 in., 74.7 in.
   c. 0.9979
5. a. Normal distribution.
   b. 51.2 lb
   c. Normal distribution
6. Normal approximation: 0.0436; exact value: 0.0355. Because the probability of getting only 2 women by chance is so small, it appears that the company is discriminating based on gender.
7. Yes. A histogram is roughly bell-shaped, and a normal quantile plot shows points that are reasonably close to a straight-line pattern.
8. Yes. A histogram is roughly bell-shaped, and a normal quantile plot shows points that are very close to a straight-line pattern.

Chapter 5 Cumulative Review Exercises

1. a. 63.0 mm
   b. 64.5 mm
   c. 66 mm
   d. 4.2 mm
   e. -0.95
   f. 75%
   g. 82.89%
   h. ratio
   i. continuous
2. a. 0.001
   b. 0.271
   c. The requirement that $np \geq 5$ is not satisfied, indicating that the normal approximation would result in errors that are too large.
   d. 5.0
   e. 2.1
   f. No, 8 is within two standard deviations of the mean and is within the range of values that could easily occur by chance.

Chapter 6 Answers

Section 6-2

1. 2.575
3. 2.33
5. $p = 0.250 \pm 0.030$
7. $p = 0.654 \pm 0.050$
9. $\hat{p} = 0.464$; $E = 0.020$
11. $\hat{p} = 0.655$; $E = 0.023$
13. 0.0300
15. 0.0405
17. $0.708 < p < 0.792$
19. $0.0887 < p < 0.124$
21. 461
23. 232
25. a. There is 95% confidence that the limits of 0.0489 and 0.0531 contain the population proportion.
   b. Yes, about 5% of males aged 18–20 drive while impaired.
   c. 5.31%
27. a. $0.731 < p < 0.786$
   b. The results do not contradict Mendel's theory. The confidence interval limits contain 75%, so the population proportion could easily be 75%.
29. a. $15.1\% < p < 21.6\%$
   b. Yes
31. a. 473
   b. 982
   c. The sample would be a voluntary response sample, so the results would not be valid.
33. a. $0.0267\% < p < 0.0376\%$
   b. No, because 0.0340% is in the confidence interval.

35. a. $0.612 < p < 0.918$
    b. Yes. The confidence interval limits do not contain the proportion of 0.5 that would be expected with an ineffective treatment. Because the proportion of boys appears to be well above the proportion of 0.5, it is highly unlikely that the results occurred by chance.
37. a. $1.07\% < p < 8.68\%$
    b. $70.1\% < p < 75.3\%$
    c. Yes, if wearing orange had no effect, we would expect that the percentage of injured hunters wearing orange should be between 70.1% and 75.3%, but it is much lower.
39. 895

Section 6-3

1. 2.33
3. 2.05
5. Yes
7. Yes
9. $2419.62; $92,580 < \mu < $97,420$
11. 0.823 sec; 4.42 sec $< \mu < 6.06$ sec
13. 62
15. 250
17. 318.1
19. $\mu = 318.10 \pm 56.01$
21. $30.0°C < \mu < 30.8°C$; it is unrealistic to know σ.
23. $141.4 < \mu < 203.6$; it is unrealistic to know σ.
25. 217
27. Answer varies, but using 0 and 8 for minimum and maximum yields 174.
29. The range is 61, so σ is estimated to be 15.25 by the range rule of thumb, and the sample size is 100. The sample standard deviation is $s = 11.6$, which results in a sample size of 58. The sample size of 58 is likely to be better because s is a better estimate of σ than range/4.
31. 105

Section 6-4

1. $t_{\alpha/2} = 2.776$
3. Neither the normal nor the t distribution applies.
5. $t_{\alpha/2} = 1.662$
7. $z_{\alpha/2} = 2.33$
9. 60; $436 < \mu < 556$
11. $112.84 < \mu < 121.56$; there is 95% confidence that the interval from 112.84 to 121.56 contains the true value of the population mean μ.
13. a. $1651.6 < \mu < 2098.8$
    b. The two confidence intervals have considerable overlap, and they do not appear to be dramatically different.
15. a. 1.7 in. $< \mu < 7.1$ in.
    b. The CI does not include 0 in. The claim is supported because the CI shows that the difference is very likely to be positive, indicating that the husbands tend to be taller than their wives.

17. a. $164 < \mu < 186$
    b. $111 < \mu < 137$
    c. 186
    d. The confidence intervals do not overlap at all, suggesting that the two population means are likely to be significantly different, and the mean heart rate for those who manually shovel snow appears to be higher than the mean heart rate for those who use the electric snow thrower.
19. 95% CI for 4000 BC: $125.7 < \mu < 131.6$;
    95% CI for 150 AD: $130.1 < \mu < 136.5$;
    The two CIs overlap, so it is possible that the two population means are the same, and we cannot conclude that head sizes appear to have changed.
21. a. $24.53 < \mu < 27.47$
    b. $23.10 < \mu < 28.38$
    c. Men and women appear to have very similar mean body mass indexes.
23. Setosa: $3.31 < \mu < 3.53$; Versicolor: $2.68 < \mu < 2.86$; Virginica: $2.88 < \mu < 3.07$. There are dramatic differences. The confidence intervals do not overlap at all, suggesting that the means are all different.
25. The question requires careful thought. The confidence interval is $4.5 < \mu < 4.9$, which seems to suggest that the sample is acceptable because the confidence interval suggests that it is very likely that the *mean* cell count is between 4 and 5. However, 244 of the 400 individual cell counts are not between 4 and 5, which suggests that the sample is not acceptable. The sample is acceptable only if the *mean* must fall between 4 and 5, but it is not acceptable if the individual values must be between 4 and 5.
27. The CI limits will be closer together than they should be.

Section 6-5

1. 6.262, 27.488
3. 51.172, 116.321
5. $9388 < \sigma < $18,030$
7. 2.06 sec $< \sigma < 3.20$ sec
9. 191
11. 133,448; no
13. $239.5 < \sigma < 601.5$
15. $1.195 < \sigma < 4.695$; yes, the confidence interval is likely to be a poor estimate because the value of 5.40 appears to be an outlier, suggesting that the assumption of a normally distributed population is not correct.
17. a. $9.2 < \sigma < 14.3$
    b. $10.1 < \sigma < 15.8$
    c. The population standard deviations do not appear to be significantly different.
19. a. $2.62 < \sigma < 4.71$
    b. $4.71 < \sigma < 8.46$
    c. With a 99% confidence level, the CIs just barely overlap at the value of 4.71, so it appears that the two populations have different amounts of variation.

Chapter 6 Review Exercises

1. a. 26.0%
   b. $23.4\% < p < 28.7\%$
   c. 2653
2. a. $4.28\ m < \mu < 5.62\ m$
   b. $3.65\ m < \mu < 5.31\ m$
   c. The confidence intervals overlap, suggesting that the two population means are not dramatically different.
3. a. $1.08\ m < \sigma < 2.08\ m$
   b. $1.36\ m < \sigma < 2.60\ m$
   c. The confidence intervals overlap, suggesting that the two population standard deviations are not dramatically different.
4. a. 50.4%
   b. $45.7\% < p < 55.1\%$
   c. No, perhaps respondents are trying to impress the pollsters, or perhaps their memories have a tendency to indicate that they voted for the winner.
5. 2944
6. 221
7. $126.388 < \mu < 139.807$; $121.114 < \mu < 132.691$. The confidence intervals overlap, suggesting that the means might be equal or close in value. It does not appear that exercise has an effect on the mean systolic blood pressure reading taken during the time of stress caused by the math test.

Chapter 6 Cumulative Review Exercises

1. a. 121.0 lb
   b. 123.0 lb
   c. 119 lb, 128 lb
   d. 116.5 lb
   e. 23.0 lb
   f. $56.8\ lb^2$
   g. 7.5 lb
   h. 119.0 lb
   i. 123.0 lb
   j. 127.0 lb
   k. ratio
   l.

   m. $112.6\ lb < \mu < 129.4\ lb$
   n. $4.5\ lb < \sigma < 18.4\ lb$
   o. 95
   p. The individual supermodel weights do not appear to be considerably different from weights of randomly selected women, because they are all within 1.31 standard deviations of the mean of 143 lb. However, when considered as a group, their mean is significantly less than the mean of 143 lb (see part m).

2. a. 0.0089
   b. $0.260 < p < 0.390$
   c. Because the confidence interval limits do not contain 0.25, it is unlikely that the expert is correct.
3. a. 39.0%
   b. $36.1\% < p < 41.9\%$
   c. Yes, because the entire confidence interval is below 50%.
   d. The required sample size depends on the confidence level and the sample proportion, not the population size.

Chapter 7 Answers

Section 7-2

1. There is not sufficient evidence to support the claim that the gender selection method is effective.
3. There does appear to be sufficient evidence to support the claim that the majority of Americans are opposed to the cloning of humans.
5. $H_0: \mu = \$50,000$. $H_1: \mu > \$50,000$.
7. $H_0: p = 0.5$. $H_1: p > 0.5$.
9. $H_0: \sigma = 2.8$. $H_1: \sigma < 2.8$.
11. $H_0: \mu = 12$. $H_1: \mu < 12$.
13. $z = \pm1.96$
15. $z = 2.33$
17. $z = \pm1.645$
19. $z = -2.05$
21. -13.45
23. 1.85
25. 0.2912
27. 0.0512
29. 0.0244
31. 0.4412
33. There is sufficient evidence to support the claim that the proportion of married women is greater than 0.5.
35. There is not sufficient evidence to support the claim that the proportion of fatal commercial aviation crashes is different from 0.038.
37. Type I: Support $p > 0.5$ when in reality $p = 0.5$. Type II: Fail to reject $p = 0.5$ (and therefore fail to support $p > 0.5$) when in reality $p > 0.5$.
39. Type I: Support $p \neq 0.038$ when in reality $p = 0.038$. Type II: Fail to reject $p = 0.038$ (and therefore fail to support $p \neq 0.038$) when in reality $p \neq 0.038$.
41. P-value $= 0.9999$. With an alternative hypothesis of $p > 0.5$, it is impossible for a sample statistic of 0.27 to fall in the critical region. No sample proportion less than 0.5 can ever support a claim that $p > 0.5$.
43. a. 0.9977 is very high, indicating that for this test, there is a very high chance of rejecting $p = 0.5$ when the true value of p is actually 0.45.
    b. 0.0023

Section 7-3

1. a. $z = -0.12$
   b. $z = \pm 1.96$
   c. 0.9044
   d. There is not sufficient evidence to warrant rejection of the claim that green-flowered peas occur at a rate of 25%.
   e. No, a hypothesis test cannot be used to prove that a proportion is equal to some claimed value.

3. H_0: $p = 0.5$. H_1: $p > 0.5$. Test statistic: $z = 3.00$. Critical value: $z = 2.33$. P-value: 0.0013. Reject H_0. There is sufficient evidence to support the claim that for couples using this method, the proportion of girls is greater than 0.5.

5. H_0: $p = 0.5$. H_1: $p < 0.5$. Test statistic: $z = -2.83$. Critical value: $z = -1.645$. P-value: 0.0023. Reject H_0. There is sufficient evidence to support the claim that for couples using this method, the proportion of boys is less than 0.5.

7. H_0: $p = 0.10$. H_1: $p < 0.10$. Test statistic: $z = -1.06$. Critical value: $z = -1.645$. P-value: 0.1446. Fail to reject H_0. There is not sufficient evidence to support the claim that less than 10% of all adults say that cloning of humans should be allowed. Newspapers can (and do) run any headlines they want, but it would be misleading to claim that less than 10% of adults approve of cloning.

9. H_0: $p = 0.000340$. H_1: $p \neq 0.000340$. Test statistic: $z = -0.66$. Critical values: $z = \pm 2.81$. P-value: 0.5092. Fail to reject H_0. There is not sufficient evidence to support the claim that the rate is different from 0.0340%. Cell phone users should not be concerned about cancer of the brain or nervous system.

11. H_0: $p = 0.5$. H_1: $p > 0.5$. Test statistic: $z = 0.83$. Critical value: $z = 1.28$. P-value: 0.2033. Fail to reject H_0. There is not sufficient evidence to support the claim that the majority are smoking a year after treatment. The nicotine patch therapy does appear to help around half of all subjects.

13. H_0: $p = 0.05$. H_1: $p < 0.05$. Test statistic: $z = -1.25$ (using $x = 7$). Critical value: $z = -2.33$. P-value: 0.1056. Fail to reject H_0. There is not sufficient evidence to support the claim that fewer than 5% of all Ziac users experience dizziness.

15. H_0: $p = 0.078$. H_1: $p < 0.078$. Test statistic: $z = -2.35$. Critical value: $z = -2.33$. P-value: 0.0094. Reject H_0. There is sufficient evidence to support the claim that the air-bag hospitalization rate is lower than 7.8%.

17. H_0: $p = 0.5$. H_1: $p > 0.5$. Test statistic: $z = 15.62$. Critical value: $z = 1.645$. P-value: 0.0001. Reject H_0. There is sufficient evidence to support the claim that males are depicted in a majority of nonreproductive anatomy illustrations.

19. a. H_0: $p = 0.10$. H_1: $p \neq 0.10$. Test statistic: $z = 2.00$. Critical values: $z = \pm 1.96$. Reject H_0. There is sufficient evidence to warrant rejection of the claim that the proportion of zeros is 0.1.
    b. H_0: $p = 0.10$. H_1: $p \neq 0.10$. Test statistic: $z = 2.00$. P-value: 0.0456. There is sufficient evidence to warrant rejection of the claim that the proportion of zeros is 0.1.

c. $0.0989 < p < 0.139$; because 0.1 is contained within the confidence interval, fail to reject H_0: $p = 0.10$. There is not sufficient evidence to warrant rejection of the claim that the proportion of zeros is 0.1.
d. The traditional and P-value methods both lead to rejection of the claim, but the confidence interval method does not lead to rejection.

21. 0.9909; power is 0.0091. The probability of making a type II error is very high (0.9909), and the power of 0.0091 is very low. Under these conditions, it is very likely that the hypothesis test will result in a wrong conclusion.

Section 7-4

1. Yes
3. No
5. $z = 1.18$, P-value: 0.1190; critical value: $z = 1.645$. There is not sufficient evidence to support the claim that the mean is greater than 118.
7. $z = 0.89$, P-value: 0.3734; critical values: $z = \pm 2.575$. There is not sufficient evidence to warrant rejection of the claim that the mean equals 5.00 sec.
9. H_0: $\mu = 30.0$. H_1: $\mu > 30.0$. Test statistic: $z = 1.84$. P-value: 0.0329. (Critical value: $z = 1.645$.) Reject H_0. There is sufficient evidence to support the claim that the mean is greater than 30.0°C.
11. H_0: $\mu = 200.0$. H_1: $\mu \neq 200.0$. Test statistic: $z = -1.46$. P-value: 0.1442. (Critical values: $z = \pm 2.575$.) Fail to reject H_0. There is not sufficient evidence to warrant rejection of the claim that the mean equals 200.0.
13. a. It is not likely that σ is known.
    b. 2.10
    c. No, so the assumption that $\sigma = 0.62$ is a safe assumption in the sense that if σ is actually some value other than 0.62, it is very unlikely that the result of the hypothesis test would be affected.
15. 98.45

Section 7-5

1. Student t
3. Normal
5. Between 0.005 and 0.01
7. Less than 0.01
9. $t = 0.745$. P-value is greater than 0.10. Critical value: $t = 1.729$. There is not sufficient evidence to support the claim that the mean is greater than 118.
11. $t = 0.900$; P-value is greater than 0.20. Critical values: $t = \pm 2.639$. There is not sufficient evidence to warrant rejection of the claim that the mean equals 5.00 sec.
13. H_0: $\mu = 3.39$. H_1: $\mu \neq 3.39$. Test statistic: $t = 1.734$. P-value: 0.103. Fail to reject H_0. There is not sufficient evidence to warrant rejection of the claim that the mean equals 3.39 kg (assuming a significance level of 0.05). There is not sufficient evidence to conclude that the vitamin supplement has an effect on birth weight.

15. $H_0: \mu = 0$. $H_1: \mu > 0$. Test statistic: $t = 4.685$. P-value is less than 0.005. Critical value: $t = 2.539$. Reject H_0. There is sufficient evidence to support the claim that the mean is greater than 0.

17. $H_0: \mu = 0.3$. $H_1: \mu < 0.3$. Test statistic: $t = -0.119$. P-value is greater than 0.10. Critical value: $t = -1.753$. Fail to reject H_0. There is not sufficient evidence to support the claim that the mean is less than 0.3 g.

19. $H_0: \mu = 10.5$ $H_1: \mu < 10.5$ Test statistic: $t = -0.095$. The critical values depend on the significance level, but the test statistic will not fall in the critical region for any reasonable choices. P-value is greater than 0.10. Fail to reject H_0. There is not sufficient evidence to support the claim that the mean is less than 10.5 sec. Because the data are taken in consecutive Olympic games, the population mean is changing as athletes become faster. We cannot conclude that future times should be around 10.5 sec.

21. $H_0: \mu = 3420$. $H_1: \mu > 3420$. Test statistic: $t = 2.857$. P-value is less than 0.005. Critical value: $t = 1.662$ (approximately). Reject H_0. There is sufficient evidence to support the claim that Norwegian newborn babies have weights with a mean greater than 3420 g.

23. $H_0: \mu = 0.92$ kg. $H_1: \mu < 0.92$ kg. Test statistic: $t = -2.327$. P-value is between 0.025 and 0.05. Critical value: $t = -2.132$ (assuming a 0.05 significance level). Reject H_0. There is sufficient evidence to support the claim that fertilized trees have a mean weight less than 0.92 kg. The fertilizer appears to have the effect of decreasing weights.

25. More likely to reject the null hypothesis because the critical t values are more extreme than the corresponding critical z values.

Section 7-6

1. Test statistic: $\chi^2 = 8.444$. Critical values: $\chi^2 = 8.907, 32.852$. P-value: Between 0.02 and 0.05. Reject H_0. There is sufficient evidence to support the claim that $\sigma \neq 15$.

3. Test statistic: $\chi^2 = 10.440$. Critical value: $\chi^2 = 14.257$. P-value: Less than 0.005. Reject H_0. There is sufficient evidence to support the claim that $\sigma < 50$.

5. $H_0: \sigma = 2.11$. $H_1: \sigma < 2.11$. Test statistic: $\chi^2 = 9.066$. Critical value: $\chi^2 = 67.328$ (approximately). (P-value: Less than 0.005.) Reject H_0. There is sufficient evidence to support the claim that the standard deviation is less than 2.11°F.

7. $H_0: \sigma = 2.5$. $H_1: \sigma < 2.5$. Test statistic: $\chi^2 = 3.831$. Critical value: $\chi^2 = 8.672$. (P-value: Less than 0.005.) Reject H_0. There is sufficient evidence to support the claim that the heights of supermodels vary less than the heights of women in general.

9. $H_0: \sigma = 0.15$. $H_1: \sigma < 0.15$. Test statistic: $\chi^2 = 44.800$. Critical value: $\chi^2 = 51.739$. (P-value: Between 0.005 and 0.01.) Reject H_0. There is sufficient evidence to support the claim that the variation is lower with the new machine. The purchase of the new machine should be considered.

11. $H_0: \sigma = 28.7$. $H_1: \sigma \neq 28.7$. Test statistic: $\chi^2 = 32.818$. Critical values: $\chi^2 = 24.433, 59.342$ (approximately). (P-value: Greater than 0.20.) Fail to reject H_0. There is not sufficient evidence to warrant rejection of the claim that the standard deviation is 28.7 lb.

13. a. Estimated values: 73.772, 129.070; Table A-4 values: 74.222, 129.561.

    b. 116.643, 184.199

Chapter 7 Review Exercises

1. a. No, the sample is a voluntary response sample, so the results do not necessarily apply to the population of adult Americans.

    b. No, even though there does appear to be statistically significant weight loss, the average amount of lost weight is so small that the drug is not practical.

    c. 0.001, because this P-value corresponds to results that provide the most support for the effectiveness of the cure.

    d. There is not sufficient evidence to support the claim that the mean is greater than 12 oz.

    e. rejecting a true null hypothesis.

2. a. $H_1: \mu < \$10,000$; Student t distribution.

    b. $H_1: \sigma > 1.8$ sec; chi-square distribution.

    c. $H_1: p > 0.5$; normal distribution.

    d. $H_1: \mu \neq 100$; normal distribution.

3. $H_0: p = 0.10$. $H_1: p < 0.10$. Test statistic: $z = -0.94$. Critical value: $z = -1.645$. P-value: 0.1736. Fail to reject H_0. There is not sufficient evidence to support the claim that the proportion of tree deaths for treated trees is less than 10%.

4. No, the researcher should use a hypothesis test to judge the significance of the results. $H_0: p = 0.5$. $H_1: p < 0.5$. Test statistic: $z = -1.58$. Critical value: $z = -1.645$. P-value: 0.0571. Fail to reject H_0. There is not sufficient evidence to support the claim that the proportion is less than 0.5 (or 50%).

5. $H_0: \mu = 12$ oz. $H_1: \mu < 12$ oz. Test statistic: $t = -4.741$. P-value is less than 0.005. Critical value: $t = -1.714$ (assuming $\alpha = 0.05$). Reject H_0. There is sufficient evidence to support the claim that the mean is less than 12 oz. Windsor's argument is not valid.

6. $H_0: \mu = 25$ mg. $H_1: \mu \neq 25$ mg. Test statistic: $t = -0.744$. P-value is greater than 0.10. Critical values: $t = \pm 2.145$. Fail to reject H_0. There is not sufficient evidence to warrant rejection of the claim that the pills come from a population in which the mean amount of atorvastatin is equal to 25 mg.

7. $H_0: \sigma = 0.5$. $H_1: \sigma > 0.5$. Test statistic: $\chi^2 = 19.509$. Critical value: $\chi^2 = 23.685$. (P-value: Greater than 0.10.) Fail to reject H_0. There is not sufficient evidence to support the claim that the standard deviation is greater than 0.5. If the standard deviation is too large, the variation is too large and some pills will have too much atorvastatin while others do not have enough.

8. $H_0: \mu = 62.2$. $H_1: \mu > 62.2$. Test statistic: $t = 3.164$. P-value is between 0.005 and 0.01. Critical value: $t = 2.896$. Reject H_0. There is not sufficient evidence to support the claim that the mean weight is higher with the enriched feed.

9. H_0: $\mu = 1.39$. H_1: $\mu < 1.39$. Test statistic: $t = -38.817$. P-value is less than 0.005. Critical value: $t = -2.364$ (approximately). Reject H_0. There is sufficient evidence to support the claim that the seat belt sample comes from a population with a mean of less than 1.39 days. The seat belts do seem to help.

Chapter 7 Cumulative Review Exercises

1. a. 0.0793 ng/m³
   b. 0.044 ng/m³
   c. 0.0694 ng/m³
   d. 0.0048
   e. 0.158 ng/m³
   f. $0.0259 < \mu < 0.1326$
   g. H_0: $\mu = 0.16$. H_1: $\mu < 0.16$. Test statistic: $t = -3.491$. P-value is less than 0.005. Critical value: $t = -1.860$. Reject H_0. There is sufficient evidence to support the claim that the mean is less than 0.16 ng/m³.
   h. Yes, the data are listed in order and there appears to be a trend of decreasing values. The population is changing over time.
2. a. 0.8023
   b. 0.3324
   c. 0.9713
   d. 4054 g
3. a. 124.23, 22.52
   b. H_0: $\mu = 114.8$. H_1: $\mu \neq 114.8$. Test statistic: $t = 1.675$. P-value is between 0.10 and 0.20. Critical value: $t = \pm 2.131$. Fail to reject H_0. There is not sufficient evidence to warrant rejection of the claim that the sample comes from a population with a mean blood pressure equal to 114.8.
   c. $112.23 < \mu < 136.23$. The confidence interval limits contain the value of 114.8.
   d. H_0: $\sigma = 13.1$. H_1: $\sigma \neq 13.1$. Test statistic: $\chi^2 = 44.339$. Critical values: $\chi^2 = 6.262, 27.488$. (P-value: Less than 0.01.) Reject H_0. There is sufficient evidence to reject the claim that the standard deviation is equal to 13.1
   e. There does not appear to be a significant difference between the sample mean and 114.8 for healthy women, but the amount of variation does appear to be affected. If the standard deviation for the infected group becomes large enough, there will be more women with blood pressure readings that are too low or too high, so there would be more women whose health would be adversely affected.

Chapter 8 Answers

Section 8-2

1. 30
3. 85
5. a. 0.417
   b. 2.17
   c. ± 1.96
   d. 0.0300

7. $0.00216 < p_1 - p_2 < 0.00623$; it appears that physical activity corresponds to a lower rate of coronary heart disease.
9. H_0: $p_1 = p_2$. H_1: $p_1 \neq p_2$. Test statistic: $z = 2.43$. P-value: 0.0150. Critical values for $\alpha = 0.05$: $z = \pm 1.96$. Critical values for $\alpha = 0.01$: $z = \pm 2.575$. Difference is significant at the 0.05 level, but not the 0.01 level.
11. a. H_0: $p_1 = p_2$. H_1: $p_1 > p_2$. Test statistic: $z = 12.63$. P-value: 0.0001. Critical value: $z = 2.33$. Reject H_0. There is sufficient evidence to support the claim that men have a higher rate of red/green color blindness than women.
    b. $0.0573 < p_1 - p_2 < 0.117$; because the difference in percentages is estimated to be between 5.73% and 11.7%, there does appear to be a substantial difference.
    c. We need a large sample for women so that the requirements of $np \geq 5$ and $nq \geq 5$ are both satisfied.
13. H_0: $p_1 = p_2$. H_1: $p_1 > p_2$. Test statistic: $z = 3.86$. P-value: 0.0001. Reject H_0. Critical value: $z = 2.33$. There is sufficient evidence to support the claim that the proportion of drinkers among convicted arsonists is greater than the proportion of drinkers convicted of fraud.
15. a. H_0: $p_1 = p_2$. H_1: $p_1 \neq p_2$. Test statistic: $z = -3.06$. Critical values: $z = \pm 2.575$ (assuming a 0.01 significance level). P-value: 0.0022. Reject H_0: $p_1 = p_2$. There is sufficient evidence to support the claim that the two population percentages are different.
    b. $-0.0823 < p_1 - p_2 < -0.00713$; because the confidence interval limits do not contain 0, there appears to be a significant difference.
17. a. H_0: $p_1 = p_2$. H_1: $p_1 > p_2$. Test statistic: $z = 4.48$ (using $x_1 = 51$ and $x_2 = 15$). P-value: 0.0001. Critical value: $z = 1.645$ (assuming a 0.05 significance level). Reject H_0. There is sufficient evidence to support the claim that the Viagra group has dyspepsia at a rate greater than the rate for those who do not take Viagra.
    b. $0.0277 < p_1 - p_2 < 0.0699$. The confidence interval does not contain zero, suggesting that there is a significant difference between the two population proportions. It appears that that the Viagra group has dyspepsia at a rate greater than the rate for those who do not take Viagra.
19. H_0: $p_1 = p_2$. H_1: $p_1 \neq p_2$. Test statistic: $z = 3.12$. P-value: 0.0018. Critical values: $z = \pm 1.96$ (assuming a 0.05 significance level). Reject H_0. There is sufficient evidence to support the claim that there is a difference in success rates. It appears that surgery is better.
21. a. $0.0227 < p_1 - p_2 < 0.217$; because the confidence interval limits do not contain 0, it appears that $p_1 = p_2$ can be rejected.
    b. $0.491 < p_1 < 0.629$; $0.371 < p_2 < 0.509$; because the confidence intervals do overlap, it appears that $p_1 = p_2$ cannot be rejected.
    c. H_0: $p_1 = p_2$. H_1: $p_1 \neq p_2$. Test statistic: $z = 2.40$. P-value: 0.0164. Critical values: $z = \pm 1.96$. Reject H_0. There is sufficient evidence to reject $p_1 = p_2$.
    d. Reject $p_1 = p_2$. Least effective: Using the overlap between the individual confidence intervals.

23. $H_0: p_1 - p_2 = 0.10$. $H_1: p_1 - p_2 \neq 0.10$. Test statistic: $z = 1.26$. P-value: 0.2076. Critical values: $z = \pm 1.96$. Fail to reject H_0. There is not sufficient evidence to warrant rejection of the claim that the headache rate for Viagra users is 10 percentage points more than the percentage for those who are given a placebo.

Section 8-3

1. Independent samples
3. Matched pairs
5. $H_0: \mu_1 = \mu_2$. $H_1: \mu_1 > \mu_2$. Test statistic: $t = 2.790$. Critical value: $t = 2.390$. P-value < 0.005. (Using TI-83/84 Plus calculator: df $= 122$, P-value $= 0.003$.) Reject H_0. There is sufficient evidence to support the claim that the population of heavy marijuana users has a lower mean than the light users. Heavy users of marijuana should be concerned about deteriorating cognitive abilities.
7. $-0.61 < \mu_1 - \mu_2 < 0.71$ (TI-83/84 Plus calculator: df $= 34$ and $-0.59 < \mu_1 - \mu_2 < 0.69$.) Because the confidence interval does contain zero, we should not conclude that the two population means are different. There is not sufficient evidence to support the claim that the magnets are effective in reducing pain.
9. $-0.01 < \mu_1 - \mu_2 < 0.23$; because this CI contains zero, there does not appear to be a significant difference between the two population means, so it does not appear that obsessive–compulsive disorders have a biological basis. (With a TI-83/84 Plus calculator, df $= 18$ and $0.01 < \mu_1 - \mu_2 < 0.21$, which does not contain zero, suggesting that there is a significant difference so that obsessive–compulsive disorders appear to have a biological basis. This is a rare case where the simple and conservative estimate of df leads to a different conclusion than the more accurate Formula 8-1.)
11. $1.46 < \mu_1 - \mu_2 < 3.52$ (TI-83/84 Plus calculator: df $= 25$ and $1.47 < \mu_1 - \mu_2 < 3.51$.) Because the confidence interval does not contain zero, there appears to be a significant difference between the two population means. It does appear that there are significantly more errors made by those treated with alcohol.
13. $-2.226 < \mu_1 - \mu_2 < -0.070$ (TI-83/84 Plus calculator: df $= 4$ and $-2.211 < \mu_1 - \mu_2 < -0.089$.)
15. $H_0: \mu_1 = \mu_2$. $H_1: \mu_1 \neq \mu_2$. Test statistic: $t = -39.614$. Critical values: $t = \pm 2.009$ (approximately). P-value: Less than 0.01. (Using TI-83/84 Plus calculator: df $= 62$, P-value $= 0.0000$.) Reject H_0. There is sufficient evidence to warrant rejection of the claim that setosa and versicolor irises have the same mean petal length.
17. Men: $n_1 = 40$, $\bar{x}_1 = 25.9975$, $s_1 = 3.430742$. Women: $n_2 = 40$, $\bar{x}_2 = 25.7400$, $s_2 = 6.16557$. $H_0: \mu_1 = \mu_2$. $H_1: \mu_1 \neq \mu_2$. Test statistic: $t = 0.231$. Critical values: $t = \pm 2.024$ (assuming a 0.05 significance level). P-value: Greater than 0.20. (Using TI-83/84 Plus calculator: df $= 61$, P-value $= 0.818$.) Fail to reject H_0. There is not sufficient evidence to warrant rejection of the claim that the mean BMI of men is equal to the mean BMI of women.
19. $H_0: \mu_1 = \mu_2$. $H_1: \mu_1 \neq \mu_2$. Test statistic: $t = 3.343$. Critical values: $t = \pm 2.009$ (assuming a 0.05 significance level). P-value: Less than

0.01. (Using TI-83/84 Plus calculator: df $= 97$, P-value $= 0.0012$.) Reject H_0. There is sufficient evidence to support the claim that two-month-old boys and two-month-old girls have different mean head circumferences.
21. $H_0: \mu_1 = \mu_2$. $H_1: \mu_1 > \mu_2$. Test statistic: $t = 2.330$. Critical value: $t = 2.364$ (approximately). P-value: Greater than 0.01. (Using TI-83/84 Plus calculator: df $= 374$, P-value $= 0.0102$.) Fail to reject H_0. There is not sufficient evidence to support the claim that the population of children not wearing seat belts has a higher mean number of days spent in ICU. Based on these particular results, there is not significant evidence in favor of seat belt use among children. However, more data have been used to provide significant evidence in favor of seat belt use among children.
23. $H_0: \mu_1 = \mu_2$. $H_1: \mu_1 < \mu_2$. Test statistic: $t = -2.908$. Critical value: $t = -1.653$ (approximately). P-value: Less than 0.005. (Using TI-83/84 Plus calculator: df $= 374$, P-value $= 0.0019$.) Reject H_0. There is sufficient evidence to support the claim that prenatal cocaine exposure is associated with lower scores of four-year-old children on the test of object assembly.
25. $H_0: \mu_1 = \mu_2$. $H_1: \mu_1 > \mu_2$. Test statistic: $t = 2.785$. Critical value: $t = 2.364$ (approximately). P-value < 0.005. (Using TI-83/84 Plus calculator: df $= 127$, P-value $= 0.0031$.) Reject H_0. There is sufficient evidence to support the claim that the population of heavy marijuana users has a lower mean than the light users. Heavy users of marijuana should be concerned about deteriorating cognitive abilities.
27. $-0.59 < \mu_1 - \mu_2 < 0.69$ (TI-83/84 Plus calculator: df $= 34$ and $-0.59 < \mu_1 - \mu_2 < 0.69$.) Because the confidence interval does contain zero, we should not conclude that the two population means are different. There is not sufficient evidence to support the claim that the magnets are effective in reducing pain.
29. a. 50/3
    b. 2/3
    c. 50/3 + 2/3 = 52/3
    d. The range of the x-y values equals the range of the x values plus the range of the y values.

Section 8-4

1. a. -0.2
   b. 2.8
   c. $t = -0.161$
   d. ± 2.776
3. $-3.6 < \mu_d < 3.2$
5. a. $H_0: \mu_d = 0$. $H_1: \mu_d \neq 0$. Test statistic: $t = -1.532$. Critical values: $t = \pm 2.228$. P-value: 0.157. Fail to reject H_0. There is not sufficient evidence to warrant rejection of the claim that there is no difference between the yields from the two types of seed.
   b. $-2.678 < \mu_d < 0.496$
   c. No

7. a. $H_0: \mu_d = 0$. $H_1: \mu_d \neq 0$. Test statistic: $t = -0.984$. Critical values: $t = \pm 2.201$. P-value: 0.346 approximately. Fail to reject H_0. There is not sufficient evidence to support the claim that there is a difference between self-reported heights and measured heights.

   b. $-3.2 < \mu_d < 1.2$; because the confidence interval limits contain 0, there is not sufficient evidence to support the claim that there is a difference between self reported heights and measured heights.

9. a. $0.69 < \mu_d < 5.56$

   b. $H_0: \mu_d = 0$. $H_1: \mu_d > 0$. Test statistic: $t = 3.036$. Critical value: $t = 1.895$. P-value: 0.009. Reject H_0. There is sufficient evidence to support the claim that the sensory measurements are lower after hypnosis.

   b. Yes

11. a. $H_0: \mu_d = 0$. $H_1: \mu_d \neq 0$. Test statistic: $t = -0.41$. P-value: 0.691. Fail to reject H_0. There is not sufficient evidence to support the claim that astemizole has an effect. Don't take astemizole for motion sickness.

   b. 0.3455; there is not sufficient evidence to support the claim that astemizole prevents motion sickness.

13. $H_0: \mu_d = 0$. $H_1: \mu_d \neq 0$. Test statistic: $t = -0.501$. Critical values: $t = \pm 2.201$. P-value: 0.626. Fail to reject H_0. There is not sufficient evidence to support the claim that there is a difference between self-reported weights and measured weights of males aged 12–16.

15. a. $-1.40 < \mu_d < -0.17$

   b. $H_0: \mu_d = 0$. $H_1: \mu_d \neq 0$. Test statistic: $t = -2.840$. Critical values: $t = \pm 2.228$. P-value < 0.02. Reject H_0. There is sufficient evidence to warrant rejection of the claim that the mean difference is 0. Morning and night body temperatures do not appear to be about the same.

17. a. Test statistic: $t = 1.861$. Critical value: $t = 1.833$. P-value: 0.045 Reject H_0. There is sufficient evidence to support $\mu_d > 0$.

   b. Test statistic: $t = 1.627$. Critical value: $t = 1.833$. P-value: 0.072. Fail to reject H_0. There is not sufficient evidence to support $\mu_1 > \mu_2$.

   c. Yes, the conclusion is affected by the test that is used.

Section 8-5

1. 4.55; $2.99 < OR < 6.93$. The confidence interval does not contain 1, indicating that Viagra appears to affect headaches.

3. 0.47; $0.26 < OR < 0.86$

5. 21.39; $13.39 < OR < 34.19$

7. 0.04; $0.01 < OR < 0.16$

9. a.

	Mouth or throat soreness	No mouth or throat soreness
Nicorette treatment	43	109
Placebo	35	118

   b. 1.33

c. $0.79 < OR < 2.23$. We are 95% confident that the limits of 0.79 and 2.23 contain the true odds ratio. The Nicorette treatment group had a soreness rate of 28%, compared to 23% for the placebo group. It appears that Nicorette affects soreness of the mouth or throat, and Nicorette users are more likely to experience that adverse reaction.

11. The results are the same.

13. a. Division by 0 causes a calculation error.

   b. Division by 0 causes a calculation error.

   c. $4.07 < OR < 1133.00$. The result makes sense, because it indicates that the treatment affects disease. Among those treated, 25% are diseased, compared to 0% for those in the placebo group.

Section 8-6

1. $H_0: \sigma_1^2 = \sigma_2^2$. $H_1: \sigma_1^2 \neq \sigma_2^2$. Test statistic: $F = 2.2500$. Upper critical value: $F = 2.1540$. Reject H_0. There is sufficient evidence to support the claim that the treatment and placebo populations have different variances.

3. $H_0: \sigma_1^2 = \sigma_2^2$. $H_1: \sigma_1^2 > \sigma_2^2$. Test statistic: $F = 2.1267$. The critical F value is between 2.1555 and 2.2341. Fail to reject H_0. There is not sufficient evidence to support the claim that the pain reductions for the sham treatment group vary more than the pain reductions for the magnet treatment group.

5. $H_0: \sigma_1^2 = \sigma_2^2$. $H_1: \sigma_1^2 > \sigma_2^2$. Test statistic: $F = 3.7539$. Critical value: $F = 3.4445$. Reject H_0. There is sufficient evidence to support the claim that king-size cigarettes with filters have amounts of nicotine that vary more than the amounts of nicotine in nonfiltered king-size cigarettes.

7. $H_0: \sigma_1 = \sigma_2$. $H_1: \sigma_1 > \sigma_2$. Test statistic: $F = 1.1842$. Critical value of F is between 1.0000 and 1.3519, but it is much closer to 1.3519. (The P-value from a TI-83/84 Plus calculator is 0.0755.) Fail to reject H_0. There is not sufficient evidence to support the claim that birth weights for the placebo population have a greater variance than birth weights for the population treated with zinc supplements.

9. $H_0: \sigma_1^2 = \sigma_2^2$. $H_1: \sigma_1^2 \neq \sigma_2^2$. Test statistic: $F = 2.59$. P-value: 0.0599. Fail to reject H_0. There is not sufficient evidence to warrant rejection of the claim that the two samples come from populations with the same variance.

11. $0.41 < \sigma_1^2/\sigma_2^2 < 3.96$ (approximately)

Chapter 8 Review Exercises

1. a. $H_0: p_1 = p_2$. $H_1: p_1 < p_2$. Test statistic: $z = -2.82$. Critical value: $z = -1.645$. P-value: 0.0024. Reject H_0. There is sufficient evidence to support the stated claim. It appears that surgical patients should be routinely warmed.

   b. 90%

   c. $-0.205 < p_1 - p_2 < -0.0543$

   d. No, the conclusions may be different.

2. a. $-27.80 < \mu_1 - \mu_2 < 271.04$; TI-83/84 Plus calculator uses df $= 17.7$ to get limits of -17.32 and 260.56.

  b. $H_0: \mu_1 = \mu_2$. $H_1: \mu_1 \neq \mu_2$. Test statistic: $t = 1.841$. Critical values: $t = \pm 2.262$. P-value: between 0.05 and 0.10. Fail to reject H_0. There is not sufficient evidence to warrant rejection of the claim of no difference.

  c. No

3. $H_0: \sigma_1 = \sigma_2$. $H_1: \sigma_1 \neq \sigma_2$. Test statistic: $F = 1.2922$. Upper critical value: $F = 4.0260$. Fail to reject H_0. There is not sufficient evidence to support the claim that the two populations have different amounts of variation.

4. $H_0: \mu_1 = \mu_2$. $H_1: \mu_1 \neq \mu_2$. Test statistic: $t = -3.500$. Critical values: $t = \pm 2.365$. P-value: 0.010. Reject H_0. There is sufficient evidence to warrant rejection of the claim that the means are equal. Filters appear to be effective in reducing carbon monoxide.

5. $H_0: \mu_1 = \mu_2$. $H_1: \mu_1 > \mu_2$. Test statistic: $t = 2.169$. Critical value: $t = 1.650$ approximately. P-value is between 0.01 and 0.025. Reject H_0. There is sufficient evidence to support the claim that zinc supplementation is associated with increased birth weights.

6. a. $H_0: \mu_d = 0$. $H_1: \mu_d \neq 0$. Test statistic: $t = 2.301$. Critical values: $t = \pm 2.262$ (assuming a 0.05 significance level). P-value: 0.046. Reject H_0 (assuming a 0.05 significance level). There is sufficient evidence to conclude that there is a difference between pretraining and post-training weights.

  b. $0.0 < \mu_d < 4.0$

7. a. $H_0: \mu_d = 0$. $H_1: \mu_d < 0$. Test statistic: $t = -3.847$. Critical value: $t = -1.796$. P-value: 0.001. Reject H_0. There is sufficient evidence to support the claim that the Dozenol tablets are more soluble after the storage period.

  b. $-45.6 < \mu_d < -12.4$

Chapter 8 Cumulative Review Exercises

1. a. 0.0707

  b. 0.369

  c. 0.104

  d. 0.0540

  e. $H_0: p_1 = p_2$. $H_1: p_1 < p_2$. Test statistic: $z = -2.52$. Critical value: $z = -1.645$. P-value: 0.0059. Reject H_0. There is sufficient evidence to support the claim that the percentage of women ticketed for speeding is less than the percentage for men.

2. There must be an error, because the rates of 13.7% and 10.6% are not possible with sample sizes of 100.

3. a. $0.0254 < p < 0.0536$ (using $x = 29$)

  b. $0.0103 < p < 0.0311$ (using $x = 15$)

  c. $0.00133 < p_1 - p_2 < 0.0363$

  d. method iii

Chapter 9 Answers

Section 9-2

1. a. Yes, because the absolute value of the test statistic exceeds the critical values $r = \pm 0.707$.

  b. 0.986

3. a. Yes, because the absolute value of the test statistic exceeds the critical values $r = \pm 0.312$.

  b. 0.876

5. The scatterplot suggests that there is a correlation, but it is not linear. With $r = 0$ and critical values of $r = \pm 0.878$ (for a 0.05 significance level), there is not a significant linear correlation.

7. a. There appears to be a linear correlation.

  b. $r = 0.906$. Critical values: $r = \pm 0.632$ (for a 0.05 significance level). There is a significant linear correlation.

  c. $r = 0$. Critical values: $r = \pm 0.666$ (for a 0.05 significance level). There does not appear to be a significant linear correlation.

  d. The effect from a single pair of values can be very substantial, and it can change the conclusion.

9. $r = 0.796$. Critical values: $r = \pm 0.666$. There is a significant linear correlation. No, the population of all women is different from the population of supermodels, so inferences about supermodels do not necessarily apply to all women.

11. $r = 0.658$. Critical values: $r = \pm 0.532$. There is a significant linear correlation. Another issue is the accuracy of the measurements, which appear to vary widely. A study might be conducted to determine whether the subject's blood pressure really does vary considerably, or whether the measurements are in error because of other factors.

13. $r = 0.262$. Critical values: $r = \pm 0.576$. There is a not significant linear correlation. Because the cigarette counts were reported, subjects may have given wrong values. Subjects may have been exposed to varying levels of secondhand smoke.

15. $r = 0.304$. Critical values: $r = \pm 0.361$ (approximately). There is no significant linear correlation. A significant linear correlation would not suggest that the drug is effective in treating pain. If the drug is totally ineffective and each before score is equal to the after score, the linear correlation coefficient would be 1, indicating that there is a significant linear correlation.

17. $r = 0.493$. Critical values: $r = \pm 0.312$. There is a significant linear correlation.

19. $r = 0.802$. Critical values: $r = \pm 0.444$ There is a significant linear correlation.

21. $r = 0.787$. Critical values: $r = \pm 0.279$. There is a significant linear correlation. There is a significant linear correlation, as in Exercise 20, but the correlation appears to be much stronger with Versicolor irises.

23. The presence of a significant linear correlation does not necessarily mean that one of the variables *causes* the other. Correlation does not imply causality.

25. Averages tend to suppress variation among individuals, so a correlation among *averages* does not necessarily mean that there is a correlation among *individuals*.

27. a. 0.972
    b. 0.905
    c. 0.999 (largest)
    d. 0.992
    e. −0.984

29. a. There appears to be a correlation.
    b. $r = 0.915$. Critical values: $r = \pm0.497$. There is a significant linear correlation.
    c. $r = -0.213$. Critical values: $r = \pm0.707$. There is not a significant linear correlation.
    d. $r = -0.164$. Critical values: $r = \pm0.707$. There is not a significant linear correlation.
    e. Combining different groups in the same data set could make it appear that there is a significant linear correlation between two variables, even though there is no correlation

Section 9-3

1. a. 18.00
   b. 5.00
3. 401 lb
5. $\hat{y} = 2 + 0x$ (or $\hat{y} = 2$)
7. a. $\hat{y} = 0.264 + 0.906x$
   b. $\hat{y} = 2 + 0x$ (or $\hat{y} = 2$)
   c. The results are very different, indicating that one point can dramatically affect the regression equation.
9. $\hat{y} = -152 + 3.88x$; with a significant linear correlation, the best predicted value is found by using the regression equation: 116 lb.
11. $\hat{y} = -14.4 + 0.769x$; 79
13. $\hat{y} = 139 + 2.48x$; 175.2
15. $\hat{y} = 1.91 + 0.297x$; equation of regression line for a drug with no effect: $\hat{y} = x$.
17. $\hat{y} = 122 + 0.0998x$; 142
19. $\hat{y} = 16.8 + 0.736x$; 63.9 in.
21. $\hat{y} = -0.0843 + 0.331x$; 0.41 mm; given the small size of the data set, the results are not dramatically different.
23. Yes; yes. The point is far away from the others, and it does have a dramatic effect on the regression line.
25. The equation $\hat{y} = -49.9 + 27.2x$ is better because it has $r = 0.997$, which is higher than $r = 0.963$ for $\hat{y} = -103.2 + 134.9 \ln x$.
27. No.

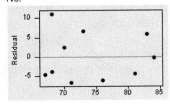

Section 9-4

1. 0.64; 64%
3. 0.253; 25.3%
5. $r = 0.934$; critical value: ± 0.279 (approximately, assuming a 0.05 significance level). Yes, significant linear correlation.
7. 273 lb
9. a. 287.37026
   b. 166.62974
   c. 454
   d. 0.63297415
   e. 4.8789597
11. a. 3696.9263
    b. 1690.5830
    c. 5387.5093
    d. 0.68620324
    e. 11.869369
13. a. 116 lb
    b. 103.7 lb $< y < 128.8$ lb
15. a. 44 ft
    b. 6.2 ft $< y < 81.4$ ft
17. -55 lb $< y < 317$ lb
19. -20 lb $< y < 277$ lb
21. $-663 < \beta_0 < -39.9$; $4.91 < \beta_1 < 14.41$

Section 9-5

1. $\hat{y} = -271.711 - 0.870x_1 + 0.554x_2 + 12.153x_3$
3. Yes, because the *P*-value is 0.000 and the adjusted R^2 value is 0.924.
5. Waist circumference, because it has the smallest *P*-value and highest adjusted R^2.
7. $\hat{y} = -206 + 2.66\text{HT} + 2.15\text{WAIST}$. That equation has the best combination of lowest *P*-value and highest value of adjusted R^2. (Using the three independent variables of height, waist, and cholesterol results in the same *P*-value and adjusted R^2, but it is better to use the equation with fewer independent variables.)
9. a. $\hat{y} = 21.6 + 6.90x$
   b. $\hat{y} = 45.7 + 0.293x$
   c. $\hat{y} = 9.80 + 0.658x_1 + 0.200x_2$ (where x_1 represents mother's height).
   d. The regression equation in part (a), because it has the best combination of small *P*-value, high adjusted R^2, and uses fewer variables. The *P*-value is 0.000 for parts (a) and (c), and the adjusted R^2 is 0.347 for part (a) and 0.366 for part (c), so part (c) might seem slightly better, but the small difference in adjusted R^2 values does not justify using three variables instead of two.
   e. The regression equation from part (a) is a good model, but the scatterplot shows points that vary considerably from the regression line, so the predicted values could be off by considerable amounts.

11. $\hat{y} = 3.06 + 2.91x_1 + 82.3x_2$ (where x_1 represents age and x_2 represents sex). The sex does appear to have a substantial effect on the weight of the bear.
    a. 61.3 lb
    b. 143.6 lb

Chapter 9 Review Exercises

1. a. $r = 0.874$. Critical values: $r = \pm 0.707$ (assuming a 0.05 significance level). There is a significant linear correlation. There is sufficient evidence to conclude that there is an association between the number of chirps in a minute and the temperature.
   b. $\hat{y} = 27.6 + 0.0523x$
   c. 92°F
   d. 76.3%
   e. No, we can only show that there is an association between two variables. We cannot show a cause/effect relationship.
2. $r = -0.069$. Critical values: $r = \pm 0.707$ (assuming a 0.05 significance level). There is no significant linear correlation. The BAC level does not appear to be related to the age of the person tested.
3. a. $r = 0.340$; critical values: $r = \pm 0.632$; no significant linear correlation.
   b. $\hat{y} = 826 + 17.1x$
   c. 1405.9 (mean)
4. a. $r = -0.157$; critical values: $r = \pm 0.632$; no significant linear correlation.
   b. $\hat{y} = 2990 - 32.5x$
   c. 1405.9 (mean)
5. a. $r = 0.892$; critical values: $r = \pm 0.632$; significant linear correlation.
   b. $\hat{y} = -724 + 0.178x$
   c. 1643 million bushels
6. $\hat{y} = -3145 + 10.1x_1 + 40.4x_2 + 0.187x_3$; 0.874; 0.811; 0.004; yes.
7. a. $r = 0.954$; critical values: $r = \pm 0.811$; significant linear correlation.
   b. $\hat{y} = 14.4 + 0.977x$
   c. No. A significant linear correlation shows only that there is an association between the two variables, but it does not show that the stress from the math test has an effect on systolic blood pressure. For example, if math test has absolutely no effect, the paired values should be roughly the same, so r would be close to 1. If the math test has the strong effect of doubling all readings, r would again be close to 1. Instead of correlation, the effect of the test should be investigated using other methods, such as those in Chapter 8.

Chapter 9 Cumulative Review Exercises

1. a. $\bar{x} = 99.1$, $s = 8.5$
   b. $\bar{x} = 102.8$, $s = 8.7$

c. No, but a better comparison would involve treating the data as matched pairs instead of two independent samples.
d. $H_0: \mu = 100$. $H_1: \mu \neq 100$. Test statistic: $t = 0.546$. Critical values: $t = \pm 2.069$. P-value is greater than 0.20. Fail to reject H_0. There is not sufficient evidence to support the claim that the mean IQ score of twins reared apart is different from the mean IQ of 100.
e. Yes. $r = 0.702$ and the critical values are $r = \pm 0.576$ (assuming a 0.05 significance level). There is a significant linear correlation. It appears that there is an association between IQ and factors of heredity.
2. a. No. Correlation can be used to investigate an association between the two variables, not whether differences between values of the two variables are significant.
   b. Test statistic: $t = 1.185$. Critical values: $t = \pm 2.262$. Fail to reject H_0: $\mu_d = 0$. There is not sufficient evidence to warrant rejection of the claim that position has no effect. Position does not appear to have a significant effect.

Chapter 10 Answers

Section 10-2

1. a. $H_0: p_1 = p_2 = p_3 = p_4$
   b. 8, 8, 8, 8
   c. $\chi^2 = 4.750$
   d. $\chi^2 = 7.815$
   e. There is not sufficient evidence to warrant rejection of the claim that the four categories are equally likely.
3. a. df = 37, so $\chi^2 = 51.805$ (approximately).
   b. $0.10 < $ P-value < 0.90
   c. There is not sufficient evidence to warrant rejection of the claim that the roulette slots are equally likely.
5. Test statistic: $\chi^2 = 2.632$. Critical value: $\chi^2 = 16.919$ (assuming a 0.05 significance level). There is not sufficient evidence to support the claim that the outcomes are not equally likely. The results appear to be as expected from an effective weighing procedure.
7. Test statistic: $\chi^2 = 9.233$. Critical value: $\chi^2 = 12.592$. There is not sufficient evidence to warrant rejection of the claim that accidents occur with equal frequency on the different days.
9. Test statistic $\chi^2 = 80.750$. Critical value: $\chi^2 = 16.919$ (assuming a 0.05 significance level). There is sufficient evidence to warrant rejection of the claim that the digits occur with the same frequency. All of the digits are even numbers, suggesting that pulse rates were recorded for 30 seconds, then doubled.
11. Test statistic: $\chi^2 = 11.552$. Critical value: $\chi^2 = 5.991$. There is sufficient evidence to warrant rejection of the claim that the offspring frequencies fit the expected distribution.
13. Test statistic: $\chi^2 = 9.658$. Critical value: $\chi^2 = 5.991$. There is sufficient evidence to warrant rejection of the claim that the actual frequencies correspond to the predicted distribution.

15. Test statistic: $\chi^2 = 28.607$. Critical value: $\chi^2 = 5.991$. There is sufficient evidence to warrant rejection of the claim that the distribution of concussions fits the distribution of exposures. Previous concussions appear to be associated with a greater likelihood of a concussion.

17. Instead of a right-tailed test, use a left-tailed test. If the sample data fit the claimed distribution almost perfectly, the test statistic will be located near the extreme left position of zero.

19. a. 0.0853, 0.2968, 0.3759, 0.1567, 0.0853
    b. 17.06, 59.36, 75.18, 31.34, 17.06
    c. Test statistic: $\chi^2 = 60.154$. Critical value: $\chi^2 = 13.277$. Reject the null hypothesis that the IQ scores come from a normally distributed population with the given mean and standard deviation. There is sufficient evidence to warrant rejection of the claim that the IQ scores were randomly selected from a normally distributed population with mean 100 and standard deviation 15.

Section 10-3

1. Test statistic: $\chi^2 = 0.413$. P-value: 0.521. There is not sufficient evidence to warrant rejection of the claim that race and ethnicity is independent of whether someone is stopped by police. There is not sufficient evidence to support a claim of racial profiling.

3. Test statistic: $\chi^2 = 32.273$. Critical value: $\chi^2 = 3.841$. There is sufficient evidence to warrant rejection of the claim that whether a subject lies is independent of the polygraph indication. It appears that the polygraph is effective, but it is not 100% accurate.

5. Test statistic: $\chi^2 = 63.908$. Critical value: $\chi^2 = 3.841$. There is sufficient evidence to warrant rejection of the claim that gender is independent of the fear of flying.

7. Test statistic: $\chi^2 = 3.062$. Critical value: $\chi^2 = 5.991$. There is not sufficient evidence to warrant rejection of the claim that success is independent of the method used. The evidence does not suggest that any method is significantly better than the others.

9. Test statistic: $\chi^2 = 65.524$. Critical value: $\chi^2 = 7.815$ (assuming a 0.05 significance level). There is sufficient evidence to warrant rejection of the claim that occupation is independent of whether the cause of death was homicide. Cashiers appear to be most vulnerable to homicide.

11. Test statistic: $\chi^2 = 19.645$. Critical value: $\chi^2 = 3.841$. There is sufficient evidence to warrant rejection of the claim that formal training is independent of how firearms are stored.

13. Test statistic: $\chi^2 = 1.199$. Critical value: $\chi^2 = 7.815$. There is not sufficient evidence to warrant rejection of the claim that getting a headache is independent of the amount of atorvastatin used as a treatment.

15. Test statistic: $\chi^2 = 1.174$. Critical value: $\chi^2 = 3.841$. There is not sufficient evidence to warrant rejection of the claim that the treatment (drug or placebo) is independent of the reaction (whether or not mouth or throat soreness was experienced). If you are thinking about using Nicorette, there should not be concern about mouth or throat soreness, but any concerns should also be weighed against the adverse effects of smoking.

17. Without Yates's correction: $\chi^2 = 0.413$. With Yates's correction: $\chi^2 = 0.270$. Yates's correction decreases the test statistic so that sample data must be more extreme to be considered significant.

Section 10-4

1. 144

3. 130

5. b and c are discordant pairs.

7. $\chi^2 = 6.635$

9. Test statistic: $\chi^2 = 2.382$. Critical value: $\chi^2 = 3.841$. Fail to reject the null hypothesis that the following two proportions are the same: (1) The proportion of subjects with no cure from the fungicide and a cure from the placebo; (2) the proportion of subjects with a cure from the fungicide and no cure from the placebo. The fungicide does not appear to be effective.

11. Test statistic: $\chi^2 = 6.750$. Critical value: $\chi^2 = 3.841$ (assuming a 0.05 significance level). Reject the null hypothesis that the following two proportions are the same: (1) The proportion of tumors with incorrect staging from MRI and correct staging from PET/CT; (2) the proportion of tumors with correct staging from MRI and incorrect staging from PET/CT. The PET/CT technology appears to be more accurate.

13. Test statistic with continuity correction: $\chi^2 = 20.021$. Test statistic without continuity correction: $\chi^2 = 21.333$. The test statistic is larger without the continuity correction. The test statistic without the continuity correction is generally larger (except when $b = c$), so the results appear to have more significant differences.

15. $P(6$ or more successes in 8 trials) $= P(6$ successes$) + P(7$ successes$) + P(8$ successes$) = 0.109375 + 0.03125 + 0.00390625 = 0.1445$, so that P-value $= 2(0.1445) = 0.289$. This is the same P-value found using the chi-square approximation, so the approximation would have been quite good here, even though the condition of $b + c \geq 10$ is not satisfied. Based on the results, there is not sufficient evidence to conclude that either treatment is better than the other.

Chapter 10 Review Exercises

1. Test statistic: $\chi^2 = 6.780$. Critical value: $\chi^2 = 12.592$. There is not sufficient evidence to support the theory that more gunfire deaths occur on weekends.

2. Test statistic: $\chi^2 = 49.731$. Critical value: $\chi^2 = 11.071$ (assuming a 0.05 significance level). There is sufficient evidence to support the claim that the type of crime is related to whether the criminal drinks or abstains.

3. Test statistic: $\chi^2 = 5.297$. Critical value: $\chi^2 = 3.841$. There is sufficient evidence to warrant rejection of the claim that whether a newborn is discharged early or late is independent of whether the newborn was rehospitalized within a week of discharge. The conclusion changes if the significance level is changed to 0.01.

4. Test statistic: $\chi^2 = 13.225$. Critical value: $\chi^2 = 3.841$. Reject the null hypothesis that the two given proportions are the same. The treatment appears to be effective.

Chapter 10 Cumulative Review Exercises

1. $\bar{x} = 80.9$; median: 81.0; range: 28.0; $s^2 = 74.4$; $s = 8.6$; 5-number summary: 66, 76.0, 81.0, 86.5, 94.
2. a. 0.272
   b. 0.468
   c. 0.614
   d. 0.282
3. Contingency table; see Section 10-3. Test statistic: $\chi^2 = 0.055$. Critical value: $\chi^2 = 7.815$ (assuming a 0.05 significance level). There is not sufficient evidence to warrant rejection of the claim that men and women choose the different answers in the same proportions.
4. Use correlation; see Section 9-2. Test statistic: $r = 0.978$. Critical values: $r = \pm 0.950$ (assuming a 0.05 significance level). There is sufficient evidence to support the claim that there is a relationship between the memory and reasoning scores.
5. Use the test for matched pairs; see Section 8-4. $\bar{d} = -10.25$; $s_d = 1.5$. Test statistic: $t = -13.667$. Critical value: $t = -2.353$ (assuming a 0.05 significance level). Reject H_0. There is sufficient evidence to support the claim that the training session is effective in raising scores.
6. Test for the difference between two independent samples; see Section 8-3. Test statistic: $t = -2.014$. Critical values: $t = \pm 3.182$ (assuming a 0.05 significance level). Fail to reject H_0: $\mu_1 = \mu_2$. There is not sufficient evidence to warrant rejection of the claim that men and women have the same mean score.

Chapter 11 Answers

Section 11-2

1. a. H_0: $\mu_1 = \mu_2 = \mu_3 = \mu_4$
   b. At least one of the four population means is different from the others.
   c. 8.448
   d. 3.2874
   e. 0.002
   f. Reject the null hypothesis that the four population means are equal.
   g. The outlier of 1.34 does not have a dramatic effect on the results. The conclusion is the same with or without the outlier.
3. a. $\mu_1 = \mu_2 = \mu_3$
   b. At least one of the three means is different from the others.
   c. $F = 0.1887$
   d. 3.0804
   e. 0.8283
   f. No
5. Test statistic: $F = 2.5423$. Critical value: $F = 3.2389$. P-value: 0.0929. Fail to reject H_0: $\mu_1 = \mu_2 = \mu_3 = \mu_4$. There is not suffi-

cient evidence to reject the claim that the weights are from populations with the same mean. It does not appear that the weights are different for the various treatments.
7. Test statistic: $F = 0.9922$. Critical value: $F = 3.2389$. P-value: 0.4216. Fail to reject H_0: $\mu_1 = \mu_2 = \mu_3 = \mu_4$. There is not sufficient evidence to support the claim that larger cars are safer.
9. Test statistic: $F = 4.0497$. Critical value: $F = 3.4028$. P-value: 0.0305. Reject H_0: $\mu_1 = \mu_2 = \mu_3$. There is sufficient evidence to support the claim that the mean breadth is not the same for the different epochs.
11. Test statistic: $F = 119.2645$. Critical value: $F = 2.9957$ approximately. P-value: 0.0000. Reject H_0: $\mu_1 = \mu_2 = \mu_3$. There is sufficient evidence to reject the claim that the three populations have different mean sepal lengths.
13. Test statistic: $F = 2.4749$. Critical value: $F = 3.0984$ (assuming a 0.05 significance level. P-value: 0.0911. Fail to reject H_0: $\mu_1 = \mu_2 = \mu_3 = \mu_4$. There is not sufficient evidence to warrant rejection of the claim that the four gender/race combinations have the same mean. There does not appear to be a serious problem with the sample because of significant differences among the four strata.
15. The results are similar in the sense that both tests identify sample 4 (mean weights of poplar trees treated with fertilizer and irrigation) as being significantly different from the mean weights of poplar trees with the other three treatment categories. Although the displayed levels of significance are different for the two tests, they do not differ by large amounts.

Section 11-3

1. "Two-way" refers to the inclusion of two different factors, which are properties or characteristics used to distinguish different populations from one another. "Analysis of variance" refers to the method used, which is based on two different estimates of the assumed common population variance.
3. If there is an interaction between factors, we should not consider the effects of either factor without considering those of the other.
5. Test statistic: $F = 1.28$. P-value: 0.313. Fail to reject the null hypothesis of no interaction. There does not appear to be a significant effect from the interaction between site and age.
7. Test statistic: $F = 249.85$. P-value: 0.000. Reject the null hypothesis that age has no effect on the amount of DDT. There is sufficient evidence to support the claim that age has an effect on the amount of DDT.
9. Test statistic: $F = 1.686$. P-value: 0.206. Fail to reject the null hypothesis that gender has no effect on running time. There is not sufficient evidence to support the claim that gender has an effect on running times.
11. Test statistic: $F = 3.87$. P-value: 0.000. Reject the null hypothesis that the choice of subject has no effect on the hearing test score. There is sufficient evidence to support the claim that the choice of subject has an effect on the hearing test score.

13. For interaction, the test statistic is $F = 0.36$ and the P-value is 0.701, so there is no significant interaction effect. For gender, the test statistic is $F = 0.09$ and the P-value is 0.762, so there is no significant effect from gender. For age, the test statistic is $F = 0.36$ and the P-value is 0.701, so there is no significant effect from age.

15. Test statistic: $F = 1.4330$. P-value: 0.2401. Fail to reject the null hypothesis that site has no effect on weight. Weight does not appear to be affected by the site.

17. Test statistic: $F = 0.0061$. P-value: 0.9428. Fail to reject the null hypothesis that site has no effect on weight. Weight does not appear to be affected by the site.

Chapter 11 Review Exercises

1. Test statistic: $F = 46.900$. P-value: 0.000. Reject H_0: $\mu_1 = \mu_2 = \mu_3$. There is sufficient evidence to warrant rejection of the claim of equal population means.

2. Test statistic: $F = 1179.034$. P-value: 0.000. Reject H_0: $\mu_1 = \mu_2 = \mu_3$. There is sufficient evidence to warrant rejection of the claim of equal petal length means.

3. Test statistic: $F = 47.3645$. P-value: 0.000. Reject H_0: $\mu_1 = \mu_2 = \mu_3$. There is sufficient evidence to warrant rejection of the claim of equal sepal width means.

4. Test statistic: $F = 0.19$. P-value: 0.832. Fail to reject the null hypothesis of no interaction. There does not appear to be a significant effect from the interaction between gender and major.

5. Test statistic: $F = 0.78$. P-value: 0.395. Fail to reject the null hypothesis that gender has no effect on SAT scores. There is not sufficient evidence to support the claim that estimated length is affected by gender.

6. Test statistic: $F = 0.13$. P-value: 0.876. Fail to reject the null hypothesis that major has no effect on SAT scores. There is not sufficient evidence to support the claim that estimated length is affected by major.

Chapter 11 Cumulative Review Exercises

1. a. 0.100 in.
   b. 0.263 in.
   c. 0.00, 0.00, 0.00, 0.010, 1.41
   d. 0.92 in., 1.41 in.
   e. Answer varies, depending on the number of classes used, but the histogram should depict a distribution that is skewed to the right.
   f. No, because the data do not appear to come from a normally distributed population.
   g. 19/52 or 0.365

2. a. 960.5, 980.0, 1046.0; no
   b. 914.5, 1010.5, 1008.5; no
   c. 174.6, 239.6, 226.8; no
   d. Test statistic: $t = -0.294$. Critical values: $t = \pm 2.093$ (assuming a 0.05 significance level). Fail to reject H_0: $\mu_1 = \mu_2$.

   e. $878.8 < \mu < 1042.2$
   f. Test statistic: $F = 0.8647$. P-value: 0.4266. Fail to reject H_0: $\mu_1 = \mu_2 = \mu_3$. There is not sufficient evidence to warrant rejection of the claim that the three populations have the same mean SAT score.

3. a. 0.3372
   b. 0.0455
   c. 1/8

Chapter 12 Answers

Section 12-2

1. The test statistic of $x = 5$ is not less than or equal to the critical value of 3. There is not sufficient evidence to warrant rejection of the claim of no difference.

3. The test statistic of $z = -0.95$ is not less than or equal to the critical value of -1.96. There is not sufficient evidence to warrant rejection of the claim of no difference.

5. The test statistic of $x = 3$ is not less than or equal to the critical value of 1. There is not sufficient evidence to reject the claim of no difference between the yields from the two types of seed. Neither type of seed appears to be better.

7. The test statistic of $x = 5$ is not less than or equal to the critical value of 2. There is not sufficient evidence to support the claim that there is a difference between self-reported heights and measured heights.

9. The test statistic of $x = 0$ is less than or equal to the critical value of 0. There is sufficient evidence to warrant rejection of the claim that there is no significant difference between the measurements taken in the two positions. There appears to be a difference.

11. The test statistic of $x = 1$ is less than or equal to the critical value of 3. There is sufficient evidence to support a claim that the company is cheating consumers.

13. Convert $x = 30$ to the test statistic $z = -1.19$. Critical value: $z = -1.645$ (assuming a 0.05 significance level). There is not sufficient evidence to support the claim that among smokers who try to quit with nicotine patch therapy, the majority are smoking a year after the treatment.

15. First approach: $z = -1.90$; reject H_0.
    Second approach: $z = -1.73$; reject H_0.
    Third approach: $z = 0$; fail to reject H_0.

17. Convert $x = 18$ to the test statistic $z = -2.31$. Critical value: $z = -2.33$. There is not sufficient evidence to support a charge of gender bias. If the binomial distribution is used instead of the normal approximation, the P-value is 0.0099, which is less than 0.01, so there is sufficient evidence to support a charge of gender bias. Using the normal approximation, the test statistic is just barely outside of the critical region; using the binomial distribution, the test statistic is just barely inside the critical region.

Section 12-3

1. Test statistic: $T = 1$. Critical value: $T = 2$. Reject the null hypothesis that both samples come from populations having differences with a median equal to zero.
3. Test statistic: $T = 13.5$. Critical value: $T = 8$. Fail to reject the null hypothesis that both samples come from populations having differences with a median equal to zero.
5. Test statistic: $T = 34$. Critical value: $T = 14$. Fail to reject the null hypothesis that both samples come from populations having differences with a median equal to zero.
7. Test statistic: $T = 0$. Critical value: $T = 8$. Reject the null hypothesis that both samples come from populations having differences with a median equal to zero.
9. Convert $T = 661$ to test statistic $z = -5.67$. Critical values: $z = 1.96$. There is sufficient evidence to warrant rejection of the claim that healthy adults have a median body temperature that is equal to 98.6°F.

Section 12-4

1. $R_1 = 120.5$, $R_2 = 155.5$, $\mu_R = 132$, $\sigma_R = 16.248$, test statistic: $z = -0.71$. Critical values: $z = \pm1.96$. Fail to reject the null hypothesis that the populations have equal medians.
3. $\mu_R = 150$, $\sigma_R = 17.321$, $R = 96.5$, $z = -3.09$. Test statistic: $z = -3.09$. Critical values: $z = \pm2.575$. There is sufficient evidence to warrant rejection of the claim that the two samples come from populations with equal medians.
5. $\mu_R = 169$, $\sigma_R = 18.385$, $R = 196$, $z = 1.47$. Test statistic: $z = 1.47$. Critical values: $z = \pm1.96$. There is not sufficient evidence to warrant rejection of the claim that the two samples come from populations with equal medians.
7. $\mu_R = 1620$, $\sigma_R = 103.923$, $R = 1339.5$, $z = -2.70$. Test statistic: $z = -2.70$. Critical values: $z = \pm1.96$. There is sufficient evidence to warrant rejection of the claim that the two samples come from populations with equal medians.
9. $\mu_R = 1620$, $\sigma_R = 103.923$, $R = 2003.5$, $z = 3.69$. Test statistic: $z = 3.69$. Critical values: $z = \pm1.96$. There is sufficient evidence to warrant rejection of the claim that the two samples come from populations with equal medians.
11. $z = -0.98$; the test statistic is the same number with opposite sign.

Section 12-5

1. a. The four treatments result in weights with equal medians.
   b. The four treatments result in weights with at least one of the medians different from the others.
   c. $H = 8.37$
   d. 7.815
   e. 0.039

f. Reject the null hypothesis that the populations have equal medians. The different treatments do not appear to result in the same median.
3. Test statistic: $H = 4.269$. Critical value: $\chi^2 = 7.815$. P-value: 0.234. Fail to reject the null hypothesis of equal medians. There is not sufficient evidence to support the claim that the medians are not all equal. It does not appear that the weights are different for the various treatments at the drier site.
5. Test statistic: $H = 6.631$. Critical value: $\chi^2 = 5.991$. P-value: 0.036. There is sufficient evidence to support the claim that the different epochs do not all have the same median skull breadth.
7. Test statistic: $H = 1.1914$. Critical value: $\chi^2 = 7.815$. P-value: 0.755. There is not sufficient evidence to support the claim that the measurements of head injuries for the four weight categories are not all the same. The given data do not provide sufficient evidence to conclude that heavier cars are safer in a crash.
9. Test statistic: $H = 59.155$. Critical value: $\chi^2 = 5.991$. P-value: 0.000. There is sufficient evidence to support the claim that the median cotinine levels are not all equal for the three groups. The results suggest that second-hand smoke affects those who are exposed to it.
11. Test statistic: $H = 15.158$. Critical value: $\chi^2 = 5.991$. P-value: 0.0005. There is sufficient evidence to support the claim that the median petal widths are not all equal for the three species.

Section 12-6

1. $r_s = 1$ and there appears to be a correlation between x and y.
3. $r_s = 0$ and there does not appear to be a correlation between x and y.
5. ±0.280
7. ±0.654
9. $r_s = 0.261$. Critical values: $r_s = \pm0.648$. No significant correlation. There does not appear to be a correlation between salary and physical demand.
11. $r_s = 0.557$. Critical values: $r_s = \pm0.700$. No significant correlation. There does not appear to be a correlation between height and weight.
13. $r_s = 0.364$. Critical values: $r_s = \pm0.738$. No significant correlation. Based on the given data, there does not appear to be a correlation between the systolic and diastolic blood pressure measurements.
15. $r_s = 0.624$. Critical values: $r_s = \pm0.587$. Significant correlation. Based on the given data, there does appear to be a correlation between the number of cigarettes smoked and the cotinine level.
17. $r_s = 0.520$. Critical values: $r_s = \pm0.314$. Significant correlation. Based on the given data, there does appear to be a correlation between cholesterol and body mass index.
19. $r_s = 0.984$. Critical values: $r_s = \pm0.269$. Significant correlation. Based on the given data, there does appear to be a correlation between chest size and weight.

21. a. ± 0.707
    b. ± 0.514
    c. ± 0.361
    d. ± 0.463
    e. ± 0.834

Chapter 12 Review Exercises

1. $r_s = 0.0952$. Critical values: $r_s = \pm 0.738$ (assuming a 0.05 significance level). No significant correlation. Based on the given data, there does not appear to be a correlation between the temperature and the winning time.
2. Convert $x = 22$ to the test statistic $z = -2.585$. Critical values: $z = \pm 2.575$. There is sufficient evidence to support the claim that the method has an effect on gender.
3. Wilcoxon rank-sum test: $\mu_R = 162$, $\sigma_R = 19.442$, $R = 89.5$, $z = -3.73$. Test statistic: $z = -3.73$. Critical values: $z = \pm 1.96$. Reject the null hypothesis that the two samples come from populations with the same median. There is sufficient evidence to warrant rejection of the claim that beer drinkers and liquor drinkers have the same median BAC levels. The liquor drinkers appear to be more dangerous.
4. Kruskal-Wallis test: Test statistic: $H = 4.234$. Critical value: $\chi^2 = 7.815$. Fail to reject equality of the population medians. There is not sufficient evidence to support the claim that the injury measurements are not the same for the four categories. There is not sufficient evidence to support the claim that heavier cars are safer.
5. The test statistic $x = 2$ is not less than or equal to the critical value of 1. There is not sufficient evidence to warrant rejection of the claim that the course has no effect. The course does not appear to have an effect.
6. Test statistic: $T = 9.5$. Critical value: $T = 6$. There is not sufficient evidence to warrant rejection of the claim that the course has no effect. The course does not appear to have an effect.

Chapter 12 Cumulative Review Exercises

1. Test statistic: $r = -0.515$. Critical values: $r = \pm 0.707$. No significant linear correlation.
2. Test statistic: $r_s = -0.463$. Critical values: $r_s = \pm 0.738$. No significant correlation.
3. The test statistic $x = 3$ is not less than or equal to the critical value of 0. There is not sufficient evidence to support the claim that there is a difference between the heights of the winning and losing candidates.
4. Test statistic: $T = 10$. Critical value: $T = 2$. There is not sufficient evidence to support the claim that there is a difference between the heights of the winning and losing candidates.
5. Test statistic: $t = 0.851$. Critical values: $t = \pm 2.365$. Fail to reject H_0: $\mu_1 = \mu_2$. There is not sufficient evidence to support the claim that there is a difference between the heights of the winning and losing candidates.

Chapter 13 Answers

Section 13-2

1. A cohort life table is a record of the actual observed mortality experience for a particular group, whereas a period life table describes mortality and longevity data for a *hypothetical* group that would have lived with the same mortality conditions throughout their lives.
3. The values in columns 3 through 6 would be halved, but the other values would remain the same.
5. 0.000412
7. 0.99446
9. 0.999112; 249.778
11. 0.816753
13. a. 0.008230
    b. 0.000827
       The results are so different because of the exceptionally high mortality rate at or very near birth.
15. 0 – 5; 0.008230; 100,000; 822 (or 823); 496,307 (or 496,308); 7,686,810; 76.9

Section 13-3

1. 4,405,932; expected numbers of drivers are helpful in determining future traffic needs, or resources required by state drivers license issuing sites.
3. 3,896,286
5. Using H_1: $p > 0.000888$, the test statistic is $z = 1.17$, the P-value is 0.121, and the critical value is $z = 1.645$ (assuming a 0.05 significance level), so fail to reject the null hypothesis of $p = 0.000888$ and conclude that there is not sufficient evidence to support the claim that the number of deaths is unusually high. (If the probability of dying is calculated from the third column of Table 13-1, use H_1: $p > 0.000881870$ to get a test statistic of $z = 1.20$, a P-value of 0.115, and the same critical value and conclusions.)
7. Using H_1: $p > 0.012603$, the test statistic is $z = 3.49$, the P-value is 0.0002, and the critical value is $z = 1.645$ (assuming a 0.05 significance level), so reject the null hypothesis of $p = 0.012603$ and conclude that there is sufficient evidence to support the claim that the number of deaths is unusually high.
9. $415.52
11. $8675.45
13. $586.32; because this result is higher than $415.52, it appears that 49-year-old black females have a higher death rate than those in the general population.

Chapter 13 Review Exercises

1. 0.000790
2. 53
3. 0.98655
4. 0–2; 0.013450; 100,000; 1345; 197,584; 7,493,035; 74.9

5. 74.9; 74.9; both values are the same. A possible explanation is that among black females who do not survive to the first birthday, a disproportionately large number of deaths occur at or near birth.

Chapter 13 Cumulative Review Exercises

1. Using H_1: $p < 0.012672$, the test statistic is $z = -1.90$, the P-value is 0.0287, and the critical value is $z = -1.645$ (assuming a 0.05 significance level), so reject the null hypothesis of $p = 0.012672$ and conclude that there is sufficient evidence to support the claim that the proportion of deaths is less than 0.012672.

2. $0.98994 < p < 0.99989$; the confidence interval limits do not contain 0.987328, which suggests that the mortality rate has been lowered. A hypothesis test might also be used to justify such a conclusion.

3. 0.962

4. 0.0497; no, because being in the same family causes the events to be dependent, instead of being independent, as required by the multiplication rule.

Credits

Photographs

Chapter Openers

Chapter 1: Child with Doctor © Sandy King/Getty Images
Chapter 2: Person Smoking in Elevator © Getty
Chapter 3: Pregnancy Test © nestonik—Fotolia
Chapter 4: Couple on Beach with Child © Corbis
Chapter 5: Gondola on Snowy Mountain © Corbis
Chapter 6: Peas © goodween123—Fotolia
Chapter 7: Peas © unikat—Fotolia
Chapter 8: Nasal Spray for Child © AP Wideworld
Chapter 9: Bear © byrdyak—Fotolia
Chapter 10: Doctor Giving Vaccine © Corbis Bettmann
Chapter 11: Poplar Trees © Denis Tabler—Fotolia
Chapter 12: Poplar Trees © Corbis RF
Chapter 13: Group of Teens © Corbis RF

Chapter 1
p. 4, The State of Statistics; Copyright Pavel L Photo and Video/Shutterstock
p. 7, Should You Believe a Statistical Study?; Copyright Jeff Greenberg/Photo Edit
p. 11, Clinical Trials vs. Observational Studies; Copyright Keith Brofsky/Getty Images

Chapter 2
p. 43, Class Size Paradox; Copyright Alexander Raths/Shutterstock
p. 47, Just an Average Guy; Copyright Brand New Images/Shutterstock
p. 69, Process of Drug Approval; Copyright Corbis RF/BrandX Pictures

Chapter 3
p. 118, Redundancy; Copyright Jim Ross, NASA DFRC
p. 124, Composite Sampling; Copyright Eugene Sim/Shutterstock
p. 145, The Random Secretary; Copyright sil63/Shutterstock

Chapter 4
p. 168, The Power of a Graph; Copyright James Leynse/Corbis
p. 170, Do Boys or Girls Run in the Family; Copyright Anthony Redpath/Corbis
p. 182, Not at Home; Copyright tmc_photos—Fotolia

Chapter 5
p. 192, Changing Populations; Copyright Taxi/Getty Images
p. 193, Reliability and Validity; Copyright Colorblind/Shutterstock
p. 205, Clinical Trial Cut Short; Copyright Seits/Photo Researchers, Inc.
p. 213, Growth Charts Updated; Copyright Kelly/Mooney Photography/Corbis
p. 242, Mannequins/Reality; Copyright Alexey Kuznetsov—Fotolia

Chapter 6
p. 274, Estimating Wildlife Population Sizes; Copyright Josh Anon/Shutterstock
p. 285, Publication Bias; Copyright Ryan McVay/Getty Images

Chapter 7
p. 342, Test of Touch Therapy; Copyright Alexey Klementiev/Fotolia
p. 348, Does Aspirin Help Prevent Heart Attacks? Copyright PhotoDisc

Chapter 8
p. 387, Gender Gap in Drug Testing; Copyright PhotoDisc
p. 401, Research in Twins; Copyright Felix Mizioznikov/Fotolia
p. 404, Twins in Twinsburg; Copyright Cohen/Ostrow/Getty Images

Chapter 9
p. 430, Direct Link Between Smoking and Cancer; Copyright Hedgehog/Fotolia
p. 450, Cell Phones and Crashes; Copyright Galina Barskaya/Fotolia
p. 465, Do Air Bags Save Lives? Copyright hero—Fotolia

Chapter 10
p. 493, Safest Airplane Seats; Copyright PhotoDisc
p. 522, Expensive Diet Pill; Copyright Lance Manning/Corbis

Chapter 12
p. 570, Lie Detectors; Copyright Monica Vinella/Getty Images
p. 597, Boys and Girls Are Not Equally Likely; Copyright Matt Antonino/Shutterstock

Chapter 13
p. 612, Is Parachuting Safe? Copyright Franck Vantomme - Fotolia

Illustrations

The following margin illustrations were rendered by Bob Giuliani.

Chapter 2
p. 35, Florence Nightingale

Chapter 3
p. 93, Probabilities That Challenge Intuition
p. 139, Monkey Typist

Chapter 6
p. 257, Poll Resistance

Chapter 7
p. 366, Delaying Death

Chapter 8
p. 406, Crest and Dependent Samples
p. 410, Hawthorne and Experimenter Effects

Chapter 10
p. 507, Detecting Phony Data

Chapter 11
p. 534, Ethics in Experiments

Chapter 13
p. 613, Sensitive Surveys

The following margin illustrations were rendered by James Bryant.

Chapter 3
p. 103, Convicted by Probability
p. 113, Independent Jet Engines

Chapter 5
p. 222, The Fuzzy Central Limit Theorem
p. 232, Survey Medium Affects Results

Chapter 7
p. 355, Palm Reading

The following illustration was rendered by Darwen Hennings.

Chapter 11
p. 532, The Placebo Effect

Index

TABLE A-3   t Distribution: Critical t Values

Degrees of Freedom	0.005	0.01	Area in One Tail 0.025	0.05	0.10
	0.01	0.02	Area in Two Tails 0.05	0.10	0.20
1	63.657	31.821	12.706	6.314	3.078
2	9.925	6.965	4.303	2.920	1.886
3	5.841	4.541	3.182	2.353	1.638
4	4.604	3.747	2.776	2.132	1.533
5	4.032	3.365	2.571	2.015	1.476
6	3.707	3.143	2.447	1.943	1.440
7	3.499	2.998	2.365	1.895	1.415
8	3.355	2.896	2.306	1.860	1.397
9	3.250	2.821	2.262	1.833	1.383
10	3.169	2.764	2.228	1.812	1.372
11	3.106	2.718	2.201	1.796	1.363
12	3.055	2.681	2.179	1.782	1.356
13	3.012	2.650	2.160	1.771	1.350
14	2.977	2.624	2.145	1.761	1.345
15	2.947	2.602	2.131	1.753	1.341
16	2.921	2.583	2.120	1.746	1.337
17	2.898	2.567	2.110	1.740	1.333
18	2.878	2.552	2.101	1.734	1.330
19	2.861	2.539	2.093	1.729	1.328
20	2.845	2.528	2.086	1.725	1.325
21	2.831	2.518	2.080	1.721	1.323
22	2.819	2.508	2.074	1.717	1.321
23	2.807	2.500	2.069	1.714	1.319
24	2.797	2.492	2.064	1.711	1.318
25	2.787	2.485	2.060	1.708	1.316
26	2.779	2.479	2.056	1.706	1.315
27	2.771	2.473	2.052	1.703	1.314
28	2.763	2.467	2.048	1.701	1.313
29	2.756	2.462	2.045	1.699	1.311
30	2.750	2.457	2.042	1.697	1.310
31	2.744	2.453	2.040	1.696	1.309
32	2.738	2.449	2.037	1.694	1.309
34	2.728	2.441	2.032	1.691	1.307
36	2.719	2.434	2.028	1.688	1.306
38	2.712	2.429	2.024	1.686	1.304
40	2.704	2.423	2.021	1.684	1.303
45	2.690	2.412	2.014	1.679	1.301
50	2.678	2.403	2.009	1.676	1.299
55	2.668	2.396	2.004	1.673	1.297
60	2.660	2.390	2.000	1.671	1.296
65	2.654	2.385	1.997	1.669	1.295
70	2.648	2.381	1.994	1.667	1.294
75	2.643	2.377	1.992	1.665	1.293
80	2.639	2.374	1.990	1.664	1.292
90	2.632	2.368	1.987	1.662	1.291
100	2.626	2.364	1.984	1.660	1.290
200	2.601	2.345	1.972	1.653	1.286
300	2.592	2.339	1.968	1.650	1.284
400	2.588	2.336	1.966	1.649	1.284
500	2.586	2.334	1.965	1.648	1.283
750	2.582	2.331	1.963	1.647	1.283
1000	2.581	2.330	1.962	1.646	1.282
2000	2.578	2.328	1.961	1.646	1.282
Large	2.576	2.326	1.960	1.645	1.282

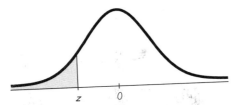

NEGATIVE z Scores

TABLE A-2	Standard Normal (z) Distribution: Cumulative Area from the LEFT									
z	.00	.01	.02	.03	.04	.05	.06	.07	.08	.09
−3.50 and lower	.0001									
−3.4	.0003	.0003	.0003	.0003	.0003	.0003	.0003	.0003	.0003	.0002
−3.3	.0005	.0005	.0005	.0004	.0004	.0004	.0004	.0004	.0004	.0003
−3.2	.0007	.0007	.0006	.0006	.0006	.0006	.0006	.0005	.0005	.0005
−3.1	.0010	.0009	.0009	.0009	.0008	.0008	.0008	.0008	.0007	.0007
−3.0	.0013	.0013	.0013	.0012	.0012	.0011	.0011	.0011	.0010	.0010
−2.9	.0019	.0018	.0018	.0017	.0016	.0016	.0015	.0015	.0014	.0014
−2.8	.0026	.0025	.0024	.0023	.0023	.0022	.0021	.0021	.0020	.0019
−2.7	.0035	.0034	.0033	.0032	.0031	.0030	.0029	.0028	.0027	.0026
−2.6	.0047	.0045	.0044	.0043	.0041	.0040	.0039	.0038	.0037	.0036
−2.5	.0062	.0060	.0059	.0057	.0055	.0054	.0052	.0051 *	.0049	.0048
−2.4	.0082	.0080	.0078	.0075	.0073	.0071	.0069	.0068	.0066	.0064
−2.3	.0107	.0104	.0102	.0099	.0096	.0094	.0091	.0089	.0087	.0084
−2.2	.0139	.0136	.0132	.0129	.0125	.0122	.0119	.0116	.0113	.0110
−2.1	.0179	.0174	.0170	.0166	.0162	.0158	.0154	.0150	.0146	.0143
−2.0	.0228	.0222	.0217	.0212	.0207	.0202	.0197	.0192	.0188	.0183
−1.9	.0287	.0281	.0274	.0268	.0262	.0256	.0250	.0244	.0239	.0233
−1.8	.0359	.0351	.0344	.0336	.0329	.0322	.0314	.0307	.0301	.0294
−1.7	.0446	.0436	.0427	.0418	.0409	.0401	.0392	.0384	.0375	.0367
−1.6	.0548	.0537	.0526	.0516	.0505 *	.0495	.0485	.0475	.0465	.0455
−1.5	.0668	.0655	.0643	.0630	.0618	.0606	.0594	.0582	.0571	.0559
−1.4	.0808	.0793	.0778	.0764	.0749	.0735	.0721	.0708	.0694	.0681
−1.3	.0968	.0951	.0934	.0918	.0901	.0885	.0869	.0853	.0838	.0823
−1.2	.1151	.1131	.1112	.1093	.1075	.1056	.1038	.1020	.1003	.0985
−1.1	.1357	.1335	.1314	.1292	.1271	.1251	.1230	.1210	.1190	.1170
−1.0	.1587	.1562	.1539	.1515	.1492	.1469	.1446	.1423	.1401	.1379
−0.9	.1841	.1814	.1788	.1762	.1736	.1711	.1685	.1660	.1635	.1611
−0.8	.2119	.2090	.2061	.2033	.2005	.1977	.1949	.1922	.1894	.1867
−0.7	.2420	.2389	.2358	.2327	.2296	.2266	.2236	.2206	.2177	.2148
−0.6	.2743	.2709	.2676	.2643	.2611	.2578	.2546	.2514	.2483	.2451
−0.5	.3085	.3050	.3015	.2981	.2946	.2912	.2877	.2843	.2810	.2776
−0.4	.3446	.3409	.3372	.3336	.3300	.3264	.3228	.3192	.3156	.3121
−0.3	.3821	.3783	.3745	.3707	.3669	.3632	.3594	.3557	.3520	.3483
−0.2	.4207	.4168	.4129	.4090	.4052	.4013	.3974	.3936	.3897	.3859
−0.1	.4602	.4562	.4522	.4483	.4443	.4404	.4364	.4325	.4286	.4247
−0.0	.5000	.4960	.4920	.4880	.4840	.4801	.4761	.4721	.4681	.4641

NOTE: For values of z below −3.49, use 0.0001 for the area.
*Use these common values that result from interpolation:

z score	Area
−1.645	0.0500
−2.575	0.0050